BRIEF CONTENTS

D0224500

CONTENTS

CHAPTER **4**   **Variability**   **109**

clivewa/Shutterstock.com

CHAPTER **5**   **z-Scores: Location of Scores
and Standardized Distributions**   **149**

clivewa/Shutterstock.com

CHAPTER **6**   **Probability**   **177**

clivewa/Shutterstock.com

CHAPTER **7**   **Probability and Samples: The Distribution
of Sample Means**   **213**

clivewa/Shutterstock.com

CHAPTER **8**   # Introduction to Hypothesis Testing   243

clivewa/Shutterstock.com

CHAPTER **9**   # Introduction to the *t* Statistic   291

clivewa/Shutterstock.com

CHAPTER **10**

The *t* Test for Two Independent Samples   **323**

clivewa/Shutterstock.com

CHAPTER **11**

The *t* Test for Two Related Samples   **359**

clivewa/Shutterstock.com

CHAPTER **15**

The Chi-Square Statistic: Tests for Goodness of Fit and Independence   533

clivewa/Shutterstock.com

APPENDIXES

PREFACE

M any students in the behavioral sciences view the required statistics course as an intimidating obstacle that has been placed in the middle of an otherwise interesting curriculum. They want to learn about psychology and human behavior—not about math and science. As a result, the statistics course is seen as irrelevant to their education and career goals. However, as long as psychology and the behavioral sciences in general are founded in science, knowledge of statistics will be necessary. Statistical procedures provide researchers with objective and systematic methods for describing and interpreting their research results. Scientific research is the system that we use to gather information, and statistics are the tools that we use to distill the information into sensible and justified conclusions. The goal of this book is not only to teach the methods of statistics, but also to convey the basic principles of objectivity and logic that are essential for the behavioral sciences and valuable for decision making in everyday life.

Essentials of Statistics for the Behavioral Sciences, *Tenth Edition*, is intended for an undergraduate statistics course in psychology or any of the related behavioral sciences. The overall learning objectives of this book include the following, which correspond to some of the learning goals identified by the American Psychological Association (Noland and the Society for the Teaching of Psychology Statistical Literacy Taskforce, 2012).

1. Calculate and interpret the meaning of basic measures of central tendency and variability.
2. Distinguish between causal and correlational relationships.
3. Interpret data displayed as statistics, graphs, and tables.
4. Select and implement an appropriate statistical analysis for a given research design, problem, or hypothesis.
5. Identify the correct strategy for data analysis and interpretation when testing hypotheses.
6. Select, apply, and interpret appropriate descriptive and inferential statistics.
7. Produce and interpret reports of statistical analyses using APA style.
8. Distinguish between statistically significant and chance findings in data.
9. Calculate and interpret the meaning of basic tests of statistical significance.
10. Calculate and interpret the meaning of confidence intervals.
11. Calculate and interpret the meaning of basic measures of effect size statistics.
12. Recognize when a statistically significant result may also have practical significance.

The book chapters are organized in the sequence that we use for our own statistics courses. We begin with descriptive statistics (Chapters 1–4), then lay the foundation for inferential statistics (Chapters 5–8), and then we examine a variety of statistical procedures focused on sample means and variance (Chapters 9–13) before moving on to correlational methods and nonparametric statistics (Chapters 14 and 15). Information about modifying this sequence is presented in the "To the Instructor" section for individuals who prefer a different organization. Each chapter contains numerous examples (many based on actual research studies), learning objectives and learning checks for each section, a summary and list of key terms, instructions for using SPSS®, detailed problem-solving tips and demonstrations, and a set of end-of-chapter problems.

Those of you who are familiar with previous editions of *Statistics for the Behavioral Sciences* and *Essentials of Statistics for the Behavioral Sciences* will notice that some changes have been made. These changes are summarized in the "To the Instructor" section. Students who are using this edition should read the section of the preface entitled "To the Student." In revising this text, our students have been foremost in our minds. Over the years, they have provided honest and useful feedback, and their hard work and perseverance has made our writing and teaching most rewarding. We sincerely thank them.

To the Instructor

Previous users of any of the Gravetter-franchise textbooks should know that we have maintained all the hallmark features of our *Statistics* and *Essentials of Statistics* textbooks: the organization of chapters and content within chapters; the student-friendly, conversational tone; and the variety of pedagogical aids, including, Tools You Will Need, chapter outlines, and section-by-section Learning Objectives and Learning Checks, as well as end-of-chapter Summaries, Key Terms lists, Focus on Problem Solving tips, Demonstrations of problems solved, SPSS sections, and end-of-chapter Problems (with solutions to odd-numbered problems provided to students in Appendix C).

■ New to This Edition

Those of you familiar with the previous edition of *Statistics for the Behavioral Sciences* will be pleased to see that *Essentials of Statistics for the Behavioral Sciences* has the same "look and feel" and includes much of its content. For those of you familiar with *Essentials*, the following are highlights of the changes that have been made:

- Every chapter begins with a Preview, which highlights an example of a published study. These have been selected for level of interest so that they will draw the student in. The studies are used to illustrate the purpose and rationale of the statistical procedure presented in the chapter.

- There has been extensive revision of the end-of-chapter Problems. Many old problems have been replaced with new examples that cite research studies. As an enhanced instructional resource for students, the odd-numbered solutions in Appendix C now show the work for intermediate answers for problems that require more than one step. The even-numbered solutions are available online in the instructor's resources.

- The sections on research design and methods in Chapter 1 have been revised to be consistent with Gravetter and Forzano, *Research Methods for the Behavioral Sciences, Sixth Edition*. The interval and ratio scales discussion in Chapter 1 has been refined and includes a new table distinguishing scales of measurement.

- In Chapter 2, a new section on stem and leaf displays describes this exploratory data analysis as a simple alternative to a frequency distribution table or graph. A basic presentation of percentiles and percentile ranks has been added to the coverage of frequency distribution tables in Chapter 2. The topic is revisited in Chapter 6 (Section 6-4, Percentiles and Percentile Ranks), showing how percentiles and percentile ranks can be determined with normal distributions.

- Chapter 3 (Central Tendency) has added coverage for the median when there are tied scores in the middle of the distribution. It includes a formula for determining the median with interpolation.

- The coverage of degrees of freedom in Chapter 4 (Variability) has been revised, including a new box feature (Degrees of Freedom, Cafeteria-Style) that provides an analogy for the student. Rounding and rounding rules are discussed in a new paragraph in Section 4-2, Defining Variance and Standard Deviation. It was presented in this section because Example 4.2 is the first instance where the answer is an irrational number. A section on quartiles and the interquartile range has been added.

- Coverage of the distribution of sample means (Chapter 7) has been revised to provide more clarification. The topic is revisited in Chapter 9, where the distribution of sample means is more concretely compared and contrasted with the distribution of z-scores, along with a comparison between the unit normal table and the t distribution table. Chapter 7 also includes a new box feature that depicts the law of large numbers using an illustration of online shopping (The Law of Large Numbers and Online Shopping).

- In Chapter 8 (Introduction to Hypothesis Testing), the section on statistical power has been completely rewritten. It is now organized and simplified into steps that the student can follow. The figures for this section have been improved as well.

- A new box feature has been added to Chapter 10 demonstrating how the t statistic for an independent-measures study can be calculated from sample means, standard deviations, and sample sizes in a published research paper. There is an added section describing the role of individual differences in the size of standard error.

- The comparison of independent- and repeated-measures designs has been expanded in Chapter 11, and includes the issue of power.

- In Chapter 12 the section describing the numerator and denominator in the F-ratio has been expanded to include a description of the sources of the random and unsystematic differences.

- Chapter 13 now covers only the two-factor, independent-measures ANOVA. The single-factor, repeated-measures ANOVA was dropped because repeated-measures designs are typically performed in a mixed design that also includes one (or more) between-subject factors. As a result, Chapter 13 now has expanded coverage of the two-factor, independent-measures ANOVA.

- For Chapter 14, three graphs have been redrawn to correct minor inaccuracies and improve clarity. As with other chapters, there is a new SPSS section with figures and end-of-chapter Problems have been updated with current research examples.

- Chapter 15 has minor revisions and an updated SPSS section with four figures. As with other chapters, the end-of-chapter Problems have been extensively revised and contain current research examples.

- Many research examples have been updated with an eye toward selecting examples that are of particular interest to college students and that cut across the domain of the behavioral sciences.

- Learning Checks have been revised.

- All SPSS sections have been revised using SPSS® 25 and new examples. New screenshots of analyses are presented. Appendix D, General Instructions for Using SPSS®, has been significantly expanded.

- A summary of statistics formulas has been added.

- This edition of *Essentials of Statistics for the Behavioral Sciences* has been edited to align with Gravetter and Forzano, *Research Methods*, providing a more seamless transition from statistics to research methods in its organization and terminology. Taken together, the two books provide a smooth transition for a two-semester sequence of Statistics and Methods, or, even an integrated Statistics/Methods course.

■ Matching the Text to Your Syllabus

The book chapters are organized in the sequence that we use for our own statistics courses. However, different instructors may prefer different organizations and probably will choose to omit or deemphasize specific topics. We have tried to make separate chapters, and even sections of chapters, completely self-contained, so that they can be deleted or reorganized to fit the syllabus for nearly any instructor. Instructors using MindTap® can easily control the inclusion and sequencing of chapters to match their syllabus exactly. Following are some common examples:

- It is common for instructors to choose between emphasizing analysis of variance (Chapters 12 and 13) or emphasizing correlation/regression (Chapter 14). It is rare for a one-semester course to complete coverage of both topics.

- Although we choose to complete all the hypothesis tests for means and mean differences before introducing correlation (Chapter 14), many instructors prefer to place correlation much earlier in the sequence of course topics. To accommodate this, Sections 14-1, 14-2, and 14-3 present the calculation and interpretation of the Pearson correlation and can be introduced immediately following Chapter 4 (Variability). Other sections of Chapter 14 refer to hypothesis testing and should be delayed until the process of hypothesis testing (Chapter 8) has been introduced.

- It is also possible for instructors to present the chi-square tests (Chapter 15) much earlier in the sequence of course topics. Chapter 15, which presents hypothesis tests for proportions, can be presented immediately after Chapter 8, which introduces the process of hypothesis testing. If this is done, we also recommend that the Pearson correlation (Sections 14-1, 14-2, and 14-3) be presented early to provide a foundation for the chi-square test for independence.

To the Student

A primary goal of this book is to make the task of learning statistics as easy and painless as possible. Among other things, you will notice that the book provides you with a number of opportunities to practice the techniques you will be learning in the form of Examples, Learning Checks, Demonstrations, and end-of-chapter Problems. We encourage you to take advantage of these opportunities. Read the text rather than just memorizing the formulas. We have taken care to present each statistical procedure in a conceptual context that explains why the procedure was developed and when it should be used. If you read this material and gain an understanding of the basic concepts underlying a statistical formula, you will find that learning the formula and how to use it will be much easier. In the "Study Hints" that follow, we provide advice that we give our own students. Ask your instructor for advice as well; we are sure that other instructors will have ideas of their own.

■ Study Hints

You may find some of these tips helpful, as our own students have reported.

- The key to success in a statistics course is to keep up with the material. Each new topic builds on previous topics. If you have learned the previous material, then the new topic is just one small step forward. Without the proper background, however, the new topic can be a complete mystery. If you find that you are falling behind, get help immediately.

- You will learn (and remember) much more if you study for short periods several times a week rather than try to condense all of your studying into one long session. Distributed practice is best for learning. For example, it is far more effective to study and do problems for half an hour every night than to have a single three-and-a-half-hour study session once a week. We cannot even work on *writing* this book without frequent rest breaks.

- Do some work before class. Stay a little bit ahead of the instructor by reading the appropriate sections before they are presented in class. Although you may not fully understand what you read, you will have a general idea of the topic, which will make the lecture easier to follow. Also, you can identify material that is particularly confusing and then be sure the topic is clarified in class.

- Pay attention and think during class. Although this advice seems obvious, often it is not practiced. Many students spend so much time trying to write down every example presented or every word spoken by the instructor that they do not actually understand and process what is being said. Check with your instructor—there may not be a need to copy every example presented in class, especially if there are many examples like it in the text. Sometimes, we tell our students to put their pens and pencils down for a moment and just listen.

- Test yourself regularly. Do not wait until the end of the chapter or the end of the week to check your knowledge. As you are reading the textbook, stop and do the examples. Also, stop and do the Learning Checks at the end of each section. After each lecture, work on solving some of the end-of-chapter Problems and check your work for odd-numbered problems in Appendix C . Review the Demonstration problems, and be sure you can define the Key Terms. If you are having trouble, get your questions answered *immediately*—reread the section, go to your instructor, or ask questions in class. By doing so, you will be able to move ahead to new material.

- Do not kid yourself! Avoid denial. Many students observe their instructor solving problems in class and think to themselves, "This looks easy, I understand it." Do you really understand it? Can you really do the problem on your own without having to read through the pages of a chapter? Although there is nothing wrong with using examples in the text as models for solving problems, you should try working a problem with your book closed to test your level of mastery.

- We realize that many students are embarrassed to ask for help. It is our biggest challenge as instructors. You must find a way to overcome this aversion. Perhaps contacting the instructor directly would be a good starting point, if asking questions in class is too anxiety-provoking. You could be pleasantly surprised to find that your instructor does not yell, scold, or bite! Also, your instructor might know of another student who can offer assistance. Peer tutoring can be very helpful.

■ Contact Us

Over the years, the students in our classes and other students using our book have given us valuable feedback. If you have any suggestions or comments about this book, you can write to Professor Emeritus Larry Wallnau, Professor Lori-Ann Forzano, or Associate Professor James Witnauer at the Department of Psychology, The College at Brockport, SUNY, 350 New Campus Drive, Brockport, New York 14420. You can also contact us directly at: lforzano@brockport.edu or jwitnaue@brockport.edu or lwallnau@brockport.edu.

Ancillaries

Ancillaries for this edition include the following.

- **MindTap® Psychology**   *MindTap® Psychology for Gravetter/Wallnau/Forzano/ Witnauer's Essentials of Statistics for the Behavioral Sciences, Tenth Edition,* is the digital learning solution that helps instructors engage and transform today's students into critical thinkers. Through paths of dynamic assignments and applications that you can personalize, real-time course analytics, and an accessible reader, MindTap helps you turn cookie cutter into cutting edge, apathy into engagement, and memorizers into higher-level thinkers. As an instructor using MindTap, you have at your fingertips the right content and unique set of tools curated specifically for your course, such as video tutorials that walk students through various concepts and interactive problem tutorials that provide students opportunities to practice what they have learned, all in an interface designed to improve workflow and save time when planning lessons and course structure. The control to build and personalize your course is all yours, focusing on the most relevant material while also lowering costs for your students. Stay connected and informed in your course through real-time student tracking that provides the opportunity to adjust the course as needed based on analytics of interactivity in the course.

- **Online Instructor's Manual**   The manual includes learning objectives, key terms, a detailed chapter outline, a chapter summary, lesson plans, discussion topics, student activities, "What If" scenarios, media tools, a sample syllabus, and an expanded test bank. The learning objectives are correlated with the discussion topics, student activities, and media tools.

- **Online PowerPoints**   Helping you make your lectures more engaging while effectively reaching your visually oriented students, these handy Microsoft PowerPoint® slides outline the chapters of the main text in a classroom-ready presentation. The PowerPoint slides are updated to reflect the content and organization of the new edition of the text.

- **Cengage Learning Testing, powered by Cognero®**   Cengage Learning Testing, powered by Cognero®, is a flexible online system that allows you to author, edit, and manage test bank content. You can create multiple test versions in an instant and deliver tests from your LMS in your classroom.

Acknowledgments

It takes a lot of good, hard-working people to produce a book. Our friends at Cengage have made enormous contributions to this textbook. We thank: Laura Ross, Product Director; Josh Parrott, Product Manager; Kat Wallace, Product Assistant; and Bethany Bourgeois, Art Director. Special thanks go to Brian Pierce and Tangelique Williams-Grayer, our Content Managers, and to Lori Hazzard, who led us through production at MPS Limited.

Reviewers play an important role in the development of a manuscript. Accordingly, we offer our appreciation to the following colleagues for their assistance: Kara Moore, Knox College; Tom Williams, Mississippi College; Stacey Todaro, Adrian College; Dave Matz, Augsburg University; Barbara Friesth, Indiana University-Purdue University Indianapolis; Bethany Jurs, Transylvania University; Ben Denkinger, Augsburg University; Sara Festini,

University of Tampa; Lindsey Johnson, University of Southern Mississippi; Lawrence Preiser, York College CUNY; Stephen Blessing, University of Tampa; Pamela A MacLaughlin, Indiana University.

We must give our heartfelt thanks to our families: Naomi and Nico Wallnau; Charlie, Ryan, and Alex Forzano; and Beth, JJ, Nate, and Ben Witnauer. This book could not have been written without their patience and support.

Finally, it is with great sorrow that we acknowledge Fred Gravetter's passing. His expertise in statistics, teaching experience, and years of assisting students are woven into the fabric of every edition of this book. His students had the utmost praise for his courses and his teaching ability. Fred was appreciated as a mentor to students and faculty alike, including his fellow authors. Yet, he was modest despite his accomplishments, and he was approachable and engaging. We were reminded of his contributions as we worked on each chapter during the course of this revision and were guided by his vision during this process. He has, no doubt, left a lasting legacy for his students and colleagues. We were most fortunate to benefit from his friendship, and he is sorely missed.

Larry B. Wallnau
Lori-Ann B. Forzano
James E. Witnauer

FREDERICK J GRAVETTER was Professor Emeritus of Psychology at The College at Brockport, State University of New York. While teaching at Brockport, Dr. Gravetter specialized in statistics, experimental design, and cognitive psychology. He received his bachelor's degree in mathematics from M.I.T. and his Ph.D. in psychology from Duke University. In addition to publishing this textbook and several research articles, Dr. Gravetter coauthored all editions of the best-selling *Statistics for the Behavioral Sciences, Essentials of Statistics for the Behavioral Sciences,* and *Research Methods for the Behavioral Sciences.* Dr. Gravetter passed away in November 2017.

LARRY B. WALLNAU is Professor Emeritus of Psychology at The College at Brockport, State University of New York. While teaching at Brockport, his research has been published in journals such as *Pharmacology Biochemistry and Behavior, Physiology and Behavior, Journal of Human Evolution, Folia Primatologica,* and *Behavior Research Methods and Instrumentation.* He also has provided editorial consultation. His courses have included statistics, bio-psychology, animal behavior, psychopharmacology, and introductory psychology. With Dr. Gravetter, he coauthored all previous editions of *Statistics for the Behavioral Sciences* and *Essentials of Statistics for the Behavioral Sciences.* Dr. Wallnau received his bachelor's degree from the University of New Haven and his Ph.D. in psychology from the State University of New York at Albany. In his leisure time, he is an avid runner with his canine companion, Gracie.

LORI-ANN B. FORZANO is Professor of Psychology at The College at Brockport, State University of New York, where she regularly teaches undergraduate and graduate courses in research methods, statistics, learning, animal behavior, and the psychology of eating. She earned a Ph.D. in experimental psychology from the State University of New York at Stony Brook, where she also received her B.S. in psychology. Dr. Forzano's research examines impulsivity and self-control in adults and children. Her research has been published in the *Journal of the Experimental Analysis of Behavior, Learning and Motivation,* and *The Psychological Record.*

Dr. Forzano has also coauthored *Essentials of Statistics for the Behavioral Sciences, Ninth Edition,* and all previous editions of *Research Methods for the Behavioral Sciences,* now in its sixth edition.

JAMES E. WITNAUER is Associate Professor of Psychology at The College at Brockport, State University of New York, where he teaches undergraduate courses in experimental psychology and graduate courses in statistics and biopsychology. He earned a Ph.D. in cognitive psychology from State University of New York, Binghamton, and a B.A. in psychology from State University of New York, Buffalo State College. Dr. Witnauer's research aims to test mathematical models of learning and behavior and has been published in *Behavioural Processes*, *Journal of Experimental Psychology: Animal Behavior Processes*, and *Neurobiology of Learning and Memory*.

Introduction to Statistics

clivewa/Shutterstock.com

PREVIEW

PREVIEW

Before we begin our discussion of statistics, we ask you to take a few moments to read the following paragraph, which has been adapted from a classic psychology experiment reported by Bransford and Johnson (1972).

> The procedure is actually quite simple. First you arrange things into different groups depending on their makeup. Of course, one pile may be sufficient, depending on how much there is to do. If you have to go somewhere else due to lack of facilities, that is the next step; otherwise you are pretty well set. It is important not to overdo any particular endeavor. That is, it is better to do too few things at once than too many. In the short run this may not seem important, but complications from doing too many can easily arise. A mistake can be expensive as well. The manipulation of the appropriate mechanisms should be self-explanatory, and we need not dwell on it here. At first the whole procedure will seem complicated. Soon, however, it will become just another facet of life. It is difficult to foresee any end to the necessity for this task in the immediate future, but then one never can tell.

You probably find the paragraph a little confusing, and most of you probably think it is describing some obscure statistical procedure. Actually, the paragraph describes the everyday task of doing laundry. Now that you know the topic (or context) of the paragraph, try reading it again—it should make sense now.

Why did we begin a statistics textbook with a paragraph about washing clothes? Our goal is to demonstrate the importance of context—when not in the proper context, even the simplest material can appear difficult and confusing. In the Bransford and Johnson (1972) experiment, people who knew the topic before reading the paragraph were able to recall 73% more than people who did not know that it was about laundry. When you are given the appropriate background, it is much easier to fit new material into your memory and recall it later. In this book each chapter begins with a preview that provides the background context for the new material in the chapter. As you read each preview section, you should gain a general overview of the chapter content. Similarly, we begin each section within each chapter with clearly stated learning objectives that prepare you for the material in that section. Finally, we introduce each new statistical procedure by explaining its purpose. Note that all statistical methods were developed to serve a purpose. If you understand why a new procedure is needed, you will find it much easier to learn and remember the procedure.

The objectives for this first chapter are to provide an introduction to the topic of statistics and to give you some background for the rest of the book. We will discuss the role of statistics in scientific inquiry, and we will introduce some of the vocabulary and notation that are necessary for the statistical methods that follow. In some respects, this chapter serves as a preview section for the rest of the book.

As you read through the following chapters, keep in mind that the general topic of statistics follows a well-organized, logically developed progression that leads from basic concepts and definitions to increasingly sophisticated techniques. Thus, the material presented in the early chapters of this book will serve as a foundation for the material that follows, even if those early chapters seem basic. The content of the first seven chapters provides an essential background and context for the statistical methods presented in Chapter 8. If you turn directly to Chapter 8 without reading the first seven chapters, you will find the material incomprehensible. However, if you learn the background material and practice the statistics procedures and methods described in early chapters, you will have a good frame of reference for understanding and incorporating new concepts as they are presented in each new chapter.

Finally, we cannot promise that learning statistics will be as easy as washing clothes. But if you begin each new topic with the proper context, you should eliminate some unnecessary confusion.

1-1 Statistics and Behavioral Sciences

LEARNING OBJECTIVES

1. Define the terms population, sample, parameter, and statistic, and describe the relationships between them; identify examples of each.

2. Define the two general categories of statistics, descriptive and inferential statistics, and describe how they are used to summarize and make decisions about data.

3. Describe the concept of sampling error and explain how sampling error creates the fundamental problem that inferential statistics must address.

■ Definitions of Statistics

By one definition, *statistics* consist of facts and figures such as the average annual snowfall in Buffalo or the average yearly income of recent college graduates. These statistics are usually informative and time-saving because they condense large quantities of information into a few simple figures. Later in this chapter we return to the notion of calculating statistics (facts and figures) but, for now, we concentrate on a much broader definition of statistics. Specifically, we use the term statistics to refer to a general field of mathematics. In this case, we are using the term *statistics* as a shortened version of *statistical methods* or *statistical procedures*. For example, you are probably using this book for a statistics course in which you will learn about the statistical procedures that are used to summarize and evaluate research results in the behavioral sciences.

Research in the behavioral sciences (and other fields) involves gathering information. To determine, for example, whether college students learn better by reading material on printed pages or on a computer screen, you would need to gather information about students' study habits and their academic performance. When researchers finish the task of gathering information, they typically find themselves with pages and pages of measurements such as preferences, personality scores, opinions, and so on. In this book, we present the statistics that researchers use to analyze and interpret the information that they gather. Specifically, statistics serve two general purposes:

1. Statistics are used to organize and summarize the information so that the researcher can see what happened in the study and can communicate the results to others.

2. Statistics help the researcher to answer the questions that initiated the research by determining exactly what general conclusions are justified based on the specific results that were obtained.

> The term **statistics** refers to a set of mathematical procedures for organizing, summarizing, and interpreting information.

Statistical procedures help ensure that the information or observations are presented and interpreted in an accurate and informative way. In somewhat grandiose terms, statistics help researchers bring order out of chaos. In addition, statistics provide researchers with a set of standardized techniques that are recognized and understood throughout the scientific community. Thus, the statistical methods used by one researcher will be familiar to other researchers, who can accurately interpret the statistical analysis with a full understanding of how it was done and what the results signify.

■ Populations and Samples

Research in the behavioral sciences typically begins with a general question about a specific group (or groups) of individuals. For example, a researcher may want to know what factors are associated with academic dishonesty among college students. Or a researcher may want to determine the effect of lead exposure on the development of emotional problems in school-age children. In the first example, the researcher is interested in the group of college students. In the second example the researcher is studying school-age children. In statistical terminology, a *population* consists of all possible members of the group a researcher wishes to study.

> A **population** is the set of all the individuals of interest in a particular study.

As you can well imagine, a population can be quite large—for example, the entire set of all registered voters in the United States. A researcher might be more specific, limiting the study's population to people in their twenties who are registered voters in the United States. A smaller population would be first-time voter registrants in Burlington, Vermont. Populations can be extremely small too, such as those for people with a rare disease or members of an endangered species. The Siberian tiger, for example, has a population of roughly only 500 animals.

Thus, populations can obviously vary in size from extremely large to very small, depending on how the investigator identifies the population to be studied. The researcher should always specify the population being studied. In addition, the population need not consist of people—it could be a population of laboratory rats, North American corporations, engine parts produced in an automobile factory, or anything else an investigator wants to study. In practice, however, populations are typically very large, such as the population of college sophomores in the United States or the population of coffee drinkers that patronize a major national chain of cafés.

Because populations tend to be very large, it usually is impossible for a researcher to examine every individual in the population of interest. Therefore, researchers typically select a smaller, more manageable group from the population and limit their studies to the individuals in the selected group. In statistical terms, a set of individuals selected from a population is called a *sample*. A sample is intended to be representative of its population, and a sample should always be identified in terms of the population from which it was selected. We shall see later that one way to ensure that a sample is representative of a population is to select a *random sample*. In random sampling every individual has the same chance of being selected from the population.

> A **sample** is a set of individuals selected from a population, usually intended to represent the population in a research study. In a **random sample** everyone in the population has an equal chance of being selected.

Just as we saw with populations, samples can vary in size. For example, one study might examine a sample of only 20 middle-school students in an experimental reading program, and another study might use a sample of more than 2,000 people who take a new cholesterol medication.

So far, we have talked about a sample being selected from a population. However, this is actually only half of the full relationship between a sample and its population. Specifically, when a researcher finishes examining the sample, the goal is to generalize the results

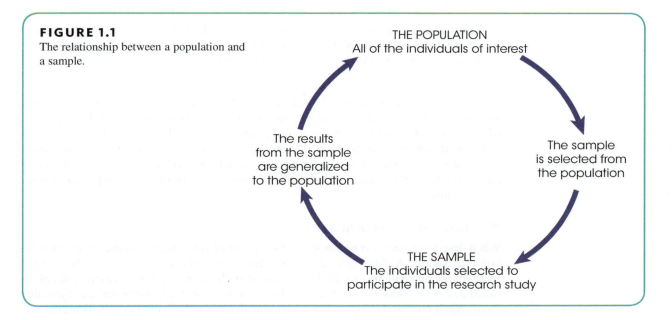

FIGURE 1.1
The relationship between a population and a sample.

back to the entire population. Remember that the researcher started with a general question about the population. To answer the question, a researcher studies a sample and then generalizes the results from the sample to the population. The full relationship between a sample and a population is shown in Figure 1.1.

■ Variables and Data

Typically, researchers are interested in specific characteristics of the individuals in the population (or in the sample), or they are interested in outside factors that may influence behavior of the individuals. For example, Bakhshi, Kanuparthy, and Gilbert (2014) wanted to determine if the weather is related to online ratings of restaurants. As the weather changes, do people's reviews of restaurants change too? Something that can change or have different values is called a *variable*.

> A **variable** is a characteristic or condition that changes or has different values for different individuals.

In the case of the previous example, both weather and people's reviews of restaurants are variables. By the way, in case you are wondering, the authors did find a relationship between weather and online reviews of restaurants. Reviews were worse during bad weather (for example, during extremely hot or cold days).

Once again, variables can be characteristics that differ from one individual to another, such as weight, gender identity, personality, or motivation and behavior. Also, variables can be environmental conditions that change, such as temperature, time of day, or the size of the room in which the research is being conducted.

To demonstrate changes in variables, it is necessary to make measurements of the variables being examined. The measurement obtained for each individual is called a *datum*, or more commonly, a *score* or *raw score*. The complete set of scores is called the *data set* or simply the *data*.

> **Data** (plural) are measurements or observations. A **data set** is a collection of measurements or observations. A **datum** (singular) is a single measurement or observation and is commonly called a **score** or **raw score**.

Before we move on, we should make one more point about samples, populations, and data. Earlier, we defined populations and samples in terms of *individuals*. For example, we previously discussed a population of registered voters and a sample of middle-school children. Be forewarned, however, that we will also refer to populations or samples of *scores*. Research typically involves measuring each individual to obtain a score, therefore every sample (or population) of individuals produces a corresponding sample (or population) of scores.

■ Parameters and Statistics

When describing data it is necessary to distinguish whether the data come from a population or a sample. A characteristic that describes a population—for example, the average score for the population—is called a *parameter*. A characteristic that describes a sample is called a *statistic*. Thus, the average score for a sample is an example of a statistic. Typically, the research process begins with a question about a population parameter. However, the actual data come from a sample and are used to compute sample statistics.

> A **parameter** is a value, usually a numerical value, that describes a population. A parameter is usually derived from measurements of the individuals in the population.
>
> A **statistic** is a value, usually a numerical value, that describes a sample. A statistic is usually derived from measurements of the individuals in the sample.

Every population parameter has a corresponding sample statistic, and most research studies involve using statistics from samples as the basis for answering questions about population parameters. As a result, much of this book is concerned with the relationship between sample statistics and the corresponding population parameters. In Chapter 7, for example, we examine the relationship between the mean obtained for a sample and the mean for the population from which the sample was obtained.

■ Descriptive and Inferential Statistical Methods

Although researchers have developed a variety of different statistical procedures to organize and interpret data, these different procedures can be classified into two general categories. The first category, *descriptive statistics*, consists of statistical procedures that are used to simplify and summarize data.

> **Descriptive statistics** are statistical procedures used to summarize, organize, and simplify data.

Descriptive statistics are techniques that take raw scores and organize or summarize them in a form that is more manageable. Often the scores are organized in a table or graph so that it is possible to see the entire set of scores. Another common technique is to summarize a set of scores by computing an average. Note that even if the data set has hundreds of scores, the average provides a single descriptive value for the entire set.

The second general category of statistical techniques is called *inferential statistics*. Inferential statistics are methods that use sample data to make general statements about a population.

> **Inferential statistics** consist of techniques that allow us to study samples and then make generalizations about the populations from which they were selected.

Because populations are typically very large, it usually is not possible to measure everyone in the population. Therefore, researchers select a sample that represents the population. By analyzing the data from the sample, we hope to make general statements about the population. Typically, researchers use sample statistics as the basis for drawing conclusions about population parameters or relationships between variables that might exist in the population. One problem with using samples, however, is that a sample provides only limited information about the population. Although samples are generally *representative* of their populations, a sample is not expected to give a perfectly accurate picture of the whole population. There usually is some discrepancy between a sample statistic and the corresponding population parameter. This discrepancy is called *sampling error*, and it creates the fundamental problem inferential statistics must always address.

> **Sampling error** is the naturally occurring discrepancy, or error, that exists between a sample statistic and the corresponding population parameter.

The concept of sampling error is illustrated in Figure 1.2. The figure shows a population of 1,000 college students and two samples, each with five students who were selected from the population. Notice that each sample contains different individuals who have different characteristics. Because the characteristics of each sample depend on the specific people in the sample, statistics will vary from one sample to another. For example, the five students in sample 1 have an average age of 19.8 years and the students in sample 2 have an average age of 20.4 years. It is unlikely that the statistics for a sample will be identical to the parameter for the entire population. Both of the statistics in the example vary slightly from the population parameter (21.3 years) from which the samples were drawn. The difference between these sample statistics and the population parameter illustrate sampling error.

You should also realize that Figure 1.2 shows only two of the hundreds of possible samples. Each sample would contain different individuals and would produce different statistics. This is the basic concept of sampling error: sample statistics vary from one sample to another and typically are different from the corresponding population parameters.

One common example of sampling error is the error associated with a sample proportion (or percentage). For instance, in newspaper articles reporting results from political polls, you frequently find statements such as this:

Candidate Brown leads the poll with 51% of the vote. Candidate Jones has 42% approval, and the remaining 7% are undecided. This poll was taken from a sample of registered voters and has a margin of error of plus or minus 4 percentage points.

The "margin of error" is the sampling error. In this case, the reported percentages were obtained from a sample and are being generalized to the whole population of potential voters.

FIGURE 1.2

A demonstration of sampling error. Two samples are selected from the same population. Notice that the sample statistics are different from one sample to another, and all of the sample statistics are different from the corresponding population parameters. The natural differences that exist, by chance, between a sample statistic and a population parameter are called sampling error.

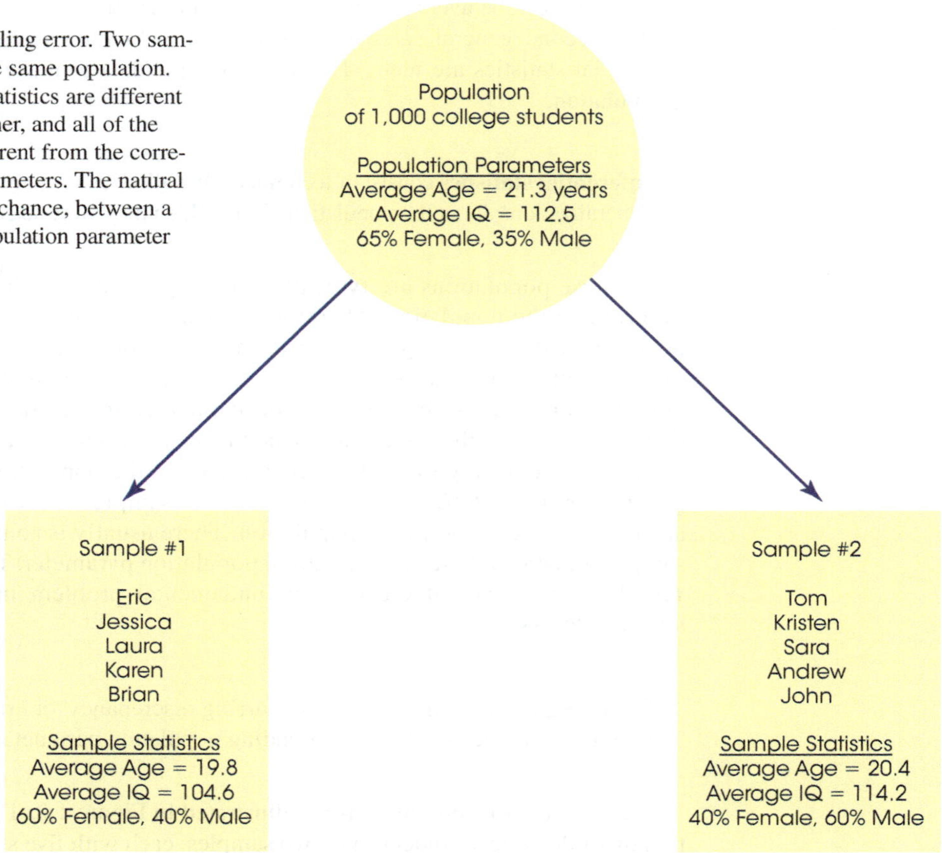

Population
of 1,000 college students

Population Parameters
Average Age = 21.3 years
Average IQ = 112.5
65% Female, 35% Male

Sample #1

Eric
Jessica
Laura
Karen
Brian

Sample Statistics
Average Age = 19.8
Average IQ = 104.6
60% Female, 40% Male

Sample #2

Tom
Kristen
Sara
Andrew
John

Sample Statistics
Average Age = 20.4
Average IQ = 114.2
40% Female, 60% Male

As always, you do not expect the statistics from a sample to be a perfect reflection of the population. There always will be some "margin of error" when sample statistics are used to represent population parameters.

As a further demonstration of sampling error, imagine that your statistics class is separated into two groups by drawing a line from front to back through the middle of the room. Now imagine that you compute the average age (or height, or GPA) for each group. Will the two groups have exactly the same average? Almost certainly they will not. No matter what you choose to measure, you will probably find some difference between the two groups. However, the difference you obtain does not necessarily mean that there is a systematic difference between the two groups. For example, if the average age for students on the right-hand side of the room is higher than the average for students on the left, it is unlikely that some mysterious force has caused the older people to gravitate to the right side of the room. Instead, the difference is probably the result of random factors such as chance. The unpredictable, unsystematic differences that exist from one sample to another are an example of sampling error. Inferential statistics tell us whether the differences between samples (e.g., a difference in age, height, or GPA) are the result of random factors (sampling error) or the result of some meaningful relationship in the population.

■ Statistics in the Context of Research

The following example shows the general stages of a research study and demonstrates how descriptive statistics and inferential statistics are used to organize and interpret the data. At the end of the example, note how sampling error can affect the interpretation of experimental results, and consider why inferential statistical methods are needed to deal with this problem.

EXAMPLE 1.1   Figure 1.3 shows an overview of a general research situation and demonstrates the roles that descriptive and inferential statistics play. The purpose of the research study is to address a question that we posed earlier: do college students learn better by studying text on printed pages or on a computer screen? Two samples of six students each are selected from the population of college students. The students in sample A read text on a computer

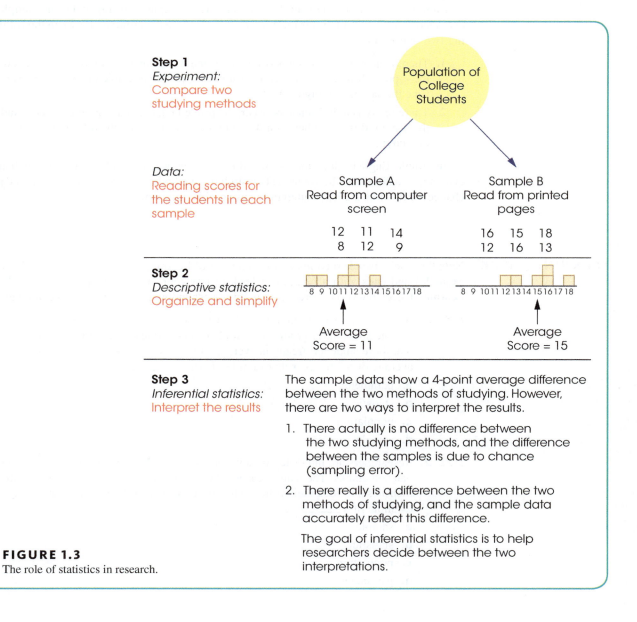

Step 1
Experiment:
Compare two studying methods

Population of College Students

Data:
Reading scores for the students in each sample

Sample A
Read from computer screen

Sample B
Read from printed pages

| 12 | 11 | 14 | | 16 | 15 | 18 |
| 8 | 12 | 9 | | 12 | 16 | 13 |

Step 2
Descriptive statistics:
Organize and simplify

8 9 10 11 12 13 14 15 16 17 18

8 9 10 11 12 13 14 15 16 17 18

Average
Score = 11

Average
Score = 15

Step 3
Inferential statistics:
Interpret the results

The sample data show a 4-point average difference between the two methods of studying. However, there are two ways to interpret the results.

1. There actually is no difference between the two studying methods, and the difference between the samples is due to chance (sampling error).

2. There really is a difference between the two methods of studying, and the sample data accurately reflect this difference.

The goal of inferential statistics is to help researchers decide between the two interpretations.

FIGURE 1.3
The role of statistics in research.

screen to study for 30 minutes and the students in sample B are given printed pages. Next, all of the students are given a multiple-choice test to evaluate their knowledge of the material. At this point, the researcher has two groups of data: the scores for sample A and the scores for sample B (see the figure). Now is the time to begin using statistics.

First, descriptive statistics are used to simplify the pages of data. For example, the researcher could draw a graph showing the scores for each sample or compute the average score for each group. Note that descriptive methods provide a simplified, organized description of the scores. In this example, the students who studied text on the computer screen averaged 11 on the test, and the students who studied printed pages had an average score of 15. These descriptive statistics efficiently summarize—with only two values—the two samples containing six scores each.

Once the researcher has described the results, the next step is to interpret the outcome. This is the role of inferential statistics. In this example, the researcher has found a difference of 4 points between the two samples (sample A averaged 11 and sample B averaged 15). The problem for inferential statistics is to differentiate between the following two interpretations:

1. There is no real difference between the printed page and a computer screen, and the 4-point difference between the samples is just an example of sampling error (like the samples in Figure 1.2).

2. There really is a difference between the printed page and a computer screen, and the 4-point difference between the samples was caused by the different methods of studying.

In simple English, does the 4-point difference between samples provide convincing evidence of a difference between the two studying methods, or is the 4-point difference just chance? Inferential statistics attempt to answer this question. ■

LEARNING CHECK   Note that each chapter section begins with a list of Learning Objectives (see page 3 for an example) and ends with a Learning Check to test your mastery of the objectives. Each Learning Check question is preceded by its corresponding Learning Objective number.

LO1 **1.** A researcher is interested in the Netflix binge-watching habits of American college students. A group of 50 students is interviewed and the researcher finds that these students stream an average of 6.7 hours per week. For this study, the average of 6.7 hours is an example of a(n) _____.

    a. parameter

    b. statistic

    c. population

    d. sample

LO2 **2.** Researchers are interested in how robins in New York State care for their newly hatched chicks. The team measures how many times per day the adults visit their nests to feed their young. The entire group of robins in the state is an example of a _____.

    a. sample

    b. statistic

    c. population

    d. parameter

LO2 3. Statistical techniques that use sample data to draw conclusions about the population are _____.

    a. population statistics

    b. sample statistics

    c. descriptive statistics

    d. inferential statistics

LO3 4. The SAT is standardized so that the population average score on the verbal test is 500 each year. In a sample of 100 graduating seniors who have taken the verbal SAT, what value would you expect to obtain for their average verbal SAT score?

    a. 500

    b. Greater than 500

    c. Less than 500

    d. Around 500 but probably not equal to 500

ANSWERS   **1. b   2. c   3. d   4. d**

1-2   Observations, Measurement, and Variables

LEARNING OBJECTIVES

4. Explain why operational definitions are developed for constructs and identify the two components of an operational definition.

5. Describe discrete and continuous variables and identify examples of each.

6. Define real limits and explain why they are needed to measure continuous variables.

7. Compare and contrast the four scales of measurement (nominal, ordinal, interval, and ratio) and identify examples of each.

■ Observations and Measurements

Science is *empirical*. This means it is based on observation rather than intuition or conjecture. Whenever we make a precise observation we are taking a measurement, either by assigning a numerical value to observations or by classifying them into categories. Observation and measurement are part and parcel of the scientific method. In this section, we take a closer look at the variables that are being measured and the process of measurement.

■ Constructs and Operational Definitions

The scores that make up the data from a research study are the result of observing and measuring variables. For example, a researcher may obtain a set of memory recall scores, personality scores, or reaction-time scores when conducting a study. Some variables, such as height, weight, and eye color are well-defined, concrete entities that can be observed and measured directly. On the other hand, many variables studied by behavioral scientists are internal characteristics that people use to help describe and explain behavior. For example, we say that a student does well in school because the student has strong *motivation* for achievement. Or we say that someone is *anxious* in social situations, or that someone seems to be *hungry*. Variables like motivation, anxiety, and hunger are called *constructs*, and because they are intangible and cannot be directly observed, they are often called *hypothetical constructs*.

Although constructs such as intelligence are internal characteristics that cannot be directly observed, it is possible to observe and measure behaviors that are representative of the construct. For example, we cannot "see" high self-esteem but we can see examples of behavior reflective of a person with high self-esteem. The external behaviors can then be used to create an operational definition for the construct. An *operational definition* defines a construct in terms of external behaviors that can be observed and measured. For example, your self-esteem is measured and operationally defined by your score on the Rosenberg Self-Esteem Scale, or hunger can be measured and defined by the number of hours since last eating.

> **Constructs** are internal attributes or characteristics that cannot be directly observed but are useful for describing and explaining behavior.
>
> An **operational definition** identifies a measurement procedure (a set of operations) for measuring an external behavior and uses the resulting measurements as a definition and a measurement of a hypothetical construct. Note that an operational definition has two components. First, it describes a set of operations for measuring a construct. Second, it defines the construct in terms of the resulting measurements.

■ Discrete and Continuous Variables

The variables in a study can be characterized by the type of values that can be assigned to them and, as we will discuss in later chapters, the type of values influences the statistical procedures that can be used to summarize or make inferences about those values. A *discrete variable* consists of separate, indivisible categories. For this type of variable, there are no intermediate values between two adjacent categories. Consider the number of questions that each student answers correctly on a 10-item multiple-choice quiz. Between neighboring values—for example, seven correct and eight correct—no other values can ever be observed.

> A **discrete variable** consists of separate, indivisible categories. No values can exist between two neighboring categories.

Discrete variables are commonly restricted to whole, countable numbers (i.e., integers)—for example, the number of children in a family or the number of students attending class. If you observe class attendance from day to day, you may count 18 students one day and 19 students the next day. However, it is impossible ever to observe a value between 18 and 19. A discrete variable may also consist of observations that differ qualitatively. For example, people can be classified by birth order (first-born or later-born), by occupation (nurse, teacher, lawyer, etc.), and college students can be classified by academic major (art, biology, chemistry, etc.). In each case, the variable is discrete because it consists of separate, indivisible categories.

On the other hand, many variables are not discrete. Variables such as time, height, and weight are not limited to a fixed set of separate, indivisible categories. You can measure time, for example, in hours, minutes, seconds, or fractions of seconds. These variables are called *continuous* because they can be divided into an infinite number of fractional parts.

FIGURE 1.4
When measuring weight to the nearest whole pound, 149.6 and 150.3 are assigned the value of 150 (top). Any value in the interval between 149.5 and 150.5 is given the value of 150.

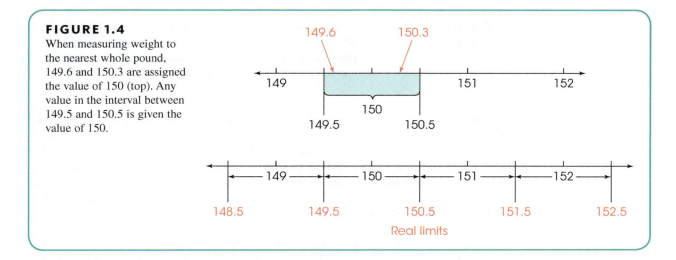

For a **continuous variable**, there are an infinite number of possible values that fall between any two observed values. A continuous variable is divisible into an infinite number of fractional parts.

Suppose, for example, that a researcher is measuring weights for a group of individuals participating in a diet study. Because weight is a continuous variable, it can be pictured as a continuous line (Figure 1.4). Note that there are an infinite number of possible points on the line without any gaps or separations between neighboring points. For any two different points on the line, it is always possible to find a third value that is between the two points.

Two other factors apply to continuous variables:

1. When measuring a continuous variable, it should be very rare to obtain identical measurements for two different individuals. Because a continuous variable has an infinite number of possible values, it should be almost impossible for two people to have exactly the same score. If the data show a substantial number of tied scores, then you should suspect either the variable is not really continuous or that the measurement procedure is very crude—meaning the continuous variable is divided into widely separated discrete numbers.

2. When measuring a continuous variable, researchers must first identify a series of measurement categories on the scale of measurement. Measuring weight to the nearest pound, for example, would produce categories of 149 pounds, 150 pounds, and so on. However, each measurement category is actually an *interval* that must be defined by boundaries. To differentiate a weight of 150 pounds from the surrounding values of 149 and 151, we must set up boundaries on the scale of measurement. These boundaries are called *real limits* and are positioned exactly halfway between adjacent scores. Thus, a score of 150 pounds is actually an interval bounded by a *lower real limit* of 149.5 at the bottom and an *upper real limit* of 150.5 at the top. Any individual whose weight falls between these real limits will be assigned a score of 150. As a result, two people who both claim to weigh 150 pounds are probably not *exactly* the same weight. One person may actually weigh 149.6 and the other 150.3, but they are both assigned a weight of 150 pounds (see Figure 1.4).

Students often ask whether a measurement of exactly 150.5 should be assigned a value of 150 or a value of 151. The answer is that 150.5 is the *boundary* between the two intervals and is not necessarily in one or the other. Instead, the placement of 150.5 depends on the rule that you are using for rounding numbers. If you are rounding up, then 150.5 goes in the higher interval (151) but if you are rounding down, then it goes in the lower interval (150).

> **Real limits** are the boundaries of intervals for scores that are represented on a continuous number line. The real limit separating two adjacent scores is located exactly halfway between the scores. Each score has two real limits. The **upper real limit** is at the top of the interval, and the **lower real limit** is at the bottom.

The concept of real limits applies to any measurement of a continuous variable, even when the score categories are not whole numbers. For example, if you were measuring time to the nearest tenth of a second, the measurement categories would be 31.0, 31.1, 31.2, and so on. Each of these categories represents an interval on the scale that is bounded by real limits. For example, a score of $X = 31.1$ seconds indicates that the actual measurement is in an interval bounded by a lower real limit of 31.05 and an upper real limit of 31.15. Remember that the real limits are always halfway between adjacent categories.

Later in this book, real limits are used for constructing graphs and for various calculations with continuous scales. For now, however, you should realize that real limits are a necessity whenever you make measurements of a continuous variable.

Finally, we should warn you that the terms *continuous* and *discrete* apply to the variables that are being measured and not to the scores that are obtained from the measurement. For example, measuring people's heights to the nearest inch produces scores of 60, 61, 62, and so on. Although the scores may appear to be discrete numbers, the underlying variable is continuous. One key to determining whether a variable is continuous or discrete is that a continuous variable can be divided into any number of fractional parts. Height can be measured to the nearest inch, the nearest 0.5 inch, or the nearest 0.1 inch. Similarly, a professor evaluating students' knowledge could use a pass/fail system that classifies students into two broad categories. However, the professor could choose to use a 10-point quiz that divides student knowledge into 11 categories corresponding to quiz scores from 0 to 10. Or the professor could use a 100-point exam that potentially divides student knowledge into 101 categories from 0 to 100. Whenever you are free to choose the degree of precision or the number of categories for measuring a variable, the variable must be continuous.

■ Scales of Measurement

It should be obvious by now that data collection requires that we make measurements of our observations. Measurement involves assigning individuals or events to categories. The categories can simply be names such as introvert/extrovert or employed/unemployed, or they can be numerical values such as 68 inches or 175 pounds. The categories used to measure a variable make up a *scale of measurement*, and the relationships between the categories determine different types of scales. The distinctions among the scales are important because they identify the limitations of certain types of measurements and because certain statistical procedures are appropriate for scores that have been measured on some scales but not on others. If you were interested in people's heights, for example, you could measure a group of individuals by simply classifying them into three categories: tall, medium, and short. However, this simple classification would not tell you much about the actual heights of the individuals, and these measurements would not give you enough information to calculate an average height for the group. Although the simple classification would be adequate for some purposes, you would need more sophisticated measurements before you could answer more detailed questions. In this section, we examine four different scales of measurement, beginning with the simplest and moving to the most sophisticated.

The Nominal Scale   The word *nominal* means "having to do with names." Measurement on a nominal scale involves classifying individuals into categories that have different names but are not quantitative or numerically related to each other. For example, if you were measuring the academic majors for a group of college students, the categories would be art, biology, business, chemistry, and so on. Each student would be classified in one category according to his or her major. The measurements from a nominal scale allow us to determine whether two individuals are different, but they do not identify either the direction or the size of the difference. If one student is an art major and another is a biology major we can say that they are different, but we cannot say that art is "more than" or "less than" biology and we cannot specify how much difference there is between art and biology. Other examples of nominal scales include classifying people by race, gender, or occupation.

A **nominal scale** consists of a set of categories that have different names. Measurements on a nominal scale label and categorize observations, but do not make any quantitative distinctions between observations.

Although the categories on a nominal scale are not quantitative values, they are occasionally represented by numbers. For example, the rooms or offices in a building may be identified by numbers. You should realize that the room numbers are simply names and do not reflect any quantitative information. Room 109 is not necessarily bigger than Room 100 and certainly not 9 points bigger. It also is fairly common to use numerical values as a code for nominal categories when data are entered into computer programs for analysis. For example, the data from a political opinion poll may code Democrats with a 0 and Republicans with a 1 as a group identifier. Again, the numerical values are simply names and do not represent any quantitative difference. The scales that follow do reflect an attempt to make quantitative distinctions.

The Ordinal Scale   The categories that make up an *ordinal scale* not only have different names (as in a nominal scale) but also are organized in a fixed order corresponding to differences of magnitude.

An **ordinal scale** consists of a set of categories that are organized in an ordered sequence. Measurements on an ordinal scale rank observations in terms of size or magnitude.

Often, an ordinal scale consists of a series of ranks (first, second, third, and so on) like the order of finish in a horse race. Occasionally, the categories are identified by verbal labels (like small, medium, and large drink sizes at a fast-food restaurant). In either case, the fact that the categories form an ordered sequence means that there is a directional relationship between categories. With measurements from an ordinal scale you can determine whether two individuals are different, and you can determine the direction of difference. However, ordinal measurements do not allow you to determine the size of the difference between two individuals. For example, suppose in the Winter Olympics you watch the medal ceremony for the women's downhill ski event. You know that the athlete receiving the gold medal had the fastest time, the silver medalist had the second fastest time, and the bronze medalist had the third fastest time. This represents an ordinal scale of measurement and reflects no more information than first, second, and third place. Note that it does not provide information about how much time difference there was between competitors. The first-place skier might have won the event by a mere one one-hundredth of a second—or perhaps by as much as one second. Other examples of ordinal scales include socioeconomic class (upper, middle, lower) and T-shirt sizes (small, medium, large). In addition, ordinal scales are often used to measure variables for which it is difficult to assign numerical scores. For example, people can rank their food preferences but might have trouble explaining "how much more" they prefer chocolate ice cream to cheesecake.

The Interval and Ratio Scales   Both an *interval scale* and a *ratio scale* consist of a series of ordered categories (like an ordinal scale) with the additional requirement that the categories form a series of intervals that are all exactly the same size. Thus, the scale of measurement consists of a series of equal intervals, such as inches on a ruler. Examples of interval scales are the temperature in degrees Fahrenheit or Celsius and examples of ratio scales are the measurement of time in seconds or weight in pounds. Note that, in each case, the difference between two adjacent values (1 inch, 1 second, 1 pound, 1 degree) is the same size, no matter where it is located on the scale. The fact that the differences between adjacent values are all the same size makes it possible to determine both the size and the direction of the difference between two measurements. For example, you know that a measurement of 80° Fahrenheit is higher than a measure of 60°, and you know that it is exactly 20° higher.

The factor that differentiates an interval scale from a ratio scale is the nature of the zero point. An interval scale has an arbitrary zero point. That is, the value 0 is assigned to a particular location on the scale simply as a matter of convenience or reference. In particular, a value of zero does not indicate a total absence of the variable being measured. The two most common examples are the Fahrenheit and Celsius temperature scales. For example, a temperature of 0° Fahrenheit does not mean that there is no temperature, and it does not prohibit the temperature from going even lower. Interval scales with an arbitrary zero point are not common in the physical sciences or with physical measurements.

A ratio scale is anchored by a zero point that is not arbitrary but rather is a meaningful value representing none (a complete absence) of the variable being measured. The existence of an absolute, non-arbitrary zero point means that we can measure the absolute amount of the variable; that is, we can measure the distance from 0. This makes it possible to compare measurements in terms of ratios. For example, a fuel tank with 10 gallons of gasoline has twice as much gasoline as a tank with only 5 gallons because there is a true absolute zero value. A completely empty tank has 0 gallons of fuel. Ratio scales are used in the behavioral sciences, too. A reaction time of 500 milliseconds is exactly twice as long as a reaction time of 250 milliseconds and a value of 0 milliseconds is a true absolute zero. To recap, with a ratio scale, we can measure the direction and the size of the difference between two measurements and we can describe the difference in terms of a ratio. Ratio scales are common and include physical measurements such as height and weight, as well as measurements of variables such as reaction time or the number of errors on a test. The distinction between an interval scale and a ratio scale is demonstrated in Example 1.2 and in Table 1.1.

An **interval scale** consists of ordered categories that are all intervals of exactly the same size. Equal differences between numbers on a scale reflect equal differences in magnitude. However, the zero point on an interval scale is arbitrary and does not indicate a zero amount of the variable being measured.

A **ratio scale** is an interval scale with the additional feature of an absolute zero point. With a ratio scale, ratios of numbers do reflect ratios of magnitude.

TABLE 1.1

Scales of Measurement for a Marathon

Scale	Information	Example
Nominal	Category only	Country of athlete (U.S., U.K., Ethiopia, Japan, Kenya, etc.)
Ordinal	*Ordered* category	Finishing position in a race (1st, 2nd, 3rd, etc.)
Interval	Ordered category *with equal intervals separating adjacent scores and arbitrary (not absolute) zero*	Time difference (above or below) from the course record, an arbitrary zero point (Example: a person who finishes the Boston Marathon 4 minutes slower than the course record takes 3 minutes longer to finish the race than a person who was 1 minute slower than the course record, but does not take four times longer.)
Ratio	Ordered category with equal amounts separating adjacent scores, *and a true absolute zero*	Amount of time to complete a marathon (Example: a person who finishes the Boston Marathon in 4 hours, 30 minutes takes 2 times longer than one who finishes in 2 hours, 15 minutes.)

EXAMPLE 1.2

A researcher obtains measurements of height for a group of 8-year-old boys. Initially, the researcher simply records each child's height in inches, obtaining values such as 44, 51, 49, and so on. These initial measurements constitute a ratio scale. A value of zero represents no height (absolute zero). Also, it is possible to use these measurements to form ratios. For example, a child who is 60 inches tall is one-and-a-half times taller than a child who is 40 inches tall.

Now suppose that the researcher converts the initial measurement into a new scale by calculating the difference between each child's actual height and the average height for this age group. A child who is 1 inch taller than average now gets a score of $+1$; a child 4 inches taller than average gets a score of $+4$. Similarly, a child who is 2 inches shorter than average gets a score of -2. On this scale, a score of zero corresponds to average height. Because zero no longer indicates a complete absence of height, the new scores constitute an interval scale of measurement.

Notice that original scores and the converted scores both involve measurement in inches, and you can compute differences, or distances, on either scale. For example, there is a 6-inch difference in height between two boys who measure 57 and 51 inches tall on the first scale. Likewise, there is a 6-inch difference between two boys who measure $+9$ and $+3$ on the second scale. However, you should also notice that ratio comparisons are not possible on the second scale. For example, a boy who measures $+9$ is not three times taller than a boy who measures $+3$. ■

Statistics and Scales of Measurement   For our purposes, scales of measurement are important because they help determine the statistics that are used to evaluate the data. Specifically, there are certain statistical procedures that are used with numerical scores from interval or ratio scales and other statistical procedures that are used with non-numerical scores from nominal or ordinal scales. The distinction is based on the fact that numerical scores are compatible with basic arithmetic operations (adding, multiplying, and so on) but non-numerical scores are not. For example, in a memory experiment a researcher might record how many words participants can recall from a list they previously studied. It is possible to add the recall scores together to find a total and then calculate the average score for the group. On the other hand, if you measure the academic major for each student, you cannot add the scores to obtain a total. (What is the total for three psychology majors plus an English major plus two chemistry majors?) The vast majority of the statistical techniques presented in this book are designed for numerical scores from interval or ratio scales. For most statistical applications, the distinction between an interval scale and a ratio scale is not important because both scales produce numerical values that permit us to compute differences between scores, add scores, and calculate mean scores. On the other hand, measurements from nominal or ordinal scales are typically not numerical values, do not measure distance, and are not compatible with many basic arithmetic operations. Therefore, alternative statistical techniques are necessary for data from nominal or ordinal scales of measurement (for example, the median and the mode in Chapter 3, the Spearman correlation in Chapter 14, and the chi-square tests in Chapter 15).

LEARNING CHECK   **LO4 1.** An operational definition is used to _____ a hypothetical construct.
   a. define
   b. measure
   c. measure and define
   d. None of the other choices is correct.

LO5 2. A researcher studies the factors that determine the length of time a consumer stays on a website before clicking off. The variable, length of time, is an example of a _____ variable.
   a. discrete
   b. continuous
   c. nominal
   d. ordinal

LO5 3. A researcher records the number of bites a goat takes of different plants. The variable, number of bites, is an example of a _____ variable.
   a. discrete
   b. continuous
   c. nominal
   d. ordinal

LO6 4. When measuring height to the nearest inch, what are the real limits for a score of 68.0 inches?
   a. 67 and 69
   b. 67.5 and 68.5
   c. 67.75 and 68.75
   d. 67.75 and 68.25

LO7 5. The professor in a communications class asks students to identify their favorite reality television show. The different television shows make up a _____ scale of measurement.
   a. nominal
   b. ordinal
   c. interval
   d. ratio

LO7 **6.** Ranking jobs, taking into account growth potential, work-life balance, and salary, would be an example of measurement on a(n) _____ scale.

    a. nominal

    b. ordinal

    c. interval

    d. ratio

ANSWERS   **1. c** **2. b** **3. a** **4. b** **5. a** **6. b**

1-3 Three Data Structures, Research Methods, and Statistics

LEARNING OBJECTIVES

8. Describe, compare, and contrast correlational, experimental, and nonexperimental research, and identify the data structures associated with each.

9. Define independent, dependent, and quasi-independent variables and recognize examples of each.

■ Data Structure 1. One Group with One or More Separate Variables Measured for Each Individual: Descriptive Research

Some research studies are conducted simply to describe individual variables as they exist naturally. For example, a college official may conduct a survey to describe the eating, sleeping, and study habits of a group of college students. Table 1.2 shows an example of data from this type of research. Although the researcher might measure several different variables, the goal of the study is to describe each variable separately. In particular, this type of research is not concerned with relationships between variables.

A study that produces the kind of data shown in Table 1.2 and is focused on describing individual variables rather than relationships is an example of *descriptive research* or the *descriptive research strategy*.

> **Descriptive research** or the **descriptive research strategy** involves measuring one or more separate variables for each individual with the intent of simply describing the individual variables.

TABLE 1.2

Three separate variables measured for each individual in a group of eight students.

	Weekly Number of Student Fast-Food Meals	Number of Hours Sleeping Each Day	Number of Hours Studying Each Day
A	0	9	3
B	4	7	2
C	2	8	4
D	1	10	3
E	0	11	2
F	0	7	4
G	5	7	3
H	3	8	2

When the results from a descriptive research study consist of numerical scores—such as the number of hours spent studying each day—they are typically described by the statistical techniques that are presented in Chapters 3 and 4. For example, a researcher may want to know the average number of meals eaten at fast-food restaurants each week for students at the college. Non-numerical scores are typically described by computing the proportion or percentage in each category. For example, a recent newspaper article reported that 34.9% of American adults are obese, which is roughly 35 pounds over a healthy weight.

■ Relationships Between Variables

Most research, however, is intended to examine relationships between two or more variables. For example, is there a relationship between the amount of violence in the video games played by children and the amount of aggressive behavior they display? Is there a relationship between vocabulary development in childhood and academic success in college? To establish the existence of a relationship, researchers must make observations— that is, measurements of the two variables. The resulting measurements can be classified into two distinct data structures that also help to classify different research methods and different statistical techniques. In the following section we identify and discuss these two data structures.

■ Data Structure 2. One Group with Two Variables Measured for Each Individual: The Correlational Method

One method for examining the relationship between variables is to observe the two variables as they exist naturally for a set of individuals. That is, simply measure the two variables for each individual. For example, research results tend to find a relationship between Facebook™ use and academic performance, especially for freshmen (Junco, 2015). Figure 1.5 shows an example of data obtained by measuring time on Facebook and academic performance for eight students. The researchers then look for consistent patterns in the data to provide evidence for a relationship between variables. For example, as Facebook time changes from one student to another, is there also a tendency for academic performance to change?

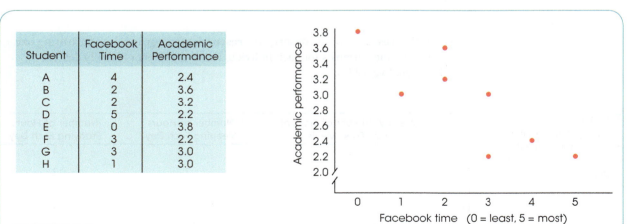

FIGURE 1.5

One of two data structures for studies evaluating the relationship between variables. Note that there are two separate measurements for each individual (Facebook time and academic performance). The same scores are shown in a table and a graph.

Consistent patterns in the data are often easier to see if the scores are presented in a graph. Figure 1.5 also shows the scores for the eight students in a graph called a scatter plot. In the scatter plot, each individual is represented by a point so that the horizontal position corresponds to the student's Facebook time and the vertical position corresponds to the student's academic performance score. The scatter plot shows a clear relationship between Facebook time and academic performance: as Facebook time increases, academic performance decreases.

A research study that simply measures two different variables for each individual and produces the kind of data shown in Figure 1.5 is an example of the *correlational method*, or the *correlational research strategy*.

> In the **correlational method**, two different variables are observed to determine whether there is a relationship between them.

Statistics for the Correlational Method   When the data from a correlational study consist of numerical scores, the relationship between the two variables is usually measured and described using a statistic called a correlation. Correlations and the correlational method are discussed in detail in Chapter 14. Occasionally, the measurement process used for a correlational study simply classifies individuals into categories that do not correspond to numerical values. For example, a researcher could classify study participants by age (40 years of age and over, or under 40 years) and by preference for smartphone use (talk or text). Note that the researcher has two scores for each individual (age category and phone use preference) but neither of the scores is a numerical value. These types of data are typically summarized in a table showing how many individuals are classified into each of the possible categories. Table 1.3 shows an example of this kind of summary table. The table shows, for example, that 15 of the people 40 and over in the sample preferred texting and 35 preferred talking. The pattern is quite different for younger participants—45 preferred texting and only 5 preferred talking. Note that by presenting the data in a table, one can see the difference in preference for age at a quick glance. The relationship between categorical variables (such as the data in Table 1.3) is usually evaluated using a statistical technique known as a *chi-square test*. Chi-square tests are presented in Chapter 15.

Limitations of the Correlational Method   The results from a correlational study can demonstrate the existence of a relationship between two variables, but they do not provide an explanation for the relationship. In particular, a correlational study cannot demonstrate a cause-and-effect relationship. For example, the data in Figure 1.5 show a systematic relationship between Facebook time and academic performance for a group of college students; those who spend more time on Facebook tend to have lower grades. However, there are many possible explanations for the relationship and we do not know exactly what factor (or factors) is responsible for Facebook users having lower grades. For example,

TABLE 1.3

Correlational data consisting of non-numerical scores. Note that there are two measurements for each individual: age and smartphone preference. The numbers indicate how many people fall into each category.

	Smartphone Preference		
	Text	Talk	
40 years and over	15	35	50
Under 40	45	5	50

many students report that they multitask with Facebook while they are studying. In this case, their lower grades might be explained by the distraction of multitasking while studying. Another possible explanation is that there is a third variable involved that produces the relationship. For example, perhaps level of interest in the course material accounts for the relationship. That is, students who have less interest in the course material might study it less and spend more time on interesting pursuits like Facebook. In particular, we cannot conclude that simply reducing time on Facebook would cause their academic performance to improve. To demonstrate a cause-and-effect relationship between two variables, researchers must use the experimental method, which is discussed next.

■ Data Structure 3. Comparing Two (or More) Groups of Scores: Experimental and Nonexperimental Methods

The second method for examining the relationship between two variables compares two or more groups of scores. In this situation, the relationship between variables is examined by using one of the variables to define the groups, and then measuring the second variable to obtain scores for each group. For example, Polman, de Castro, and van Aken (2008) randomly divided a sample of 10-year-old boys into two groups. One group then played a violent video game and the second played a nonviolent game. After the game-playing session, the children went to a free play period and were monitored for aggressive behaviors (hitting, kicking, pushing, frightening, name-calling, fighting, quarreling, or teasing another child). An example of the resulting data is shown in Figure 1.6. The researchers then compared the scores for the violent-video group with the scores for the nonviolent-video group. A systematic difference between the two groups provides evidence for a relationship between playing violent video games and aggressive behavior for 10-year-old boys.

Statistics for Comparing Two (or More) Groups of Scores   Most of the statistical procedures presented in this book are designed for research studies that compare groups of scores like the study in Figure 1.6. Specifically, we examine descriptive statistics that summarize and describe the scores in each group and we use inferential statistics to determine whether the differences between the groups can be generalized to the entire population.

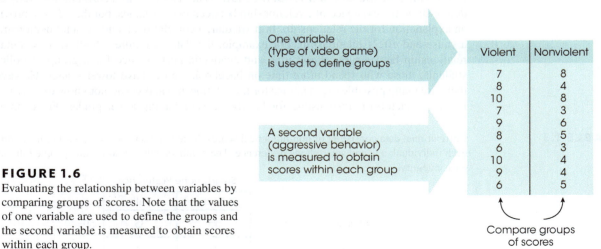

FIGURE 1.6
Evaluating the relationship between variables by comparing groups of scores. Note that the values of one variable are used to define the groups and the second variable is measured to obtain scores within each group.

When the measurement procedure produces numerical scores, the statistical evaluation typically involves computing the average score for each group and then comparing the averages. The process of computing averages is presented in Chapter 3, and a variety of statistical techniques for comparing averages are presented in Chapters 8–13. If the measurement process simply classifies individuals into non-numerical categories, the statistical evaluation usually consists of computing proportions for each group and then comparing proportions. Previously, in Table 1.3, we presented an example of non-numerical data examining the relationship between age and smartphone preference. The same data can be used to compare the proportions for participants age 40 and over with the proportions for those under 40 years of age. For example, 30% of people 40 and over prefer to text compared to 90% of those under 40. As before, these data are evaluated using a chi-square test, which is presented in Chapter 15.

■ Experimental and Nonexperimental Methods

There are two distinct research methods that both produce groups of scores to be compared: the experimental and the nonexperimental strategies. These two research methods use exactly the same statistics and they both demonstrate a relationship between two variables. The distinction between the two research strategies is how the relationship is interpreted. The results from an experiment allow a cause-and-effect explanation. For example, we can conclude that changes in one variable are responsible for causing differences in a second variable. A nonexperimental study does not permit a cause-and effect explanation. We can say that changes in one variable are accompanied by changes in a second variable, but we cannot say why. Each of the two research methods is discussed in the following sections.

■ The Experimental Method

One specific research method that involves comparing groups of scores is known as the *experimental method* or the *experimental research strategy*. The goal of an experimental study is to demonstrate a cause-and-effect relationship between two variables. Specifically, an experiment attempts to show that changing the value of one variable causes changes to occur in the second variable. To accomplish this goal, the experimental method has two characteristics that differentiate experiments from other types of research studies:

In more complex experiments, a researcher may systematically manipulate more than one variable and may observe more than one variable. Here we are considering the simplest case, in which only one variable is manipulated and only one variable is observed.

1. **Manipulation** The researcher manipulates one variable by changing its value from one level to another. In the Polman et al. (2008) experiment examining the effect of violence in video games on aggressive behavior (Figure 1.6), the researchers manipulate the amount of violence by giving one group of boys a violent game to play and giving the other group a nonviolent game. A second variable is observed (measured) to determine whether the manipulation causes changes to occur. In the Polman et al. (2008) experiment, aggressive behavior was measured.

2. **Control** The researcher must exercise control over the research situation to ensure that other, extraneous variables do not influence the relationship being examined. Control usually involves matching different groups as closely as possible on those variables that we don't want to manipulate.

To demonstrate these two characteristics, consider the Polman et al. (2008) study examining the effect of violence in video games on aggression (see Figure 1.6). To be able to say that the difference in aggressive behavior is caused by the amount of violence in the game, the researcher must rule out any other possible explanation for the difference. That is, any other variables that might affect aggressive behavior must be controlled. Two of the general categories of variables that researchers must consider:

1. **Environmental Variables** These are characteristics of the environment such as lighting, time of day, and weather conditions. A researcher must ensure that the

individuals in treatment A are tested in the same environment as the individuals in treatment B. Using the video game violence experiment (see Figure 1.6) as an example, suppose that the individuals in the nonviolent condition were all tested in the morning and the individuals in the violent condition were all tested in the evening. It would be impossible to determine if the results were due to the type of video game the children played or the time of day they were tested because an uncontrolled environmental variable (time of day) is allowed to vary with the treatment conditions. Whenever a research study allows more than one explanation for the results, the study is said to be *confounded* because it is impossible to reach an unambiguous conclusion.

According to APA convention, the term *participants* is used when referring to research with humans and the term *subjects* is used when referring to research with animals.

2. **Participant Variables**   These are characteristics such as age, gender, motivation, and personality that vary from one individual to another. Because no two people (or animals) are identical, the individuals who participate in research studies will be different on a wide variety of participant variables. These differences, known as *individual differences*, are a part of every research study. Whenever an experiment compares different groups of participants (one group in treatment A and a different group in treatment B), the concern is that there may be consistent differences between groups for one or more participant variables. For the experiment shown in Figure 1.6, for example, the researchers would like to conclude that the violence in the video game causes a change in the participants' aggressive behavior. In the study, the participants in both conditions were 10-year-old boys. Suppose, however, that the participants in the violent video game condition, just by chance, had more children who were bullies. In this case, there is an alternative explanation for the difference in aggression that exists between the two groups. Specifically, the difference between groups may have been caused by the amount of violence in the game, but it also is possible that the difference was caused by preexisting differences between the groups. Again, this would produce a confounded experiment.

Researchers typically use three basic techniques to control other variables. First, the researcher could use *random assignment*, which means that each participant has an equal chance of being assigned to each of the treatment conditions. The goal of random assignment is to distribute the participant characteristics evenly between the two groups so that neither group is noticeably smarter (or older, or faster) than the other. Random assignment can also be used to control environmental variables. For example, participants could be assigned randomly for testing either in the morning or in the afternoon. A second technique for controlling variables is to use *matching* to ensure groups are equivalent in terms of participant variables and environmental variables. For example, the researcher could match groups by ensuring that each group has exactly 60% females and 40% males. Finally, the researcher can control variables by *holding them constant*. For example, in the video game violence study discussed earlier (Polman et al., 2008), the researchers used only 10-year-old boys as participants (holding age and gender constant). In this case the researchers can be certain that one group is not noticeably older or has a larger proportion of females than the other.

The matched-subject is a method to prevent preexisting participant differences between groups and is covered in Chapter 11.

In the **experimental method**, one variable is manipulated while another variable is observed and measured. To establish a cause-and-effect relationship between the two variables, an experiment attempts to control all other variables to prevent them from influencing the results.

The individuals in a research study differ on a variety of participant variables such as age, weight, skills, motivation, and personality. The differences from one participant to another are known as **individual differences**.

Terminology in the Experimental Method   Specific names are used for the two variables that are studied by the experimental method. The variable that is manipulated by the experimenter is called the *independent variable*. It can be identified as the treatment conditions to which participants are assigned. For the example in Figure 1.6, the amount of violence in the video game is the independent variable. The variable that is observed and measured to obtain scores within each condition is the *dependent variable*. In Figure 1.6, the level of aggressive behavior is the dependent variable.

> The **independent variable** is the variable that is manipulated by the researcher. In behavioral research, the independent variable usually consists of the two (or more) treatment conditions to which subjects are exposed. The independent variable is manipulated *prior* to observing the dependent variable.
>
> The **dependent variable** is the one that is observed to assess the effect of the treatment. The dependent variable is the variable that is measured in the experiment and its value changes in a way that depends on the status of the independent variable.

An experimental study evaluates the relationship between two variables by manipulating one variable (the independent variable) and measuring one variable (the dependent variable). Note that in an experiment only one variable is actually measured. You should realize that this is different from a correlational study, in which all variables are measured and the data consist of at least two separate scores for each individual.

Control Conditions in an Experiment   Often an experiment will include a condition in which the participants do not receive any experimental treatment. The scores from these individuals are then compared with scores from participants who do receive the treatment. The goal of this type of study is to demonstrate that the treatment has an effect by showing that the scores in the treatment condition are substantially different from the scores in the no-treatment condition. In this kind of research, the no-treatment condition is called the *control condition*, and the treatment condition is called the *experimental condition*.

> Individuals in a **control condition** do not receive the experimental treatment. Instead, they either receive no treatment or they receive a neutral, placebo treatment. The purpose of a control condition is to provide a baseline for comparison with the experimental condition.
>
> Individuals in the **experimental condition** do receive the experimental treatment.

Note that the independent variable always consists of at least two values. (Something must have at least two different values before you can say that it is "variable.") For the video game violence experiment (see Figure 1.6), the independent variable is the amount of violence in the video game. For an experiment with an experimental group and a control group, the independent variable is treatment versus no treatment.

■ Nonexperimental Methods: Nonequivalent Groups and Pre-Post Studies

In informal conversation, there is a tendency for people to use the term *experiment* to refer to any kind of research study. You should realize, however, that the term applies only to studies that satisfy the specific requirements outlined earlier. In particular, a real

FIGURE 1.7

Two examples of nonexperimental studies that involve comparing two groups of scores. In (a), the study uses two preexisting groups (suburban/rural) and measures a dependent variable (verbal scores) in each group. In (b), the study uses time (before/after) to define the two groups and measures a dependent variable (depression) in each group.

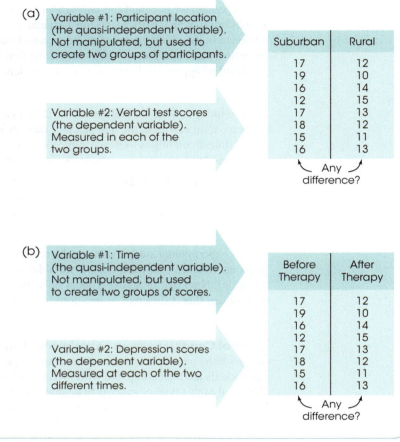

(a)

Variable #1: Participant location (the quasi-independent variable). Not manipulated, but used to create two groups of participants.

Variable #2: Verbal test scores (the dependent variable). Measured in each of the two groups.

Suburban	Rural
17	12
19	10
16	14
12	15
17	13
18	12
15	11
16	13

Any difference?

(b)

Variable #1: Time (the quasi-independent variable). Not manipulated, but used to create two groups of scores.

Variable #2: Depression scores (the dependent variable). Measured at each of the two different times.

Before Therapy	After Therapy
17	12
19	10
16	14
12	15
17	13
18	12
15	11
16	13

Any difference?

experiment must include manipulation of an independent variable and rigorous control of other, extraneous variables. As a result, there are a number of other research designs that are not true experiments but still examine the relationship between variables by comparing groups of scores. Two examples are shown in Figure 1.7 and are discussed in the following paragraphs. This type of research study is classified as nonexperimental.

The top part of Figure 1.7 shows an example of a *nonequivalent groups study* comparing third-grade students from suburban communities to those from rural communities. Notice that this study involves comparing two groups of scores (like an experiment). However, the researcher has no ability to control which participants go into which group—group assignment for the children is determined by where they live, not by the researcher. Because this type of research compares preexisting groups, the researcher cannot control the assignment of participants to groups and cannot ensure equivalent groups. Other examples of nonequivalent group studies include comparing 8-year-old children and 10-year-old children, people diagnosed with an eating disorder and those not diagnosed with a disorder, and comparing children from a single-parent home and those from a two-parent home. Because it is impossible to use techniques like random assignment to control participant variables and ensure equivalent groups, this type of research is not a true experiment.

Correlational studies are also examples of nonexperimental research. In this section, however, we are discussing nonexperimental studies that compare two or more groups of scores.

The bottom part of Figure 1.7 shows an example of a *pre-post study* comparing depression scores before therapy and after therapy. A pre-post study uses the passage of time (before/after) to create the groups of scores. In Figure 1.7 the two groups of scores are obtained by measuring the same variable (depression) twice for each participant; once before therapy and again after therapy. In a pre-post study, however, the researcher has no control over the passage of time. The "before" scores are always measured earlier than the "after" scores. Although a difference between the two groups of scores may be caused by the treatment, it is always possible that the scores simply change as time goes by. For example, the depression scores may decrease over time in the same way that the symptoms of a cold disappear over time. In a pre-post study the researcher also has no control over other variables that change with time. For example, the weather could change from dark and gloomy before therapy to bright and sunny after therapy. In this case, the depression scores could improve because of the weather and not because of the therapy. Because the researcher cannot control the passage of time or other variables related to time, this study is not a true experiment.

Terminology in Nonexperimental Research   Although the two research studies shown in Figure 1.7 are not true experiments, you should notice that they produce the same kind of data that are found in an experiment (see Figure 1.6). In each case, one variable is used to create groups, and a second variable is measured to obtain scores within each group. In an experiment, the groups are created by manipulation of the independent variable, and the participants' scores are the dependent variable. The same terminology is often used to identify the two variables in nonexperimental studies. That is, the variable that is used to create groups is the independent variable and the scores are the dependent variable. For example, the top part of Figure 1.7, the child's location (suburban/rural), is the independent variable and the verbal test scores are the dependent variable. However, you should realize that location (suburban/rural) is not a true independent variable because it is not manipulated. For this reason, the "independent variable" in a nonexperimental study is often called a *quasi-independent variable*.

In a nonexperimental study, the "independent variable" that is used to create the different groups of scores is often called the **quasi-independent variable**.

LEARNING CHECK   **LO8**   **1.** Which of the following is most likely to be a purely correlational study?
a. One variable and one group
b. One variable and two groups
c. Two variables and one group
d. Two variables and two groups

LO8   **2.** A research study comparing alcohol use for college students in the United States and Canada reports that more Canadian students drink but American students drink more (Kuo, Adlaf, Lee, Gliksman, Demers, & Wechsler, 2002). What research design did this study use?
a. Correlational
b. Experimental
c. Nonexperimental
d. Noncorrelational

LO9 3. Stephens, Atkins, and Kingston (2009) found that participants were able to tolerate more pain when they shouted their favorite swear words over and over than when they shouted neutral words. For this study, what is the independent variable?

   a. The amount of pain tolerated

   b. The participants who shouted swear words

   c. The participants who shouted neutral words

   d. The kind of word shouted by the participants

ANSWERS   1. c   2. c   3. d

1-4 Statistical Notation

LEARNING OBJECTIVES

10. Identify what is represented by each of the following symbols: X, Y, N, n, and Σ.

11. Perform calculations using summation notation and other mathematical operations following the correct order of operations.

The measurements obtained in research studies provide the data for statistical analysis. Most of the statistical analyses use the same general mathematical operations, notation, and basic arithmetic that you have learned during previous years of schooling. In case you are unsure of your mathematical skills, there is a mathematics review section in Appendix A at the back of this book. The appendix also includes a skills assessment exam (p. 570) to help you determine whether you need the basic mathematics review. In this section, we introduce some of the specialized notation that is used for statistical calculations. In later chapters, additional statistical notation is introduced as it is needed.

■ Scores

Quiz Scores	Height	Weight
X	X	Y
37	72	165
35	68	151
35	67	160
30	67	160
25	68	146
17	70	160
16	66	133

Measuring a variable in a research study yields a value or a score for each individual. Raw scores are the original, unchanged scores obtained in the study. Scores for a particular variable are typically represented by the letter X. For example, if performance in your statistics course is measured by tests and you obtain a 35 on the first test, then we could state that $X = 35$. A set of scores can be presented in a column that is headed by X. For example, a list of quiz scores from your class might be presented as shown in the margin (the single column on the left).

When observations are made for two variables, there will be two scores for each individual. The data can be presented as two lists labeled X and Y for the two variables. For example, observations for people's height in inches (variable X) and weight in pounds (variable Y) can be presented as shown in the double column in the margin. Each pair X, Y represents the observations made of a single participant.

The letter N is used to specify how many scores are in a set. An uppercase letter N identifies the number of scores in a population and a lowercase letter n identifies the number of scores in a sample. Throughout the remainder of the book you will notice that we often use notational differences to distinguish between samples and populations. For the height and weight data in the preceding table, $n = 7$ for both variables. Note that by using a lowercase letter n, we are implying that these data are a sample.

■ Summation Notation

Many of the computations required in statistics involve adding a set of scores. Because this procedure is used so frequently, a special notation is used to refer to the sum of a set of scores. The Greek letter *sigma*, or Σ, is used to stand for summation. The expression ΣX means to add all the scores for variable X. The summation sign Σ can be read as "the sum of." Thus, ΣX is read "the sum of the scores." For the following set of quiz scores, 10, 6, 7, 4,

$$\Sigma X = 27 \quad \text{and} \quad N = 4.$$

To use summation notation correctly, keep in mind the following two points:

1. The summation sign, Σ, is always followed by a symbol or mathematical expression. The symbol or expression identifies exactly which values are to be added. To compute ΣX, for example, the symbol following the summation sign is X, and the task is to find the sum of the X values. On the other hand, to compute $\Sigma(X - 1)$, the summation sign is followed by a relatively complex mathematical expression, so your first task is to calculate all the $(X - 1)$ values and then add those results.

2. The summation process is often included with several other mathematical operations, such as multiplication or squaring. To obtain the correct answer, it is essential that the different operations be done in the correct sequence. Following is a list showing the correct *order of operations* for performing mathematical operations. Most of this list should be familiar, but you should note that we have inserted the summation process as the fourth operation in the list.

Order of Mathematical Operations

More information on the order of operations for mathematics is available in the Math Review Appendix A, Section A.1.

1. Any calculation contained within parentheses is done first.

2. Squaring (or raising to other exponents) is done second.

3. Multiplying and/or dividing is done third. A series of multiplication and/or division operations should be done in order from left to right.

4. Summation using the Σ notation is done next.

5. Finally, any other addition and/or subtraction is done.

The following examples demonstrate how summation notation is used in most of the calculations and formulas we present in this book. Notice that whenever a calculation requires multiple steps, we use a computational table to help demonstrate the process. The table simply lists the original scores in the first column and then adds columns to show the results of each successive step. Notice that the first three operations in the order-of-operations list all create a new column in the computational table. When you get to summation (number 4 in the list), you simply add the values in the last column of your table to obtain the sum.

EXAMPLE 1.3

X	X^2
3	9
1	1
7	49
4	16

A set of four scores consists of values 3, 1, 7, and 4. We will compute ΣX, ΣX^2, and $(\Sigma X)^2$ for these scores. To help demonstrate the calculations, we will use a computational table showing the original scores (the X values) in the first column. Additional columns can then be added to show additional steps in the series of operations. You should notice that the first three operations in the list (parentheses, squaring, and multiplying) all create a new column of values. The last two operations, however, produce a single value corresponding to the sum.

The table to the left shows the original scores (the X values) and the squared scores (the X^2 values) that are needed to compute ΣX^2.

The first calculation, ΣX, does not include any parentheses, squaring, or multiplication, so we go directly to the summation operation. The X values are listed in the first column of the table, and we simply add the values in this column:

$$\Sigma X = 3 + 1 + 7 + 4 = 15$$

To compute ΣX^2, the correct order of operations is to square each score and then find the sum of the squared values. The computational table shows the original scores and the results obtained from squaring (the first step in the calculation). The second step is to find the sum of the squared values, so we simply add the numbers in the X^2 column:

$$\Sigma X^2 = 9 + 1 + 49 + 16 = 75$$

The final calculation, $(\Sigma X)^2$, includes parentheses, so the first step is to perform the calculation inside the parentheses. Thus, we first find ΣX and then square this sum. Earlier, we computed $\Sigma X = 15$, so

$$(\Sigma X)^2 = (15)^2 = 225 \qquad \blacksquare$$

EXAMPLE 1.4

Use the same set of four scores from Example 1.3 and compute $\Sigma(X - 1)$ and $\Sigma(X - 1)^2$. The following computational table will help demonstrate the calculations.

X	$(X - 1)$	$(X - 1)^2$	
3	2	4	The first column lists the
1	0	0	original scores. A second column lists the $(X - 1)$
7	6	36	values, and a third column shows the $(X - 1)^2$ values.
4	3	9	

To compute $\Sigma(X - 1)$, the first step is to perform the operation inside the parentheses. Thus, we begin by subtracting one point from each of the X values. The resulting values are listed in the middle column of the table. The next step is to add the $(X - 1)$ values, so we simply add the values in the middle column.

$$\Sigma(X - 1) = 2 + 0 + 6 + 3 = 11$$

The calculation of $\Sigma(X - 1)^2$ requires three steps. The first step (inside parentheses) is to subtract 1 point from each X value. The results from this step are shown in the middle column of the computational table. The second step is to square each of the $(X - 1)$ values. The results from this step are shown in the third column of the table. The final step is to add the $(X - 1)^2$ values, so we add the values in the third column to obtain

$$\Sigma(X - 1)^2 = 4 + 0 + 36 + 9 = 49$$

Notice that this calculation requires squaring before adding. A common mistake is to add the $(X - 1)$ values and then square the total. Be careful! $\qquad \blacksquare$

EXAMPLE 1.5

In both the preceding examples, and in many other situations, the summation operation is the last step in the calculation. According to the order of operations, parentheses, exponents, and multiplication all come before summation. However, there are situations in which extra addition and subtraction are completed after the summation. For this example, use the same scores that appeared in the previous two examples, and compute $\Sigma X - 1$.

With no parentheses, exponents, or multiplication, the first step is the summation. Thus, we begin by computing ΣX. Earlier we found $\Sigma X = 15$. The next step is to subtract one point from the total. For these data,

$$\Sigma X - 1 = 15 - 1 = 14 \qquad \blacksquare$$

EXAMPLE 1.6

For this example, each individual has two scores. The first score is identified as X, and the second score is Y. With the help of the following computational table, compute ΣX, ΣY, $\Sigma X \Sigma Y$, and ΣXY.

Person	X	Y	XY
A	3	5	15
B	1	3	3
C	7	4	28
D	4	2	8

To find ΣX, simply add the values in the X column.

$$\Sigma X = 3 + 1 + 7 + 4 = 15$$

Similarly, ΣY is the sum of the Y values in the middle column.

$$\Sigma Y = 5 + 3 + 4 + 2 = 14$$

To find $\Sigma X \Sigma Y$ you must add the X values and add the Y values. Then you multiply these sums.

$$\Sigma X \Sigma Y = 15(14) = 210$$

To compute ΣXY, the first step is to multiply X times Y for each individual. The resulting products (XY values) are listed in the third column of the table. Finally, we add the products to obtain

$$\Sigma XY = 15 + 3 + 28 + 8 = 54$$ ■

The following example is an opportunity for you to test your understanding of summation notation.

EXAMPLE 1.7

Calculate each value requested for the following scores: 5, 2, 4, 2

    a. ΣX^2 **b.** $\Sigma(X + 1)$ **c.** $\Sigma(X + 1)^2$

You should obtain answers of 49, 17, and 79 for a, b, and c, respectively. Good luck. ■

LEARNING CHECK

LO10 1. What value is represented by the lowercase letter n?
    a. The number of scores in a population
    b. The number of scores in a sample
    c. The number of values to be added in a summation problem
    d. The number of steps in a summation problem

LO11 2. What is the value of $\Sigma(X - 2)$ for the following scores: 6, 2, 4, 2?
    a. 12
    b. 10
    c. 8
    d. 6

LO11 3. What is the first step in the calculation of $(\Sigma X)^2$?
    a. Square each score.
    b. Add the scores.
    c. Subtract 2 points from each score.
    d. Add the $X - 2$ values.

ANSWERS   **1. b**   **2. d**   **3. b**

SUMMARY

1. The term *statistics* is used to refer to methods for organizing, summarizing, and interpreting data.

2. Scientific questions usually concern a population, which is the entire set of individuals one wishes to study. Usually, populations are so large that it is impossible to examine every individual, so most research is conducted with samples. A sample is a group selected from a population, usually for purposes of a research study.

3. A characteristic that describes a sample is called a statistic, and a characteristic that describes a population is called a parameter. Although sample statistics are usually representative of corresponding population parameters, there is typically some discrepancy between a statistic and a parameter. The naturally occurring difference between a statistic and a parameter is called sampling error.

4. Statistical methods can be classified into two broad categories: descriptive statistics, which organize and summarize data, and inferential statistics, which use sample data to draw inferences about populations.

5. A construct is a variable that cannot be directly observed. An operational definition defines a construct in terms of external behaviors that are representative of the construct.

6. A discrete variable consists of indivisible categories, often whole numbers that vary in countable steps. A continuous variable consists of categories that are infinitely divisible, with each score corresponding to an interval on the scale. The boundaries that separate intervals are called real limits and are located exactly halfway between adjacent scores.

7. A measurement scale consists of a set of categories that are used to classify individuals. A nominal scale consists of categories that differ only in name and are not differentiated in terms of magnitude or direction. In an ordinal scale, the categories are differentiated in terms of direction, forming an ordered series. An interval scale consists of an ordered series of categories that are all equal-sized intervals. With an interval scale, it is possible to differentiate direction and distance between categories. Finally, a ratio scale is an interval scale for which the zero point indicates none of the variable being measured. With a ratio scale, ratios of measurements reflect ratios of magnitude.

8. The correlational method examines relationships between variables by measuring two different variables for each individual. This method allows researchers to measure and describe relationships, but cannot produce a cause-and-effect explanation for the relationship.

9. The experimental method examines relationships between variables by manipulating an independent variable to create different treatment conditions and then measuring a dependent variable to obtain a group of scores in each condition. The groups of scores are then compared. A systematic difference between groups provides evidence that changing the independent variable from one condition to another also caused a change in the dependent variable. All other variables are controlled to prevent them from influencing the relationship. The intent of the experimental method is to demonstrate a cause-and-effect relationship between variables.

10. Nonexperimental studies also examine relationships between variables by comparing groups of scores, but they do not have the rigor of true experiments and cannot produce cause-and-effect explanations. Instead of manipulating a variable to create different groups, a nonexperimental study uses a preexisting participant characteristic (such as older/younger) or the passage of time (before/after) to create the groups being compared.

11. In an experiment, the independent variable is manipulated by the researcher and the dependent variable is the one that is observed to assess the effect of the treatment. The variable that is used to create the groups in a nonexperiment is a quasi-independent variable.

12. The letter X is used to represent scores for a variable. If a second variable is used, Y represents its scores. The letter N is used as the symbol for the number of scores in a population; n is the symbol for a number of scores in a sample.

13. The Greek letter sigma (Σ) is used to stand for summation. Therefore, the expression ΣX is read "the sum of the scores." Summation is a mathematical operation (like addition or multiplication) and must be performed in its proper place in the order of operations; summation occurs after parentheses, exponents, and multiplying/dividing have been completed.

KEY TERMS

Statistics, statistical methods,
   statistical procedures (3)

population (4)

sample (4)

random sample (4)

variable (5)

data (6)

data set (6)

datum (6)

score, raw score (6)

parameter (6)

statistic (6)

descriptive statistics (6)

inferential statistics (7)

sampling error (7)

constructs (12)

operational definition (12)

discrete variable (12)

continuous variable (13)

real limits (14)

upper real limit (14)

lower real limit (14)

nominal scale (15)

ordinal scale (15)

interval scale (16)

ratio scale (16)

descriptive research, descriptive
   research strategy (19)

correlational method (21)

experimental method (24)

individual differences (24)

independent variable (25)

dependent variable (25)

control condition (25)

experimental condition (25)

nonequivalent groups study (26)

pre–post study (27)

quasi-independent variable (27)

FOCUS ON PROBLEM SOLVING

It may help to simplify summation notation if you observe that the summation sign is always followed by a symbol or symbolic expression—for example, ΣX or $\Sigma(X + 3)$. This symbol specifies which values you are to add. If you use the symbol as a column heading and list all the appropriate values in the column, your task is simply to add up the numbers in the column. To find $\Sigma(X + 3)$ for example, start a column headed with $(X + 3)$ next to the column of Xs. List all the $(X + 3)$ values; then find the total for the column.

Often, summation notation is part of a relatively complex mathematical expression that requires several steps of calculation. The series of steps must be performed according to the order of mathematical operations (see page 29). The best procedure is to use a computational table that begins with the original X values listed in the first column. Except for summation, each step in the calculation creates a new column of values. For example, computing $\Sigma(X + 1)^2$ involves three steps and produces a computational table with three columns. The final step is to add the values in the third column (see Example 1.4).

DEMONSTRATION 1.1

SUMMATION NOTATION

A set of scores consists of the following values:

$$7 \quad 3 \quad 9 \quad 5 \quad 4$$

For these scores, compute each of the following:

ΣX

$(\Sigma X)^2$

ΣX^2

$\Sigma X + 5$

$\Sigma(X - 2)$

Compute ΣX To compute ΣX, we simply add all of the scores in the group.

$$\Sigma X = 7 + 3 + 9 + 5 + 4 = 28$$

X	X^2
7	49
3	9
9	81
5	25
4	16

Compute $(\Sigma X)^2$ The first step, inside the parentheses, is to compute ΣX. The second step is to square the value for ΣX.

$$\Sigma X = 28 \quad \text{and} \quad (\Sigma X)^2 = (28)^2 = 784$$

Compute ΣX^2 The first step is to square each score. The second step is to add the squared scores. The computational table shows the scores and squared scores. To compute ΣX^2 we add the values in the X^2 column.

$$\Sigma X^2 = 49 + 9 + 81 + 25 + 16 = 180$$

X	$X - 2$
7	5
3	1
9	7
5	3
4	2

Compute $\Sigma X + 5$ The first step is to compute ΣX. The second step is to add 5 points to the total.

$$\Sigma X = 28 \quad \text{and} \quad \Sigma X + 5 = 28 + 5 = 33$$

Compute $\Sigma(X - 2)$ The first step, inside parentheses, is to subtract 2 points from each score. The second step is to add the resulting values. The computational table shows the scores and the $(X - 2)$ values. To compute $\Sigma(X - 2)$, add the values in the $(X - 2)$ column

$$\Sigma(X - 2) = 5 + 1 + 7 + 3 + 2 = 18$$

SPSS®

*Note: The Statistical Package for the Social Sciences, known as SPSS, is a computer program that performs most of the statistical calculations that are presented in this book, and is commonly available on college and university computer systems. Appendix D contains a general introduction to SPSS. In the SPSS section at the end of each chapter for which SPSS is applicable, there are step-by-step instructions for using SPSS to perform the statistical operations presented in the chapter.

Following are detailed instructions for using SPSS to calculate the number of scores in a data set (N or n) and the sum of the scores (ΣX).

Demonstration Example

Suppose that a researcher measures participants' reaction times to a verbal prompt (in seconds) and observes the following scores:

Participant	Reaction Time
A	30
B	19
C	15
D	24
E	15
F	21
G	13
H	26
I	26
J	13
K	17
L	6
M	17

(continued)

N	15
O	13
P	14
Q	20
R	20
S	14
T	19

We can use SPSS to find the number and sum of scores.

Data Entry

1. Enter information in the **Variable View.** In the **Name** field, enter a short, descriptive name for the variable that does not include spaces. Here, "RT" (for reaction time) is used. The default settings for **Type, Width, Values, Missing, Align, and Role** are acceptable.
2. For **Decimals**, enter "0" because reaction time was measured to the nearest whole second.
3. In the **Label** field, a descriptive title for the variable should be used. Here, we used "Reaction Time to Verbal Prompt (seconds)."
4. In the **Measure** field, select **Scale** because time is a ratio scale. The **Variable View** should now look similar to the SPSS figure below.

	Name	Type	Width	Decimals	Label	Values	Missing	Columns	Align	Measure	Role
1	RT	Numeric	8	0	Reaction Time to Verbal Prompt (seconds)	None	None	8	🔲 Right	🖉 Scale	↘ Input
2											
3											

5. Select the **Data View** in the bottom-left corner of the screen and enter the values from the reaction time measurement in the table above. When you have finished, the table should be similar to the figure below.

	RT	var	var
1	30		
2	19		
3	15		
4	24		
5	15		
6	21		
7	13		
8	26		
9	26		
10	13		
11	17		
12	6		
13	17		
14	15		
15	13		
16	14		
17	20		
18	20		
19	14		
20	19		

Data Analysis

1. Click **Analyze** on the tool bar, select **Descriptive Statistics,** and click on **Descriptives** as below.

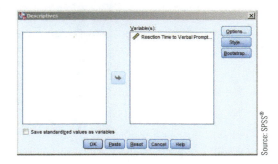

Source: SPSS®

2. Highlight the column label "Reaction Time to . . ." and click the arrow to move it to the **Variables** box.

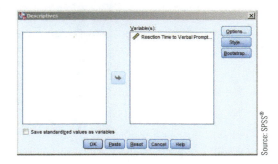

Source: SPSS®

3. Click **Options**. On the following screen, check the **Sum** box and uncheck the others (mean and standard deviation will be covered in later chapters). Your Options window should be as below.

Click the **Continue** button in the Options window and click the **OK** button in the Descriptives window.

SPSS Output

Your SPSS output includes a summary table with the number of scores and the sum of scores. Notice that SPSS always symbolizes the number of scores with an upper-case "N," even when you are analyzing a sample. Don't worry about this—SPSS uses computations that are appropriate for samples. Also, SPSS identifies your variable in the table based on the text that you entered in the Label field of the Variable View.

Try It Yourself

For the following set of scores, use SPSS to find the number of scores and ΣX.

Participant	Reaction Time
A	7
B	9
C	11
D	8
E	13
F	12
G	8
H	14
I	8
J	8
K	6
L	10
M	8
N	12
O	7
P	9
Q	18
R	14

Your output table should report that $\Sigma X = 182$ and $N = 18$.

PROBLEMS

Solutions to odd-numbered problems are provided in Appendix C.

1. A researcher is interested in the texting habits of high school students in the United States. The researcher selects a group of 100 students, measures the number of text messages that each individual sends each day, and calculates the average number for the group.
   a. Identify the population for this study.
   b. Identify the sample for this study.
   c. The average number that the researcher calculated is an example of a _____ .

2. Define the terms population and sample, and explain the role of each in a research study.

3. A researcher conducted an experiment on the effect of caffeine on memory in college students in the United States. The researcher randomly assigned each of 100 students to one of two groups. One group received caffeinated coffee followed by a memory test. The second group received decaffeinated coffee followed by a memory test. The researcher calculated the average number of items correctly recalled in each group.
   a. What is the population?
   b. What is the sample?
   c. The group that received decaffeinated coffee is a(n) _____ .

d. The group that received caffeinated coffee is a(n) _____ .
   e. The sample contains _____ participants. The population contains _____ .
   f. The averages calculated after the memory test is a _____ .

4. Statistical methods are classified into two major categories: descriptive and inferential. Describe the general purpose for the statistical methods in each category.

5. We know that the average IQ of everyone in the United States is 100. We randomly select 10 people and observe that their average IQ is 105.
   a. The value of 105 is a _____ .
   b. The value of 100 is a _____ .

6. Define the terms statistic and parameter and explain how these terms are related to the concept of sampling error.

7. A professor is interested in whether student performance on exams is better in the afternoon than in the morning. One sample of students was randomly assigned to receive the exam in the morning and another sample was randomly assigned to receive the exam in the afternoon. The following data were collected:

Participant	Time of Exam	Exam Score
1	Morning	65
2	Morning	73
3	Morning	90
4	Afternoon	70
5	Afternoon	75
6	Afternoon	95

The average score for morning students was 76 and the average score for afternoon students was 80. The professor concludes that the afternoon is the best time for students to complete the exam and that the difference in average scores reveals an important difference between afternoon and morning classes in college.
   a. Describe how sampling error could account for this difference.
   b. What type of statistic would the professor use to determine if the difference in exam averages between the samples provides convincing evidence of a difference between the time of day, or if the difference is just chance?

8. Explain why *honesty* is a hypothetical construct instead of a concrete variable. Describe how honesty might be measured and defined using an operational definition.

9. A tax form asks people to identify their age, annual income, number of dependents, and social security number. For each of these four variables, identify the scale of measurement that probably is used and identify whether the variable is continuous or discrete.

10. In your most recent checkup, your physician listed that your height is 70 inches, rounded to the nearest whole inch. Why is it unlikely that your height is exactly 70 inches? What are the upper and lower real limits of your height?

11. Four scales of measurement were introduced in this chapter, from simple classification on a nominal scale to the more informative measurements from a ratio scale.
   a. What additional information is obtained from measurements on an ordinal scale compared to measurements on a nominal scale?
   b. What additional information is obtained from measurements on an interval scale compared to measurements on an ordinal scale?
   c. What additional information is obtained from measurements on a ratio scale compared to measurements on an interval scale?

12. Your friend measures the temperature of her coffee to be 70° Celsius. Your friend also notices that the temperature outside is 35° Celsius. Why is it incorrect to say that the coffee is twice as warm as the temperature outside?

13. Describe the data for a correlational research study and explain how these data are different from the data obtained in experimental and nonexperimental studies, which also evaluate relationships between two variables.

14. Describe how the goal of an experimental research study is different from the goal for nonexperimental or correlational research. Identify the two elements that are necessary for an experiment to achieve its goal.

15. The results of a recent study showed that children who routinely drank reduced fat milk (1% or skim) were more likely to be overweight or obese at ages 2 and 4 compared to children who drank whole or 2% milk (Scharf, Demmer, & DeBoer, 2013).
   a. Is this an example of an experimental or a nonexperimental study?
   b. Explain how individual differences could provide an alternative explanation for the difference in weight between the groups.
   c. Create a research study that would be able to differentiate among those interpretations of the results.

16. Gentile, Lynch, Linder, and Walsh (2004) surveyed more than 600 eighth- and ninth-grade students regarding their gaming habits and other behaviors. Their results showed that the adolescents who experienced more video game violence were also more hostile and had more frequent arguments with teachers. Is this an experimental or a nonexperimental study? Explain your answer.

17. Deters and Mehl (2013) studied the effect of Facebook status updates on feelings of loneliness. Eighty-six participants were randomly assigned to two groups. One group was instructed to post more social media status updates and the other group was not. The researchers measured participants' loneliness using the UCLA Loneliness Scale, which consists of 10 items that ask participants to rate from 1 ("Never feel this way") to 4 ("I often feel this way") how often they experience specific feelings of loneliness (for example, "How often do you feel shut out and excluded by others?"). Participants who were instructed to post status updates had lower loneliness scores.
   a. For the measurement in this study, identify whether it is discrete or continuous and list the scale of measurement.
   b. What is the value of n?
   c. Is this an experimental or nonexperimental study? Explain.
   d. The group that was instructed to post more status updates is a(n) _____.

18. A research study comparing alcohol use for college students in the United States and Canada reports that more Canadian students drink but American students drink more (Kuo, Adlaf, Lee, Gliksman, Demers, & Wechsler, 2002). Is this study an example of an experiment? Explain why or why not.

19. Ackerman and Goldsmith (2011) compared learning performance for students who studied material printed on paper versus students who studied the same material presented on a computer screen. All students were then given a test on the material and the researchers recorded the number of correct answers.
   a. Identify the dependent variable for this study.
   b. Is the dependent variable discrete or continuous?
   c. What scale of measurement (nominal, ordinal, interval, or ratio) is used to measure the dependent variable?

20. Dwyer, Figuerooa, Gasalla, and Lopez (2018) showed that learning of flavor preferences depends on the relative value of the reward with which a flavor is paired. In their experiment, rats received pairings of a cherry flavor with 8% sucrose solution after exposure to 32% sucrose solution, which made the 8% solution a relatively low value. On other trials, a grape flavor was paired with 8% sucrose solution after exposure to a 2% sucrose solution, which made the 8% solution a relatively high value. Thus, cherry was paired with a relatively low-value reward and grape was paired with a relatively high-value reward. They observed that rats consumed more in ounces of cherry flavor than grape flavor at a later test.
   a. Identify the independent and dependent variables for this study.
   b. What scale of measurement is the dependent variable?
   c. Is the dependent variable discrete or continuous?
   d. Imagine that the researcher reported that subject number 4 consumed 2.5 ounces of cherry-flavored water. Consumption of the solution was rounded to the nearest tenth of an ounce. What are the lower and upper real limits of subject 4's score?

21. Doebel and Munakata (2018) discovered that delay of gratification by children is influenced by social context. All children were told that they were in the "green group" and were placed in a room with a single marshmallow. Participants were told that they could either eat the single marshmallow now or wait for the experimenter to return with two marshmallows. Before choosing between one marshmallow now or two later, children were randomly assigned to one of two conditions. They were told that either (1) other children in the green group waited and kids in the orange group didn't wait or (2) other children in the green group didn't wait and kids in the orange group waited. Children were more likely to choose to wait after being told that other members of their group waited.

a. Did this study use experimental or nonexperimental methods?
b. Identify the variables in this study.

22. Ford and Torok (2008) found that motivational signs were effective in increasing physical activity on a college campus. Signs such as "Step up to a healthier lifestyle" and "An average person burns 10 calories a minute walking up the stairs" were posted by the elevators and stairs in a college building. Students and faculty increased their use of the stairs during times that the signs were posted compared to times when there were no signs.
   a. Identify the independent and dependent variables for this study.
   b. What scale of measurement is used for the independent variable?

23. For the following scores, find the value of each expression:
   a. ΣX
   b. $(\Sigma X)^2$
   c. $\Sigma X - 3$
   d. $\Sigma(X - 3)$

X
4
2
6
3

24. For the following set of scores, find the value of each expression:
   a. $n\Sigma(X - 1)$
   b. $\Sigma X - 3^2$
   c. $\dfrac{\Sigma(X - 2)}{n}$
   d. $\Sigma(X - 4)^2$

X
3
5
4
2
1

25. For the following set of scores, find the value of each expression:
   a. $\Sigma(X - 4)^2$
   b. $(\Sigma X)^2$
   c. ΣX^2
   d. $\Sigma(X + 3)$

X
-1
-3
6
-4
0

26. Two scores, X and Y, are recorded for each of $n = 5$ participants. For these scores, find the value of each expression.
   a. ΣX
   b. ΣY
   c. $\Sigma(X + Y)$
   d. ΣXY

Participant	X	Y
A	3	1
B	1	5
C	-2	2
D	-4	2
E	2	4

27. For the following set of scores, find the value of each expression:

a. ΣXY
b. $\Sigma X \Sigma Y$
c. ΣY
d. $n = ?$

Participant	X	Y
A	6	1
B	3	0
C	0	2
D	−1	4

28. Use summation notation to express the following calculations.

a. Multiply scores X and Y and then add each product.
b. Sum the scores X and sum the scores Y and then multiply the sums.
c. Subtract X from Y and sum the differences.
d. Sum the X scores.

29. Use summation notation to express each of the following calculations:

a. Add the scores and then square the sum.
b. Square each score and then add the squared values.
c. Subtract 2 points from each score and then add the resulting values.
d. Subtract 1 point from each score and square the resulting values. Then add the squared values.

30. For the following set of scores, find the value of each expression:

a. ΣX^2
b. $(\Sigma X)^2$
c. $\Sigma(X - 3)$
d. $\Sigma(X - 3)^2$

X
6
1
4
5
2

31. For the following set of scores, find the value of each expression:

a. $n \Sigma X^2$
b. $(\Sigma Y)^2$
c. ΣXY
d. $\Sigma X \Sigma Y$

Participant	X	Y
A	3	2
B	1	6
C	5	0
D	2	5
E	0	6

32. For the following set of scores, find the value of each expression:

a. $n \Sigma X^2$
b. $(\Sigma Y)^2$
c. ΣXY
d. $\Sigma X \Sigma Y$

Participant	X	Y
A	5	1
B	3	3
C	0	5
D	−3	7
E	−5	9

Frequency Distributions

Tools You Will Need

The following items are considered essential background material for this chapter. If you doubt your knowledge of any of these items, you should review the appropriate chapter or section before proceeding.

- Proportions (Appendix A)
  - Fractions
  - Decimals
  - Percentages
- Scales of measurement (Chapter 1): Nominal, ordinal, interval, and ratio
- Continuous and discrete variables (Chapter 1)
- Real limits (Chapter 1)

clivewa/Shutterstock.com

PREVIEW

PREVIEW

Behavioral scientists have observed effects of watching television shows and other media on behavior in laboratory settings. Jena, Jain, and Hicks (2018) wanted to know if a movie that glorifies reckless behavior and risk-taking has an effect on its viewers in real life settings. *The Fast and the Furious* movie franchise has produced eight movies as of 2017 and a ninth release is expected in 2020. The series emphasizes, among other things, powerful modified cars, reckless driving, and street racing. Researchers compared the speeding tickets during the three weeks before each movie release to the three weeks afterward over a six-year period in Montgomery County, MD. They found the speeding tickets in the weeks prior to the release of each *The Fast and the Furious* movie averaged 16 mph above the posted speed limit. During the weeks afterward, tickets averaged 19 mph above the speed limit. This represents a nearly 20% change in amount of speed above the posted limit.

Table 2.1 lists hypothetical data similar to those of the study, showing miles per hour (mph) above the speed limit for each ticket.

You probably find it difficult to see a clear pattern simply by looking at an unorganized list of numbers. Can you tell how much difference, if any, there is between the two groups in speeding? One way to address this question is to organize each group of scores into a frequency distribution, which provides a clearer picture of any differences between the groups.

For example, the same data in Table 2.1 have been organized in a frequency distribution graph in Figure 2.1. In the figure each individual is represented as a block

TABLE 2.1

Speeding tickets during the three weeks before and after release of *The Fast and the Furious* movies. The scores for the hypothetical data reflect miles per hour above the posted speed limit.

Before Movie Release	After Movie Release
15	17
16	20
18	20
14	19
15	22
16	15
16	19
19	20
15	22
16	16

that is placed above the individual's score on the horizontal line. The resulting pile of blocks shows a picture of how individual scores are distributed. The distribution makes it clear that during the three weeks following the movie release, tickets are generally for speeds that are higher than during the three weeks before the movie release. Before the movie release most tickets were approximately 16 mph above the speed limit. After its release, most tickets were 19 mph or more above the posted limit.

In this chapter we present techniques for organizing data into tables and graphs so that an entire set of scores can be presented in an organized display or illustration.

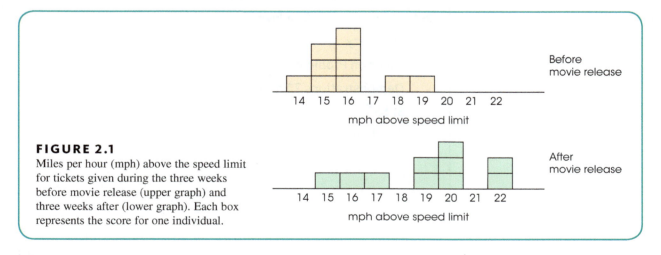

FIGURE 2.1

Miles per hour (mph) above the speed limit for tickets given during the three weeks before movie release (upper graph) and three weeks after (lower graph). Each box represents the score for one individual.

2-1   Frequency Distributions and Frequency Distribution Tables

LEARNING OBJECTIVES

1. Use and create frequency distribution tables and explain how they are related to the original set of scores.

2. Calculate the following from a frequency table: ΣX, ΣX^2, and the proportion and percentage of the group associated with each score.

3. Define percentiles and percentile ranks.

4. Determine percentiles and percentile ranks for values corresponding to real limits in a frequency distribution table.

The results from a research study usually consist of pages of numbers like those listed in Table 2.1, or in large spreadsheets in a computer file, corresponding to the measurements or scores collected during the study. The immediate problem for the researcher is to organize the scores into some comprehensible form so that any patterns in the data can be seen easily and communicated to others. This is the job of descriptive statistics: to simplify the organization and presentation of data. One of the most common procedures for organizing a set of data is to place the scores in a *frequency distribution*.

A **frequency distribution** is an organized tabulation of the number of individuals located in each category on the scale of measurement.

A frequency distribution takes a disorganized set of scores and places them in order from highest to lowest, grouping together individuals who all have the same score. If the highest score is $X = 10$, for example, the frequency distribution groups together all the 10s, then all the 9s, then the 8s, and so on. Thus, a frequency distribution allows the researcher to see "at a glance" the entire set of scores. It shows whether the scores are generally high or low, whether they are concentrated in one area or spread out across the entire range, and generally provides an organized picture of the data. In addition to providing a picture of the entire set of scores, a frequency distribution allows you to see the location of any individual score relative to all the other scores in the set.

A frequency distribution can be structured either as a table or a graph, but in both cases, the distribution presents the same two elements:

1. The set of categories that make up the original measurement scale.

2. A record of the frequency, or number of individuals in each category.

It is customary to list categories from highest to lowest, but this is an arbitrary arrangement. Some computer programs list categories from lowest to highest, while others provide an option for using either descending or ascending order for X.

Thus, a frequency distribution presents a picture of how the individual scores are distributed on the measurement scale—hence the name *frequency distribution*.

■ Frequency Distribution Tables

The simplest *frequency distribution table* presents the measurement scale by listing the different measurement categories (X values) in a column from highest to lowest. Beside each X value, we indicate the frequency, or the number of times that particular measurement occurred in the data. It is customary to use X as the column heading for the scores and f as the column heading for the frequencies. An example of a frequency distribution table follows (Example 2.1).

EXAMPLE 2.1

The following set of $N = 20$ scores was obtained from a 10-point statistics quiz. We will organize these scores by constructing a frequency distribution table.

Scores:

8	9	8	7	10	9	6	4	9	8
7	8	10	9	8	6	9	7	8	8

X	f
10	2
9	5
8	7
7	3
6	2
5	0
4	1

1. The highest score is $X = 10$, and the lowest score is $X = 4$. Therefore, the first column of the table lists the categories that make up the scale of measurement (X values) from 10 down to 4. Notice that all the possible values are listed in the table. For example, no one had a score of $X = 5$, but this value is included. With an ordinal, interval, or ratio scale, the categories are listed in order (usually highest to lowest). For a nominal scale, the categories can be listed in any order.

2. The frequency associated with each score is recorded in the second column. For example, two people had scores of $X = 10$, so there is a 2 in the f column beside $X = 10$.

Because the table organizes the scores, it is possible to see very quickly the general quiz results. For example, there were only two perfect scores, but most of the class had high grades (8s and 9s). With one exception (the score of $X = 4$), it appears that the class has learned the material fairly well.

Notice that the X values in a frequency distribution table represent the scale of measurement, *not* the actual set of scores. For example, the X column lists the value 10 only one time, but the frequency column indicates that there are actually two values of $X = 10$. Also, the X column lists a value of $X = 5$, but the frequency column indicates that no one actually had a score of $X = 5$.

You also should notice that the frequencies can be used to find the total number of scores in the distribution. By adding up the frequencies, you obtain the total number of individuals:

$$\Sigma f = N$$

■

Obtaining ΣX from a Frequency Distribution Table   There may be times when you need to compute the sum of the scores, ΣX, or perform other computations for a set of scores that has been organized into a frequency distribution table. To complete these calculations correctly, you must use all the information presented in the table. That is, it is essential to use the information in the f column as well as the X column to obtain the full set of scores.

When it is necessary to perform calculations for scores that have been organized into a frequency distribution table, the safest procedure is to use the information in the table to recover the complete list of individual scores before you begin any computations. This process is demonstrated in the following example.

EXAMPLE 2.2

X	f
5	1
4	2
3	3
2	3
1	1

Consider the frequency distribution table shown in the margin. The table shows that the distribution has one 5, two 4s, three 3s, three 2s, and one 1, for a total of 10 scores. If you simply list all 10 scores, you can safely proceed with calculations such as finding ΣX or ΣX^2. For example, to compute ΣX you must add all 10 scores:

$$\Sigma X = 5 + 4 + 4 + 3 + 3 + 3 + 2 + 2 + 2 + 1$$

For the distribution in this table, you should obtain $\Sigma X = 29$. Try it yourself.

Similarly, to compute ΣX^2 you square each of the 10 scores and then add the squared values.

$$\Sigma X^2 = 5^2 + 4^2 + 4^2 + 3^2 + 3^2 + 3^2 + 2^2 + 2^2 + 2^2 + 1^2$$

This time you should obtain $\Sigma X^2 = 97$.

■

An alternative way to get ΣX from a frequency distribution table is to multiply each X value by its frequency and then add these products. This sum may be expressed in symbols as ΣfX. The computation is summarized as follows for the data in Example 2.2:

Caution: Doing calculations within the table works well for ΣX but can lead to errors for more complex formulas such as ΣfX^2.

X	f	fx	
5	1	5	(the one 5 totals 5)
4	2	8	(the two 4s total 8)
3	3	9	(the three 3s total 9)
2	3	6	(the three 2s total 6)
1	1	1	(the one 1 totals 1)
		$\Sigma X = 29$	

No matter which method you use to find ΣX, the important point is that you must use the information given in the frequency column in addition to the information in the X column.

Similarly, one can compute ΣX^2 from a frequency distribution table; however, it is necessary to perform an operation (squaring) on each of the values for X before multiplying those Xs by their corresponding frequencies. Squared values for X are placed in a column headed X^2. Then the frequency is multiplied by each X^2 value and placed in a column labeled fX^2. Finally, the values in column fX^2 are summed.

X	f	X^2	fX^2	
5	1	25	25	(5 squared times 1 is 25)
4	2	16	32	(4 squared times 2 is 32)
3	3	9	27	(3 squared times 3 is 27)
2	3	4	12	(2 squared times 3 is 12)
1	1	1	1	(1 squared times 1 is 1)

$$\Sigma fX^2 = 25 + 32 + 27 + 12 + 1 = 97$$

Remember, to compute ΣX^2 for the entire distribution by this alternate method you must use the information given in both the X and frequency columns and find ΣfX^2.

The following example is an opportunity for you to test your understanding by computing ΣX and ΣX^2 for scores in a frequency distribution table.

EXAMPLE 2.3 Calculate ΣX and ΣX^2 for scores shown in the frequency distribution table in Example 2.1 (p. 46). You should obtain $\Sigma X = 158$ and $\Sigma X^2 = 1,288$. Good luck. ■

■ Proportions and Percentages

In addition to the two basic columns of a frequency distribution, there are other measures that describe the distribution of scores and can be incorporated into the table. The two most common are proportion and percentage.

Proportion measures the fraction of the total group that is associated with each score. In Example 2.2, there were two individuals with $X = 4$. Thus, 2 out of 10 people had $X = 4$, so the proportion would be $\frac{2}{10} = 0.20$. In general, the proportion associated with each score is

$$\text{proportion} = p = \frac{f}{N}$$

Because proportions describe the frequency (f) in relation to the total number (N), they often are called *relative frequencies*. Although proportions can be expressed as fractions (for example, $\frac{2}{10}$), they more commonly appear as decimals. A column of proportions, headed with a p, can be added to the basic frequency distribution table (see Example 2.4).

In addition to using frequencies (f) and proportions (p), researchers often describe a distribution of scores with percentages. For example, an instructor might describe the results of an exam by saying that 15% of the class earned As, 23% Bs, and so on. To compute the *percentage* associated with each score, you first find the proportion (p) and then multiply by 100:

$$\text{percentage} = p(100) = \frac{f}{N}(100)$$

Percentages can be included in a frequency distribution table by adding a column headed with %. Example 2.4 demonstrates the process of adding proportions and percentages to a frequency distribution table.

EXAMPLE 2.4   The frequency distribution table from Example 2.2 is repeated here. This time we have added columns showing the proportion (p) and the percentage (%) associated with each score.

X	f	p = f/N	% = p(100)
5	1	1/10 = 0.10	10%
4	2	2/10 = 0.20	20%
3	3	3/10 = 0.30	30%
2	3	3/10 = 0.30	30%
1	1	1/10 = 0.10	10%

■ Percentile and Percentile Ranks

Although the primary purpose of a frequency distribution is to provide a description of an entire set of scores, it also can be used to describe the position of an individual within the set. Individual scores, or X values, are called raw scores. By themselves, raw scores do not provide much information. For example, if you are told that your score on an exam is $X = 43$, you cannot tell how well you did relative to other students in the class. To evaluate your score, you need more information, such as the average score or the number of people who had scores above and below you. With this additional information, you would be able to determine your relative position in the class. Because raw scores do not provide much information, it is desirable to transform them into a more meaningful form. One transformation that we will consider changes raw scores into percentiles.

Suppose, for example, that you have a score of $X = 43$ on an exam and you know that exactly 60% of the class had scores of 43 or lower. Then your score $X = 43$ has a percentile rank of 60%, and your score would be called the 60th percentile. Notice that *percentile rank* refers to a percentage and that *percentile* refers to a score. Also notice that your rank or percentile describes your exact position within the distribution.

The **percentile rank** of a particular score is defined as the percentage of individuals in the distribution with scores at or below the particular value.

When a score is identified by its percentile rank, the score is called a **percentile**.

■ Cumulative Frequency and Cumulative Percentage

To determine percentiles or percentile ranks, the first step is to find the number of individuals who are located at or below each point in the distribution. This can be done most easily with a frequency distribution table by simply counting the number of scores that are in or below each category on the scale. The resulting values are called *cumulative frequencies* because they represent the accumulation of individuals as you move up the scale.

EXAMPLE 2.5

In the following frequency distribution table, we have included a cumulative frequency column headed by *cf*. For each row, the cumulative frequency value is obtained by adding up the frequencies in and below that category. For example, the score $X = 3$ has a cumulative frequency of 14 because exactly 14 individuals had scores of $X = 3$ or less.

X	f	cf	
5	1	20	$cf = 1 + 5 + 8 + 4 + 2 = 20$
4	5	19	$cf = 5 + 8 + 4 + 2 = 19$
3	8	14	$cf = 8 + 4 + 2 = 14$
2	4	6	$cf = 4 + 2 = 6$
1	2	2	$cf = 2$

■

The cumulative frequencies show the number of individuals located at or below each score. To find percentiles, we must convert these frequencies into percentages. The resulting values are called *cumulative percentages* because they show the percentage of individuals who are accumulated as you move up the scale.

EXAMPLE 2.6

This time we have added a cumulative percentage column (*c%*) to the frequency distribution table from Example 2.5. The values in this column represent the percentage of individuals who are located in and below each category. For example, 70% of the individuals (14 out of 20) had scores of $X = 3$ or lower. Cumulative percentages can be computed by

$$c\% = \frac{cf}{N}(100\%)$$

X	f	cf	c%
5	1	20	100%
4	5	19	95%
3	8	14	70%
2	4	6	30%
1	2	2	10%

■

> It is possible to estimate the X value for a percentile that does not exist in the $c\%$ column of the table using a method called interpolation. Interpolation is covered in Chapter 3 for determining the 50th percentile.

The cumulative percentages in a frequency distribution table give the percentage of individuals with scores at or below each X value. However, you must remember that the X values in the table are usually measurements of a continuous variable and, therefore, represent intervals on the scale of measurement (see page 13). A score of $X = 2$, for example, means that the measurement was somewhere between the real limits of 1.5 and 2.5. Thus, when a table shows that a score of $X = 2$ has a cumulative percentage of 30%, you should interpret this as meaning that 30% of the individuals have been accumulated by the time you reach the top of the interval for $X = 2$. Notice that each cumulative percentage value is associated with the upper real limit of its interval; in this case X_{URL} is 2.5. This point also is demonstrated in Figure 2.2. Note the shaded area in the graph is the section

FIGURE 2.2

A frequency distribution histogram with shaded area in the graph below the upper real limit for $X = 2$. This corresponds to 30% of the distribution, the same as the percentile rank shown in the table for Example 2.6.

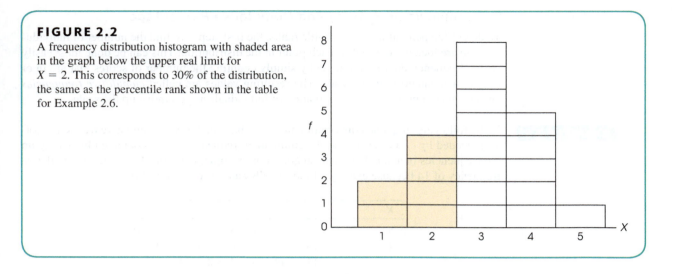

of the distribution below the upper real limit for $X = 2$. In terms of "blocks" there are 6 of 20 blocks in the shaded area of the graph. This corresponds to 30% of the distribution, the same as the percentile rank shown in the table for Example 2.6.

LEARNING CHECK   **LO1**   **1.** If the following scores are placed in a frequency distribution table, then what is the frequency value corresponding to $X = 3$? Scores: 2, 3, 1, 1, 3, 3, 2, 4, 3, 1

 a. 1
 b. 2
 c. 3
 d. 4

LO1   **2.** For the following distribution that reports the number of smiles displayed by a childcare worker to a baby in a 20-minute time frame, how many smiles were observed?

 a. 5
 b. 10
 c. 15
 d. 21

X	f
5	6
4	5
3	5
2	3
1	2

LO2   **3.** For the following frequency distribution, what is the value of ΣX^2?

 a. 50
 b. 55
 c. 74
 d. 225

X	f
5	1
4	0
3	2
2	1
1	3

LO3 4. In a distribution of exam scores, which of the following would be the highest score?

    a. The 20th percentile.

    b. The 80th percentile.

    c. A score with a percentile rank of 15%.

    d. A score with a percentile rank of 75%.

LO4 5. Following are three rows from a frequency distribution table. For this distribution, what is the 90th percentile?

    a. $X = 24.5$

    b. $X = 25$

    c. $X = 29$

    d. $X = 29.5$

X	$c\%$
30–34	100%
25–29	90%
20–24	60%

ANSWERS   **1. d   2. d   3. a   4. b   5. d**

2-2 Grouped Frequency Distribution Tables

LEARNING OBJECTIVE

5. Choose when it is useful to set up a grouped frequency distribution table, and use and create this type of table for a set of scores.

When a set of data covers a wide range of values, it is unreasonable to list all the individual scores in a frequency distribution table. Consider, for example, a set of exam scores that range from a low of $X = 41$ to a high of $X = 96$. These scores cover a *range* of more than 50 points.

If we were to list all the individual scores from $X = 96$ down to $X = 41$, it would take 56 rows to complete the frequency distribution table. Although this would organize the data, the table would be long and cumbersome. Remember: The purpose for constructing a table is to obtain a relatively simple, organized picture of the data. This can be accomplished by grouping the scores into intervals and then listing the intervals in the table instead of listing each individual score. For example, we could construct a table showing the number of students who had scores in the 90s, the number with scores in the 80s, and so on. The result is called a *grouped frequency distribution table* because we are presenting groups of scores rather than individual values. The groups, or intervals, are called *class intervals*.

There are several guidelines that help guide you in the construction of a grouped frequency distribution table. Note that these are simply guidelines, rather than absolute requirements, but they do help produce a simple, well-organized, and easily understood table.

> When the scores are whole numbers, the total number of rows for a regular table can be obtained by finding the difference between the highest and lowest scores and adding 1:
>
> rows = highest − lowest + 1

GUIDELINE 1   The grouped frequency distribution table should have about 10 class intervals. If a table has many more than 10 intervals, it becomes cumbersome and defeats the purpose of a frequency distribution table. On the other hand, if you have too few intervals, you begin to lose information about the distribution of the scores. At the extreme, with only one interval, the table would not tell you anything about how the scores are distributed. Remember that the purpose of a frequency distribution is to help a researcher see the data. With too few or too many intervals, the table will not provide a clear picture. You should note that 10 intervals is a general guide. If you are constructing a table on a blackboard, for example, you probably want only 5 or 6 intervals. If the table is to be printed in a scientific report, you may want 12 or 15 intervals. In each case, your goal is to present a table that is relatively easy to see and understand.

GUIDELINE 2   The width of each interval should be a relatively simple number. For example, 2, 5, 10, or 20 would be a good choice for the interval width. Notice that it is easy to count by 5s or 10s. These numbers are easy to understand because one can readily see how you have divided the range of scores.

GUIDELINE 3   The bottom score in each class interval should be a multiple of the width. If you are using a width of 10 points, for example, the intervals should start with 10, 20, 30, 40, and so on. Again, this makes it easier for someone to understand how the table has been constructed.

GUIDELINE 4   All intervals should be the same width. They should cover the range of scores completely with no gaps and no overlaps, so that any particular score belongs in exactly one interval.

The application of these rules is demonstrated in Example 2.7.

EXAMPLE 2.7   An instructor has obtained the set of $N = 25$ exam scores shown here. To help organize these scores, we will place them in a frequency distribution table. The scores are:

82	75	88	93	53	84	87	58	72	94	69	84	61
91	64	87	84	70	76	89	75	80	73	78	60	

The first step is to determine the range of scores. For these data, the smallest score is $X = 53$ and the largest score is $X = 94$, so a total of 42 rows would be needed for a table that lists each individual score. Because 42 rows would not provide a simple table, we have to group the scores into class intervals. The best method for finding a good interval width is a systematic trial-and-error approach that uses guidelines 1 and 2 simultaneously. Specifically, we want about 10 intervals and we want the interval width to be a simple number. For this example, the scores cover a range of 42 points, so we will try several different interval widths to see how many intervals are needed to cover this range. For example, if each interval were 2 points wide, it would take 21 intervals to cover a range of 42 points. This is too many, so we move on to an interval width of 5 or 10 points. The following table shows how many intervals would be needed for these possible widths:

Notice that an interval width of 5 will result in about 10 intervals, which is exactly what we want.

Width	Number of Intervals Needed to Cover a Range of 42 Points	
2	21	(too many)
5	9	(OK)
10	5	(too few)

The next step is to actually identify the intervals. The lowest score for these data is $X = 53$, so the lowest interval should contain this value. Because the interval should have a multiple of 5 as its bottom score, the interval should begin at 50. The interval has a width of 5, so it should contain 5 values: 50, 51, 52, 53, and 54. Thus, the bottom interval is 50–54. The next interval would start at 55 and go to 59. Note that this interval also has a

TABLE 2.2

This grouped frequency distribution table shows the data from Example 2.7. The original scores range from a high of $X = 94$ to a low of $X = 53$. This range has been divided into 9 intervals with each interval exactly 5 points wide. The frequency column (f) lists the number of individuals with scores in each of the class intervals.

X	f
90–94	3
85–89	4
80–84	5
75–79	4
70–74	3
65–69	1
60–64	3
55–59	1
50–54	1

bottom score that is a multiple of 5, and contains exactly 5 scores (55, 56, 57, 58, and 59). The complete frequency distribution table showing all of the class intervals is presented in Table 2.2.

Once the class intervals are listed, you complete the table by adding a column of frequencies. The values in the frequency column indicate the number of individuals who have scores located in that class interval. For this example, there were three students with scores in the 60–64 interval, so the frequency for this class interval is $f = 3$ (see Table 2.2). The basic table can be extended by adding columns showing the proportion and percentage associated with each class interval.

Finally, you should note that after the scores have been placed in a grouped table, you lose information about the specific value for any individual score. For example, Table 2.2 shows that one person had a score between 65 and 69, but the table does not identify the exact value for the score. In general, the wider the class intervals are, the more information is lost. In Table 2.2 the interval width is 5 points, and the table shows that there are three people with scores in the lower 60s and one person with a score in the upper 60s. This information would be lost if the interval width were increased to 10 points. With an interval width of 10, all of the 60s would be grouped together into one interval labeled 60–69. The table would show a frequency of four people in the 60–69 interval, but it would not tell whether the scores were in the upper 60s or the lower 60s. ■

■ Real Limits and Frequency Distributions

Recall from Chapter 1 that a continuous variable has an infinite number of possible values and can be represented by a number line that is continuous and contains an infinite number of points. However, when a continuous variable is measured, the resulting measurements correspond to *intervals* on the number line rather than single points. If you are measuring time in seconds, for example, a score of $X = 8$ seconds actually represents an interval bounded by the real limits 7.5 seconds and 8.5 seconds. Thus, a frequency distribution table showing a frequency of $f = 3$ individuals all assigned a score of $X = 8$ does not mean that all three individuals had exactly the same measurement. Instead, you should realize that the three measurements are simply located in the same interval between 7.5 and 8.5.

The concept of real limits also applies to the class intervals of a grouped frequency distribution table. For example, a class interval of 40–49 contains scores from $X = 40$ to $X = 49$. These values are called the *apparent limits* of the interval because it appears that they form the upper and lower boundaries for the class interval. If you are measuring a continuous variable, however, a score of $X = 40$ is actually an interval from 39.5 to 40.5. Similarly, $X = 49$ is an interval from 48.5 to 49.5. Therefore, the real limits of the interval are 39.5 (the lower real limit) and 49.5 (the upper real limit). Notice that the next higher-class interval is 50–59, which has a lower real limit of 49.5. Thus, the two intervals meet at the real limit 49.5, so there are no gaps in the scale. You also should notice that the width of each class interval becomes easier to understand when you consider the real limits of an interval. For example, the interval 50–59 has real limits of 49.5 and 59.5. The distance between these two real limits (10 points) is the width of the interval.

LEARNING CHECK   **LO5 1.** A set of scores ranges from a high of $X = 86$ to a low of $X = 17$. If these scores are placed in a grouped frequency distribution table with an interval width of 10 points, the top interval in the table would be _____.
  a. 80–89
  b. 80–90
  c. 81–90
  d. 77–86

LO5 2. What is the highest score in the following distribution?
  a. $X = 16$
  b. $X = 17$
  c. $X = 1$
  d. Cannot be determined.

X	f
24–25	2
22–23	4
20–21	6
18–19	3
16–17	1

LO5 3. Which of the following statements is *false* regarding grouped frequency distribution tables?
  a. An interval width should be used that yields about 10 intervals.
  b. Intervals are listed in descending order, starting with the highest value at the top of the X column.
  c. The bottom score for each interval is a multiple of the interval width.
  d. The value for N can be determined by counting the number of intervals in the X column.

ANSWERS   **1. a   2. d   3. d**

2-3 Frequency Distribution Graphs

LEARNING OBJECTIVES

6. Describe how the three types of frequency distribution graphs—histograms, polygons, and bar graphs—are constructed and identify when each is used.

7. Use and create frequency distribution graphs and explain how they are related to the original set of scores.

8. Explain how frequency distribution graphs for populations differ from the graphs used for samples.

9. Identify the shape of a distribution—symmetrical, positively or negatively skewed—by looking at a frequency distribution table or graph.

A frequency distribution graph is basically a picture of the information available in a frequency distribution table. We will consider several different types of graphs, but all start with two perpendicular lines called *axes*. The horizontal line is the *X*-axis, or the abscissa (ab-SIS-uh). The vertical line is the *Y*-axis, or the ordinate. The measurement scale (set of *X* values) is listed along the *X*-axis with values increasing from left to right. The frequencies are listed on the *Y*-axis with values increasing from bottom to top. As a general rule, the point where the two axes intersect should have a value of zero for both the scores and the frequencies. A final general rule is that the graph should be constructed so that its height (*Y*-axis) is approximately two-thirds to three-quarters of its length (*X*-axis). Violating these guidelines can result in graphs that give a misleading picture of the data (see Box 2.1, page 60).

■ Graphs for Interval or Ratio Data

When the data consist of numerical scores that have been measured on an interval or ratio scale, there are two options for constructing a frequency distribution graph. The two types of graphs are called *histograms* and *polygons*.

Histograms   To construct a histogram, you first list the numerical scores or class intervals (the categories of measurement) along the *X*-axis. Then you draw a bar above each *X* value so that

  a. the height of the bar corresponds to the frequency for that category.

  b. for continuous variables, the width of the bar extends to the real limits of the category. For discrete variables, each bar extends exactly half the distance to the adjacent category on each side.

For both continuous and discrete variables, each bar in a histogram extends to the midpoint between adjacent categories. As a result, adjacent bars touch and there are no spaces or gaps between bars. An example of a histogram is shown in Figure 2.3.

   When data have been grouped into class intervals, you can construct a frequency distribution histogram by drawing a bar above each interval so that the width of the bar extends exactly half the distance to the adjacent category on each side. This process is demonstrated in Figure 2.4.

   For the two histograms shown in Figures 2.3 and 2.4, notice that the values on both the vertical and horizontal axes are clearly marked and that both axes are labeled. Also note that, whenever possible, the units of measurement are specified; for example, Figure 2.4 shows a distribution of heights measured in inches. Finally, notice that the horizontal axis in Figure 2.4 does not list all the possible heights starting from zero and going up to 45 inches. Instead, the graph clearly shows a break between zero and 30, indicating that some scores have been omitted.

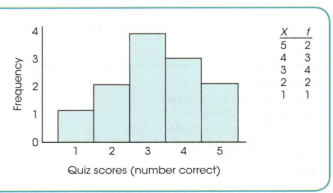

FIGURE 2.3
An example of a frequency distribution histogram. The same set of quiz scores is presented in a frequency distribution table and in a histogram.

FIGURE 2.4

An example of a frequency distribution histogram for grouped data. The same set of children's heights is presented in a frequency distribution table and in a histogram.

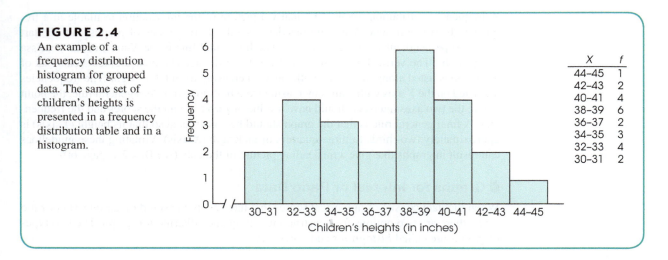

X	f
44–45	1
42–43	2
40–41	4
38–39	6
36–37	2
34–35	3
32–33	4
30–31	2

An Informal Histogram A slight modification to the traditional histogram produces an easily drawn and simple to understand sketch of a frequency distribution. Instead of drawing a bar above each score, the informal sketch consists of drawing a stack of blocks. Each block represents one individual, so the number of blocks above each score corresponds to the frequency for that score. An example is shown in Figure 2.5.

Note that the number of blocks in each stack makes it very easy to see the absolute frequency for each category. In addition, it is easy to see the exact difference in frequency from one category to another. In Figure 2.5, for example, there are exactly two more people with scores of $X = 2$ than with scores of $X = 1$. Because the frequencies are clearly displayed by the number of blocks, this type of display eliminates the need for a vertical line (the Y-axis) showing frequencies. In general, this kind of graph provides a simple picture of the distribution for a sample of scores. Note that we often will use this kind of graph to show sample data throughout the book. You should also note, however, that this kind of display simply provides a quick, informal sketch of the distribution. For formal presentations, such as a paper in a scientific journal or a presentation at a conference, a histogram with bars and the labeled axis for frequencies should be used.

Polygons The second option for graphing a distribution of numerical scores from an interval or ratio scale of measurement is called a polygon. To construct a polygon, you begin by listing the numerical scores (the categories of measurement) along the X-axis. Then:

a. A dot is centered above each score so that the vertical position of the dot corresponds to the frequency for the category.

b. A continuous line is drawn from dot to dot to connect the series of dots.

FIGURE 2.5

A frequency distribution graph in which each individual is represented by a block placed directly above the individual's score. For example, three people had scores of $X = 2$.

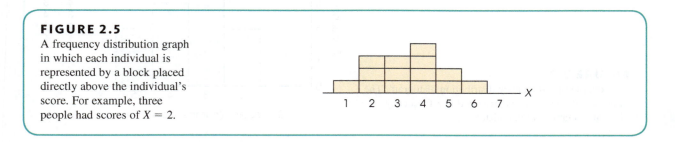

FIGURE 2.6

An example of a frequency distribution polygon. The same set of data is presented in a frequency distribution table and in a polygon.

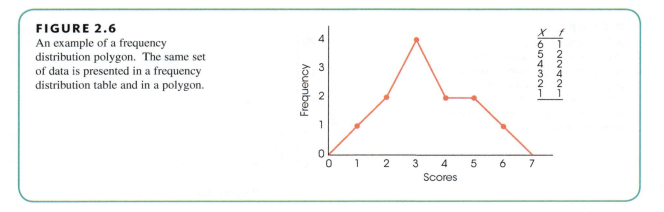

X	f
6	1
5	2
4	2
3	4
2	2
1	1

c. The graph is completed by drawing a line down to the *X*-axis (zero frequency) at each end of the range of scores. The final lines are usually drawn so that they reach the *X*-axis at a point that is one category below the lowest score on the left side and one category above the highest score on the right side. An example of a polygon is shown in Figure 2.6.

A polygon also can be used with data that have been grouped into class intervals. For a grouped distribution, you position each dot directly above the midpoint of the class interval. The midpoint can be found by averaging the highest and the lowest scores in the interval. For example, a class interval that is listed as 20–29 would have a midpoint of 24.5.

$$\text{midpoint} = \frac{20 + 29}{2} = \frac{49}{2} = 24.5$$

An example of a frequency distribution polygon with grouped data is shown in Figure 2.7.

■ Graphs for Nominal or Ordinal Data

When the scores are measured on a nominal or ordinal scale (usually non-numerical values), the frequency distribution can be displayed in a *bar graph*.

Bar Graphs   A bar graph is essentially the same as a histogram, except that spaces are left between adjacent bars. For a nominal scale, the space between bars emphasizes that

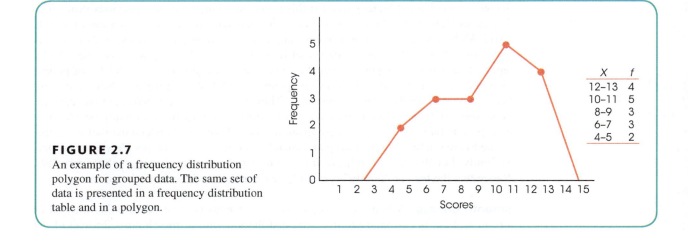

X	f
12–13	4
10–11	5
8–9	3
6–7	3
4–5	2

FIGURE 2.7

An example of a frequency distribution polygon for grouped data. The same set of data is presented in a frequency distribution table and in a polygon.

FIGURE 2.8

A bar graph showing the distribution of personality types in a sample of college students. Because personality type is a discrete variable measured on a nominal scale, the graph is drawn with space between the bars.

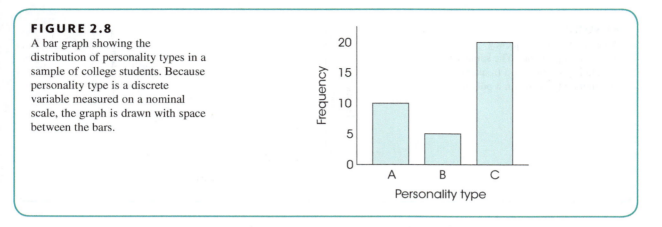

the scale consists of separate, distinct categories. For ordinal scales, separate bars are used because you cannot assume that the categories are all the same size.

To construct a bar graph, list the categories of measurement along the X-axis and then draw a bar above each category so that the height of the bar corresponds to the frequency for the category. An example of a bar graph is shown in Figure 2.8.

■ Graphs for Population Distributions

When you can obtain an exact frequency for each score in a population, you can construct frequency distribution graphs that are exactly the same as the histograms, polygons, and bar graphs that are typically used for samples. For example, if a population is defined as a specific group of $N = 50$ people, we could easily determine how many have IQs of $X = 110$. However, if we were interested in the entire population of adults in the United States, it would be impossible to obtain an exact count of the number of people with an IQ of 110. Although it is still possible to construct graphs showing frequency distributions for extremely large populations, the graphs usually involve two special features: relative frequencies and smooth curves.

Relative Frequencies   Sometimes samples are so large that reporting absolute frequencies does not sufficiently simplify the data. A common alternative is using *relative frequencies*. For example, the American Pet Products Association estimated that in 2017–18 there were nearly 85 million households with a pet (as reported by the Humane Society of the United States) and that this reflected an increase over previous years. The American Veterinary Medical Association has studied how pet owners view their pets (2012 AVMA Sourcebook) by using a sample of more than 50,000 households from this population. Rather than report the actual frequencies, which are quite large, the AVMA reported the findings as percentages. For example, it was observed that 63.2% of people view their pets as family, 35.8% as companions, and 1.0% as property. Note that these percentages are not the actual frequencies. They are *relative frequencies*, but one can still make some statements about these data. For example, almost twice as many people view their pets as family compared to companions. You should also understand that these frequencies are relative to 100. So, approximately 63 out of every 100 people view their pets as family. Finally, data for relative frequencies can be displayed in a graph (Figure 2.9). Notice that the bar for "family" is roughly twice as tall as the one for "companion."

Smooth Curves   When a population consists of numerical scores from an interval or a ratio scale, it is customary to draw the distribution with a smooth curve instead of the

FIGURE 2.9
An example of a relative frequency distribution. Percentage of owners who view their pets as family members, companions, or property.

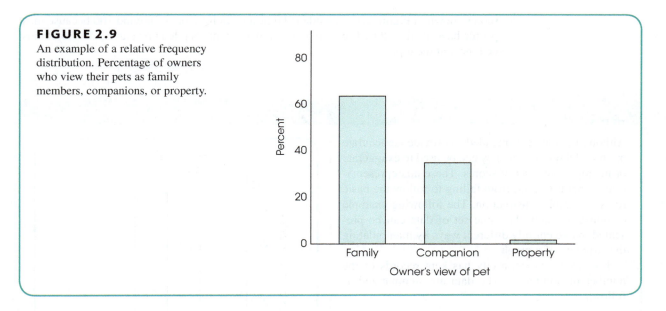

jagged, step-wise shapes that occur with histograms and polygons. The smooth curve indicates that you are not connecting a series of dots (real frequencies) but instead are showing the relative changes that occur from one score to the next. One commonly occurring population distribution is the normal curve. The word *normal* refers to a specific shape that can be precisely defined by an equation. Less precisely, we can describe a *normal distribution* as being symmetrical, with the greatest frequency in the middle and relative frequencies decreasing as you approach either extreme. A good example of a normal distribution is the population distribution for IQ scores shown in Figure 2.10. Because normal-shaped distributions occur commonly and because this shape is mathematically guaranteed in certain situations, we give it extensive attention throughout this book.

In the future, we will be referring to *distributions of scores*. Whenever the term *distribution* appears, you should conjure up an image of a frequency distribution graph. The graph provides a picture showing exactly where the individual scores are located. To make this concept more concrete, you might find it useful to think of the graph as showing a pile of individuals just like we showed a pile of blocks in Figure 2.4. For the population of IQ

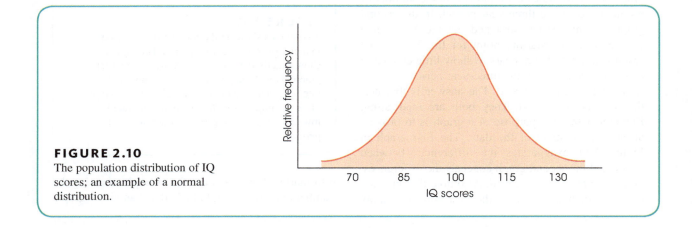

FIGURE 2.10
The population distribution of IQ scores; an example of a normal distribution.

scores shown in Figure 2.10, the pile is highest at an IQ score of around 100 because most people have average IQs. There are only a few individuals piled up at an IQ of 130; it must be lonely at the top.

BOX 2.1 The Use and Misuse of Graphs

Although graphs are intended to provide an accurate picture of a set of data, they can be used to exaggerate or misrepresent a set of scores. These misrepresentations generally result from failing to follow the basic rules for graph construction. The following example demonstrates how the same set of data can be presented in two entirely different ways by manipulating the structure of a graph.

For several years, a city has kept records of the number of homicides. The data are summarized as follows:

Year	Number of Homicides
2016	218
2017	225
2018	229

In an election year, a candidate for mayor posts a graph on Facebook that suggests the incumbent mayor has done a poor job of addressing the homicide problem in the city. Not to be upstaged, the current mayor posts her own graph on Facebook to support the claim she has a strong track record of preventing homicides from getting worse. Their graphs of the homicide numbers are shown in Figure 2.11. In the first graph, the candidate has exaggerated the height of the Y-axis for frequency and started numbering the Y-axis at 215 rather than at zero. As a result, the graph seems to indicate a rapid rise in the number of homicides over the three-year period. In the second graph, the mayor has stretched out the X-axis and used zero as the starting point for the Y-axis. The result is a graph that appears to show little change in the homicide rate over the three-year period.

Which graph is correct? The answer is that neither one is very good. They both are misleading. Remember that the purpose of a graph is to provide an accurate display of the data. The first graph in Figure 2.11 exaggerates the differences between years, and the second graph conceals the differences. Some compromise is needed. Also note that in some cases a graph may not be the best way to display

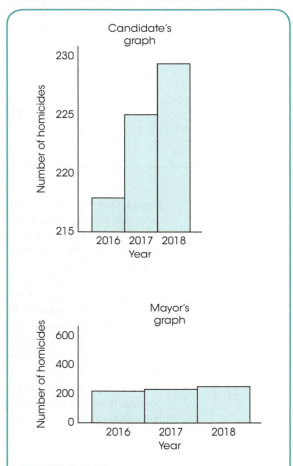

FIGURE 2.11
Two graphs showing the number of homicides in a city over a three-year period. Both graphs show exactly the same data. However, the first graph gives the appearance that the homicide rate is high and rising rapidly. The second graph gives the impression that the homicide rate is low and has not changed over the three-year period.

information. For these data, for example, showing the numbers in a table would be better than either graph.

■ The Shape of a Frequency Distribution

Rather than drawing a complete frequency distribution graph, researchers often simply describe a distribution by listing its characteristics. There are three characteristics that completely describe any distribution: shape, central tendency, and variability. In simple terms, central tendency measures where the center of the distribution is located, and variability measures the degree to which the scores are spread over a wide range or clustered together. Central tendency and variability are covered in detail in Chapters 3 and 4. Technically, the shape of a distribution is defined by an equation that prescribes the exact relationship between each X and Y value on the graph. However, we will rely on a few less-precise terms that serve to describe the shape of most distributions.

Nearly all distributions can be classified as being either symmetrical or skewed.

In a **symmetrical distribution**, it is possible to draw a vertical line through the middle so that one side of the distribution is a mirror image of the other (see Figure 2.12).

In a **skewed distribution**, the scores tend to pile up toward one end of the scale and taper off gradually at the other end (see Figure 2.12).

The section where the scores taper off toward one end of a distribution is called the **tail of the distribution**.

A skewed distribution with the tail on the right-hand side is **positively skewed** because the tail points toward the positive (above-zero) end of the X-axis. If the tail points to the left, the distribution is **negatively skewed** (see Figure 2.12).

For a very difficult exam, most scores tend to be low, with only a few individuals earning high scores. This produces a positively skewed distribution. Similarly, a very easy exam tends to produce a negatively skewed distribution, with most of the students earning high scores and only a few with low values.

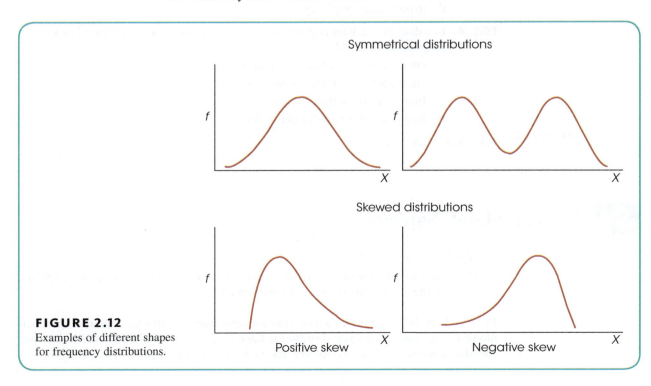

FIGURE 2.12

Examples of different shapes for frequency distributions.

Not all distributions are perfectly symmetrical or obviously skewed in one direction. Therefore, it is common to modify these descriptions of shape with phrases like "roughly symmetrical" or "tends to be positively skewed." The goal is to provide a general idea of the appearance of the distribution.

LEARNING CHECK   **LO6** **1.** Which of the following measurement scales are displayed by frequency distribution polygons?

   a. Either interval or ratio scales.

   b. Only ratio scales.

   c. Either nominal or ordinal scales.

   d. Only nominal scales.

LO7 **2.** A group of quiz scores is shown in a histogram. If the bars in the histogram gradually increase in height from left to right, what can you conclude about the set of quiz scores?

   a. There are more high scores than there are low scores.

   b. There are more low scores than there are high scores.

   c. The height of the bars always decreases as the scores increase.

   d. None of the above.

LO8 **3.** Instead of showing the actual number of individuals in each category, a population frequency distribution graph usually shows a(n)_____.

   a. estimated frequency

   b. grouped frequency

   c. relative frequency

   d. hypothetical frequency

LO9 **4.** In a distribution with negative skew, where are the scores with the highest frequencies located?

   a. On the right side of the distribution.

   b. On the left side of the distribution.

   c. In the middle of the distribution.

   d. Represented at two distinct peaks.

ANSWERS   **1. a   2. a   3. c   4. a**

2-4   Stem and Leaf Displays

LEARNING OBJECTIVE

10. Construct and describe the basic elements of a stem and leaf display and explain how the display shows the entire distribution of scores.

In 1977, J. W. Tukey presented a technique for organizing data that provides a simple alternative to a grouped frequency distribution table or graph (Tukey, 1977). This technique, called a *stem and leaf display*, requires that each score be separated into two parts: The first digit (or digits) is called the *stem*, and the last digit is called the *leaf*. For example, $X = 85$ would be separated into a stem of 8 and a leaf of 5. Similarly, $X = 42$ would have a stem of 4 and a leaf of 2. To construct a stem and leaf display for a set of scores, the first step is to list all the stems in a column. For the data in Table 2.3, for example, the lowest scores are in the 30s and the highest scores are in the 90s, so the list of stems would be

Stems
3
4
5
6
7
8
9

The next step is to go through the scores, one score at a time, and write the leaf for each score beside its stem. For the data in Table 2.3, the first score is $X = 83$, so you would write 3 (the leaf) beside the 8 in the column of stems. This process is continued for the entire set of scores. The complete stem and leaf display is shown with the original data in Table 2.3. When constructing a stem and leaf display by hand, the leaves for each stem do not have to be sorted in ascending order. However, when using statistical software for this task, the leaves will be sorted by magnitude. For example, in Table 2.3 the leaves for stem 7 would be displayed as 1344668.

■ Comparing Stem and Leaf Displays with Grouped Frequency Distributions

Notice that the stem and leaf display is similar to a grouped frequency distribution. Each of the stem values corresponds to a class interval. For example, the stem 3 represents all scores in the 30s—that is, all scores in the interval 30–39. The number of leaves in the display shows the frequency associated with each stem. It also should be clear that the stem and leaf display has one important advantage over a traditional grouped frequency distribution. Specifically, the stem and leaf display allows you to identify every individual score in the data. In the display shown in Table 2.3, for example, you know that there were three scores in the 60s and that the specific values were 62, 68, and 63. A grouped frequency distribution would tell you only the number of scores in a class interval. It will not tell you the specific values. This advantage can be very valuable, especially if you need to do any calculations with the original scores. For example, if you need to add all the scores, you can recover the actual values from the stem and leaf display and compute the total. With a grouped frequency distribution, however, the individual scores are not available.

TABLE 2.3

A set of $N = 24$ scores presented as raw data and organized in a stem and leaf display.

	Data		Stem	and Leaf Display
83	82	63	3	23
62	93	78	4	26
71	68	33	5	6279
76	52	97	6	283
85	42	46	7	1643846
32	57	59	8	3521
56	73	74	9	37
74	81	76		

LO10 1. For the scores shown in the following stem and leaf display, what is the lowest score in the distribution?

a. 7

b. 15

c. 50

d. 51

9	374
8	945
7	7042
6	68
5	14

LO10 2. For the scores shown in the following stem and leaf display, how many people had scores in the 70s?

a. 1

b. 2

c. 3

d. 4

9	374
8	945
7	7042
6	68
5	14

ANSWERS   **1. d 2. d**

SUMMARY

1. The goal of descriptive statistics is to simplify the organization and presentation of data. One descriptive technique is to place the data in a frequency distribution table or graph that shows exactly how many individuals (or scores) are located in each category on the scale of measurement.

2. A frequency distribution table lists the categories that make up the scale of measurement (the X values) in one column. Beside each X value, in a second column, is the frequency or number of individuals in that category. The table may include a proportion column showing the relative frequency for each category:

$$\text{proportion} = p = \frac{f}{n}$$

The table may include a percentage column showing the percentage associated with each X value:

$$\text{percentage} = p(100) = \frac{f}{n}(100)$$

3. The cumulative percentage is the percentage of individuals with scores at or below a particular point in the distribution. The cumulative percentage values are associated with the upper real limits of the corresponding scores or intervals.

4. Percentiles and percentile ranks are used to describe the position of individual scores within a distribution.

Percentile rank gives the cumulative percentage associated with a particular score. A score that is identified by its rank is called a percentile.

5. It is recommended that a frequency distribution table have a maximum of 10–15 rows to keep it simple. If the scores cover a range that is wider than this suggested maximum, it is customary to divide the range into sections called class intervals. These intervals are then listed in the frequency distribution table along with the frequency or number of individuals with scores in each interval. The result is called a grouped frequency distribution. The guidelines for constructing a grouped frequency distribution table are as follows:

a. There should be about 10 intervals.

b. The width of each interval should be a simple number (e.g., 2, 5, or 10).

c. The bottom score in each interval should be a multiple of the width.

d. All intervals should be the same width, and they should cover the range of scores with no gaps.

6. A frequency distribution graph lists scores on the horizontal axis and frequencies on the vertical axis. The type of graph used to display a distribution depends on the scale of measurement used. For interval or ratio scales, you should use a histogram or a polygon. For a histogram, a bar is drawn above each score so that the height of the bar corresponds to the frequency. Each

bar extends to the real limits of the score, so that adjacent bars touch. For a polygon, a dot is placed above the midpoint of each score or class interval so that the height of the dot corresponds to the frequency; then lines are drawn to connect the dots. Bar graphs are used with nominal or ordinal scales. Bar graphs are similar to histograms except that gaps are left between adjacent bars.

7. Shape is one of the basic characteristics used to describe a distribution of scores. Most distributions can be classified as either symmetrical or skewed. A skewed distribution that tails off to the right is positively skewed. If it tails off to the left, it is negatively skewed.

8. A stem and leaf display is an alternative procedure for organizing data. Each score is separated into a stem (the first digit or digits) and a leaf (the last digit). The display consists of the stems listed in a column with the leaf for each score written beside its stem. A stem and leaf display is similar to a grouped frequency distribution table; however, the stem and leaf display identifies the exact value of each score and the grouped frequency distribution does not.

KEY TERMS

frequency distribution (45)

frequency distribution table (45)

proportion (p) (47)

percentage (48)

percentile (48)

percentile rank (48)

cumulative frequency (cf) (49)

cumulative percentage (c%) (49)

range (51)

grouped frequency distribution (51)

class interval (51)

apparent limits (53)

histogram (55)

polygon (56)

bar graph (57)

relative frequency (58)

normal distribution (59)

symmetrical distribution (61)

skewed distribution (61)

tail(s) of a distribution (61)

positively skewed distribution (61)

negatively skewed distribution (61)

stem and leaf display (62)

FOCUS ON PROBLEM SOLVING

1. When constructing or working with a grouped frequency distribution table, a common mistake is to calculate the interval width by using the highest and lowest values that define each interval. For example, some students are tricked into thinking that an interval identified as 20–24 is only 4 points wide. To determine the correct interval width, you can
   a. Count the individual scores in the interval. For this example, the scores are 20, 21, 22, 23, and 24 for a total of 5 values. Thus, the interval width is 5 points.
   b. Use the real limits to determine the real width of the interval. For example, an interval identified as 20–24 has a lower real limit of 19.5 and an upper real limit of 24.5 (halfway to the next score). Using the real limits, the interval width is

$$24.5 - 19.5 = 5 \text{ points}$$

DEMONSTRATION 2.1

A GROUPED FREQUENCY DISTRIBUTION TABLE

For the following set of $N = 20$ scores, construct a grouped frequency distribution table using an interval width of 5 points. The scores are:

14	8	27	16	10	22	9	13	16	12
10	9	15	17	6	14	11	18	14	11

STEP 1   **Set up the class intervals.**   The largest score in this distribution is $X = 27$, and the lowest is $X = 6$. Therefore, a frequency distribution table for these data would have 22 rows and would be too large. A grouped frequency distribution table would be better. We have asked specifically for an interval width of five points, and the resulting table has five rows.

X
25–29
20–24
15–19
10–14
5–9

Remember that the interval width is determined by the real limits of the interval. For example, the class interval 25–29 has an upper real limit of 29.5 and a lower real limit of 24.5. The difference between these two values is the width of the interval—namely, 5.

STEP 2   **Determine the frequencies for each interval.**   Examine the scores, and count how many scores fall into the class interval of 25–29. Cross out each score that you have already counted. Record the frequency for this class interval. Now repeat this process for the remaining intervals. The result is the following table:

X	f	
25–29	1	(the score $X = 27$)
20–24	1	($X = 22$)
15–19	5	(the scores $X = 16, 16, 15, 17,$ and 18)
10–14	9	($X = 14, 10, 13, 12, 10, 14, 11, 14,$ and 11)
5–9	4	($X = 8, 9, 9,$ and 6)

DEMONSTRATION 2.2

FINDING PERCENTILES AND PERCENTILE RANKS

Find the 50th percentile for the following frequency distribution table.

X	f
15	1
14	1
13	1
12	2
11	3
10	2

STEP 1   **Find the cumulative frequency (*cf*) and cumulative percentage values and add these values to the basic frequency table.**   Cumulative frequencies indicate the number of individuals located in or below each score. To find these frequencies, begin with the bottom score, then add the frequencies as you move up the column. For this example, there are 2 individuals with a score of 10 (*cf* = 2). Moving up the column, the score of 11 contains an additional 3 individuals, so the cumulative value for this score is 2 + 3 = 5 (simply add the 3 individuals that received a score of 11 to the number of individuals who received scores below 11). Continue moving up the column, cumulating frequencies for each interval.

Cumulative percentages are determined from the cumulative frequencies by the relationship

$$c\% = \left(\frac{cf}{N}\right)100\%$$

For example, the *cf* column shows that 2 individuals (out of the total set of $N = 10$) have scores of 10. The corresponding cumulative percentage is

$$c\% = \left(\frac{2}{10}\right)100\% = \left(\frac{1}{5}\right) = 20\%$$

The complete set of cumulative frequencies and cumulative percentages is shown in the following table:

X	f	cf	c%
15	1	10	100%
14	1	9	90%
13	1	8	80%
12	2	7	70%
11	3	5	50%
10	2	2	20%

STEP 2 **Locate the score that corresponds to the percentile that you are asked to find.** In this example, 50% is listed for a score of $X = 11$. However, the cumulative percentages in the *c%* column are associated with the upper real limits of the scores listed in the first column.

STEP 3 **Identify the upper real limit of the score.** In this example, the upper real limit of 11 is 11.5. Thus, the 50th percentile is 11.5.

SPSS®

General instructions for using SPSS are presented in Appendix D. Following are detailed instructions for using SPSS to produce **Frequency Distribution Tables and Graphs**.

Demonstration Example

Suppose that an instructor is interested describing the distribution of quiz scores from her class. The instructor records the following quiz scores:

Student	Quiz scores
A	19
B	22
C	22
D	25
E	23
F	16
G	19
H	22
I	21
J	24

Student	Quiz scores
K	21
L	18
M	21
N	22
O	23
P	24
Q	23
R	20
S	20
T	20

Here, we will use SPSS to summarize the distribution of scores with a frequency distribution table and a graph.

Data Entry

1. Enter information in the **Variable View.** In the **Name** field, enter a short, descriptive name for the variable that does not include spaces. Here, "score" is used. The default settings for **Type, Width, Values, Missing, Align, and Role** are acceptable.
2. For **Decimals**, enter "0."
3. In the **Label** field, a descriptive title for the variable should be used. Here, we used "Score on Quiz 7 in PSY 101."
4. In the **Measure** field, select **Scale** because quiz score is a ratio scale.
5. In the **Data View** section, enter the quiz scores in the "score" column.

Data Analysis

1. Click **Analyze** on the tool bar, select **Descriptive Statistics**, and click on **Frequencies**.
2. Highlight the column label for the set of scores (score) in the left box and click the arrow to move it into the **Variable** box.
3. Click **Charts.**
4. Select either **Bar Graphs** or **Histogram**.
5. Click **Continue**.
6. Be sure that the option to **Display Frequency Table** is selected.
7. Click **OK**.

SPSS Output

It is not uncommon for a SPSS output to contain multiple sections. This output has three. The first section ("Statistics") reports the number of scores.

Statistics

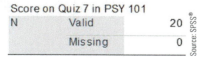

Score on Quiz 7 in PSY 101		
N	Valid	20
	Missing	0

The second section reports the frequency distribution table. The frequency distribution table will list the score values in the left-most column. Scores are sorted from smallest to largest, which is different from the largest to smallest arrangement of frequency tables that you have seen in this text. Score values that do not occur (zero frequencies) are not included in the table, and the program does not group scores into class intervals (all values are listed). SPSS also reports percentage and cumulative percentage also listed for each score.

Score on Quiz 7 in PSY 101

		Frequency	Percent	Valid Percent	Cumulative Percent
Valid	16	1	5.0	5.0	5.0
	18	1	5.0	5.0	10.0
	19	2	10.0	10.0	20.0
	20	3	15.0	15.0	35.0
	21	3	15.0	15.0	50.0
	22	4	20.0	20.0	70.0
	23	3	15.0	15.0	85.0
	24	2	10.0	10.0	95.0
	25	1	5.0	5.0	100.0
	Total	20	100.0	100.0	

Source: SPSS®

You will notice that in our quiz score example, SPSS reports that $X = 22$ occurred four times in the dataset (i.e., the value in the "Frequency" column reports the absolute frequency of the score). Relatedly, SPSS reports a measure of relative frequency—percent. $X = 22$ consisted of 20% of the scores in the dataset because $Percent = 100\left(\frac{f}{N}\right) = 100\left(\frac{4}{20}\right)$.

Moreover, the Cumulative Percent column reports a value of 70.0 for $X = 22$. This means that the upper real limit of $X = 22$ is the 70th percentile.

The third section displays a histogram of quiz scores. SPSS will display a frequency distribution table and a graph. Note that SPSS often produces a histogram that groups the scores in unpredictable intervals. A bar graph usually produces a clearer picture of the actual frequency associated with each score.

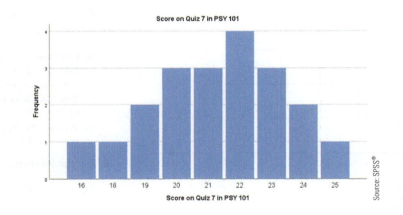

Score on Quiz 7 in PSY 101

Source: SPSS®

Try It Yourself

For the following set of scores, use SPSS to summarize the distribution with a frequency table and a histogram.

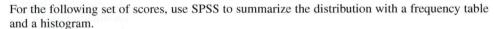

10   9   6   9   9   9   8   11   11   10   14   10   12   9   11   13

What is the cumulative percent value corresponding to the upper real limit of $X = 13$? You should come up with 93.8%.

PROBLEMS

1. For the following set of scores:

9	10	7	8	15	11	13	12	9	10
14	13	10	10	11	7	12	12	14	13

   a. Place the scores in a frequency distribution table. Include columns for proportion and percentage in your table.

   b. $n = ?$

2. For the following set of scores:

2	6	4	4	3	6	7	5	4	8
4	5	8	3	5	5	7	6	1	4

   a. Place the scores in a frequency distribution table. Include columns for proportion and percentage in your table.

   b. $n = ?$

3. Using the following informal histogram, calculate each of the following.

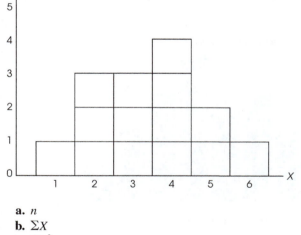

   a. n
   b. ΣX
   c. ΣX^2

4. Based on the following absolute frequency table, calculate each of the following.

X	f
15	1
14	1
13	2
12	3
11	5
10	4

   a. n
   b. ΣX
   c. ΣX^2

5. Find each of the following values for the distribution shown in the following polygon.

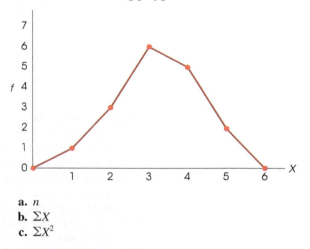

   a. n
   b. ΣX
   c. ΣX^2

6. For the following set of scores:

6	3	7	6	1	2	7	3	6	4
1	2	4	7	4	4	5	5	6	6

   a. Construct a frequency distribution table to organize the scores. Include cumulative frequency and cumulative percent.

   b. What is the percentile rank of the upper real limit of $X = 5$?

   c. What is the upper real limit of the score that corresponds to the 50th percentile?

7. For the following set of scores:

18	15	16	18	17	15	13	17	14	19
16	13	16	14	15	17	20	16	17	19

   a. Construct a frequency distribution table to organize the scores. Include cumulative frequency and cumulative percent.

   b. What is the percentile rank of the upper real limit of $X = 15$?

   c. What is the upper real limit of the score that corresponds to the 75th percentile?

8. For each of the following, list the class intervals that would be best for a grouped frequency distribution.
   a. Lowest $X = 3$, highest $X = 84$
   b. Lowest $X = 17$, highest $X = 32$
   c. Lowest $X = 52$, highest $X = 97$

9. You are interested in how much time you spend on Instagram™ so you recorded the number of minutes spent browsing your newsfeed each day for three weeks. You obtain the following data:

35	62	25	29	37	27	64
17	24	46	14	29	28	54
17	39	32	39	73	41	23

a. Create a grouped frequency distribution table that (i) has the best possible class interval width and (ii) an appropriate number of class intervals.
b. Describe the shape of the distribution.
c. For each class interval, identify the upper and lower real limits.

10. The international affective picture system is a collection of images that differ in their emotional content. The system contains some images that evoke fear in participants (e.g., a photograph of a spider), some images that have little emotional content (e.g., a pair of shoes), and some images have emotionally appealing images (e.g., a beautiful landscape). These images are useful to researchers who study emotion because participants' responses to these images are distributed in well-understood ways (Marchewka, Zurawski, Jenorog, & Grabowska, 2014). In one study, participants rated the pleasantness of a set of landscape images that were judged to be relaxing on a continuous scale, ranging between one being "very negative" and nine being "very positive". The following are a set of scores similar to those obtained by the researchers.

8	6	8	6	6	7	3
9	4	6	9	5	7	7
6	5	5	6	5	7	9

a. Construct a frequency distribution table to organize the scores.
b. Draw a frequency distribution histogram for these data.

11. Describe the difference in appearance between a bar graph and a histogram and describe the circumstances in which each type of graph is used to represent sample data. How would the same variables be represented in a population?

12. The following scores are the ages for a random sample of $n = 32$ drivers who were issued parking tickets in Chicago during 2019. Determine the best interval width and place the scores in a grouped frequency distribution table. From looking at your table, does it appear that tickets are issued equally across age groups?

57	30	45	59	39	53	28	19
34	21	34	38	52	29	64	39
22	44	46	26	56	20	33	58
32	25	48	22	51	26	63	51

13. What information is available about the scores in a regular frequency distribution table that you cannot obtain for the scores in a grouped table?

14. Draw a polygon for the distribution of scores shown in the following table.

X	f
6	2
5	5
4	3
3	2
2	1

15. For the following set of scores:

12	13	8	14	10	8	9	13	9
9	14	8	12	8	13	13	7	12

a. Organize the scores in a frequency distribution table.
b. Based on the frequencies, identify the shape of the distribution.

16. Place the following scores in a frequency distribution table. Based on the frequencies, what is the shape of the distribution?

15	14	9	10	15	12	14	11	13
14	13	14	12	14	13	13	12	11

17. A survey given to a sample of college students contained questions about the following variables. For each variable, identify the kind of graph that should be used to display the distribution of scores (histogram, polygon, or bar graph).
a. Age
b. Birth-order position among siblings (oldest = first)
c. Academic major
d. Registered voter (yes/no)

18. For the following set of scores:

7	5	6	4	4	3	8	9	4	7	5	5	6
9	4	7	5	10	6	8	5	6	3	4	8	5

a. Construct a frequency distribution table.
b. Sketch a histogram showing the distribution.
c. What is the shape of the distribution?

19. A local fast-food restaurant normally sells coffee in three sizes—small, medium, and large—at three different prices. Recently they had a special sale, charging only $1 for any sized coffee. During the sale, an employee recorded the number of each coffee size that was purchased on Wednesday morning. The following Wednesday, when prices had returned to normal, she again recorded the number of coffees sold for each size. The results are shown in the following table.

Regular Prices		All Sizes for $1	
X	f	X	f
Large	12	Large	41
Medium	25	Medium	27
Small	31	Small	11

a. What kind of graph would be appropriate for showing the distribution of coffee sizes for each of the two time periods?
b. Draw the two frequency distribution graphs.
c. Based on your two graphs, did the sale have an influence on the size of coffee that customers ordered?

20. Weinstein, McDermott, and Roediger (2010) published an experimental study examining different techniques that students use to prepare for a test. Students read a passage, knowing that they would have a quiz on the material. After reading the passage, students in one condition were asked to continue studying by simply reading the passage again. In a second condition, students answered a series of prepared questions about the material. Then all students took the quiz. The following table shows quiz scores similar to the results obtained in the study.

Quiz Scores for Two Groups of Students	
Simply Reread	Answer Questions
8, 5, 7, 9, 8	9, 7, 8, 9, 9
9, 9, 8, 6, 9	8, 10, 9, 5, 10
7, 7, 4, 6, 5	7, 9, 8, 7, 8

Sketch a polygon showing the frequency distribution for students who reread the passage. In the same graph,

sketch a polygon showing the scores for the students who answered questions. (Use two different colors or use a solid line for one polygon and a dashed line for the other.) Does it look like there is a difference between the two groups?

21. Your instructors, your parents, and your feelings of stress during finals week all tell you that cramming is a bad way to prepare for exams. Participants in Kornell's (2009) research study received two sets of flash cards with vocabulary questions that they studied multiple times. One stack of flash cards was studied with a long amount of time between consecutive presentations of the same question. The other stack of flash cards was crammed—participants studied those questions with only a short amount of time between consecutive presentations. After studying the flash cards, participants were tested for the number of correctly remembered answers from all flash cards. The following represent data like those observed by Kornell (2009):

Number of Correctly Remembered Questions	
Short time between flash cards	Long time between flash cards
0, 1, 3, 2, 2, 3, 2, 1, 3, 3	3, 4, 3, 3, 2, 3, 1, 4, 3, 3

Create a frequency table for each of the two conditions. Does there appear to be a difference between the two groups?

22. For the following set of scores:

| 30 | 69 | 41 | 51 | 36 | 53 | 60 | 24 | 55 | 44 |
| 61 | 25 | 74 | 63 | 55 | 13 | 42 | 56 | 54 | 49 |

a. Construct a stem and leaf plot.
b. What is the shape of the distribution?

23. For the following set of scores:

| 37 | 68 | 55 | 52 | 83 | 72 | 67 | 69 | 76 | 65 |
| 87 | 96 | 62 | 67 | 63 | 25 | 94 | 38 | 78 | 60 |

a. Construct a stem and leaf plot.
b. What is the shape of the distribution?

Central Tendency

clivewa/Shutterstock.com

Tools You Will Need

The following items are considered essential background material for this chapter. If you doubt your knowledge of any of these items, you should review the appropriate chapter or section before proceeding.

- Summation notation (Chapter 1)
- Frequency distributions (Chapter 2)

PREVIEW

Dyslexia is a learning disability that affects language skills, especially reading and writing. Children with dyslexia might have difficulty learning to read and problems with word and letter interpretation. Today much reading is done on a screen, such as a computer monitor, tablet, or smartphone, rather than on a printed paper page. One advantage of material presented on a computer monitor, for example, is the ability to modify the background color and the color of the letters. Rello and Bigham (2017) did a study that looked at reading speed in adults with and without dyslexia when the background color was varied. The participants were given short passages to read with different pale-colored-screen backgrounds. The time taken to complete the passage was recorded. The researchers also measured comprehension to make sure that participants were actually reading and understanding the text. It is not surprising they found that participants with dyslexia were slower readers than those without this diagnosis. However, the interesting finding is that both groups read material faster when the background was a warm color (peach, orange, or yellow) and slower when the background color was cool (blue, blue-gray, or green). The findings have implications for making written material on screens more accessible for those with dyslexia. The following hypothetical data compares reading time in seconds for warm versus cool color screen backgrounds for adults with dyslexia.

> Warm background color: 11, 13, 15, 11, 12, 10, 14, 12, 10, 12
>
> Cool background color: 17, 16, 18, 16, 15, 20, 17, 17, 20, 14

The purpose of the study is to determine if background screen color has an effect on reading performance. Just glancing at the listed data does not give us a clear idea about the results. The results also are presented in a frequency distribution graph (see Figure 3.1).

Although it seems obvious that participants reading the passage with warm color backgrounds read faster, this conclusion is based on a general impression, or a subjective interpretation, of the figure. In fact, this conclusion is not always true. For example, there is overlap between the two groups—some of the reading scores with cool background colors were faster. What we need is a method to precisely summarize each group as a whole so that we can objectively describe how much difference exists between the two groups.

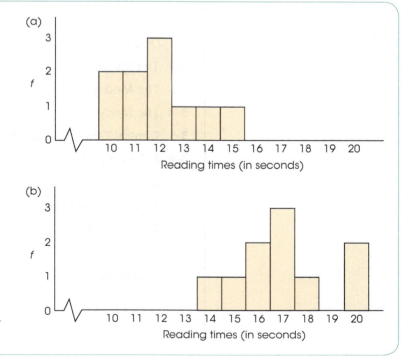

FIGURE 3.1

Frequency distribution for reading time in seconds for those who read material when the background was a warm color (a) and for those who read material when the background was a cool color (b).

The solution to this problem is to identify the typical, or average reading time for each group. Then the research results can be described by saying that the typical reading time for the warm color group is faster than the typical reading time for the cool color group.

In this chapter we introduce the statistical techniques used to identify the typical, or average score for a distribution. Although there are several reasons for defining the average score, the primary advantage of an average is that it provides a single number that describes an entire distribution and can be compared with other distributions.

3-1 Overview

The general purpose of descriptive statistical methods is to organize and summarize a set of scores. Perhaps the most common method for summarizing and describing a distribution is to find a single value that defines the average score and that serves as a typical example to represent the entire distribution. In statistics, the concept of an average or representative score is called *central tendency*. The goal in measuring central tendency is to describe a distribution of scores by determining a single value that identifies the center of the distribution. Ideally, this central value will be the score that is the best representative value for all of the individuals in the distribution.

> **Central tendency** is a statistical measure to determine a single score that defines the center of a distribution. The goal of central tendency is to find the single score that is most typical or most representative of the entire group.

In everyday language, central tendency attempts to identify the "average" or "typical" individual. This average value can then be used to provide a simple description of either an entire population or a sample. In addition to describing an entire distribution, measures of central tendency are also useful for making comparisons between groups of individuals or between sets of data. For example, weather data indicate that for Seattle, Washington, the average yearly temperature is 53° Fahrenheit and the average annual precipitation is 34 inches. By comparison, the average temperature in Phoenix, Arizona, is 71° and the average precipitation is 7.4 inches. The point of these examples is to demonstrate the great advantage of being able to describe a large set of data with a single, representative number. Central tendency characterizes what is typical for a large population, and in doing so makes large amounts of data more digestible. Statisticians sometimes use the expression "number crunching" to illustrate this aspect of data description. That is, we take a distribution consisting of many scores and "crunch" them down to a single value that describes them all.

Unfortunately, there is no single, standard procedure for determining central tendency. The problem is that no single measure produces a central, representative value in every situation. The three distributions shown in Figure 3.2 should help demonstrate this fact. Before we discuss the three distributions, take a moment to look at the figure and try to identify the "center" or the "most representative score" for each distribution.

1. The first distribution [Figure 3.2(a)] is symmetrical, with the scores forming a distinct pile centered around $X = 5$. For this type of distribution, it is easy to identify the "center," and most people would agree that the value $X = 5$ is an appropriate measure of central tendency.

2. In the second distribution [Figure 3.2(b)], however, problems begin to appear. Now the scores form a negatively skewed distribution, piling up at the high end of the scale around $X = 8$, but tapering off to the left all the way down to $X = 1$. Where

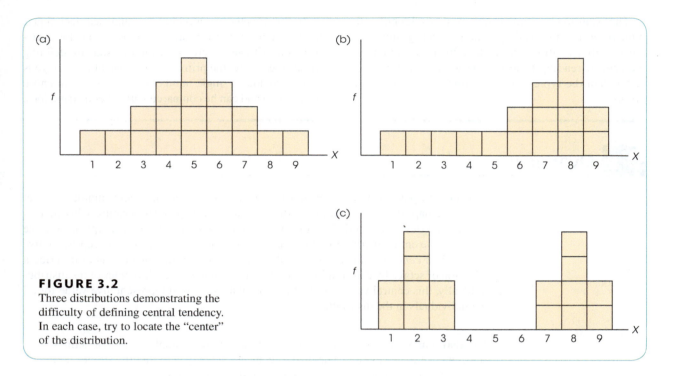

FIGURE 3.2

Three distributions demonstrating the difficulty of defining central tendency. In each case, try to locate the "center" of the distribution.

is the "center" in this case? Some people might select $X = 8$ as the center because more individuals had this score than any other single value. However, $X = 8$ is clearly not in the middle of the distribution. In fact, the majority of the scores (10 out of 16) have values less than 8, so it seems reasonable that the "center" should be defined by a value that is less than 8.

3. Now consider the third distribution [Figure 3.2(c)]. Again, the distribution is symmetrical, but now there are two distinct piles of scores. Because the distribution is symmetrical with $X = 5$ as the midpoint, you may choose $X = 5$ as the "center." However, none of the scores is located at $X = 5$ (or even close), so this value is not particularly good as a representative score. On the other hand, because there are two separate piles of scores with one group centered at $X = 2$ and the other centered at $X = 8$, it is tempting to say that this distribution has two centers. But can one distribution have two centers?

Clearly, there can be problems defining the "center" of a distribution. Occasionally, you will find a nice, neat distribution like the one shown in Figure 3.2(a), for which everyone will agree on the center. But you should realize that other distributions are possible and that there may be different opinions concerning the definition of the center. To deal with these problems, statisticians have developed three different methods for measuring central tendency: the mean, the median, and the mode. They are computed differently and have different characteristics. To decide which of the three measures is best for any particular distribution, you should keep in mind that the general purpose of central tendency is to find the single most representative score. Each of the three measures we present has been developed to work best in a specific situation. We examine this issue in more detail after we introduce the three measures.

3-2 The Mean

LEARNING OBJECTIVES

1. Define the mean, and calculate both the population mean and the sample mean.
2. Explain the alternative definitions of the mean as the amount each individual receives when the total is divided equally and as a balancing point.
3. Calculate a weighted mean.
4. Find n, ΣX, and M using scores in a frequency distribution table.
5. Describe the effect on the mean and calculate the outcome for each of the following: changing a score, adding or removing a score, adding or subtracting a constant from each score, and multiplying or dividing each score by a constant.

The *mean*, also known as the arithmetic average, is computed by adding all the scores in the distribution and dividing by the number of scores. The mean for a population is identified by the Greek letter mu, μ (pronounced "mew"), and the mean for a sample is identified by M or $\overline{X}$ (read "x-bar").

The convention in many statistics textbooks is to use $\overline{X}$ to represent the mean for a sample. However, in manuscripts and in published research reports the letter M is the standard notation for a sample mean. Because you will encounter the letter M when reading research reports and because you should use the letter M when writing research reports, we have decided to use the same notation in this text. Keep in mind that the $\overline{X}$ notation is still appropriate for identifying a sample mean, and you may find it used on occasion, especially in textbooks.

The **mean** for a distribution is the sum of the scores divided by the number of scores.

The formula for the *population mean* is

$$\mu = \frac{\Sigma X}{N} \tag{3.1}$$

First, add all the scores in the population, and then divide by N. For a sample, the computation is exactly the same, but the formula for the *sample mean* uses symbols that signify sample values:

$$\text{sample mean} = M = \frac{\Sigma X}{n} \tag{3.2}$$

In general, we use Greek letters to identify characteristics of a population (parameters) and letters of our own alphabet to stand for sample values (statistics). If a mean is identified with the symbol M, you should realize that we are dealing with a sample. Also note that the equation for the sample mean uses a lowercase n as the symbol for the number of scores in the sample.

EXAMPLE 3.1 For the following population of $N = 4$ scores,

$$3, 7, 4, 6$$

the mean is

$$\mu = \frac{\Sigma X}{N} = \frac{20}{4} = 5$$

■

■ Alternative Definitions for the Mean

Although the procedure of adding the scores and dividing by the number of scores provides a useful definition of the mean, there are two alternative definitions that may give you a better understanding of this important measure of central tendency.

Dividing the Total Equally   The first alternative is to think of the mean as the amount each individual receives when the total (ΣX) is divided equally among all the individuals (N) in the distribution. Consider the following example.

EXAMPLE 3.2	A group of $n = 6$ children buys a box of baseball cards at a garage sale and discovers that the box contains a total of 180 cards. If the children divide the cards equally among themselves, how many cards will each child get? You should recognize that this problem represents the standard procedure for computing the mean. Specifically, the total (ΣX) is divided by the number (n) to produce the mean, $\frac{180}{6} = 30$ cards for each child. ■

The previous example demonstrates that it is possible to define the mean as the amount that each individual gets when the total is distributed equally. This somewhat socialistic technique is particularly useful in problems for which you know the mean and must find the total. Consider the following example.

EXAMPLE 3.3	Now suppose that the $n = 6$ children from Example 3.2 decide to sell their baseball cards on eBay. If they make an average of $M = \$5$ per child, what is the total amount of money for the whole group? Although you do not know exactly how much money each child has, the new definition of the mean tells you that if they pool their money together and then distribute the total equally, each child will get \$5. For each of $n = 6$ children to get \$5, the total must be $6(\$5) = \$30$. To check this answer, use the formula for the mean:

$$M = \frac{\Sigma X}{n} = \frac{\$30}{6} = \$5 \qquad\blacksquare$$

The Mean as a Balance Point   The second alternative definition of the mean describes the mean as a balance point for the distribution. Consider a population consisting of $N = 5$ scores (1, 2, 6, 6, 10). For this population, $\Sigma X = 25$ and $\mu = \frac{25}{5} = 5$. Figure 3.3 shows this population drawn as a histogram, with each score represented as a box that is sitting on a seesaw. If the seesaw is positioned so that it pivots at a point equal to the mean, then it will be balanced and will rest level.

The reason the seesaw is balanced over the mean becomes clear when we measure the distance of each box (score) from the mean:

Score	Distance from the Mean
$X = 1$	4 points below the mean
$X = 2$	3 points below the mean
$X = 6$	1 point above the mean
$X = 6$	1 point above the mean
$X = 10$	5 points above the mean

Notice that the mean balances the distances. That is, the total distance below the mean is the same as the total distance above the mean:

below the mean: $4 + 3 = 7$ points
above the mean: $1 + 1 + 5 = 7$ points

FIGURE 3.3

The frequency distribution shown as a seesaw balanced at the mean. Based on Weinberg, G. H., Schumaker, J. A., and Oltman, D. (1981). *Statistics: An intuitive approach* (p. 14). Belmont, CA: Wadsworth.

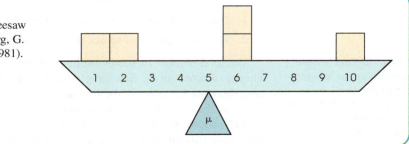

Because the mean serves as a balance point, the value of the mean will always be located somewhere between the highest score and the lowest score; that is, the mean can never be outside the range of scores. If the lowest score in a distribution is $X = 8$ and the highest is $X = 15$, then the mean *must* be between 8 and 15. If you calculate a value that is outside this range, then you have made an error.

The image of a seesaw with the mean at the balance point is also useful for determining how a distribution is affected if a new score is added or if an existing score is removed. For the distribution in Figure 3.3, for example, what would happen to the mean (balance point) if a new score were added at $X = 10$? For another view, see Box 3.1.

■ The Weighted Mean

Often it is necessary to combine two sets of scores and then find the overall mean for the combined group. Suppose, for example, that we begin with two separate samples. The first sample has $n = 12$ scores and a mean of $M = 6$. The second sample has $n = 8$ and $M = 7$. If the two samples are combined, what is the mean for the total group?

BOX 3.1 Another Look at the Mean as the Balance Point

Determining the distance of a score from the mean is a simple matter of subtracting the mean from a score. In a notational expression, distance from the mean is $X - \mu$ for a population (and $X - M$ for a sample). Once again, consider the population used to demonstrate the mean as the balance point of a distribution. For $N = 5$ scores with a mean of $\mu = 5$, the scores in the population are:

$$1, 2, 6, 6, 10$$

Using $X - \mu$, we can find the distances of each score from the mean:

For $X = 1$ $\quad$ $X - \mu = 1 - 5 = -4$

For $X = 2$ $\quad$ $X - \mu = 2 - 5 = -3$

For $X = 6$ $\quad$ $X - \mu = 6 - 5 = +1$

For $X = 6$ $\quad$ $X - \mu = 6 - 5 = +1$

For $X = 10$ $\quad$ $X - \mu = 10 - 5 = +5$

Notice the signs of these differences. A negative sign tells you the score is below the mean by a certain distance and a positive sign tells you the score is above the mean. The sum of the negative distances is -7 and the sum of the positive distances is $+7$. Thus the sum of all $N = 5$ distances equals zero because the mean is the balance point of the distribution. Using the notation for distance from the mean, $X - \mu$, the sum of the distances will always equal zero, or

$$\Sigma(X - \mu) = 0$$

To calculate the overall mean, we need two values:

1. The overall sum of the scores for the combined group (ΣX).
2. The total number of scores in the combined group (n).

When the data involve more than one sample (or population), we use subscripts to identify the sample. For example, n_1 refers to the number of scores in sample 1.

The total number of scores in the combined group can be found easily by adding the number of scores in the first sample (n_1) and the number in the second sample (n_2). In this case, there are 12 scores in the first sample and 8 in the second, for a total of $12 + 8 = 20$ scores in the combined group. Similarly, the overall sum for the combined group can be found by adding the sum for the first sample (ΣX_1) and the sum for the second sample (ΣX_2). With these two values, we can compute the mean using the basic equation

$$\text{overall mean} = M = \frac{\Sigma X \text{ (overall sum for the combined group)}}{n \text{ (total number in the combined group)}}$$

$$= \frac{\Sigma X_1 + \Sigma X_2}{n_1 + n_2} \tag{3.3}$$

To find the sum of the scores for each sample, remember that the mean can be defined as the amount each person receives when the total (ΣX) is distributed equally. The first sample has $n = 12$ and $M = 6$. (Expressed in dollars instead of scores, this sample has $n = 12$ people and each person gets \$6 when the total is divided equally.) For each of 12 people to get $M = 6$, the total must be $\Sigma X = 12 \times 6 = 72$. In the same way, the second sample has $n = 8$ and $M = 7$ so the total must be $\Sigma X = 8 \times 7 = 56$. Using these values, we obtain an overall mean of

$$\text{overall mean} = M = \frac{\Sigma X_1 + \Sigma X_2}{n_1 + n_2} = \frac{72 + 56}{12 + 8} = \frac{128}{20} = 6.4$$

The following table summarizes the calculations.

First Sample	Second Sample	Combined Sample
$n_1 = 12$	$n_2 = 8$	$n = 20 \ (12 + 8)$
$\Sigma X = 72$	$\Sigma X = 56$	$\Sigma X = 128 \ (72 + 56)$
$M_1 = 6$	$M_2 = 7$	$M = 6.4$

Note that the overall mean is not halfway between the original two sample means. That is, you shouldn't simply add up the sample means ($6 + 7$) and divide by the number of means (2). Because the samples are not the same size, one sample makes a larger contribution to the total group and therefore carries more weight in determining the overall mean. For this reason, the overall mean we have calculated is called the *weighted mean*. In this example, the overall mean of $M = 6.4$ is closer to the value of $M = 6$ (the larger sample) than it is to $M = 7$ (the smaller sample). When sample sizes are not equal, the weighted mean will always be closer to the mean of the larger sample.

The following example is an opportunity for you to test your understanding by computing a weighted mean yourself.

EXAMPLE 3.4

One sample has $n = 4$ scores with a mean of $M = 8$ and a second sample has $n = 8$ scores with a mean of $M = 5$. If the two samples are combined, what is the mean for the combined group? For this example, you should obtain a mean of $M = 6$. Good luck and remember that you can use the example in the text as a model. ∎

■ Computing the Mean from a Frequency Distribution Table

When a set of scores has been organized in a frequency distribution table, the calculation of the mean is usually easier if you first remove the individual scores from the table. Table 3.1 shows a distribution of scores organized in a frequency distribution table. To compute the mean for this distribution you must be careful to use both the X values in the first column and the frequencies in the second column. The values in the table show that the distribution consists of one 10, two 9s, four 8s, and one 6, for a total of $n = 8$ scores.

To find the sum of the scores, you must add all eight scores:

$$\Sigma X = 10 + 9 + 9 + 8 + 8 + 8 + 8 + 6 = 66$$

Note that you can also find the sum of the scores by computing ΣfX as we demonstrated in Chapter 2 (page 47). For the data in Table 3.1,

$$\Sigma X = \Sigma fX = 10 + 18 + 32 + 0 + 6 = 66$$

Remember that you also can determine the number of scores by adding the frequencies, $n = \Sigma f$. For the data in Table 3.1,

$$n = \Sigma f = 1 + 2 + 4 + 0 + 1 = 8$$

Once you have found ΣX and n, you compute the mean as usual. For these data,

$$M = \frac{\Sigma X}{n} = \frac{66}{8} = 8.25$$

■ Characteristics of the Mean

The mean has many characteristics that will be important in future discussions. In general, these characteristics result from the fact that every score in the distribution contributes to the value of the mean. Specifically, every score adds to the total (ΣX) and every score contributes one point to the number of scores (n). These two values (ΣX and n) determine the value of the mean. We now discuss four of the more important characteristics of the mean.

Changing a Score   Changing the value of any score will change the mean. For example, a sample of quiz scores for a psychology lab section consists of 9, 8, 7, 5, and 1. Note that the sample consists of $n = 5$ scores with $\Sigma X = 30$. The mean for this sample is

$$M = \frac{\Sigma X}{n} = \frac{30}{5} = 6.00$$

Now suppose that the score of $X = 1$ is changed to $X = 8$. Note that we have added 7 points to this individual's score, which will also add 7 points to the total (ΣX). After changing the score, the new distribution consists of

$$9, 8, 7, 5, 8$$

.

TABLE 3.1

Statistics quiz scores for a section of $n = 8$ students.

Quiz Score (X)	f	fX
10	1	10
9	2	18
8	4	32
7	0	0
6	1	6

There are still $n = 5$ scores, but now the total is $\Sigma X = 37$. Thus, the new mean is

$$M = \frac{\Sigma X}{n} = \frac{37}{5} = 7.40$$

Notice that changing a single score in the sample has produced a new mean. You should recognize that changing any score also changes the value of ΣX (the sum of the scores), and thus always changes the value of the mean.

Introducing a New Score or Removing a Score   Adding a new score to a distribution, or removing an existing score, will usually change the mean. The exception is when the new score (or the removed score) is exactly equal to the mean. It is easy to visualize the effect of adding or removing a score if you remember that the mean is defined as the balance point for the distribution. Figure 3.4 shows a distribution of scores represented as boxes on a seesaw that is balanced at the mean, $\mu = 7$. Imagine what would happen if we added a new score (a new box) at $X = 10$. Clearly, the seesaw would tip to the right and we would need to move the pivot point (the mean) to the right to restore balance.

Now imagine what would happen if we removed the score (the box) at $X = 9$. This time the seesaw would tip to the left and, once again, we would need to change the mean to restore balance.

Finally, consider what would happen if we added a new score of $X = 7$, exactly equal to the mean. It should be clear that the seesaw would not tilt in either direction, so the mean would stay in exactly the same place. Also note that if we remove the new score at $X = 7$, the seesaw will remain balanced and the mean will not change. In general, adding a new score or removing an existing score will cause the mean to change unless the new score (or existing score) is located exactly at the mean.

The following example demonstrates exactly how the new mean is computed when a new score is added to an existing sample.

EXAMPLE 3.5   Adding a score (or removing a score) has the same effect on the mean whether the original set of scores is a sample or a population. To demonstrate the calculation of the new mean, we will use the set of scores that is shown in Figure 3.4 (below). This time, however, we will treat the scores as a sample with $n = 5$ and $M = 7$. Note that this sample must have $\Sigma X = 35$. What will happen to the mean if a new score of $X = 13$ is added to the sample?

To find the new sample mean, we must determine how the values for n and ΣX will be changed by a new score. We begin with the original sample and then consider the effect of adding the new score. The original sample had $n = 5$ scores, so adding one new score will produce $n = 6$. Similarly, the original sample had $\Sigma X = 35$. Adding a score of $X = 13$ will increase the sum by 13 points, producing a new sum of $\Sigma X = 35 + 13 = 48$. Finally, the new mean is computed using the new values for n and ΣX.

$$M = \frac{\Sigma X}{n} = \frac{48}{6} = 8$$

FIGURE 3.4
A distribution of $N = 5$ scores that is balanced at the mean, $\mu = 7$.

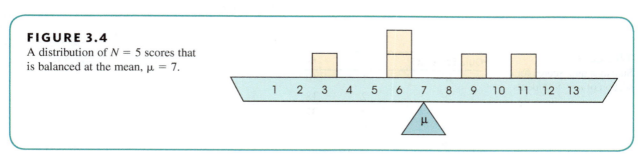

The entire process can be summarized as follows:

Original Sample	New Sample, Adding $X = 13$
$n = 5$	$n = 6$
$\Sigma X = 35$	$\Sigma X = 48$
$M = \frac{35}{5} = 7$	$M = \frac{48}{6} = 8$

■

The following example is an opportunity for you to test your understanding by determining how the mean is changed by removing a score from a distribution.

EXAMPLE 3.6

We begin with a sample of $n = 5$ scores with $\Sigma X = 35$ and $M = 7$. If one score with a value of $X = 11$ is removed from the sample, what is the mean for the remaining scores? You should obtain a mean of $M = 6$. Good luck and remember that you can use Example 3.5 as a model.

■

Adding or Subtracting a Constant from Each Score   If a constant value is added to every score in a distribution, the same constant will be added to the mean. Similarly, if you subtract a constant from every score, the same constant will be subtracted from the mean.

Consider the Rello and Bigham (2017) study in the Preview. The researchers found that reading times for material on a computer screen were quicker with warm color backgrounds than with cool colors when testing adults with dyslexia. Table 3.2 shows a sample of $n = 4$ participants and their reading times. Note that the total for the warm color column is $\Sigma X = 48$ for a sample of $n = 4$ participants, so the mean is $M = \frac{48}{4} = 12$. Now suppose that the effect of background color is to speed up reading by a constant amount of 3 points. This would add 3 points to each individual's reading score when the background is a cool color. The resulting scores with cool colors are shown in the second column of the table. For these scores, the total is $\Sigma X = 60$, so the mean is $M = \frac{60}{4} = 15$. Adding 3 points to each rating score has also added 3 points to the mean, from $M = 12$ to $M = 15$. (It is important to note that treatment effects are usually not as simple as adding or subtracting a constant amount. Nonetheless, the concept of adding a constant to every score is important and will be addressed in later chapters when we are using statistics to evaluate mean differences.)

Multiplying or Dividing Each Score by a Constant   If every score in a distribution is multiplied by (or divided by) a constant value, the mean will change in the same way.

Multiplying (or dividing) each score by a constant value is a common method for changing the unit of measurement. To change a set of measurements from minutes to seconds, for example, you multiply by 60; to change from inches to feet, you divide by 12. One common task for researchers is converting measurements into metric units to conform to international standards. For example, publication guidelines of the American Psychological Association call for metric equivalents to be reported in parentheses when most nonmetric units are used. Table 3.3 shows

TABLE 3.2

Reading speed (seconds) for different background colors.

Participant	Warm Color	Cool Color
A	11	14
B	12	15
C	14	17
D	11	14
	$\Sigma X = 48$	$\Sigma X = 60$
	$M = 12$	$M = 15$

TABLE 3.3
Measurements transformed from inches to centimeters.

Original Measurement in Inches	Conversion to Centimeters (Multiply by 2.54)
10	25.40
9	22.86
12	30.48
8	20.32
11	27.94
$\Sigma X = 50$	$\Sigma X = 127.00$
$M = 10$	$M = 25.40$

how a sample of $n = 5$ scores measured in inches would be transformed into a set of scores measured in centimeters. (Note that 1 inch equals 2.54 centimeters.) The first column shows the original scores that total $\Sigma X = 50$ with $M = 10$ inches. In the second column, each of the original scores has been multiplied by 2.54 (to convert from inches to centimeters) and the resulting values total $\Sigma X = 127$, with $M = 25.4$. Multiplying each score by 2.54 has also caused the mean to be multiplied by 2.54. You should realize, however, that although the numerical values for the individual scores and the sample mean have changed, the actual measurements are not changed.

LEARNING CHECK   **LO1 1.** A population of $N = 5$ scores has a mean of $\mu = 12$. What is ΣX for this sample?

    a. $\frac{12}{5} = 2.40$

    b. $\frac{5}{12} = 0.417$

    c. $5(12) = 60$

    d. Cannot be determined from the information given.

LO2 2. A sample has a mean of $M = 72$. If one person with a score of $X = 98$ is removed from the sample, what effect will it have on the sample mean?

    a. The sample mean will increase.

    b. The sample mean will decrease.

    c. The sample mean will remain the same.

    d. Cannot be determined from the information given.

LO3 3. One sample of $n = 4$ scores has a mean of $M = 10$, and a second sample of $n = 10$ scores has a mean of $M = 20$. If the two samples are combined, then what value will be obtained for the mean of the combined sample?

    a. Equal to 15

    b. Greater than 15 but less than 20

    c. Less than 15 but more than 10

    d. None of the other choices is correct.

LO4 4. For the following frequency distribution table, what are the values for ΣX and n?

    a. 20; 4

    b. 10; 10

    c. 20; 10

    d. 10; 2.0

X	f
4	1
3	2
2	3
1	4

LO5 **5.** A population of $N = 10$ scores has a mean of 30. If every score in the distribution is multiplied by 3, then what is the value of the new mean?

    a. Still 30

    b. 33

    c. 60

    d. 90

ANSWERS   **1. c** **2. b** **3. b** **4. c** **5. d**

3-3 | The Median

> **LEARNING OBJECTIVE**
>
> **6.** Define the median, identify the median for discrete scores, and calculate the precise median for a continuous variable.

The second measure of central tendency we will consider is called the *median*. The goal of the median is to locate the midpoint of the distribution. Unlike the mean, there are no specific symbols or notation to identify the median. Instead, the median is simply identified by the word *median*. In addition, the definition and the computations for the median are identical for a sample and for a population.

> If the scores in a distribution are listed in order from smallest to largest, the **median** is the midpoint of the list. More specifically, the median is the point on the measurement scale below which 50% of the scores in the distribution are located.

■ Finding the Median for Simple Distributions

Defining the median as the *midpoint* of a distribution means that the scores are being divided into two equal-sized groups. We are not locating the midpoint between the highest and lowest X values. To find the median, list the scores in order from smallest to largest. Begin with the smallest score and count the scores as you move up the list. The median is the first point you reach that is greater than 50% of the scores in the distribution. The median can be equal to a score in the list or it can be a point between two scores. Notice that the median is not algebraically defined in this section (that is, we are not presenting an equation for computing the median of scores).

EXAMPLE 3.7   This example demonstrates the calculation of the median when N (or n) is an odd number. With an odd number of scores, you list the scores in order (lowest to highest), and the median is the middle score in the list. Consider the following set of $N = 5$ scores, which have been listed in order:

$$3, 5, 8, 10, 11$$

The middle score is $X = 8$, so the median is equal to 8. Using the counting method, with $N = 5$ scores, the 50% point would be $2\frac{1}{2}$ scores. Starting with the smallest scores, we must count the 3, the 5, and the 8 before we reach the target of at least 50%. Again, for this distribution, the median is the middle score, $X = 8$. ■

EXAMPLE 3.8

This example demonstrates the calculation of the median when N (or n) is an even number. With an even number of scores in the distribution, you list the scores in order (lowest to highest) and then locate the median by finding the average of the middle two scores. Consider the following population:

$$1, 1, 4, 5, 7, 8$$

Now we select the middle pair of scores (4 and 5), add them together, and divide by 2:

$$\text{median} = \frac{4 + 5}{2} = \frac{9}{2} = 4.5$$

Using the counting procedure, with $N = 6$ scores, the 50% point is 3 scores. Starting with the smallest scores, we must count the first 1, the second 1, and the 4 before we reach the target of at least 50%. Again, the median for this distribution is 4.5, which is the first point on the scale beyond $X = 4$. For this distribution, exactly 3 scores (50%) are located below 4.5. Note: If there is a gap between the middle two scores, the convention is to define the median as the midpoint between the two scores. For example, if the middle two scores are $X = 4$ and $X = 6$, the median would be defined as 5. ■

The simple technique of listing and counting scores is sufficient to determine the median for many simple distributions and is always appropriate for discrete variables. Notice that this technique will always produce a median that is either a whole number or is halfway between two whole numbers. With a continuous variable, however, it is possible to divide a distribution precisely in half so that *exactly* 50% of the distribution is located below (and above) a specific point. The procedure for locating the precise median is discussed in the following section.

■ Finding the Precise Median for a Continuous Variable

Recall from Chapter 1 that a continuous variable consists of categories that can be split into an infinite number of fractional parts. For example, time can be measured in seconds, tenths of a second, hundredths of a second, and so on. When the scores in a distribution are measurements of a continuous variable, it is possible to split one of the categories into fractional parts and find the median by locating the precise point that separates the bottom 50% of the distribution from the top 50%. The following example demonstrates this process.

EXAMPLE 3.9

For this example, we will find the precise median for the following sample of $n = 8$ scores:

$$1, 2, 3, 4, 4, 4, 4, 6$$

The frequency distribution for this sample is shown in Figure 3.5(a). With an even number of scores, you normally would compute the average of the middle two scores to find the median. This process produces a median of $X = 4$. For a discrete variable, $X = 4$ is the correct value for the median. Recall from Chapter 1 that a discrete variable consists of indivisible categories such as the number of children in a family. Some families have 4 children and some have 5, but none have 4.31 children. For a discrete variable, the category $X = 4$ cannot be divided and the whole number 4 is the median.

However, if you look at the distribution histogram, the value $X = 4$ does not appear to divide the distribution exactly in half. The problem comes from the tendency to interpret a score of $X = 4$ as meaning exactly 4.00. However, if the scores are measurements of a continuous variable, then the score $X = 4$ actually corresponds to an interval from 3.5 to 4.5, and the median corresponds to a point within this interval.

To find the precise median, we first observe that the distribution contains $n = 8$ scores represented by eight blocks in the graph. The median is the point that has exactly four

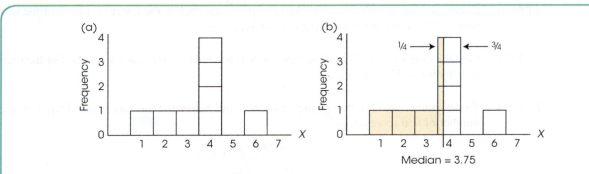

FIGURE 3.5

A distribution with several scores clustered at the median. The median for this distribution is positioned so that each of the four boxes at $X = 4$ is divided into two sections with $\frac{1}{4}$ of each box below the median (to the left) and $\frac{3}{4}$ of each box above the median (to the right). As a result, there are exactly 4 boxes, 50% of the distribution, on each side of the median.

blocks (50%) on each side. Starting at the left-hand side and moving up the scale of measurement, we accumulate a total of three blocks when we reach a value of 3.5 on the X-axis [see Figure 3.5(a)]. What is needed is one more block to reach the goal of four blocks (50%). The problem is that the next interval contains four blocks. The solution is to take a fraction of each block so that the fractions combine to give you one block. For this example, if we take $\frac{1}{4}$ of each block, the four quarters will combine to make one whole block. This solution is shown in Figure 3.5(b). The fraction is determined by the number of blocks needed to reach 50% and the number that exists in the interval

$$\text{fraction} = \frac{\text{number needed to reach 50\%}}{\text{number in the interval}}$$

For this example, we needed one out of the four blocks in the interval, so the fraction is $\frac{1}{4}$. To obtain $\frac{1}{4}$ of each block, the median is the point that is located exactly $\frac{1}{4}$ of the way into the interval. The interval for $X = 4$ extends from 3.5 to 4.5. The interval width is 1 point, so $\frac{1}{4}$ of the interval corresponds to 0.25 points. Starting at the bottom of the interval and moving up 0.25 points produces a value of $3.50 + 0.25 = 3.75$. This is the median, with exactly 50% of the distribution (four boxes) on each side. Notice that the median divides the area in the distribution in half so that 50% of the area is below and above the median. ■

■ A Formula for the Median with Continuous Variables

Example 3.9 was an example of a process called *interpolation*. We can use interpolation more generally to estimate an intermediate value between any two other X values. For example, suppose your new puppy weighed 10 pounds when it was 20 weeks old. You weighed it again when it was 30 weeks old and it was 20 pounds. What would you estimate it weighed at 25 weeks old? The estimate would be 15 pounds, halfway between 10 and 20 pounds, because 25 weeks is halfway between 20 and 30 weeks. Interpolation can be used when the median falls somewhere within the upper and lower real limits of several tied scores. It is summarized in the following five steps.

STEP 1   Determine how many scores should fall above and below the median by taking one-half of N, or $0.5N$.

STEP 2   Count the number of scores (or blocks in the graph) below the lower real limit of the tied values. The notation for this frequency is $f_{\text{BELOW LRL}}$.

STEP 3   Find the number of additional scores (blocks) needed to make exactly one-half of the total distribution, $0.5N - f_{\text{BELOW LRL}}$.

STEP 4   To determine what fraction of the tied scores fall below the median, divide Step 3 by the number of tied scores, f_{TIED}.

$$\frac{0.5N - f_{\text{BELOW LRL}}}{f_{\text{TIED}}}$$

STEP 5   Add the fraction (Step 4) to the lower real limit, X_{LRL}, of the interval containing the tied scores.

When these steps are incorporated into a formula, we obtain

$$\text{median} = X_{\text{LRL}} + \left(\frac{0.5N - f_{\text{BELOW LRL}}}{f_{\text{TIED}}}\right) \tag{3.4}$$

Where X_{LRL} is the lower real limit of the tied values, $f_{\text{BELOW LRL}}$ is the frequency of scores with values below X_{LRL}, and f_{TIED} is the frequency of tied values. Thus, for Example 3.9,

$$\text{median} = 3.5 + \frac{(0.5(8) - 3)}{4}$$

$$= 3.5 + \frac{(4 - 3)}{4}$$

$$= 3.5 + \tfrac{1}{4}$$

$$= 3.5 + .25 = 3.75$$

Remember, finding the precise midpoint by dividing scores into fractional parts is sensible for a continuous variable; however, it is not appropriate for a discrete variable. For example, a median time of 3.75 seconds is reasonable, but a median family size of 3.75 children is not.

■ The Median, the Mean, and the Middle

Earlier, we defined the mean as the "balance point" for a distribution because the distances above the mean must have the same total as the distances below the mean. One consequence of this definition is that the mean is always located inside the group of scores, somewhere between the smallest score and the largest score. You should notice, however, that the concept of a balance point focuses on distances rather than scores. In particular, it is possible to have a distribution in which the vast majority of the scores are located on one side of the mean. Figure 3.6 shows a distribution of $N = 6$ scores in which 5 out of 6 scores have values less than the mean. In this figure, the total of the distances above the mean is 8 points and the total of the distances below the mean is 8 points. Thus, the mean is located in the middle of the distribution if you use the concept of distance to define the "middle." However, you should realize that the mean is not necessarily located at the exact center of the group of scores.

FIGURE 3.6

A population of $N = 6$ scores with a mean of $\mu = 4$. Notice that the mean does not necessarily divide the scores into two equal groups. In this example, 5 out of the 6 scores have values less than the mean.

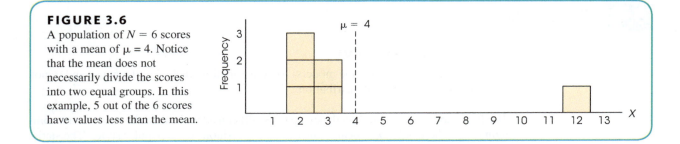

The median, on the other hand, defines the middle of the distribution in terms of scores. In particular, the median is located so that half of the scores are on one side and half are on the other side. For the distribution in Figure 3.6, for example, the median is located at $X = 2.5$, with exactly 3 scores above this value and exactly 3 scores below. Thus, it is possible to claim that the median is located in the middle of the distribution, provided that the term "middle" is defined by the number of scores.

In summary, the mean and the median are both methods for defining and measuring central tendency. However, it is important to point out that although they both define the middle of the distribution, they use different definitions of the term "middle."

LEARNING CHECK   **LO6 1.** What is the median for the following set of scores?

Scores: 1, 6, 8, 19

    a. 6
    b. 6.5
    c. 7
    d. 7.5

LO6 2. What is the median for the sample presented in the following frequency distribution table?

    a. 1.5
    b. 2.0
    c. 2.5
    d. 3.0

X	f
4	1
3	2
2	2
1	3

LO6 3. Find the precise median for the following scores measuring a continuous variable.

Scores: 1, 4, 5, 5, 5, 6, 7, 8

    a. 5
    b. 5.17
    c. 5.67
    d. 6

ANSWERS   **1. c   2. b   3. b**

3-4 The Mode

LEARNING OBJECTIVE

7. Define and determine the mode(s) for a distribution, including the major and minor modes for a bimodal distribution.

The final measure of central tendency that we will consider is called the *mode*. In its common usage, the word *mode* means "the customary fashion" or "a popular style." The statistical definition is similar in that the mode is the most common observation among a group of scores.

> In a frequency distribution, the **mode** is the score or category that has the greatest frequency.

As with the median, there are no symbols or special notation used to identify the mode or to differentiate between a sample mode and a population mode. In addition, the definition of the mode is the same for a population and for a sample distribution.

The mode is a useful measure of central tendency because it can be used to determine the typical or most frequent value for any scale of measurement, including a nominal scale (see Chapter 1). Consider, for example, the data shown in Table 3.4. These data were obtained by asking a sample of 100 students to name their favorite restaurants in town. The result is a sample of $n = 100$ scores with each score corresponding to the restaurant that the student named.

Caution: **The mode is a score or category, not a frequency. For this example, the mode is Luigi's, not $f = 42$.**

For these data, the mode is Luigi's, the restaurant (score) that was named most frequently as a favorite place. Although we can identify a modal response for these data, you should notice that it would be impossible to compute a mean or a median. Specifically, you cannot add restaurants to obtain ΣX and you cannot list the scores (named restaurants) in order.

The mode also can be useful because it is the only measure of central tendency that must correspond to an actual score in the data; by definition, the mode is the most frequently occurring score. The mean and the median, on the other hand, are both calculated values and often produce an answer that does not equal any score in the distribution. For example, in Figure 3.6 (page 89) we presented a distribution with a mean of 4 and a median of 2.5. Note that none of the scores is equal to 4 and none of the scores is equal to 2.5. However, the mode for this distribution is $X = 2$ and there are three individuals who actually have scores of $X = 2$.

TABLE 3.4

Favorite restaurants named by a sample of $n = 100$ students.

Restaurant	f
College Grill	5
George & Harry's	16
Luigi's	42
Oasis Diner	18
Roxbury Inn	7
Sutter's Mill	12

In a frequency distribution graph, the greatest frequency will appear as the tallest part of the figure. To find the mode, you simply identify the score located directly beneath the highest point in the distribution.

Although a distribution will have only one mean and only one median, it is possible to have more than one mode. Specifically, it is possible to have two or more scores that have the same highest frequency. In a frequency distribution graph, the different modes will correspond to distinct, equally high peaks. A distribution with two modes is said to be *bimodal,* and a distribution with more than two modes is called *multimodal.* Occasionally, a distribution with several equally high points is said to have no mode.

Incidentally, a bimodal distribution is often an indication that two separate and distinct groups of individuals exist within the same population (or sample). For example, if you measured height for each person in a set of 100 college students, the resulting distribution would probably have two modes, one corresponding primarily to the males in the group and one corresponding primarily to the females.

Technically, the mode is the score with the absolute highest frequency. However, the term *mode* is often used more casually to refer to scores with relatively high frequencies—that is, scores that correspond to peaks in a distribution even though the peaks are not the absolute highest points. For example, Sibbald (2014) looked at frequency distribution graphs of student achievement scores for individual classrooms in Ontario, Canada. The goal of the study was to identify bimodal distributions, which would suggest two different levels of student achievement within a single class. For this study, bimodal was defined as a distribution having two or more significant local maximums. Figure 3.7 shows a distribution of scores that is similar to a graph presented in the study. There are two distinct peaks in the distribution, one located at $X = 17$ and the other located at $X = 22$. Each of these values is a mode in the distribution. Note, however, that the two modes do not have identical frequencies. Seven students had scores of $X = 22$ and only six had scores of $X = 17$. Nonetheless, both of these points are called modes. When two modes have unequal frequencies, researchers occasionally differentiate the two values by calling the taller peak the *major mode,* and the shorter one the *minor mode.* By the way, the author interpreted a bimodal distribution as a suggestion for the teacher to consider using two different teaching strategies; one for the high achievers and one designed specifically to help low-achieving students.

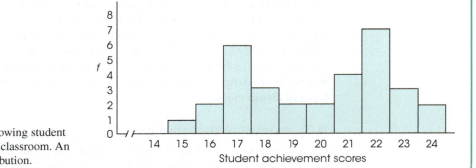

FIGURE 3.7

A frequency distribution showing student achievement scores for one classroom. An example of a bimodal distribution.

LO7 **1.** For the sample shown in the frequency distribution table, what is the mode?

   a. 4

   b. 2

   c. 2.5

   d. 1

X	f
5	1
4	4
3	3
2	4
1	5

LO7 **2.** If the mean, median, and mode are all computed for a distribution of scores, which of the following statements cannot be true?

   a. No one had a score equal to the mean.

   b. No one had a score equal to the median.

   c. No one had a score equal to the mode.

   d. All of the other three statements cannot be true.

LO7 **3.** What is the mode for the following set of $n = 8$ scores? Scores: 2, 4, 4, 5, 7, 8, 8, 8

   a. 4

   b. 5

   c. 5.5

   d. 8

ANSWERS   **1. d   2. c   3. d**

3-5 Central Tendency and the Shape of the Distribution

LEARNING OBJECTIVE

8. Explain how the three measures of central tendency—mean, median, and mode—are related to each other for symmetrical and skewed distributions, and predict their relative values based on the shape of the distribution.

We have identified three different measures of central tendency, and often a researcher calculates all three for a single set of data. Because the mean, the median, and the mode are all trying to measure the same thing, it is reasonable to expect that these three values should be related. In fact, there are some consistent and predictable relationships among the three measures of central tendency. Specifically, there are situations in which all three measures will have exactly the same value. On the other hand, there are situations in which the three measures are guaranteed to be different. In part, the relationships among the mean, median, and mode are determined by the shape of the distribution. We will consider two general types of distributions.

■ Symmetrical Distributions

For a *symmetrical distribution*, the right-hand side of the graph is a mirror image of the left-hand side. If a distribution is perfectly symmetrical, the median is exactly at the center because exactly half of the area in the graph will be on either side of the center. The mean also is exactly at the center of a perfectly symmetrical distribution because each score on the left side of the distribution is balanced by a corresponding score (the mirror image) on the right side. As a result, the mean (the balance point) is located at the center of the distribution. Thus, for a perfectly symmetrical distribution, the mean and the median are the same [Figure 3.8(a)]. If a distribution is roughly symmetrical, but not perfect, the mean and median will be close together in the center of the distribution.

If a symmetrical distribution has only one mode, it will also be in the center of the distribution. Thus, for a perfectly symmetrical distribution with one mode, all three measures of central tendency—the mean, the median, and the mode—have the same value. For a roughly symmetrical distribution, the three measures are clustered together in the center of the distribution. On the other hand, a bimodal distribution that is symmetrical [see Figure 3.8(b)] will have the mean and median together in the center with the modes on each side. A rectangular distribution [see Figure 3.8(c)] has no mode because all *X* values occur with the same frequency. Still, the mean and the median are in the center of the distribution.

■ Skewed Distributions

> The positions of the mean, median, and mode are not as consistently predictable in distributions of discrete variables (see Von Hippel, 2005).

In *skewed distributions,* especially distributions for continuous variables, there is a strong tendency for the mean, median, and mode to be located in predictably different positions. Figure 3.9(a), for example, shows a *positively skewed distribution* with the peak (highest frequency) on the left-hand side. This is the position of the mode. However, it should be clear that the vertical line drawn at the mode does not divide the distribution into two equal parts. To have exactly 50% of the distribution on each side, the median must be located to the right of the mode. Finally, the mean is typically located to the right of the median because it is influenced most by the extreme scores in the tail and is displaced farthest to the right toward the tail of the distribution. Therefore, in a positively skewed distribution, the most likely order of the three measures of central tendency from smallest to largest (left to right) is the mode, the median, and the mean.

> Notice that the mean is always displaced toward the tail of the distribution. In this situation, the "tail wags the dog."

Negatively skewed distributions are lopsided in the opposite direction, with the scores piling up on the right-hand side and the tail tapering off to the left. The grades on an easy exam, for example, tend to form a negatively skewed distribution [see Figure 3.9(b)]. For a distribution with negative skew, the mode is on the right-hand side (with the peak), while the mean is displaced toward the left by the extreme scores in the tail. As before, the median is usually located between the mean and the mode. Therefore, in a negatively skewed distribution, the most probable order for the three measures of central tendency from smallest value to largest value (left to right), is the mean, the median, and the mode.

FIGURE 3.8
Measures of central tendency for three symmetrical distributions: normal, bimodal, and rectangular.

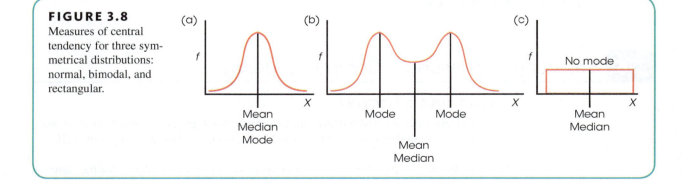

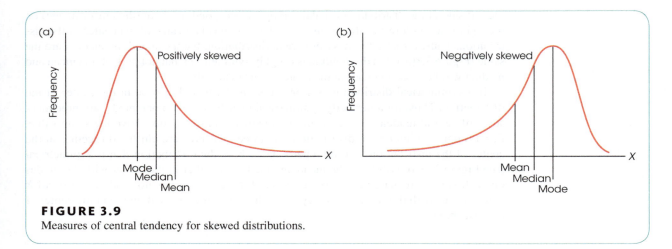

FIGURE 3.9
Measures of central tendency for skewed distributions.

LEARNING CHECK   **LO8**   **1.** For a distribution of scores, the mean is equal to the median. What is the most likely shape of this distribution?

    a. Symmetrical

    b. Positively skewed

    c. Negatively skewed

    d. Impossible to determine the shape

LO8   **2.** For a positively skewed distribution with a mode of $X = 20$ and a median of $X = 25$, what is the most likely value for the mean?

    a. Greater than 25

    b. Less than 20

    c. Between 20 and 25

    d. Cannot be determined from the information given

LO8   **3.** For a positively skewed distribution, what is the most probable order for the three measures of central tendency from smallest to largest?

    a. Mean, median, mode

    b. Mean, mode, median

    c. Mode, mean, median

    d. Mode, median, mean

ANSWERS   **1. a   2. a   3. d**

3-6   Selecting a Measure of Central Tendency

LEARNING OBJECTIVE

9. Explain when each of the three measures of central tendency—mean, median, and mode—should be used, and identify the advantages and disadvantages of each.

You usually can compute two or even three measures of central tendency for the same set of data. Although the three measures often produce similar results, there are situations in which they are predictably different (see Section 3.5). Deciding which measure of central tendency is best to use depends on several factors. Before we discuss these factors, however, note that whenever the scores are numerical values (interval or ratio scale) the mean is usually the preferred measure of central tendency. Because the mean uses every score in the distribution, it typically produces a good representative value. Remember that the goal of central tendency is to find the single value that best represents the entire distribution. Besides being a good representative, the mean has the added advantage of being closely related to variance and standard deviation, the most common measures of variability (Chapter 4). This relationship makes the mean a valuable measure for purposes of inferential statistics. For these reasons, and others, the mean generally is considered to be the best of the three measures of central tendency. But there are specific situations in which it is impossible to compute a mean or in which the mean is not particularly representative. It is in these situations that the mode and the median are used.

■ When to Use the Median

We will consider four situations in which the median serves as a valuable alternative to the mean. In the first three cases, the data consist of numerical values (interval or ratio scales) for which you would normally compute the mean. However, each case also involves a special problem so that it is either impossible to compute the mean, or the calculation of the mean produces a value that is not central or not representative of the distribution. The fourth situation involves measuring central tendency for ordinal data.

Extreme Scores or Skewed Distributions   As noted in the previous section, when a distribution is skewed or has a few extreme scores—scores that are very different in value from most of the others—then the mean may not be a good representative of the majority of the distribution. The problem comes from the fact that the extreme values can have a large influence and cause the mean to be displaced. In this situation, the fact that the mean uses all of the scores equally can be a disadvantage. Consider, for example, the distribution of $n = 10$ scores in Figure 3.10. For this sample, the mean is

$$M = \frac{\Sigma X}{n} = \frac{203}{10} = 20.3$$

Notice that the mean is not very representative of any score in this distribution. Although most of the scores are clustered between 10 and 13, the extreme score of $X = 100$ inflates the value of ΣX and distorts the mean.

  The median, on the other hand, is not easily affected by extreme scores. For this sample, $n = 10$, there should be five scores on either side of the median. The median is 11.50. Notice that this is a very representative value. Also note that the median would be unchanged even if the extreme score were 1,000 instead of only 100. Because it is relatively unaffected by extreme scores, the median commonly is used when reporting the average value for a skewed distribution. For example, the distribution of personal incomes is very skewed, with a small segment of the population earning incomes that are astronomical. These extreme values distort the mean, so that it is not very representative of the salaries that most of us earn. As in the previous example, the median is the preferred measure of central tendency when extreme scores exist.

Undetermined Values   Occasionally, you will encounter a situation in which an individual has an unknown or undetermined score. This often occurs when you are measuring the number of errors (or amount of time) required for an individual to complete a task. For example, suppose that preschool children are asked to assemble a wooden puzzle as quickly as possible. The experimenter records how long (in minutes) it takes each child

FIGURE 3.10

A frequency distribution with one extreme score. Notice that the graph shows two breaks in the X-axis. Rather than listing all of the scores for 0–100, the graph skips directly to the lowest score, which is $X = 10$, and then breaks again between $X = 15$ and $X = 100$. The breaks in the X-axis are the conventional way of notifying the reader that some values have been omitted.

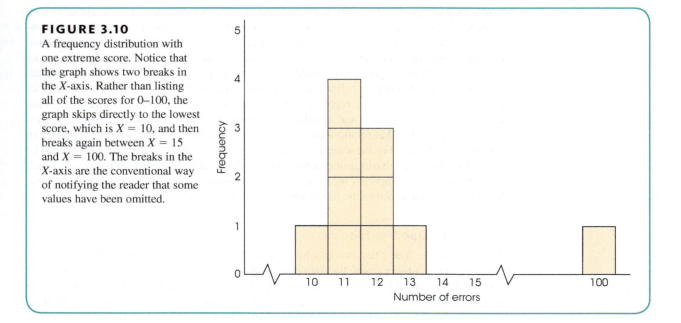

to arrange all the pieces to complete the puzzle. Table 3.5 presents results for a sample of $n = 6$ children.

Notice that one child never completed the puzzle. After an hour, this child still showed no sign of solving the puzzle, so the experimenter stopped him or her. This participant has an undetermined score. (There are two important points to be noted. First, the experimenter should not throw out this individual's score. The whole purpose for using a sample is to gain a picture of the population, and this child tells us that part of the population cannot solve the puzzle. Second, this child should not be given a score of $X = 60$ minutes. Even though the experimenter stopped the individual after one hour, the child did not finish the puzzle. The score that is recorded is the amount of time needed to finish. For this individual, we do not know how long this is.)

It is impossible to compute the mean for these data because of the undetermined value. We cannot calculate the ΣX part of the formula for the mean. However, it is possible to determine the median. For these data, the median is 12.5. Three scores are below the median, and three scores (including the undetermined value) are above the median.

Open-Ended Distributions   A distribution is said to be *open-ended* when there is no upper limit (or lower limit) for one of the categories. The table in the margin of the next page provides an example of an open-ended distribution, showing the number of pizzas eaten

TABLE 3.5

Number of minutes needed to assemble a wooden puzzle.

Person	Time (Min.)
1	8
2	11
3	12
4	13
5	17
6	Never finished

Number of Pizzas (X)	f
5 or more	3
4	2
3	2
2	3
1	6
0	4

during a one-month period for a sample of $n = 20$ high school students. The top category in this distribution shows that three of the students consumed "5 or more" pizzas. This is an open-ended category. Notice that it is impossible to compute a mean for these data because you cannot find ΣX (the total number of pizzas for all 20 students). However, you can find the median. Listing the 20 scores in order produces $X = 1$ and $X = 2$ as the middle two scores. For these data, the median is 1.5.

Ordinal Scale   Many researchers believe that it is not appropriate to use the mean to describe central tendency for ordinal data. When scores are measured on an ordinal scale, the median is always appropriate and is usually the preferred measure of central tendency.

You should recall that ordinal measurements allow you to determine direction (greater than or less than), but do not allow you to determine distance. The median is compatible with this type of measurement because it is defined by direction: half of the scores are above the median and half are below the median. The mean, on the other hand, defines central tendency in terms of distance. Remember that the mean is the balance point for the distribution, so that the distances above the mean are exactly balanced by the distances below the mean. Because the mean is defined in terms of distances, and because ordinal scales do not measure distance, it is not appropriate to compute a mean for scores from an ordinal scale.

■ When to Use the Mode

We will consider three situations in which the mode is commonly used as an alternative to the mean, or is used in conjunction with the mean to describe central tendency.

Nominal Scales   The primary advantage of the mode is that it can be used to measure and describe central tendency for data that are measured on a nominal scale. Recall that the categories that make up a nominal scale are differentiated only by name, such as classifying people by occupation or college major. Because nominal scales do not measure quantity (distance or direction), it is impossible to compute a mean or a median for data from a nominal scale. Therefore, the mode is the only option for describing central tendency for nominal data. When the scores are numerical values from an interval or ratio scale, the mode is usually not the preferred measure of central tendency.

Discrete Variables   Recall that discrete variables are those that exist only in whole, indivisible categories. Often, discrete variables are numerical values, such as the number of children in a family or the number of rooms in a house. When these variables produce numerical scores, it is possible to calculate means. However, the calculated means are usually fractional values that cannot actually exist. For example, computing means will generate results such as "the average family has 2.4 children and a house with 5.33 rooms." The mode, on the other hand, always identifies an actual score (the most typical case) and, therefore, it produces more sensible measures of central tendency. Using the mode, our conclusion would be "the typical, or modal, family has 2 children and a house with 5 rooms." In many situations, especially with discrete variables, people are more comfortable using the realistic, whole-number values produced by the mode.

Describing Shape   Because the mode requires little or no calculation, it is often included as a supplementary measure along with the mean or median as a no-cost extra. The value of the mode (or modes) in this situation is that it gives an indication of the shape of the distribution as well as a measure of central tendency. Remember that the mode identifies the location of the peak (or peaks) in the frequency distribution graph. For example, if you are told that a set of exam scores has a mean of 72 and a mode of 80, you should have a better picture of the distribution than would be available from the mean alone (see Section 3.5).

IN THE LITERATURE

Reporting Measures of Central Tendency

Measures of central tendency are commonly used in the behavioral sciences to summarize and describe the results of a research study. For example, a researcher may report the sample means from two different treatments or the median score for a large sample. These values may be reported in text describing the results, or presented in tables or in graphs.

In reporting results, many behavioral science journals use guidelines adopted by the American Psychological Association (APA), as outlined in the *Publication Manual of the American Psychological Association* (6th ed., 2010). We will refer to the APA manual from time to time in describing how data and research results are reported in the scientific literature. The APA style uses the letter *M* as the symbol for the sample mean. Thus, a study might state:

> The treatment group showed fewer errors ($M = 2.56$) on the task than the control group ($M = 11.76$).

The median can be reported using the abbreviation *Mdn,* as in "Mdn = 8.5 errors," or it can simply be reported in narrative text, as follows:

> The median number of errors for the treatment group was 8.5, compared to a median of 13 for the control group.

There is no special symbol or convention for reporting the mode. If mentioned at all, the mode is usually just reported in narrative text.

When there are many means to report, tables with headings provide an organized and more easily understood presentation. Table 3.6 illustrates this point. Here we use a simplified version of the Rello and Bigham (2017) study from the Preview showing hypothetical results.

TABLE 3.6
The mean time in seconds to read a passage for adults with or without dyslexia with warm or cool screen background colors.

	Warm Colors	Cool Colors
Adults with dyslexia	12.85	16.76
Adults without dyslexia	10.17	14.21

■ Presenting Means and Medians in Graphs

Graphs also can be used to report and compare measures of central tendency. Usually, graphs are used to display values obtained for sample means, but occasionally you will see sample medians reported in graphs (modes are rarely, if ever, shown in a graph). The value of a graph is that it allows several means (or medians) to be shown simultaneously. It is then possible to make quick comparisons between groups or treatment conditions. When preparing a graph, it is customary to list the different groups or treatment conditions on the horizontal axis. Typically, these are the different values that make up the independent variable or the quasi-independent variable. Values for the dependent variable (the scores) are listed on the vertical axis. The means (or medians) are then displayed using a *line graph, histogram,* or *bar graph*, depending on the scale of measurement used for the independent variable.

Figure 3.11 shows an example of a *line graph* displaying the relationship between drug dose (the independent variable) and food consumption (the dependent variable). In this

FIGURE 3.11
The relationship between an independent variable (drug dose) and a dependent variable (food consumption). Because drug dose is a continuous variable measured on a ratio scale, a line graph is used to show the relationship.

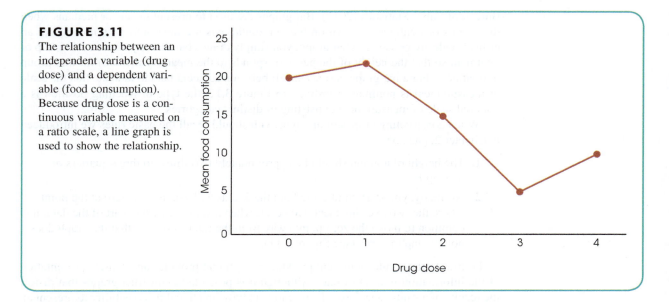

study, there were five different drug doses (treatment conditions), which are listed along the horizontal axis. The five means appear as points in the graph. To construct this graph, a point was placed above each treatment condition so that the vertical position of the point corresponds to the mean score for the treatment condition. The points are then connected with straight lines. A line graph is used when the values on the horizontal axis are measured on an interval or a ratio scale. An alternative to the line graph is a *histogram*. For this example, the histogram would show a bar above each drug dose so that the height of each bar corresponds to the mean food consumption for that group, with no space between adjacent bars.

Figure 3.12 shows a *bar graph* displaying the median weekly income for different types of teaching positions according to data from the United States Department of Labor,

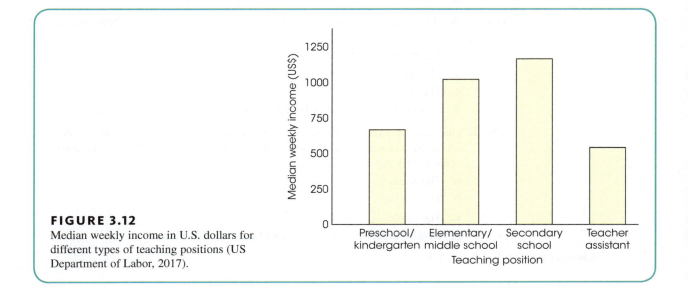

FIGURE 3.12
Median weekly income in U.S. dollars for different types of teaching positions (US Department of Labor, 2017).

Bureau of Labor Statistics (2017). Bar graphs are used to present means or medians when the groups or treatments shown on the horizontal axis are measured on a nominal or an ordinal scale. To construct a bar graph, you simply draw a bar directly above each group or treatment so that the height of the bar corresponds to the mean (or median) for that group or treatment. For a bar graph, a space is left between adjacent bars to indicate that the scale of measurement is nominal or ordinal. In Figure 3.12, the type of teaching position is a nominal scale of measurement consisting of distinct categories.

When constructing graphs of any type, you should recall the basic rules we introduced in Chapter 2, page 55:

1. The height of a graph should be approximately two-thirds to three-quarters of its length.

2. Normally, you start numbering both the X-axis and Y-axis with zero at the point where the two axes intersect. However, when a value of zero is part of the data, it is common to move the zero point away from the intersection so that the graph does not overlap the axes (see Figure 3.11).

Following these rules will help produce a graph that provides an accurate presentation of the information in a set of data. Although it is possible to construct graphs that distort the results of a study (see Box 2.1), researchers have an ethical responsibility to present an honest and accurate report of their research results.

LEARNING CHECK   **LO9**   **1.** A researcher is measuring problem-solving times for a sample of $n = 20$ laboratory rats. However, one of the rats fails to solve the problem so the researcher has an undetermined score. What is the best measure of central tendency for these data?

    a. The mean

    b. The median

    c. The mode

    d. Central tendency cannot be determined for these data.

LO9   **2.** What is the best measure of central tendency for an extremely skewed distribution of scores?

    a. The mean

    b. The median

    c. The mode

    d. Central tendency cannot be determined for a skewed distribution.

LO9   **3.** One item on a questionnaire asks students to identify their preferred animal for the school mascot from three different choices. What is the best measure of central tendency for the data from this question?

    a. The mean

    b. The median

    c. The mode

    d. Central tendency cannot be determined for these data.

ANSWERS   **1. b   2. b   3. c**

SUMMARY

1. The purpose of central tendency is to determine the single value that identifies the center of the distribution and best represents the entire set of scores. The three standard measures of central tendency are the mean, the median, and the mode.

2. The mean is the arithmetic average. It is computed by adding all the scores and then dividing by the number of scores. Conceptually, the mean is obtained by dividing the total (ΣX) equally among the number of individuals (N or n). The mean can also be defined as the balance point for the distribution. The distances above the mean are exactly balanced by the distances below the mean. Although the calculation for a population mean is the same as the calculation for a sample mean, a population mean is identified by the symbol μ, and a sample mean is identified by M. In most situations with numerical scores from an interval or a ratio scale, the mean is the preferred measure of central tendency.

3. Changing any score in the distribution causes the mean to be changed. When a constant value is added to (or subtracted from) every score in a distribution, the same constant value is added to (or subtracted from) the mean. If every score is multiplied by a constant, the mean is multiplied by the same constant.

4. The median is the midpoint of a distribution of scores. The median is the preferred measure of central tendency when a distribution has a few extreme scores that displace the value of the mean. The median also is used when there are undetermined (infinite) scores that make it impossible to compute a mean. Finally, the median is the preferred measure of central tendency for data from an ordinal scale.

5. The mode is the most frequently occurring score in a distribution. It is easily located by finding the peak in a frequency distribution graph. For data measured on a nominal scale, the mode is the appropriate measure of central tendency. It is possible for a distribution to have more than one mode.

6. For symmetrical distributions, the mean will equal the median. If there is only one mode, then it will have the same value, too.

7. For skewed distributions, the mode is located toward the side where the scores pile up, and the mean is pulled toward the extreme scores in the tail. The median is usually located between these two values.

KEY TERMS

central tendency (75)

mean (77)

population mean (μ) (77)

sample mean (M) (77)

weighted mean (80)

median (85)

interpolation (87)

mode (90)

bimodal (91)

multimodal (91)

major mode (91)

minor mode (91)

symmetrical distribution (92)

skewed distribution (93)

positive skew (93)

negative skew (93)

line graph (98)

FOCUS ON PROBLEM SOLVING

X	f
4	1
3	4
2	3
1	2

1. Although the three measures of central tendency appear to be very simple to calculate, there is always a chance for errors. The most common sources of error follow.
   a. Many students find it very difficult to compute the mean for data presented in a frequency distribution table. They tend to ignore the frequencies in the table and simply average the score values listed in the X column. You must use the frequencies *and* the scores! Remember that the number of scores is found by $N = \Sigma f$, and the sum of all N scores is found by ΣfX. For the distribution shown in the margin, the mean is $\frac{24}{10} = 2.40$.

b. The median is the midpoint of the distribution of scores, not the midpoint of the scale of measurement. For a 100-point test, for example, many students incorrectly assume that the median must be $X = 50$. To find the median, you must have the *complete set* of individual scores. The median separates the individuals into two equal-sized groups.

c. The most common error with the mode is for students to report the highest frequency in a distribution rather than the score with the highest frequency. Remember that the purpose of central tendency is to find the most representative score. For the distribution in the margin, the mode is $X = 3$, not $f = 4$.

DEMONSTRATION 3.1

COMPUTING MEASURES OF CENTRAL TENDENCY

For the following sample, find the mean, median, and mode. The scores are:

$$5, 6, 9, 11, 5, 11, 8, 14, 2, 11$$

Compute the Mean   The calculation of the mean requires two pieces of information: the sum of the scores, ΣX; and the number of scores, n. For this sample, $n = 10$ and

$$\Sigma X = 5 + 6 + 9 + 11 + 5 + 11 + 8 + 14 + 2 + 11 = 82$$

Therefore, the sample mean is

$$M = \frac{\Sigma X}{n} = \frac{82}{10} = 8.2$$

See Example 3.9 (page 86) if you are computing the precise median for continuous data.

Find the Median   To find the median, first list the scores in order from smallest to largest. With an even number of scores, the median is the average of the middle two scores in the list. Listed in order, the scores are:

$$2, 5, 5, 6, 8, 9, 11, 11, 11, 14$$

The middle two scores are 8 and 9, so the median is 8.5.

Find the Mode   For this sample, $X = 11$ is the score that occurs most frequently. The mode is $X = 11$.

SPSS®

General instructions for using SPSS are presented in Appendix D. Following are detailed instructions for using SPSS to compute the **Mean, Median, Number of Scores, and ΣX** for two groups of scores.

Demonstration Example

Working in a noisy environment is associated with a reduced ability to hear high-frequency sounds (like the sound of a high-pitched whistle). The ability to hear these sounds is especially important for understanding speech. Below is a hypothetical dataset of middle-age adults from two groups. The top 21 rows of scores come from a group of participants that worked in a noisy environment. The bottom 21 rows of scores are from a group that worked in a quiet environment. Each score represents the maximum frequency sound (in thousands of Hertz) that the person can reliably hear.

Participant	Score	Work Environment	Participant	Score	Work Environment
1	13	noisy	22	16	quiet
2	10	noisy	23	16	quiet
3	7	noisy	24	15	quiet
4	10	noisy	25	14	quiet
5	10	noisy	26	13	quiet
6	5	noisy	27	12	quiet
7	9	noisy	28	14	quiet
8	12	noisy	29	10	quiet
9	12	noisy	30	14	quiet
10	6	noisy	31	16	quiet
11	6	noisy	32	13	quiet
12	15	noisy	33	15	quiet
13	8	noisy	34	15	quiet
14	8	noisy	35	12	quiet
15	11	noisy	36	15	quiet
16	12	noisy	37	12	quiet
17	9	noisy	38	12	quiet
18	10	noisy	39	16	quiet
19	12	noisy	40	20	quiet
20	9	noisy	41	18	quiet
21	10	noisy	42	15	quiet

We will use SPSS to compute the mean, median, number of scores, and sum of scores for each group of participants.

Data Entry

1. You will create two variables in the **Variable View.** In the **Name** field for the first variable, enter "maxFrequency" for the measurement. In the **Name** field for the second variable, enter "workEnv." The default settings for **Width, Values, Missing, Align, and Role** are acceptable. Be sure that **Type** is numeric for the first variable and string for the second variable.

2. For **Decimals** of both variables, enter "0."

3. In the **Label** field, a descriptive title for the variable should be used. Here, we used "Maximum Audible Sounds (in thousands of Hertz)" for the first variable and "Work Environment (Noisy vs. Quiet)" for the second variable.

4. In the **Measure** field, select **Scale** for the first variable because frequency is a ratio scale of measurement. For the second variable, select **Nominal.** When you have finished entering information about the variables, your Variable View should be similar to the figure below.

	Name	Type	Width	Decimals	Label	Values	Missing	Columns	Align	Measure	Role
1	maxFreque..	Numeric	8	0	Maximum Audible Sounds (in thousands of Hertz)	None	None	8	Right	Scale	Input
2	workEnv	String	8	0	Work Environment (Noisy vs. Quiet)	None	None	8	Left	Nominal	Input

5. Click **Data View** to return to the table where you will enter values. The data format for this problem is like Data Format 2 described in Appendix D. Enter the scores above in the "maxFrequency" column. Enter the work environment for each participant in the "workEnv" column by typing "noisy" or "quiet" in each cell. When you are finished, your Data View should be like the figure below.

	maxFrequency	workEnv
1	13.00	noisy
2	10.00	noisy
3	7.00	noisy
4	10.00	noisy
5	10.00	noisy
6	5.00	noisy
7	9.00	noisy
8	12.00	noisy
9	12.00	noisy
10	6.00	noisy
11	6.00	noisy
12	15.00	noisy
13	8.00	noisy
14	8.00	noisy
15	11.00	noisy
16	12.00	noisy
17	9.00	noisy
18	10.00	noisy
19	12.00	noisy
20	9.00	noisy
21	10.00	noisy
22	16.00	quiet
23	16.00	quiet
24	15.00	quiet
25	14.00	quiet
26	13.00	quiet
27	12.00	quiet
28	14.00	quiet
29	10.00	quiet
30	14.00	quiet
31	16.00	quiet
32	13.00	quiet
33	15.00	quiet
34	15.00	quiet
35	12.00	quiet
36	15.00	quiet
37	12.00	quiet
38	12.00	quiet
39	16.00	quiet
40	20.00	quiet
41	18.00	quiet
42	15.00	quiet

Source: SPSS®

Data Analysis

1. Click **Analyze** on the tool bar, select **Compare Means**, and click on **Means**.
2. Highlight the column label for the set of scores ("maxFrequency") in the left box and click the arrow to move it into the **Dependent List** box. Highlight the column label variable where you recorded the work environment ("workEnv") and click the arrow to move it into the **Independent List** box.
3. Click on the **Options** box, and use the arrow to move statistics between the **Statistics** box and the **Cell Statistics** box. SPSS will compute all of the statistics listed in the Cell Statistics box. Be sure that your list includes Mean, Number of Cases, Median, and Sum. Some of the statistics that are selected by default (e.g., Std. Deviation) are covered in later chapters. You can deselect those by clicking the arrow to remove them from the Cell Statistics box.
4. Check that the Means Options window is similar to the figure below and click **Continue.** Check that the Means window is as seen in the image below and click **OK**.

SPSS Output

The SPSS output will contain two sections, as seen in the image below.

Case Processing Summary

	Cases					
	Included		Excluded		Total	
	N	Percent	N	Percent	N	Percent
Maximum Audible Sounds (in thousands of Hertz) * Work Environment (Noisy vs. Quiet)	42	100.0%	0	0.0%	42	100.0%

Report

Maximum Audible Sounds (in thousands of Hertz)

Work Environment (Noisy vs. Quiet)	Mean	N	Median	Sum
noisy	9.71	21	10.00	204
quiet	14.43	21	15.00	303
Total	12.07	42	12.00	507

The Case Processing Summary section of the SPSS Output reports the total number of scores that were included in the analysis ($N = 42$) and excluded from the analysis ($N = 0$). The Report section of the output lists the mean, number of scores, median, and sum of scores in three ways: (1) for participants that worked in noisy environments only, (2) for participants that worked in quiet environments only, and (3) for participants from both groups. You should notice that the scores from the noisy work environment had a lower mean ($M = 9.71$) and median ($Mdn = 10.00$) than scores from the quiet work environment ($M = 14.43$ and $Mdn = 15.00$). You will also find that SPSS uses the sorting method to find the median.

Try It Yourself

For the following set of scores, use SPSS to compute the mean, median, sum of scores, and number of scores in each group.

Participant	Score	Group
1	15	Group 1
2	8	Group 1
3	9	Group 1
4	14	Group 1
5	14	Group 1
6	6	Group 1
7	12	Group 1
8	9	Group 1
9	7	Group 2
10	−1	Group 2
11	10	Group 2
12	9	Group 2
13	4	Group 2
14	2	Group 2
15	7	Group 2
16	4	Group 2

SPSS will report the following statistics:

	Mean	N	Median	Sum
Group 1	10.88	8	10.50	87
Group 2	5.25	8	5.50	42
Total	8.06	16	8.50	129

PROBLEMS

1. A sample of $n = 9$ scores has $\Sigma X = 108$. What is the sample mean?

2. A sample of $n = 12$ scores has $\Sigma X = 72$. What is the sample mean?

3. Find the mean for the following set of scores: 2, 7, 9, 4, 5, 3, 0, 6

4. Find the mean for the following set of scores: 8, 2, 5, 7, 12, 9, 11, 3, 6

5. Using the following informal histogram, what is the value of the mean? Explain your answer.

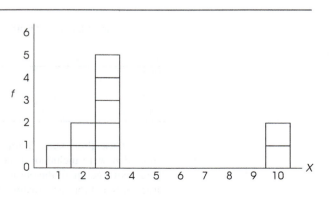

6. In a sample of $n = 6$ scores, five of the scores are each above the mean by one point. Where is the sixth score located relative to the mean?

7. Which statistic is equivalent to dividing the sum of scores equally across all members of a sample?

8. A sample with a mean of $M = 8$ has $\Sigma X = 56$. How many scores are in the sample?

9. A population of $N = 7$ scores has a mean of $\mu = 13$. What is the value of ΣX for this population?

10. One sample of $n = 10$ scores has a mean of 8, and a second sample of $n = 5$ scores has a mean of 2. If the two samples are combined, what is the mean for the combined sample?

11. One sample has a mean of $M = 6$, and a second sample has a mean of $M = 12$. The two samples are combined into a single set of scores.
    a. What is the mean for the combined set if both the original samples have $n = 4$ scores?
    b. What is the mean for the combined set if the first sample has $n = 3$ and the second sample has $n = 6$?
    c. What is the mean for the combined set if the first sample has $n = 6$ and the second sample has $n = 3$?

12. Find the mean for the scores in the following frequency distribution table:

X	f
6	1
5	4
4	2
3	2
2	1

13. A sample of $n = 10$ scores has a mean of $M = 7$. If one score is changed from $X = 21$ to $X = 11$, what is the value of the new sample mean?

14. A sample of $n = 6$ scores has a mean of $M = 10$. If one score is changed from $X = 12$ to $X = 0$, what is the value of the new sample mean?

15. A sample of $n = 6$ scores has a mean of $M = 10$. If one score with a value of $X = 12$ is removed from the sample, then what is the value of the new sample mean?

16. A sample of $n = 5$ scores has a mean of $M = 12$. If one new score with a value of $X = 17$ is added to the sample, then what is the mean for the new sample?

17. A population of $N = 10$ scores has a mean of $\mu = 12$. If one score with a value of $X = 21$ is removed from the population, then what is the value of the new population mean?

18. A sample of scores has a mean of $M = 6$. Calculate the mean for each of the following.
    a. A constant value of 3 is added to each score.
    b. A constant value of 1 is subtracted from each score.
    c. Each score is multiplied by a constant value of 6.
    d. Each score is divided by a constant value of 2.

19. A population of scores has a mean of $\mu = 50$. Calculate the mean for each of the following.
    a. A constant value of 50 is added to each score.
    b. A constant value of 50 is subtracted from each score.
    c. Each score is multiplied by a constant value of 2.
    d. Each score is divided by a constant value of 50.

20. In 2016, the 50th percentile score for household income in the United States was \$59,039. What statistic for central tendency is this?

21. Find the median for the following set of scores: 1, 9, 3, 6, 4, 3, 11, 10

22. Find the median for the following set of scores: 1, 4, 8, 7, 13, 26, 6

23. For the following sample of $n = 10$ scores, 6, 5, 4, 3, 3, 3, 2, 2, 2, 1
    a. Assume that the scores are measurements of a discrete variable and find the median.
    b. Assume that the scores are measurements of a continuous variable and find the precise median by locating the precise midpoint of the distribution.

24. For the following sample of $n = 10$ scores: 2, 3, 4, 4, 5, 5, 5, 6, 6, 7
    a. Assume that the scores are measurements of a discrete variable and find the median.
    b. Assume that the scores are measurements of a continuous variable and find the precise median by locating the precise midpoint of the distribution.

25. Find the mean, median, and mode for the distribution of scores in the following frequency distribution table.

X	f
9	1
8	1
7	3
6	4
5	1

26. Find the mean, median, and mode for the following scores: 8, 7, 5, 7, 0, 10, 2, 4, 11, 7, 8, 7

27. What shape is the distribution displayed in the following informal histogram? Identify the major and minor modes.

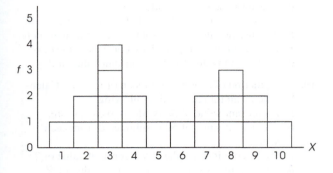

28. For the following frequency distribution table, identify the shape of the distribution.

X	f
5	1
6	2
7	5
8	3
9	1
10	1
11	2
12	3
13	1

29. Solve the following problems.
   a. Find the mean, median, and mode for the following scores.

9	6	7	10	7	9	9	7
9	4	9	8	3	6	8	9

   b. Based on the relative values for the mean, median, and mode, what is the most likely shape for this distribution of scores (symmetrical, positively skewed, or negatively skewed)?

30. Anderson (1999) was interested in the effects of attention load on reaction time. Participants in her study received a dual-task procedure in which they needed to respond as quickly as possible to a stimulus while simultaneously paying attention to the sounds of spoken words. She recorded reaction time (in hundreds of milliseconds). Below are data like those observed by Anderson:

$$3, 4, 4, 4, 5, 6, 8, 12, 20, 25$$

   a. Find the mean, median, and mode.
   b. Based on the relative values of those statistics, what is the shape of the distribution?
   c. Anderson (1999) reported median reaction times. Why?

31. Solve the following problems.
   a. Find the mean, median, and mode for the scores in the following frequency distribution table.

X	f
5	2
4	5
3	2
2	3
1	0
0	2

   b. Based on the three values for central tendency, what is the most likely shape for this distribution of scores (symmetrical, positively skewed, or negatively skewed)?

32. Identify the circumstances in which the median may be better than the mean as a measure of central tendency and explain why.

Variability

Tools You Will Need

The following items are considered essential background material for this chapter. If you doubt your knowledge of any of these items, you should review the appropriate chapter or section before proceeding.

- Summation notation (Chapter 1)
- Central tendency (Chapter 3)
  - Mean
  - Median

clivewa/Shutterstock.com

PREVIEW

PREVIEW

Mazes have been used for many years with laboratory animals to study learning, memory, and motivation, including the function of the brain and drug effects. Researchers have used mazes in a variety of configurations (for example, see Tolman, 1948). Typically, a laboratory rat is placed in the start box and allowed to run down the maze alleys, making sequences of left and right turns until it reaches the goal box, where a food reward awaits. The researcher might record the number of wrong turns the rat makes or the time it takes the rat to reach the goal box. Consider the following basic example of rats learning to solve a maze. The experimenter takes a sample of six rats and tests them in a multiple T-maze (Figure 4.1).

Each rat is placed at the start of the maze and left inside until it finds the food reward at the end of one alley. This consists of one learning trial. The rats are tested for 10 trials, and the amount of time it takes to find the food reward is recorded for each rat on every trial. In the first few trials the rats explore the maze, sometimes backtracking their path and revisiting the same incorrect alley. Occasionally they might rear up and sniff the walls of the maze, or even hesitate when they get to a choice point before making a turn. Eventually they happen to find the food reward. Over the course of the trials the rats solve the maze faster as they hesitate less, make fewer errors, and learn the location of the reward. Hypothetical data (time in minutes) are presented in Table 4.1 for the rats' learning performance on the first and tenth trials. The mean amount of time it takes the sample of rats to solve the maze on Trial 1 and Trial 10 is also shown.

If you compare these data, you will notice that the scores on the first trial are more spread out than the scores on the tenth trial. This greater spread of the scores reflects more variability in behavior on the first trial. The rats show more individual differences in how they

TABLE 4.1

Performance of rats in a multiple T-maze.

Rat	Time to Solve on Trial 1	Time to Solve on Trial 10
A	8	1
B	18	3
C	5	1
D	19	4
E	13	2
F	9	1
	$M = 12$	$M = 2$

respond on their very first exposure to the maze, with scores ranging from 5 to 19. As they learn the location of the reward and the correct sequence of responses to get to the goal box, their behavior becomes more uniform and less variable. By the tenth trial their scores range from 1 to 4. Another way to look at the variability of the data in this study is to use the mean as a reference point. First, look at the data on the tenth trial. The scores are clustered close to the mean. One score equals the mean, $M = 2$, and the others vary no more than 1 or 2 points from the mean. Now look at the scores on the first trial. They are spread farther from the mean, $M = 12$. Two scores are as many as 5 points from the mean.

In this chapter we introduce the statistical concept of variability. We will describe the methods that are used to measure and objectively describe the differences that exist from one score to another within a distribution. In addition to describing distributions of scores, variability also helps us determine which outcomes are likely and which are very unlikely to be obtained. This aspect of variability will play an important role in inferential statistics, which is covered in later chapters.

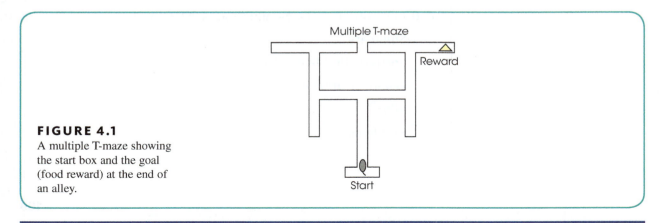

FIGURE 4.1

A multiple T-maze showing the start box and the goal (food reward) at the end of an alley.

4-1 Introduction to Variability

LEARNING OBJECTIVES

1. Define variability and explain its use and importance as a statistical measure.
2. Define and calculate the range as a simple measure of variability and explain its limitations.
3. Define and calculate the interquartile range and explain its advantages over the simple range.

The term *variability* has much the same meaning in statistics as it has in everyday language; to say that things are variable means that they are not all the same. In statistics, our goal is to measure the amount of variability for a particular set of scores, a distribution. In simple terms, if the scores in a distribution are all the same, then there is no variability. If there are small differences between scores, then the variability is small, and if there are large differences between scores, then the variability is large.

In this chapter we introduce variability as a statistical concept. We will describe the methods that are used to measure and objectively describe the differences that exist from one score to another within a distribution. In addition to describing distributions of scores, variability also helps us determine which outcomes are likely and which are very unlikely to be obtained. This aspect of variability will play an important role in inferential statistics.

Variability provides a quantitative measure of the differences between scores in a distribution and describes the degree to which the scores are spread out or clustered together.

Figure 4.2 shows two distributions of familiar values for the population of adult males: Part (a) shows the distribution of men's heights (in inches), and part (b) shows the distribution of men's weights (in pounds). Notice that the two distributions differ in terms of central tendency. The mean height is 70 inches (5 feet, 10 inches) and the mean weight is 170 pounds. In addition, notice that the distributions differ in terms of variability. For example, most heights are clustered close together, within 5 or 6 inches of the mean. Weights, on the other hand, are spread over a much wider range. In the weight distribution it is not unusual to find two men whose weights differ by 40 or 50 pounds.

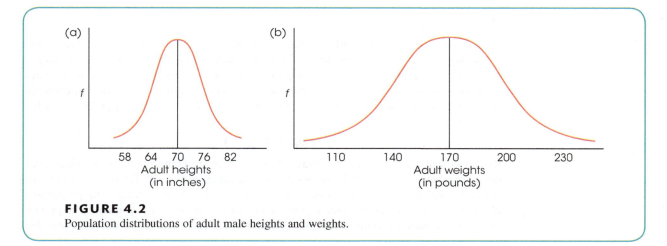

FIGURE 4.2
Population distributions of adult male heights and weights.

Variability can also be viewed as measuring predictability, consistency, or even diversity. If your morning commute to work or school always takes between 15 and 17 minutes, then your commuting time is very *predictable* and you do not need to leave home 60 minutes early just to be sure that you arrive on time. Similarly, *consistency* in performance from trial to trial is viewed as a skill. For example, the ability to hit a target time after time is an indication of skilled performance in many sports. Finally, corporations, colleges, and government agencies often make attempts to increase the *diversity* of their students or employees. Once again, they are referring to the differences from one individual to the next. Thus, predictability, consistency, and diversity are all concerned with the differences between scores or between individuals, which is exactly what is measured by variability.

In general, a good measure of variability serves two purposes:

1. Variability describes the distribution of scores. Specifically, it tells whether the scores are clustered close together or spread out over a large distance. Usually, variability is defined in terms of *distance*. It tells how much distance to expect between one score and another, or how much distance to expect between an individual score and the mean. For example, we know that the heights for most adult males are clustered close together, within five or six inches of the average. Although more extreme heights exist, they are relatively rare.

2. Variability measures how well an individual score (or group of scores) represents the entire distribution. This aspect of variability is very important for inferential statistics, in which relatively small samples are used to answer questions about populations. For example, suppose that you selected a sample of one adult male to represent the entire population. Because most men have heights that are within a few inches of the population average (the distances between scores and the population mean are small), there is a very good chance that you would select someone whose height is within six inches of the population mean. For men's weights, on the other hand, there are relatively large differences from one individual to another. For example, it would not be unusual to select an individual whose weight differs from the population average by more than 30 pounds. Thus, when using a sample to represent a population, variability provides information about how much error to expect between the sample data and the population mean.

In this chapter, we consider four different measures of variability: the range, the interquartile range, the standard deviation, and the variance. Of these four, the standard deviation and the related measure of variance are by far the most important because they play a central role in inferential statistics.

■ The Range

The obvious first step toward defining and measuring variability is the *range*, which is the distance covered by the scores in a distribution, from the smallest to the largest score. Although the concept of the range is fairly straightforward, there are several distinct methods for computing the numerical value. One commonly used definition of the range simply measures the difference between the largest score (X_{max}) and the smallest score (X_{min}):

$$\text{range} = X_{max} - X_{min}$$

By this definition, scores having values from 1 to 5 cover a range of 4 points. Many computer programs, such as SPSS, use this definition, which works well for variables with precisely defined upper and lower boundaries. For example, if you are measuring proportions of an object, like pieces of a pizza, you can obtain values such as $\frac{1}{8}$, $\frac{1}{4}$, $\frac{1}{2}$, and $\frac{3}{4}$. Expressed as decimal values, the proportions range from 0 to 1. You can never have a value less than

0 (none of the pizza) and you can never have a value greater than 1 (all of the pizza). Thus, the complete set of proportions is bounded by 0 at one end and by 1 at the other. As a result, the proportions cover a range of 1 point.

Continuous and discrete variables are discussed in Chapter 1 on pages 12–14.

An alternative definition of the range is often used when the scores are measurements of a continuous variable. In this case, the range can be defined as the difference between the upper real limit (URL) for the largest score (X_{max}) and the lower real limit (LRL) for the smallest score (X_{min}).

$$\text{range} = \text{URL for } X_{max} - \text{LRL for } X_{min} \qquad (4.1)$$

According to this definition, scores having values from 1 to 5 cover a range of $5.5 - 0.5 = 5$ points.

When the scores are whole numbers, the range can also be defined as the number of measurement categories. If every individual is classified as either 1, 2, or 3, then there are three measurement categories and the range is 3 points. Defining the range as the number of measurement categories also works for discrete variables that are measured with numerical scores. For example, suppose a study measures the number of children in participating families and the following scores are obtained:

Remember, a discrete variable consists of separate and indivisible categories so that there are no values between neighboring categories (or scores).

$$2 \quad 2 \quad 3 \quad 1 \quad 0 \quad 4 \quad 3 \quad 1$$

The data consist of values from 0 to 4; therefore, there are five measurement categories (0, 1, 2, 3, and 4) and the range is 5 points. By this definition, when the scores are all whole numbers based on a discrete variable, the range can be obtained by

$$X_{max} - X_{min} + 1 \qquad (4.2)$$

Using any of these definitions, the range is probably the most obvious way to describe how spread out the scores are—simply find the distance between the maximum and the minimum scores. The problem with using the range as a measure of variability is that it is completely determined by those two extreme values and ignores the other scores in the distribution. Thus, a distribution with one unusually large (or small) score will have a large range even if the other scores are all clustered close together.

Note that the range is determined by only the most extreme high and extreme low scores in the distribution. The range does not consider all the scores in the distribution; therefore, it often does not give an accurate description of the variability for the entire distribution. For these reasons, the range is considered to be a crude and unreliable measure of variability. The range is seldom used in formal descriptions of variability because of its failure to reveal the typical distance among common scores.

■ The Interquartile Range

In basic terms, the interquartile range is the range of scores that make up the middle 50% of the distribution. The 25% extreme low and 25% extreme high scores are not used for this measure of variability. The interquartile range is based on *quartiles*, which are a type of percentile rank. As the name implies, a quartile is one-fourth of the distribution. The first quartile, Q1, corresponds to the score that has a percentile rank of 25%. That is, 25% of the scores fall below it. The second quartile has a percentile rank of 50%, which we saw in Chapter 3 is the median. The third quartile, Q3, is a score with a rank of 75%, or 75% of the scores fall below it. The fourth quartile, Q4, is the highest score in the distribution and its rank is 100%. While a percentile divides a distribution into 100 equal parts, each corresponding to 1% of the distribution, a quartile divides the distribution into four equal parts, each corresponding to 25% of the distribution.

Quartiles are the scores having percentile ranks of 25%, 50%, 75%, and 100%, which are termed the first, second, third, and fourth quartile, respectively. Quartiles divide the distribution into four equal parts such that each quartile section corresponds to 25% of the distribution.

Consider the following set of scores from a continuous variable:

17	14	12	11	11	14	9	13	6	10
15	11	12	13	11	10	11	10	15	12

These data are displayed in Figure 4.3. For $N = 20$ scores, the quartiles divide the distribution into four equal parts containing five scores each and are labeled along with the X values. Notice that quartiles are associated with their upper real limits. For example, the first quartile, Q1, equals 10.5. For Q1, 25% of the scores in the distribution fall below 10.5. Similarly, for Q3, 75% of the distribution falls below 13.5.

The *interquartile range (IQR)* is the distance between the 25th and 75th percentile, or between Q1 and Q3. Note that the bottom 25% and top 25% of the distribution are excluded so that the interquartile range spans the scores in the middle 50% of the distribution (Figure 4.3). To compute the interquartile range, you first identify Q3 (the 75th percentile) and Q1 (the 25th percentile). This can be done with a histogram like Figure 4.3 or with a frequency distribution table that includes columns for cf and $c\%$. Finally, find the difference between the first and third quartiles:

$$\text{Interquartile range} = \text{IQR} = Q3 - Q1 \qquad \text{(4.3)}$$

For the data in Figure 4.3,

$$\text{IQR} = Q3 - Q1 = 13.5 - 10.5 = 3$$

The shaded area of Figure 4.3 represents the scores within the interquartile range.

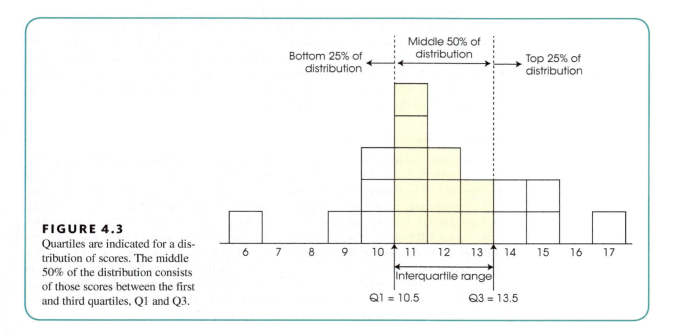

FIGURE 4.3

Quartiles are indicated for a distribution of scores. The middle 50% of the distribution consists of those scores between the first and third quartiles, Q1 and Q3.

> The **interquartile range (IQR)** is the distance between the X values that correspond to the first (Q1) and third (Q3) quartiles. It reflects the range for the scores that fall in the middle 50% of the distribution.

Notice that the interquartile range is a better description of variability than the range. For these data the range is as follows:

$$\text{range} = \text{URL for } X_{max} - \text{LRL for } X_{min}$$
$$= 17.5 - 5.5 = 12$$

Notice that the extreme scores, $X = 6$ and $X = 17$, have a large influence on the value of the range, even though most scores are clustered in the center of the distribution. By excluding the extreme 25% of scores at the top and bottom of the distribution, the interquartile range is not influenced by extreme values. In these instances, the interquartile range is more descriptive than the basic range because it trims the extreme scores and provides a range that reflects the cluster of scores in the center of the distribution.

When to Use the Interquartile Range   The interquartile range typically is used when measuring central tendency with the median. Both measures are related to percentiles. The median is the 50th percentile, and Q1 and Q3 are the 25th and 75th percentile, respectively. The interquartile range is used in the same set of circumstances where the median is preferred (Chapter 3, pages 95–97), especially with distributions that have extreme scores or are skewed. It often can be used when there are undetermined values and open-ended distributions as well. Finally, like the median it can be used for data measured with an ordinal scale of measurement.

LEARNING CHECK   **LO1**   **1.** Which of the following is a consequence of increasing variability?

   a. The distance from one score to another tends to increase and a single score tends to provide a more accurate representation of the entire distribution.

   b. The distance from one score to another tends to increase and a single score tends to provide a less accurate representation of the entire distribution.

   c. The distance from one score to another tends to decrease and a single score tends to provide a more accurate representation of the entire distribution.

   d. The distance from one score to another tends to decrease and a single score tends to provide a less accurate representation of the entire distribution.

LO2   **2.** What is the range for the following set of scores? Scores: 5, 7, 9, 15

   a. 4 points

   b. 5 points

   c. 10 or 11 points

   d. 15 points

LO2   **3.** For the following scores, which of the following actions will increase the range? Scores: 3, 7, 10, 15

   a. Add 4 points to the score $X = 3$

   b. Add 4 points to the score $X = 7$

   c. Add 4 points to the score $X = 10$

   d. Add 4 points to the score $X = 15$

LO3 **4.** For the following scores, find the interquartile range. Scores:

3, 4, 4, 1, 7, 3, 2, 6, 4, 2

1, 6, 3, 4, 5, 2, 5, 4, 3, 4

a. 7

b. 2

c. 3.5

d. 1

ANSWERS **1. b** **2. c** **3. d** **4. b**

4-2 | Defining Variance and Standard Deviation

LEARNING OBJECTIVES

4. Define variance and standard deviation and describe what is measured by each.

5. Calculate variance and standard deviation for a simple set of scores.

6. Estimate the standard deviation for a set of scores based on a visual examination of a frequency distribution graph of the distribution.

The standard deviation is the most commonly used and the most important descriptive measure of variability. Standard deviation uses the mean of the distribution as a reference point and measures variability by considering the distance between each score and the mean.

In simple terms, the standard deviation provides a measure of the standard, or average, distance from the mean, and describes whether the scores are clustered closely around the mean or are widely scattered.

Although the concept of standard deviation is straightforward, the actual equations tend to be more complex and lead us to the related concept of *variance* before we finally reach the standard deviation. Therefore, we begin by looking at the logic that leads to these equations. If you remember that our goal is to measure the standard, or typical, distance from the mean, then this logic and the equations that follow should be easier to remember.

STEP 1   The first step in finding the standard distance from the mean is to determine the *deviation*, or distance from the mean, for each individual score. By definition, the deviation for each score is the difference between the score and the mean.

> A **deviation** or **deviation score** is the difference between a score and the mean, and is calculated as
>
> $$\text{deviation} = X - \mu$$

For a distribution of scores with $\mu = 50$, if your score is $X = 53$, then your deviation score is

$$X - \mu = 53 - 50 = 3 \text{ points}$$

If your score is $X = 45$, then your deviation score is

$$X - \mu = 45 - 50 = -5 \text{ points}$$

In some sources you might see a lowercase *x* used as notation for a deviation score.

Notice that there are two parts to a deviation score: the sign (+ or −) and the number. The sign (+ or −) tells the direction from the mean—that is, whether the score is located above (+) or below (−) the mean, and the number gives the actual distance from the mean. For example, a deviation score of −6 corresponds to a score that is below the mean by a distance of 6 points.

STEP 2   Because our goal is to compute a measure of the standard distance from the mean, you might be tempted to calculate the mean of the deviation scores. To compute this mean, you first add up the deviation scores and then divide by *N*. This process is demonstrated in the following example.

EXAMPLE 4.1   We start with the following set of *N* = 4 scores. These scores add up to $\Sigma X = 12$, so the mean is $\mu = \frac{12}{4} = 3$. For each score, we have computed the deviation.

X	$X - \mu$
8	+5
1	−2
3	0
0	−3
	$0 = \Sigma(X - \mu)$

∎

Note that the deviation scores add up to zero. This should not be surprising if you remember that the mean serves as a balance point for the distribution. The total of the distances above the mean is exactly equal to the total of the distances below the mean (page 78). Thus, the total for the positive deviations is exactly equal to the total for the negative deviations, and the complete set of deviations always adds up to zero (Box 3.1).

Because the sum of the deviations is *always* zero, the mean of the deviations is also zero and is of no value as a measure of variability. Specifically, the mean of the deviations is zero if the scores are closely clustered and it is zero if the scores are widely scattered. (You should note, however, that the constant value of zero is useful in other ways. Whenever you are working with deviation scores, you can check your calculations by making sure that the deviation scores add up to zero.)

STEP 3   The average of the deviation scores will not work as a measure of variability because it is always zero. Clearly, this problem results from the positive and negative values canceling each other out. The solution is to get rid of the signs (+ and −). The standard procedure for accomplishing this is to square each deviation score. Using the squared values, you then compute the average of the squared deviations, or the *mean squared deviation*, which is called *variance*.

Variance equals the mean of the squared deviations. Variance is the average squared distance from the mean.

Note that the process of squaring deviation scores does more than simply get rid of plus and minus signs. It results in a measure of variability based on *squared* distances. Although variance is valuable for some of the *inferential* statistical methods covered later, the concept of squared distance is not an intuitive or easy-to-understand *descriptive* measure. For example, it is not particularly useful to know that the squared distance from New York City to Boston is 26,244 miles squared. The squared value becomes meaningful, however, if you take the square root. Therefore, we continue the process with one more step.

STEP 4   Remember that our goal is to compute a measure of the standard distance from the mean. Variance, which measures the average squared distance from the mean, is not exactly what we want. The final step simply takes the square root of the variance to obtain the *standard deviation*, which measures the standard distance from the mean.

> **Standard deviation** is the square root of the variance and provides a measure of the standard, or average distance from the mean.
>
> $$\text{Standard deviation} = \sqrt{\text{variance}}$$

Figure 4.4 shows the overall process of computing variance and standard deviation. Remember that our goal is to measure variability by finding the standard distance from the mean. However, we cannot simply calculate the average of the distances because this value will always be zero. Therefore, we begin by squaring each distance, then we find the average of the squared distances, and finally we take the square root to obtain a measure of the standard distance. Technically, the standard deviation is the square root of the average squared deviation. Conceptually, however, the standard deviation provides a measure of the average distance from the mean.

Although we still have not presented any formulas for variance or standard deviation, you should be able to compute these two statistical values from their definitions. The following example demonstrates this process.

EXAMPLE 4.2   We will calculate the variance and standard deviation for the following population of $N = 5$ scores:

$$1, \quad 9, \quad 5, \quad 8, \quad 7$$

Remember that the purpose of standard deviation is to measure the standard distance from the mean, so we begin by computing the population mean. These five scores add up to

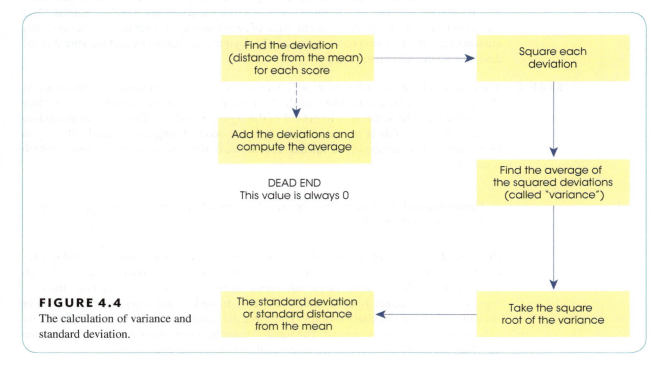

FIGURE 4.4

The calculation of variance and standard deviation.

$\Sigma X = 30$ so the mean is $\mu = \frac{30}{5} = 6$. Next, we find the deviation (distance from the mean) for each score and then square the deviations. Using the population mean $\mu = 6$, these calculations are shown in the following table.

Score X	Deviation X − μ	Squared Deviation (X − μ)²
1	−5	25
9	3	9
5	−1	1
8	2	4
7	1	1
		40 = the sum of the squared deviations

For this set of $N = 5$ scores, the squared deviations add up to 40. The mean of the squared deviations, the variance, is $\frac{40}{5} = 8$, and the standard deviation is $\sqrt{8} = 2.83$. ■

You should note that a standard deviation of 2.83 is a sensible answer for this distribution. The five scores in the population are shown in a histogram in Figure 4.5 so that you can see the distances more clearly. Note that the scores closest to the mean are only 1 point away. Also, the score farthest from the mean is 5 points away. For this distribution, the largest distance from the mean is 5 points and the smallest distance is 1 point. Thus, the standard distance should be somewhere between 1 and 5. By looking at a distribution in this way, you should be able to make a rough estimate of the standard deviation. In this case, the standard deviation should be between 1 and 5, probably around 3 points. The value we calculated for the standard deviation is in excellent agreement with this estimate.

Making a quick estimate of the standard deviation can help you avoid errors in calculation. For example, if you calculated the standard deviation for the scores in Figure 4.5 and obtained a value of 12, you should realize immediately that you have made an error. (If the biggest deviation is only 5 points, then it is impossible for the standard deviation to be 12.)

The following example is an opportunity for you to test your understanding by computing variance and standard deviation yourself.

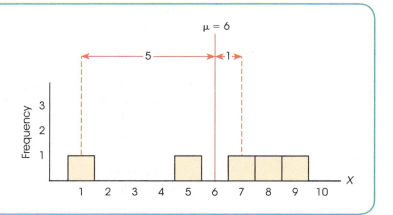

FIGURE 4.5

A frequency distribution histogram for a population of $N = 5$ scores. The mean for this population is $\mu = 6$. The smallest distance from the mean is 1 point and the largest distance is 5 points. The standard distance (or standard deviation) should be between 1 and 5 points.

EXAMPLE 4.3   Compute the variance and standard deviation for the following set of $N = 6$ scores: 12, 0, 1, 7, 4, and 6. You should obtain a variance of 16 and a standard deviation of 4. Good luck.   ■

Because the standard deviation and variance are defined in terms of distance from the mean, these measures of variability are used only with numerical scores that are obtained from measurements on an interval or a ratio scale. Recall from Chapter 1 (page 16) that these two scales are the only ones that provide information about distance; nominal and ordinal scales do not. Also, recall from Chapter 3 (page 97) that it is inappropriate to compute a mean for ordinal data and impossible to compute a mean for nominal data. Because the mean is a critical component in the calculation of standard deviation and variance, the same restrictions that apply to the mean also apply to these two measures of variability. Specifically, the mean, the standard deviation, and the variance should be used only with numerical scores from interval or ordinal scales of measurement.

A Note about Rounding   For Example 4.2, your calculator or computer program might give an answer for standard deviation like 2.828427125, depending on how many digits and decimal places it reports. However, for Example 4.2 we have, for convenience, reported just two decimal places. The rule we have used is as follows. If the third decimal place is 5 or higher, we drop it and any decimal values on the right, then make the second decimal place one point higher. Thus, we reported a standard deviation of 2.83. However, if the third decimal place is less than 5, then we simply drop that decimal and all to the right, and leave the second decimal place unchanged. We will usually round to two decimal places. You might check with your instructor, who might have a preference for how you round off answers.

LEARNING CHECK   **LO4** **1.** Which of the following sets of scores has the largest variance?
   a. 1, 3, 8, 12
   b. 12, 13, 14, 15
   c. 2, 2, 2, 2
   d. 22, 24, 25, 27

LO5 **2.** What is the variance for the following set of scores? Scores: 4, 1, 7
   a. $\frac{66}{3} = 22$
   b. 18
   c. 9
   d. 6

LO6 **3.** A set of scores ranges from a high of $X = 24$ to a low of $X = 12$ and has a mean of 18. Which of the following is the most likely value for the standard deviation for these scores?
   a. 3 points
   b. 6 points
   c. 12 points
   d. 24 points

ANSWERS   **1. a   2. d   3. a**

LEARNING OBJECTIVES

7. Calculate *SS*, the sum of the squared deviations, for a population using either the definitional or the computational formula and describe the circumstances in which each formula is appropriate.

8. Calculate the variance and the standard deviation for a population.

The concepts of variance and standard deviation are the same for both samples and populations. However, the details of the calculations differ slightly, depending on whether you have data from a sample or from a complete population. We first consider the formulas for populations and then look at samples in Section 4-4.

■ The Sum of Squared Deviations (SS)

Recall that variance is defined as the mean of the squared deviations. This mean is computed exactly the same way you compute any mean: first find the sum, and then divide by the number of scores.

$$\text{Variance} = \text{mean squared deviation} = \frac{\text{sum of squared deviations}}{\text{number of scores}}$$

The value in the numerator of this equation, the sum of the squared deviations, is a basic component of variability, and we will focus on it. To simplify things, it is identified by the notation *SS* (for sum of squared deviations), and it generally is referred to as the *sum of squares*.

> *SS*, or **sum of squares**, is the sum of the squared deviation scores.

You need to know two formulas to compute *SS*. These formulas are algebraically equivalent (they always produce the same answer), but they look different and are used in different situations.

The first of these formulas is called the definitional formula because the symbols in the formula literally define the process of adding up the squared deviations:

$$\text{Definitional formula: } SS = \Sigma(X - \mu)^2 \qquad \text{(4.4)}$$

To find the sum of the squared deviations, the formula instructs you to perform the following sequence of calculations:

1. Find each deviation score $(X - \mu)$.

2. Square each deviation score $(X - \mu)^2$.

3. Add the squared deviations.

The result is *SS*, the sum of the squared deviations. The following example demonstrates using this formula.

EXAMPLE 4.4   We will compute *SS* for the following set of $N = 4$ scores. These scores have a sum of $\Sigma X = 8$, so the mean is $\mu = \frac{8}{4} = 2$. The following table shows the deviation and the squared deviation for each score. The sum of the squared deviations is $SS = 22$.

Score X	Deviation $X - \mu$	Squared Deviation $(X - \mu)^2$	
1	-1	1	$\Sigma X = 8$
0	-2	4	$\mu = 2$
6	$+4$	16	
1	-1	$\dfrac{1}{22}$	$\Sigma(X - \mu)^2 = 22$

∎

Notice that the value of SS is always greater than or equal to zero because it is based on squared deviation scores. If you obtain an SS value of less than zero, you have made a mistake.

Although the definitional formula is the most direct method for computing SS, it can be awkward to use. In particular, when the mean is not a whole number, the deviations all contain decimals or fractions, and the calculations become difficult. In addition, calculations with decimal values introduce the opportunity for rounding error, which can make the result less accurate. For these reasons, an alternative formula has been developed for computing SS. The alternative, known as the computational formula, performs calculations with the scores (not the deviations) and therefore minimizes the complications of decimals and fractions.

$$\text{Computational formula: } SS = \Sigma X^2 - \frac{(\Sigma X)^2}{N} \tag{4.5}$$

The first part of this formula directs you to square each score and then add the squared values, ΣX^2. In the second part of the formula, you find the sum of the scores, ΣX, then square this total and divide the result by N. Finally, subtract the second part from the first. The use of this formula is shown in Example 4.5 with the same scores that we used to demonstrate the definitional formula.

EXAMPLE 4.5

The computational formula is used to calculate SS for the same set of $N = 4$ scores we used in Example 4.4. Note that the formula requires the calculation of two sums: first, compute ΣX, and then square each score and compute ΣX^2. These calculations are shown in the following table. The two sums are used in the formula to compute SS.

X	X^2	
1	1	$SS = \Sigma X^2 - \dfrac{(\Sigma X)^2}{N}$
0	0	$= 38 - \dfrac{(8)^2}{4}$
6	36	$= 38 - \dfrac{64}{4}$
1	1	$= 38 - 16$
$\Sigma X = 8$	$\Sigma X^2 = 38$	$= 22$

∎

Remember, it is impossible to get an SS value of less than zero unless a mistake is made. By definition, sum of squares is based on the squared deviations.

Note that the two formulas produce exactly the same value for SS. Although the formulas look different, they are in fact equivalent. The definitional formula provides the most direct representation of the concept of SS; however, this formula can be awkward to use, especially if the mean includes a fraction or decimal value. If you have a small group of scores and the mean is a whole number, then the definitional formula is fine; otherwise the computational formula is usually easier to use.

In the same way that sum of squares, or *SS*, is used to refer to the sum of squared deviations, the term *mean square*, or *MS*, is often used to refer to variance, which is the mean squared deviation.

■ Final Formulas and Notation

With the definition and calculation of *SS* behind you, the equations for variance and standard deviation become relatively simple. Remember that variance is defined as the mean squared deviation. The mean is the sum of the squared deviations divided by *N*, so the equation for the *population variance* is

$$\text{variance} = \frac{SS}{N}$$

Standard deviation is the square root of variance, so the equation for the *population standard deviation* is

$$\text{standard deviation} = \sqrt{\frac{SS}{N}}$$

There is one final bit of notation before we work completely through an example computing *SS*, variance, and standard deviation. Like the mean (μ), variance and standard deviation are parameters of a population and are identified by Greek letters. To identify the standard deviation, we use the Greek letter sigma (the Greek letter *s*, standing for standard deviation). The capital letter sigma (Σ) has been used already, so we now use the lowercase sigma, σ, as the symbol for the population standard deviation. To emphasize the relationship between standard deviation and variance, we use σ^2 as the symbol for population variance (standard deviation is the square root of the variance). Thus,

$$\text{population standard deviation} = \sigma = \sqrt{\sigma^2} = \sqrt{\frac{SS}{N}} \tag{4.6}$$

$$\text{population variance} = \sigma^2 = \frac{SS}{N} \tag{4.7}$$

Population variance is represented by the symbol $\boldsymbol{\sigma^2}$ and equals the mean squared distance from the mean. Population variance is obtained by dividing the sum of squares (*SS*) by *N*.

Population standard deviation is represented by the symbol $\boldsymbol{\sigma}$ and equals the square root of the population variance.

Earlier, in Examples 4.3 and 4.4, we computed the sum of squared deviations for a simple population of *N* = 4 scores (1, 0, 6, 1) and obtained *SS* = 22. For this population, the variance is

$$\sigma^2 = \frac{SS}{N} = \frac{22}{4} = 5.50$$

and the standard deviation is

$$\sigma = \sqrt{5.50} = 2.35$$

LEARNING CHECK   **LO7**   **1.** What is *SS*, the sum of the squared deviations, for the following population of *N* = 5 scores? Scores: 1, 9, 0, 2, 3

a. 10

b. 41

    c. 50

    d. 95

LO8 **2.** What is the standard deviation for the following population of scores? Scores: 1, 3, 9, 3

    a. 36

    b. 9

    c. 6

    d. 3

LO8 **3.** A population of $N = 8$ scores has a standard deviation of $\sigma = 3$. What is the value of SS, the sum of the squared deviations, for this population?

    a. 72

    b. 24

    c. $8\sqrt{3}$

    d. $\frac{9}{8} = 1.125$

ANSWERS   **1. c**   **2. d**   **3. a**

4-4   Measuring Variance and Standard Deviation for a Sample

LEARNING OBJECTIVES

  9. Explain why it is necessary to make a correction to the formulas for variance and standard deviation when computing these statistics for a sample.

  10. Calculate SS, the sum of the squared deviations, for a sample using either the definitional or the computational formula and describe the circumstances in which each formula is appropriate.

  11. Calculate the variance and the standard deviation for a sample.

■ The Problem with Sample Variability

The goal of inferential statistics is to use the limited information from samples to draw general conclusions about populations. The basic assumption of this process is that samples should be representative of the populations from which they come. This assumption poses a special problem for variability because samples consistently tend to be less variable than their populations. The mathematical explanation for this fact is beyond the scope of this book, but a simple demonstration of this general tendency is shown in Figure 4.6. Notice that a few extreme scores in the population tend to make the population variability relatively large. However, these extreme values are unlikely to be obtained when you are selecting a sample, which means that the sample variability is relatively small. The fact that a sample tends to be less variable than its population means that sample variability gives a *biased* estimate of population variability. This bias is in the direction of underestimating the population value rather than being right on the mark. (The concept of a biased statistic is discussed in more detail in Section 4-5.)

    Fortunately, the bias in sample variability is consistent and predictable, which means it can be corrected. For example, if the speedometer in your car consistently shows speeds that are 5 mph slower than you are actually driving, it does not mean that the speedometer is useless. It simply means that you must make an adjustment to the speedometer reading

A sample statistic is said to be *biased* if, on average, it consistently overestimates or underestimates the corresponding population parameter.

FIGURE 4.6

The population of adult heights forms a normal distribution. If you select a sample from this population, you are most likely to obtain individuals who are near average in height. As a result, the variability for the scores in the sample is smaller than the variability for the scores in the population.

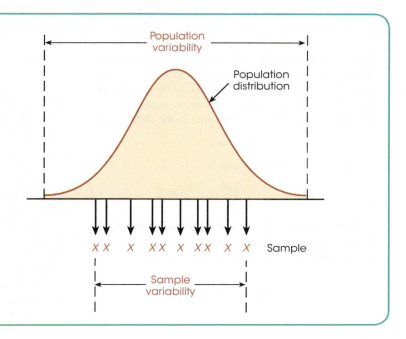

to get an accurate speed. In the same way, we will make an adjustment in the calculation of sample variance. The purpose of the adjustment is to make the resulting value for sample variance an accurate and unbiased representative of the population variance.

■ Formulas for Sample Variance and Standard Deviation

The calculations of variance and standard deviation for a sample follow the same steps that were used to find population variance and standard deviation. First, calculate the sum of squared deviations (*SS*). Second, calculate the variance. Third, find the square root of the variance, which is the standard deviation.

Except for minor changes in notation, calculating the sum of squared deviations, *SS*, is exactly the same for a sample as it is for a population. The changes in notation involve using *M* for the sample mean instead of μ, and using *n* (instead of *N*) for the number of scores. Thus, the definitional formula for *SS* for a sample is

$$\text{Definitional formula: } SS = \Sigma(X - M)^2 \tag{4.8}$$

Note that the sample formula has exactly the same structure as the population formula (Equation 4.4 on page 121) and instructs you to find the sum of the squared deviations using the following three steps:

1. Find the deviation from the mean for each score: deviation = $X - M$
2. Square each deviation: squared deviation = $(X - M)^2$
3. Add the squared deviations: $SS = \Sigma(X - M)^2$

The value of *SS* also can be obtained using a computational formula. Except for one minor difference in notation (using *n* in place of *N*), the computational formula for *SS* is the same for a sample as it was for a population (see Equation 4.5, page 122). Using sample notation, this formula is:

$$\text{Computational formula: } SS = \Sigma X^2 - \frac{(\Sigma X)^2}{n} \tag{4.9}$$

BOX 4.1 A Note about the Computational Formula for *SS*

When comparing the definitional and computational formulas for *SS*, you might wonder where the *N* came from in the computational formula. In the definitional formula, you first compute deviation scores by subtracting the mean from every score in the distribution,

$$(X - \mu)$$

The formula for the mean is

$$\frac{\Sigma X}{N}$$

If we substitute the formula for the mean into the definitional formula, we obtain an expression that looks similar to the computational formula:

$$\Sigma\left(X - \frac{\Sigma X}{N}\right)^2$$

There is an *N* in the definitional formula because it is based on deviations from the mean. Thus, in the definitional formula, $\Sigma(X - \mu)^2$, the *N* is "hidden" but it is there in the formula for μ. This also applies to *SS* for sample data. The *n* in the computational formula comes from the fact that the definitional formula uses the sample mean, $\frac{\Sigma X}{N}$, to compute deviations scores.

While it is possible to algebraically prove that the computational and definitional formulas are equivalent, it is beyond the scope of this book. What you should remember, however, is that the *N* and *n* are *always* used in the computational formulas for populations and samples, respectively. This is especially important for samples. *Never use* n − *1 in the computational formula for SS with samples.* The term *n* − 1 is *only* used for computing sample variance and standard deviation.

Again, calculating *SS* for a sample is exactly the same as for a population, except for minor changes in notation. Notice that *n*, the number scores in the sample, is used in place of *N* (see Box 4.1).

Formulas for sample variance, s^2, and standard deviation, *s*, divide *SS* by *n* − 1, unlike population formulas which divide by *N*. This is the adjustment that is necessary to correct for the bias in sample variability. The effect of the adjustment is to increase the value you will obtain. Dividing by a smaller number (*n* − 1 instead of *n*) produces a larger result and makes sample variance an accurate and unbiased estimator of population variance. The following example demonstrates the calculation of variance and standard deviation for a sample.

> **Remember, sample variability tends to underestimate population variability unless some correction is made.**

EXAMPLE 4.6   We have selected a sample of *n* = 8 scores from a population. The scores are 4, 6, 5, 11, 7, 9, 7, and 3. The frequency distribution histogram for this sample is shown in Figure 4.7.

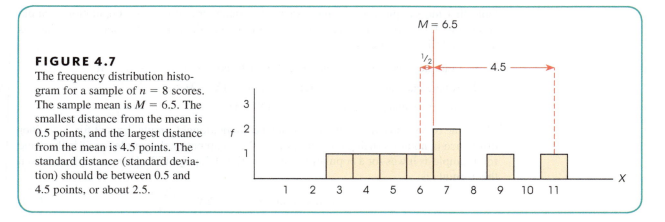

FIGURE 4.7
The frequency distribution histogram for a sample of *n* = 8 scores. The sample mean is *M* = 6.5. The smallest distance from the mean is 0.5 points, and the largest distance from the mean is 4.5 points. The standard distance (standard deviation) should be between 0.5 and 4.5 points, or about 2.5.

Before we begin any calculations, you should be able to look at the sample distribution and make a preliminary estimate of the outcome. Remember that standard deviation measures the standard distance from the mean. For this sample the mean is $M = \frac{52}{8} = 6.5$. The scores closest to the mean are $X = 6$ and $X = 7$, both of which are exactly 0.50 points away. The score farthest from the mean is $X = 11$, which is 4.50 points away. With the smallest distance from the mean equal to 0.50 and the largest distance equal to 4.50, we should obtain a standard distance (standard deviation) somewhere between 0.50 and 4.50, probably around 2.5.

We begin the calculations by finding the value of SS for this sample. Because the mean is not a whole number ($M = 6.5$), the computational formula is easier to use. The scores, and the squared scores, needed for this formula are shown in the following table.

Scores X	Squared Scores X^2
4	16
6	36
5	25
11	121
7	49
9	81
7	49
3	9
$\Sigma X = 52$	$\Sigma X^2 = 386$

A common error is to use $n - 1$ in the computational formula for SS when you have scores from a sample. Remember, the SS formula always uses n (or N). After you compute SS for a sample, you must correct for the sample bias by using $n - 1$ in the formulas for s^2 and s.

Using the two sums,

$$SS = \Sigma X^2 - \frac{(\Sigma X)^2}{n} = 386 - \frac{(52)^2}{8}$$
$$= 386 - 338$$
$$= 48$$

the sum of squared deviations for this sample is $SS = 48$. Continuing the calculations,

$$\text{sample variance} = s^2 = \frac{SS}{n - 1} = \frac{48}{8 - 1} = 6.86$$

Finally, the standard deviation is

$$s = \sqrt{s^2} = \sqrt{6.86} = 2.62$$

Note that the value we obtained is in excellent agreement with our preliminary prediction (see Figure 4.7). ■

The following example is an opportunity for you to test your understanding by computing sample variance and standard deviation yourself.

EXAMPLE 4.7   For the following sample of $n = 5$ scores, compute the variance and standard deviation: 1, 5, 5, 1, and 8. You should obtain $s^2 = 9$ and $s = 3$. Good luck. ■

Remember that the formulas for sample variance and standard deviation were constructed so that the sample variability provides a good estimate of population variability. For this reason, the *sample variance* is often called *estimated population variance*, and the *sample standard deviation* is called *estimated population standard deviation*. When you have only a sample to work with, the variance and standard deviation for the sample provide the best possible estimates of the population variability.

■ Sample Variability and Degrees of Freedom

Why, in particular, is $n - 1$ used for sample variance to make it an unbiased estimate of population variance? Why not some other value? Although the concept of a deviation score and the calculation of SS are almost exactly the same for samples and populations, the minor differences in notation are really very important. Specifically, with a population, you find the deviation for each score by measuring its distance from the population mean. With a sample, on the other hand, the value of μ is unknown and you must measure distances from the sample mean. Because the value of the sample mean varies from one sample to another, you must first compute the sample mean before you can begin to compute deviations. However, calculating the value of M places a restriction on the variability of the scores in the sample. This restriction is demonstrated in the following example.

EXAMPLE 4.8   Suppose we select a sample of $n = 3$ scores and compute a mean of $M = 5$. The first two scores in the sample have no restrictions; they are independent of each other and they can have any values. For this demonstration, we will assume that we obtained $X = 2$ for the first score and $X = 9$ for the second. At this point, however, the third score in the sample is restricted.

X	A sample of $n = 3$ scores with a mean of $M = 5$.
2	
9	
—	← What is the third score?

For this example, the third score must be $X = 4$. The reason that the third score is restricted to $X = 4$ is that the entire sample of $n = 3$ scores has a mean of $M = 5$. For 3 scores to have a mean of 5, the scores must have a total of $\Sigma X = 15$. Because the first two scores add up to 11 $(9 + 2)$, the third score must be $X = 4$.

Similarly, you can look at the restriction imposed by the sample means by considering deviations from the mean. As previously noted, you must first compute the sample mean before you can begin to compute deviations. The mean places a restriction on $n - 1$ deviations. For these data of $n = 3$ scores, $M = 5$ and we assumed the first two scores are $X = 2$ and $X = 9$. Their deviation scores are as follows:

X	$X - M$
2	−3
9	+4
—	← What is the third deviation?

Because the sum of the deviations always equals zero (see Example 4.1, page 117), the third deviation must be −1. Note that one point below the mean is a score of $X = 4$. ■

In Example 4.8, the first two out of three scores were free to have any values, but the final score was dependent on the values chosen for the first two. In general, with a sample of n scores, the first $n - 1$ scores are free to vary, but the final score is restricted. As a result, the sample is said to have $n - 1$ *degrees of freedom.*

> For a sample of n scores, the **degrees of freedom**, or *df*, for the sample variance are defined as $df = n - 1$. The degrees of freedom determine the number of scores in the sample that are independent and free to vary.

The $n - 1$ degrees of freedom for a sample is the same $n - 1$ that is used in the formulas for sample variance and standard deviation. Remember that variance is defined as the mean squared deviation. To calculate sample variance (mean squared deviation), we find the sum of the squared deviations (*SS*) and divide by the number of scores that are free to vary. This number is $n - 1 = df$. Thus, the formula for sample variance is

$$s^2 = \frac{\text{sum of squared deviations}}{\text{number of scores free to vary}} = \frac{SS}{df} = \frac{SS}{n-1} \qquad (4.10)$$

The formula for sample standard deviation is

$$s = \sqrt{s^2} = \sqrt{\frac{SS}{df}} = \sqrt{\frac{SS}{n-1}} \qquad (4.11)$$

Later in this book, we use the concept of degrees of freedom in other situations. For now, remember that knowing the sample mean places a restriction on sample variability. Only $n - 1$ of the scores are free to vary; $df = n - 1$. See Box 4.2 for an analogy.

BOX 4.2 Degrees of Freedom, Cafeteria-Style

The cafeteria is an unlikely place to start a discussion of statistics, yet we can make a food-based analogy for degrees of freedom and talk about food for a bit. This analogy, while not truly what we mean by degrees of freedom in statistics, gives you some idea of the general notion that sample variability is restricted.

There is an enormous population of desserts that might be prepared and served—seemingly countless recipes for pies, cakes, pastries, puddings, ice cream flavors, cookies, and so on. On a given day, the cafeteria will offer a choice of desserts. These choices are just a sample from the large population of different desserts that could be offered. On this particular afternoon, the cafeteria has five desserts (Figure 4.8). To keep this simple, imagine that the cafeteria has only one of each dessert and only five customers. These people work their way down the cafeteria line, piling food on their trays, when they arrive at the display of desserts. We will observe their responses to

the desserts, which may vary from person to person. The number of observations (the selection of a dessert) are free to vary. Consider the following scenario:

- The first person in line gets to select from five desserts and chooses the apple pie.

- The second person in lines gets to select from four remaining desserts and chooses the chocolate cake.

- The third person gets to select from the three remaining desserts and chooses the sundae.

- The fourth person has a choice between the two remaining desserts and takes the fruit.

Each of these observations is free to vary—that is, until we get to the last person, who must settle for the stale cookie. Thus, just as $n - 1$ scores are free to vary for a sample, $n - 1$ dessert choices are free to vary in the cafeteria.

FIGURE 4.8
Of the $n = 5$ deserts, $n - 1$ selections are free to vary.

LEARNING CHECK   **LO9** **1.** Which of the following explains why it is necessary to make a correction to the formula for sample variance?

     a. If sample variance is computed by dividing by n instead of $n - 1$, the resulting values will tend to underestimate the population variance.

     b. If sample variance is computed by dividing by n instead of $n - 1$, the resulting values will tend to overestimate the population variance.

     c. If sample variance is computed by dividing by $n - 1$ instead of n, the resulting values will tend to underestimate the population variance.

     d. If sample variance is computed by dividing by $n - 1$ instead of n, the resulting values will tend to underestimate the population variance.

LO10 **2.** Under what circumstances is the computational formula preferred over the definitional formula when computing SS, the sum of the squared deviations, for a sample?

     a. When the sample mean is a whole number.

     b. When the sample mean is not a whole number.

     c. When the sample variance is a whole number.

     d. When the sample variance is not a whole number.

LO11 **3.** What is the variance for the following sample of $n = 5$ scores? Scores: 2, 0, 8, 2, 3

     a. $\frac{81}{4} = 20.25$

     b. $\frac{36}{4} = 9$

     c. $\frac{36}{5} = 7.2$

     d. $\sqrt{9} = 3$

ANSWERS   **1. a   2. b   3. b**

4-5   Sample Variance as an Unbiased Statistic

LEARNING OBJECTIVES

12. Define biased and unbiased statistics.

13. Explain why the sample mean and the sample variance (dividing by $n - 1$) are unbiased statistics.

■ Biased and Unbiased Statistics

Earlier we noted that sample variability tends to underestimate the variability in the corresponding population; as a result, it is a *biased statistic*. To correct for this problem we adjusted the formula for sample variance by dividing by $n - 1$ instead of dividing by n. The result of the adjustment is that sample variance provides a much more accurate representation of the population variance. Specifically, dividing by $n - 1$ produces a sample variance that provides an *unbiased* estimate of the corresponding population variance. This does not mean that each individual sample variance will be exactly equal to its population variance. In fact, some sample variances will overestimate the population value and some will underestimate it. However, the average of all the sample variances will produce an accurate estimate of the population variance. This is the idea behind the concept of an *unbiased statistic*.

A sample statistic is **unbiased** if the average value of the statistic is equal to the population parameter. (The average value of the statistic is obtained from all the possible samples for a specific sample size, n.)

A sample statistic is **biased** if the average value of the statistic either underestimates or overestimates the corresponding population parameter.

The following example demonstrates the concept of biased and unbiased statistics.

EXAMPLE 4.9

We have structured this example assuming "sampling with replacement." In this type of random sampling, scores are replaced after every selection is made. Thus, the same score can be selected more than once in a sample. You will learn about sampling with replacement in Chapter 6.

We begin with a population that consists of exactly $N = 3$ scores: 0, 3, 9. With a few calculations you should be able to verify that this population has a mean of $\mu = 4$ and a variance of $\sigma^2 = 14$.

Next, we select samples of $n = 2$ scores from this population. In fact, we obtain every single possible sample with $n = 2$. The complete set of samples is listed in Table 4.2. Notice that the samples are listed systematically to ensure that every possible sample is included. We begin by listing all the samples that have $X = 0$ as the first score, then all the samples with $X = 3$ as the first score, and so on. Notice that the table shows a total of 9 samples.

Finally, we have computed the mean and the variance for each sample. Note that the sample variance has been computed two different ways. First, we make no correction for bias and compute each sample variance as the average of the squared deviations by simply dividing SS by n. Second, we compute the correct sample variances for which SS is divided by $n - 1$ to produce an unbiased measure of variance. You should verify our calculations by computing one or two of the values for yourself. The complete set of sample means and sample variances is presented in Table 4.2. ∎

First, consider the column of biased sample variances, which were calculated dividing by n. These 9 sample variances add up to a total of 63, which produces an average value of $\frac{63}{9} = 7$. The original population variance, however, is $\sigma^2 = 14$. Note that the average of the sample variances is *not* equal to the population variance. If the sample variance is computed by dividing by n, the resulting values will not produce an accurate estimate of

TABLE 4.2   The set of all the possible random samples for $n = 2$ is selected from the population described in Example 4.9. The mean is computed for each sample, and the variance is computed two different ways: (1) dividing SS by n, which is incorrect and produces a biased statistic; and (2) dividing SS by $n - 1$, which is correct and produces an unbiased statistic.

			Sample Statistics		
Sample	First Score	Second Score	Mean M	Biased Variance (Using n)	Unbiased Variance (Using $n - 1$)
1	0	0	0.00	0.00	0.00
2	0	3	1.50	2.25	4.50
3	0	9	4.50	20.25	40.50
4	3	0	1.50	2.25	4.50
5	3	3	3.00	0.00	0.00
6	3	9	6.00	9.00	18.00
7	9	0	4.50	20.25	40.50
8	9	3	6.00	9.00	18.00
9	9	9	9.00	0.00	0.00
		Totals	36.00	63.00	126.00

the population variance. On average, these sample variances underestimate the population variance and, therefore, are biased statistics.

Next, consider the column of sample variances that are computed using $n - 1$. Although the population has a variance of $\sigma^2 = 14$, you should notice that none of the samples has a variance exactly equal to 14. However, if you consider the complete set of sample variances, you will find that the 9 values add up to a total of 126, which produces an average value of $\frac{126}{9} = 14.00$. Thus, the average of the sample variances is exactly equal to the original population variance. On average, the sample variance (computed using $n - 1$) produces an accurate, unbiased estimate of the population variance.

Finally, direct your attention to the column of sample means. For this example, the original population has a mean of $\mu = 4$. Although none of the samples has a mean exactly equal to 4, if you consider the complete set of sample means, you will find that the 9 sample means add up to a total of 36, so the average of the sample means is $\frac{36}{9} = 4$. Note that the average of the sample means is exactly equal to the population mean. Again, this is what is meant by the concept of an unbiased statistic. On average, the sample values provide an accurate representation of the population. In this example, the average of the 9 sample means is exactly equal to the population mean.

In summary, both the sample mean and the sample variance (using $n - 1$) are examples of unbiased statistics. This fact makes the sample mean and sample variance extremely valuable for use as inferential statistics. Although no individual sample is likely to have a mean and variance exactly equal to the population values, both the sample mean and the sample variance, on average, do provide accurate estimates of the corresponding population values.

LEARNING CHECK   **LO12**   **1.** A researcher takes a sample from a population and computes a statistic for the sample. Which of the following statements is correct?

    a. If the sample statistic overestimates the corresponding population parameter, then the statistic is biased.

    b. If the sample statistic underestimates the corresponding population parameter, then the statistic is biased.

    c. If the sample statistic is equal to the corresponding population parameter, then the statistic is unbiased.

    d. None of the above.

LO12   **2.** A researcher takes all of the possible samples of $n = 4$ from a population. Next, the researcher computes a statistic for each sample and calculates the average of all the statistics. Which of the following statements is the most accurate?

    a. If the average statistic overestimates the corresponding population parameter, then the statistic is biased.

    b. If the average statistic underestimates the corresponding population parameter, then the statistic is biased.

    c. If the average statistic is equal to the corresponding population parameter, then the statistic is unbiased.

    d. All of the above.

LO13   **3.** All the possible samples of $n = 3$ scores are selected from a population with $\mu = 30$ and $\sigma = 5$, and the mean is computed for each of the samples. If the average is calculated for all of the sample means, what value will be obtained?

a. 30

b. Greater than 30

c. Less than 30

d. Near 30 but not exactly equal to 30

ANSWERS   **1. d   2. d   3. a**

4-6 | More about Variance and Standard Deviation

LEARNING OBJECTIVES

14. Describe how the mean and standard deviation are represented in a frequency distribution graph of a population or sample distribution.

15. Describe the effect on the mean and standard deviation and calculate the outcome for each of the following: adding or subtracting a constant from each score, and multiplying or dividing each score by a constant.

16. Describe how the mean and standard deviation are reported in research journals.

17. Determine the general appearance of a distribution based on the values for the mean and standard deviation.

18. Explain how patterns in sample data are affected by sample variance.

■ Presenting the Mean and Standard Deviation in a Frequency Distribution Graph

In frequency distribution graphs, we identify the position of the mean by drawing a vertical line and labeling it with μ or M. Because the standard deviation measures distance from the mean, it is represented by a horizontal line or an arrow drawn from the mean outward for a distance equal to the standard deviation and labeled with σ or an s. Figure 4.9(a) shows an example of a population distribution with a mean of $\mu = 80$ and a standard deviation of $\sigma = 8$, and Figure 4.9(b) shows the frequency distribution for a sample with a mean of $M = 16$ and a standard deviation of $s = 2$. For rough sketches, you can identify the mean with a vertical line in the middle of the distribution. The standard deviation line should extend approximately halfway from the mean to the most extreme score. [*Note:* In Figure 4.9(a), we show the standard deviation as a line to the right of the mean. You should realize that we could have drawn the line pointing to the left, or we could have drawn two lines (or arrows), with one pointing to the right and one pointing to the left, as in Figure 4.9(b). In each case, the goal is to show the standard distance from the mean.]

■ Transformations of Scale

Occasionally a set of scores is transformed by adding a constant to each score or by multiplying each score by a constant value. This happens, for example, when exposure to a treatment adds a fixed amount to each participant's score or when you want to change the unit of measurement (for example, to convert from minutes to seconds, multiply each

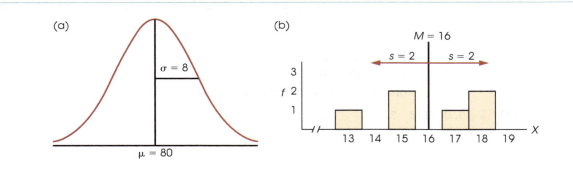

FIGURE 4.9

Showing means and standard deviations in frequency distribution graphs. (a) A population distribution with a mean of $\mu = 80$ and a standard deviation of $\sigma = 8$. (b) A sample with a mean of $M = 16$ and a standard deviation of $s = 2$.

score by 60). What happens to the standard deviation when the scores are transformed in this manner?

The easiest way to determine the effect of a transformation is to remember that the standard deviation is a measure of distance. If you select any two scores and see what happens to the distance between them, you also will find out what happens to the standard deviation.

1. **Adding a constant to each score does not change the standard deviation.** If you begin with a distribution that has $\mu = 40$ and $\sigma = 10$, what happens to the standard deviation if you add 5 points to every score? Consider any two scores in this distribution: Suppose, for example, that these are exam scores and that you had a score of $X = 41$ and your friend had $X = 43$. The distance between these two scores is $43 - 41 = 2$ points. After adding the constant, 5 points, to each score, your score would be $X = 46$, and your friend would have $X = 48$. The distance between scores is still 2 points. Adding a constant to every score does not affect any of the distances and, therefore, does not change the standard deviation. This fact can be seen clearly if you imagine a frequency distribution graph. If, for example, you add 10 points to each score, then every score in the graph is moved 10 points to the right. The result is that the entire distribution is shifted to a new position 10 points up the scale. Note that the mean moves along with the scores and is increased by 10 points. However, the variability does not change because each of the deviation scores $(X - \mu)$ does not change.

2. **Multiplying each score by a constant causes the standard deviation to be multiplied by the same constant.** Consider the same distribution of exam scores we looked at earlier. If $\mu = 40$ and $\sigma = 10$, what would happen to the standard deviation if each score were multiplied by 2? Again, we will look at two scores, $X = 41$ and $X = 43$, with a distance between them equal to 2 points. After all the scores have been multiplied by 2, these scores become $X = 82$ and $X = 86$. Now the distance between scores is 4 points, twice the original distance. Multiplying each score causes each distance to be multiplied, so the standard deviation also is multiplied by the same amount.

IN THE LITERATURE

The dependent variables in psychology research are often numerical values obtained from measurements on interval or ratio scales. Thus, when reporting the results of a study, the researcher will provide descriptive information for both central tendency and variability. When the mean is reported in a study as a descriptive statistic of central tendency, the standard deviation is reported with it to describe the amount of variability. Similarly, when the median is reported in the study, the interquartile range accompanies it.

Reporting the Standard Deviation

In many journals, especially those following APA style, the symbol *SD* is used for the sample standard deviation. For example, the results might state:

> Children who viewed the violent cartoon displayed more aggressive responses ($M = 12.45$, $SD = 3.70$) than those who viewed the control cartoon ($M = 4.22$, $SD = 1.04$).

When reporting the descriptive measures for several groups, the findings may be summarized in a table. Table 4.3 illustrates the results of hypothetical data.

Sometimes the table also indicates the sample size, *n,* for each group. You should remember that the purpose of the table is to present the data in an organized, concise, and accurate manner. The mean also may be presented with a measure of variability in a graph. This will be demonstrated in Chapter 7 with standard error, another measure of variability.

Reporting the Interquartile Range

Just as the mean and standard deviation are reported together as descriptive measures of central tendency and variability, so too are the median and interquartile range. This makes sense because both measures are related to percentile ranks. Remember, the median is the 50th percentile (Q2) and the interquartile range is based on the range of scores between the 25th percentile (Q1) and 75th percentile (Q3). These measures can be presented in a table or in a graph called a Box Plot.

Nitzschner, Melis, Kaminski, and Tomasello (2012) examined whether dogs could evaluate people by watching other dogs interact with them. A test dog would be in a separate enclosure while watching a demonstrator dog in a room with two people—a "nice" person who gave the demonstrator dog attention and an "ignoring" person who did not interact with the dog. Later, the test dogs were placed in the room with the two people. The "nice" person and "ignoring" person were in opposite corners, and how

TABLE 4.3

The number of aggressive behaviors for male and female adolescents after playing a violent or nonviolent video game.

	Type of Video Game	
	Violent	Nonviolent
Males	$M = 7.72$	$M = 4.34$
	$SD = 2.43$	$SD = 2.16$
Females	$M = 2.47$	$M = 1.61$
	$SD = 0.92$	$SD = 0.68$

FIGURE 4.10
The box plot shows the interquartile range (height of the box), median (horizontal bar within the box), and the range (vertical whiskers).

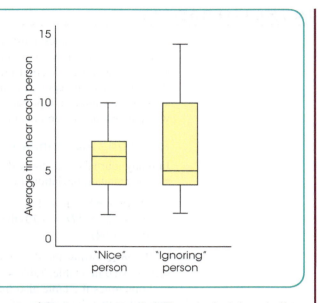

much time the test dogs spent near each person was recorded. Hypothetical data similar to those observed in this study may be reported as follows:

	Response of Observing Dogs	
	"Nice" Person	"Ignoring" Person
Median time (seconds)	6	5
Interquartile range	3	6

The median and interquartile range are often presented in a graph called a *box plot*. A basic box plot also includes the range of scores from the minimum *X* to maximum *X* values. Figure 4.10 shows a box plot for the data presented in the table. The plot has three basic features: a box, a horizontal line through the box, and vertical lines that are often called "whiskers." The height of the box reflects the interquartile range, so that the bottom of the box corresponds to the *X* value for Q1 and the top of the box corresponds to the *X* value for Q3. The horizontal bar through the box is placed at the value for the median. The vertical whiskers signify the range of scores, so that the bottom and top show the lowest and highest scores, respectively, in the distribution.

The advantage of a box plot is that a quick glance provides a summary of the data. For this study, dogs did not show a preference for people simply by observing demonstrator dogs. Dogs that do have direct experience, however, exhibit strong preference for the "nice" person, as you might guess.

■ Standard Deviation and Descriptive Statistics

Because standard deviation requires extensive calculations, there is a tendency to get lost in the arithmetic and forget what standard deviation is and why it is important. Standard deviation is primarily a descriptive measure; it describes how variable, or how spread out, the scores are in a distribution. Behavioral scientists must deal with the variability that comes from studying people and animals. People are not all the same; they have different attitudes, opinions, talents, IQs, and personalities. Although we can calculate the average value for any of these variables, it is equally important to describe the variability. Standard deviation describes variability by measuring *distance from the mean*. In any distribution, some individuals will be close to the mean, and others will be relatively far from the mean. Standard deviation provides a measure of the typical, or standard, distance from the mean.

Describing an Entire Distribution   Rather than listing all the individual scores in a distribution, research reports typically summarize the data by reporting only the mean and the standard deviation. When you are given these two descriptive statistics, however, you should be able to visualize the entire set of data. For example, consider a sample with a mean of $M = 36$ and a standard deviation of $s = 4$. Although there are several different ways to picture the data, one simple technique is to imagine (or sketch) a histogram in which each score is represented by a box in the graph. For this sample, the data can be pictured as a pile of boxes (scores) with the center of the pile located at a value of $M = 36$. The individual scores or boxes are scattered on both sides of the mean with some of the boxes relatively close to the mean and some farther away. As a rule of thumb, roughly 70% of the scores in a distribution are located within a distance of one standard deviation from the mean, and almost all of the scores (roughly 95%) are within two standard deviations of the mean. In this example, the standard distance from the mean is $s = 4$ points, so your image should have most of the boxes within 4 points of the mean, and nearly all the boxes within 8 points. One possibility for the resulting image is shown in Figure 4.11.

Describing the Location of Individual Scores   Notice that Figure 4.11 not only shows the mean and the standard deviation, but also uses these two values to reconstruct the underlying scale of measurement (the X values along the horizontal line). The scale of measurement helps complete the picture of the entire distribution and helps to relate each individual score to the rest of the group. In this example, you should realize that a score of $X = 34$ is located near the center of the distribution, only slightly below the mean. On the other hand, a score of $X = 45$ is an extremely high score, located far out in the right-hand tail of the distribution.

Notice that the relative position of a score depends in part on the size of the standard deviation. For example, in Figure 4.9 (page 134), we show a population distribution with a mean of $\mu = 80$ and a standard deviation of $\sigma = 8$, and a sample distribution with a mean of $M = 16$ and a standard deviation of $s = 2$. In the population distribution, a score that is 4 points above the mean is slightly above average but is certainly not an extreme value. In the sample distribution, however, a score that is 4 points above the mean is an extremely high score. In each case, the relative position of the score depends on the size of the standard deviation. For the population, a deviation of 4 points from the mean is relatively small, corresponding to only one-half of the standard deviation. For the sample, on the other hand, a 4-point deviation is very large, equaling twice the size of the standard deviation.

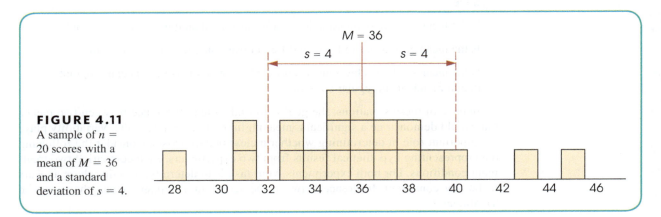

FIGURE 4.11
A sample of $n = 20$ scores with a mean of $M = 36$ and a standard deviation of $s = 4$.

> ### BOX 4.3 An Analogy for the Mean and the Standard Deviation
>
> Although the basic concepts of the mean and the standard deviation are not overly complex, the following analogy often helps students gain a more complete understanding of these two statistical measures.
>
> In our local community, the site for a new high school was selected because it provides a central location. An alternative site on the western edge of the community was considered, but this site was rejected because it would require extensive busing for students living on the east side. In this example, the location of the high school is analogous to the concept of the mean: just as the high school is located in the center of the community, the mean is located in the center of the distribution of scores.
>
> For each student in the community, it is possible to measure the distance between home and the new high school. Some students live only a few blocks from the new school and others live as much as three miles away. The average distance that a student must travel to the school was calculated to be 0.80 miles. The average distance from the school is analogous to the concept of the standard deviation; that is, the standard deviation measures the standard distance from an individual score to the mean.

The general point of this discussion is that the mean and standard deviation are not simply abstract concepts or mathematical equations. Instead, these two values should be concrete and meaningful, especially in the context of a set of scores. The mean and standard deviation are central concepts for most of the statistics that are presented in the following chapters. A good understanding of these two statistics will help you with the more complex procedures that follow (see Box 4.3).

■ Variance and Inferential Statistics

In very general terms, the goal of inferential statistics is to detect meaningful and significant patterns in research results. The basic question is whether the patterns observed in the sample data reflect corresponding patterns that exist in the population, or if they are simply random fluctuations that occur by chance. Variability plays an important role in the inferential process because the variability in the data influences how easy it is to see patterns. In general, low variability means that existing patterns can be seen clearly, whereas high variability tends to obscure any patterns that might exist. The following example provides a simple demonstration of how variance can influence the perception of patterns.

EXAMPLE 4.10 In many research studies the goal is to compare means for two (or more) sets of data. For example:

Is the mean level of depression lower after therapy than it was before therapy?

Is the mean attitude score for men different from the mean score for women?

Is the mean reading achievement score higher for students in a special program than for students in regular classrooms?

In each of these situations, the goal is to find a clear difference between two means that would demonstrate a significant, meaningful pattern in the results. Variability plays an important role in determining whether a clear pattern exists. Consider the following data representing hypothetical results from two experiments, each comparing two treatment conditions. For both experiments, your task is to determine whether there appears to be any consistent difference between the scores in Treatment 1 and the scores in Treatment 2.

Experiment A		Experiment B	
Treatment 1	Treatment 2	Treatment 1	Treatment 2
10	14	8	17
9	16	15	20
11	15	12	13
10	15	5	10

For each experiment, the data have been constructed so that there is a 5-point mean difference between the two treatments—on average, the scores in Treatment 2 are 5 points higher than the scores in Treatment 1. The 5-point difference is relatively easy to see in Experiment A, where the variability is low, but the same 5-point difference is difficult to see in Experiment B, where the variability is large. Again, high variability tends to obscure any patterns in the data. This general fact is perhaps even more convincing when the data are presented in a graph. Figure 4.12 shows the two sets of data from Experiments A and B. Notice that the results from Experiment A clearly show the 5-point difference between treatments. One group of scores piles up around 10 and the second group piles up around 15. On the other hand, the scores from Experiment B seem to be mixed together randomly with no clear difference between the two treatments.   ■

In the context of inferential statistics, the variance that exists in a set of sample data is often classified as *error variance*. This term is used to indicate that the sample variance represents unexplained and uncontrolled differences between scores. As the error variance increases, it becomes more difficult to see any systematic differences or patterns that might exist in the data. An analogy is to think of variance as the static that occurs on a radio station or a smartphone when you enter an area of poor reception. In general, variance makes it difficult to get a clear signal from the data. High variance can make it difficult or impossible to see a mean difference between two sets of scores, or to see any other meaningful patterns in the results from a research study.

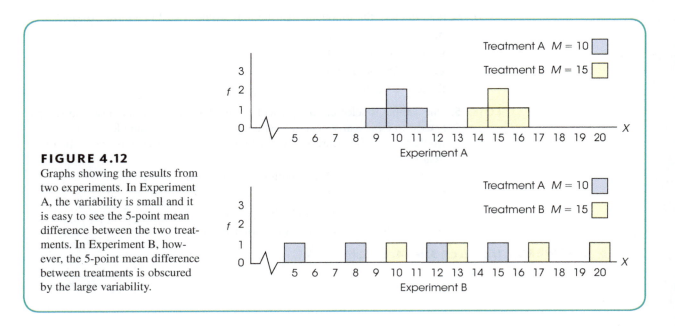

FIGURE 4.12
Graphs showing the results from two experiments. In Experiment A, the variability is small and it is easy to see the 5-point mean difference between the two treatments. In Experiment B, however, the 5-point mean difference between treatments is obscured by the large variability.

LEARNING CHECK   **LO14**   **1.** If a normal-shaped population with $\mu = 40$ and $\sigma = 5$ is shown in a frequency distribution graph, how would the mean and standard deviation be represented?

    a. The mean is represented by a vertical line drawn at $X = 40$ and the standard deviation is represented by a vertical line drawn at $X = 45$.

    b. The mean is represented by an arrow under the graph pointing up to $X = 40$ and the standard deviation is represented by a vertical line drawn at $X = 45$.

    c. The mean is represented by a vertical line drawn at $X = 40$ and the standard deviation is represented by a horizontal line drawn from $X = 40$ to $X = 45$.

    d. The mean is represented by an arrow under the graph pointing up to $X = 40$ and the standard deviation is represented by a horizontal line drawn from $X = 40$ to $X = 45$.

LO15   **2.** If 5 points are added to every score in a population with a mean of $\mu = 45$ and a standard deviation of $\sigma = 6$, what are the new values for μ and σ?

    a. $\mu = 45$ and $\sigma = 6$

    b. $\mu = 45$ and $\sigma = 11$

    c. $\mu = 50$ and $\sigma = 6$

    d. $\mu = 50$ and $\sigma = 11$

LO16   **3.** A research study obtains a mean of 12.7 and a standard deviation of 2.3 for a sample of $n = 25$ participants. How would the sample mean and standard deviation be reported in a research journal report?

    a. $M = 12.7$ and $s = 2.3$

    b. $M = 12.7$ and $SD = 2.3$

    c. $Mn = 12.7$ and $s = 2.3$

    d. $Mn = 12.7$ and $SD = 2.3$

LO17   **4.** For which of the following distributions would $X = 35$ be most extreme?

    a. $\mu = 30$ and $\sigma = 5$

    b. $\mu = 30$ and $\sigma = 10$

    c. $\mu = 25$ and $\sigma = 5$

    d. $\mu = 25$ and $\sigma = 10$

LO18   **5.** One sample is selected to represent scores in treatment 1 and a second sample is used to represent scores in treatment 2. Which set of sample statistics would present the clearest picture of a real mean difference between the two treatments?

    a. $M_1 = 40$, $M_2 = 45$, and both variances $= 15$

    b. $M_1 = 40$, $M_2 = 45$, and both variances $= 3$

    c. $M_1 = 40$, $M_2 = 42$, and both variances $= 15$

    d. $M_1 = 40$, $M_2 = 42$, and both variances $= 3$

ANSWERS   **1. c   2. c   3. b   4. c   5. b**

SUMMARY

1. The purpose of variability is to measure and describe the degree to which the scores in a distribution are spread out or clustered together. There are four basic measures of variability: the range, interquartile range, the variance, and the standard deviation.

   The range is the distance covered by the set of scores, from the smallest score to the largest score. The range is completely determined by the two extreme scores and is considered to be a relatively crude measure of variability.

   The interquartile range is more descriptive than the basic range because it trims the extreme scores and provides a range that reflects the middle 50% of scores in the center of the distribution.

   Standard deviation and variance are the most commonly used measures of variability. Both of these measures are based on the idea that each score can be described in terms of its deviation or distance from the mean. The variance is the mean of the squared deviations. The standard deviation is the square root of the variance and provides a measure of the standard distance from the mean.

2. To calculate variance or standard deviation, you first need to find the sum of the squared deviations, SS. Except for minor changes in notation, the calculation of SS is identical for samples and populations. There are two methods for calculating SS:

   I. By definition, you can find SS using the following steps:
   a. Find the deviation $(X - \mu)$ for each score.
   b. Square each deviation.
   c. Add the squared deviations.
   This process can be summarized in a formula as follows:

   $$\text{Definitional formula: } SS = \Sigma(X - \mu)^2$$

   II. The sum of the squared deviations can also be found using a computational formula, which is especially useful when the mean is not a whole number:

   $$\text{Computational formula: } SS = \Sigma X^2 - \frac{(\Sigma X)^2}{N}$$

3. Variance is the mean squared deviation and is obtained by finding the sum of the squared deviations and then dividing by the number of scores. For a population, variance is

   $$\sigma^2 = \frac{SS}{N}$$

   For a sample, only $n - 1$ of the scores are free to vary (degrees of freedom or $df = n - 1$), so sample variance is

   $$s^2 = \frac{SS}{n - 1} = \frac{SS}{df}$$

   Using $n - 1$ in the sample formula makes the sample variance an accurate and unbiased estimate of the population variance.

4. Standard deviation is the square root of the variance. For a population, this is

   $$\sigma = \sqrt{\frac{SS}{N}}$$

   Sample standard deviation is

   $$s = \sqrt{\frac{SS}{n - 1}} = \sqrt{\frac{SS}{df}}$$

5. Adding a constant value to every score in a distribution does not change the standard deviation. Multiplying every score by a constant, however, causes the standard deviation to be multiplied by the same constant.

6. Because the mean identifies the center of a distribution and the standard deviation describes the average distance from the mean, these two values should allow you to create a reasonably accurate image of the entire distribution. Knowing the mean and standard deviation should also allow you to describe the relative location of any individual score within the distribution.

7. Large variance can obscure patterns in the data and, therefore, can create a problem for inferential statistics.

KEY TERMS

variability (111)

range (112)

quartile (114)

interquartile range (115)

deviation or deviation score (116)

variance or mean squared deviation (*MS*) (117)

standard deviation (118)

sum of squares or sum of squared deviations (*SS*) (121)

population variance (σ^2) (123)

population standard deviation
   (σ) (123)
sample variance or estimated
   population variance (s^2) (127)

sample standard deviation or
   estimated population standard
   deviation (s) (127)
degrees of freedom (df) (128)

biased statistic (131)
unbiased statistic (131)
box plot (136)

FOCUS ON PROBLEM SOLVING

1. The purpose of variability is to provide a measure of how spread out the scores are in a distribution. Usually this is described by the standard deviation. Because the calculations are relatively complicated, it is wise to make a preliminary estimate of the standard deviation before you begin. Remember that standard deviation provides a measure of the typical, or standard, distance from the mean. Therefore, the standard deviation must have a value somewhere between the largest and the smallest deviation scores. As a rule of thumb, the standard deviation should be about one-fourth of the range.

2. Rather than trying to memorize all the formulas for SS, variance, and standard deviation, you should focus on the definitions of these values and the logic that relates them to each other:

   SS is the sum of squared deviations.
   Variance is the mean squared deviation.
   Standard deviation is the square root of variance.

   The only formula you should need to memorize is the computational formula for SS.

3. A common error is to use $n - 1$ in the computational formula for SS when you have scores from a sample. Remember that the SS formula always uses n (or N). After you compute SS for a sample, you must correct for the sample bias by using $n - 1$ in the formulas for variance and standard deviation.

DEMONSTRATION 4.1

COMPUTING MEASURES OF VARIABILITY

For the following sample data, compute the variance and standard deviation. The scores are:

$$10, \quad 7, \quad 6, \quad 10, \quad 6, \quad 15$$

STEP 1   **Compute SS, the sum of squared deviations.**   We will use the computational formula. For this sample, $n = 6$ and

$$\Sigma X = 10 + 7 + 6 + 10 + 6 + 15 = 54$$
$$\Sigma X^2 = 10^2 + 7^2 + 6^2 + 10^2 + 6^2 + 15^2 = 546$$
$$SS = \Sigma X^2 - \frac{(\Sigma X)^2}{N} = 546 - \frac{(54)^2}{6}$$
$$= 546 - 486$$
$$= 60$$

STEP 2   **Compute the sample variance.**   For sample variance, SS is divided by the degrees of freedom, $df = n - 1$.

$$s^2 = \frac{SS}{n - 1} = \frac{60}{5} = 12$$

STEP 3 **Compute the sample standard deviation.** Standard deviation is simply the square root of the variance.

$$s = \sqrt{12} = 3.46$$

SPSS®

General instructions for using SPSS are presented in Appendix D. Following are detailed instructions for using SPSS to compute the **Range, Standard Deviation, IQR,** and **Variance** for a sample of scores. These steps will also be used to produce a **Box plot** of the data.

Demonstration Example

Suppose that a college admissions office is interested in the economic diversity of a pool of college applicants. They receive the annual household income of a sample of $n = 30$ applicants. The table below lists the annual incomes of those applicants.

$34,863 $73,633 $ 91,625 $67,317 $54,457 $80,673 $32,487 $32,191 $6,236 $42,729
$67,922 $55,959 $103,594 $ 2,990 $35,539 $65,953 $53,403 $43,685 $85,985 $47,346
$54,062 $91,879 $101,336 $31,319 $31,688 $37,414 $51,717 $41,943 $78,754 $52,015

We will use SPSS to summarize the variability of the scores listed above.

Data Entry

1. Use the **Variable View** of the data editor to create a new variable. Enter "income" in the **Name** field. Select **Numeric** in the **Type** field and **Scale** in the **Measure** field. Enter a brief, descriptive title for the variable in the **Label** field (here, "Annual Household Income of Applicant" was used).

2. Click on **Data View** and enter all the scores in the "Income" column of the data editor.

Data Analysis

1. Click **Analyze** on the tool bar, select **Descriptive Statistics**, and click on **Explore**.
2. Highlight the column label for the set of scores ("Annual Household Income . . .") in the left box and click the arrow to move it into the **Dependent List** box.
3. Click **Plots** and uncheck **Stem-and-leaf** in the "Explore: Plots" window. Click **Continue.**
4. Click **OK**.

SPSS Output

The SPSS Output contains three sections. The **Case Processing Summary** section reports the number of scores that was included in the analysis ($N = 30$). The **Descriptives** section lists statistics for the sample in a table. Most of the statistics in this report should be familiar to you. For example, the Mean row reports that the mean income was $55,023.80. Note that measures for the 95% Confidence Interval and Std. Error will be covered in later chapters. Also note, 5% Trimmed Mean, Skewness, and Kurtosis are not discussed in this textbook. Importantly, the Descriptives section lists several measurements of variability. The range = $100,604, IQR = $39,543, and $s = $25,609.07. Notice that the value for variance (s^2) is very large because the variance statistic reports the mean *squared* deviation between each score and the mean. SPSS uses the formula for sample variance deviation and not the formula for population variance.

The **Annual Household Income of Applicant** section contains a box plot for the data. The thick black line in the center of the box displays the median ($Mdn = $52,709). The line that creates the top of the box is the 75th percentile score (i.e., Q3 = $74,913) and the bottom of the box is the 25th percentile (i.e., Q1 = $35,370). The horizontal line at the top of the figure represents the maximum score ($103,594) and the horizontal line at the bottom of the figure represents the minimum score ($2990).

Case Processing Summary

	Cases					
	Valid		Missing		Total	
	N	Percent	N	Percent	N	Percent
Annual Household Income of Applicant	30	100.0%	0	0.0%	30	100.0%

Descriptives

			Statistic	Std. Error
Annual Household Income of Applicant	Mean		55023.80	4675.554
	95% Confidence Interval for Mean	Lower Bound	45461.22	
		Upper Bound	64586.38	
	5% Trimmed Mean		55197.93	
	Median		52709.00	
	Variance		655824245.200	
	Std. Deviation		25609.066	
	Minimum		2990	
	Maximum		103594	
	Range		100604	
	Interquartile Range		39543	
	Skewness		0.113	0.427
	Kurtosis		-0.351	0.833

Annual Household Income of Applicant

Source: SPSS®

Try It Yourself

Use SPSS to summarize the variability of the following set of scores:

665	542	496	564	452	413	524	455	311	604
456	510	445	602	617	323	419	501	506	408

SPSS will summarize those scores with the following statistics:

Mean	490.65
Median	498.50
Variance	8732.03
Std. Deviation	93.45
Minimum	311
Maximum	665
Range	354
Interquartile Range	133

PROBLEMS

1. Briefly explain the goal for defining and measuring variability.

2. What is the range for the following set of scores? (You may have more than one answer.) Scores: 6, 12, 9, 17, 11, 4, 14

3. Calculate the range and interquartile range for the following set of scores from a continuous variable: 5, 1, 6, 5, 4, 6, 7, 12. Identify the score that corresponds to the 75th percentile and the score that corresponds to the 25th percentile. Why is the interquartile range a better description of variability in the data than the range?

4. Calculate the range and interquartile range for the following set of scores from a continuous variable: 23, 13, 10, 8, 10, 9, 11, 12.

5. In words, explain what is measured by variance and standard deviation.

6. Is it possible to obtain a negative value for SS (sum of squared deviations), variance, and standard deviation?

7. Calculate SS, σ^2, and σ for the following population of $N = 4$ scores: 0, 6, 6, 8.

8. Calculate SS, σ^2, and σ for the following population of $N = 5$ scores: 6, 0, 4, 2, 3.

9. Describe the scores in a sample that has a standard deviation of zero.

10. There are two different formulas or methods that can be used to calculate SS.
    a. Under what circumstances is the definitional formula easy to use?
    b. Under what circumstances is the computational formula preferred?

11. Calculate the mean for both of the following sets of scores. Use both the computational and definitional formulas to compute SS for both sets of scores. Round to two decimal places for each calculation. Why is there a difference in the calculated SS for Set A and not Set B? Which one is correct?

Set A:	8	9	11	9	12	9
Set B:	8	9	11	9	12	5

12. For the following population of $N = 10$ scores: 5, 12, 14, 6, 14, 8, 11, 8, 12, 10
    a. Sketch a histogram showing the population distribution.
    b. Locate the value of the population mean in your sketch, and make an estimate of the standard deviation (as seen in Example 4.2, page 118).

c. Compute SS, variance, and standard deviation for the population. (How well does your estimate compare with the actual value of σ?)

13. For the following population of $N = 8$ scores: 1, 3, 1, 10, 1, 0, 1, 3
    a. Calculate SS, σ^2, and σ.
    b. Which formula should be used to calculate SS? Explain.

14. Calculate SS, variance, and standard deviation for the following population of $N = 7$ scores: 8, 1, 4, 3, 5, 3, 4.

15. For the following set of scores: 6, 2, 3, 0, 4
    a. If the scores are a population, what are the variance and standard deviation?
    b. If the scores are a sample, what are the variance and standard deviation?

16. Explain why the formula for sample variance is different from the formula for population variance. Why is it inappropriate to use the formula for population variance in calculating the variance of a sample?

17. For a sample of $n = 12$ scores, what value should be used in the denominator of the formula for variance? What value should be used in the denominator of the formula for the mean? Explain why the two formulas use different values in the denominator.

18. For the following sample of $n = 6$ scores: 0, 11, 5, 10, 5, 5
    a. Sketch a histogram showing the sample distribution.
    b. Locate the value of the sample mean in your sketch, and make an estimate of the standard deviation (as seen in Example 4.6, page 126).
    c. Compute SS, variance, and standard deviation for the sample. (How well does your estimate compare with the actual value of s?)

19. Calculate SS, variance, and standard deviation for the following sample of $n = 9$ scores: 4, 16, 5, 15, 12, 9, 10, 10, 9.

20. Calculate SS, variance, and standard deviation for the following sample of $n = 5$ scores: 2, 9, 5, 5, 9.

21. For the following population of scores: 1, 4, 7
    a. Calculate the population mean and population variance.
    b. Complete the following table that lists all possible samples of $n = 2$ scores from the population. Use the first three rows (Samples a–c) as examples.

Sample	Score 1	Score 2	$M = \dfrac{\Sigma X}{n}$	SS	$\dfrac{SS}{n-1}$	$\dfrac{SS}{n}$
a	1	1	1.00	0.00	0.00	0.00
b	1	4	2.50	4.50	4.50	2.25
c	1	7	4.00	18.00	18.00	9.00
d	4	1	2.50			
e	4	4	4.00			
f	4	7	5.50			
g	7	1	4.00			
h	7	4	5.50			
i	7	7	7.00			

c. Calculate the mean for all values in the M column. Calculate the mean for all values in the $\frac{SS}{n-1}$ column. Calculate the mean for all values in the $\frac{SS}{n}$ column. Which values match the parameters of the population? Identify those statistics that were biased. Identify those statistics that were unbiased.

22. $\frac{SS}{n} = 4$ for the following set of sample scores: 2, 8, 4, 6, 5. What is your best guess about the actual value of variance in the population?

23. A sample of $n = 12$ scores has a sample mean of $M = 60$ and a sample standard deviation of $s = 3$. What are the values of ΣX and SS?

24. A sample of $n = 10$ scores has a sample mean of $M = 25$ and a sample standard deviation of $s = 4$. What are the values of ΣX and SS?

25. A population has a mean of $\mu = 100$ and a standard deviation of $\sigma = 20$. Sketch a frequency distribution for the population and label the mean and standard deviation.

26. A population has a mean of $\mu = 50$ and a standard deviation of $\sigma = 10$.
   a. If 3 points were added to every score in the population, what would be the new values for the mean and standard deviation?
   b. If every score in the population were multiplied by 2, then what would be the new values for the mean and standard deviation?

27. Solve the following problems.
   a. After 6 points have been added to every score in a sample, the mean is found to be $M = 70$ and the standard deviation is $s = 13$. What were the values for the mean and standard deviation for the original sample?
   b. After every score in a sample is multiplied by 3, the mean is found to be $M = 48$ and the standard deviation is $s = 18$. What were the values for the mean and standard deviation for the original sample?

28. Compute the mean and standard deviation for the following sample of $n = 5$ scores: 70, 72, 71, 80, and

72. *Hint*: To simplify the arithmetic, you can subtract 70 points from each score to obtain a new sample. Then, compute the mean and standard deviation for the new sample. Finally, make the correction for having added 70 points to each score to find the mean and standard deviation for the original sample.

29. For the following sample of $n = 8$ scores: 0, 1, $\frac{1}{2}$, 0, 3, $\frac{1}{2}$, 0, 1
   a. Simplify the arithmetic by first multiplying each score by 2 to obtain a new sample. Then, compute the mean and standard deviation for the new sample.
   b. Starting with the values you obtained in part a, make the correction for having multiplied by 2 to obtain the values for the mean and standard deviation for the original sample.

30. For the following population of $N = 6$ scores: 2, 9, 6, 8, 9, 8
   a. Calculate the range and the standard deviation. (Use either definition for the range—see page 113.)
   b. Add 2 points to each score and compute the range and standard deviation again.
   c. Describe how adding a constant to each score influences measures of variability.

31. The range is completely determined by the two extreme scores in a distribution. The standard deviation, on the other hand, uses every score.
   a. Compute the range (choose either definition), the variance, and the standard deviation for the following sample of $n = 4$ scores. Note that there are two scores located in the center of the distribution and two extreme values. Scores: 0, 6, 6, 12.
   b. Now we will increase the variability by moving the two central scores out to the extremes. Once again compute the range, variance, and standard deviation. New scores: 0, 0, 12, 12.
   c. According to the range, how do the two distributions compare in variability? How do they compare according to the variance and standard deviation?

32. For the data in the following sample: 1, 1, 9, 1
   a. Find the mean, SS, variance, and standard deviation.
   b. Now change the score of $X = 9$ to $X = 3$, and find the new values for SS, variance, and standard deviation.
   c. Describe how one extreme score influences the mean and standard deviation.

33. Luhmann, Schimmack, and Eid (2011) were interested in the relationship between income and subjective well-being. The researchers measured participants' income across time between 1991 and 2006 and observed that both the central tendency and variability of income tended to increase across time. The following annual income scores (in thousands of dollars) are similar to the data observed by Luhmann et al.

1991	2006
8	15
18	13
3	11
8	27
8	5
3	21
13	3
3	23
8	17

a. Compute M, SS, s^2, and s for these samples.
b. Report these descriptive statistics in a format that is appropriate for a scientific journal.

34. Everyone experiences "ups and downs" in life satisfaction. Boehm, Winning, Segerstrom, and Kubzansky (2015) studied whether such variability in life satisfaction is correlated with mortality rate. Over a nine-year period, 4,458 Australian older adult participants answered surveys about life satisfaction, and the researchers recorded whether the participants were deceased at the time of planned follow-up interviews. Participants rated their life satisfaction on a scale of 0 (dissatisfied) to 10 (satisfied). The researchers observed data similar to the following:

Low Mortality Sample	High Mortality Sample
1	0
6	10
6	10
7	4
6	9
5	3
5	0
3	10
8	2
3	3
5	4

a. Compute M, SS, s^2, and s for these samples.
b. Report the mean and standard deviation in a format that is appropriate for a scientific journal.

35. If you have ever tried to learn a new mechanical skill, you probably noticed that the hand-eye coordination needed to perform the skill is learned with practice. To demonstrate the effects of such practice, Li (2008) studied the effect of prism goggles on the accuracy of pointing at a target by participants. Prism goggles shift the image in the eye to one side. Participants in this experiment served in three phases: a baseline phase without prism goggles, a pre-adaptation phase with prism goggles, and a post-adaptation phase with prism goggles after participants practiced a visual task while wearing the goggles. The researcher measured how far away from the target the participant pointed in centimeters. The researcher observed data like those following:

Pre-adaptation	Post-adaptation
5	0
23	10
5	0
20	7
9	7
11	3
11	8

Describe the effect of adaptation on distance between the participants' points and the target. Be sure that your description includes some discussion of central tendency and variability.

36. On an exam with a mean of $M = 40$, you obtain a score of $X = 35$.
    a. Relative to other students, would your performance on the exam be better with a standard deviation of $s = 2$ or with a standard deviation of $s = 8$? (*Hint*: Sketch each distribution and find the location of your score.)
    b. If your score were $X = 46$, would you prefer $s = 2$ or $s = 8$? Explain your answer.

37. One population has a mean of $\mu = 50$ and a standard deviation of $\sigma = 15$, and a different population has a mean of $\mu = 50$ and a standard deviation of $\sigma = 5$.
    a. Sketch both distributions, labelling μ and σ.
    b. Would a score of $X = 65$ be considered an extreme value (out in the tail) in one of these distributions? Explain your answer.

38. A teacher is interested in the effect of a study session on quiz performance. Two different classes receive a pretest (before the study session) and a posttest (after the study session). Thus, the teacher records the following four sets of scores:

Class 1 Pretest	Class 1 Posttest
14	18
5	20
12	20
12	20
8	24
9	18
10	20

Class 2 Pretest	Class 2 Posttest
20	10
2	21
12	12
8	28
12	21
4	20
12	28

a. For each of the four sets of scores above, calculate the sample mean and standard deviation.
b. In which class is the effect of the study session most obvious in the pattern of data? Explain.

z-Scores: Location of Scores and Standardized Distributions

clivewa/Shutterstock.com

Tools You Will Need

The following items are considered essential background material for this chapter. If you doubt your knowledge of any of these items, you should review the appropriate chapter and section before proceeding.

- The mean (Chapter 3)
- The standard deviation (Chapter 4)
- Basic algebra (math review, Appendix A)

A common test of cognitive ability requires participants to search through a visual display and respond to specific targets as quickly as possible. This kind of test is called a perceptual-speed test. Measures of perceptual speed are commonly used for predicting performance on jobs that demand a high level of speed and accuracy. Although many different tests are used, a typical example is shown in Figure 5.1. This task requires the participant to search through the display of digit pairs as quickly as possible and circle each pair that adds to 10. Your score is determined by the amount of time required to complete the task with a correction for the number of errors you make. One complaint about this kind of paper-and-pencil test is that it is tedious and time-consuming to score because a researcher must also search through the entire display to identify errors to determine the participant's level of accuracy. An alternative, proposed by Ackerman and Beier (2007), is a computerized version of the task. The computer version presents a series of digit pairs that participants respond to on a touch-sensitive monitor. The computerized test is very reliable and the scores are equivalent to the paper-and-pencil tests in terms of assessing cognitive skill. The advantage of the computerized test is that the computer produces a test score immediately when a participant finishes the test.

Suppose that you took Ackerman and Beier's test and your combined time and errors produced a score of 92. How did you do? Are you faster than average, fairly normal in perceptual speed, or does your score indicate a serious deficit in cognitive skill? The answer is that you have no idea how your score of 92 compares with scores for others who took the same test. Now suppose that you are also told that the distribution of perceptual speed scores has a mean of $\mu = 86.75$ and a standard deviation of $\sigma = 10.50$. With this additional information, you should realize that your score ($X = 92$) is somewhat higher than average but not an exceptionally high score.

In this chapter we introduce a statistical procedure for converting individual scores into z-scores, so that each z-score is a meaningful value that identifies exactly where the original score is located in the distribution. As you will see, z-scores use the mean as a reference point to determine whether the score is above or below average. A z-score also uses the standard deviation as a yardstick for describing how much an individual score differs from the average. For example, a z-score will tell you if your score is above the mean by a distance equal to two standard deviations, or below the mean by one-half of a standard deviation. A good understanding of the mean and the standard deviation will be a valuable background for learning about z-scores.

64	23	19	31	19	46	31	91	83	82	82	46	19	87
11	42	94	87	64	44	19	55	82	46	57	98	39	46
78	73	72	66	63	71	67	42	62	73	45	22	62	99
73	91	52	37	55	97	91	51	44	23	46	64	97	62
97	31	21	49	93	91	89	46	73	82	55	98	12	56
73	82	37	55	89	83	73	27	83	82	73	46	97	62
57	96	46	55	46	19	13	67	73	26	58	64	32	73
23	94	66	55	91	73	67	73	82	55	64	62	46	39
87	11	99	73	56	73	63	73	91	82	63	33	16	88
19	42	62	91	12	82	32	92	73	46	68	19	11	64
93	91	32	82	63	91	46	46	36	55	19	92	62	71

FIGURE 5.1

An example of a perceptual speed task. The participant is asked to search through the display as quickly as possible and circle each pair of digits that add up to 10.

5-1 | Introduction

In the previous two chapters, we introduced the concepts of the mean and standard deviation as methods for describing an entire distribution of *raw scores*. Now we shift attention to the individual scores within a distribution. In this chapter, we introduce a statistical technique that uses the mean and the standard deviation to transform each score (X value) into a *z-score* or a *standard score*. The purpose of z-scores, or standard scores, is to identify and describe the exact location of each score in a distribution.

The following example demonstrates why z-scores are useful and introduces the general concept of transforming X values into z-scores.

EXAMPLE 5.1

Suppose you received a score of $X = 26$ on a statistics exam. How did you do compared to other students? It should be clear that you need more information to answer this question. Your score of $X = 26$ could be one of the best scores in the class, or it might be the lowest score in the distribution. To find the location of your score, you must have information about the other scores in the distribution. It would be useful, for example, to know the mean for the class. If the mean were $M = 20$, you would be in a much better position than if the mean were $M = 35$. Obviously, your position relative to the rest of the class depends on the mean. However, the mean by itself is not sufficient to tell you the exact location of your score. Suppose you know that the mean for the statistics exam is $M = 20$ and your score is $X = 26$. At this point, you know that your score is 6 points above the mean, but you still do not know exactly where it is located. Six points may be a relatively big distance and you may have one of the highest scores in the class, or 6 points may be a relatively small distance and you are only slightly above the average. Figure 5.2 shows two possible distributions. Both distributions have a mean of $M = 20$, but for one distribution, the standard deviation is $s = 2$, and for the other, $s = 10$. The location of $X = 26$ is identified by the shaded box in each of the two distributions. When the standard deviation is $s = 2$, your score of $X = 26$ is in the extreme right-hand tail, the highest score in the distribution. However, in the other distribution, where $s = 10$, your score is only slightly above average. Thus, the amount of variability in the distribution is important too. The location of your score within the distribution depends on the standard deviation as well as the mean. ■

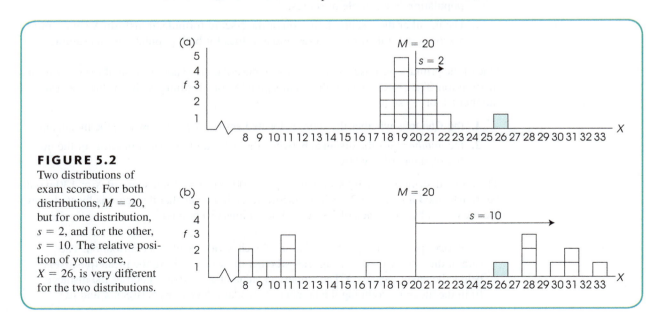

FIGURE 5.2

Two distributions of exam scores. For both distributions, $M = 20$, but for one distribution, $s = 2$, and for the other, $s = 10$. The relative position of your score, $X = 26$, is very different for the two distributions.

The intent of the preceding example is to demonstrate that a score *by itself* does not necessarily provide much information about its position within a distribution. These original, unchanged scores that are the direct result of measurement are called *raw scores*. To make raw scores more meaningful, they are often transformed into new values that contain more information. This transformation is one purpose for z-scores. In particular, we transform X values into z-scores so that the resulting z-scores tell exactly where within a distribution the original scores are located.

A second purpose for z-scores is to *standardize* an entire distribution. A common example of a standardized distribution is the distribution of IQ scores. Although there are several different tests for measuring IQ, the tests usually are standardized so that they have a mean of 100 and a standard deviation of 15. Because most of the different tests are standardized, it is possible to understand and compare IQ scores even though they come from different tests. For example, we all understand that an IQ score of 95 is a little below average, *no matter which IQ test was used*. Similarly, an IQ of 145 is extremely high, *no matter which IQ test was used*. In general terms, the process of standardizing takes different distributions and makes them equivalent. The advantage of this process is that it is possible to compare distributions even though they may have been quite different before standardization.

In summary, the process of transforming X values into z-scores serves two useful purposes:

1. Each z-score tells the exact location of the original X value within the distribution.

2. The z-scores form a standardized distribution that can be directly compared to other distributions that also have been transformed into z-scores.

Each of these purposes is discussed in the following sections.

5-2 z-Scores and Locations in a Distribution

LEARNING OBJECTIVES

1. Explain how a z-score identifies a precise location in a distribution for either a population or a sample of scores.

2. Using either the z-score definition or the z-score formula, transform X values into z-scores and transform z-scores into X values for both populations and samples.

One of the primary purposes of a z-score is to describe the exact location of a score within a distribution. The z-score accomplishes this goal by transforming each X value into a signed number (+ or −) so that

1. the *sign* tells whether the score is located above (+) or below (−) the mean, and

2. the *number* tells the distance between the score and the mean in terms of the number of standard deviations.

Thus, in a distribution of IQ scores with $\mu = 100$ and $\sigma = 15$, a score of $X = 130$ would be transformed into $z = +2.00$. The z-score value indicates that the score is located above the mean (+) by a distance of 2 standard deviations (30 points).

> A **z-score** specifies the precise location of each X value within a distribution. The sign of the z-score (+ or −) signifies whether the score is above the mean (positive) or below the mean (negative). The numerical value of the z-score specifies the distance from the mean by counting the number of standard deviations between X and μ.

FIGURE 5.3
The relationship between *z*-score values and locations in a population distribution.

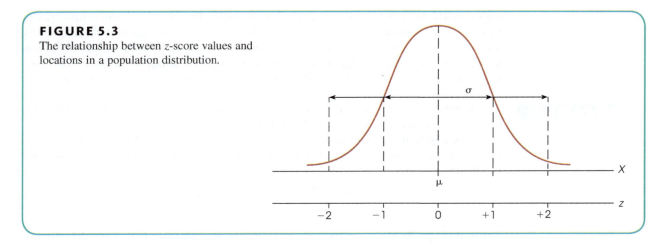

Notice that a *z*-score always consists of two parts: a sign (+ or −) and a magnitude (the numerical value). Both parts are necessary to describe completely where a raw score is located within a distribution.

Figure 5.3 shows a population distribution with various positions identified by their *z*-score values. Notice that all *z*-scores above the mean are positive and all *z*-scores below the mean are negative. The sign of a *z*-score tells you immediately whether the score is located above or below the mean. Also, note that a *z*-score of $z = +1.00$ corresponds to a position exactly one standard deviation above the mean. A *z*-score of $z = +2.00$ is always located exactly two standard deviations above the mean. The numerical value of the *z*-score tells you the number of standard deviations from the mean. Finally, you should notice that Figure 5.3 does not give any specific values for the population mean or the standard deviation. The locations identified by *z*-scores are the same for *all distributions*, no matter what mean or standard deviation the distributions may have.

Now we can return to the two distributions shown in Figure 5.2 and use a *z*-score to describe the position of $X = 26$ within each distribution as follows:

> In Figure 5.2(a), with a standard deviation of $s = 2$, the score $X = 26$ corresponds to a *z*-score of $z = +3.00$. That is, the score is located above the mean by exactly three standard deviations.

> In Figure 5.2(b), with $s = 10$, the score $X = 26$ corresponds to a *z*-score of $z = +0.60$. In this distribution, the score is located above the mean by exactly six-tenths of a standard deviation.

> *Whenever you are working with z-scores, you should imagine or draw a picture similar to that shown in Figure 5.3. Although you should realize that not all distributions are normal, we will use the normal shape as an example when showing z-scores for populations.*

■ The *z*-Score Formula for a Population

The *z*-score definition is adequate for transforming back and forth between X values and *z*-scores as long as the arithmetic is easy to do in your head. For more complicated values, it is best to have an equation to help structure the calculations. Fortunately, the relationship between X values and *z*-scores is easily expressed in a formula. The formula for transforming scores into *z*-scores is

$$z = \frac{X - \mu}{\sigma} \tag{5.1}$$

The numerator of the equation, $X - \mu$, is a *deviation score* (Chapter 4, page 116). The deviation measures the distance in points between X and μ and the sign of the deviation indicates whether X is located above or below the mean. The deviation score is then divided by σ because we want the z-score to measure distance in terms of standard deviation units. The formula performs exactly the same arithmetic that is used with the z-score definition, and it provides a structured equation to organize the calculations when the numbers are more difficult. The following examples demonstrate the use of the z-score formula.

EXAMPLE 5.2

A distribution of scores has a mean of $\mu = 100$ and a standard deviation of $\sigma = 10$. What z-score corresponds to a score of $X = 130$ in this distribution?

According to the definition, the z-score will have a value of $+3$ because the score is located above the mean by exactly three standard deviations. Using the z-score formula, we obtain

$$z = \frac{X - \mu}{\sigma} = \frac{130 - 100}{10} = \frac{30}{10} = +3.00$$

The formula produces exactly the same result that is obtained using the z-score definition. ■

EXAMPLE 5.3

A distribution of scores has a mean of $\mu = 86$ and a standard deviation of $\sigma = 7$. What z-score corresponds to a score of $X = 95$ in this distribution?

Note that this problem is not particularly easy, especially if you try to use the z-score definition and perform the calculations in your head. However, the z-score formula organizes the numbers and allows you to finish the final arithmetic with your calculator. Using the formula, we obtain

$$z = \frac{X - \mu}{\sigma} = \frac{95 - 86}{7} = \frac{9}{7} = +1.29$$

According to the formula, a score of $X = 95$ corresponds to $z = 1.29$. The z-score indicates a location that is above the mean (positive) by slightly more than one standard deviation. ■

When you use the z-score formula, it can be useful to pay attention to the definition of a z-score as well. For example, we used the formula in Example 5.3 to calculate the z-score corresponding to $X = 95$, and obtained $z = 1.29$. Using the z-score definition, we note that $X = 95$ is located above the mean by 9 points, which is slightly more than one standard deviation ($\sigma = 7$). Therefore, the z-score should be positive and have a value slightly greater than 1.00. In this case, the answer predicted by the definition is in perfect agreement with the calculation. However, if the calculations produce a different value, for example $z = 0.78$, you should realize that this answer is not consistent with the definition of a z-score. In this case, an error has been made and you should check the calculations.

■ Determining a Raw Score (*X*) from a z-Score

Although the z-score equation (Equation 5.1, page 153) works well for transforming X values into z-scores, it can be awkward when you are trying to work in the opposite direction and change z-scores back into X values. In general it is easier to use the definition of a z-score, rather than a formula, when you are changing z-scores into X values. Remember, the z-score describes exactly where the score is located by identifying the direction and distance from the mean. It is possible, however, to express this definition as a formula, and we will use a simple problem to demonstrate how the formula can be created:

For a distribution with a mean of $\mu = 60$ and $\sigma = 8$, what X value corresponds to a z-score of $z = -1.50$?

To solve this problem, we will use the z-score definition and carefully monitor the step-by-step process. The value of the z-score indicates that X is located below the mean by a distance equal to 1.5 standard deviations. Thus, the first step in the calculation is to determine the distance corresponding to 1.5 standard deviations. For this problem, the standard deviation is $\sigma = 8$ points, so 1.5 standard deviations is $1.5(8) = 12$ points. The next step is to find the value of X that is located below the mean by 12 points. With a mean of $\mu = 60$, the score is

$$X = \mu - 12 = 60 - 12 = 48$$

The two steps can be combined to form a single formula:

$$X = \mu + z\sigma \qquad (5.2)$$

In the formula, the value of $z\sigma$ is the *deviation* of X and determines both the direction and the size of the distance from the mean. In this problem, $z\sigma = (-1.5)(8) = -12$, or 12 points below the mean. Equation 5.2 simply combines the mean and the deviation from the mean to determine the exact value of X. Notice the importance of the sign of the *z*-score. If the *z*-score is positive, then $z\sigma$ is added to the mean. However, if the *z*-score is negative, then $z\sigma$ is subtracted from the mean. For negative *z*-scores the computation for the value of X looks like:

$$X = \mu + (-z\sigma)$$

Because a positive (+) value times a negative (−) value equals a negative value, the computation is the same as

$$X = \mu - z\sigma$$

Finally, you should realize that Equations 5.1 and 5.2 are actually two different versions of the same equation. If you begin with Equation 5.1 and use basic algebra, solving the equation for X will result in Equation 5.2. We will leave this as an exercise for those who want to try it.

■ Computing *z*-Scores for Samples

The definition and the purpose of a *z*-score is the same for a sample as for a population, provided that you use the sample mean and the sample standard deviation to specify each *z*-score location. Thus, for a sample, each X value is transformed into a *z*-score so that

1. the sign of the *z*-score indicates whether the X value is above (+) or below (−) the sample mean, and

2. the numerical value of the *z*-score identifies the distance from the sample mean by measuring the number of sample standard deviations between the score (X) and the sample mean (M).

Expressed as a formula, each X value in a sample can be transformed into a *z*-score as follows:

$$z = \frac{X - M}{s} \qquad (5.3)$$

Similarly, each *z*-score can be transformed back into an X value, as follows:

$$X = M + zs \qquad (5.4)$$

You should recognize that these two equations are identical to the population equations (5.1 and 5.2), except that we are now using sample statistics, M and s, in place of the population parameters μ and σ. The following example demonstrates the transformation of X values and *z*-scores for a sample.

EXAMPLE 5.4

In a sample with a mean of $M = 40$ and a standard deviation of $s = 10$, what is the z-score corresponding to $X = 35$ and what is the X value corresponding to $z = +2.00$?

The score, $X = 35$, is located below the mean by 5 points, which is exactly half of the standard deviation. According to the z-score definition, the corresponding z-score is $z = -0.50$. Using Equation 5.3, the z-score for $X = 35$ is

$$z = \frac{X - M}{s}$$

$$= \frac{35 - 40}{10} = \frac{-5}{10} = -0.50$$

If the z-score is negative, do not forget to include the sign in the formula, $X = M + (-zs)$.

Using the z-score definition, $z = +2.00$ corresponds to a location above the mean by two standard deviations. With a standard deviation of $s = 10$, this is a distance of 20 points. The score that is located 20 points above the mean is $X = 60$. Using Equation 5.4 we obtain

$$X = M + zs$$

$$= 40 + 2.00(10)$$

$$= 40 + 20$$

$$= 60$$

LEARNING CHECK   **LO1**   **1.** What location in a distribution corresponds to $z = -3.00$?

  a. Above the mean by 3 points.

  b. Above the mean by a distance equal to three standard deviations.

  c. Below the mean by 3 points.

  d. Below the mean by a distance equal to three standard deviations.

LO2   **2.** For a population with $\mu = 90$ and $\sigma = 12$, what is the z-score corresponding to $X = 102$?

  a. $+0.50$

  b. $+1.00$

  c. $+1.20$

  d. $+12.00$

LO2   **3.** For a sample with $M = 72$ and $s = 4$, what is the X value corresponding to $z = -2.00$?

  a. $X = 70$

  b. $X = 68$

  c. $X = 64$

  d. $X = 60$

ANSWERS   **1. d   2. b   3. c**

| 5-3 | Other Relationships between *z*, *X*, the Mean, and the Standard Deviation |

LEARNING OBJECTIVE

3. Explain how *z*-scores establish a relationship among *X*, the mean, the standard deviation, and the value of *z*, and use that relationship to find an unknown mean when given a *z*-score, a score, and the standard deviation; or find an unknown standard deviation when given a *z*-score, a score, and the mean.

In most cases, we simply transform scores (*X* values) into *z*-scores, or change *z*-scores back into *X* values. However, you should realize that a *z*-score establishes a relationship between the score, the mean, and the standard deviation. This relationship can be used to answer a variety of different questions about scores and the distributions in which they are located. The following two examples demonstrate some possibilities.

EXAMPLE 5.5   In a population with a mean of $\mu = 65$, a score of $X = 59$ corresponds to $z = -2.00$. What is the standard deviation for the population?

To answer the question, we begin with the *z*-score value. A *z*-score of -2.00 indicates that the corresponding score is located below the mean by a distance of two standard deviations. You also can determine that the score ($X = 59$) is located below the mean ($\mu = 65$) by a distance of 6 points. Thus, two standard deviations correspond to a distance of 6 points, which means that one standard deviation must be $\sigma = 3$ points. ■

The same relationships exist for samples, as demonstrated in the following example.

EXAMPLE 5.6   In a sample with a standard deviation of $s = 6$, a score of $X = 33$ corresponds to $z = +1.50$. What is the mean for the sample?

Again, we begin with the *z*-score value. In this case, a *z*-score of $+1.50$ indicates that the score is located above the mean by a distance corresponding to 1.50 standard deviations. With a standard deviation of $s = 6$, this distance is $(1.50)(6) = 9$ points. Thus, the score is located 9 points above the mean. The score is $X = 33$, so the mean must be $M = 24$. ■

Many students find problems like those in Examples 5.5 and 5.6 easier to understand if they draw a picture showing all the information presented in the problem. For the problem in Example 5.5, the picture would begin with a distribution that has a mean of $\mu = 65$ (we use a normal distribution that is shown in Figure 5.4, page 158). The value of the standard deviation is unknown, but you can add arrows to the sketch pointing outward from the mean for a distance corresponding to one standard deviation. Finally, use standard deviation arrows to identify the location of $z = -2.00$ (two standard deviations below the mean) and add $X = 59$ at that location. All of these factors are shown in Figure 5.4. In the figure, it is easy to see that $X = 59$ is located 6 points below the mean, and that the 6-point distance corresponds to exactly two standard deviations. Again, if two standard deviations equal 6 points, then one standard deviation must be $\sigma = 3$ points.

A slight variation on Examples 5.5 and 5.6 is demonstrated in the following example. This time you must use the *z*-score information to find both the population mean and the standard deviation.

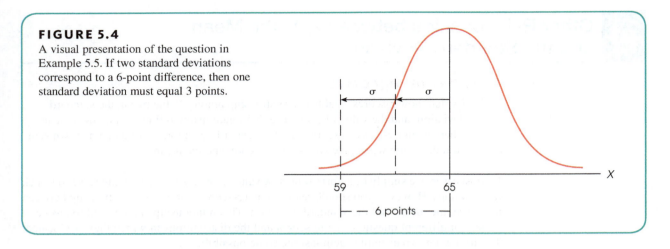

FIGURE 5.4

A visual presentation of the question in Example 5.5. If two standard deviations correspond to a 6-point difference, then one standard deviation must equal 3 points.

EXAMPLE 5.7   In a population distribution, a score of $X = 54$ corresponds to $z = +2.00$ and a score of $X = 42$ corresponds to $z = -1.00$. What are the values for the mean and the standard deviation for the population? Once again, many students find this kind of problem easier to understand if they can see it in a picture, so we have sketched this example in Figure 5.5.

The key to solving this kind of problem is to focus on the distance between the two scores. Notice that the distance can be measured in points and in standard deviations. In points, the distance from $X = 42$ to $X = 54$ is 12 points. According to the two z-scores, $X = 42$ is located one standard deviation below the mean and $X = 54$ is located two standard deviations above the mean (see Figure 5.5). Thus, the total distance between the two scores is equal to three standard deviations. We have determined that the distance between the two scores is 12 points, which is equal to three standard deviations. As an equation,

$$3\sigma = 12 \text{ points}$$

Dividing both sides by 3, we obtain

$$\sigma = 4 \text{ points}$$

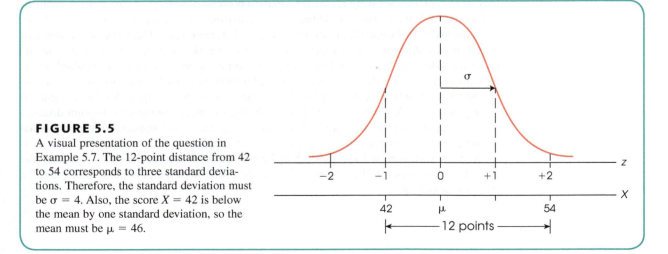

FIGURE 5.5

A visual presentation of the question in Example 5.7. The 12-point distance from 42 to 54 corresponds to three standard deviations. Therefore, the standard deviation must be $\sigma = 4$. Also, the score $X = 42$ is below the mean by one standard deviation, so the mean must be $\mu = 46$.

Finally, note that $X = 42$ corresponds to $z = -1.00$, which means that $X = 42$ is located one standard deviation below the mean. With a standard deviation of $\sigma = 4$, the mean must be $\mu = 46$. Thus, the population has a mean of $\mu = 46$ and a standard deviation of $\sigma = 4$. ■

The following example is an opportunity for you to test your understanding by solving a problem similar to the demonstration in Example 5.7.

EXAMPLE 5.8   In a sample distribution, a score of $X = 64$ corresponds to $z = 0.50$ and a score of $X = 72$ has a z-score of $z = 1.50$. What are the values for the sample mean and standard deviation? Remember, a problem is easier to solve if you draw a picture showing all the information presented in the problem. You should obtain $M = 60$ and $s = 8$. Good luck. ■

LEARNING CHECK   **LO3** **1.** In a population with $\mu = 70$, a score of $X = 68$ corresponds to a z-score of $z = -0.50$. What is the population standard deviation?

    a. 1

    b. 2

    c. 4

    d. Cannot be determined without additional information.

LO3 **2.** In a sample with a standard deviation of $s = 4$, a score of $X = 64$ corresponds to $z = -0.50$. What is the sample mean?

    a. $M = 62$

    b. $M = 60$

    c. $M = 66$

    d. $M = 68$

LO3 **3.** In a population of scores, $X = 50$ corresponds to $z = +2.00$ and $X = 35$ corresponds to $z = -1.00$. What is the population mean?

    a. 35

    b. 40

    c. 37.5

    d. 45

LO3 **4.** In a sample, $X = 70$ corresponds to $z = +2.00$ and $X = 65$ corresponds to $z = +1.00$. What are the sample mean and standard deviation?

    a. $M = 60$ and $s = 5$

    b. $M = 60$ and $s = 10$

    c. $M = 50$ and $s = 10$

    d. $M = 50$ and $s = 5$

ANSWERS   **1. c** **2. c** **3. b** **4. a**

5-4 Using z-Scores to Standardize a Distribution

LEARNING OBJECTIVE

4. Describe the effects of standardizing a distribution by transforming the entire set of scores into z-scores, and explain the advantages of this transformation.

■ Population Distributions

It is possible to transform every X value in a population into a corresponding z-score. The result of this process is that the entire distribution of X values is transformed into a distribution of z-scores (see Figure 5.6). The new distribution of z-scores has characteristics that make the *z-score transformation* a very useful tool. Specifically, if every X value is transformed into a z-score, then the distribution of z-scores will have the following properties:

1. **Shape** The distribution of z-scores will have exactly the same shape as the original distribution of scores. If the original distribution is negatively skewed, for example, then the z-score distribution will also be negatively skewed. If the original distribution is normal, the distribution of z-scores will also be normal. Transforming raw scores into z-scores does not change anyone's position in the distribution. For example, any raw score that is above the mean by one standard deviation will be transformed to a z-score of +1.00, which is still above the mean by one standard deviation. Transforming a distribution from X values to z values does not move scores from one position to another; the procedure simply relabels each score (see Figure 5.6). Because each individual score stays in its same position within the distribution, the overall shape of the distribution does not change.

2. **The Mean** The z-score distribution will *always* have a mean of zero. In Figure 5.6, the original distribution of X values has a mean of $\mu = 100$. When this value, $X = 100$, is transformed into a z-score, the result is

$$z = \frac{X - \mu}{\sigma} = \frac{100 - 100}{10} = 0$$

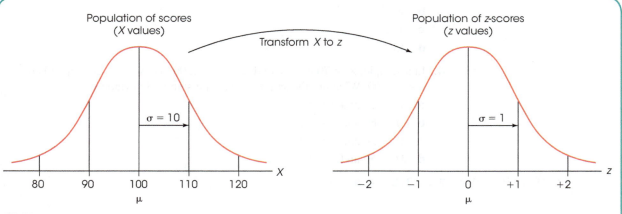

FIGURE 5.6

An entire population of scores is transformed into z-scores. The transformation does not change the shape of the distribution, but the mean is transformed into a value of 0 and the standard deviation is transformed to a value of 1.

Thus, the original population mean is transformed into a value of zero in the z-score distribution. The fact that the z-score distribution has a mean of zero makes the mean a convenient reference point. Recall from the definition of z-scores that all positive z-scores are above the mean and all negative z-scores are below the mean. In other words, for z-scores, $\mu = 0$.

3. **The Standard Deviation**   The distribution of z-scores will *always* have a standard deviation of $\sigma = 1$. In Figure 5.6, the original distribution of X values has $\mu = 100$ and $\sigma = 10$. In this distribution, a value of $X = 110$ is above the mean by exactly 10 points or 1 standard deviation. When $X = 110$ is transformed, it becomes $z = +1.00$, which is above the mean by exactly 1 point in the z-score distribution. Thus, the standard deviation corresponds to a 10-point distance in the X distribution and is transformed into a 1-point distance in the z-score distribution. The advantage of having a standard deviation of 1 is that the numerical value of a z-score is exactly the same as the number of standard deviations from the mean. For example, a z-score of $z = 1.50$ is exactly 1.50 standard deviations from the mean.

In Figure 5.6, we showed the z-score transformation as a process that changed a distribution of X values into a new distribution of z-scores. In fact, there is no need to create a whole new distribution. Instead, you can think of the z-score transformation as simply *relabeling* the values along the X-axis. That is, after a z-score transformation, you still have the same distribution, but now each individual is labeled with a z-score instead of an X value. Figure 5.7 demonstrates this concept with a single distribution that has two sets of labels: the X values along one line and the corresponding z-scores along another line. Notice that the mean for the distribution of z-scores is zero and the standard deviation is 1.

■ Sample Distributions

If all the scores in a sample are transformed into z-scores, the result is a sample distribution of z-scores. The transformed distribution of z-scores will have the same properties that exist when a population of X values is transformed into z-scores. Specifically,

1. the distribution for the sample of z-scores will have the same shape as the original sample of scores.

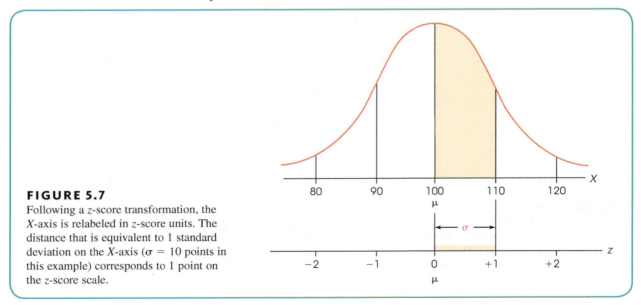

FIGURE 5.7
Following a z-score transformation, the X-axis is relabeled in z-score units. The distance that is equivalent to 1 standard deviation on the X-axis ($\sigma = 10$ points in this example) corresponds to 1 point on the z-score scale.

2. the sample of z-scores will have a mean of $M_z = 0$.

3. the sample of z-scores will have a standard deviation of $s_z = 1$.

Note that the set of z-scores is still considered to be a sample (just like the set of X values) and the sample formulas must be used to compute variance and standard deviation. The following example demonstrates the process of transforming the scores from a sample into z-scores.

EXAMPLE 5.9

We begin with a sample of $n = 5$ scores: 0, 2, 4, 4, 5. With a few simple calculations, you should be able to verify that the sample mean is $M = 3$, the sample variance is $s^2 = 4$, and the sample standard deviation is $s = 2$. Using the sample mean and sample standard deviation, we can convert each X value into a z-score. For example, $X = 5$ is located above the mean by 2 points. Thus, $X = 5$ is above the mean by exactly one standard deviation and has a z-score of $z = +1.00$. The z-scores for the entire sample are shown in the following table.

X	z
0	−1.50
2	−0.50
4	+0.50
4	+0.50
5	+1.00

Again, a few simple calculations demonstrate that the sum of the z-score values is $\Sigma_z = 0$, so the mean is $M_z = 0$. Figure 5.8 shows the z-score transformation.

Because the mean is zero, each z-score value is its own deviation from the mean. Therefore, the sum of the squared z-scores is equal to the sum of the squared deviations. For this sample of z-scores,

$$SS = \Sigma z^2 = (-1.50)^2 + (-0.50)^2 + (+0.50)^2 + (0.50)^2 + (+1.00)^2$$
$$= 2.25 + 0.25 + 0.25 + 0.25 + 1.00$$
$$= 4.00$$

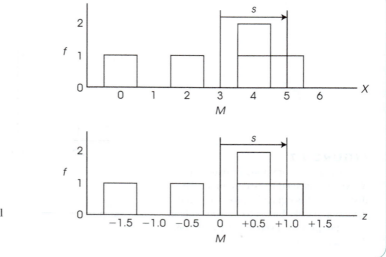

FIGURE 5.8

Transforming a distribution of raw scores (top) into z-scores (bottom) will not change the shape of the distribution. The z-scores will have a mean of $M = 0$ and a standard deviation of $s = 1$.

Notice that this set of z-scores is a sample and the variance and standard deviation are computed using the sample formula with $df = n - 1$.

The variance for the sample of z-scores is

$$s_z^2 = \frac{SS}{n-1} = \frac{4}{4} = 1.00$$

Finally, the standard deviation for the sample of z-scores is $s_z = \sqrt{1.00} = 1.00$. As always, the distribution of z-scores has a mean of 0 and a standard deviation of 1. ■

■ Using z-Scores for Making Comparisons

When *any* distribution (with any mean or standard deviation) is transformed into z-scores, the resulting distribution will always have a mean of $\mu = 0$ and a standard deviation of $\sigma = 1$. Because all z-score distributions have the same mean and the same standard deviation, the z-score distribution is called a *standardized distribution*.

> A **standardized distribution** is composed of scores that have been transformed to create predetermined values for μ and σ. Standardized distributions are used to make dissimilar distributions comparable.

A z-score distribution is an example of a standardized distribution with $\mu = 0$ and $\sigma = 1$. That is, when any distribution (with any mean or standard deviation) is transformed into z-scores, the transformed distribution will always have $\mu = 0$ and $\sigma = 1$. One advantage of standardizing distributions is that it makes it possible to compare different scores or different individuals even though they come from completely different distributions. Normally, if two scores come from different distributions, it is impossible to make any direct comparison between them. Suppose, for example, Dave received a score of $X = 60$ on a psychology exam and a score of $X = 56$ on a biology test. For which course should Dave expect the better grade?

Because the scores come from two different distributions, you cannot make any direct comparison. Without additional information, it is even impossible to determine whether Dave is above or below the mean in either distribution. Before you can begin to make comparisons, you must know the values for the mean and standard deviation for each distribution. Suppose the biology scores had $\mu = 48$ and $\sigma = 4$, and the psychology scores had $\mu = 50$ and $\sigma = 10$. With this new information, you could sketch the two distributions, locate Dave's score in each distribution, and compare the two locations.

Instead of drawing the two distributions to determine where Dave's two scores are located, we simply can compute the two z-scores to find the two locations. For psychology, Dave's z-score is

$$z = \frac{X - \mu}{\sigma} = \frac{60 - 50}{10} = \frac{10}{10} = +1.0$$

Be sure to use the μ and σ values for the distribution to which X belongs.

For biology, Dave's z-score is

$$z = \frac{56 - 48}{4} = \frac{8}{4} = +2.0$$

Note that Dave's z-score for biology is $+2.0$, which means that his test score is two standard deviations above the class mean. On the other hand, his z-score is $+1.0$ for psychology, or one standard deviation above the mean. In terms of relative class standing, Dave is doing much better in the biology class.

Notice that we cannot compare Dave's two exam scores ($X = 60$ and $X = 56$) because the scores come from different distributions with different means and standard deviations. However, we can compare the two z-scores because all distributions of z-scores have the same mean ($\mu = 0$) and the same standard deviation ($\sigma = 1$).

LEARNING CHECK   **LO4 1.** A population with $\mu = 90$ and $\sigma = 20$ is transformed into *z*-scores. After the transformation, what is the mean for the population of *z*-scores?

    a. $\mu = 80$

    b. $\mu = 1.00$

    c. $\mu = 0$

    d. Cannot be determined from the information given.

LO4 2. A sample with a mean of $M = 70$ and a standard deviation of $s = 15$ is being transformed into *z*-scores. After the transformation, what is the standard deviation for the sample of *z*-scores?

    a. 0

    b. 1

    c. $n - 1$

    d. n

LO4 3. Which of the following is an advantage of transforming X values into *z*-scores?

    a. All negative numbers are eliminated.

    b. The distribution is transformed to a normal shape.

    c. All scores are moved closer to the mean.

    d. Dissimilar distributions can be compared.

LO4 4. Last week Sarah had exams in math and Spanish. On the math exam, the mean was $\mu = 30$ with $\sigma = 5$, and Sarah had a score of $X = 45$. On the Spanish exam, the mean was $\mu = 60$ with $\sigma = 6$, and Sarah had a score of $X = 65$. For which class should Sarah expect the better grade?

    a. Math

    b. Spanish

    c. The grades should be the same because the two exam scores are in the same location.

    d. There is not enough information to determine which is the better grade.

ANSWERS   1. c   2. b   3. d   4. a

5-5 Other Standardized Distributions Based on *z*-Scores

LEARNING OBJECTIVE

5. Use *z*-scores to transform any distribution into a standardized distribution with a predetermined mean and a predetermined standard deviation.

■ Transforming *z*-Scores to a Distribution with a Predetermined Mean and Standard Deviation

Although *z*-score distributions have distinct advantages, many people find them cumbersome because they contain negative values and decimals. For this reason, it is common to standardize a distribution by transforming the scores into a new distribution with a predetermined mean and standard deviation that are positive whole numbers. The goal is to create a new (standardized) distribution that has "simple" values for the mean and standard deviation but does not change any individual's location within the distribution. Standardized scores of this type are frequently used in psychological or educational testing. For example, raw scores for the SAT are transformed to a standardized distribution that has $\mu = 500$ and $\sigma = 100$. For intelligence tests, raw scores are frequently converted to standard scores that have a mean of 100 and a standard deviation of 15. Because most IQ tests are standardized so that they have the same mean and standard deviation, it is possible to compare IQ scores even though they may come from different tests.

The procedure for standardizing a distribution to create new values for the mean and standard deviation is a two-step process that can be used either with a population or a sample:

1. The original scores are transformed into z-scores.

2. The z-scores are then transformed into new X values so that the specific mean and standard deviation are attained.

This process ensures that each individual has exactly the same z-score location in the new distribution as in the original distribution. The following example demonstrates the standardization procedure for a population.

EXAMPLE 5.10   An instructor gives an exam to a psychology class. For this exam, the distribution of raw scores has a mean of $\mu = 57$ with $\sigma = 14$. The instructor would like to simplify the distribution by transforming all scores into a new, standardized distribution with $\mu = 50$ and $\sigma = 10$. To demonstrate this process, we will consider what happens to two specific students: Maria, who has a raw score of $X = 64$ in the original distribution, and Joe, whose original raw score is $X = 43$.

STEP 1   Transform each of the original raw scores into z-scores.

For Maria, $X = 64$, so her z-score is

$$z = \frac{X - \mu}{\sigma} = \frac{64 - 57}{14} = +0.5$$

For Joe, $X = 43$, and his z-score is

$$z = \frac{X - \mu}{\sigma} = \frac{43 - 57}{14} = -1.0$$

Remember: the values of μ and σ are for the distribution from which X was taken.

STEP 2   Change each z-score into an X value in the new standardized distribution that has a mean of $\mu = 50$ and a standard deviation of $\sigma = 10$.

Maria's z-score, $z = +0.50$, indicates that she is located above the mean by one-half of a standard deviation. In the new, standardized distribution, this location corresponds to $X = 55$ (above the mean by 5 points).

Joe's z-score, $z = -1.00$, indicates that he is located below the mean by exactly one standard deviation. In the new distribution, this location corresponds to $X = 40$ (below the mean by 10 points).

The results of this two-step transformation process are summarized in Table 5.1. Note that Joe, for example, has exactly the same z-score ($z = -1.00$) in both the original distribution and the new standardized distribution. This means that Joe's position relative to the other students in the class has not changed.

TABLE 5.1

A demonstration of how two individual scores are changed when a distribution is standardized, using the data from Example 5.10.

	Original Scores $\mu = 57$ and $\sigma = 14$		z-Score Location		Standardized Scores $\mu = 50$ and $\sigma = 10$
Maria	$X = 64$	$\rightarrow$	$z = +0.50$	$\rightarrow$	$X = 55$
Joe	$X = 43$	$\rightarrow$	$z = -1.00$	$\rightarrow$	$X = 40$

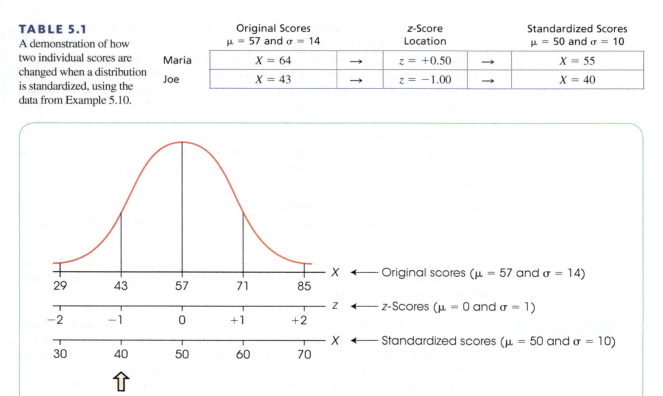

FIGURE 5.9

The distribution of exam scores from Example 5.10. The original distribution was standardized to produce a distribution with $\mu = 50$ and $\sigma = 10$. Note that each individual is identified by an original score, a z-score, and a new, standardized score. For example, Joe has an original score of 43, a z-score of -1.00, and a standardized score of 40.

Figure 5.9 provides another demonstration of the concept that standardizing a distribution does not change the individual positions within the distribution. The figure shows the original exam scores from Example 5.10, with a mean of $\mu = 57$ and a standard deviation of $\sigma = 14$. In the original distribution, Joe is located at a score of $X = 43$. In addition to the original scores, we have included a second scale showing the z-score value for each location in the distribution. In terms of z-scores, Joe is located at a value of $z = -1.00$. Finally, we have added a third scale showing the *standardized scores* where the mean is $\mu = 50$ and the standard deviation is $\sigma = 10$. For the standardized scores, Joe is located at $X = 40$. Note that Joe is always in the same place in the distribution. The only thing that changes is the number that is assigned to Joe: for the original scores, Joe is at 43; for the z-scores, Joe is at -1.00; and for the standardized scores, Joe is at 40.

LEARNING CHECK   LO5 1. A set of scores has a mean of $\mu = 63$ and a standard deviation of $\sigma = 8$. If these scores are standardized so that the new distribution has $\mu = 50$ and $\sigma = 10$, what new value would be obtained for a score of $X = 59$ from the original distribution?

   a. The score would still be $X = 59$.

   b. 45

    c. 46

    d. 55

LO5 2. A distribution with $\mu = 35$ and $\sigma = 8$ is being standardized so that the new mean and standard deviation will be $\mu = 50$ and $\sigma = 10$. When the distribution is standardized, what value will be obtained for a score of $X = 39$ from the original distribution?

    a. $X = 54$

    b. $X = 55$

    c. $X = 1.10$

    d. Impossible to determine without more information.

LO5 3. Using z-scores, a sample with $M = 37$ and $s = 6$ is standardized so that the new mean is $M = 50$ and $s = 10$. How does an individual's z-score in the new distribution compare with his/her z-score in the original sample?

    a. New $z =$ old $z + 13$

    b. New $z = (10/6)$(old z)

    c. New $z =$ old z

    d. Cannot be determined with the information given.

ANSWERS   **1. b**   **2. b**   **3. c**

5-6 | Looking Ahead to Inferential Statistics

LEARNING OBJECTIVE

6. Explain how z-scores can help researchers use the data from a sample to draw inferences about populations.

Recall that inferential statistics are techniques that use the information from samples to answer questions about populations. In later chapters, we will use inferential statistics to help interpret the results from research studies. A typical research study begins with a question about how a treatment will affect the individuals in a population. Because it is usually impossible to study an entire population, the researcher selects a sample and administers the treatment to the individuals in the sample. This general research situation is shown in Figure 5.10. To evaluate the effect of the treatment, the researcher simply compares the treated sample with the original population. If the individuals in the sample are noticeably different from the individuals in the original population, the researcher has evidence that the treatment has had an effect. On the other hand, if the sample is not noticeably different from the original population, it would appear that the treatment has no effect.

    Notice that the interpretation of the research results depends on whether the sample is *noticeably different* from the population. One technique for deciding whether a sample is noticeably different is to use z-scores. For example, an individual with a z-score near 0 is located in the center of the population and would be considered to be a fairly typical or representative individual. However, an individual with an extreme z-score, beyond $+2.00$ or -2.00 for example, would be considered "noticeably different" from most of the individuals in the population. Thus, we can use z-scores to help decide whether the treatment has caused a change. Specifically, if the individuals who receive the treatment finish the

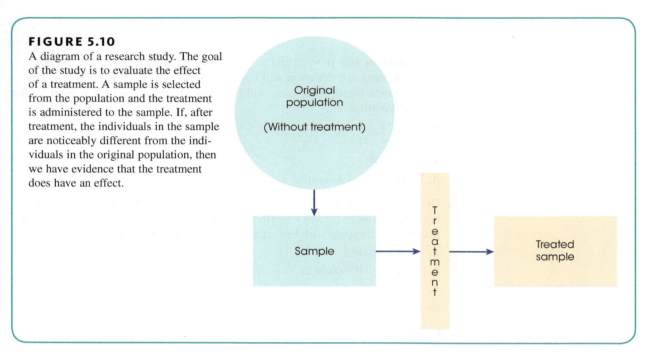

FIGURE 5.10

A diagram of a research study. The goal of the study is to evaluate the effect of a treatment. A sample is selected from the population and the treatment is administered to the sample. If, after treatment, the individuals in the sample are noticeably different from the individuals in the original population, then we have evidence that the treatment does have an effect.

research study with extreme *z*-scores, we can conclude that the treatment does appear to have an effect. The following example demonstrates this process.

EXAMPLE 5.11

A researcher is evaluating the effect of a new growth hormone. It is known that regular adult rats weigh an average of $\mu = 400$ grams. The weights vary from rat to rat, and the distribution of weights is normal with a standard deviation of $\sigma = 20$ grams. The population distribution is shown in Figure 5.11. Note that this is the distribution of weight for regular rats that have not received any special treatment. Next, the researcher selects one newborn rat and injects the rat with the growth hormone. When the rat reaches maturity, it is weighed to determine whether there is any evidence that the hormone has had an effect.

First, assume that the hormone-injected rat weighs $X = 418$ grams. Although this is more than the average nontreated rat ($\mu = 400$ grams), is it convincing evidence that the hormone has had an effect? If you look at the distribution in Figure 5.11, you should realize that a rat weighing 418 grams is not noticeably different from the regular rats that did not receive any hormone injection. Specifically, our injected rat would be located near the center of the distribution for regular rats with a *z*-score of

$$z = \frac{X - \mu}{\sigma} = \frac{418 - 400}{20} = \frac{18}{20} = 0.90$$

Because the injected rat still looks the same as a regular, nontreated rat, the conclusion is that the growth hormone does not appear to have an effect.

Now, assume that our injected rat weighs $X = 450$ grams. In the distribution of regular rats (see Figure 5.11), this animal would have a *z*-score of

$$z = \frac{X - \mu}{\sigma} = \frac{450 - 400}{20} = \frac{50}{20} = 2.50$$

In this case, the hormone-injected rat is substantially bigger than most ordinary rats, and it would be reasonable to conclude that the hormone does have an effect on weight.   ∎

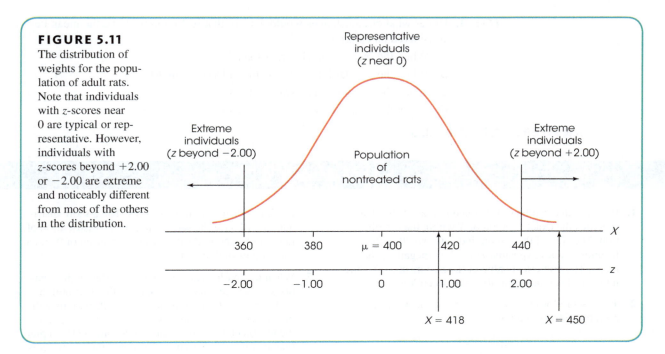

FIGURE 5.11

The distribution of weights for the population of adult rats. Note that individuals with z-scores near 0 are typical or representative. However, individuals with z-scores beyond +2.00 or −2.00 are extreme and noticeably different from most of the others in the distribution.

In the preceding example, we used z-scores to help interpret the results obtained from a sample. Specifically, if the individuals who receive the treatment in a research study have extreme z-scores compared to those who do not receive the treatment, we can conclude that the treatment does appear to have an effect. The example, however, used an arbitrary definition to determine which z-score values are noticeably different. Although it is reasonable to describe individuals with z-scores near 0 as "highly representative" of the population, and individuals with z-scores beyond ±2.00 as "extreme," you should realize that these z-score boundaries were not determined by any mathematical rule. In the following chapter we introduce *probability*, which gives us a rationale for deciding exactly where to set the boundaries.

LEARNING CHECK   **LO6 1.** For the past 20 years, the high temperature on April 15 has averaged $\mu = 60$ degrees with a standard deviation of $\sigma = 4$. Last year, the high temperature was 75 degrees. Based on this information, last year's temperature on April 15 was _____.

   a. a little above average

   b. far above average

   c. above average, but it is impossible to describe how much above average

   d. There is not enough information to compare last year with the average.

LO6 2. A score of $X = 75$ is obtained from a population. Which set of population parameters would make $X = 75$ an extreme, unrepresentative score?

   a. $\mu = 65$ and $\sigma = 8$

   b. $\mu = 65$ and $\sigma = 3$

   c. $\mu = 70$ and $\sigma = 8$

   d. $\mu = 70$ and $\sigma = 3$

LO6 3. Under what circumstances would a score that is 20 points above the mean be considered an extreme score?

  a. When the mean is much larger than 20.

  b. When the standard deviation is much larger than 20.

  c. When the mean is much smaller than 20.

  d. When the standard deviation is much smaller than 20.

ANSWERS **1. b** **2. b** **3. d**

SUMMARY

1. Each X value can be transformed into a z-score that specifies the exact location of X within the distribution. The sign of the z-score indicates whether the location is above (positive) or below (negative) the mean. The numerical value of the z-score specifies the number of standard deviations between X and μ.

2. The z-score formula is used to transform X values into z-scores. For a population:

$$z = \frac{X - \mu}{\sigma}$$

   For a sample:

$$z = \frac{X - M}{s}$$

3. To transform z-scores back into X values, it usually is easier to use the z-score definition rather than a formula. However, the z-score formula can be transformed into a new equation. For a population:

$$X = \mu + z\sigma$$

   For a sample:

$$X = M + zs$$

4. When an entire distribution of scores (either a population or a sample) is transformed into z-scores, the result is a distribution of z-scores. The z-score distribution will have the same shape as the distribution of raw scores, and it always will have a mean of 0 and a standard deviation of 1.

5. When comparing raw scores from different distributions, it is necessary to standardize the distributions with a z-score transformation. The distributions will then be comparable because they will have the same mean (0) and the same standard deviation (1). In practice, it is necessary to transform only those raw scores that are being compared.

6. In certain situations, such as psychological testing, a distribution may be standardized by converting the original X values into z-scores and then converting the z-scores into a new distribution of scores with predetermined values for the mean and the standard deviation.

7. In inferential statistics, z-scores provide an objective method for determining how well a specific score represents its population. A z-score near 0 indicates that the score is close to the population mean and therefore is representative. A z-score beyond $+2.00$ (or -2.00) indicates that the score is extreme and is noticeably different from the other scores in the distribution.

KEY TERMS

raw score (152)

z-score (152)

deviation score (153)

z-score transformation (160)

standardized distribution (163)

standardized score (166)

FOCUS ON PROBLEM SOLVING

1. When you are converting an X value to a z-score (or vice versa), do not rely entirely on the formula. You can avoid careless mistakes if you use the definition of a z-score (sign and numerical value) to make a preliminary estimate of the answer before you begin

computations. For example, a z-score of $z = -0.85$ identifies a score located *below* the mean by almost one standard deviation. When computing the X value for this z-score, be sure that your answer is smaller than the mean, and check that the distance between X and μ is slightly less than the standard deviation.

2. When comparing scores from distributions that have different means and standard deviations, it is important to be sure that you use the correct values in the z-score formula. Use the mean and the standard deviation for the distribution from which the score was taken.

3. Remember that a z-score specifies a relative position within the context of a specific distribution. A z-score is a relative value, not an absolute value. For example, a z-score of $z = -2.0$ does not necessarily suggest a very low raw score—it simply means that the raw score is among the lowest within that specific group.

DEMONSTRATION 5.1

TRANSFORMING X VALUES INTO z-SCORES

A distribution of scores has a mean of $\mu = 60$ with $\mu = 12$. Find the z-score for $X = 75$.

STEP 1   **Determine the sign of the z-score.**   First, determine whether X is above or below the mean. This will determine the sign of the z-score. For this demonstration, X is larger than (above) μ, so the z-score will be positive.

STEP 2   **Convert the distance between X and μ into standard deviation units.**   For $X = 75$ and $\mu = 60$, the distance between X and μ is 15 points. With $\sigma = 12$ points, this distance corresponds to $\frac{15}{12} = 1.25$ standard deviations.

STEP 3   **Combine the sign from Step 1 with the numerical value from Step 2.**   The score is above the mean ($+$) by a distance of 1.25 standard deviations. Thus,

$$z = +1.25$$

STEP 4   **Confirm the answer using the z-score formula.**   For this example, $X = 75$, $\mu = 60$, and $\sigma = 12$.

$$z = \frac{X - \mu}{\sigma} = \frac{75 - 60}{12} = \frac{+15}{12} = +1.25$$

DEMONSTRATION 5.2

CONVERTING z-SCORES TO X VALUES

For a sample with $M = 60$ and $s = 12$, what is the X value corresponding to $z = -0.50$? Notice that in this situation we know the z-score and must find X.

STEP 1   **Locate X in relation to the mean.**   A z-score of -0.50 indicates a location below the mean by half of a standard deviation.

STEP 2   **Convert the distance from standard deviation units to points.**   With $s = 12$, half of a standard deviation is 6 points.

STEP 3   **Identify the X value.**   The value we want is located below the mean by 6 points. The mean is $M = 60$, so the score must be $X = 54$.

General instructions for using SPSS are presented in Appendix D. Following are detailed instructions for using SPSS to **Transform X Values into z-Scores for a Sample**.

Demonstration Example

An employer is interested in identifying the most extroverted people in a pool of job applicants. Each applicant submitted the results of either Extraversion Assessment 1 or Extraversion Assessment 2. The data from the applicants are below.

Extraversion Assessment 1

Applicant	Score
1	17
2	8
3	9
4	9
5	16
6	4
7	5

Extraversion Assessment 2

Applicant	Score
8	149
9	99
10	192
11	61
12	62
13	138
14	184

The employer is tasked with identifying the top five most extraverted individuals from either assessment in the pool of applicants. Importantly, the employer can't select the top five *raw scores* because the raw scores come from different assessments. We will use SPSS to transform extraversion scores to z-scores for the two different assessments.

Data Entry

1. Click the **Variable View** tab to enter information about the variables.
2. In the first row, enter "extScore" (for extraversion score). Fill in the remaining information about your variable where necessary. Be sure that **Type** = "Numeric", **Width** = "8", **Decimals** = "0", **Label** = "Extraversion Score", **Values** = "None", **Missing** = "None", **Columns** = "8", **Align** = "Right", and **Measure** = "Scale".
3. In the second row, enter "assessNum" (for Assessment Number). Fill in the remaining information about your variable where necessary. Be sure that **Type** = "Numeric", **Width** = "8", **Decimals** = "0", **Label** = "Extraversion Score", **Values** = "None", **Missing** = "None", **Columns** = "10", **Align** = "Right", and **Measure** = "Nominal". When you are finished entering information about your variables, the Variable View should be similar to the figure below.

	Name	Type	Width	Decimals	Label	Values	Missing	Columns	Align	Measure	Role
1	extScore	Numeric	8	0	Extraversion Sc...	None	None	8	≣ Right	◆ Scale	↘ Input
2	assessNum	Numeric	8	0	Assessment N...	None	None	8	≣ Right	⬩ Nominal	↘ Input

Source: SPSS®

4. Click the **Data View** tab to return to the data editor and enter your scores. Each row represents a single applicant. Enter the extraversion score in the first column and the assessment type in the second column. The Data View should be as below.

	extScore	assessNum
1	17	1
2	8	1
3	9	1
4	9	1
5	16	1
6	4	1
7	5	1
8	149	2
9	99	2
10	192	2
11	61	2
12	62	2
13	138	2
14	184	2

Source: SPSS®

Data Analysis

1. To program SPSS to treat the scores as coming from different assessments, click **Data** on the tool bar and select **Split File.** Click the "Organize output by groups" button. Select "Assessment Number" in the box and use the arrow to move it to the "Groups Based on:" box. Click **OK.** The output window should confirm that the file has been split into two different groups.

2. Click **Analyze** on the tool bar, select **Descriptive Statistics**, and click on **Descriptives**.

3. Highlight the column label for the set of extraversion scores in the left box and click the arrow to move it into the **Variable** box.

4. Click the box to **Save standardized values as variables** at the bottom of the **Descriptives** screen.

5. Click **OK**.

SPSS Output

The program will produce the usual output display listing the number of scores (N), the maximum and minimum scores, the mean, and the standard deviation for each assessment. However, if you return to the Data Editor by clicking the **Window** tool bar and click the entry that ends with "Data Editor," you will find that SPSS has produced a new column showing the z-score corresponding to each of the original X values (i.e., "ZextScore"). The z-scores will allow you to compare between the two personality assessments and should appear like the following figure.

	extScore	assessNum	ZextScore
1	17	1	1.45025
2	8	1	-.34124
3	9	1	-.14218
4	9	1	-.14218
5	16	1	1.25120
6	4	1	-1.13745
7	5	1	-.93840
8	149	2	.41864
9	99	2	-.50873
10	192	2	1.21617
11	61	2	-1.21352
12	62	2	-1.19498
13	138	2	.21462
14	184	2	1.06780

Source: SPSS®

You should notice that Applicants 1, 5, 8, 10, and 14 had the highest z-scores from their respective assessments.

Caution: The SPSS program computes the z-scores using the sample standard deviation instead of the population standard deviation. If your set of scores is intended to be a population, SPSS will not produce the correct z-score values. You can convert the SPSS values into population z-scores by multiplying each z-score value by $\sqrt{\frac{n}{n-1}}$.

Try It Yourself

Suppose that an instructor wants to identify the five students with the most extremely low or extremely high grades (that is, the five students with grades that are farthest from the mean) in two different statistics classes. Below are the distributions of total points earned from the two classes.

Scores for Statistics Class 1

529   421   485   491   487   558   483   407   651   430   682   637   511

Scores for Statistics Class 2

374   388   199   278   315   303   395   315   293   347   277   335   232

Use SPSS to find the five most extreme values from both classes. You should find that $X = 199$ (Class 2, $z = -1.94$), $X = 682$ (Class 1, $z = +1.82$), $X = 651$ (Class 1, $z = +1.47$), $X = 395$ (Class 2, $z = +1.44$), and $X = 232$ (Class 2, $z = -1.37$) are the five most extreme scores in the two distributions.

PROBLEMS

1. Explain how a z-score identifies an exact location in a distribution with a single number.

2. You are told that the results of an extraversion assessment gave you a z-score of +2.00. What does that mean about your level of extraversion relative to the mean?

3. Identify the letters in the following distribution that correspond to the following z-scores.
   a. $z = +2.00$
   b. $z = 0.00$
   c. $z = -2.00$
   d. $z = -0.50$

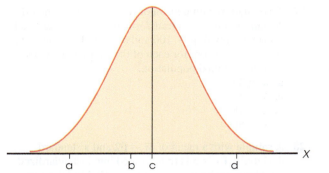

4. A score of $X = 75$ is measured in a population with a mean of $\mu = 100$. A z-score of $z = +1.50$ is calculated. Without knowing the standard deviation, explain why the z-score of $z = +1.50$ is incorrect.

5. For a population with a standard deviation of $\sigma = 10$, find the z-score for each of the following locations in the distribution.
   a. Above the mean by 10 points
   b. Above the mean by 5 points
   c. Below the mean by 20 points
   d. Below the mean by 6 points

6. For a sample with a standard deviation of $s = 15$, describe the location of each of the following z-scores in terms of its position relative to the mean. For example, $z = +1.00$ is a location that is 15 points above the mean.
   a. $z = -1.20$
   b. $z = +0.80$
   c. $z = +2.00$
   d. $z = -1.60$

7. For a population with $\mu = 50$ and $\sigma = 6$:
   a. Find the z-score for each of the following X values.

$X = 50$	$X = 62$	$X = 53$
$X = 44$	$X = 47$	$X = 38$

   b. Find the score (X value) that corresponds to each of the following z-scores.

$z = +1.00$	$z = +2.50$	$z = +1.50$
$z = -1.50$	$z = -3.00$	$z = -2.50$

8. For a population with $\mu = 80$ and $\sigma = 9$, find the z-score for each of the following X values.

$X = 83$	$X = 75$	$X = 91$
$X = 67$	$X = 85$	$X = 68$

9. A sample has a mean of $M = 90$ and a standard deviation of $s = 20$.
   a. Find the z-score for each of the following X values.

$X = 95$	$X = 98$	$X = 105$
$X = 80$	$X = 88$	$X = 76$

b. Find the X value for each of the following z-scores.

$z = -1.00$	$z = +0.50$	$z = -1.50$
$z = +0.75$	$z = -1.25$	$z = +2.60$

10. A sample has a mean of $M = 53$ and a standard deviation of $s = 11$. For this sample, find the z-score for each of the following X values.

$X = 30$	$X = 39$	$X = 70$
$X = 48$	$X = 57$	$X = 64$

11. You receive your midterm exam scores for four classes. In all four classes, your midterm score was $X = 70$. Below is the mean and standard deviation for each of your classes. Compute the z-score for each midterm exam score and summarize in words how you performed relative to other students in the class.
   a. Class 1: $\mu = 70$ and $\sigma = 10$
   b. Class 2: $\mu = 50$ and $\sigma = 20$
   c. Class 3: $\mu = 80$ and $\sigma = 5$
   d. Class 4: $\mu = 80$ and $\sigma = 20$

12. Find the X value corresponding to $z = 0.75$ for each of the following distributions.
   a. $\mu = 90$ and $\sigma = 4$
   b. $\mu = 90$ and $\sigma = 8$
   c. $\mu = 90$ and $\sigma = 12$
   d. $\mu = 90$ and $\sigma = 20$

13. Find the z-score corresponding to $X = 105$ and the X value corresponding to $z = +0.40$ for each of the following populations.
   a. $\mu = 100$ and $\sigma = 12$
   b. $\mu = 100$ and $\sigma = 4$
   c. $\mu = 80$ and $\sigma = 14$
   d. $\mu = 80$ and $\sigma = 6$

14. Find the z-score corresponding to $X = 24$ and the X value corresponding to $z = +1.50$ for each of the following samples.
   a. $M = 20$ and $s = 12$
   b. $M = 20$ and $s = 4$
   c. $M = 30$ and $s = 8$
   d. $M = 30$ and $s = 10$

15. A score that is 20 points below the mean corresponds to a z-score of $z = -0.50$. What is the population standard deviation?

16. A score that is 10 points above the mean corresponds to a z-score of $z = +1.20$. What is the sample standard deviation?

17. For a population with a standard deviation of $\sigma = 4$, a score of $X = 24$ corresponds to $z = -1.50$. What is the population mean?

18. For a population with a standard deviation of $\sigma = 12$, a score of $X = 115$ corresponds to $z = +1.25$. What is the population mean?

19. For a population with a mean of $\mu = 45$, a score of $X = 54$ corresponds to $z = +1.50$. What is the population standard deviation?

20. For a sample with a standard deviation of $s = 4$, a score of $X = 35$ corresponds to $z = -1.25$. What is the sample mean?

21. For a sample with a mean of $M = 63$, a score of $X = 54$ corresponds to $z = -0.75$. What is the sample standard deviation?

22. In a population distribution, a score of $X = 56$ corresponds to $z = -0.40$ and a score of $X = 70$ corresponds to $z = +1.00$. Find the mean and standard deviation for the population. (*Hint*: Sketch the distribution and locate the two scores on your sketch.)

23. In a sample distribution, a score of $X = 21$ corresponds to $z = -1.00$ and a score of $X = 12$ corresponds to $z = -2.50$. Find the mean and standard deviation for the sample.

24. The Graduate Record Exam is a standardized test taken by many graduating college seniors. GRE scores are often included in applications to masters or doctoral programs. Suppose that a psychology major received a score of $X = 160$ on the verbal section of the GRE and a score of $X = 159$ on the quantitative section of the GRE. The mean verbal GRE score for the years 2014–2017 among psychology majors was $M = 152$, $s = 7$. The mean quantitative score for the same years was $M = 149$, $s = 7$. Use z-scores to describe the student's performance relative to other psychology majors. On which section, verbal or quantitative, was the student's performance better?

25. For each of the following, identify the exam score that should lead to the better grade. In each case, explain your answer.
 a. A score of $X = 70$ on an exam with $\mu = 82$ and $\sigma = 8$; or a score of $X = 60$ on an exam with $\mu = 72$ and $\sigma = 12$.
 b. A score of $X = 58$ on an exam with $\mu = 49$ and $\sigma = 6$; or a score of $X = 85$ on an exam with $\mu = 70$ and $\sigma = 10$.
 c. A score of $X = 32$ on an exam with $\mu = 24$ and $\sigma = 4$; or a score of $X = 26$ on an exam with $\mu = 20$ and $\sigma = 2$.

26. Your professor tells you that all exam scores were transformed to z-scores for the midterm examination. Describe the mean and standard deviation of the resulting distribution of z-scores.

27. A population with a mean of $\mu = 41$ and a standard deviation of $\sigma = 4$ is transformed into a standardized distribution with $\mu = 100$ and $\sigma = 20$. Find the new, standardized score for each of the following values from the original population.
 a. $X = 39$
 b. $X = 36$
 c. $X = 45$
 d. $X = 50$

28. A sample with a mean of $M = 62$ and a standard deviation of $s = 5$ is transformed into a standardized distribution with $\mu = 50$ and $\sigma = 10$. Find the new, standardized score for each of the following values from the original population.
 a. $X = 61$
 b. $X = 55$
 c. $X = 65$
 d. $X = 74$

29. A population consists of the following $N = 7$ scores: 6, 1, 0, 7, 4, 13, 4.
 a. Compute μ and σ for the population.
 b. Find the z-score for each score in the population.
 c. Transform the original population into a new population of $N = 7$ scores with a mean of $\mu = 50$ and a standard deviation of $\sigma = 20$.

30. A sample consists of the following $n = 5$ scores: 8, 4, 10, 0, 3.
 a. Compute the mean and standard deviation for the sample.
 b. Find the z-score for each score in the sample.
 c. Transform the original sample into a new sample with a mean of $M = 100$ and $s = 20$.

31. A researcher is interested in the effects of a "smart drug" on performance on a standardized intelligence test with a mean of $\mu = 200$ and a standard deviation of $\sigma = 50$. Suppose that a participant who receives the smart drug subsequently earns a score of $X = 220$ on the intelligence test. Should the researcher be convinced that the participant's performance is appreciably different from the mean? Explain your answer.

32. For each of the following populations, would a score of $X = 85$ be considered a central score (near the middle of the distribution) or an extreme score (far out in the tail of the distribution)?
 a. $\mu = 75$ and $\sigma = 15$
 b. $\mu = 80$ and $\sigma = 2$
 c. $\mu = 90$ and $\sigma = 20$
 d. $\mu = 93$ and $\sigma = 3$

Probability

Tools You Will Need

The following items are considered essential background material for this chapter. If you doubt your knowledge of any of these items, you should review the appropriate chapter or section before proceeding.

- Proportions (math review, Appendix A)
  - Fractions
  - Decimals
  - Percentages
- Basic algebra (math review, Appendix A)
- z-Scores (Chapter 5)

clivewa/Shutterstock.com

PREVIEW

The insurance industry and government agencies gather data on injuries suffered by U.S. citizens, including those that result in accidental deaths. This might seem like a somewhat morbid way to begin a discussion of probability, but some basic points about probability and risk can be made from the data that have been collected. Table 6.1 shows the number of accidental deaths for several causes.

From these data one can determine the probability of accidental death for the reported causes. The data in Table 6.1 are based on the 2014 population of the United States of almost 318.9 million people. The probability, or risk, of death for an accidental cause is determined by a fraction:

$$\text{probability} = \frac{\text{number of fatalities}}{\text{total number in relevant population}}$$

Consider the probability of death by all motor vehicle accidents (including motorcyclists, passengers, pedestrians, and so on). For example, there were 35,398 motor vehicle fatalities in 2014. When divided by the number of people in the United States, we obtain

$$\text{probability} = \frac{35,398}{318,857,050} = .000111015265$$

Notice that probabilities are often expressed as a proportion but may also be described as a fraction and a percentage. As a percentage, one multiplies the preceding proportion by 100 and it would be reported as a 0.0111% chance of dying by a motor vehicle mishap. As a fraction, the reciprocal of the proportion, 1/proportion, results in a yearly risk of 1 in 9,008 for motor vehicle fatalities.

Many people are under the impression that air travel seems like a dangerous way to get to a destination. Airplane crashes get lots of media attention, sometimes with graphic news reports. Adding to this perception,

the Smithsonian Channel has a series, *Air Disasters*, which dramatizes official investigations of airplane crashes. Yet, Table 6.1 shows that there were only 412 fatalities in 2014, and these include small private airplanes. However, is the U.S. population the appropriate number to use in determining the probability of dying by air travel? One might argue that more people travel to and from their destinations by motor vehicles than by airplane. Consider the fact that there were nearly 912.5 million passengers on all U.S. domestic flights in 2015, according to the Federal Aviation Administration (2017). Yes, this number is much more than the U.S. population—but many passengers are frequent flyers, many might have taken one or more connecting flights on each trip, and most have to fly back to their original destination. Each time they board the airplane, they are counted as a passenger. Additionally, this figure is not surprising when you consider that every day in 2015 there were approximately 43,000 domestic flights and 2.5 million passengers. So, the probability of death by airline accident on domestic flights might more accurately be based on the population of airplane passengers per year, not on the U.S. population. Based on number of passengers, the yearly risk of death by airplane accident would be

$$\text{probability} = \frac{412}{912,485,113} = 0.0000004515 \text{ or}$$
$$0.000045\%$$

or a yearly risk of 1 in 2,214,839.

If you now realize that accidental death by air travel is highly unlikely, consider the probability of death by lightning strikes for the U.S. population. According to Table 6.1 there were 25 deaths by lightning strikes in the United States for 2014. Given the size of the U.S. population, the probability would be

$$\text{probability} = \frac{25}{318,857,050} = 0.0000000784$$

or a 1 in 12,754,282 yearly risk. This does not mean it is safe to stand under a tree in a thunderstorm. If everyone did that, the probability of death by lightning would be much greater. According to *Live Science* (2011), Florida leads the way, with an average of 1.45 million lightning strikes per year. Now imagine if everyone in the Sunshine State stood outside during thunderstorms. It is best to be safe and stay indoors.

TABLE 6.1

The number of deaths in the United States by cause of injury for 2014.

Cause of Death	Number of Deaths
All motor vehicle accidents	35,398
Air transport accidents	412
Lightning strikes	25

Source: Insurance Information Institute (2015).

In this chapter, we will introduce the concept of probability, examine how it is applied in several different situations, and discuss its general role in the field of statistics. We will also examine the normal distribution and how it is used to answer questions about proportion and probability. Finally, we will take a look ahead at the role of probability in inferential statistics.

6-1 Introduction to Probability

> **LEARNING OBJECTIVE**
>
> **1.** Define probability and calculate (from information provided or from a frequency distribution graph) the probability of a specific outcome as a proportion, decimal, and percentage.

In Chapter 1, we introduced the idea that research studies begin with a general question about an entire population, but the actual research is conducted using a sample. In this situation, the role of inferential statistics is to use the sample data as the basis for answering questions about the population. To accomplish this goal, inferential procedures are typically built around the concept of probability. Specifically, the relationships between samples and populations are usually defined in terms of probability.

For example, suppose you are selecting a single marble from a jar that contains 50 black marbles and 50 white marbles. (In this example, the jar of marbles is the *population* and the single marble to be selected is the *sample*.) Although you cannot guarantee the exact outcome of your sample, it is possible to talk about the potential outcomes in terms of probabilities. In this case, you have a 50-50 chance of getting either color. Now consider another jar (population) that has 90 black marbles and only 10 white marbles. Again, you cannot specify the exact outcome of a sample, but now you know that the sample probably will be a black marble. By knowing the makeup of a population, we can determine the probability of obtaining specific samples. In this way, probability gives us a connection between populations and samples, and this connection is the foundation for the inferential statistics to be presented in the chapters that follow.

You may have noticed that the preceding examples begin with a population and then use probability to describe the samples that could be obtained. This is exactly backward from what we want to do with inferential statistics. Remember that the goal of inferential statistics is to begin with a sample and then answer a general question about the population. We reach this goal in a two-stage process. In the first stage, we develop probability as a bridge from populations to samples. This stage involves identifying the types of samples that probably would be obtained from a specific population. Once this bridge is established, we simply reverse the probability rules to allow us to move from samples to populations (Figure 6.1). The process of reversing the probability relationship can be demonstrated by considering again the two jars of marbles we looked at earlier (Jar 1 has 50 black and 50 white marbles; Jar 2 has 90 black and only 10 white marbles). This time, suppose you are blindfolded when the sample is selected, so you do not know which jar is being used. Your task is to look at the sample that you obtain and then decide which jar is most likely. If you select a sample of $n = 4$ marbles and all are black, which jar would you choose? It should be clear that it would be relatively unlikely (low probability) to obtain this sample from Jar 1; in four draws, you almost certainly would get at least one white marble. On the other hand, this sample would have a high probability of coming from Jar 2, where nearly all the marbles are black. Your decision therefore is that the sample probably came from Jar 2. Note that you are now using the sample to make an inference about the population.

FIGURE 6.1

The role of probability in inferential statistics. Probability is used to predict the type of samples that are likely to be obtained from a population. Thus, probability establishes a connection between samples and populations. Inferential statistics rely on this connection when they use sample data as the basis for making conclusions about populations.

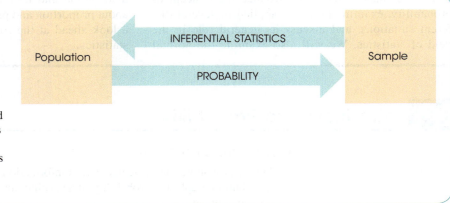

You should also notice that it is not *impossible* to obtain a sample of $n = 4$ black marbles from Jar 1. Thus, there is some uncertainty in your decision. In statistics, uncertainty is usually expressed as a statement about probability.

It may appear that selecting marbles from a jar has nothing to do with interpreting research results in the behavioral sciences, but the same principles apply. For example, suppose that a psychologist gives an anxiety questionnaire to a sample of students during final exams and obtains a sample mean of $M = 20$. Based on this result, we can conclude that the sample is more likely to have come from a population with a mean near $\mu = 20$ than from a population with a mean that is not near 20.

■ Defining Probability

Probability is a huge topic that extends far beyond the limits of introductory statistics, and we will not attempt to examine it all here. Instead, we concentrate on the few concepts and definitions that are needed for an introduction to inferential statistics. We begin with a relatively simple definition of *probability*.

> For a situation in which several different outcomes are possible, the **probability** for any specific outcome is defined as a fraction or a proportion of all the possible outcomes. If the possible outcomes are identified as A, B, C, D, and so on, then
>
> $$\text{probability of } A = \frac{\text{number of outcomes classified as } A}{\text{total number of possible outcomes}}$$

For example, if you are selecting a card from a complete deck, there are 52 possible outcomes. The probability of selecting the king of hearts is $p = \frac{1}{52}$. The probability of selecting an ace is $p = \frac{4}{52}$ because there are 4 aces in the deck.

To simplify the discussion of probability, we use a notation system that eliminates a lot of the words. The probability of a specific outcome is expressed with a p (for probability), followed by the specific outcome in parentheses. For example, the probability of selecting a king from a deck of cards is written as $p(\text{king})$. The probability of obtaining heads for a coin toss is written as $p(\text{heads})$.

Note that probability is defined as a proportion, or a part of the whole. This definition makes it possible to restate any probability problem as a proportion problem. For example, the probability problem "What is the probability of selecting a king from a deck of cards?"

Typically, we use proportions to summarize previous observations, and probability is used to predict future, uncertain outcomes.

can be restated as "What proportion of the whole deck consists of kings?" In each case, the answer is $\frac{4}{52}$, or "4 out of 52." This translation from probability to proportion may seem trivial now, but it will be a great aid when the probability problems become more complex. In most situations, we are concerned with the probability of obtaining a particular sample from a population. The terminology of *sample* and *population* will not change the basic definition of probability. For example, the whole deck of cards can be considered as a population, and the single card we select is the sample.

Probability Values   The definition we are using identifies probability as a fraction or a proportion. If you work directly from this definition, the probability values you obtain are expressed as fractions. For example, if you are selecting a card at random,

$$p(\text{spade}) = \frac{13}{52} = \frac{1}{4}$$

Or if you are tossing a coin,

$$p(\text{heads}) = \frac{1}{2}$$

If you are unsure how to convert from fractions to decimals or percentages, you should review the section on proportions in the math review, Appendix A.

You should be aware that these fractions can be expressed equally well as either decimals or percentages:

$$p = \frac{1}{4} = 0.25 = 25\%$$

$$p = \frac{1}{2} = 0.50 = 50\%$$

By convention, probability values most often are expressed as decimal values. But you should realize that any of these three forms is acceptable.

You also should note that all the possible probability values are contained in a limited range. At one extreme, when an event never occurs, the probability is zero, or 0%. At the other extreme, when an event always occurs, the probability is 1, or 100%. Thus, all probability values are contained in a range from 0 to 1. For example, suppose you have a jar containing 10 white marbles. The probability of randomly selecting a marble of any color other than white is

$$p(\text{any other color}) = \frac{0}{10} = 0$$

The probability of selecting a white marble is

$$p(\text{white}) = \frac{10}{10} = 1$$

■ Random Sampling

For the preceding definition of probability to be accurate, it is necessary that the outcomes be obtained by a process called *random sampling*.

> **Random sampling** requires that each individual in the population has an *equal chance* of being selected. A sample obtained by this process is called a **simple random sample**.

A second requirement, necessary for many statistical formulas, states that if more than one individual is being selected, the probabilities must *stay constant* from one selection to the next. Adding this second requirement produces what is called *independent random sampling*.

The term *independent* refers to the fact that the probability of selecting any particular individual is not influenced by the individuals already selected for the sample. For example, the probability that you will be selected is constant and does not change even when other individuals are selected before you are.

Because an independent random sample is usually a required component for most statistical applications, we will always assume that this is the sampling method being used. To simplify discussion, we will typically omit the word "independent" and simply refer to this sampling technique as *random sampling*. However, you should always assume that both requirements (equal chance and constant probability) are part of the selection process. Samples that are obtained using this technique are called *independent random samples* or simply *random samples*.

> **Independent random sampling** requires that each individual has an equal chance of being selected and that the probability of being selected stays constant from one selection to the next if more than one individual is selected. A sample obtained with this technique is called an **independent random sample**, or simply a **random sample**.

Each of the two requirements for random sampling has some interesting consequences. The first assures that there is no bias in the selection process. For a population with N individuals, each individual must have the same probability, $p = \frac{1}{N}$, of being selected. This means, for example, that you would not get a random sample of people in your city by selecting names from a yacht club membership list. Similarly, you would not get a random sample of college students by selecting individuals from your psychology classes. You also should note that the first requirement of random sampling prohibits you from applying the definition of probability to situations in which the possible outcomes are not equally likely. Consider, for example, the question of whether you will win $1 million in the lottery tomorrow. There are only two possible alternatives:

1. You will win.
2. You will not win.

According to our simple definition, the probability of winning would be one out of two, or $p = \frac{1}{2}$. However, the two alternatives are not equally likely, so the simple definition of probability does not apply.

The second requirement also is more interesting than may be apparent at first glance. Consider, for example, the selection of $n = 2$ cards from a complete deck. For the first draw, the probability of obtaining the jack of diamonds is

$$p(\text{jack of diamonds}) = \frac{1}{52}$$

After selecting one card for the sample, you are ready to draw the second card. What is the probability of obtaining the jack of diamonds this time? Assuming that you still are holding the first card, there are two possibilities:

$$p(\text{jack of diamonds}) = \frac{1}{51} \text{ if the first card was not the jack of diamonds}$$

or

$$p(\text{jack of diamonds}) = 0 \text{ if the first card was the jack of diamonds}$$

In either case, the probability is different from its value for the first draw. This contradicts the requirement for random sampling, which states that the probability must stay constant. To keep the probabilities from changing from one selection to the next, it is necessary to return each individual to the population before you make the next selection. This process is called *sampling with replacement*. The second requirement for random samples (constant probability) demands that you sample with replacement.

(*Note*: We are using a definition of random sampling that requires equal chance of selection and constant probabilities. This kind of sampling is also known as independent random sampling, and often is called *random sampling with replacement*. Many of the statistics we will encounter later are founded on this kind of sampling. However, you should realize that other definitions exist for the concept of random sampling. In particular, it is very common to define random sampling without the requirement of constant probabilities—that is, *random sampling without replacement*. In addition, there are many different sampling techniques that are used when researchers are selecting individuals to participate in research studies.)

■ Probability and Frequency Distributions

The situations in which we are concerned with probability usually involve a population of scores that can be displayed in a frequency distribution graph. If you think of the graph as representing the entire population, then different portions of the graph represent different portions of the population. Because probabilities and proportions are equivalent, a particular portion of the graph corresponds to a particular probability in the population. Thus, whenever a population is presented in a frequency distribution graph, it will be possible to represent probabilities as proportions of the graph. The relationship between graphs and probabilities is demonstrated in the following example.

EXAMPLE 6.1   We will use a very simple population that contains only $N = 10$ scores with values 1, 1, 2, 3, 3, 4, 4, 4, 5, 6. This population is shown in the frequency distribution graph in Figure 6.2. If you are taking a random sample of $n = 1$ score from this population, what is the probability of obtaining an individual with a score greater than 4? In probability notation,

$$p(X > 4) = ?$$

Using the definition of probability, there are 2 scores that meet this criterion out of the total group of $N = 10$ scores, so the answer would be $p = \frac{2}{10}$. This answer can be obtained directly from the frequency distribution graph if you recall that probability and proportion measure the same thing. Looking at the graph (see Figure 6.2), what proportion of the population consists of scores greater than 4? The answer is the shaded part of the distribution—that is, 2 squares out of the total of 10 squares in the distribution. Notice that we now are defining probability as a proportion of *area* in the frequency distribution graph. This provides a very concrete and graphic way of representing probability.

Using the same population once again, what is the probability of selecting an individual with a score of less than 5? In symbols,

$$p(X < 5) = ?$$

Going directly to the distribution in Figure 6.2, we now want to know what part of the graph is not shaded. The unshaded portion consists of 8 out of the 10 blocks ($\frac{8}{10}$ of the area of the graph), so the answer is $p = \frac{8}{10}$. ■

FIGURE 6.2

A frequency distribution histogram for a population of $N = 10$ scores. The shaded part of the figure indicates the portion of the whole population that corresponds to scores greater than $X = 4$. The shaded portion is two-tenths $\left(\frac{2}{10}\right)$ of the whole distribution.

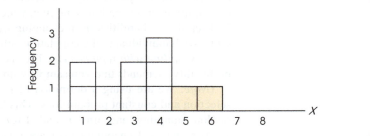

LEARNING CHECK   **LO1**   **1.** An introductory psychology class with $n = 44$ students has 20 freshmen, 14 sophomores, 2 juniors, and 8 seniors. If one student is randomly selected from this class, what is the probability of getting a sophomore?

    a. $\frac{8}{24}$

    b. $\frac{20}{24}$

    c. $\frac{20}{44}$

    d. $\frac{14}{44}$

    LO1   **2.** A jar contains 10 Snickers bars and 20 Hershey bars. If one candy bar is selected from this jar, what is the probability that it will be a Snickers bar?

    a. $\frac{1}{30}$

    b. $\frac{1}{20}$

    c. $\frac{10}{30}$

    d. $\frac{10}{20}$

    LO1   **3.** Random sampling requires sampling with replacement. What is the goal of sampling with replacement?

    a. It ensures that every individual has an equal chance of selection.

    b. It ensures that the probabilities stay constant from one selection to the next.

    c. It ensures that the same individual is not selected twice.

    d. All of the other options are goals of sampling with replacement.

ANSWERS   **1. d   2. c   3. b**

6-2   Probability and the Normal Distribution

LEARNING OBJECTIVE

2. Use the unit normal table to find the following: (1) proportions/probabilities for specific z-score values, and (2) z-score locations that correspond to specific proportions.

The normal distribution was first introduced in Chapter 2 as an example of a commonly occurring shape for population distributions. An example of a normal distribution is shown in Figure 6.3.

FIGURE 6.3
The normal distribution. The exact shape of the normal distribution is specified by an equation relating each *X* value (score) with each *Y* value (frequency). The equation is

$$Y = \frac{1}{\sqrt{2\pi\sigma^2}}e^{-(X-\mu)^2/2\sigma^2}$$

(π and *e* are mathematical constants). In simpler terms, the normal distribution is symmetrical with a single mode in the middle. The frequency tapers off as you move farther from the middle in either direction.

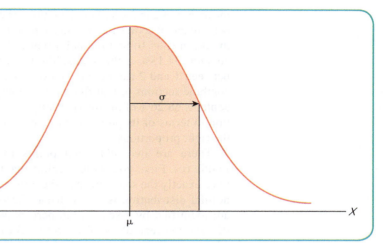

Note that the normal distribution is symmetrical, with the highest frequency in the middle and frequencies tapering off as you move toward either extreme. Although the exact shape for the normal distribution is defined by an equation (see Figure 6.3), the normal shape can also be described by the proportions of area contained in each section of the distribution. Statisticians often identify sections of a normal distribution by using *z*-scores. Figure 6.4 shows a normal distribution with several sections marked in *z*-score units. You should recall that *z*-scores measure positions in a distribution in terms of standard deviations from the mean. (Thus, *z* = +1 is 1 standard deviation above the

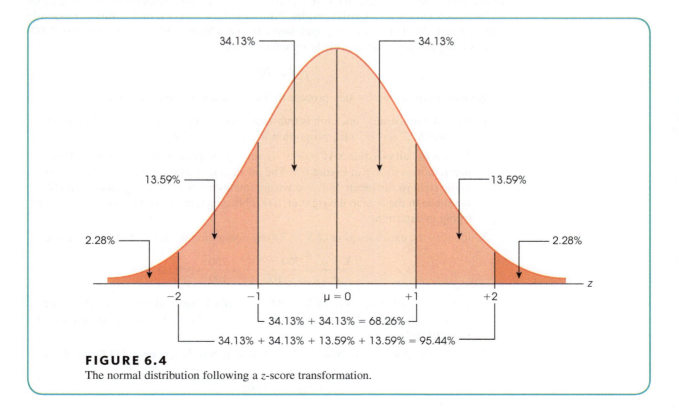

FIGURE 6.4
The normal distribution following a *z*-score transformation.

mean, $z = +2$ is 2 standard deviations above the mean, and so on.) The graph shows the percentage of scores that fall in each of these sections. For example, the section between the mean ($z = 0$) and the point that is 1 standard deviation above the mean ($z = +1$) contains 34.13% of the scores. Similarly, 13.59% of the scores are located in the section between 1 and 2 standard deviations above the mean. We also see that it is possible to combine sections of that figure. For example, the sections between $z = -1$ and $z = +1$ contains 68.26% of the distribution. In this way it is possible to define a normal distribution in terms of its proportions; that is, a distribution is normal if and only if it has all the right proportions.

There are two additional points to be made about the distribution shown in Figure 6.4. First, you should realize that the sections on the left side of the distribution have exactly the same areas as the corresponding sections on the right side because the normal distribution is symmetrical. Second, because the locations in the distribution are identified by z-scores, the percentages shown in the figure apply to *any normal distribution* regardless of the values for the mean and standard deviation. Remember: when any distribution is transformed into z-scores, the mean becomes zero and the standard deviation becomes one.

Because the normal distribution is a good model for many naturally occurring distributions and because this shape is guaranteed in some circumstances (as you will see in Chapter 7), we devote considerable attention to this particular distribution. The process of answering probability questions about a normal distribution is introduced in the following example.

EXAMPLE 6.2   Suppose the population distribution of Math SAT scores is normal with a mean of $\mu = 500$ and a standard deviation of $\sigma = 100$. Given this information about the population and the known proportions for a normal distribution (see Figure 6.4), we can determine the probabilities associated with specific samples. For example, what is the probability of randomly selecting an individual from this population who has a Math SAT score greater than 700?

Restating this question in probability notation, we get

$$p(X > 700) = ?$$

We will follow a step-by-step process to find the answer to this question.

1. First, the probability question is translated into a proportion question: Out of all possible SAT scores, what proportion is greater than 700?

2. The set of "all possible SAT scores" is simply the population distribution. This population is shown in Figure 6.5. The mean is $\mu = 500$, so the score $X = 700$ is to the right of the mean. Because we are interested in all scores greater than 700, we shade in the area to the right of 700. This area represents the proportion we are trying to determine.

3. Identify the exact position of $X = 700$ by computing a z-score. For this example,

$$z = \frac{X - \mu}{\sigma} = \frac{700 - 500}{100} = \frac{200}{100} = +2.00$$

That is, a Math SAT score of $X = 700$ is exactly 2 standard deviations above the mean and corresponds to a z-score of $z = +2.00$. We have also located this z-score in Figure 6.5.

4. The proportion we are trying to determine may now be expressed in terms of its z-score:

$$p(z > +2.00) = ?$$

FIGURE 6.5
The distribution of Math SAT
scores described in Example 6.2.

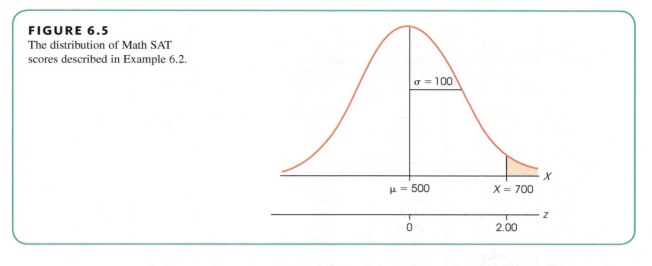

According to the proportions shown in Figure 6.4, all normal distributions, regardless of the values for μ and σ, will have 2.28% of the scores in the tail beyond $z = +2.00$. Thus, for the population of SAT scores,

$$p(X > 700) = p(z > +2.00) = 2.28\%$$

■ The Unit Normal Table

Before we attempt any more probability questions, we must introduce a more useful tool than the graph of the normal distribution shown in Figure 6.4. The graph shows proportions for only a few selected z-score values. A more complete listing of z-scores and proportions is provided in the *unit normal table*. This table lists proportions of the normal distribution for a full range of possible z-score values.

The complete unit normal table is provided in Appendix B, Table B.1, and part of the table is reproduced in Figure 6.6. Notice that the table is structured in a four-column format. The first column (A) lists z-score values corresponding to different positions in a normal distribution. If you imagine a vertical line drawn through a normal distribution, then the exact location of the line can be described by one of the z-score values listed in column A. You should also realize that a vertical line separates the distribution into two sections: a larger section called the *body* and a smaller section called the *tail*. Columns B and C in the table identify the proportion of the distribution in each of the two sections. Column B presents the proportion in the body (the larger portion), and column C presents the proportion in the tail. Finally, we have added a fourth column, column D, which identifies the proportion of the distribution that is located *between* the mean and the z-score.

We will use the distribution in Figure 6.7(a) to help introduce the unit normal table. The figure shows a normal distribution with a vertical line drawn at $z = +0.25$. Using the portion of the table shown in Figure 6.6, find the row in the table that contains $z = 0.25$ in column A. Reading across the row, you should find that the line at $z = +0.25$ separates the distribution into two sections, with the larger section (the body) containing 0.5987 or 59.87% of the distribution and the smaller section (the tail) containing 0.4013 or 40.13% of the distribution. Also, there is exactly 0.0987 or 9.87% of the distribution between the mean and $z = +0.25$.

To make full use of the unit normal table, there are a few facts to keep in mind:

1. The *body* always corresponds to the larger part of the distribution whether it is on the right-hand or left-hand side. Similarly, the *tail* is always the smaller section whether it is on the right or left.

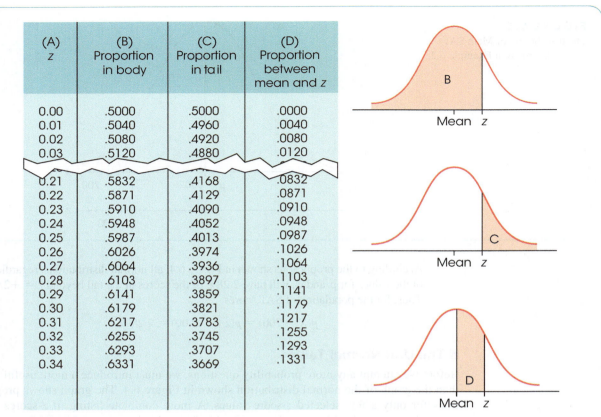

(A) z	(B) Proportion in body	(C) Proportion in tail	(D) Proportion between mean and z
0.00	.5000	.5000	.0000
0.01	.5040	.4960	.0040
0.02	.5080	.4920	.0080
0.03	.5120	.4880	.0120
0.21	.5832	.4168	.0832
0.22	.5871	.4129	.0871
0.23	.5910	.4090	.0910
0.24	.5948	.4052	.0948
0.25	.5987	.4013	.0987
0.26	.6026	.3974	.1026
0.27	.6064	.3936	.1064
0.28	.6103	.3897	.1103
0.29	.6141	.3859	.1141
0.30	.6179	.3821	.1179
0.31	.6217	.3783	.1217
0.32	.6255	.3745	.1255
0.33	.6293	.3707	.1293
0.34	.6331	.3669	.1331

FIGURE 6.6
A portion of the unit normal table. This table lists proportions of the normal distribution corresponding to each z-score value. Column A of the table lists z-scores. Column B lists the proportion in the body of the normal distribution for that z-score value, and Column C lists the proportion in the tail of the distribution. Column D lists the proportion between the mean and the z-score.

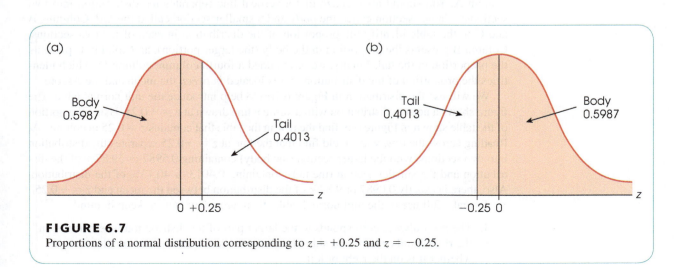

FIGURE 6.7
Proportions of a normal distribution corresponding to z = +0.25 and z = −0.25.

2. Because the normal distribution is symmetrical, the proportions on the right-hand side are exactly the same as the corresponding proportions on the left-hand side. Earlier, for example, we used the unit normal table to obtain proportions for $z = +0.25$. Figure 6.7(b) shows the same proportions for $z = -0.25$. For a negative z-score, however, notice that the tail of the distribution is on the left side and the body is on the right. For a positive z-score [Figure 6.7(a)], the positions are reversed. However, the proportions in each section are exactly the same, with 0.5987 in the body and 0.4013 in the tail. Once again, the table does not list negative z-score values. To find proportions for negative z-scores, you must look up the corresponding proportions for the positive value of z.

3. Although the z-score values change signs ($+$ and $-$) from one side to the other, the proportions are always positive. Thus, column C in the table always lists the proportion in the tail whether it is the right-hand or left-hand tail.

■ Probabilities, Proportions, and z-Scores

The unit normal table lists relationships between z-score locations and proportions in a normal distribution. For any z-score location, you can use the table to look up the corresponding proportions. Similarly, if you know the proportions, you can use the table to find the specific z-score location. Because we have defined probability as equivalent to proportion, you can also use the unit normal table to look up probabilities for normal distributions. The following examples demonstrate a variety of different ways that the unit normal table can be used.

Finding Proportions/Probabilities for Specific z-Score Values   For each of the following examples, we begin with a specific z-score value and then use the unit normal table to find probabilities or proportions associated with the z-score.

EXAMPLE 6.3A

What proportion of the normal distribution corresponds to z-score values greater than $z = 1.00$? First, you should sketch the distribution and shade in the area you are trying to determine. This is shown in Figure 6.8(a). In this case, the shaded portion is the tail of the distribution beyond $z = 1.00$. To find this shaded area, you simply look for $z = 1.00$ in column A to find the appropriate row in the unit normal table. Then scan across the row to column C (tail) to find the proportion. Using the table in Appendix B, you should find that the answer is 0.1587.

You also should notice that this same problem could have been phrased as a probability question. Specifically, we could have asked, "For a normal distribution, what is the probability of selecting a z-score value greater than $z = +1.00$?" Again, the answer is $p(z > 1.00) = 0.1587$ (or 15.87%).   ■

EXAMPLE 6.3B

For a normal distribution, what is the probability of selecting a z-score less than $z = 1.50$? In symbols, $p(z < 1.50) = ?$ Our goal is to determine what proportion of the normal distribution corresponds to z-scores less than 1.50. A normal distribution is shown in Figure 6.8(b) and $z = 1.50$ is located in the distribution. Note that we have shaded all the values to the left of (less than) $z = 1.50$. This is the portion we are trying to find. Clearly the shaded portion is more than 50%, so it corresponds to the body of the distribution. Therefore, find $z = 1.50$ in column A of the unit normal table and read across the row to obtain the proportion from column B. The answer is $p(z < 1.50) = 0.9332$ (or 93.32%).   ■

EXAMPLE 6.3C

Many problems require that you find proportions for negative z-scores. For example, what proportion of the normal distribution is contained in the tail beyond $z = -0.50$? That is,

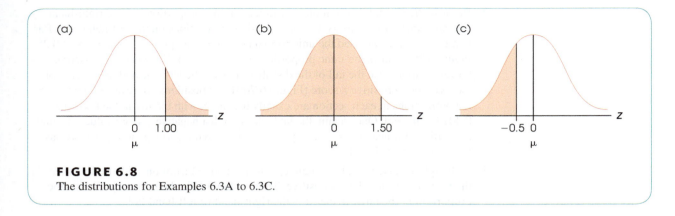

FIGURE 6.8
The distributions for Examples 6.3A to 6.3C.

$p(z < -0.50) = ?$ This portion has been shaded in Figure 6.8(c). To answer questions with negative z-scores, simply remember that the normal distribution is symmetrical with a z-score of zero at the mean, positive values to the right, and negative values to the left. The proportion in the left tail beyond $z = -0.50$ is identical to the proportion in the right tail beyond $z = +0.50$. To find this proportion, look up $z = 0.50$ in column A, and read across the row to find the proportion in column C (tail). You should get an answer of 0.3085 (30.85%). ∎

The following example is an opportunity for you to test your understanding by finding proportions in a normal distribution yourself.

EXAMPLE 6.4

Find the proportion of a normal distribution corresponding to each of the following sections:

a. $z > 0.80$

b. $z > -0.75$

You should obtain answers of 0.2119 and 0.7734 for a and b, respectively. Good luck. ∎

Finding the z-Score Location that Corresponds to Specific Proportions The preceding examples all involved using a z-score value in column A to look up proportions in columns B or C. You should realize, however, that the table also allows you to begin with a known proportion and then look up the corresponding z-score. The following examples demonstrate this process.

EXAMPLE 6.5A

For a normal distribution, what z-score separates the top 10% from the remainder of the distribution? To answer this question, we have sketched a normal distribution [Figure 6.9(a)] and drawn a vertical line that separates the highest 10% (approximately) from the rest. The problem is to locate the exact position of this line. For this distribution, we know that the tail contains 0.1000 (10%) and the body contains 0.9000 (90%). To find the z-score value, you simply locate the row in the unit normal table that has 0.1000 in column C or 0.9000 in column B. For example, you can scan down the values in column C (tail) until you find a proportion of 0.1000. Note that you probably will not find the exact proportion, but you can use the closest value listed in the table. For this example, a proportion of 0.1000 is not listed in column C but you can use 0.1003, which is listed. Once you have found the correct proportion in the table, simply read across the row to find the corresponding z-score value in column A.

For this example, the z-score that separates the extreme 10% in the tail is $z = 1.28$. At this point you must be careful because the table does not differentiate between

FIGURE 6.9

The distributions for
Examples 6.5A and 6.5B.

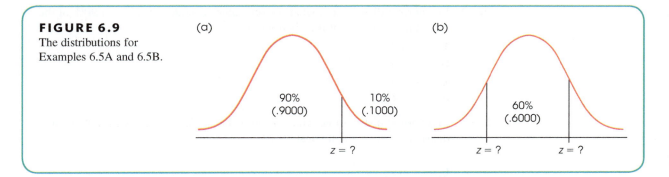

(a)

90%
(.9000)   10%
          (.1000)

z = ?

(b)

60%
(.6000)

z = ?   z = ?

the right-hand and left-hand tail of the distribution. Specifically, the final answer could be either $z = +1.28$, which separates 10% in the right-hand tail, or $z = -1.28$, which separates 10% in the left-hand tail. For this problem we want the right-hand tail (the highest 10%), so the z-score value is $z = +1.28$. ∎

EXAMPLE 6.5B For a normal distribution, what z-score values form the boundaries that separate the middle 60% of the distribution from the rest of the scores?

Again, we have sketched a normal distribution [Figure 6.9(b)] and drawn vertical lines so that roughly 60% of the distribution is in the central section, with the remainder split equally between the two tails. The problem is to find the z-score values that define the exact locations for the lines. To find the z-score values, we begin with the known proportions: 0.6000 in the center and 0.4000 divided equally between the two tails. Although these proportions can be used in several different ways, this example provides an opportunity to demonstrate how column D in the table can be used to solve problems. For this problem, the 0.6000 in the center can be divided in half with exactly 0.3000 to the right of the mean and exactly 0.3000 to the left. Each of these sections corresponds to the proportion listed in column D. Begin by scanning down column D, looking for a value of 0.3000. Again, this exact proportion is not in the table, but the closest value is 0.2995. Reading across the row to column A, you should find a z-score value of $z = 0.84$. Looking again at the sketch [Figure 6.9(b)], the right-hand line is located at $z = +0.84$ and the left-hand line is located at $z = -0.84$. ∎

You may have noticed that we have sketched distributions for each of the preceding problems. As a general rule, you should always sketch a distribution, locate the mean with a vertical line, and shade in the portion you are trying to determine. Look at your sketch. It will help you determine which columns to use in the unit normal table. If you make a habit of drawing sketches, you will avoid careless errors when using the table.

LEARNING CHECK   **LO2**   **1.** What is the probability of randomly selecting a z-score greater than $z = 0.25$ from a normal distribution?

    a. 0.5987

    b. 0.4013

    c. −0.5987

    d. −0.4013

LO2 2. In a normal distribution, what z-score value separates the highest 90% of the scores from the rest of the distribution?

    a. $z = 1.28$

    b. $z = -1.28$

    c. $z = 0.13$

    d. $z = -0.13$

LO2 3. In a normal distribution, what z-score value separates the lowest 20% of the distribution from the highest 80%?

    a. $z = 0.20$

    b. $z = 0.80$

    c. $z = 0.84$

    d. $z = -0.84$

ANSWERS   1. b   2. b   3. d

6-3 Probabilities and Proportions for Scores from a Normal Distribution

LEARNING OBJECTIVES

3. Calculate the probability for a specific X value.

4. Calculate the score (X value) corresponding to a specific proportion in a distribution.

In the preceding section, we used the unit normal table to find probabilities and proportions corresponding to specific z-score values. In most situations, however, it is necessary to find probabilities for specific X values. Consider the following example:

> It is known that IQ scores form a normal distribution with $\mu = 100$ and $\sigma = 15$. Given this information, what is the probability of randomly selecting an individual with an IQ score of less than 120?

This problem is asking for a specific probability or proportion of a normal distribution. However, before we can look up the answer in the unit normal table, we must first transform the IQ scores (X values) into z-scores. Thus, to solve this new kind of probability problem, we must add one new step to the process. Specifically, to answer probability questions about scores (X values) from a normal distribution, you must use the following two-step procedure:

Caution: The unit normal table can be used only with normal-shaped distributions. If a distribution is not normal, transforming X values to z-scores will not make it normal.

1. Transform the X values into z-scores.

2. Use the unit normal table to look up the proportions corresponding to the z-score values.

This process is demonstrated in the following examples. Once again, we suggest that you sketch the distribution and shade the portion you are trying to find in order to avoid careless mistakes.

EXAMPLE 6.6 We will now answer the probability question about IQ scores that we presented earlier. Specifically, what is the probability of randomly selecting an individual with an IQ score of less than 120? Restated in terms of proportions, we want to find the proportion of

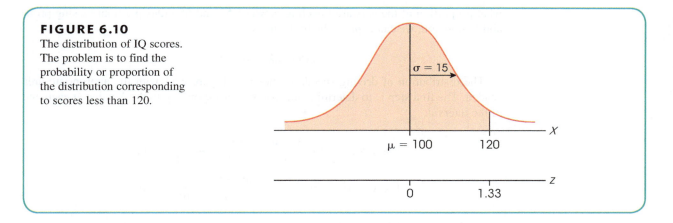

FIGURE 6.10
The distribution of IQ scores.
The problem is to find the
probability or proportion of
the distribution corresponding
to scores less than 120.

the IQ distribution that corresponds to scores less than 120. The distribution is drawn in Figure 6.10, and the portion we want has been shaded.

The first step is to change the X values into z-scores. In particular, the score of $X = 120$ is changed to

$$z = \frac{X - \mu}{\sigma} = \frac{120 - 100}{15} = \frac{20}{15} = 1.33$$

Thus, an IQ score of $X = 120$ corresponds to a z-score of $z = 1.33$, and IQ scores less than 120 correspond to z-scores less than 1.33.

Next, look up the z-score value in the unit normal table. Because we want the proportion of the distribution in the body to the left of $X = 120$ (see Figure 6.10), the answer will be found in column B. Consulting the table, we see that a z-score of 1.33 corresponds to a proportion of 0.9082. The probability of randomly selecting an individual with an IQ of less than 120 is $p = 0.9082$. In symbols,

$$p(X < 120) = p(z < 1.33) = 0.9082 \text{ (or 90.82\%)}$$

Finally, notice that we phrased this question in terms of a *probability*. Specifically, we asked, "What is the probability of selecting an individual with an IQ of less than 120?" However, the same question can be phrased in terms of a *proportion*: "What proportion of all the individuals in the population have IQ scores of less than 120?" Both versions ask exactly the same question and produce exactly the same answer. A third alternative for presenting the same question is introduced in Section 6-4. ■

Finding Proportions/Probabilities Located between Two Scores The next example demonstrates the process of finding the probability of selecting a score that is located *between* two specific values. Although these problems can be solved using the proportions of columns B and C (body and tail), they are often easier to solve with the proportions listed in column D.

EXAMPLE 6.7 The highway department conducted a study measuring driving speeds on a local section of interstate highway. They found an average speed of $\mu = 58$ miles per hour with a standard deviation of $\sigma = 10$. The distribution was approximately normal. Given this information, what proportion of the cars are traveling between 55 and 65 miles per hour? Using probability notation, we can express the problem as

$$p(55 < X < 65) = ?$$

The distribution of driving speeds is shown in Figure 6.11 with the appropriate area shaded. The first step is to determine the z-score corresponding to the X value at each end of the interval.

$$\text{For } X = 55: \quad z = \frac{X - \mu}{\sigma} = \frac{55 - 58}{10} = \frac{-3}{10} = -0.30$$

$$\text{For } X = 65: \quad z = \frac{X - \mu}{\sigma} = \frac{65 - 58}{10} = \frac{7}{10} = 0.70$$

Looking again at Figure 6.11, we see that the proportion we are seeking can be divided into two sections: (1) the area left of the mean, and (2) the area right of the mean. The first area is the proportion between the mean and $z = -0.30$, and the second is the proportion between the mean and $z = +0.70$. Using column D of the unit normal table, these two proportions are 0.1179 and 0.2580. The total proportion is obtained by adding these two sections:

$$p(55 < X < 65) = p(-0.30 < z < +0.70) = 0.1179 + 0.2580 = 0.3759 \qquad ■$$

EXAMPLE 6.8   Using the same distribution of driving speeds from the previous example, what proportion of cars are traveling between 65 and 75 miles per hour?

$$p(65 < X < 75) = ?$$

The distribution is shown in Figure 6.12 with the appropriate area shaded. Again, we start by determining the z-score corresponding to each end of the interval.

$$\text{For } X = 65: \quad z = \frac{X - \mu}{\sigma} = \frac{65 - 58}{10} = \frac{7}{10} = 0.70$$

$$\text{For } X = 75: \quad z = \frac{X - \mu}{\sigma} = \frac{75 - 58}{10} = \frac{17}{10} = 1.70$$

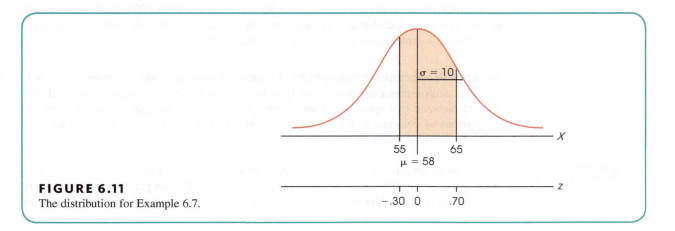

FIGURE 6.11
The distribution for Example 6.7.

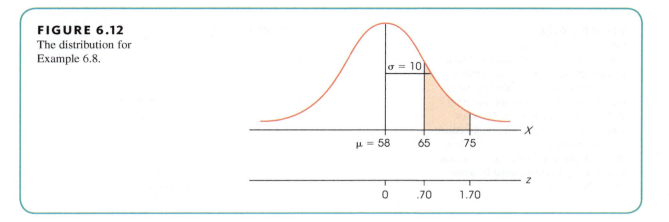

FIGURE 6.12
The distribution for
Example 6.8.

There are several different ways to use the unit normal table to find the proportion between these two z-scores. For this example, we will use the proportions in the tail of the
distribution (column C). According to column C in the unit normal table, the proportion in
the tail beyond $z = 0.70$ is $p = 0.2420$. Note that this proportion includes the section that
we want, but it also includes an extra, unwanted section located in the tail beyond $z = 1.70$.
Locating $z = 1.70$ in the table, and reading across the row to column C, we see that the
unwanted section is $p = 0.0446$. To obtain the correct answer, we subtract the unwanted
portion from the total proportion in the tail beyond $z = 0.70$.

$$p(65 < X < 75) = p(0.70 < z < 1.70) = 0.2420 - 0.0446 = 0.1974$$ ■

The following example is an opportunity for you to test your understanding by finding
probabilities for scores in a normal distribution yourself.

EXAMPLE 6.9   For a normal distribution with $\mu = 60$ and a standard deviation of $\sigma = 12$, find each probability requested.

a. $p(X > 66)$

b. $p(48 < X < 72)$

You should obtain answers of 0.3085 and 0.6826 for a and b, respectively. Good luck. ■

Finding Scores Corresponding to Specific Proportions or Probabilities   In the
previous three examples, the problem was to find the proportion or probability corresponding to specific X values. The two-step process for finding these proportions is shown in
Figure 6.13. Working with probabilities for a normal distribution involves two steps: (1)
using a z-score formula and (2) using the unit normal table. However, the order of the steps
may vary, depending on the type of probability question you are trying to answer.

In one instance you may start with a known X value and have to find a probability that
is associated with it (as in Example 6.6). First, you must convert the X value to a z-score
using Equation 5.1 (page 153). Then you consult the unit normal table to get the probability associated with the particular area of the graph. *Note:* You cannot go directly from the
X value to the unit normal table. You must find the z-score first.

However, suppose that you begin with a known probability value and want to find the X
value associated with it. In this case, you use the unit normal table first to find the z-score
that corresponds with the probability value. Then you convert the z-score into an X value
using Equation 5.2 (page 155). Figure 6.13 illustrates the steps you must take when moving

FIGURE 6.13

On this map, the solid lines are the "roads" that you must travel to find a probability value that corresponds to any specific score or to find the score that corresponds to any specific probability value. In taking these routes, you must pass through the intermediate step of finding a z-score value. [*Note:* You may not travel directly between X values and probability (that is, along the dashed line).]

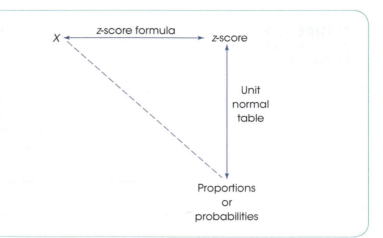

from an X value to a probability or from a probability back to an X value. This diagram is like a map and guides you through the essential steps as you "travel" between X values and probabilities. The solid lines are the "roads" you must take as you navigate between probability and X values. You must always travel through the step of finding a z-score when working with X values and probability. The following example demonstrates this process.

EXAMPLE 6.10   The U.S. Census Bureau (2017) reports that Americans spend an average of $\mu = 26.1$ minutes commuting to work each day. Assuming that the distribution of commuting times is normal with a standard deviation of $\sigma = 10$ minutes, how much time do you have to spend commuting each day to be in the highest 10% nationwide? The distribution is shown in Figure 6.14 with a portion representing approximately 10% shaded in the right-hand tail.

In this problem, we begin with a proportion (10% or 0.10), and we are looking for a score. According to the map in Figure 6.13, we can move from p (proportion) to X (score) via z-scores. The first step is to use the unit normal table to find the z-score that corresponds to a proportion of 0.10 in the tail. First, scan the values in column C to locate the row that has a proportion of 0.10 in the tail of the distribution. Note that you will not find 0.1000 exactly, but locate the closest value possible. In this case, the closest value is 0.1003. Reading across the row, we find $z = 1.28$ in column A.

The next step is to determine whether the z-score is positive or negative. Remember that the table does not specify the sign of the z-score. Looking at the distribution in

FIGURE 6.14

The distribution of commuting times for American workers. The problem is to find the score that separates the highest 10% of commuting times from the rest.

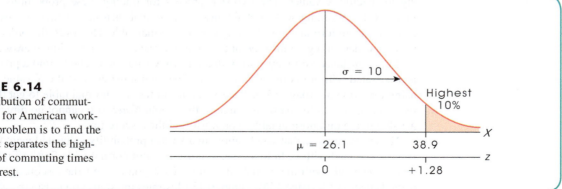

Figure 6.14, you should realize that the score we want is above the mean, so the z-score is positive, $z = +1.28$.

The final step is to transform the z-score into an X value. By definition, a z-score of $+1.28$ corresponds to a score that is located above the mean by 1.28 standard deviations. One standard deviation is equal to 10 points ($\sigma = 10$), so 1.28 standard deviations is

$$1.28\sigma = 1.28(10) = 12.8 \text{ points}$$

Thus, our score is located above the mean ($\mu = 26.1$) by a distance of 12.8 points. Therefore,

$$X = 26.1 + 12.8 = 38.9$$

The answer to our original question is that you must commute at least 38.9 minutes per day to be in the top 10% of American commuters. ■

EXAMPLE 6.11   Again, the distribution of commuting times for American workers is normal, with a mean of $\mu = 26.1$ minutes and a standard deviation of $\sigma = 10$ minutes. For this example, we will find the range of values that defines the middle 90% of the distribution. The entire distribution is shown in Figure 6.15 with the middle portion shaded.

The 90% (0.9000) in the middle of the distribution can be split in half, with 45% (0.4500) on each side of the mean. Looking up 0.4500 in column D of the unit normal table, you will find that the exact proportion is not listed. However, you will find 0.4495 and 0.4505, which are equally close. Technically, either value is acceptable, but we will use 0.4505 so that the total area in the middle is at least 90%. Reading across the row, you should find a z-score of $z = 1.65$ in column A. Thus, the z-score at the right boundary is $z = +1.65$ and the z-score at the left boundary is $z = -1.65$. In either case, a z-score of 1.65 indicates a location that is 1.65 standard deviations away from the mean. For the distribution of commuting times, one standard deviation is $\sigma = 10$, so 1.65 standard deviations is a distance of

$$1.65\sigma = 1.65(10) = 16.5 \text{ points}$$

Therefore, the score at the right-hand boundary is located above the mean by 16.5 points and corresponds to $X = 26.1 + 16.5 = 42.6$. Similarly, the score at the left-hand boundary is below the mean by 16.5 points and corresponds to $X = 26.1 - 16.5 = 9.6$. The middle 90% of the distribution corresponds to values between 9.6 and 42.6. Thus, 90% of American commuters spend between 9.6 and 42.6 minutes commuting to work each day. Only 10% of commuters spend either more time or less time commuting. ■

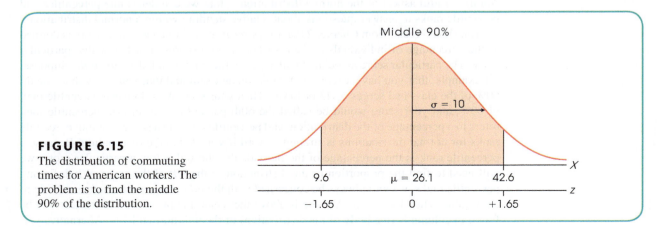

FIGURE 6.15

The distribution of commuting times for American workers. The problem is to find the middle 90% of the distribution.

LEARNING CHECK   **LO3**   **1.** The population of SAT scores forms a normal distribution with a mean of $\mu = 500$ and $\sigma = 100$. What proportion of the population consists of individuals with SAT scores higher than 400?

    a. 0.1587

    b. 0.8413

    c. 0.3413

    d. −0.1587

LO3   **2.** A normal distribution has $\mu = 100$ and $\sigma = 20$. What is the probability of randomly selecting a score of greater than 130 from this distribution?

    a. $p = 0.9032$

    b. $p = 0.9332$

    c. $p = 0.0968$

    d. $p = 0.0668$

LO4   **3.** For a normal distribution with $\mu = 70$ and $\sigma = 10$, what is the minimum score necessary to be in the top 60% of the distribution?

    a. 67.5

    b. 62.5

    c. 65.2

    d. 68.4

ANSWERS   1. b   2. d   3. a

6-4   Percentiles and Percentile Ranks

LEARNING OBJECTIVE

5. For normally distributed scores, find the percentile rank corresponding to a score and find the percentile score corresponding to a proportion of the normal distribution.

6. Find the interquartile range for a normal distribution.

Another useful aspect of the normal distribution is that we can determine percentiles and percentile ranks to answer questions about relative standing within a normal distribution.

You should recall from Chapter 2 that the *percentile rank* of a particular score is defined as the percentage of individuals in the distribution with scores at or below that particular score. The particular score associated with a percentile rank is called a *percentile*. Suppose, for example, that you have a score of $X = 43$ on an exam and that you know that exactly 60% of the class had scores of 43 or lower. Then your score $X = 43$ has a percentile rank of 60%, and your score would be called the 60th percentile. Remember, percentile rank refers to a percentage of the distribution, and percentile refers to a score. Finding percentile ranks for normal distributions is straightforward if you sketch the distribution. Because a percentile rank is the percentage of the individuals who fall below a particular score, we will need to find the proportion of the distribution to the *left* of the score. When finding percentile ranks, we will always be concerned with the percentage on the left-hand side of an X value. Therefore, if the X value is above the mean, the proportion to the left will be found in column B (the body of the distribution) of the unit normal table. Alternatively, if the X value is below the mean, everything to its left is reflected in column C (the tail). The following examples illustrate these points.

EXAMPLE 6.12

A population is normally distributed with $\mu = 100$ and $\sigma = 10$. What is the percentile rank for $X = 114$?

Because a percentile rank indicates a score's standing relative to all lower scores, we must focus on the area of the distribution to the left of $X = 114$. The distribution is shown in Figure 6.16 and the area of the curve containing all scores below $X = 114$ is shaded. The proportion for this shaded area will give us the percentile rank. Because the distribution is normal, we can use the unit normal table to find this proportion. The first step is to compute the z-score for the X value we are considering.

$$z = \frac{X - \mu}{\sigma} = \frac{114 - 100}{10} = \frac{14}{10} = +1.40$$

The next step is to consult the unit normal table. Note that the shaded area in Figure 6.16 makes up the large body of the graph. The proportion for this area is presented in column B. For $z = +1.40$, column B indicates a proportion of 0.9192. The percentile rank for $X = 114$ is 91.92%. ■

EXAMPLE 6.13

For the distribution in Example 6.12, what is the percentile rank for $X = 92$?

This example is diagrammed in Figure 6.17. The score $X = 92$ is placed in the left side of the distribution because it is below the mean. Again, percentile ranks deal with the area of the distribution below the score in question. Therefore, we have shaded the area to the left of $X = 92$.

First, the X value is transformed to a z-score:

$$z = \frac{X - \mu}{\sigma} = \frac{92 - 100}{10} = \frac{-8}{10} = -0.80$$

Now the unit normal table can be consulted. The proportion to the left-hand tail beyond $z = -0.80$ can be found in column C. According to the unit normal table, for $z = 0.80$, the proportion in the tail is $p = 0.2119$. This also is the area beyond $z = -0.80$. Thus, the percentile rank for $X = 92$ is 21.19%. That is, a score of 92 is greater than 21.19% of the scores in the distribution. ■

■ Finding Percentiles

The process of finding a particular percentile is very similar to the process used in Example 6.10. You are given a percentage (this time a percentile rank), and you must find the corresponding X value (the percentile). You should recall that finding an X value from a

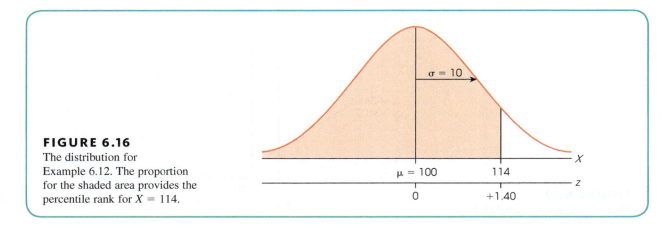

FIGURE 6.16

The distribution for Example 6.12. The proportion for the shaded area provides the percentile rank for $X = 114$.

FIGURE 6.17
The distribution for Example 6.13. The proportion for the shaded area provides the percentile rank for $X = 92$.

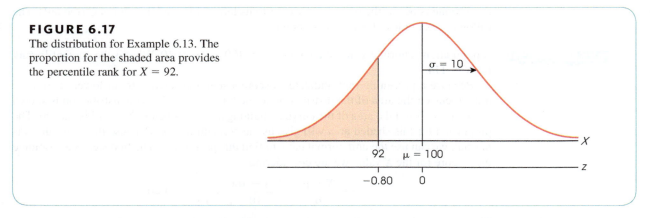

percentage requires the intermediate step of determining the z-score for that proportion of the distribution. The following example demonstrates this process for percentiles.

EXAMPLE 6.14   A population is normally distributed with $\mu = 60$ and $\sigma = 5$. For this population, what is the 34th percentile?

In this example, we are looking for an X value (percentile) that has 34% (or $p = 0.3400$) of the distribution below it. This problem is illustrated in Figure 6.18. Note that 34% is roughly equal to one-third of the distribution, so the corresponding shaded area in Figure 6.18 is located entirely on the left-hand side of the mean. In this problem, we begin with a proportion (0.3400), and we are looking for a score (the percentile). The first step in moving from a proportion to a score is to find the z-score (Figure 6.13). You must look at the unit normal table to find the z-score that corresponds to a proportion of 0.3400. Because the proportion is in the tail beyond z, you must look in column C for a proportion of 0.3400. Although the exact value of 0.3400 is not listed in the table, you may use the closest value, which is 0.3409. The z-score corresponding to this value is $z = -0.41$. Notice that we have added a negative sign to the z-score value because the position we want is located below the mean. Thus, the X value for which we are looking has a z-score of $z = -0.41$.

The next step is to convert the z-score to an X value. Using the z-score formula solved for X, we obtain

$$X = \mu + z\sigma$$
$$= 60 + (-0.41)(5)$$
$$= 60 - 2.05$$
$$= 57.95$$

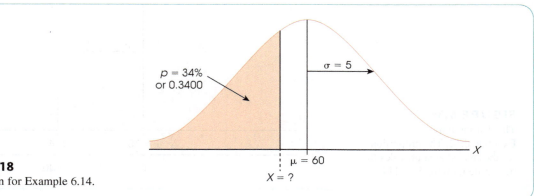

FIGURE 6.18
The distribution for Example 6.14.

The 34th percentile for this distribution is $X = 57.95$. This answer makes sense because the 50th percentile for this example is 60 (the mean and median). Therefore the 34th percentile must be a value less than 60. ■

■ Quartiles

We first looked at *quartiles* in Chapter 4 (page 113) in considering the *interquartile range* for sample data. When population data form a normal distribution there is a simple method to compute the interquartile range from the unit normal table. Recall that percentiles divide the distribution into 100 equal parts, each corresponding to 1% of the distribution. The area in a distribution can also be divided into four equal parts called quartiles, each corresponding to 25%. The first quartile (Q1) is the score that separates the lowest 25% of the distribution from the rest. Thus, the first quartile is the same as the 25th percentile. Similarly, the second quartile (Q2) is the score that has 50% (two quarters) of the distribution below it. You should recognize that Q2 is the median, or 50th percentile of the distribution. The third quartile (Q3) is the X value that has 75% (three quarters) of the distribution below it. The Q3 for a distribution is also the 75th percentile.

For a normal distribution, the first quartile always corresponds to $z = -0.67$, the second quartile corresponds to $z = 0.00$ (the mean), and the third quartile corresponds to $z = +0.67$ (Figure 6.19). These values can be found by consulting the unit normal table and are true of any normal distribution. This makes finding quartiles and the interquartile range straightforward for normal distributions. The following example demonstrates the use of quartiles.

EXAMPLE 6.15   A population is normally distributed and has a mean of $\mu = 50$ with a standard deviation of $\sigma = 10$. Find the first, second, and third quartile, and compute the interquartile range.

The first quartile, Q1, is the same as the 25th percentile, which has a corresponding z-score of $z = -0.67$. With $\mu = 50$ and $\sigma = 10$, we can determine the X value of Q1.

$$X = \mu + z\sigma$$
$$= 50 + (-0.67)(10)$$
$$= 50 - 6.70$$
$$= 43.30$$

The second quartile, Q2, is also the 50th percentile, or median. For a normal distribution, the median equals the mean, so Q2 is 50. By the formula, with a z-score of 0, we obtain

$$X = \mu + z\sigma$$
$$= 50 + (0.00)(10)$$
$$= 50 - 0.00$$
$$= 50.00$$

The third quartile, Q3, is also the 75th percentile. It has a corresponding z-score of $z = +0.67$. Using the z-score formula solved for X, we obtain

$$X = \mu + z\sigma$$
$$= 50 + (+0.67)(10)$$
$$= 50 + 6.70$$
$$= 56.70$$

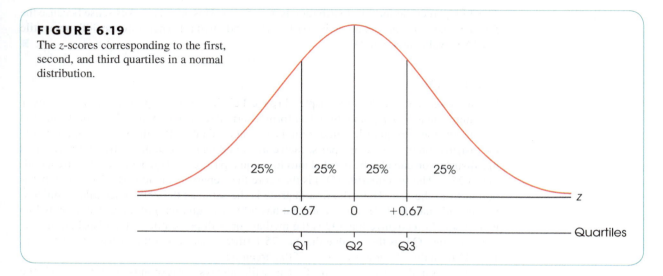

FIGURE 6.19

The z-scores corresponding to the first, second, and third quartiles in a normal distribution.

In Chapter 4 (page 114), we learned that the interquartile range was defined as the distance between the first and third quartile, or

$$IQR = Q3 - Q1$$

For this example, the interquartile range is

$$IQR = Q3 - Q1$$
$$IQR = 56.70 - 43.30$$
$$IQR = 13.40$$

Notice that Q1 and Q3 are the same distance from the mean (6.7 points in the previous example). Q1 and Q3 will always be equidistant from the mean for normal distributions because normal distributions are symmetrical. Therefore, the distance between Q1 and Q3 (the interquartile range by definition) will also equal twice the distance of Q3 from the mean (see Figure 6.19). The distance of Q3 from the mean can be obtained simply by multiplying the z-score for Q3 ($z = +0.67$) times the standard deviation. Using this shortcut greatly simplifies the computation. For a normal distribution,

$$IQR = 2(0.67\sigma) \tag{6.1}$$

Remember, this simplified formula is used *only* for a normally distributed population. ■

LEARNING CHECK   **LO 5 1.** For a population with a $\mu = 500$ and $\sigma = 100$, which of the following is the percentile rank of a score of $X = 550$?

  a. 69.15%

  b. $z = +0.50$

  c. 38.05%

  d. $z = -0.50$

LO5 2. For a population with a $\mu = 100$ and $\sigma = 10$, which of the following is the 20th percentile score?

  a. $X = 91.60$

  b. $z = -8.40$

   c. $X = 108.40$

   d. $z = -0.84$

LO6 **3.** Find the interquartile range for a normal distribution with $\sigma = 50$.

   a. Cannot be determined without more information.

   b. 67

   c. 13.40

   d. 25%

ANSWERS   **1. a**   **2. a**   **3. b**

6-5   Looking Ahead to Inferential Statistics

LEARNING OBJECTIVE

7. Explain how probability can be used to evaluate a treatment effect by identifying likely and very unlikely outcomes.

Probability forms a direct link between samples and the populations from which they come. As we noted at the beginning of this chapter, this link is the foundation for the inferential statistics in future chapters. The following example provides a brief preview of how probability is used in the context of inferential statistics.

We ended Chapter 5 with a demonstration of how inferential statistics are used to help interpret the results of a research study. A general research situation was shown in Figure 5.10 (page 168) and is repeated here in Figure 6.20. The research begins with a population that forms a normal distribution with a mean of $\mu = 400$ and a standard deviation of $\sigma = 20$. A sample is selected from the population and a treatment is administered to the sample. The goal for the study is to evaluate the effect of the treatment.

To determine whether the treatment has an effect, the researcher simply compares the treated sample with the original population. If the individuals in the sample have scores

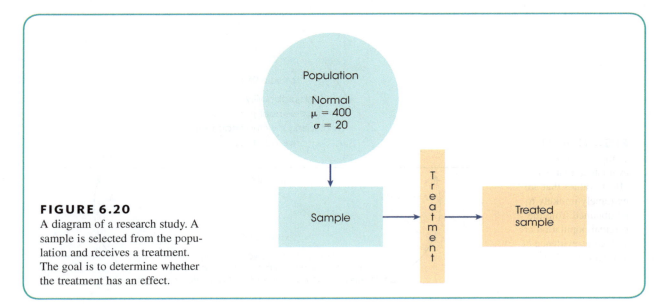

FIGURE 6.20

A diagram of a research study. A sample is selected from the population and receives a treatment. The goal is to determine whether the treatment has an effect.

around 400 (the original population mean), then we must conclude that the treatment appears to have no effect. On the other hand, if the treated individuals have scores that are noticeably different from 400, then the researcher has evidence that the treatment does have an effect. Notice that the study is using a sample to help answer a question about a population; this is the essence of inferential statistics.

The problem for the researcher is determining exactly what is meant by "noticeably different" from 400. If a treated individual has a score of $X = 415$, is that enough to say that the treatment has an effect? What about $X = 420$ or $X = 450$? In Chapter 5, we suggested that z-scores provide one method for solving this problem. Specifically, we suggested that a z-score value beyond $z = 2.00$ (or -2.00) was an extreme value and therefore noticeably different. However, the choice of $z = \pm 2.00$ was purely arbitrary. Now we have another tool, *probability,* to help us decide exactly where to set the boundaries.

Figure 6.21 shows the original population from our hypothetical research study. Note that most of the scores are located close to $\mu = 400$. Also note that we have added boundaries separating the middle 95% of the distribution from the extreme 5% or 0.0500 in the two tails. Dividing the 0.0500 in half produces a proportion of 0.0250 in the right-hand tail and 0.0250 in the left-hand tail. Using column C of the unit normal table, the z-score boundaries for the right and left tails are $z = +1.96$ and $z = -1.96$, respectively. If we are selecting an individual from the original untreated population, then it is very unlikely that we would obtain a score beyond the $z = \pm 1.96$ boundaries.

The boundaries set at $z = \pm 1.96$ provide objective criteria for deciding whether our sample provides evidence that the treatment has an effect. Specifically, if our sample is located in the tail beyond one of the ± 1.96 boundaries, then we can conclude:

1. The sample is an extreme value, nearly 2 standard deviations away from the average, and therefore is noticeably different from most individuals in the original population.

2. If the treatment has no effect, then the sample is a very unlikely outcome. Specifically, the probability of obtaining a sample that is beyond the ± 1.96 boundaries is less than 5%.

Therefore, the sample provides clear evidence that the treatment has had an effect.

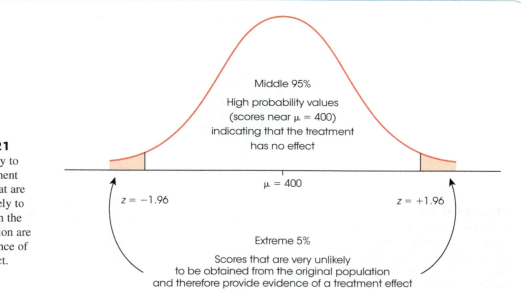

FIGURE 6.21
Using probability to evaluate a treatment effect. Values that are extremely unlikely to be obtained from the original population are viewed as evidence of a treatment effect.

Middle 95%

High probability values
(scores near $\mu = 400$)
indicating that the treatment
has no effect

$\mu = 400$

$z = -1.96$ $z = +1.96$

Extreme 5%

Scores that are very unlikely
to be obtained from the original population
and therefore provide evidence of a treatment effect

LEARNING CHECK   **LO7** **1.** Which of the following accurately describes a score of $X = 57$ or larger in a normal distribution with $\mu = 40$ and $\sigma = 5$?

  a. It is an extreme, very unlikely score.

  b. It is higher than average but not extreme or unlikely.

  c. It is a little above average.

  d. It is an average, representative score.

LO7 **2.** For a normal distribution with $\mu = 60$ and $\sigma = 10$, what X values form the boundaries between the middle 95% of the distribution and the extreme 5% in the tails?

  a. 51.6 and 68.4

  b. 47.2 and 72.8

  c. 43.5 and 65.5

  d. 40.4 and 79.6

LO7 **3.** An individual is selected from a normal population with a mean of $\mu = 80$ with $\sigma = 20$, and a treatment is administered to the individual. After treatment, the individual's score is found to be $X = 105$. How likely is it that a score this large or larger would be obtained if the treatment has no effect?

  a. $p = 0.1056$

  b. $p = 0.3944$

  c. $p = 0.8944$

  d. $p = 1.2500$

ANSWERS   **1.** a   **2.** d   **3.** a

SUMMARY

1. The probability of a particular event A is defined as a fraction or proportion:

$$p(A) = \frac{\text{number of outcomes classified as } A}{\text{total number of possible outcomes}}$$

2. Our definition of probability is accurate only for random samples. There are two requirements that must be satisfied for a random sample:
   a. Every individual in the population has an equal chance of being selected.
   b. When more than one individual is being selected, the probabilities must stay constant. This means there must be sampling with replacement.

3. All probability problems can be restated as proportion problems. The "probability of selecting a king from a deck of cards" is equivalent to the "proportion of the deck that consists of kings." For frequency distributions, probability questions can be answered by determining proportions of area. The "probability of selecting an individual with an IQ greater than 108" is equivalent to the "proportion of the whole population that consists of IQs greater than 108."

4. For normal distributions, probabilities (proportions) can be found in the unit normal table. The table provides a listing of the proportions of a normal distribution that correspond to each z-score value. With the table, it is possible to move between X values and probabilities using a two-step procedure:
   a. The z-score formula (Chapter 5) allows you to transform X to z or to change z back to X.
   b. The unit normal table allows you to look up the probability (proportion) corresponding to each z-score or the z-score corresponding to each probability.

5. Percentiles and percentile ranks measure the relative standing of a score within a distribution. Percentile rank is the percentage of individuals with scores at or below a particular X value. A percentile is an X value that is identified by its rank. The percentile rank always corresponds to the proportion to the left of the score in question.

KEY TERMS

probability (180)

random sampling (181)

simple random sample (181)

independent random sampling (182)

independent random sample (182)

random sample (182)

sampling with replacement (183)

sampling without replacement (183)

unit normal table (187–188)

percentile rank (198)

percentile (198)

quartiles (201)

interquartile range (201)

FOCUS ON PROBLEM SOLVING

1. We have defined probability as being equivalent to a proportion, which means that you can restate every probability problem as a proportion problem. This definition is particularly useful when you are working with frequency distribution graphs in which the population is represented by the whole graph and probabilities (proportions) are represented by portions of the graph. When working problems with the normal distribution, you always should start with a sketch of the distribution. You should shade the portion of the graph that reflects the proportion you are looking for.

2. Remember that the unit normal table shows only positive z-scores in column A. However, since the normal distribution is symmetrical, the proportions in the table apply to both positive and negative z-score values.

3. A common error for students is to use negative values for proportions on the left-hand side of the normal distribution. Proportions (or probabilities) are always positive: 10% is 10% whether it is in the left or right tail of the distribution.

4. The proportions in the unit normal table are accurate only for normal distributions. If a distribution is not normal, you cannot use the table.

DEMONSTRATION 6.1

FINDING PROBABILITY FROM THE UNIT NORMAL TABLE

A population is normally distributed with a mean of $\mu = 45$ and a standard deviation of $\sigma = 4$. What is the probability of randomly selecting a score that is greater than 43? In other words, what proportion of the distribution consists of scores greater than 43?

STEP 1   **Sketch the distribution.**   For this demonstration, the distribution is normal with $\mu = 45$ and $\sigma = 4$. The score of $X = 43$ is lower than the mean and therefore is placed to the left of the mean. The question asks for the proportion corresponding to scores greater than 43, so shade in the area to the right of this score. Figure 6.22 shows the sketch.

STEP 2   **Transform the X value to a z-score.**

$$z = \frac{X - \mu}{\sigma} = \frac{43 - 45}{4} = \frac{-2}{4} = -0.5$$

STEP 3   **Find the appropriate proportion in the unit normal table.**   Ignoring the negative size, locate $z = -0.50$ in column A. In this case, the proportion we want corresponds to the body of the distribution and the value is found in column B. For this example,

$$p(X > 43) = p(z > -0.50) = 0.6915$$

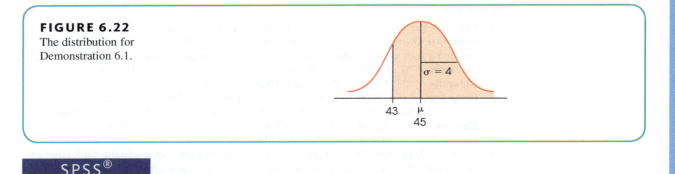

FIGURE 6.22
The distribution for
Demonstration 6.1.

SPSS®

General instructions for using SPSS are presented in Appendix D. Following are detailed instructions for using SPSS to **Transform *X* Values into percentiles for normally distributed scores**.

Demonstration Example

The Stroop procedure is a test of cognitive functioning that can help to assess neuropsychological disorders. In this task, participants are exposed to a card with the name of a color printed on it in colored ink. Participants are asked to either read the name of a color that is printed on a card or report the color ink that was used to print the name. When there is a mismatch between the color name and the color ink, participants are often slower to report the color ink than the color name that is written on the card. Thus, it would be difficult for participants to report *red* when the word "green" is printed in red colored ink. Below are hypothetical data from the Stroop procedure. Each interference score represents the difference in response time (report color ink minus report color name) in seconds across a block of trials. These data are similar to those reported by Troyer, Leach, and Strauss (2006). Notice that the mean of these scores is $M = 11$ and the standard deviation is $s = 3$.

Participant	Interference Score
1	7
2	9
3	14
4	11
5	12
6	8
7	8
8	13
9	11
10	12
11	16
12	17
13	6
14	15
15	11
16	10
17	13
18	11
19	11
20	9
21	7

We will use SPSS to convert these scores to percentile ranks based on the normal distribution.

Data Entry

1. Click the **Variable View** tab to enter information about the variables.
2. In the first row, enter "intScore" (for interference score). Fill in the remaining information about your variable where necessary. Be sure that **Type** = "Numeric", **Width** = "8", **Decimals** = "0", **Label** = "Interference Score", **Values** = "None", **Missing** = "None", **Columns** = "8", **Align** = "Right", and **Measure** = "Scale".
3. In the second row, enter "percRank" (for percentile rank). Fill in the remaining information about your variable where necessary. Be sure that **Type** = "Numeric", **Width** = "8", **Decimals** = "2", **Label** = "Percentile Rank", **Values** = "None", **Missing** = "None", **Columns** = "8", **Align** = "Right", and **Measure** = "Scale".
4. Click the **Data View** tab to return to the data editor and enter scores in the "intScore" column. Leave the "percRank" column blank because SPSS will enter the percentile ranks and place them in that column.

Data Analysis

1. Click **Transform** on the tool bar and click **Compute Variable**.
2. In the **Target Variable:** field, enter "percRank". This programs SPSS to display the computed percentiles in the "percRank" column.
3. In the **Function group** box, select "CDF & Noncentral CDF". In the **Functions and Special Variables** box, double-click "Cdf.Normal". This function will report the proportion of the normal distribution below a value. The **Numeric Expression** box should now be populated with "CDF.NORMAL(?,?,?)".
4. Enter "100*" before the expression to change proportion to percent.
5. Replace the first question mark in the expression with "intScore".
6. Replace the second question mark in the expression with the mean ("11") and the third question mark with the standard deviation ("3").
7. Check that your "Compute Variable" window is as below and click **OK**. You will be asked to confirm that you would like to change the values in the percRank column. Click **OK**.

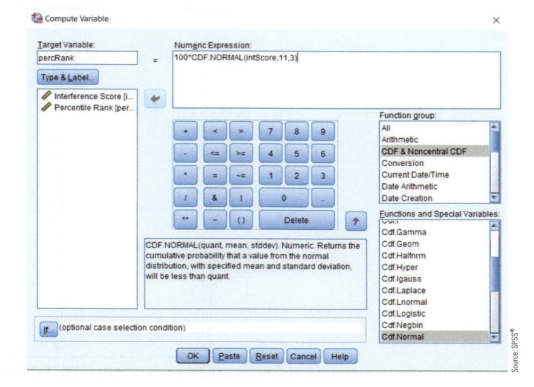

Source: SPSS®

SPSS Output

The output for this analysis is displayed in the Data View rather than in an output window. You should find that the "percRank" column is populated with percentile ranks for each of the scores in your data set. The Data View should appear as below.

	intScore	percRank
1	7	9.12
2	9	25.25
3	14	84.13
4	11	50.00
5	12	63.06
6	8	15.87
7	8	15.87
8	13	74.75
9	11	50.00
10	12	63.06
11	16	95.22
12	17	97.72
13	6	4.78
14	15	90.88
15	11	50.00
16	10	36.94
17	13	74.75
18	11	50.00
19	11	50.00
20	9	25.25
21	7	9.12

Source: SPSS®

For example, a score of $X = 17$ has a percentile rank of 97.72%. This makes good sense when you think about the distribution. $X = 17$ is 6 points greater than the mean of $M = 11$. Because the standard deviation is $s = 3$, the z-score is $z = +2.00$. The unit normal table in Appendix B reports that a z-score of $+2.00$ is greater than 97.72% of the distribution (proportion $= .9772$). Thus, the percentile ranks computed by SPSS agree with the unit normal table in Appendix B.

Try It Yourself

For the following data set, identify all scores in the data set that are above the 70th percentile. The mean of those scores is $M = 10$ and the standard deviation is $s = 4$. If you have done this correctly, you should find that four scores ($X = 14$, $X = 14$, $X = 16$, and $X = 18$) are greater than the 70th percentile.

Participant	Interference Score
1	12
2	2
3	14
4	11
5	14
6	10
7	8
8	12
9	12
10	4
11	16
12	6
13	11
14	8
15	8
16	18
17	11
18	8
19	4
20	9
21	12

PROBLEMS

1. What are the two requirements for a random sample?

2. Define sampling with replacement and explain why it is used.

3. A psychology class consists of 32 freshmen and 48 sophomores. If the professor selects names from the class list using *random sampling*,
   a. what is the probability that the first student selected will be a freshman?
   b. and if a random sample of $n = 6$ students is selected and the first five selected are sophomores, what is the probability that the sixth student selected will be a freshman?
   c. Repeat question a (above) after 10 sophomore students join the class.

4. Bags of Skittles™ candies include six different colors: red, orange, yellow, green, blue, and purple. If the bag has an equal number of each of the six colors, what are the probabilities for each of the following?
   a. Randomly selecting a green candy?
   b. Randomly selecting either a green or a yellow candy?
   c. Randomly selecting something other than a green candy?

5. Draw a vertical line through a normal distribution for each of the following *z*-score locations. Determine whether the body is on the right or left side of the line and find the proportion in the body.
   a. $z = +2.00$
   b. $z = +0.50$
   c. $z = -1.50$
   d. $z = -1.67$

6. Draw a vertical line through a normal distribution for each of the following *z*-score locations. Determine whether the tail is on the right or left side of the line and find the proportion in the tail.
   a. $z = +1.00$
   b. $z = +0.33$
   c. $z = -0.10$
   d. $z = -0.67$

7. Draw a vertical line through a normal distribution for each of the following *z*-score locations. Find the proportion of the distribution located between the mean and the *z*-score.
   a. $z = +1.60$
   b. $z = +0.90$
   c. $z = -1.50$
   d. $z = -0.40$

8. Find each of the following probabilities for a normal distribution.

a. $p(z > +2.00)$
b. $p(z > -1.00)$
c. $p(z < +0.50)$
d. $p(z < +1.75)$

9. What proportion of a normal distribution is located between each of the following z-score boundaries?
   a. $z = -1.64$ and $z = +1.64$
   b. $z = -1.96$ and $z = +1.96$
   c. $z = -1.00$ and $z = +1.00$

10. Find each of the following probabilities for a normal distribution.
    a. $p(-1.80 < z < 0.20)$
    b. $p(-0.40 < z < 1.40)$
    c. $p(0.25 < z < 1.25)$
    d. $p(-0.90 < z < -0.60)$

11. Find the z-score location of a vertical line that separates a normal distribution as described in each of the following.
    a. 5% in the tail on the right
    b. 20% in the tail on the left
    c. 90% in the body on the right
    d. 50% on each side of the distribution

12. Find the z-score boundaries that separate a normal distribution as described in each of the following.
    a. The middle 20% from the 80% in the tails
    b. The middle 25% from the 75% in the tails
    c. The middle 70% from the 30% in the tails
    d. The middle 90% from the 10% in the tails

13. Find the z-score boundaries that separate a normal distribution as described in each of the following.
    a. The middle 95% from the 5% in the tails
    b. The middle 50% from the 50% in the tails
    c. The middle 75% from the 25% in the tails
    d. The middle 60% from the 40% in the tails

14. A normal distribution has a mean of $\mu = 50$ and a standard deviation of $\sigma = 5$. For each of the following scores, indicate whether the tail is to the right or left of the score and find the proportion of the distribution located in the tail.
    a. $X = 45$
    b. $X = 35$
    c. $X = 55$
    d. $X = 60$

15. A normal distribution has a mean of $\mu = 70$ and a standard deviation of $\sigma = 12$. For each of the following scores, indicate whether the body is to the right or left of the score and find the proportion of the distribution located in the body.
    a. $X = 74$
    b. $X = 84$
    c. $X = 54$
    d. $X = 58$

16. For a normal distribution with a mean of $\mu = 85$ and a standard deviation of $\sigma = 20$, find the proportion of the population corresponding to each of the following.
    a. Scores greater than 89
    b. Scores less than 72
    c. Scores between 70 and 100

17. IQ test scores are standardized to produce a normal distribution with a mean of $\mu = 100$ and a standard deviation of $\sigma = 15$. Find the proportion of the population in each of the following IQ categories.
    a. Genius or near genius: IQ over 140
    b. Very superior intelligence: IQ from 120 to 140
    c. Average or normal intelligence: IQ from 90 to 109
    d. $p(X > ?) = .05$
    e. $p(X < ?) = .75$

18. Suppose that the distribution of scores on the Graduate Record Exam (GRE) is approximately normal, with a mean of $\mu = 150$ and a standard deviation of $\sigma = 5$. For the population of students who have taken the GRE:
    a. What proportion have GRE scores less than 145?
    b. What proportion have GRE scores greater than 157?
    c. What is the minimum GRE score needed to be in the highest 20% of the population?
    d. If a graduate school accepts only students from the top 10% of the GRE distribution, what is the minimum GRE score needed to be accepted?

19. An important reason that students struggle in college is that they are sometimes unaware that they have not yet mastered a new skill. Struggling students often overestimate their level of mastery in part because the skills needed to master a topic are the same skills needed to identify weaknesses in understanding. For example, students in a psychology class who had just submitted an exam with a mean of $\mu = 35$ and a standard deviation of $\sigma = 6$ were asked to rate how well they performed on the exam. Below are scores like those observed by Dunning, Johnson, Ehrlinger, and Kruger (2003).

Student	Actual exam score	Perceived exam score
1	33	35
2	30	35
3	36	40
4	36	32
5	26	34
6	35	36
7	40	37
8	38	39
9	44	40
10	42	41
11	21	35
12	35	37
13	41	43
14	29	32
15	36	37

a. Identify Q1 and Q3 and calculate the interquartile range of actual exam scores based on μ and σ.

b. Compute z-scores and use the unit normal table (Appendix B) to identify the percentile rank of each student's actual exam score.

c. For students who earned actual exam scores in the bottom 25%, compute the mean perceived exam score. For students in the top 25%, compute the mean perceived exam score. Which group has greater accuracy in estimating their exam grades?

20. According to a recent report, the average American consumes 22.7 teaspoons of sugar each day (Cohen, August 2013). Assuming that the distribution is approximately normal with a standard deviation of $\sigma = 4.5$, find each of the following values.

a. What percent of people consume more than 32 teaspoons of sugar a day?

b. What percent of people consume more than 18 teaspoons of sugar a day?

c. How much daily sugar intake corresponds to the top 5% of the population?

21. A report in 2016 indicates that Americans between the ages of 8 and 18 spend an average of $\mu = 10$ hours per day using some sort of electronic device such as a smartphone computer, or tablet. Assume that the distribution of times is normal with a standard deviation of $\sigma = 2.5$ hours and find the following values.

a. What is the probability of selecting an individual who uses electronic devices more than 9 hours a day?

b. What proportion of 8- to 18-year-old Americans spend between 8 and 12 hours per day using electronic devices? In symbols, $p(8 < X < 12) = ?$

c. What is the interquartile range in the distribution of time spent using electronic devices?

22. Seattle, Washington, averages $\mu = 34$ inches of annual precipitation. Assuming that the distribution of precipitation amounts is approximately normal with a standard deviation of $\sigma = 6.5$ inches, determine whether each of the following represents a fairly typical year, an extremely wet year, or an extremely dry year.

a. Annual precipitation of 41.8 inches

b. Annual precipitation of 49.6 inches

c. Annual precipitation of 28.0 inches

23. Suppose that a researcher is interested in the effect of new smart drug on IQ. Scores from the IQ test are normally distributed with a mean of $\mu = 100$ and $\sigma = 15$. A participant receives the smart drug and completes the IQ assessment.

a. If the treatment has no effect on IQ, what is the probability that $X > 145$?

b. If the treatment has no effect on IQ, what is the probability that $X > 110$?

Probability and Samples: The Distribution of Sample Means

Tools You Will Need

The following items are considered essential background material for this chapter. If you doubt your knowledge of any of these items, you should review the appropriate chapter and section before proceeding.

- Random sampling (Chapter 6)
- Probability and the normal distribution (Chapter 6)
- z-Scores (Chapter 5)

clivewa/Shutterstock.com

PREVIEW

PREVIEW

Now that you have some understanding of probability, consider the following problem presented to research participants by Nobel Laureates Amos Tversky and Daniel Kahneman:

> Imagine an urn filled with balls. Two-thirds of the balls are one color, and the remaining one-third are a second color. One individual selects 5 balls from the urn and finds that 4 are red and 1 is white. Another individual selects 20 balls and finds that 12 are red and 8 are white. Which of these two individuals should feel more confident that the urn contains two-thirds red balls and one-third white balls, rather than the opposite?*

When Tversky and Kahneman (1974) presented this problem to a group of participants in their now classic study, they found that most people felt that the first sample (4 out of 5) provided much stronger evidence and, therefore, should give more confidence. At first glance, it may appear that this is the correct decision. After all, the first sample contained 4/5 = 80% red balls, and the second sample contained only 12/20 = 60% red balls. However, one sample contains only $n = 5$ balls, and the other sample contains $n = 20$. The correct answer to the problem is that the larger sample (12 out of 20) gives much stronger justification for concluding that the balls in the urn are predominantly red. It appears that most people tend to focus on the sample proportion and pay little attention to the sample size.

The importance of sample size may be easier to appreciate if you approach the urn problem from a different perspective. Suppose that you are the individual assigned the responsibility for selecting a sample and then deciding which color is in the majority. Before you select your sample, you are offered a choice between selecting either a sample of 5 balls or a sample of 20 balls. Which would you prefer? It should be clear that a large sample would be better. With a small number, you risk obtaining an unrepresentative sample. By chance, you could end up with 3 white balls and 2 red balls even though the reds outnumber the whites 2 to 1. The larger sample is much more likely to provide an accurate representation of the population. It provides more information about the population. This is an example of the *law of large numbers*, which states that large samples will be representative of the population from which they are selected. One final example should help demonstrate this law. If you were tossing a coin, you probably would not be surprised to obtain 3 heads in a row. However, if you obtained a series of 20 heads in a row, you most certainly would suspect a trick coin. The large sample has more authority.

In this chapter, we will examine the relationship between samples and populations. More specifically, we will consider the relationship between sample means and the population mean. As you will see, sample size is one of the primary considerations in determining how well a sample mean represents the population mean.

*Adapted from Tversky, A. and Kahneman, D. (1974). Judgments under uncertainty: Heuristics and biases. *Science*, *185*, 1124-1131. Copyright 1974 by the AAAS.

7-1   Samples, Populations, and the Distribution of Sample Means

LEARNING OBJECTIVE

1. Define the distribution of sample means, describe the logically predictable characteristics of the distribution, and use this information to determine characteristics of the distribution of sample means for a specific population and sample size.

The preceding two chapters presented the topics of *z*-scores and probability. Whenever a score is selected from a population, you should be able to compute a *z*-score that describes exactly where the score is located in the distribution. If the population is normal, you also should be able to determine the probability value for obtaining any individual score. In a normal distribution, for example, any score located in the tail of the distribution beyond $z = +2.00$ is an extreme value, and a score this large has a probability of only $p = 0.0228$.

However, the z-scores and probabilities that we have considered so far are limited to situations in which the sample consists of a single score. Most research studies involve much larger samples, such as $n = 20$ laboratory rats in a study of memory, or $n = 100$ school-age children in a study of moral judgment. In these situations, the sample mean, rather than a single score, is used to answer questions about the population. In this chapter we extend the concepts of z-scores and probability to cover situations with samples greater than $n = 1$. In particular, we introduce a procedure for transforming a sample mean into a z-score. Thus, a researcher is able to compute a z-score that describes an entire sample. As always, a z-score value near zero indicates a central, representative sample; a z-value beyond $+2.00$ or -2.00 indicates an extreme sample. Thus, it is possible to describe how any specific sample is related to all the other possible samples. In most situations, we also can use the z-score value to find the probability of obtaining a specific sample, no matter how many scores the sample contains.

In general, the difficulty of working with samples is that a sample provides an incomplete picture of the population. Suppose, for example, a researcher randomly selects a sample of $n = 25$ students from a state college. Although the sample should be representative of the entire student population at that state college, there are almost certainly some segments of the population that are not included in the sample. In addition, any statistics that are computed for the sample will not be identical to the corresponding parameters for the entire population. For example, the average IQ for the sample of 25 students will not be the same as the overall mean IQ for the entire population. This difference, or *error* between sample statistics and the corresponding population parameters, is called *sampling error* and was illustrated in Figure 1.2 (page 8).

> **Sampling error** is the natural discrepancy, or amount of error, between a sample statistic and its corresponding population parameter.

Furthermore, samples are variable; they are not all the same. If you take two separate samples from the same population, the samples will be different. They will contain different individuals, they will have different scores, and they will have different sample means. How can you tell which sample gives the best description of the population? Can you even predict how well a sample will describe its population? What is the probability of selecting a sample with specific characteristics? These questions can be answered once we establish the rules that relate samples and populations.

■ The Distribution of Sample Means

As noted, two separate samples probably will be different even though they are taken from the same population. The samples will have different individuals, different scores, different means, and so on. In most cases, especially for very large populations, it is possible to obtain many thousands, or even millions, of different samples from one population. With all these different samples coming from the same population, it may seem hopeless to try to establish some simple rules for the relationships between samples and populations. Fortunately, however, the huge set of possible samples forms a relatively simple and orderly pattern that makes it possible to predict the characteristics of a sample with some accuracy. The ability to predict sample characteristics is based on the *distribution of sample means*.

The **distribution of sample means** is the collection of sample means for all the possible random samples of a particular size (n) that can be obtained from a population.

Notice that the distribution of sample means contains *all the possible samples*. It is necessary to have all the possible values to compute probabilities. For example, if the entire set contains exactly 100 samples, then the probability of obtaining any specific sample is 1 out of 100: $p = \frac{1}{100}$.

Also, you should notice that the distribution of sample means is different from distributions we have considered before. Until now we always have discussed distributions of scores; now the values in the distribution are not scores, but statistics (sample means). Because statistics are obtained from samples, a distribution of statistics is often referred to as a *sampling distribution*.

A **sampling distribution** is a distribution of statistics obtained by selecting all the possible samples of a specific size from a population.

Thus, the distribution of sample means is an example of a sampling distribution. In fact, it often is called the sampling distribution of M.

If you actually wanted to construct the distribution of sample means, you would first select a random sample of a specific size (n) from a population, calculate the sample mean, and place the sample mean in a frequency distribution. Then you select another random sample with the same number of scores. Again, you calculate the sample mean and add it to your distribution. This process is shown in Figure 7.1. You continue selecting samples and calculating means, over and over. Remember that all the samples have the same number of scores (n). Eventually, you would have the complete set of all the possible random samples, and your frequency distribution would show the distribution of sample means.

■ Characteristics of the Distribution of Sample Means

We demonstrate the process of constructing a distribution of sample means in Example 7.1, but first we use common sense and a little logic to predict the general characteristics of the distribution.

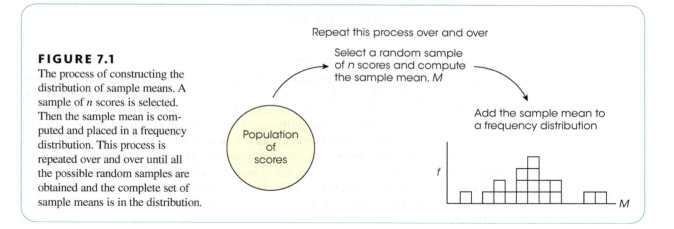

FIGURE 7.1

The process of constructing the distribution of sample means. A sample of n scores is selected. Then the sample mean is computed and placed in a frequency distribution. This process is repeated over and over until all the possible random samples are obtained and the complete set of sample means is in the distribution.

Repeat this process over and over

Select a random sample of n scores and compute the sample mean, M

Population of scores

Add the sample mean to a frequency distribution

1. The sample means should pile up around the population mean. Samples are not expected to be perfect but they are representative of the population. As a result, most of the sample means should be relatively close to the population mean.

2. The pile of sample means should tend to form a normal-shaped distribution. Logically, most of the samples should have means close to μ, and it should be relatively rare to find sample means that are substantially different from μ. As a result, the sample means should pile up in the center of the distribution (around μ) and the frequencies should taper off as the distance between M and μ increases. This describes a normal-shaped distribution.

3. In general, the larger the sample size, the closer the sample means should be to the population mean, μ. Logically, a large sample should be better than a small sample because it is more representative. Thus, the sample means obtained with a large sample size should cluster relatively close to the population mean; the means obtained from small samples should be more widely scattered.

As you will see, each of these three commonsense characteristics is an accurate description of the distribution of sample means. The following example demonstrates the process of constructing the distribution of sample means by repeatedly selecting samples from a population.

EXAMPLE 7.1

Remember that random sampling requires sampling with replacement.

We begin with a population that consists of only four scores: 2, 4, 6, 8. This population is pictured in the frequency distribution histogram in Figure 7.2.

We are going to use this population as the basis for constructing the distribution of sample means for $n = 2$. Remember: this distribution is the collection of sample means from all the possible random samples of $n = 2$ from this population. We begin by looking at all the possible samples. For this example, there are 16 different samples, and they are all listed in Table 7.1. Notice that the samples are listed systematically. First, we list all the possible samples with $X = 2$ as the first score, then all the possible samples with $X = 4$ as the first score, and so on. In this way, we are sure that we have all of the possible random samples.

Next, we compute the mean, M, for each of the 16 samples (see the last column of Table 7.1). The 16 means are then placed in a frequency distribution histogram (Figure 7.3). This is the distribution of sample means. Note that the distribution in Figure 7.3 demonstrates two of the characteristics that we predicted for the distribution of sample means.

1. The sample means pile up around the population mean. For this example, the population mean is $\mu = 5$, and the sample means are clustered around a value of 5. It should not surprise you that the sample means tend to approximate the population mean. After all, samples are supposed to be representative of the population.

FIGURE 7.2
Frequency distribution histogram for a population of $N = 4$ scores: 2, 4, 6, 8.

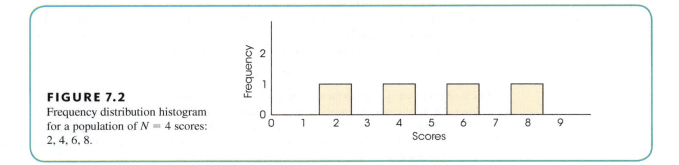

TABLE 7.1

The complete set of possible samples of $n = 2$ scores that can be obtained from the population presented in Figure 7.2. Notice that the table lists *random samples*. This requires sampling with replacement, so it is possible to select the same score twice.

Sample	Scores		Sample Mean (M)
	First	Second	
1	2	2	2
2	2	4	3
3	2	6	4
4	2	8	5
5	4	2	3
6	4	4	4
7	4	6	5
8	4	8	6
9	6	2	4
10	6	4	5
11	6	6	6
12	6	8	7
13	8	2	5
14	8	4	6
15	8	6	7
16	8	8	8

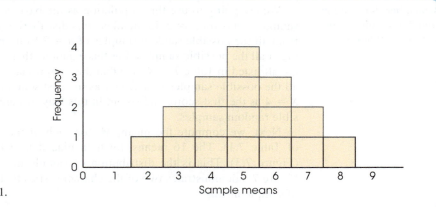

FIGURE 7.3

The distribution of sample means for $n = 2$. The distribution shows the 16 sample means from Table 7.1.

2. The distribution of sample means is approximately normal in shape. This is a characteristic that is discussed in detail later and is extremely useful because we already know a great deal about probabilities and the normal distribution (Chapter 6).

Finally, you should realize that we can use the distribution of sample means to achieve the goal for this chapter, which is to answer probability questions about sample means. For example, if you take a sample of $n = 2$ scores from the original population, what is the probability of obtaining a sample with a mean greater than 7? In symbols,

$$p(M > 7) = ?$$

Remember that our goal in this chapter is to answer probability questions about samples with $n > 1$.

Because probability is equivalent to proportion, the probability question can be restated as follows: Of all the possible sample means, what proportion have values greater than 7? In this form, the question is easily answered by looking at the distribution of sample means. All the possible sample means are pictured (see Figure 7.3), and only 1 out of the 16 means has a value greater than 7. The answer, therefore, is 1 out of 16, or $p = \frac{1}{16}$. ∎

LEARNING CHECK   **LO1**   1. If all the possible random samples, each with $n = 9$ scores, are selected from a normally distributed population with $\mu = 90$ and $\sigma = 20$, and the mean is calculated for each sample, then what is the average value for all of the sample means?

    a. 9

    b. 90

    c. $9(90) = 810$

    d. Cannot be determined without additional information.

LO1   2. All the possible random samples of size $n = 2$ are selected from a population with $\mu = 40$ and $\sigma = 10$ and the mean is computed for each sample. Then all the possible samples of size $n = 25$ are selected from the same population and the mean is computed for each sample. How will the distribution of sample means for $n = 2$ compare with the distribution for $n = 25$?

    a. The two distributions will have the same mean and variability.

    b. The mean and variability for $n = 25$ will both be larger than the mean and variability for $n = 2$.

    c. The mean and variability for $n = 25$ will both be smaller than the mean and variability for $n = 2$.

    d. The variability for $n = 25$ will be smaller than the variability for $n = 2$, but the two distributions will have the same mean.

LO1   3. If all the possible random samples of size $n = 25$ are selected from a population with $\mu = 90$ and $\sigma = 20$ and the mean is computed for each sample, then what shape is expected for the distribution of sample means?

    a. The sample means tend to form a normal-shaped distribution.

    b. The distribution of sample means will have the same shape as the sample distribution.

    c. The sample will be distributed evenly across the scale, forming a rectangular-shaped distribution.

    d. There are thousands of possible samples and it is impossible to predict the shape of the distribution.

ANSWERS   1. b   2. d   3. a

| 7-2 | # Shape, Central Tendency, and Variability for the Distribution of Sample Means |

LEARNING OBJECTIVES

2. Explain how the central limit theorem specifies the shape, central tendency, and variability for the distribution of sample means, and use this information to construct the distribution of sample means for a specific sample size from a specified population.

3. Describe how the standard error of M is calculated, explain what it measures, describe how it is related to the standard deviation for the population, and use this information to determine the standard error for samples of a specific size selected from a specified population.

Once again, the symbol for standard error is σ_M. The σ indicates that this value is a standard deviation, and the subscript M indicates that it is the standard deviation for the distribution of sample means. The standard error is an extremely valuable measure because it specifies precisely how well a sample mean estimates its population mean—that is, how much error you should expect, on average, between M and μ. Remember that one basic reason for taking samples is to use the sample data to answer questions about the population. However, you do not expect a sample to provide a perfectly accurate picture of the population. There always is some discrepancy or error between a sample statistic and the corresponding population parameter. Now we are able to calculate exactly how much error to expect. For any sample size (n), we can compute the standard error, which measures the average distance between a sample mean and the population mean.

According to the central limit theorem, the standard error is equal to $\sigma/\sqrt{n}$. Thus, the magnitude of the standard error is determined by two factors: (1) the size of the sample and (2) the standard deviation of the population from which the sample is selected. We will examine each of these factors.

The Sample Size   We previously predicted, based on common sense, that the size of a sample should influence how accurately the sample represents its population. Specifically, a large sample should be more accurate than a small sample. In general, as the sample size increases, the error between the sample mean and the population mean should decrease. This rule is also known as the *law of large numbers* (see Box 7.1).

The **law of large numbers** states that the larger the sample size (n), the more probable it is that the sample mean will be close to the population mean.

The Population Standard Deviation   As we noted earlier, the size of the standard error depends on the size of the sample. Specifically, bigger samples have smaller error,

---

BOX 7.1 The Law of Large Numbers and Online Shopping

The law of large numbers dictates that we can be more confident about the values of sample statistics when they describe larger samples than when they describe smaller samples. The law of large numbers is important in research of course, but it is also important for statistically guided decision making in the real world. A common situation you might encounter involves making decisions about online purchases based on product ratings.

Suppose that you are interested in purchasing earbuds for your smartphone. You search Amazon.com™ for "earbuds" and receive several thousand choices. You decide to sort these items by rating, hoping to find the best earbuds. At the top of the list are the EarBuddies earbuds, which received a mean rating of $M = 5.0$ stars based on $n = 2$ reviews. Next on the list, you find the Brain Tickler earbuds, which received a mean rating of $M = 4.5$ stars among $n = 10$ reviews. Next are the Skull Crusher earbuds, which received a mean rating of $M = 4.2$ stars among $n = 1{,}600$ reviews. Which earbuds should you buy? All things being equal, you should choose the earbuds with the higher *mean* rating. However, while the EarBuddies would seem to be a better choice than the Brain Ticklers, you also must consider the sample size. The EarBuddies ratings are based on a sample of only $n = 2$ customers. With such a small sample, a mean of $M = 5.0$ is based on just a very small glimpse of the population of EarBuddies users. Similarly, the Brain Ticklers are based on a sample of only 10 reviews. In contrast, the ratings for the Skull Crushers are based on a large sample $n = 1{,}600$ customers, and thus, its mean rating of $M = 4.2$ stars is likely to be representative of the population of Skull Crusher users. In this case, the Skull Crusher earbuds are the best choice.

and smaller samples have bigger error. At the extreme, the smallest possible sample (and the largest standard error) occurs when the sample consists of $n = 1$ score. At this extreme, each sample is a single score and the distribution of sample means is identical to the original population distribution of scores. In this case, the standard deviation for the distribution of sample means, which is the standard error, is identical to the standard deviation for the distribution of scores. In other words, when $n = 1$, the standard error $= \sigma_M$ is identical to the standard deviation $= \sigma$.

When $n = 1$, $\sigma_M = \sigma$ (standard error = standard deviation).

You can think of the standard deviation as the "starting point" for standard error. When $n = 1$, the standard error and the standard deviation are the same: $\sigma_M = \sigma$. As sample size increases beyond $n = 1$, the sample becomes a more accurate representative of the population, and the standard error decreases. The formula for standard error expresses this relationship between standard deviation and sample size (n).

This formula is contained in the central limit theorem.

$$\text{standard error} = \sigma_M = \frac{\sigma}{\sqrt{n}} \qquad (7.1)$$

Note that the formula satisfies all the requirements for the standard error. Specifically:

a. As sample size (n) increases, the size of the standard error decreases. (Larger samples are more accurate.)

b. When the sample consists of a single score ($n = 1$), the standard error is the same as the standard deviation ($\sigma_M = \sigma$).

Figure 7.4 illustrates the general relationship between standard error and sample size. (The calculations for the data points in Figure 7.4 are presented in Table 7.2.) Again, the basic concept is that the larger a sample is, the more accurately it represents its population. Also note that the standard error decreases in relation to the *square root* of the sample size. As a result, researchers can substantially reduce error by increasing sample size up to around $n = 30$. However, increasing sample size beyond $n = 30$ produces relatively small improvement in how well the sample represents the population.

Defining the Standard Error in Terms of Variance   In Equation 7.1 and in most of the preceding discussion, we have defined standard error in terms of the population

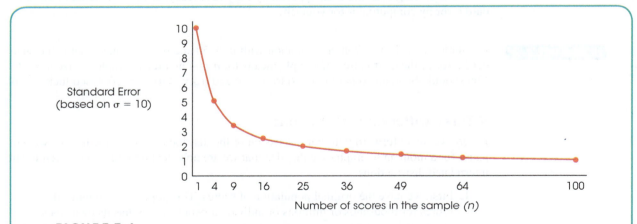

FIGURE 7.4
The relationship between standard error and sample size based on $\sigma = 10$. As the sample size is increased, there is less error between the sample mean and the population mean.

TABLE 7.2

Calculations for the points shown in Figure 7.4. Again, notice that the size of the standard error decreases as the size of the sample increases.

Sample Size (n)	Standard Error
1	$\sigma_M = \dfrac{10}{\sqrt{1}} = 10.00$
4	$\sigma_M = \dfrac{10}{\sqrt{4}} = 5.00$
9	$\sigma_M = \dfrac{10}{\sqrt{9}} = 3.33$
16	$\sigma_M = \dfrac{10}{\sqrt{16}} = 2.50$
25	$\sigma_M = \dfrac{10}{\sqrt{25}} = 2.00$
49	$\sigma_M = \dfrac{10}{\sqrt{49}} = 1.43$
64	$\sigma_M = \dfrac{10}{\sqrt{64}} = 1.25$
100	$\sigma_M = \dfrac{10}{\sqrt{100}} = 1.00$

standard deviation. However, the population standard deviation (σ) and the population variance (σ^2) are directly related, and it is easy to substitute variance into the equation for standard error. Using the simple equality $\sigma = \sqrt{\sigma^2}$, the equation for standard error can be rewritten as follows:

$$\text{standard error} = \sigma_M = \frac{\sigma}{\sqrt{n}} = \frac{\sqrt{\sigma^2}}{\sqrt{n}} = \sqrt{\frac{\sigma^2}{n}} \tag{7.2}$$

Throughout the rest of this chapter (and in Chapter 8), we will continue to define standard error in terms of the standard deviation (Equation 7.1). However, in later chapters (starting in Chapter 9) the formula based on variance (Equation 7.2) will become more useful.

The following example is an opportunity for you to test your understanding of the standard error by computing it for yourself.

EXAMPLE 7.2

If samples are selected from a population with $\mu = 50$ and $\sigma = 12$, then what is the standard error of the distribution of sample means for $n = 4$ and for a sample of size $n = 16$? You should obtain answers of $\sigma_M = 6$ for $n = 4$ and $\sigma_M = 3$ for $n = 16$. Good luck.   ∎

■ Three Different Distributions

Before we move forward with our discussion of the distribution of sample means, we will pause for a moment to emphasize the idea that we are now dealing with three different but interrelated distributions.

1. First, we have the original population of scores. This population contains the scores for thousands or millions of individual people, and it has its own shape, mean, and standard deviation. For example, suppose a population consists of millions of scores on a standardized reading comprehension test, and that these scores form a normal distribution with a mean of $\mu = 100$ and a standard deviation of $\sigma = 12$. An example of a population is shown in Figure 7.5(a).

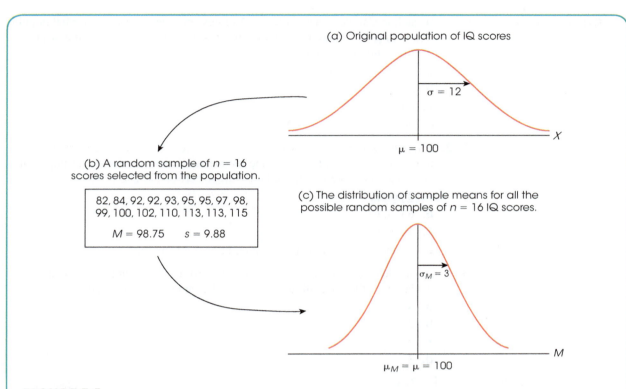

(a) Original population of IQ scores

$\sigma = 12$

$\mu = 100$

X

(b) A random sample of $n = 16$ scores selected from the population.

82, 84, 92, 92, 93, 95, 95, 97, 98, 99, 100, 102, 110, 113, 113, 115

$M = 98.75$ $s = 9.88$

(c) The distribution of sample means for all the possible random samples of $n = 16$ IQ scores.

$\sigma_M = 3$

$\mu_M = \mu = 100$

M

FIGURE 7.5

Three distributions. Part (a) shows the population of reading scores. Part (b) shows a sample of $n = 16$ reading scores. Part (c) shows the distribution of sample means for all possible samples of $n = 16$ reading scores. Note that the mean for the sample in part (b) is one of the thousands of sample means in the distribution shown in part (c).

2. Next, we have a sample that is selected from the population. The sample consists of a small set of scores for people who have been selected to represent the entire population. For example, we could select a random sample of $n = 16$ people and measure each individual's reading score. The sample mean and the sample standard deviation are calculated for these 16 scores. Note that the sample has its own mean and standard deviation. The scores for the sample are shown in Figure 7.5(b). If you sketched a frequency distribution for these data, you would see that the sample distribution also has its own shape.

3. The third distribution is the distribution of sample means. This is a theoretical distribution consisting of the sample means obtained from all the possible random samples of a specific size. For example, the distribution of sample means for samples of $n = 16$ reading scores would be normal with a mean (expected value) of $\mu = 100$ and a standard deviation (standard error) of $\sigma_M = \frac{12}{\sqrt{16}} = \frac{12}{4} = 3$. This distribution, shown in Figure 7.5(c), is narrower than the population distribution because its standard deviation ($\sigma_M = 3$) is smaller than the population standard deviation ($\sigma = 12$).

Note that the scores for the sample [Figure 7.5(b)] were taken from the original population [Figure 7.5(a)] and that the mean for the sample is one of the values contained in the distribution of sample means [Figure 7.5(c)]. Thus, the three distributions are all connected, but they are all distinct.

LEARNING CHECK

LO2 **1.** If random samples, each with $n = 4$ scores, are selected from a normal population with $\mu = 90$ and $\sigma = 20$, then what is the expected value of the mean for the distribution of sample means?

   a. 2.5

   b. 5

   c. 40

   d. 90

LO3 **2.** If random samples, each with $n = 4$ scores, are selected from a normal population with $\mu = 80$ and $\sigma = 12$, and the mean is calculated for each sample, then how much distance is expected on average between M and μ?

   a. 2 points

   b. 6 points

   c. 18 points

   d. Cannot be determined without additional information.

LO3 **3.** A sample of $n = 4$ scores has a standard error of 24. What is the standard deviation of the population from which the sample was obtained?

   a. 48

   b. 24

   c. 6

   d. 3

ANSWERS   **1. d   2. b   3. a**

7-3 z-Scores and Probability for Sample Means

LEARNING OBJECTIVES

4. Calculate the z-score for a sample mean.

5. Describe the circumstances in which the distribution of sample means is normal and, in these circumstances, find the probability associated with a specific sample.

The primary use for the distribution of sample means is to find the probability of selecting a sample with a specific mean. Recall that probability is equivalent to proportion. Because the distribution of sample means presents the entire set of all possible sample means, we can use proportions of this distribution to determine the probability of obtaining a sample with a specific mean. The following example demonstrates this process.

EXAMPLE 7.3

Caution: Whenever you have a probability question about a sample mean, you must use the distribution of sample means.

Suppose the population of scores on the Math SAT forms a normal distribution with $\mu = 500$ and $\sigma = 100$. If you take a random sample of $n = 16$ students, what is the probability that the sample mean will be greater than $M = 525$?

First, you can restate this probability question as a proportion question: Out of all the possible sample means, what proportion have values greater than 525? You know about "all the possible sample means"; this is the distribution of sample means. The problem is to find a specific portion of this distribution.

Although we cannot construct the distribution of sample means by repeatedly taking samples and calculating means (as in Example 7.1), we know exactly what the distribution looks like based on the information from the central limit theorem. Specifically, the distribution of sample means has the following characteristics:

a. The distribution is normal because the population of Math SAT scores is normal.

b. The distribution has a mean of 500 because the population mean is $\mu = 500$.

c. For $n = 16$, the distribution has a standard error of $\sigma_M = 25$:

$$\sigma_M = \frac{\sigma}{\sqrt{n}} = \frac{100}{\sqrt{16}} = \frac{100}{4} = 25$$

This distribution of sample means is shown in Figure 7.6.

We are interested in sample means greater than 525 (the shaded area in Figure 7.6), so the next step is to use a z-score to locate the exact position of $M = 525$ in the distribution. The value 525 is located above the mean by 25 points, which is exactly one standard deviation (in this case, exactly one standard error). Thus, the z-score for $M = 525$ is $z = +1.00$.

Because this distribution of sample means is normal, you can use the unit normal table to find the probability associated with $z = +1.00$. The table indicates that 0.1587 of the distribution is located in the tail of the distribution beyond $z = +1.00$. Our conclusion is that it is relatively unlikely, $p = 0.1587$ (15.87%), to obtain a random sample of $n = 16$ students with an average Math SAT score greater than 525. ■

■ A z-Score for Sample Means

As demonstrated in Example 7.3, it is possible to use a z-score to describe the exact location of any specific sample mean within the distribution of sample means. The z-score tells exactly where the sample mean is located in relation to all the other possible sample means that could have been obtained. As defined in Chapter 5, a z-score identifies the location with a signed number so that

1. the sign tells whether the location is above ($+$) or below ($-$) the mean.

2. the number tells the distance between the location and the mean in terms of the number of standard deviations.

However, we are now finding a location within the distribution of sample means. Therefore, we must use the notation and terminology appropriate for this distribution. First, we

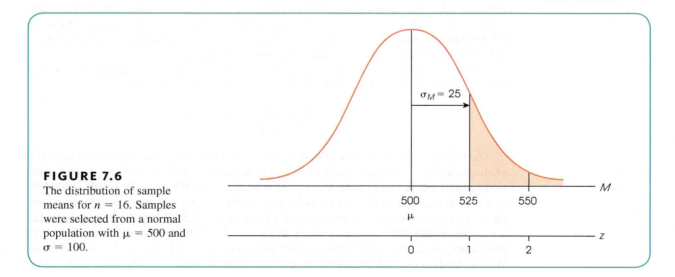

FIGURE 7.6
The distribution of sample means for $n = 16$. Samples were selected from a normal population with $\mu = 500$ and $\sigma = 100$.

are finding the location for a sample mean (M) rather than a score (X). Second, the standard deviation for the distribution of sample means is the standard error, σ_M. Therefore, the z-score for a sample mean can be defined as a signed number that identifies the location of the sample mean in the distribution of sample means so that

1. the sign tells whether the sample mean is located above ($+$) or below ($-$) the mean for the distribution (which is the population mean, μ).

2. the number tells the distance between the sample mean and μ in terms of the number of standard errors.

With these changes, the z-score formula for locating a sample mean is

$$z = \frac{M - \mu}{\sigma_M}$$
(7.3)

Caution: **When computing z for a single score, use the standard deviation, σ. When computing z for a sample mean, you must use the standard error, σ_M.**

Just as every score (X) has a z-score that describes its position in the distribution of scores, every sample mean (M) has a z-score that describes its position in the distribution of sample means. When the distribution of sample means is normal, it is possible to use z-scores and the unit normal table to find the probability associated with any specific sample mean (as in Example 7.3). The following example is an opportunity for you to test your understanding of z-scores and probability for sample means.

EXAMPLE 7.4   A sample of $n = 4$ scores is selected from a normal distribution with a mean of $\mu = 40$ and a standard deviation of $\sigma = 16$.

a. Find the z-score for a sample mean of $M = 42$.

b. Determine the probability of obtaining a sample mean larger than $M = 42$.

You should obtain $z = 0.25$ and $p = 0.4013$. Good luck.   ■

The following example demonstrates that the distribution of sample means also can be used to make quantitative predictions about the kinds of samples that should be obtained from any population.

EXAMPLE 7.5   Once again, suppose the population of Math SAT scores forms a normal distribution with a mean of $\mu = 500$ and a standard deviation of $\sigma = 100$. For this example, we are going to determine what kind of sample mean is likely to be obtained as the average SAT score for a random sample of $n = 25$ students. Specifically, we will determine the exact range of values that is expected for the sample mean 80% of the time.

We begin with the distribution of sample means for $n = 25$. This distribution is normal with an expected value of $\mu = 500$ and, with $n = 25$, the standard error is

$$\sigma_M = \frac{\sigma}{\sqrt{n}} = \frac{100}{\sqrt{25}} = \frac{100}{5} = 20$$

See Figure 7.7. Our goal is to find the range of values that make up the middle 80% of the distribution. Because the distribution is normal, we can use the unit normal table to determine the boundaries for the middle 80%. First, the 80% is split in half, with 40% (0.4000) on each side of the mean. Looking up 0.4000 in column D (the proportion between the mean and z), we find a corresponding z-score of $z = 1.28$. Thus, the z-score boundaries for the middle 80% are $z = +1.28$ and $z = -1.28$. By definition, a z-score of 1.28 represents a location that is 1.28 standard deviations (or standard errors) from the mean. With a standard error of 20 points, the distance from the mean is $1.28(20) = 25.6$ points.

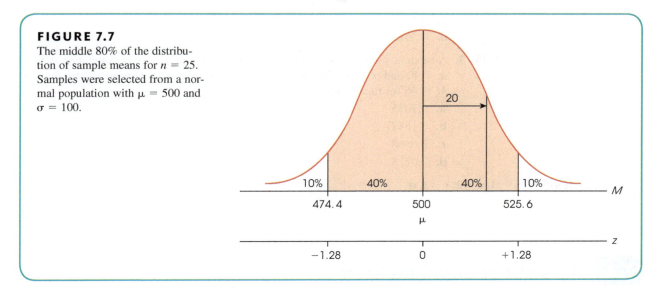

FIGURE 7.7
The middle 80% of the distribution of sample means for $n = 25$. Samples were selected from a normal population with $\mu = 500$ and $\sigma = 100$.

The mean is $\mu = 500$, so a distance of 25.6 in both directions produces a range of values from 474.4 to 525.6. A useful formula for calculating the sample means that form these boundaries is as follows:

$$M = \mu + z(\sigma_M) \tag{7.4}$$

Notice the similarity of this equation to equation 5.2 (page 155). Remember to pay attention to the sign of z.

Thus, 80% of all the possible sample means are contained in a range between 474.4 and 525.6. If we select a sample of $n = 25$ students, we can be 80% confident that the mean Math SAT score for the sample will be in this range. ∎

The point of Example 7.5 is that the distribution of sample means makes it possible to predict the value that ought to be obtained for a sample mean. Because the population mean is $\mu = 500$, we know that a sample of $n = 25$ students ought to have a mean Math SAT score around 500. More specifically, we are 80% confident that the value of the sample mean will be between 474.4 and 525.6. The ability to predict sample means in this way will be a valuable tool for the inferential statistics that follow.

LEARNING CHECK   **LO4** **1.** A sample of $n = 25$ scores is obtained from a population with $\mu = 70$ and $\sigma = 20$. If the sample mean is $M = 78$, then what is the *z*-score corresponding to the sample mean?

  a. $z = +0.25$
  b. $z = +0.50$
  c. $z = +1.00$
  d. $z = +2.00$

LO5 **2.** A random sample of $n = 4$ scores is obtained from a normal population with $\mu = 20$ and $\sigma = 4$. What is the probability of obtaining a mean greater than $M = 22$ for this sample?

  a. 0.50
  b. 1.00

   c. 0.1587

   d. 0.3085

LO5 **3.** A random sample of $n = 4$ scores is obtained from a normal population with $\mu = 40$ and $\sigma = 6$. What is the probability of obtaining a mean greater than $M = 46$ for this sample?

   a. 0.3085

   b. 0.1587

   c. 0.0668

   d. 0.0228

ANSWERS   **1. d   2. c   3. d**

7-4   More about Standard Error

LEARNING OBJECTIVE

6. Describe how the magnitude of the standard error is related to the size of the sample, and determine the sample size needed to produce a specified standard error or the new standard error produced by a specific change in the sample size.

In Chapter 5, we introduced the idea of z-scores to describe the exact location of individual scores within a distribution. In Chapter 6, we introduced the idea of finding the probability of obtaining any individual score, especially scores from a normal distribution. By now, you should realize that most of this chapter is simply repeating the same things that were covered in Chapters 5 and 6, but with two adjustments:

1. We are now using the distribution of sample means instead of a distribution of scores.

2. We are now using the standard error instead of the standard deviation.

Of these two adjustments, the primary new concept in Chapter 7 is the standard error, and the single rule that you need to remember is:

Whenever you are working with a sample mean, you must use the standard error.

This single rule encompasses essentially all the new content in Chapter 7. Therefore, this section will focus on the concept of standard error to ensure that you have a good understanding of this new concept.

■ Sampling Error and Standard Error

At the beginning of this chapter, we introduced the idea that it is possible to obtain thousands of different samples from a single population. Each sample will have its own individuals, its own scores, and its own sample mean. The distribution of sample means provides a method for organizing all of the different sample means into a single picture. Figure 7.8 shows a distribution of sample means and its relation to the normal distribution. To emphasize the fact that the distribution contains many different samples, we have constructed this figure so that it displays both a frequency distribution histogram and a normal distribution. The histogram shows a collection of samples, specifically the means

FIGURE 7.8

An example of a typical distribution of sample means. Each of the bars in the histogram represents the frequencies for different sample means. A curve for the normal distribution is superimposed on the histogram. The expected value of the distribution of sample means equals the population mean, μ.

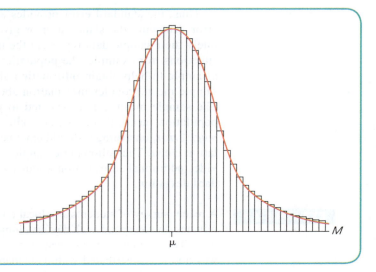

for these samples, which have been selected from a population. The height of each bar reflects the frequency for sample means of a certain value. (If each sample mean were a box, imagine these boxes stacked to create the bars.) As more samples are selected and the means are included in the histogram, the histogram begins to more closely approximate the normal distribution. Notice that the sample means tend to pile up around the population mean (μ), forming a normal-shaped distribution as predicted by the central limit theorem.

The distribution shown in Figure 7.8 provides a concrete example for reviewing the general concepts of sampling error and standard error. Although the following points may seem obvious, they are intended to provide you with a better understanding of these two statistical concepts.

1. **Sampling Error**. The general concept of sampling error is that a sample typically will not provide a perfectly accurate representation of its population. More specifically, there typically is some discrepancy (or error) between a statistic computed for a sample and the corresponding parameter for the population. As you look at Figure 7.8, notice that the individual sample means are not exactly equal to the population mean. In fact, 50% of the samples have means that are smaller than μ (the entire left-hand side of the distribution). Similarly, 50% of the samples produce means that overestimate the true population mean. In general, there will be some discrepancy, or *sampling error*, between the mean for a sample and the mean for the population from which the sample was obtained.

2. **Standard Error**. Again, looking at Figure 7.8, notice that most of the sample means are relatively close to the population mean (those in the center of the distribution). These samples provide a fairly accurate representation of the population. On the other hand, some samples produce means that are out in the tails of the distribution, relatively far from the population mean. These extreme sample means do not accurately represent the population. For each individual sample, you can measure the error (or distance) between the sample mean and the population mean. For some samples, the error will be relatively small, but for other samples, the error will be relatively large. The *standard error* provides a way to measure the "average," or standard, distance between a sample mean and the population mean.

Thus, the standard error provides a method for defining and measuring sampling error. Knowing the standard error gives researchers a good indication of how accurately their sample data represent the populations they are studying. In most research situations, for example, the population mean is unknown, and the researcher selects a sample to help obtain information about the unknown population. Specifically, the sample mean provides information about the value of the unknown population mean. The sample mean is not expected to give a perfectly accurate representation of the population mean; there will be some error, and the standard error tells *exactly how much error*, on average, should exist between the sample mean and the unknown population mean. The following example demonstrates the use of standard error and provides additional information about the relationship between standard error and standard deviation.

EXAMPLE 7.6   A recent survey of students at a local college included the following question: How many minutes do you spend each day watching electronic video (online, TV, smartphone, tablet, etc.). The average response was $\mu = 80$ minutes, and the distribution of viewing times was normally distributed with a standard deviation of $\sigma = 20$ minutes. Next, we take a sample from this population and examine how accurately the sample mean represents the population mean. More specifically, we will examine how sample size affects accuracy by considering three different samples: one with $n = 1$ student, one with $n = 4$ students, and one with $n = 100$ students.

Figure 7.9 shows the distributions of sample means based on samples of $n = 1$, $n = 4$, and $n = 100$. Each distribution shows the collection of all possible sample means that could be obtained for that particular sample size. Notice that all three sampling distributions are normal (because the original population is normal), and all three have the same mean, $\mu = 80$, which is the expected value of M. However, the three distributions differ greatly with respect to variability. We will consider each one separately.

The smallest sample size is $n = 1$. When a sample consists of a single student, the mean for the sample equals the score for the student, $M = X$. Thus, when $n = 1$, the distribution of sample means is identical to the original population of scores. In this case, the standard

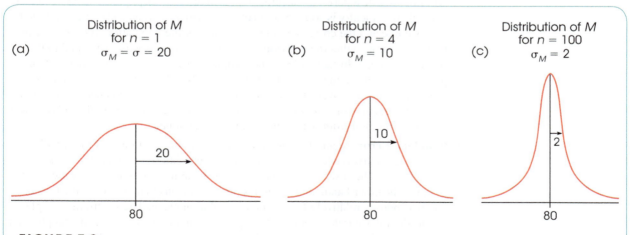

FIGURE 7.9
The distribution of sample means for (a) $n = 1$, (b) $n = 4$, and (c) $n = 100$ obtained from a normal population with $\mu = 80$ and $\sigma = 20$. Notice that the size of the standard error decreases as the sample size increases.

error for the distribution of sample means is equal to the standard deviation for the original population. Equation 7.1 confirms this observation:

$$\sigma_M = \frac{\sigma}{\sqrt{n}} = \frac{20}{\sqrt{1}} = 20$$

When the sample consists of a single student, you expect, on average, a 20-point difference between the sample mean and the mean for the population. As we noted earlier, the population standard deviation is the "starting point" for the standard error. With the smallest possible sample, $n = 1$, the standard error is equal to the standard deviation [see Figure 7.9(a)].

As the sample size increases, however, the standard error gets smaller. For a sample of $n = 4$ students, the standard error is

$$\sigma_M = \frac{\sigma}{\sqrt{n}} = \frac{20}{\sqrt{4}} = \frac{20}{2} = 10$$

That is, the typical (or standard) distance between M and μ is 10 points. Figure 7.9(b) illustrates this distribution. Notice that the sample means in this distribution approximate the population mean more closely than in the previous distribution where $n = 1$.

With a sample of $n = 100$, the standard error is still smaller.

$$\sigma_M = \frac{\sigma}{\sqrt{n}} = \frac{20}{\sqrt{100}} = \frac{20}{10} = 2$$

A sample of $n = 100$ students should produce a sample mean that represents the population much more accurately than a sample of $n = 4$ or $n = 1$. As shown in Figure 7.9(c), there is very little error between M and μ when $n = 100$. Specifically, you would expect on average only a 2-point difference between the sample mean and the population mean. ∎

In summary, this example illustrates that with the smallest possible sample ($n = 1$), the standard error and the population standard deviation are the same. As sample size increases, the standard error gets smaller, and the sample means tend to approximate μ more closely. Thus, standard error defines the relationship between sample size and the accuracy with which M represents μ.

IN THE LITERATURE

Reporting Standard Error

As we will see in future chapters, the standard error plays a very important role in inferential statistics. Because of its crucial role, the standard error for a sample mean, rather than the sample standard deviation, is often reported in scientific papers. Scientific journals vary in how they refer to the standard error, but frequently the symbols *SE* and *SEM* (for standard error of the mean) are used. The standard error is reported in two ways. Much like the standard deviation, it may be reported in a table along with the sample means (Table 7.3). Alternatively, the standard error may be reported in graphs.

Figure 7.10 illustrates the use of a bar graph to display information about the sample mean and the standard error. In this experiment, two samples (groups A and B) are given different treatments, and then the participants' scores on a dependent variable are recorded. The mean for group A is $M = 15$, and for group B it is $M = 30$. For both

(continues)

FIGURE 7.10

The mean score ($\pm SE$) for treatment groups A and B.

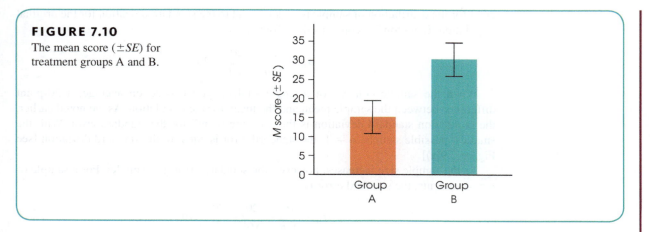

TABLE 7.3

The mean self-consciousness scores for participants who were working in front of a video camera and those who were not (controls).

	n	Mean	SE
Control	17	32.23	2.31
Camera	15	45.17	2.78

samples, the standard error of M is $\sigma_M = 4$. Note that the mean is represented by the height of the bar, and the standard error is depicted by brackets at the top of each bar. Each bracket extends for a distance equal to one standard error above and one standard error below the sample mean. Thus, the graph illustrates the mean for each group plus or minus one standard error ($M \pm SE$). When you glance at Figure 7.10, not only do you get a "picture" of the sample means, but also you get an idea of how much error you should expect for those means.

Figure 7.11 shows how sample means and standard error are displayed in a line graph. In this study, two samples representing different age groups are tested on a task for four trials. The number of errors committed on each trial is recorded for all participants. The graph shows the mean (M) number of errors committed for each group on each trial. The brackets show the size of the standard error for each sample mean. Again, the brackets extend one standard error above and below the value of the mean.

FIGURE 7.11

The mean number of mistakes ($\pm SE$) for groups A and B on each trial.

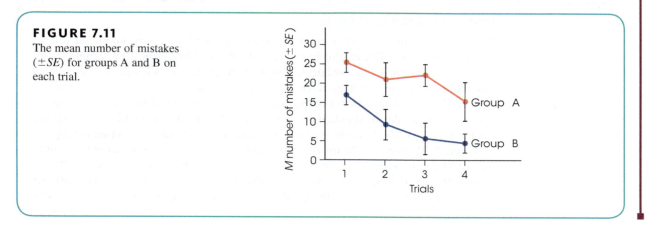

LEARNING CHECK   **LO6**   **1.** Which of the following would cause the standard error of M to get larger?

    a. Increasing both the sample size and standard deviation.

    b. Decreasing both the sample size and standard deviation.

    c. Increasing the sample size and decreasing the standard deviation.

    d. Decreasing the sample size and increasing the standard deviation.

LO6   **2.** A sample obtained from a population with $\sigma = 8$ has a standard error of 2 points. How many scores are in the sample?

    a. $n = 5$

    b. $n = 10$

    c. $n = 16$

    d. $n = 25$

LO6   **3.** A random sample is selected from a population with $\mu = 80$ and $\sigma = 20$. How large must the sample be to ensure a standard error of 2 points or less?

    a. $n = 10$

    b. $n = 25$

    c. $n = 100$

    d. It is impossible to obtain a standard error of less than 2 for any sized sample.

ANSWERS   **1. d   2. c   3. c**

7-5   Looking Ahead to Inferential Statistics

LEARNING OBJECTIVE

7. Explain how the distribution of sample means can be used to evaluate a treatment effect by identifying likely and very unlikely samples, and use this information to determine whether a specific sample suggests that a treatment effect is likely or very unlikely.

Inferential statistics are methods that use sample data as the basis for drawing general conclusions about populations. However, we have noted that a sample is not expected to give a perfectly accurate reflection of its population. In particular, there will be some error or discrepancy between a sample statistic and the corresponding population parameter. In this chapter, we focused on sample means and observed that a sample mean will not be exactly equal to the population mean. The standard error of M specifies how much difference is expected on average between the mean for a sample and the mean for the population.

The natural differences that exist between samples and populations introduce a degree of uncertainty and error into all inferential processes. Specifically, there is always a margin of error that must be considered whenever a researcher uses a sample mean as the basis for drawing a conclusion about a population mean. Remember that the sample mean is not perfect. In the next six chapters we introduce a variety of statistical methods that all use sample means to draw inferences about population means.

In each case, the distribution of sample means and the standard error will be critical elements in the inferential process. Before we begin this series of chapters, we pause briefly to demonstrate how the distribution of sample means, along with z-scores and probability, can help us use sample means to draw inferences about population means.

EXAMPLE 7.7 Suppose that a psychologist is planning a research study to evaluate the effect of a new growth hormone. It is known that regular, adult rats (with no hormone) weigh an average of $\mu = 400$ grams. Of course, not all rats are the same size, and the distribution of their weights is normal with $\sigma = 20$. The psychologist plans to select a sample of $n = 25$ newborn rats, inject them with the hormone, and then measure their weights when they become adults. The structure of this research study is shown in Figure 7.12.

The psychologist will make a decision about the effect of the hormone by comparing the sample of treated rats with the regular untreated rats in the original population. If the treated rats in the sample are noticeably different from untreated rats, then the researcher has evidence that the hormone has an effect. The problem is to determine exactly how much difference is necessary before we can say that the sample is *noticeably different*.

The distribution of sample means and the standard error can help researchers make this decision. In particular, the distribution of sample means can be used to show exactly what would be expected for a sample of rats that do not receive any hormone injections. This allows researchers to make a simple comparison between

a. the sample of treated rats (from the research study).

b. samples of untreated rats (from the distribution of sample means).

If our treated sample is noticeably different from the untreated samples, then we have evidence that the treatment has an effect. On the other hand, if our treated sample still looks like one of the untreated samples, then we must conclude that the treatment does not appear to have any effect.

We begin with the original population of untreated rats and consider the distribution of sample means for all the possible samples of $n = 25$ rats. The distribution of sample means has the following characteristics:

1. It is a normal distribution, because the population of rat weights is normal.

2. It has an expected value of 400, because the population mean for untreated rats is $\mu = 400$.

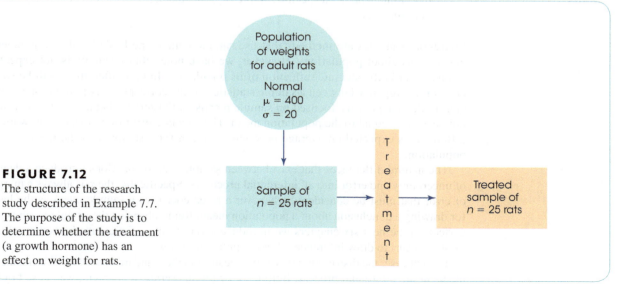

FIGURE 7.12

The structure of the research study described in Example 7.7. The purpose of the study is to determine whether the treatment (a growth hormone) has an effect on weight for rats.

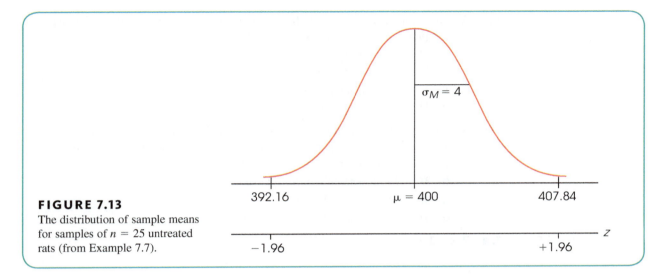

FIGURE 7.13

The distribution of sample means for samples of $n = 25$ untreated rats (from Example 7.7).

3. It has a standard error of $\sigma_M = \frac{20}{\sqrt{25}} = \frac{20}{5} = 4$, because the population standard deviation is $\sigma = 20$ and the sample size is $n = 25$.

The distribution of sample means is shown in Figure 7.13. Notice that a sample of $n = 25$ untreated rats (without the hormone) should have a mean weight of around 400 grams. To be more precise, we can use z-scores to determine the middle 95% of all the possible sample means. As demonstrated in Chapter 6 (page 204), the middle 95% of a normal distribution is located between z-score boundaries of $z = +1.96$ and $z = -1.96$ (check the unit normal table). These z-score boundaries are shown in Figure 7.13. With a standard error of $\sigma_M = 4$ points, a z-score of $z = 1.96$ corresponds to a distance of $1.96(4) = 7.84$ points from the mean. Thus, the z-score boundaries of ± 1.96 correspond to sample means of 392.16 and 407.84.

We have determined that a sample of $n = 25$ untreated rats (no growth hormone) is almost guaranteed (95% probability) to have a mean between 392.16 and 407.84. At the same time, it is very unlikely (probability of 5% or less) that a sample mean would be in the tails beyond these two boundaries without the help of a real treatment effect. Therefore, if the mean for our treated sample is beyond the boundaries, then we have evidence that the hormone does have an effect. ■

In Example 7.7 we used the distribution of sample means, together with z-scores and probability, to provide a description of what is reasonable to expect for an untreated sample. Then, we evaluated the effect of a treatment by determining whether the treated sample was noticeably different from an untreated sample. This procedure forms the foundation for the inferential technique known as *hypothesis testing* that is introduced in Chapter 8 and repeated throughout the remainder of this book.

LEARNING CHECK   **LO7** **1.** A sample is obtained from a population with $\mu = 100$ and $\sigma = 20$. Which of the following samples would produce the z-score closest to zero?

a. A sample of $n = 25$ scores with $M = 102$

b. A sample of $n = 100$ scores with $M = 102$

c. A sample of $n = 25$ scores with $M = 104$

d. A sample of $n = 100$ scores with $M = 104$

LO7 2. For a normal population with $\mu = 80$ and $\sigma = 20$, which of the following samples is *least likely* to be obtained?

    a. $M = 88$ for a sample of $n = 4$

    b. $M = 84$ for a sample of $n = 4$

    c. $M = 88$ for a sample of $n = 25$

    d. $M = 84$ for a sample of $n = 25$

LO7 3. For a sample selected from a normal population with $\mu = 100$ and $\sigma = 15$, which of the following would be the most extreme and unrepresentative?

    a. $M = 90$ for a sample of $n = 9$ scores

    b. $M = 90$ for a sample of $n = 25$ scores

    c. $M = 95$ for a sample of $n = 9$ scores

    d. $M = 95$ for a sample of $n = 25$ scores

ANSWERS   **1. a   2. c   3. b**

SUMMARY

1. The distribution of sample means is defined as the set of *M*s for all the possible random samples for a specific sample size (*n*) that can be obtained from a given population. According to the *central limit theorem*, the parameters of the distribution of sample means are as follows:

    a. *Shape*. The distribution of sample means is normal if either one of the following two conditions is satisfied:

      ■ The population from which the samples are selected is normal.

      ■ The size of the samples is relatively large (around $n = 30$ or more).

    b. *Central Tendency*. The mean of the distribution of sample means is identical to the mean of the population from which the samples are selected. The mean of the distribution of sample means is called the expected value of *M*.

    c. *Variability*. The standard deviation of the distribution of sample means is called the standard error of *M* and is defined by the formula

$$\sigma_M = \frac{\sigma}{\sqrt{n}} \quad \text{or} \quad \sigma_M = \sqrt{\frac{\sigma^2}{n}}$$

Standard error measures the standard distance between a sample mean (*M*) and the population mean (μ).

2. One of the most important concepts in this chapter is standard error. The standard error tells how much error to expect if you are using a sample mean to represent a population mean.

3. The location of each *M* in the distribution of sample means can be specified by a *z*-score:

$$z = \frac{M - \mu}{\sigma_M}$$

4. Because the distribution of sample means tends to be normal, we can use these *z*-scores and the unit normal table to find probabilities for specific sample means. In particular, we can identify which sample means are likely and which are very unlikely to be obtained from any given population. This ability to find probabilities for samples is the basis for the inferential statistics in the chapters ahead.

KEY TERMS

sampling error (215)

distribution of sample means (216)

sampling distribution (216)

central limit theorem (220)

expected value of *M* (221)

standard error of *M* (221)

law of large numbers (222)

FOCUS ON PROBLEM SOLVING

1. Whenever you are working probability questions about sample means, you must use the distribution of sample means. Remember that every probability question can be restated as a proportion question. Probabilities for sample means are equivalent to proportions of the distribution of sample means.

2. When computing probabilities for sample means, the most common error is to use standard deviation (σ) instead of standard error (σ_M) in the z-score formula. Standard deviation measures the typical deviation (or "error") for a single score. Standard error measures the typical deviation (or error) for a sample. Remember: the larger the sample is, the more accurately the sample represents the population. Thus, sample size (n) is a critical part of the standard error.

$$\text{Standard error} = \sigma_M = \frac{\sigma}{\sqrt{n}}$$

3. Although the distribution of sample means is often normal, it is not always a normal distribution. Check the criteria to be certain the distribution is normal before you use the unit normal table to find probabilities (see item 1a of the Summary). Remember that all probability problems with a normal distribution are easier if you sketch the distribution and shade in the area of interest.

DEMONSTRATION 7.1

PROBABILITY AND THE DISTRIBUTION OF SAMPLE MEANS

A population forms a normal distribution with a mean of $\mu = 60$ and a standard deviation of $\sigma = 12$. For a sample of $n = 36$ scores from this population, what is the probability of obtaining a sample mean greater than 63?

$$p(M > 63) = ?$$

STEP 1   **Rephrase the probability question as a proportion question.**   Out of all the possible sample means for $n = 36$, what proportion will have values greater than 63? *All the possible sample means* is simply the distribution of sample means, which is normal, with a mean of $\mu = 60$ and a standard error of

$$\sigma_M = \frac{\sigma}{\sqrt{n}} = \frac{12}{\sqrt{36}} = \frac{12}{6} = 2$$

STEP 2   **Compute the z-score for the sample mean.**   A sample mean of $M = 63$ corresponds to a z-score of

$$z = \frac{M - \mu}{\sigma_M} = \frac{63 - 60}{2} = \frac{3}{2} = 1.50$$

Therefore, $p(M > 63) = p(z > 1.50)$.

STEP 3   **Look up the proportion in the unit normal table.**   Find $z = 1.50$ in column A and read across the row to find $p = 0.0668$ in column C. This is the answer.

$$p(M > 63) = p(z > 1.50) = 0.0668 \text{ (or 6.68\%)}$$

SPSS®

The statistical computer package SPSS is not structured to compute the standard error or a z-score for a sample mean. In later chapters, however, we introduce new inferential statistics that are included in SPSS. When these new statistics are computed, SPSS typically includes a report of standard error that describes how accurately, on average, the sample represents its population.

PROBLEMS

1. Briefly define each of the following:
   a. Distribution of sample means
   b. Central limit theorem
   c. Expected value of M
   d. Standard error of M

2. Suppose that all possible $n = 50$ samples are selected from a population. How would the mean, standard deviation, and shape of the resulting sampling distribution compare to a sampling distribution based on all possible $n = 100$ samples?

3. Compare the following:
   a. Measures of variability, s, σ, and σ_M
   b. Measures of central tendency, M, μ, and μ_M

4. A sample is selected from a population with a mean of $\mu = 100$ and a standard deviation of $\sigma = 20$.
   a. If the sample has $n = 16$ scores, what is the expected value of M and the standard error of M?
   b. If the sample has $n = 100$ scores, what is the expected value of M and the standard error of M?

5. Describe the distribution of sample means (shape, mean, and standard error) for samples of $n = 64$ selected from a population with a mean of $\mu = 90$ and a standard deviation of $\sigma = 32$.

6. Under what circumstances is the distribution of sample means guaranteed to be a normal distribution?

7. A random sample is selected from a population with a standard deviation of $\sigma = 18$.
   a. On average, how much difference should there be between the sample mean and the population mean for a random sample of $n = 4$ scores from this population?
   b. On average, how much difference should there be for a sample of $n = 9$ scores?
   c. On average, how much difference should there be for a sample of $n = 36$ scores?

8. For a sample of $n = 36$ scores, what is the value of the population standard deviation (σ) necessary to produce each of the following standard error values?
   a. $\sigma_M = 12$ points
   b. $\sigma_M = 3$ points
   c. $\sigma_M = 2$ points

9. Suppose that a professor randomly assigns students to study groups of $n = 4$ students. The final exam in the professor's class has a mean of $\mu = 75$ and a standard deviation of $\sigma = 10$. What is the expected value of the mean and the standard deviation of the distribution of study group means?

10. For a population with a mean of $\mu = 40$ and a standard deviation of $\sigma = 12$, find the z-score corresponding to each of the following samples.
    a. $X = 52$ for a sample of $n = 1$ score
    b. $M = 52$ for a sample of $n = 9$ scores
    c. $M = 52$ for a sample of $n = 16$ scores

11. Sales representatives at a cellular phone retailer sell a mean of $\mu = 200$ and a standard deviation of $\sigma = 50$ smartphones per year. At the Rochester, New York, branch, $n = 25$ representatives sell $M = 220$. Compute the z-score for the Rochester branch.

12. A sample of $n = 64$ scores has a mean of $M = 68$. Assuming that the population mean is $\mu = 60$, find the z-score for this sample:
    a. If it was obtained from a population with $\sigma = 16$
    b. If it was obtained from a population with $\sigma = 32$
    c. If it was obtained from a population with $\sigma = 48$

13. A population forms a normal distribution with a mean of $\mu = 85$ and a standard deviation of $\sigma = 24$. For each of the following samples, compute the z-score for the sample mean.
    a. $M = 91$ for $n = 4$ scores
    b. $M = 91$ for $n = 9$ scores
    c. $M = 91$ for $n = 16$ scores
    d. $M = 91$ for $n = 36$ scores

14. Scores on a standardized reading test for fourth-grade students form a normal distribution with $\mu = 71$ and $\sigma = 24$. What is the probability of obtaining a sample mean greater than $M = 63$ for each of the following?
    a. A sample of $n = 9$ students
    b. A sample of $n = 36$ students
    c. A sample of $n = 64$ students

15. Scores from a questionnaire measuring social anxiety form a normal distribution with a mean of $\mu = 50$ and a standard deviation of $\sigma = 10$. What is the probability of obtaining a sample mean greater than $M = 53$

a. for a random sample of $n = 4$ people?
b. for a random sample of $n = 16$ people?
c. for a random sample of $n = 25$ people?

16. A normal distribution has a mean of $\mu = 58$ and a standard deviation of $\sigma = 12$.
    a. What is the probability of randomly selecting a score less than $X = 52$?
    b. What is the probability of selecting a sample of $n = 9$ scores with a mean less than $M = 52$?
    c. What is the probability of selecting a sample of $n = 16$ scores with a mean less than $M = 52$?

17. A population has a mean of $\mu = 30$ and a standard deviation of $\sigma = 8$.
    a. If the population distribution is normal, what is the probability of obtaining a sample mean greater than $M = 32$ for a sample of $n = 4$?
    b. If the population distribution is positively skewed, what is the probability of obtaining a sample mean greater than $M = 32$ for a sample of $n = 4$?
    c. If the population distribution is normal, what is the probability of obtaining a sample mean greater than $M = 32$ for a sample of $n = 64$?
    d. If the population distribution is positively skewed, what is the probability of obtaining a sample mean greater than $M = 32$ for a sample of $n = 64$?

18. For random samples of size $n = 16$ selected from a normal distribution with a mean of $\mu = 75$ and a standard deviation of $\sigma = 20$, find each of the following:
    a. The range of sample means that defines the middle 95% of the distribution of sample means.
    b. The range of sample means that defines the middle 99% of the distribution of sample means.

19. The distribution exam grades for an introductory psychology class is negatively skewed with a mean of $\mu = 71.5$ and a standard deviation of $\sigma = 12$.
    a. What is the probability of selecting a random sample of $n = 9$ students with an average grade greater than 75? (Careful: This is a trick question.)
    b. What is the probability of selecting a random sample of $n = 36$ students with an average grade greater than 75?
    c. For a sample of $n = 36$ students, what is the probability that the average grade is between 70 and 75?

20. By definition, jumbo shrimp are those that require between 10 and 15 shrimp to make a pound. Suppose that the number of jumbo shrimp in a 1-pound bag averages $\mu = 12.5$ with a standard deviation of $\sigma = 1.5$, and forms a normal distribution. What is the probability of randomly picking a sample of $n = 25$ 1-pound bags that average more than $M = 13$ shrimp per bag?

21. For a population with a mean of $\mu = 72$ and a standard deviation of $\sigma = 10$, what is the standard error

of the distribution of sample means for each of the following sample sizes?
    a. $n = 4$ scores
    b. $n = 25$ scores

22. For a population with $\sigma = 16$, how large a sample is necessary to have a standard error that is
    a. equal to 8 points?
    b. equal to 4 points?
    c. equal to 2 points?

23. If the population standard deviation is $\sigma = 24$, how large a sample is necessary to have a standard error that is
    a. equal to 6 points?
    b. equal to 3 points?
    c. equal to 2 points?

24. A normal distribution has a mean of $\mu = 60$ and a standard deviation of $\sigma = 12$. For each of the following samples, compute the z-score for the sample mean and determine whether the sample mean is a typical, representative value or an extreme value for a sample of this size.
    a. $M = 64$ for $n = 4$ scores
    b. $M = 64$ for $n = 9$ scores
    c. $M = 69$ for $n = 4$ scores
    d. $M = 69$ for $n = 9$ scores
    e. $M = 66$ for $n = 16$ scores
    f. $M = 66$ for $n = 36$ scores
    g. $M = 54$ for $n = 4$ scores
    h. $M = 54$ for $n = 36$ scores

25. Metacognition is an understanding of one's own cognitive processes, like thoughts, perceptions, and memory. In a recent study of metacognition in monkeys, a researcher presented monkeys with either a tube that was closed except at both ends, or a tube with an opening in the center that could be used to inspect the inside of the tube. In addition, subjects either watched or didn't watch a human placing food in one end of the tube that could later be retrieved by the monkey (Rosati & Santos, 2016). Apparently knowing that they needed more information to find the food, monkeys that did not observe the human were more likely to spontaneously inspect the center of the tube than monkeys that observed the human. Similarly, monkeys that did not observe the human were slower to choose the end of the tube baited with food than monkeys that observed the human. Data on amount of time to choose—similar to the results obtained in the study—are shown in the following table.

	Mean	SE
Observed human	13.4	3.0
No observation	8.5	1.5

a. Construct a bar graph that incorporates all the information in the table.

b. Looking at your graph, do you think that observing the human had an effect on the amount of time needed to choose?

26. Suppose that a researcher developed a drug that she claims increases extroversion. A sample of $n = 4$ participants has a sample mean of $M = 115$ on a personality assessment after taking the drug. The personality test has a population mean of $\mu = 100$ and $\sigma = 30$. Is the sample mean an especially unlikely result based on the population parameters for the personality test? What if the researcher increased the sample to $n = 25$ and observes the same sample mean of $M = 115$?

27. A sample of $n = 36$ scores is selected from a normal distribution with a mean of $\mu = 65$. Compute the z-score for a sample mean of $M = 59$ and determine whether the sample mean is a typical, representative value or an extreme value for each of the following:

a. A population standard deviation of $\sigma = 12$

b. A population standard deviation of $\sigma = 30$

Introduction to Hypothesis Testing

CHAPTER 8

Tools You Will Need

The following items are considered essential background material for this chapter. If you doubt your knowledge of any of these items, you should review the appropriate chapter or section before proceeding.

- z-Scores (Chapter 5)
- Distribution of sample means (Chapter 7)
  - Expected value
  - Standard error
  - Probability and sample means

clivewa/Shutterstock.com

PREVIEW

PREVIEW

Women's college basketball arguably is underappreciated among sports fans and commentators alike, especially when compared to the men's game. Television exposure and marketing of women's games lag behind those of their male counterparts. Yet at the NCAA Division I level, female athletes exhibit great skill and the competition is intense. As you would expect, the recruitment of talent from high schools by women's college programs is intense as well.

Aside from the performance statistics of female athletes on the court, let's look at another aspect of the game to illustrate the basics of hypothesis testing. Suppose that the mean height for adult women is $\mu = 64$ inches (or, 5 feet 4 inches) with a standard deviation of $\sigma = 6$ (Centers for Disease Control and Prevention, National Center for Health Statistics, 2016).[1] If a college selected a random sample of $n = 11$ women from the population to play on its basketball team, what would you expect the sample to look like, assuming that height is not an advantage in basketball and unrelated to recruitment of players? If height were an irrelevant factor, it should be clear that the sample mean, M, should be close to the population mean. That is, a sample mean of $M = 64$ inches would be reasonable.

However, recruitment of talented players is not a random process—so let's examine an actual example. Consider the roster of a Division I powerhouse in women's basketball, like the University of Connecticut. The

heights, in inches, for $n = 11$ players on the 2018–19 roster are as follows:

$$73 \quad 68 \quad 65 \quad 69 \quad 71 \quad 76 \quad 73 \quad 74 \quad 74 \quad 75 \quad 72$$

The mean height of the team is $M = 71.8$ inches, or nearly 6 feet tall. Is this a reasonable finding, or does this sample seem to be extreme compared to the general population?

Thus, we have two possibilities. On one hand the sample data might be consistent with the population. That is, the heights of players on the basketball team are what one would expect from a random sample drawn from the general population. Alternatively, the sample data might be extreme compared to the general population and suggest that the average height of the population of women basketball players is taller. These are two competing hypotheses. Because we are dealing with sample data, there is always some uncertainty about which conclusion about the population is correct. In this chapter, we will examine an inferential statistical method called hypothesis testing. It involves a series of logical steps that use sample data to test hypotheses about a population, usually in the context of assessing possible treatment effects in an experiment. Hypothesis testing takes into account the uncertainty when assessing competing hypotheses.

Finally, we should note the obvious. Height alone is no guarantee that a person will excel at basketball, let alone receive All-American honors. However, it can help.

[1]The CDC data consist of a sample of $n = 5,547$ women, for which the mean height is $M = 63.7$ inches with a standard error of 0.08. Recall from Chapter 7 that a very large sample results in a small standard error. That is, there will be very little error between sample means and the population mean. For the sake of this example, we have used reasonable approximations for μ and σ. In Chapter 9 we will introduce a method to estimate μ from sample data.

8-1 The Logic of Hypothesis Testing

LEARNING OBJECTIVES

1. Describe the purpose of a hypothesis test and explain how the test accomplishes its goal.

2. Using symbols, state the null and alternative hypotheses for a specific research example. Using words, state the null and alternative hypotheses as they relate to the independent and dependent variables in a specific research study.

3. Define the alpha level (level of significance) and the critical region for a hypothesis test and explain how the outcome of a hypothesis test is influenced by a change in alpha level.

4. Conduct a hypothesis test using the standard four-step procedure and make a statistical decision about the effect of a treatment.

It usually is impossible or impractical for a researcher to observe every individual in a population. Therefore, researchers usually collect data from a sample and then use the sample data to help answer questions about the population. Hypothesis testing is a statistical procedure that allows researchers to use sample data to draw inferences about the population of interest.

Hypothesis testing is one of the most commonly used inferential procedures. In fact, most of the remainder of this book examines hypothesis testing in a variety of different situations and applications. Although the details of a hypothesis test change from one situation to another, the general process remains constant. In this chapter, we introduce the general procedure for a hypothesis test. You should notice that we use the statistical techniques that have been developed in the preceding three chapters—that is, we combine the concepts of z-scores, probability, and the distribution of sample means to create a new statistical procedure known as *hypothesis testing*.

> **Hypothesis testing** is a statistical method that uses sample data to evaluate a hypothesis about a population.

In very simple terms, the logic underlying the hypothesis-testing procedure is as follows:

1. First, we state a hypothesis about a population. Usually the hypothesis concerns the value of a population parameter. For example, we might hypothesize that American adults gain an average of $\mu = 7$ pounds between Thanksgiving and New Year's Day each year.

2. Before we select a sample, we use the hypothesis to describe what values we should expect for the sample mean, if the hypothesis is really true. For example, if we predict that the average weight gain for the population is $\mu = 7$ pounds, then we would expect our sample to have a mean *around* 7 pounds. (Remember: the sample should be similar to the population, but you always expect a certain amount of error.)

3. Next, we obtain a random sample from the population. For example, we might select a sample of $n = 200$ American adults and measure the average weight change for the sample between Thanksgiving and New Year's Day.

4. Finally, we compare the obtained sample data with the prediction that was made from the hypothesis. If the sample mean is consistent with the prediction, we conclude that the hypothesis is reasonable. But if there is a big discrepancy between the data and the prediction, we decide that the hypothesis is probably wrong.

A hypothesis test is typically used in the context of a research study. That is, a researcher completes a research study and then uses a hypothesis test to evaluate the results. Depending on the type of research and the type of data, the details of the hypothesis test change from one research situation to another. In later chapters, we examine different versions of hypothesis testing that are used for different kinds of research. For now, however, we focus on the basic elements that are common to all hypothesis tests. To accomplish this general goal, we will examine a hypothesis test as it applies to the simplest possible situation—using a sample mean to test a hypothesis about a population mean.

In the five chapters that follow, we consider hypothesis testing in more complex research situations involving sample means and mean differences. In Chapter 14, we look at correlational research and examine how the relationships obtained for sample data are used to evaluate hypotheses about relationships in the population. Finally, in Chapter 15, we examine how the proportions that exist in a sample are used to test hypotheses about the corresponding proportions in the population.

FIGURE 8.1

The basic research situation for hypothesis testing. A population parameter (for example, μ) is known or assumed before the study. The purpose of the study is to determine whether the treatment has an effect on the population mean.

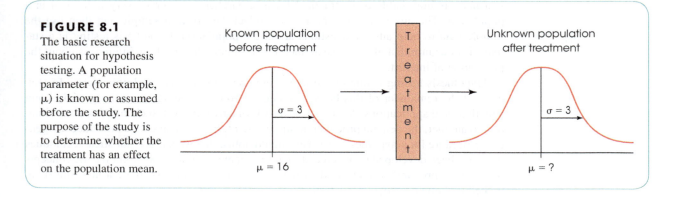

■ The Elements of a Hypothesis Test

Once again, we introduce hypothesis testing with a situation in which a researcher is using one sample mean to evaluate a hypothesis about one unknown population mean.

The Unknown Population   Figure 8.1 shows the general research situation that we will use to introduce the process of hypothesis testing. Notice that the researcher begins with a known population. This is the set of individuals as they exist *before treatment*. For this example, we are assuming that the original set of scores forms a normal distribution with $\mu = 16$ and $\sigma = 3$. The purpose of the research is to determine the effect of a treatment on the individuals in the population. That is, the goal is to determine what happens to the population *after the treatment is administered*.

To simplify the hypothesis-testing situation, one basic assumption is made about the effect of the treatment: If the treatment has any effect, it is simply to add a constant amount to (or subtract a constant amount from) each individual's score. You should recall from Chapters 3 and 4 that adding (or subtracting) a constant changes the mean but does not change the shape of the population, nor does it change the standard deviation. Thus, we assume that the population after treatment has the same shape as the original population and the same standard deviation as the original population. This assumption is incorporated into the situation shown in Figure 8.1.

Note that, after treatment, the unknown population is the focus of the research question. Specifically, the purpose of the research is to determine what would happen if the treatment were administered to every individual in the population.

The Sample in the Research Study   The goal of the hypothesis test is to determine whether the treatment has any effect on the individuals in the population (see Figure 8.1). Usually, however, we cannot administer the treatment to the entire population, so the actual research study is conducted using a sample. Figure 8.2 shows the structure of the research study from the point of view of the hypothesis test. The original population, before treatment, is shown on the left-hand side. The unknown population, after treatment, is shown on the right-hand side. Note that the unknown population is actually *hypothetical* (the treatment is never administered to the entire population). Instead, we are asking *what would happen if* the treatment were administered to the entire population. The research study involves selecting a sample from the original population, administering the treatment to the sample, and then recording scores for the individuals in the treated sample. Notice that the research study produces a treated sample. Although this sample was obtained indirectly, it is equivalent to a sample that is obtained directly from the unknown

FIGURE 8.2

From the point of view of the hypothesis test, the entire population receives the treatment and then a sample is selected from the treated population. In the actual research study, however, a sample is selected from the original population and the treatment is administered to the sample. From either perspective, the result is a treated sample that represents the treated population.

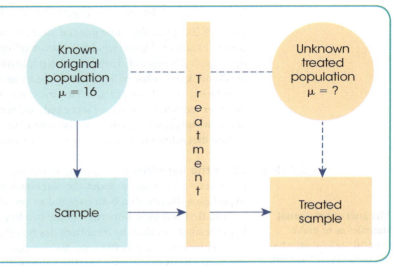

treated population. The hypothesis test uses the treated sample on the right-hand side of Figure 8.2 to evaluate a hypothesis about the unknown treated population on the right side of the figure.

A hypothesis test is a formalized procedure that follows a standard series of operations. In this way, researchers have a standardized method for evaluating the results of their research studies. Other researchers will recognize and understand exactly how the data were evaluated and how conclusions were reached. To emphasize the formal structure of a hypothesis test, we will present hypothesis testing as a four-step process that is used throughout the rest of the book. The following example provides a concrete foundation for introducing the hypothesis-testing procedure.

EXAMPLE 8.1   Previous research indicates that men rate women who are wearing red as being more attractive than when they are wearing other colors (Elliot & Niesta, 2008). Based on these results, Guéguen and Jacob (2012) reasoned that the same phenomenon might influence the way that men react to waitresses wearing red. In their study, waitresses in five different restaurants wore the same T-shirt in six different colors (red, blue, green, yellow, black, and white) on different days during a six-week period. Except for the T-shirts, the waitresses were instructed to act normally and to record each customer's gender and how much was left as a tip. The results show that male customers gave significantly bigger tips to waitresses wearing red, but that color had no effect on tipping for female customers.

A researcher decided to test this result by repeating the basic study at a local restaurant. Waitresses (and waiters) at the restaurant routinely wear white shirts with black pants, and restaurant records indicate that the waitresses' tips from male customers average $\mu = 16$ percent of the bill with a standard deviation of $\sigma = 3$ percentage points. The distribution of tip amounts is roughly normal. During the study, the waitresses are asked to wear red shirts and the researcher plans to record tips for a sample of $n = 36$ male customers.

If the mean tip for the sample is noticeably different from the baseline mean (when wearing white shirts), the researcher can conclude that wearing the color red does appear to have an effect on tipping. On the other hand, if the sample mean is still around 16 percent (the same as the baseline), the researcher must conclude that the red shirt does not appear to have any effect.

■ The Four Steps of a Hypothesis Test

Figure 8.2 depicts the same general structure as the research situation described in the preceding example. The original population before treatment (before the red shirt) has a mean tip of $\mu = 16$ percent. However, the population after treatment is unknown. Specifically, we do not know what will happen to the mean score if the waitresses wear red for the entire population of male customers. However, we do have a sample of $n = 36$ participants who were served when waitresses wore red, and we can use this sample to help draw inferences about the unknown population. The following four steps outline the hypothesis-testing procedure that allows us to use sample data to answer questions about an unknown population.

STEP 1   **State the hypotheses.** As the name implies, the process of hypothesis testing begins by stating a hypothesis about the unknown population. Actually, we state two opposing hypotheses. Notice that both hypotheses are stated in terms of population parameters.

> The goal of inferential statistics is to make general statements about the population by using sample data. Therefore, when testing hypotheses, we make our predictions about the population parameters.

The first and most important of the two hypotheses is called the *null hypothesis*. The null hypothesis states that the treatment has no effect. In general, the null hypothesis states that there is no change, no effect, no difference—nothing happened, hence the name *null*. The null hypothesis is identified by the symbol H_0. (The H stands for *hypothesis*, and the zero subscript indicates that this is the *zero-effect* hypothesis.) For the study in Example 8.1, the null hypothesis states that the red shirt has no effect on tipping behavior for the population of male customers. In symbols, this hypothesis is

$$H_0: \mu_{\text{red shirt}} = 16 \qquad \text{(Even with a red shirt, the mean tip is still 16 percent.)}$$

> The **null hypothesis** (H_0) states that in the general population there is no change, no difference, or no relationship. In the context of an experiment, H_0 predicts that the independent variable (treatment) *has no effect* on the dependent variable (scores) for the population.

The second hypothesis is simply the opposite of the null hypothesis, and it is called the *scientific*, or *alternative*, *hypothesis* (H_1). This hypothesis states that the treatment has an effect on the dependent variable.

> The **alternative hypothesis** (H_1) states that there is a change, a difference, or a relationship for the general population. In the context of an experiment, H_1 predicts that the independent variable (treatment) *does have an effect* on the dependent variable.

> The null hypothesis and the alternative hypothesis are mutually exclusive and exhaustive. They cannot both be true. The data will determine whether to reject or fail to reject the null hypothesis.

For this example, the alternative hypothesis states that the red shirt does have an effect on tipping for the population and will cause a change in the mean score. In symbols, the alternative hypothesis is represented as

$$H_1: \mu_{\text{red shirt}} \neq 16 \qquad \text{(With a red shirt, the mean tip will be different from 16 percent.)}$$

Notice that the alternative hypothesis simply states that there will be some type of change. It does not specify whether the effect will be increased or decreased tips. In some circumstances, it is appropriate for the alternative hypothesis to specify the direction of the effect. For example, the researcher might hypothesize that a red shirt will increase tips ($\mu > 16$). This type of hypothesis results in a directional hypothesis test, which is examined in detail later in this chapter. For now, we concentrate on nondirectional tests—for which the hypotheses simply state that the treatment has no effect (H_0) or has some effect (H_1).

STEP 2   **Set the criteria for a decision.**   Eventually the researcher will use the data from the sample to evaluate the credibility of the null hypothesis. The data will either be consistent with the null hypothesis or tend to refute the null hypothesis. In particular, if there is a big discrepancy between the data and the hypothesis, we will conclude that the hypothesis is wrong.

To formalize the decision process, we use the null hypothesis to predict the kind of sample mean that ought to be obtained. Specifically, we determine exactly which sample means are consistent with the null hypothesis and which sample means are at odds with the null hypothesis.

For our example, the null hypothesis states that the red shirt has no effect and the population mean is still $\mu = 16$ percent. If this is true, then the sample mean should have a value around 16. Therefore, a sample mean near 16 is consistent with the null hypothesis. On the other hand, a sample mean that is very different from 16 is not consistent with the null hypothesis. To determine exactly which values are "near" 16 and which values are "very different from" 16, we will examine all of the possible sample means that could be obtained if the null hypothesis is true. For our example, this is the distribution of sample means for $n = 36$. According to the null hypothesis, this distribution is centered at $\mu = 16$. The distribution of sample means is then divided into two sections:

1. Sample means that are likely to be obtained if H_0 is true; that is, sample means that are close to the null hypothesis

2. Sample means that are very unlikely to be obtained if H_0 is true; that is, sample means that are very different from the null hypothesis

Figure 8.3 shows the distribution of sample means divided into these two sections. Notice that the high-probability samples are located in the center of the distribution and

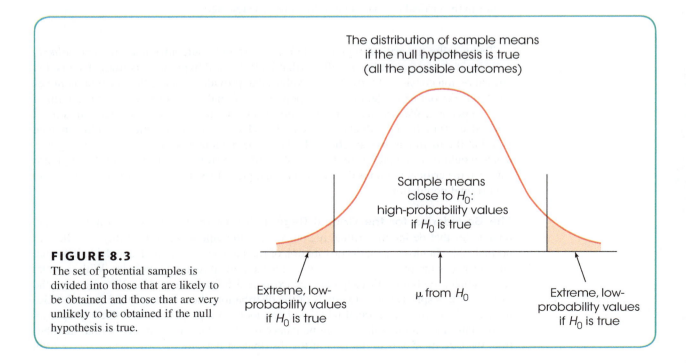

FIGURE 8.3
The set of potential samples is divided into those that are likely to be obtained and those that are very unlikely to be obtained if the null hypothesis is true.

The distribution of sample means
if the null hypothesis is true
(all the possible outcomes)

Sample means
close to H_0:
high-probability values
if H_0 is true

Extreme, low-
probability values
if H_0 is true

μ from H_0

Extreme, low-
probability values
if H_0 is true

have sample means close to the value specified in the null hypothesis. On the other hand, the low-probability samples are located in the extreme tails of the distribution. After the distribution has been divided in this way, we can compare our sample data with the values in the distribution. Specifically, we can determine whether our sample mean is consistent with the null hypothesis (like the values in the center of the distribution) or whether our sample mean is very different from the null hypothesis (like the values in the extreme tails).

The Alpha Level   To find the boundaries that separate the high-probability samples from the low-probability samples, we must define exactly what is meant by "low" probability and "high" probability. This is accomplished by selecting a specific probability value, which is known as the *level of significance*, or the *alpha level*, for the hypothesis test. The alpha (α) value is a small probability that is used to identify the low-probability samples. By convention, commonly used alpha levels are $\alpha = .05$ (5%), $\alpha = .01$ (1%), and $\alpha = .001$ (0.1%). For example, with $\alpha = .05$, we separate the most unlikely 5% of the sample means (the extreme values) from the most likely 95% of the sample means (the central values).

The extremely unlikely values, as defined by the alpha level, make up what is called the *critical region*. These extreme values in the tails of the distribution define outcomes that are not consistent with the null hypothesis; that is, they are very unlikely to occur if the null hypothesis is true. Whenever the data from a research study produce a sample mean that is located in the critical region, we conclude that the data are not consistent with the null hypothesis, and we reject the null hypothesis.

> The **alpha level**, or the **level of significance**, is a probability value that is used to define the concept of "very unlikely" in a hypothesis test.
>
> The **critical region** is composed of the extreme sample values that are very unlikely (as defined by the alpha level) to be obtained if the null hypothesis is true. The boundaries for the critical region are determined by the alpha level. If sample data fall in the critical region, the null hypothesis is rejected.

Technically, the critical region is defined by sample outcomes that are *very unlikely* to occur if the treatment has no effect (that is, if the null hypothesis is true). That is, the critical region consists of those sample values that provide evidence that the treatment has an effect. For our example, the regular population of male customers leaves a mean tip of $\mu = 16$ percent. We selected a sample from this population and administered a treatment (the red shirt) to the individuals in the sample. What kind of sample mean would convince you that the treatment has an effect? It should be obvious that the most convincing evidence would be a sample mean that is really different from $\mu = 16$ percent. In a hypothesis test, the critical region is determined by sample values that are "really different" from the original population.

The Boundaries for the Critical Region   To determine the exact location for the boundaries that define the critical region, we use the alpha-level probability and the unit normal table. In most cases, the distribution of sample means is normal, and the unit normal table provides the precise z-score location for the critical region boundaries. With $\alpha = .05$, for example, the boundaries separate the extreme 5% from the middle 95%. Because the extreme 5% is split between two tails of the distribution, there is exactly 2.5% (or 0.0250) in each tail. In the unit normal table, you can look up a proportion of 0.0250 in column C (the tail) and find that the z-score boundary is $z = 1.96$. Thus, for any normal distribution, the extreme 5% is in the tails of the distribution beyond $z = +1.96$ and $z = -1.96$.

FIGURE 8.4

The critical region (very unlikely outcomes) for $\alpha = .05$.

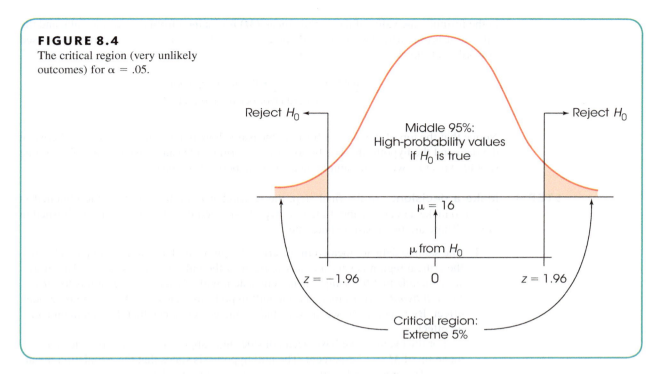

Reject H_0

Reject H_0

Middle 95%:
High-probability values
if H_0 is true

$\mu = 16$

μ from H_0

$z = -1.96$ 0 $z = 1.96$

Critical region:
Extreme 5%

These values define the boundaries of the critical region for a hypothesis test using $\alpha = .05$ (Figure 8.4). That is, the boundaries are $z = \pm 1.96$ when $\alpha = .05$.

Similarly, an alpha level of $\alpha = .01$ means that 1% or .0100 is split between the two tails. In this case, the proportion in each tail is .0050, and the corresponding z-score boundaries are $z = \pm 2.58$ (± 2.57 is acceptable as well). For $\alpha = .001$, the boundaries are located at $z = \pm 3.30$. You should verify these values in the unit normal table and be sure that you understand exactly how they are obtained.

STEP 3 **Collect data and compute sample statistics.**   At this time, we begin recording tips for male customers while the waitresses are wearing red. Notice that the data are collected *after* the researcher has stated the hypotheses and established the criteria for a decision. This sequence of events helps ensure that a researcher makes an honest, objective evaluation of the data and does not tamper with the decision criteria after the experimental outcome is known.

Next, the raw data from the sample are summarized with the appropriate statistics: For this example, the researcher would compute the sample mean. Now it is possible for the researcher to compare the sample mean (from the data) with the null hypothesis. This is the heart of the hypothesis test: comparing the data with the hypothesis.

The comparison is accomplished by computing a z-score that describes exactly where the sample mean is located relative to the hypothesized population mean from H_0. In Step 2, we constructed the distribution of sample means that would be expected if the null hypothesis were true—that is, the entire set of sample means that could be obtained if the treatment has no effect (see Figure 8.4). Now we calculate a z-score that identifies where our sample mean is located in this hypothesized distribution. The z-score formula for a sample mean is

$$z = \frac{M - \mu}{\sigma_M}$$

treats one type of error (falsely claiming that a treatment works and falsely convicting a person) as being far more serious than other errors. Here is a direct comparison between hypothesis tests and jury trials:

1. The test begins with a null hypothesis stating that there is no treatment effect. The trial begins with a null hypothesis that the accused did not commit the crime (that is, innocent until proven guilty).

2. The research study gathers evidence to test whether the treatment actually does have an effect, and the police gather evidence to test whether the accused really committed a crime.

3. If there is enough evidence, the researcher rejects the null hypothesis and concludes that there is evidence for a treatment effect. If there is enough evidence, the jury rejects the hypothesis and concludes that the defendant is guilty of a crime.

4. If there is not enough evidence, the researcher fails to reject the null hypothesis. Note that the researcher does not conclude that there is no treatment effect, simply that there is not enough evidence to conclude that there is an effect. Similarly, if there is not enough evidence, the jury fails to find the defendant guilty. Note that the jury does not conclude that the defendant is innocent, simply that there is not enough evidence for a guilty verdict.

■ A Closer Look at the *z*-Score Statistic

The *z*-score statistic that is used in the hypothesis test is the first specific example of what is called a *test statistic*. The term *test statistic* simply indicates that the sample data are converted into a single, specific statistic that is used to test hypotheses. In the chapters that follow, we introduce several other test statistics that are used in a variety of different research situations. However, most of the new test statistics have the same basic structure and serve the same purpose as the *z*-score. We have already described the *z*-score equation as a formal method for comparing the sample data and the population hypothesis. In this section, we discuss the *z*-score from two other perspectives that may give you a better understanding of hypothesis testing and the role that *z*-scores play in this inferential statistical method. In each case, keep in mind that the *z*-score serves as a general model for other test statistics that will come in future chapters.

The *z*-Score Formula as a Recipe   The *z*-score formula, like any formula, can be viewed as a recipe. If you follow instructions and use all the right ingredients, the formula produces a *z*-score. In the hypothesis-testing situation, however, you do not have all the necessary ingredients. Specifically, you do not know the value for the population mean (μ), which is one component or ingredient in the formula.

This situation is similar to trying to follow a cake recipe where one of the ingredients is not clearly listed. For example, the recipe may call for flour, but there is a grease stain on the page that makes it impossible to read how much flour. Faced with this situation, you might try the following steps:

1. Make a hypothesis about the amount of flour. For example, hypothesize that the correct amount is 2 cups.

2. To test your hypothesis, add the rest of the ingredients along with the hypothesized flour amount and bake the cake.

3. If the cake turns out to be good, you can reasonably conclude that your hypothesis was correct. But if the cake is terrible, you conclude that your hypothesis was wrong.

In a hypothesis test with z-scores, we do essentially the same thing. We have a formula (recipe) for z-scores, but one ingredient is missing. Specifically, we do not know the value for the population mean, μ. Therefore, we try the following steps:

1. Make a hypothesis about the value of μ. This is the null hypothesis.
2. Plug the hypothesized value into the formula along with the other values (ingredients).
3. If the formula produces a z-score near zero (which is where z-scores are supposed to be), we conclude that the hypothesis was correct. On the other hand, if the formula produces an extreme value (a very unlikely result), we conclude that the hypothesis was wrong.

The *z*-Score Formula as a Ratio   In the context of a hypothesis test, the z-score formula has the following structure:

$$z = \frac{M - \mu}{\sigma_M} = \frac{\text{sample mean} - \text{hypothesized population mean}}{\text{standard error between } M \text{ and } \mu}$$

Notice that the numerator of the formula involves a direct comparison between the sample data and the mean that comes from the null hypothesis. In particular, the numerator measures the obtained difference between the sample mean and the population mean according to the null hypothesis. The standard error in the denominator of the formula measures the expected amount of difference (or error) that exists between a sample mean and the population mean without any treatment effect. Thus, the z-score formula (and many other test statistics) forms a ratio:

$$z = \frac{\text{actual difference between sample } (M) \text{ and the hypothesis } (\mu)}{\text{expected difference between } M \text{ and } \mu \text{ with no treatment effect}}$$

Thus, for example, a z-score of $z = +3.00$ means that the obtained difference between the sample and the hypothesis is three times bigger than would be expected if the treatment had no effect.

In general, a large value for a test statistic like the z-score indicates a large discrepancy between the sample data that we observed and the sample data that we would expect based on the null hypothesis. Specifically, a large value indicates that the sample data are very unlikely to have occurred by chance alone. Therefore, when we obtain a large value (in the critical region), we conclude that it must have been caused by a treatment effect.

LEARNING CHECK   **LO1**   **1.** In general terms, what is a hypothesis test?
a. A descriptive technique that allows researchers to describe a sample.
b. A descriptive technique that allows researchers to describe a population.
c. An inferential technique that uses the data from a sample to draw inferences about a population.
d. An inferential technique that uses information about a population to make predictions about a sample.

LO2   **2.** A sample is selected from a population with a mean of $\mu = 75$ and a treatment is administered to the individuals in the sample. If a hypothesis test is used to evaluate the treatment effect, then what is the correct statement of the null hypothesis?
a. $\mu = 75$
b. $\mu \neq 75$

        c. $M = 75$

        d. $M \neq 75$

LO3 **3.** Which of the following accurately describes the critical region for a hypothesis test?

    a. Outcomes that have a very low probability if the null hypothesis is true.

    b. Outcomes that have a high probability if the null hypothesis is true.

    c. Outcomes that have a very low probability regardless of whether the null hypothesis is true.

    d. Outcomes that have a high probability regardless of whether the null hypothesis is true.

LO4 **4.** The psychology department is gradually changing its curriculum by increasing the number of online course offerings. To evaluate the effectiveness of this change, a random sample of $n = 36$ students who registered for Introductory Psychology is placed in the online version of the course. At the end of the semester, all students take the same final exam. The average score for the sample is $M = 76$. For the general population of students taking the traditional lecture class, the final exam scores form a normal distribution with a mean of $\mu = 71$ and a standard deviation of $\sigma = 12$. The department conducts a hypothesis test with $\alpha = .05$. Which of the following correctly describes the outcome of the hypothesis test?

    a. Reject the null hypothesis because $z = +2.50$, which is in the critical region.

    b. Accept the null hypothesis because $z = +2.50$, which is outside of the critical region.

    c. Fail to reject the null hypothesis because $z = +0.42$, which is not in the critical region.

    d. Fail to accept the alternative hypothesis because $z = +0.42$, which is inside of the critical region.

ANSWERS   **1. c**   **2. a**   **3. a**   **4. a**

8-2 Uncertainty and Errors in Hypothesis Testing

LEARNING OBJECTIVE

5. Define a Type I error and a Type II error, explain the consequences of each, and describe how a Type I error is related to the alpha level.

Hypothesis testing is an *inferential process*, which means that it uses limited information as the basis for reaching a general conclusion. Specifically, a sample provides only limited or incomplete information about the whole population, and yet a hypothesis test uses a sample to draw a conclusion about the population. In this situation, there is always the possibility that an incorrect conclusion will be made. Although sample data are usually representative of the population, there is always a chance that the sample is misleading and will cause a researcher to make the wrong decision about the research results. In a hypothesis test, there are two different kinds of errors that can be made.

■ Type I Errors

It is possible that the data will lead you to reject the null hypothesis when in fact the treatment has no effect. Remember: samples are not expected to be identical to their populations, and some extreme samples can be very different from the populations they are supposed to represent. If a researcher selects one of these extreme samples by chance, then the data from the sample may give the appearance of a strong treatment effect, even though there is no real effect. In the previous section, for example, we discussed a research study examining how the tipping behavior of male customers is influenced by a waitress wearing the color red. Suppose the researcher selects a sample of $n = 36$ men who already were good tippers. Even if the red shirt (the treatment) has no effect at all, these men will still leave higher than average tips. In this case, the researcher is likely to conclude that the treatment does have an effect, when in fact it really does not. This is an example of what is called a *Type I error*.

> A **Type I error** occurs when a researcher rejects a null hypothesis that is actually true. In a typical research situation, a Type I error means the researcher concludes that there is evidence for a treatment effect when in fact the treatment has no effect.

You should realize that a Type I error is not a careless mistake in the sense that a researcher is overlooking something that should be perfectly obvious. On the contrary, the researcher is looking at sample data that appear to show a clear treatment effect. The researcher then makes a careful decision based on the available information. The problem is that the information from the sample is misleading.

In most research situations, the consequences of a Type I error can be very serious. Because the researcher has rejected the null hypothesis and believes that the data support evidence for the treatment effect, it is likely that the researcher will report or even publish the research results. A Type I error, however, means that this is a false report of an effect. Thus, Type I errors lead to false reports in the scientific literature. Other researchers may try to build theories or develop other experiments based on the false results. A lot of precious time and resources may be wasted.

The Probability of a Type I Error   A Type I error occurs when a researcher unknowingly obtains an extreme, nonrepresentative sample. Fortunately, the hypothesis test is structured to minimize the risk that this will occur. Figure 8.4 shows the distribution of sample means and the critical region for the waitress-tipping study we have been discussing. This distribution contains all of the possible sample means for samples of $n = 36$ if the null hypothesis is true. Notice that most of the sample means are near the hypothesized population mean, $\mu = 16$, and that means in the critical region are very unlikely to occur.

With an alpha level of $\alpha = .05$, only 5% of the samples have means in the critical region. Therefore, there is only a 5% probability ($p = .05$) that one of these samples will be obtained. Thus, the alpha level determines the probability of obtaining a sample mean in the critical region when the null hypothesis is true. In other words, the alpha level determines the probability of a Type I error.

> The **alpha level** for a hypothesis test is the probability that the test will lead to a Type I error. That is, the alpha level determines the probability of obtaining sample data in the critical region even though the null hypothesis is true.

In summary, whenever the sample data are in the critical region, the appropriate decision for a hypothesis test is to reject the null hypothesis. Normally this is the correct decision because the treatment has caused the sample to be different from the original population. In this case, the hypothesis test has correctly identified evidence for a real treatment effect. Occasionally, however, sample data are in the critical region just by chance, without any treatment effect. When this occurs, the researcher will make a Type I error; that is, the researcher will conclude that a treatment effect exists when in fact it does not. Anytime the null hypothesis is rejected, there is a chance that a Type I error has been committed. Fortunately, the risk of a Type I error is under the control of the researcher. Specifically, the probability of a Type I error is equal to the alpha level. By selecting a small alpha level, the researcher can minimize the probability of a Type I error.

■ Type II Errors

Whenever a researcher rejects the null hypothesis, there is a risk of a Type I error. Similarly, whenever a researcher fails to reject the null hypothesis, there is a risk of a *Type II error*. By definition, a Type II error is the failure to reject a false null hypothesis. In other words, a Type II error means that a treatment effect really exists, but the hypothesis test fails to detect it.

> A **Type II error** occurs when a researcher fails to reject a null hypothesis that is in fact false. In a typical research situation, a Type II error means that the hypothesis test has failed to detect a real treatment effect.

A Type II error occurs when the sample mean is not in the critical region even though the treatment has an effect on the sample. Often this happens when the effect of the treatment is relatively small. In this case, the treatment does influence the sample, but the magnitude of the effect is not big enough to move the sample mean into the critical region. Because the sample is not substantially different from the original population (it is not in the critical region), the statistical decision is to fail to reject the null hypothesis and to conclude that there is not enough evidence to say there is evidence for a treatment effect.

The consequences of a Type II error are usually not as serious as those of a Type I error. In general terms, a Type II error means that the research data do not show the results that the researcher had hoped to obtain. The researcher can accept this outcome and conclude that there is no evidence of a treatment effect, or the researcher can repeat the experiment (usually with some improvement, such as a larger sample) and try to demonstrate that the treatment really does work.

Unlike a Type I error, it is impossible to determine a single, exact probability for a Type II error. Instead, the probability of a Type II error depends on a variety of factors (such as sample size and effect size) and therefore is a function of several factors, rather than a specific number, like alpha, that the researcher selects. Nonetheless, the probability of a Type II error is represented by the symbol β, the Greek letter *beta*.

In summary, a hypothesis test always leads to one of two decisions:

1. The sample data provide sufficient evidence to reject the null hypothesis and conclude that the treatment has an effect.

2. The sample data do not provide enough evidence to reject the null hypothesis. In this case, you fail to reject H_0 and conclude that the treatment does not appear to have an effect.

TABLE 8.1

Possible outcomes of a statistical decision.

		Actual Situation	
		No Effect, H_0 True	Effect Exists, H_0 False
Researcher's Decision	Reject H_0	Type I error	Decision correct
	Fail to Reject H_0	Decision correct	Type II error

In either case, there is a chance that the data are misleading and the decision is wrong. In summary, a hypothesis test always has two possibilities for error:

1. **Type I error (alpha):** rejecting a true null hypothesis. The sample data are extreme by chance and may give the appearance of a treatment effect, even though there is no real effect.

2. **Type II error (beta):** failing to reject a false null hypothesis. The sample data are not in the critical region even though the treatment has an effect on the sample.

The complete set of decisions and outcomes is shown in Table 8.1. The risk of an error is especially important in the case of a Type I error, which can lead to a false report. Fortunately, the probability of a Type I error is determined by the alpha level, which is completely under the control of the researcher. At the beginning of a hypothesis test, the researcher states the hypotheses and selects the alpha level, which immediately determines the risk of a Type I error.

■ Selecting an Alpha Level

As you have seen, the alpha level for a hypothesis test serves two very important functions. First, alpha helps determine the boundaries for the critical region by defining the concept of "very unlikely" outcomes. At the same time, alpha determines the probability of a Type I error. When you select a value for alpha at the beginning of a hypothesis test, your decision influences both of these functions.

The primary concern when selecting an alpha level is to minimize the risk of a Type I error. Thus, alpha levels tend to be very small probability values. By convention, the largest permissible value is $\alpha = .05$ (Cowles & Davis, 1982). When there is no treatment effect, an alpha level of .05 means that there is still a 5% risk, or a 1-in-20 probability, of rejecting the null hypothesis when it is actually true and committing a Type I error. Because the consequences of a Type I error can be relatively serious, many researchers and scientific publications prefer to use a more conservative alpha level such as .01 or .001 to reduce the risk that a false report is published and becomes part of the scientific literature.

At this point, it may appear that the best strategy for selecting an alpha level is to choose the smallest possible value to minimize the risk of a Type I error. However, there is a different kind of risk that develops as the alpha level is lowered. Specifically, a lower alpha level means less risk of a Type I error, but it also means that the hypothesis test demands more evidence from the research results.

The trade-off between the risk of a Type I error and the demands of the test is controlled by the boundaries of the critical region. For the hypothesis test to conclude that there is evidence for a treatment effect, the sample data must be in the critical region. If the treatment really has an effect, it should cause the sample to be different from the original population; essentially, the treatment should push the sample into the critical region. However, as the alpha level is lowered, the boundaries for the critical region move farther out and become more difficult to reach. Figure 8.5 shows how the boundaries for the critical region move farther into the tails as the alpha level decreases. Notice that $z = 0$, in the center of the distribution, corresponds to the value of μ specified in the null hypothesis. The boundaries for

FIGURE 8.5

The locations of the critical region boundaries for three different levels of significance: $\alpha = .05$, $\alpha = .01$, and $\alpha = .001$.

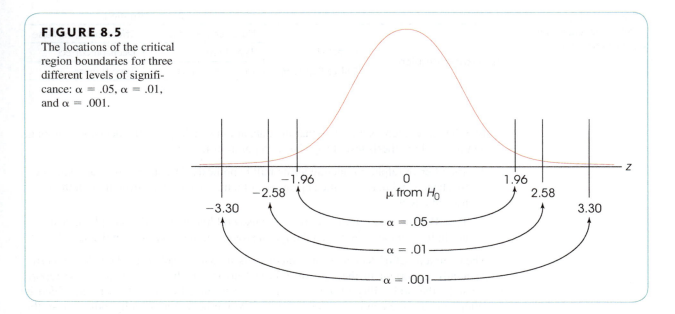

the critical region determine how much distance between the sample mean and μ is needed to reject the null hypothesis. As the alpha level gets smaller, this distance gets larger.

Thus, an extremely small alpha level, such as .000001 (one in a million), would mean almost no risk of a Type I error, but it would push the critical region so far out that it would become essentially impossible to ever reject the null hypothesis; that is, it would require an enormous treatment effect or an enormous sample size before the sample data would reach the critical boundaries.

In general, researchers try to maintain a balance between the risk of a Type I error and the demands of the hypothesis test. Alpha levels of .05, .01, and .001 are considered reasonably good values because they provide a low risk of error without placing excessive demands on the research results.

LEARNING CHECK   **LO5** **1.** What does a Type II error mean?

    a. A researcher has falsely concluded that a treatment has an effect.

    b. A researcher has correctly concluded that a treatment has no effect.

    c. A researcher has falsely concluded that a treatment has no effect.

    d. A researcher has correctly concluded that a treatment has an effect.

LO5 **2.** What does a Type I error mean?

    a. A researcher has concluded that a treatment has an effect when it really does.

    b. A researcher has concluded that a treatment has no effect when it really does not.

    c. A researcher has concluded that a treatment has no effect when it really does.

    d. A researcher has concluded that a treatment has an effect when it really does not.

LO5 **3.** What is the consequence of increasing the alpha level (for example, from .01 to .05)?

    a. It will increase the likelihood of rejecting H_0 and increase the risk of a Type I error.

    b. It will decrease the likelihood of rejecting H_0 and increase the risk of a Type I error.

    c. It will increase the likelihood of rejecting H_0 and decrease the risk of a Type I error.

    d. It will decrease the likelihood of rejecting H_0 and decrease the risk of a Type I error.

ANSWERS   **1. c** **2. d** **3. a**

8-3 | More about Hypothesis Tests

LEARNING OBJECTIVES

6. Describe how the results of a hypothesis test with a z-score test statistic are reported in the literature.

7. Explain how the outcome of a hypothesis test is influenced by the sample size, the standard deviation, and the difference between the sample mean and the hypothesized population mean.

8. Describe the assumptions underlying a hypothesis test with a z-score test statistic.

■ A Summary of the Hypothesis Test

In Example 8.1 we presented a complete example of a hypothesis test evaluating the effect of waitresses wearing red on male customers' tipping behavior. The four-step process for the hypothesis test is summarized as follows:

    Step 1: *State the hypotheses and select an alpha level.*

    Step 2: *Locate the critical region.*

    Step 3: *Compute the test statistic (the z-score).*

    Step 4: *Make a decision about the null hypothesis.*

IN THE LITERATURE

Reporting the Results of the Statistical Test

When you are writing a research report or reading a published report, a special jargon and notational system are used to discuss the outcome of a hypothesis test. If the results from the waitress-tipping study in Example 8.1 were reported in a scientific journal, for example, you would not be told explicitly that the researcher evaluated the data using a z-score as a test statistic with an alpha level of .05. Nor would you be told "the null hypothesis is rejected." Instead, you would see a statement such as:

Wearing a red shirt had a significant effect on the size of the tips left by male customers, $z = 2.40$, $p < .05$.

(continues)

hypothesis test because the tests usually work well even when the assumptions are violated. However, you should be aware of the fundamental conditions that are associated with each type of statistical test to ensure that the test is being used appropriately. The assumptions for hypothesis tests with z-scores are summarized as follows.

Random Sampling   It is assumed that the participants used in the study were selected randomly. Remember, we wish to generalize our findings from the sample to the population. Therefore, the sample must be representative of the population from which it has been drawn. Random sampling helps to ensure that it is representative.

Independent Observations   The values in the sample must consist of *independent* observations. In everyday terms, two observations are independent if there is no consistent, predictable relationship between the first observation and the second. More precisely, two events (or observations) are independent if the occurrence of the first event has no effect on the probability of the second event. Specific examples of independence and non-independence are examined in Box 8.1. Usually, this assumption is satisfied by using a

---

BOX 8.1 Independent Observations

Independent observations are a basic requirement for nearly all hypothesis tests. The critical concern is that each observation or measurement is not influenced by any other observation or measurement. An example of independent observations is the set of outcomes obtained in a series of coin tosses. Assuming that the coin is balanced, each toss has a 50–50 chance of coming up either heads or tails. More important, each toss is *independent* of the tosses that came before. On the fifth toss, for example, there is a 50% chance of heads no matter what happened on the previous four tosses; the coin does not remember what happened earlier and is not influenced by the past. (*Note*: Many people fail to believe in the independence of events. For example, after a series of four tails in a row, it is tempting to think that the probability of heads must increase because the coin is overdue to come up heads. This is a mistake, called the "gambler's fallacy." Remember that the coin does not know what happened on the preceding tosses and cannot be influenced by previous outcomes.)

In most research situations, the requirement for independent observations is typically satisfied by using a random sample of separate, unrelated individuals. Thus, the measurement obtained for each individual is not influenced by other participants in the study. The following two situations demonstrate circumstances in which the observations are *not* independent.

1. A researcher interested in the mathematical ability of new freshmen at the state college selects a sample of $n = 20$ from the group of students who attend a brief orientation describing the school's physics program. It should be obvious that the researcher does *not* have 20 independent observations. In addition to being a biased and unrepresentative sample, the students in this group probably share an unusually high level of mathematics experience. Thus, the score for each student is likely to be similar to the scores for the others in the group.

2. The principle of independent observations is violated if the sample is obtained using *sampling without replacement*. For example, if you are selecting from a group of 20 potential participants, each individual has a 1 in 20 chance of being selected first. After the first person is selected, however, there are only 19 people remaining and the probability of being selected changes to 1 in 19. Because the probability of the second selection depends on the first, the two selections are not independent.

random sample, which also helps ensure that the sample is representative of the population and that the results can be generalized to the population.

The Value of σ Is Unchanged by the Treatment   A critical part of the *z*-score formula in a hypothesis test is the standard error, σ_M. To compute the value for the standard error, we must know the sample size (*n*) and the population standard deviation (σ). In a hypothesis test, however, the sample comes from an *unknown* population (see Figure 8.2). If the population is really unknown, it would suggest that we do not know the standard deviation and, therefore, we cannot calculate the standard error. To solve this dilemma, we have made an assumption. Specifically, we assume that the standard deviation for the unknown population (after treatment) is the same as it was for the population before treatment.

Actually, this assumption is the consequence of a more general assumption that is part of many statistical procedures. This general assumption states that the effect of the treatment is to add a constant amount to (or subtract a constant amount from) every score in the population. You should recall that adding (or subtracting) a constant changes the mean but has no effect on the standard deviation. You also should note that this assumption is a theoretical ideal. In actual experiments, a treatment generally does not show a perfect and consistent additive effect.

Normal Sampling Distribution   To evaluate hypotheses with *z*-scores, we have used the unit normal table to identify the critical region. This table can be used only if the distribution of sample means is normal.

LEARNING CHECK   **LO6** **1.** A research report includes the statement, "$z = 1.18, p > .05$." What happened in the hypothesis test?

    a. The obtained sample mean was very unlikely if the null hypothesis is true, so H_0 was rejected.

    b. The obtained sample mean was very likely if the null hypothesis is true, so H_0 was rejected.

    c. The obtained sample mean was very unlikely if the null hypothesis is true, and the test failed to reject H_0.

    d. The obtained sample mean was very likely if the null hypothesis is true, and the test failed to reject H_0.

LO7 **2.** A researcher uses a hypothesis test to evaluate H_0: $\mu = 90$. Which combination of factors is most likely to result in rejecting the null hypothesis?

    a. $M = 95$ and $\sigma = 10$

    b. $M = 95$ and $\sigma = 20$

    c. $M = 100$ and $\sigma = 10$

    d. $M = 100$ and $\sigma = 20$

LO8 **3.** What assumptions are required for a *z*-score hypothesis test?

    a. The scores are obtained by random sampling.

    b. The scores in the sample are independent observations.

    c. The distribution of sample means is normal.

    d. all of the above

ANSWERS   **1. d**   **2. c**   **3. d**

8-4 Directional (One-Tailed) Hypothesis Tests

LEARNING OBJECTIVE

9. Describe the hypotheses and the critical region for a directional (one-tailed) hypothesis test.

The hypothesis-testing procedure presented in Sections 8-2 and 8-3 is the standard, *nondirectional,* or *two-tailed*, *test* format. The term *two-tailed* comes from the fact that the critical region is divided between the two tails of the distribution. This format is by far the most widely accepted procedure for hypothesis testing. Nonetheless, there is an alternative that is discussed in this section.

Usually a researcher begins an experiment with a specific prediction about the direction of the treatment effect. For example, a special training program is expected to *increase* student performance, or alcohol consumption is expected to *slow* reaction times. In these situations, it is possible to state the statistical hypotheses in a manner that incorporates the directional prediction into the statement of H_0 and H_1. The result is a directional test, or what commonly is called a *one-tailed test*.

In a **directional hypothesis test**, or a **one-tailed test**, the statistical hypotheses (H_0 and H_1) specify either an increase or a decrease in the population mean. That is, they make a statement about the direction of the effect.

The following example demonstrates the elements of a one-tailed hypothesis test.

EXAMPLE 8.3

Earlier, in Example 8.1, we discussed a research study that examined the effect of waitresses wearing red on the tips given by male customers. In the study, each participant in a sample of $n = 36$ was served by a waitress wearing a red shirt and the size of the tip was recorded. For the general population of male customers (without a red shirt), the distribution of tips was roughly normal with a mean of $\mu = 16$ percent and a standard deviation of $\sigma = 3$ percentage points. For this example, the expected effect is that the color red will increase tips. If the researcher obtains a sample mean of $M = 16.9$ percent for the $n = 36$ participants, is the result sufficient to conclude that the red shirt really increases tips? ∎

■ The Hypotheses for a Directional Test

Because a specific direction is expected for the treatment effect, it is possible for the researcher to perform a directional test. The first (and most critical) step is to state the statistical hypotheses. Remember that the null hypothesis states that there is no treatment effect and that the alternative hypothesis states that there is an effect. For this example, the predicted effect is that the red shirt will increase tips. Thus, the two hypotheses would state:

H_0: Tips are not increased. (The treatment does not work.)

H_1: Tips are increased. (The treatment works as predicted.)

To express directional hypotheses in symbols, it usually is easier to begin with the alternative hypothesis (H_1). Again, we know that the general population has an average of $\mu = 16$, and H_1 states that this value will be increased with the red shirt. Therefore, expressed in symbols, H_1 states,

H_1: $\mu > 16$ (With the red shirt, the average tip is greater than 16 percent.)

The null hypothesis states that the red shirt does not increase tips. In symbols,

H_0: $\mu \leq 16$ (With the red shirt, the average tip is not greater than 16 percent.)

Note again that the two hypotheses are mutually exclusive and cover all the possibilities. Also note that the two hypotheses concern the general population of male customers, not just the 36 men in the study. We are asking what would happen if all male customers were served by a waitress wearing a red shirt.

■ The Critical Region for Directional Tests

The critical region is defined by sample outcomes that are very unlikely to occur if the null hypothesis is true (that is, if the treatment has no effect). Earlier (page 250), we noted that the critical region can also be defined in terms of sample values that provide *convincing evidence* that the treatment really does have an effect. For a directional test, the concept of "convincing evidence" is the simplest way to determine the location of the critical region. We begin with all the possible sample means that could be obtained if the null hypothesis is true. This is the distribution of sample means and it will be normal (because the population of tips given by customers is normal), have an expected value of $\mu = 16$ (from H_0), and, for a sample of $n = 36$, will have a standard error of $\sigma_M = \frac{3}{\sqrt{36}} = 0.5$ points. The distribution is shown in Figure 8.6.

> If the prediction is that the treatment will produce a *decrease* in scores, the critical region is located entirely in the left-hand tail of the distribution.

For this example, the treatment is expected to increase tips given by customers. If the regular population of male customers has an average tip of $\mu = 16$ percent, then a sample mean that is substantially more than 16 would provide convincing evidence that the red shirt worked. Thus, the critical region is located entirely in the right-hand tail of the distribution corresponding to sample means much greater than $\mu = 16$ (Figure 8.6). Because the critical region is contained in one tail of the distribution, a directional test is commonly called a *one-tailed* test. Also note that the proportion specified by the alpha level is not divided between two tails, but rather is contained entirely in one tail. Using $\alpha = .05$ for example, the whole 5% is located in one tail. In this case, the z-score boundary for the critical region is $z = +1.65$, which is obtained by looking up a proportion of .05 in column C (the tail) of the unit normal table.

Notice that a directional (one-tailed) test requires two changes in the step-by-step hypothesis-testing procedure.

1. In the first step of the hypothesis test, the directional prediction is incorporated into the statement of the hypotheses.

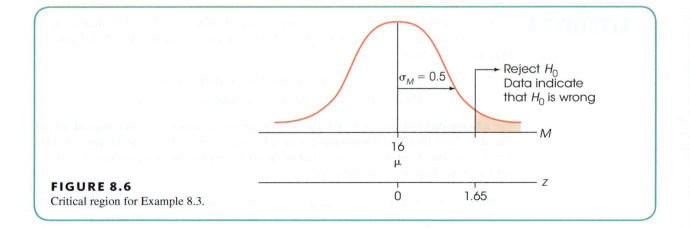

FIGURE 8.6
Critical region for Example 8.3.

2. In the second step of the process, the critical region is located entirely in one tail of the distribution.

After these two changes, a one-tailed test continues exactly the same as a regular two-tailed test. Specifically, you calculate the z-score statistic and then make a decision about H_0 depending on whether the z-score is in the critical region.

For this example, the researcher obtained a mean of $M = 16.9$ percent for the 36 participants who were served by a waitress in a red shirt. This sample mean corresponds to a z-score of

$$z = \frac{M - \mu}{\sigma_M} = \frac{16.9 - 16.0}{0.5} = \frac{0.9}{0.5} = 1.80$$

A z-score of $z = +1.80$ is in the critical region for a one-tailed test (see Figure 8.6). This is a very unlikely outcome if H_0 is true. Therefore, we reject the null hypothesis and conclude that the red shirt produces a significant increase in tips from male customers. In the literature, this result would be reported as follows:

Wearing a red shirt produced a significant increase in tips, $z = 1.80$, $p < .05$, one-tailed.

Note that the report clearly acknowledges that a one-tailed test was used.

■ Comparison of One-Tailed versus Two-Tailed Tests

The general goal of hypothesis testing is to determine whether a treatment has an effect on a population. The test is performed by selecting a sample, administering the treatment to the sample, and then comparing the result with the original population. If the treated sample is noticeably different from the original population, then we conclude that there is evidence for a treatment effect, and we reject H_0. On the other hand, if the treated sample is still similar to the original population, then we conclude that there is no convincing evidence for a treatment effect, and we fail to reject H_0. The critical factor in this decision is the *size of the difference* between the treated sample and the original population. A large difference is evidence that the treatment worked; a small difference is not sufficient to say that the treatment has any effect.

The major distinction between one-tailed and two-tailed tests is in the criteria they use for rejecting H_0. A one-tailed test allows you to reject the null hypothesis when the difference between the sample and the population is relatively small, provided the difference is in the specified direction. A two-tailed test, on the other hand, requires a relatively large difference independent of direction. This point is illustrated in the following example.

EXAMPLE 8.4

Consider again the one-tailed test in Example 8.3 evaluating the effect of waitresses wearing red on the tips from male customers. If we had used a standard two-tailed test, the hypotheses would be

H_0: $\mu = 16$ (The red shirt has no effect on tips.)

H_1: $\mu \neq 16$ (The red shirt does have an effect on tips.)

For a two-tailed test with $\alpha = .05$, the critical region consists of z-scores beyond ± 1.96. The data from Example 8.3 produced a sample mean of $M = 16.9$ percent and $z = 1.80$. For the two-tailed test, this z-score is not in the critical region, and we conclude that the red shirt does not have a significant effect. ■

With the two-tailed test in Example 8.4, the 0.9-point difference between the sample mean and the hypothesized population mean ($M = 16.9$ and $\mu = 16$) is not big enough to reject the null hypothesis. However, with the one-tailed test in Example 8.3, the same 0.9-point difference is large enough to reject H_0 and conclude that the treatment had a significant effect.

All researchers agree that one-tailed tests are different from two-tailed tests. However, there are several ways to interpret the difference. One group of researchers contends that a two-tailed test is more rigorous and, therefore, more convincing than a one-tailed test. Other researchers feel that one-tailed tests are preferable because they are more sensitive. That is, a relatively small treatment effect may be significant with a one-tailed test but fail to reach significance with a two-tailed test.

In general, two-tailed tests should be used in research situations when there is no strong directional expectation or when there are two competing predictions. For example, a two-tailed test would be appropriate for a study in which one theory predicts an increase in scores but another theory predicts a decrease. One-tailed tests should be used only in situations when the directional prediction is made before the research is conducted and there is a strong justification for making the directional prediction, including results from previous research. In particular, if a two-tailed test fails to reach significance, you should never follow up with a one-tailed test as a second attempt to salvage a significant result for the same data.

LEARNING CHECK   **LO9 1.** A population is known to have a mean of $\mu = 45$. A treatment is expected to *increase* scores for individuals in this population. If the treatment is evaluated using a one-tailed hypothesis, then which of the following is the correct statement of the null hypothesis?

   a. $\mu \geq 45$
   b. $\mu > 45$
   c. $\mu \leq 45$
   d. $\mu < 45$

LO9 2. A researcher is conducting an experiment to evaluate a treatment that is expected to *decrease* the scores for individuals in a population that is known to have a mean of $\mu = 95$. The results will be examined using a one-tailed hypothesis test. Which of the following is the correct statement of the alternative hypothesis (H_1)?

   a. $\mu > 95$
   b. $\mu \geq 95$
   c. $\mu < 95$
   d. $\mu \leq 95$

LO9 3. A researcher expects a treatment to produce an *increase* in the population mean. Assuming a normal distribution, what is the critical z-score for a one-tailed test with $\alpha = .01$?

   a. $+2.33$
   b. ± 2.58
   c. $+1.65$
   d. ± 2.33

ANSWERS   **1. c   2. c   3. a**

Cohen's *d* measures the distance between two means and is typically reported as a positive number even when the formula produces a negative value.

In this equation, $\mu_{no\ treatment}$ is the value for μ from the null hypothesis. The standard deviation is included in the calculation to standardize the size of the mean difference in much the same way that *z*-scores standardize locations in a distribution. For example, a 15-point mean difference can be a relatively large treatment effect or a relatively small effect depending on the size of the standard deviation. This principle is demonstrated in Figure 8.7. The top portion of the figure (part a) shows the results of a treatment that produces a 15-point mean difference in Math SAT scores; before treatment, the average Math SAT score is $\mu = 500$, and after treatment the average is 515. Notice that the standard deviation for SAT scores is $\sigma = 100$, so the 15-point difference appears to be small. For this example, Cohen's *d* is

$$\text{Cohen's } d = \frac{\text{mean difference}}{\text{standard deviation}} = \frac{15}{100} = 0.15$$

Now consider the treatment effect shown in Figure 8.7(b). This time, the treatment produces a 15-point mean difference in IQ scores; before treatment the average IQ is 100,

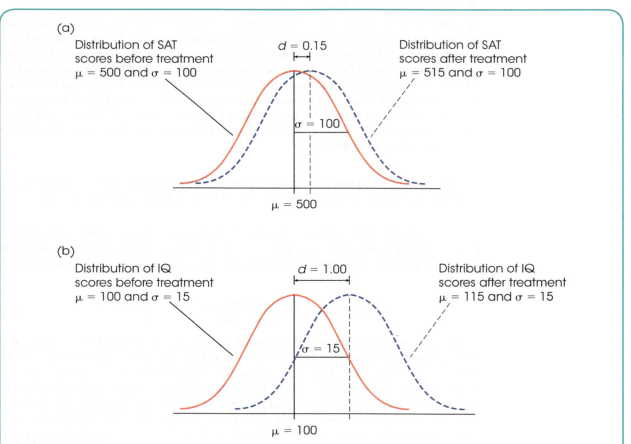

(a)

Distribution of SAT scores before treatment $\mu = 500$ and $\sigma = 100$

$d = 0.15$

Distribution of SAT scores after treatment $\mu = 515$ and $\sigma = 100$

$\sigma = 100$

$\mu = 500$

(b)

Distribution of IQ scores before treatment $\mu = 100$ and $\sigma = 15$

$d = 1.00$

Distribution of IQ scores after treatment $\mu = 115$ and $\sigma = 15$

$\sigma = 15$

$\mu = 100$

FIGURE 8.7

The appearance of a 15-point treatment effect in two different situations. In part (a), the standard deviation is $\sigma = 100$ and the 15-point effect is relatively small. In part (b), the standard deviation is $\sigma = 15$ and the 15-point effect is relatively large. Cohen's *d* uses the standard deviation to help measure effect size.

and after treatment the average is 115. Because IQ scores have a standard deviation of $\sigma = 15$, the 15-point mean difference now appears to be large. For this example, Cohen's d is

$$\text{Cohen's } d = \frac{\text{mean difference}}{\text{standard deviation}} = \frac{15}{15} = 1.00$$

Notice that Cohen's d measures the size of the treatment effect in terms of the standard deviation. For example, a value of $d = 0.50$ indicates that the treatment changed the mean by half of a standard deviation; similarly, a value of $d = 1.00$ indicates that the size of the treatment effect is equal to one whole standard deviation.

Cohen (1988) also suggested criteria for evaluating the size of a treatment effect, as shown in Table 8.2.

As one final demonstration of Cohen's d, consider the two hypothesis tests in Example 8.5. For each test, the original population had a mean of $\mu = 50$ with a standard deviation of $\sigma = 10$. For each test, the mean for the treated sample was $M = 51$. Although one test used a sample of $n = 25$ and the other test used a sample of $n = 400$, the sample size is not considered when computing Cohen's d. Therefore, both of the hypothesis tests would produce the same value:

$$\text{Cohen's } d = \frac{\text{mean difference}}{\text{standard deviation}} = \frac{1}{10} = 0.10$$

Notice that Cohen's d simply describes the size of the treatment effect and is not influenced by the number of scores in the sample. For both hypothesis tests, the original population mean was $\mu = 50$ and, after treatment, the sample mean was $M = 51$. Thus, treatment appears to have increased the scores by 1 point, which is equal to one-tenth of a standard deviation (Cohen's $d = 0.1$). Referring to Table 8.2, the treatment has a small effect.

We can now return to Example 8.1 (page 247) to demonstrate the measurement of effect size with Cohen's d after completing a hypothesis test. Recall, in Example 8.1 a researcher decided to replicate the study of Guéguen and Jacob (2012), which studied the effect of shirt color on the size of tips received by waitresses from male customers. The average tip received by waitresses wearing white shirts is known to be $\mu = 16$ percent with a standard deviation of $\sigma = 3$. A sample of $n = 36$ customers were served by the waitresses who were wearing red shirts. In one hypothetical scenario (Step 4, Number 1), the researcher found that the average tip given by the sample of customers was $M = 17.2$ percent. The results were significant with $\alpha = .05$, two tails. Although the hypothesis test provides evidence for a statistically significant treatment effect, it does not provide information about the absolute size of the effect. In this study, mean difference for Cohen's d is the difference between the sample mean, M, and the hypothesized value for μ according to the null hypothesis. For these data, Cohen's d is computed as follows:

$$\text{Cohen's } d = \frac{\text{mean difference}}{\text{standard deviation}} = \frac{17.2 - 16}{3} = \frac{1.2}{3} = 0.4$$

Based on Table 8.2, this result is a medium effect size.

TABLE 8.2
Evaluating effect size with Cohen's d.

Magnitude of d	Evaluation of Effect Size
$d = 0.2$	Small effect (mean difference around 0.2 standard deviation)
$d = 0.5$	Medium effect (mean difference around 0.5 standard deviation)
$d = 0.8$	Large effect (mean difference around 0.8 standard deviation)

LEARNING CHECK   **LO10**   **1.** Statistical significance tells us what?

    a. That the treatment effect is substantial.

    b. About the size of the treatment effect.

    c. That the results are unlikely to have occurred if there is no treatment effect.

    d. That the results are likely to have occurred if there is a treatment effect.

LO11   **2.** A sample of $n = 9$ scores is selected from a population with a mean of $\mu = 80$ and $\sigma = 12$, and a treatment is administered to the sample. After the treatment, the researcher measures effect size with Cohen's d and obtains $d = 0.25$. What was the sample mean?

    a. $M = 81$

    b. $M = 82$

    c. $M = 83$

    d. $M = 84$

LO12   **3.** If other factors are held constant, then how does sample size affect the likelihood of rejecting the null hypothesis and the value for Cohen's d?

    a. A larger sample increases the likelihood of rejecting the null hypothesis and increases the value of Cohen's d.

    b. A larger sample increases the likelihood of rejecting the null hypothesis but decreases the value of Cohen's d.

    c. A larger sample increases the likelihood of rejecting the null hypothesis but has no effect on the value of Cohen's d.

    d. A larger sample decreases the likelihood of rejecting the null hypothesis but has no effect on the value of Cohen's d.

LO12   **4.** Under what circumstances is a very small treatment effect most likely to be statistically significant?

    a. With a large sample and a large standard deviation.

    b. With a large sample and a small standard deviation.

    c. With a small sample and a large standard deviation.

    d. With a small sample and a small standard deviation.

ANSWERS   **1. c   2. c   3. c   4. b**

8-6 | Statistical Power

LEARNING OBJECTIVES

13. Define the power of a hypothesis test and explain how power is related to the probability of a Type II error.

14. Perform a power analysis.

15. Explain how power analysis is used in planning research and in interpreting the results of a hypothesis test.

16. Identify the factors that influence power and explain how power is affected by each.

As we have seen thus far, the researcher performs a hypothesis test and measures the effect size after data have been collected. However, planning must be done before any data are collected. The researcher must choose an alpha level for the hypothesis test and whether to use a one- or two-tailed test. Equally important, the researcher must decide how many participants to use in the study. Measuring the power of a statistical test assists in selecting a sample of sufficient size to detect a treatment effect when one exists. We will also see that effect size, and measures like Cohen's *d*, are related to power. The *power* of a hypothesis test is defined as the probability that the test will correctly reject the null hypothesis if the treatment really has an effect.

> The **power** of a statistical test is the probability that the test will correctly reject a false null hypothesis. That is, power is the probability that the test will identify a treatment effect if one really exists.

Whenever a treatment has an effect, there are only two possible outcomes for a hypothesis test:

1. The first outcome is failing to reject H_0 when there is a real effect, which was defined earlier (page 258) as a Type II error.
2. The second outcome is rejecting H_0 when there is a real effect.

Because there are only two possible outcomes when there is a treatment effect, the probability for the first and the probability for the second must add up to 100%, or $p = 1.00$. Therefore, if the probability of committing a Type II error is 20%, then the probability of rejecting a false H_0 is 80%. We have already identified the probability of a Type II error (the first outcome) as:

$$p \text{ (Type II error)} = \beta.$$

Therefore, the power of the test (the second outcome) must be:

$$p \text{ (rejecting a false } H_0\text{)} = 1 - \beta.$$

In the examples that follow, we demonstrate the calculation of power for hypothesis tests; that is, the probability that the test will correctly reject the null hypothesis. At the same time, however, we are computing the probability that the test will result in a Type II error. For example, if the power of the test is calculated to be 70% $(1 - \beta)$, then the probability of a Type II error must be 30% (β).

■ Calculating Power

Researchers typically calculate power as a means of determining whether an experiment is likely to be sensitive to detect a treatment effect when one exists. As we shall see, power analysis is especially helpful to the researcher in selecting a sample size that will increase the power of the hypothesis test. Thus, researchers may calculate the power of a hypothesis test *before* they actually conduct the experiment. In this way, they can determine the probability that the results will be significant (reject H_0) if the treatment really produces an effect before investing time and resources to conduct the study. To calculate power, however, it is first necessary to make assumptions about a variety of factors that influence the outcome of a hypothesis test. Factors such as the sample size, the size of the treatment effect, and the value chosen for the alpha level can all influence a hypothesis test as well as its power. The following example demonstrates the calculation of power for a specific research situation.

decided to conduct the study using a larger sample, then the power would be dramatically different. The following example demonstrates this point using a sample size of $n = 25$.

EXAMPLE 8.7   We are assuming that the treatment will produce an 8-point effect and $\alpha = .05$ for two tails. These are the same assumptions used in Example 8.6; however, now the sample size is increased from $n = 4$ to $n = 25$.

STEP 1   **Sketch the distributions for the null and alternative hypotheses.**   Figure 8.9 shows the two distributions of sample means with $n = 25$. Again, the null distribution on the left has an expected value of $\mu_{null} = 80$ and shows the set of all possible sample means if H_0 is true. With $n = 25$, the standard error for the sample means is

$$\sigma_M = \frac{\sigma}{\sqrt{n}} = \frac{10}{\sqrt{25}} = \frac{10}{5} = 2$$

The alternative distribution is on the right. It has an expected value of $\mu_{alternative} = 88$ and shows all possible sample means when there is an 8-point treatment effect. It also has a standard error of $\sigma_M = 2$. Notice that with a larger sample the standard error is smaller than in Example 8.6 when $n = 4$. This can be seen by comparing Figure 8.9 to Figure 8.8.

STEP 2   **Locate the critical regions and compute $M_{critical}$.**   As before, $\alpha = .05$ with two tails, and the null distribution has critical boundaries for the hypothesis test of $z = -1.96$ and $z = +1.96$. Note that almost all of the treated sample means in the alternative distribution are now located beyond the $+1.96$ boundary. Thus, with a sample of $n = 25$, there is a high probability (power) that you will detect the treatment effect.

With $\alpha = .05$ the critical region is defined by z values of ± 1.96. With an 8-point increase we must determine the value of the critical mean for a $z = +1.96$.

$$M_{critical} = \mu_{null} + z(\sigma_M)$$
$$M_{critical} = 80 + 1.96(\sigma_M) = 80 + 1.96(2) = 83.92 \text{ points}$$

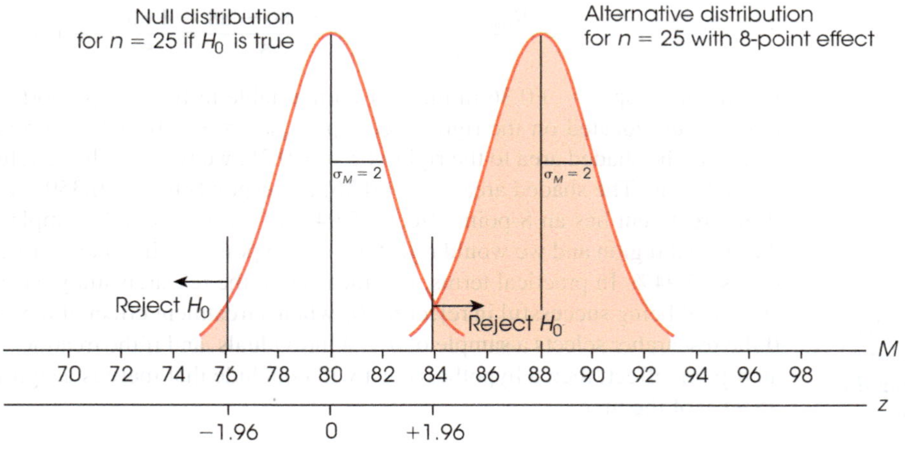

FIGURE 8.9

A demonstration of how sample size affects power for a hypothesis test. The left-hand side shows the distribution of sample means that would occur if the null hypothesis were true. The critical region is defined for this distribution. The right-hand side shows the distribution of sample means that would be obtained if there were an 8-point treatment effect. Notice that nearly all of the alternative distribution is shaded. Thus, increasing the sample size to $n = 25$ has increased the power to almost 100% compared to nearly 36% for a sample of $n = 4$ in Figure 8.8.

STEP 3 **Compute z-score for the alternative distribution and find power.** The boundary corresponds to a sample mean of $M = 83.92$. The expected value for the treated distribution is $\mu_{\text{alternative}} = 88$. The z-score for the critical mean is as follows:

$$z = \frac{M_{\text{critical}} - \mu_{\text{alternative}}}{\sigma_M} = \frac{83.92 - 88}{2} = \frac{-4.08}{2} = -2.04$$

The z-score is negative and we are interested in the proportion for the shaded area to the right of it. This proportion corresponds to column B of the unit normal table. Power equals .9793, or 97.93%. With a sample of $n = 25$, there is a high likelihood that the hypothesis test would detect a treatment effect if one exists. ■

Earlier, in Example 8.6, we found power to equal almost 36% for a sample of $n = 4$. However, when the sample size is increased to $n = 25$, power increases to nearly 98%. Holding other factors constant (such as effect size and alpha), a larger sample produces greater power for a hypothesis test. Because power is directly related to sample size, one of the primary reasons for doing a power analysis is to determine what sample size is necessary to achieve a reasonable probability for a successful research study. Before a study is conducted, researchers can compute power using different sample sizes for any given assumption about effect size to determine the probability that their research will successfully reject the null hypothesis. If the probability (power) is too small, they always have the option of increasing sample size prior to conducting the study to increase power. In this way, power analysis plays an important role in planning research.

■ Power and Effect Size

The size of an effect is another factor that is related to power. If you examine Figure 8.8, you might see how power and effect size are related. Figure 8.8 shows the calculation of power for an 8-point treatment effect based on a sample of $n = 4$ and $\alpha = .05$, two tails. Now consider what would happen if the treatment effect instead were 16 points. With a 16-point treatment effect, the alternative distribution (right side) would shift farther to the right so that its expected value would be $\mu_{\text{alternative}} = 96$. The separation between the two distributions increases. Figure 8.10 shows this larger effect size. In this new position, approximately

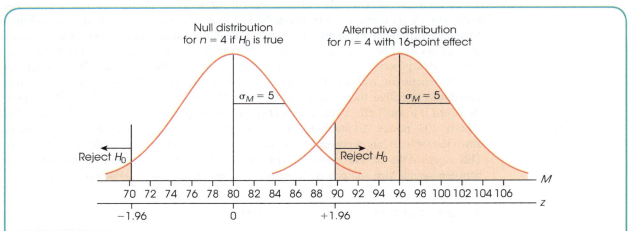

FIGURE 8.10
The null and alternative distributions for a 16-point effect when $n = 4$ and $\alpha = .05$ two tails. Compared to Example 8.6 and Figure 8.8, a 16-point effect has a power of $p = 89.25\%$, much larger than an 8-point effect which has a power of 35.94%.

90% of the treated sample means would be beyond the $z = +1.96$ boundary. Thus, with a 16-point treatment effect, there is about a 90% probability of selecting a sample that leads to rejecting the null hypothesis. In other words, the power of the test is around 90% for a 16-point effect compared to only 36% with an 8-point effect (Example 8.6). Again, it is possible to find the z-score corresponding to the exact location of the critical boundary and to look up the probability value for power in the unit normal table. For a 16-point treatment effect, you should find that the critical boundary ($M = 89.80$) corresponds to $z = -1.24$. The z-score is negative, so the proportion in the shaded area to the right is in column B of the unit normal table. The table shows the exact power of the test is $p = 0.8925$, or 89.25%.

In general, as the effect size increases, the distribution of sample means for the alternative distribution moves even farther from the null distribution so that more and more of its samples are beyond the $z = +1.96$ boundary. Thus, as the effect size increases, the probability of rejecting H_0 also increases, which means that the power of the test increases. Measures of effect size such as Cohen's d and measures of power are related. Cohen's d and power are complementary in that they both provide information about the treatment effect. As the magnitude of measures of effect size, such as Cohen's d, increase, so does statistical power. For this reason, power analysis will sometimes accompany Cohen's d and similar measures of effect size, after a study is completed and a hypothesis test has found evidence for an effect.

■ Another Look at Sample Size and Effect Size

When the effect size is small, the statistical power will be low—provided that other factors are held constant (such as alpha level and sample size). When power is low, it will be less likely that the null hypothesis will be rejected when a treatment effect exists. The study will be less likely to detect a treatment effect. We have seen that when a treatment effect is small, one way to increase power is to increase the sample size. Thus, it is important for researchers to select a sample size that is sufficiently large to detect a treatment effect. Power analysis can be performed using the assumption of a small treatment effect and repeated for different sample sizes, n, to find a sample size that results in more power. Because power analysis plays a role in planning research, power tables are often used as guides for selecting sample size.

Table 8.3 is a power table that shows the sample size required to achieve a specific level of power for medium effect size ($d = 0.50$) and small effect size ($d = 0.20$), for both $\alpha = .05$ and $\alpha = .01$, two tails. For now, we will consider only the two columns for medium and small effects when $\alpha = .05$, two tails. The level of power is listed in the first column. Notice the difference in sample size between medium and small effect sizes for any given level of power. For example, suppose a researcher is planning a study and wants it to achieve 70% power. That is, when a treatment effect exists there will be a 70% probability that the experiment will detect the effect (H_0 will be rejected). If the researcher is expecting a medium effect size, then a sample size of $n = 25$ would be needed to achieve 70% power. However, if a small effect size is expected, then a sample of $n = 155$ would be needed to achieve 70% power. Additionally, in any column there is a general pattern. As you move down the table to larger power levels, the size of the sample that is needed also increases. This aspect of the table is consistent with Example 8.7, where we looked at the relationship between sample size and power. As sample size increases, statistical power increases too.

■ Other Factors That Affect Power

Although the power of a hypothesis test is directly influenced by the size of the treatment effect, power is not meant to be a pure measure of effect size. Instead, power is influenced by several factors other than effect size that are related to the hypothesis test. Some of these factors are considered in the following subsections.

TABLE 8.3
Sample size required to achieve a level of power.

Power*	Medium effect size, $d = 0.50$ with $\alpha = .05$, two tails	Small effect size, $d = 0.20$ with $\alpha = .05$, two tails	Medium effect size, $d = 0.50$ with $\alpha = .01$, two tails	Small effect size, $d = 0.20$ with $\alpha = .01$, two tails
20%	5	32	13	76
30%	9	52	17	106
40%	12	73	22	135
50%	16	97	27	166
60%	20	123	33	201
70%	25	155	39	241
80%	32	197	47	292
90%	43	263	60	372

*Matlab™ was programmed to produce the values in this table. For each treatment effect size and alpha level in the table, power was computed for sample sizes ranging from 1 to 1,000, and the smallest sample size that exceeded the target level of power was recorded. A normal z distribution was assumed in the calculations.

Alpha Level   Reducing the alpha level for a hypothesis test also reduces the power of the test. For example, lowering α from .05 to .01 lowers the power of the hypothesis test. The effect of reducing the alpha level can be seen by looking at Figure 8.8. In this figure, the boundaries for the critical region are drawn using $\alpha = .05$. Specifically, the critical region on the right-hand side of the null distribution begins at $z = +1.96$. If α were changed to .01, the boundary would be moved farther to the right, out to $z = +2.58$. It should be clear that moving the critical boundary to the right means that a smaller portion of the alternative distribution will be in the critical region. Thus, there would be a lower probability of rejecting the null hypothesis and a lower value for the power of the test.

The relationship between alpha level and power is shown in Table 8.3 as well. For example, compare the columns for medium effect size for $\alpha = .05$, two tails, and medium effect size for $\alpha = .01$, two tails. For the same size effect, when $\alpha = .01$, a larger sample size is necessary to achieve a given level of power. To achieve 70% power for a medium-size effect ($d = 0.50$), a sample of $n = 25$ is required for $\alpha = .05$, but a sample of $n = 39$ is needed when $\alpha = .01$.

One-Tailed versus Two-Tailed Tests   Changing from a regular two-tailed test to a one-tailed test increases the power of the hypothesis test. Again, this effect can be seen by referring to Figure 8.9. The figure shows the boundaries for the critical region using a two-tailed test with $\alpha = .05$ so that the critical region on the right-hand side begins at $z = +1.96$. Changing to a one-tailed test would move the critical boundary closer to the center of the null distribution to a value of $z = +1.65$. Moving the boundary to the left would cause a larger proportion of the alternative distribution to be in the critical region and, therefore, would increase the power of the test.

■ A Note on the Direction of the Treatment Effect

We have used examples in which the assumption is that the treatment would cause an increase in mean difference. It should be clear that there are experiments in which the expected outcome is a decrease in the dependent variable. For example, perhaps the researcher assumes the treatment will cause a 20-point decrease. In this case, the null distribution will be on the right side and the alternative distribution on the left. This is shown in Figure 8.11. The calculation of power would follow the same steps. Notice that when computing $M_{critical}$ in step 2, a negative z-score (-1.96) is multiplied by the standard error, σ_M.

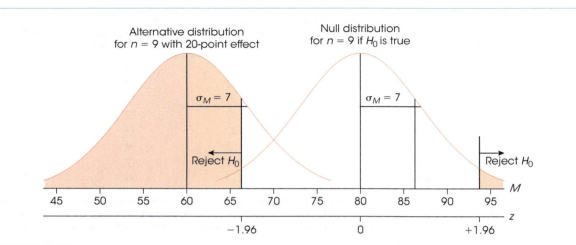

FIGURE 8.11

The distributions for a power analysis when the researcher assumes the treatment will cause a 20-point decrease. The null distribution is on the right side and the alternative distribution is on the left.

LEARNING CHECK   **LO13**   **1.** If the power of a hypothesis test is found to be $p = 0.80$, then what is the probability of a Type II error for the same test?

    a. $p = 0.20$

    b. $p = 0.80$

    c. The probability of a Type II error is not related to power.

    d. It is impossible to determine without knowing the alpha level for the test.

LO 14 and 15   **2.** Suppose that a researcher is planning a study and would like to estimate the study's power before collecting data. Assuming that the null distribution has a mean of $\mu = 80$ and a standard deviation of $\sigma = 21$, what is the power of a $\alpha = .05$, two-tailed, hypothesis test with an expected treatment effect of -20. Sample size is $n = 9$. Notice that Step 1 of this power analysis is depicted in Figure 8.11.

    a. 0.1841

    b. 0.8159

    c. 0.3821

    d. 0.6179

LO16   **3.** How does the sample size influence the likelihood of rejecting the null hypothesis and the power of the hypothesis test?

    a. Increasing sample size increases both the likelihood of rejecting H_0 and the power of the test.

    b. Increasing sample size decreases both the likelihood of rejecting H_0 and the power of the test.

    c. Increasing sample size increases the likelihood of rejecting H_0, but the power of the test is unchanged.

    d. Increasing sample size decreases the likelihood of rejecting H_0, but the power of the test is unchanged.

LO16 **4.** How is the power of a hypothesis test related to sample size and the alpha level?

    a. A larger sample and a larger alpha level will both increase power.

    b. A larger sample and a larger alpha level will both decrease power.

    c. A larger sample will increase power, but a larger alpha will decrease power.

    d. A larger sample will decrease power, but a larger alpha will increase power.

ANSWERS 1. a   2. b   3. a   4. a

SUMMARY

1. Hypothesis testing is an inferential procedure that uses the data from a sample to draw a general conclusion about a population. The procedure begins with a hypothesis about an unknown population. Then a sample is selected, and the sample data provide evidence that either supports or refutes the hypothesis.

2. In this chapter, we introduced hypothesis testing using the simple situation in which a sample mean is used to test a hypothesis about an unknown population mean. The goal for the test is to determine whether a treatment has an effect on the population mean (see Figure 8.2).

3. Hypothesis testing is structured as a four-step process that is used throughout the remainder of the book.
   a. State the hypotheses, and select an alpha level. The null hypothesis (H_0) states that there is no effect or no change. In this case, H_0 states that the mean for the population after treatment is the same as the mean before treatment. The alpha level, usually $\alpha = .05$ or $\alpha = .01$, provides a definition of the term *very unlikely* and determines the risk of a Type I error. Also state an alternative hypothesis (H_1), which is the exact opposite of the null hypothesis.
   b. Locate the critical region. The critical region is defined as sample outcomes that would be very unlikely to occur if the null hypothesis is true. The alpha level defines "very unlikely."
   c. Collect the data and compute the test statistic. The sample mean is transformed into a z-score by the formula

   $$z = \frac{M - \mu}{\sigma_M}$$

   The value of μ is obtained from the null hypothesis. The z-score test statistic identifies the

   location of the sample mean in the distribution of sample means. Expressed in words, the z-score formula is

   $$z = \frac{\text{sample mean} - \substack{\text{hypothesized} \\ \text{population mean}}}{\text{standard error}}$$

   d. Make a decision. If the obtained z-score is in the critical region, reject H_0 because it is very unlikely that these data would be obtained if H_0 were true. In this case, conclude that the treatment has changed the population mean. If the z-score is not in the critical region, fail to reject H_0 because the data are not significantly different from the null hypothesis. In this case, the data do not provide sufficient evidence to indicate that the treatment has had an effect.

4. Whatever decision is reached in a hypothesis test, there is always a risk of making the incorrect decision. There are two types of errors that can be committed:

   - A Type I error is defined as rejecting a true H_0. This is a serious error because it results in falsely reporting a treatment effect. The risk of a Type I error is determined by the alpha level and therefore is under the experimenter's control.

   - A Type II error is defined as the failure to reject a false H_0. In this case, the experiment fails to detect an effect that actually occurred. The probability of a Type II error cannot be specified as a single value and depends in part on the size of the treatment effect. It is identified by the symbol β (beta).

5. When a researcher expects that a treatment will change scores in a particular direction (increase or decrease), it is possible to do a directional, or one-tailed, test. The first step in this procedure is to incorporate

the directional prediction into the hypotheses. To locate the critical region, you must determine what kind of data would refute the null hypothesis by demonstrating that the treatment worked as predicted. These outcomes will be located entirely in one tail of the distribution.

6. In addition to using a hypothesis test to evaluate the *significance* of a treatment effect, it is recommended that you also measure and report the *effect size*. One measure of effect size is Cohen's *d*, which is a standardized measure of the mean difference. Cohen's *d* is computed as

$$\text{Cohen's } d = \frac{\text{mean difference}}{\text{standard deviation}}$$

7. The size of the sample influences the outcome of the hypothesis test, but has little or no effect on measures of effect size. As sample size increases, the likelihood of rejecting the null hypothesis also increases. The variability of the scores influences both the outcome of the hypothesis test and measures of effect size. Increased variability reduces the likelihood of

rejecting the null hypothesis and reduces measures of effect size.

8. The power of a hypothesis test is defined as the probability that the test will correctly reject the null hypothesis.

9. To determine the power for a hypothesis test, you must first identify the boundaries for the critical region. Then, you must specify the magnitude of the treatment effect, the size of the sample, and the alpha level. With these assumptions, the power of the hypothesis test is the probability of obtaining a sample mean in the critical region.

10. As the size of the treatment effect increases, statistical power increases. Also, power is influenced by several factors that can be controlled by the experimenter:
    - Increasing the alpha level increases power.
    - A one-tailed test has greater power than a two-tailed test.
    - A large sample results in more power than a small sample.

KEY TERMS

hypothesis testing (245)

null hypothesis (248)

scientific or alternative hypothesis (248)

alpha level or level of significance (250, 257)

critical region (250)

test statistic (254)

Type I error (257)

Type II error (258)

beta (258)

significant or statistically significant (262)

nondirectional hypothesis test or two-tailed test (266)

directional hypothesis test or one-tailed test (266)

effect size (271)

Cohen's *d* (271)

power (275)

FOCUS ON PROBLEM SOLVING

1. Hypothesis testing involves a set of logical procedures and rules that enable us to make general statements about a population when all we have are sample data. This logic is reflected in the four steps that have been used throughout this chapter. Hypothesis-testing problems will become easier to tackle when you learn to follow the steps.

STEP 1 State the hypotheses and set the alpha level.

STEP 2 Locate the critical region.

STEP 3 Compute the test statistic (in this case, the *z*-score) for the sample.

STEP 4 Make a decision about H_0 based on the result of Step 3.

2. Take time to consider the implications of your decision about the null hypothesis. The null hypothesis states that there is no effect. If your decision is to reject H_0, you should conclude that the sample data provide evidence for a treatment effect. If your decision

is to fail to reject H_0, you conclude that there is not enough evidence to conclude that an effect exists.

3. When you are doing a directional hypothesis test, read the problem carefully and watch for key words (such as increase or decrease, raise or lower, and more or less) that tell you which direction the researcher is predicting. The predicted direction will determine the alternative hypothesis (H_1) and the critical region.

DEMONSTRATION 8.1

HYPOTHESIS TEST WITH z

Suppose that it is known that the scores on a standardized reading test are normally distributed with $\mu = 100$ and $\sigma = 30$. A researcher suspects that special training in reading skills will produce a change in the scores for the individuals in the population. A sample of $n = 36$ individuals is selected, and the treatment is given to this sample. Following treatment, the average score for this sample is $M = 110$. Is this enough evidence to conclude that the training has an effect on test scores?

STEP 1 **State the hypotheses and select an alpha level.**   The null hypothesis states that the special training has no effect. In symbols,

$$H_0: \mu = 100 \text{ (After special training, the mean is still 100.)}$$

The alternative hypothesis states that the treatment does have an effect.

$$H_1: \mu \neq 100 \text{ (After training, the mean is different from 100.)}$$

At this time, you also select the alpha level. For this demonstration, we will use $\alpha = .05$. Thus, there is a 5% risk of committing a Type I error if we reject H_0.

STEP 2 **Locate the critical region.**   With $\alpha = .05$, the critical region consists of sample means that correspond to z-scores beyond the critical boundaries of $z = \pm 1.96$.

STEP 3 **Obtain the sample data, and compute the test statistic.**   For this example, the distribution of sample means, according to the null hypothesis, will be normal with an expected value of $\mu = 100$ and a standard error of

$$\sigma_M = \frac{\sigma}{\sqrt{n}} = \frac{30}{\sqrt{36}} = \frac{30}{6} = 5$$

In this distribution, our sample mean of $M = 110$ corresponds to a z-score of

$$z = \frac{M - \mu}{\sigma_M} = \frac{110 - 100}{5} = \frac{10}{5} = +2.00$$

STEP 4 **Make a decision about H_0, and state the conclusion.**   The z-score we obtained is in the critical region. This indicates that our sample mean of $M = 110$ is an extreme or unusual value to be obtained from a population with $\mu = 100$. Therefore, our statistical decision is to *reject* H_0. Our conclusion for the study is that the data provide sufficient evidence that the special training changes test scores.

DEMONSTRATION 8.2

EFFECT SIZE USING COHEN'S d

We will compute Cohen's d using the research situation and the data from Demonstration 8.1. Again, the original population mean was $\mu = 100$ and, after treatment (special training), the sample mean was $M = 110$. Thus, there is a 10-point mean difference. Using the population standard deviation, $\sigma = 30$, we obtain an effect size of

$$\text{Cohen's } d = \frac{\text{mean difference}}{\text{standard deviation}} = \frac{10}{30} = 0.33$$

According to Cohen's evaluation standards (see Table 8.2), this is a medium treatment effect.

DEMONSTRATION 8.3

POWER

Suppose that a researcher is interested in replicating the study described in Demonstrations 8.1 and 8.2. She proposes to use the same standardized reading test with $\mu = 100$ and $\sigma = 30$. She expects a +10-point treatment effect and will use a sample size of $n = 9$ participants. What is the power of the planned replication?

STEP 1   Sketch the distributions for the null and alternative hypotheses.   Notice that distributions for the null and alternative hypotheses have a 10-point difference between their means. Standard error is calculated as before except that we must use $n = 9$. Thus,

$$\sigma_M = \frac{\sigma}{\sqrt{n}} = \frac{30}{\sqrt{9}} = \frac{30}{3} = 10$$

The null and alternative distributions are shown in Figure 8.12.

STEP 2   Locate the critical regions and compute M_{critical}.   In this example, power is the probability of rejecting the null hypothesis when we assume a 10-point treatment effect. In order to calculate power, we must identify the criteria for rejecting the null hypothesis. As in Demonstration 8.1, our researcher will use $\alpha = .05$, two-tailed. Thus, the critical region consists of sample means beyond the critical boundaries of $z = \pm 1.96$. Because we are

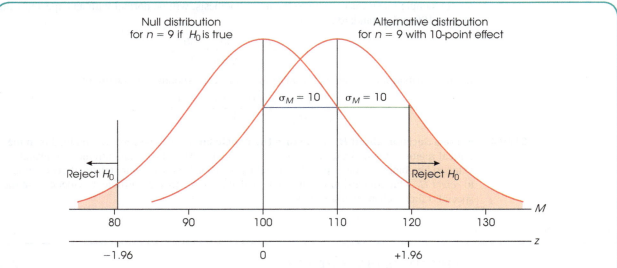

FIGURE 8.12
The null and alternative distributions for a 10-point effect when $n = 9$ and $\alpha = .05$, two tails.

expecting the treatment to increase reading scores, we calculate the value of the critical sample mean in the right tail of the null distribution:

$$M_{critical} = \mu_{null} + 1.96(\sigma_M) = 100 + 1.96(10) = 119.60$$

STEP 3   **Compute the z-Score for the alternative distribution and find power.**   The next step is to identify the position of the mean of the alternative distribution relative to $M_{critical}$. To do this, we compute a z-score as follows:

$$z = \frac{M_{critical} - \mu_{alternative}}{\sigma_M} = \frac{119.60 - 110}{10} = \frac{9.60}{10} = +0.96$$

To identify the power of the proposed study, we use the unit normal table in Appendix B. You should notice that the critical mean is *more* extreme than the mean of the alternative distribution (see Figure 8.12). Thus, we consult column C (proportion in the tail) and find that power = .1685, or 16.85%. You should immediately recognize that the proposed replication has a fairly low probability of rejecting the null hypothesis.

SPSS®

The statistical computer package SPSS is not structured to conduct hypothesis tests using z-scores. In truth, the z-score test presented in this chapter is rarely used in actual research situations. The problem with the z-score test is that it requires that you know the value of the population standard deviation, and this information is usually not available. Researchers rarely have detailed information about the populations that they wish to study. Instead, they must obtain information entirely from samples. In the following chapters we introduce new hypothesis-testing techniques that are based entirely on sample data. These new techniques are included in SPSS.

PROBLEMS

1. Does a hypothesis test allow a researcher to claim that an alternative hypothesis is true? Explain your answer.

2. Identify the four steps of a hypothesis test as presented in this chapter.

3. Suppose that a researcher is interested in the effect of a new college preparation course on scores for a standardized critical thinking test with a population mean of $\mu = 20$. Students receive training in the course and later receive the standardized test. The researcher wants to test the hypothesis that the course affected test scores.
   a. In words, state the null and alternative hypotheses as they relate to the treatment in this example.
   b. Use symbols to state the null and alternative hypothesis.

4. Define the alpha level and the critical region for a hypothesis test.

5. Define a Type I error and a Type II error and explain the consequences of each. Which type of error is worse? Why?

6. If the alpha level is changed from $\alpha = .05$ to $\alpha = .01$:
   a. What happens to the boundaries for the critical region?
   b. What happens to the probability of a Type I error?

7. Explain how each of the following influences the value of the z-score in a hypothesis test.
   a. Increasing the size of the treatment effect.
   b. Increasing the population standard deviation.
   c. Increasing the number of scores in the sample.

8. According to the CDC (2016), the average life expectancy of someone with diabetes is $\mu = 72$ years, $\sigma = 14$. Suppose that a sample of $n = 64$ people diagnosed with diabetes who received a blood glucose monitoring implant had an average life expectancy of $M = 76$ years. Test the hypothesis that the glucose monitoring implant changes life expectancy. Assume a two-tailed test, $\alpha = .05$.

9. The National Study of Student Engagement (Indiana University, 2018) reports that the average, full-time college senior in the United States spends

only $\mu = 15$, $\sigma = 9$, hours per week preparing for classes by reading, doing homework, studying, etc. A state university develops a program that is designed to increase student motivation to study. A sample of $n = 36$ students completes the program and later reports that they spend $M = 18$ hours per week studying. The university would like to test whether the program increased time spent preparing for class.

a. Assuming a two-tailed test, state the null and alternative hypotheses in a sentence that includes the two variables being examined.

b. Using the standard four-step procedure, conduct a two-tailed hypothesis test with $\alpha = .05$ to evaluate the effect of the program.

10. The personality characteristics of business leaders (e.g., CEOs) are related to the operations of the businesses that they lead (Oreg & Berson, 2018). Traits like openness to experience are related to positive financial outcomes and other traits are related to negative financial outcomes for their businesses. Suppose that a board of directors is interested in evaluating the personality of their leadership. Among a sample of $n = 16$ managers, the sample mean of the openness to experiences dimension of personality was $M = 4.50$. Assuming that $\mu = 4.24$ and $\sigma = 1.05$ (Cobb-Clark & Schurer, 2012), use a two-tailed hypothesis test with $\alpha = .05$ to test the hypothesis that this company's business leaders' openness to experience is different from the population.

11. Ackerman and Goldsmith (2011) report that students who study from a screen (smartphone, tablet, or computer) tended to have lower quiz scores than students who studied the same material from printed pages. To test this finding, a professor identifies a sample of $n = 16$ students who used the electronic version of the course textbook and determines that this sample had an average score of $M = 72.5$ on the final exam. During the previous three years, the final exam scores for the general population of students taking the course averaged $\mu = 77$ with a standard deviation of $\sigma = 8$ and formed a roughly normal distribution. The professor would like to use the sample to determine whether students studying from an electronic screen had exam scores that are significantly different from those for the general population.

a. Assuming a two-tailed test, state the null and alternative hypotheses in a sentence that includes the two variables being examined.

b. Using the standard four-step procedure, conduct a two-tailed hypothesis test with $\alpha = .05$ to evaluate the effect of studying from an electronic screen.

12. Childhood participation in sports, cultural groups, and youth groups appears to be related to improved self-esteem for adolescents (McGee, Williams, Howden-Chapman, Martin, & Kawachi, 2006). In a representative study, a sample of $n = 100$ adolescents with a history of group participation is given a standardized self-esteem questionnaire. For the general population of adolescents, scores on this questionnaire form a normal distribution with a mean of $\mu = 50$ and a standard deviation of $\sigma = 15$. The sample of group-participation adolescents had an average of $M = 53.8$.

a. Does this sample provide enough evidence to conclude that self-esteem scores for these adolescents are significantly different from those of the general population? Use a two-tailed test with $\alpha = .05$.

b. Compute Cohen's d to measure the size of the difference.

c. Write a sentence describing the outcome of the hypothesis test and the measure of effect size as it would appear in a research report.

13. A random sample is selected from a normal population with a mean of $\mu = 20$ and a standard deviation of $\sigma = 10$. After a treatment is administered to the individuals in the sample, the sample mean is found to be $M = 25$.

a. If the sample consists of $n = 25$ scores, is the sample mean sufficient to conclude that the treatment has a significant effect? Use a two-tailed test with $\alpha = .05$.

b. If the sample consists of $n = 4$ scores, is the sample mean sufficient to conclude that the treatment has a significant effect? Use a two-tailed test with $\alpha = .05$.

c. Comparing your answers for parts a and b, explain how the size of the sample influences the outcome of a hypothesis test.

14. A random sample of $n = 9$ scores is selected from a normal population with a mean of $\mu = 100$. After a treatment is administered to the individuals in the sample, the sample mean is found to be $M = 106$.

a. If the population standard deviation is $\sigma = 10$, is the sample mean sufficient to conclude that the treatment has a significant effect? Use a two-tailed test with $\alpha = .05$.

b. Repeat part a, assuming a one-tailed test with $\alpha = .05$.

c. If the population standard deviation is $\sigma = 12$, is the sample mean sufficient to conclude that the treatment has a significant effect? Use a two-tailed test with $\alpha = .05$.

d. Repeat part c, assuming a one-tailed test with $\alpha = .05$.

e. Comparing your answers for parts a through d, explain how the magnitude of the standard deviation and the number of tails in the hypothesis influence the outcome of a hypothesis test.

15. A random sample is selected from a normal population with a mean of $\mu = 40$ and a standard deviation of $\sigma = 10$. After a treatment is administered to the individuals in the sample, the sample mean is found to be $M = 46$.

    a. How large a sample is necessary for this sample mean to be statistically significant? Assume a two-tailed test with $\alpha = .05$.

    b. If the sample mean were $M = 43$, what sample size is needed to be significant for a two-tailed test with $\alpha = .05$?

16. Researchers at a weather center in the northeastern United States recorded the number of 90° Fahrenheit days each year since records first started in 1875. The numbers form a normal-shaped distribution with a mean of $\mu = 9.6$ and a standard deviation of $\sigma = 1.9$. To see if the data showed any evidence of global warming, they also computed the mean number of 90° days for the most recent $n = 4$ years and obtained $M = 12.25$. Do the data indicate that the past four years have had significantly more 90° days than would be expected for a random sample from this population? Use a one-tailed test with $\alpha = .05$.

17. A high school teacher has designed a new course intended to help students prepare for the mathematics section of the SAT. A sample of $n = 20$ students is recruited for the course and, at the end of the year, each student takes the SAT. The average score for this sample is $M = 562$. For the general population, scores on the SAT are standardized to form a normal distribution with $\mu = 500$ and $\sigma = 100$.

    a. Can the teacher conclude that students who take the course score significantly higher than the general population? Use a one-tailed test with $\alpha = .01$.

    b. Compute Cohen's d to estimate the size of the effect.

    c. Write a sentence demonstrating how the results of the hypothesis test and the measure of effect size would appear in a research report.

18. Screen time and use of social media are related to negative mental health outcomes, including suicidal thoughts (Twenge, Joiner, Rogers, & Martin, 2018). In a national survey of adolescents, the mean number of depressive symptoms was $\mu = 2.06$, $\sigma = 1.00$. Suppose that a researcher recruits $n = 25$ participants and instructs them to visit friends or family instead of using social media. The researcher observes that the average number of depressive symptoms is $M = 1.66$ after the intervention.

    a. Test the hypothesis that the treatment reduced the number of depressive symptoms. Use a one-tailed test with $\alpha = .05$.

    b. If the researcher wants to reduce the likelihood of a Type 1 error, what should they do?

    c. Compute Cohen's d to estimate the size of the effect.

    d. Write a sentence demonstrating how the results of the hypothesis test and the measure of effect size would appear in a research report.

19. Suppose that a treatment effect increases both the mean and the standard deviation of a measurement. Can a hypothesis test with z be conducted? Explain your answer.

20. After examining over one million online restaurant reviews and the associated weather conditions, Bakhshi, Kanuparthy, and Gilbert (2014) reported significantly higher ratings during moderate weather compared to very hot or very cold conditions. To verify this result, a researcher collected a sample of $n = 25$ reviews of local restaurants over an unusually hot period during July and August and obtained an average rating of $M = 7.29$. The complete set of reviews during the previous year averaged $\mu = 7.52$ with a standard deviation of $\sigma = 0.60$.

    a. Can the researcher conclude that reviews during hot weather are significantly lower than the general population average? Use a one-tailed test with $\alpha = .05$.

    b. Compute Cohen's d to measure effect size for this study.

    c. Write a sentence demonstrating how the outcome of the hypothesis test and the measure of effect size would appear in a research report.

21. A researcher is evaluating the influence of a treatment using a sample selected from a normally distributed population with a mean of $\mu = 50$ and a standard deviation of $\sigma = 10$. The researcher expects a +5-point treatment effect and plans to use a two-tailed hypothesis test with $\alpha = .05$.

    a. Compute the power of the test if the researcher uses a sample of $n = 4$ individuals (see Example 8.6).

    b. Compute the power of the test if the researcher uses a sample of $n = 25$ individuals.

22. Telles, Singh, and Balkrishna (2012) reported that yoga training improves finger dexterity. Suppose that a researcher conducts an experiment evaluating the effect of yoga on standardized O'Conner finger dexterity test scores. A sample of $n = 4$ participants is selected, and each person receives yoga training before being tested on a standardized dexterity task. Suppose that for the regular population, scores on the dexterity task form a normal distribution with $\mu = 50$ and $\sigma = 8$. The treatment is expected to increase scores on the test by an average of 3 points.

    a. If the researcher uses a two-tailed test with $\alpha = .05$, what is the power of the hypothesis test?

    b. Again, assuming a two-tailed test with $\alpha = .05$, what is the power of the hypothesis test if the sample size were increased to $n = 64$?

23. Research has shown that IQ scores have been increasing for years (Flynn, 1984, 1999). The phenomenon is called the Flynn effect and the data indicate that the increase appears to average about 7 points per decade. To examine this effect, a researcher obtains an IQ test with instructions for scoring from 10 years ago and plans to administer the test to a sample of $n = 25$ of today's high school students. Ten years ago, the scores on this IQ test produced a standardized distribution with a mean of $\mu = 100$ and a standard deviation $\sigma = 15$. If there actually has been a 7-point increase in the average IQ during the past ten years, then find the power of the hypothesis test for each of the following.
    a. The researcher uses a two-tailed hypothesis test with $\alpha = .05$ to determine whether the data indicate a significant change in IQ over the past 10 years.
    b. The researcher uses a two-tailed hypothesis test with $\alpha = .01$ to determine whether the data indicate a significant increase in IQ over the past 10 years.

24. Explain how the power of a hypothesis test is influenced by each of the following. Assume that all other factors are held constant.
    a. Increasing the alpha level from .01 to .05.
    b. Changing from a one-tailed test to a two-tailed test.
    c. Increasing effect size.

25. Suppose that a researcher is interested in the effect of an exercise program on body weight among men. The researcher expects a treatment effect of 3 pounds after 15 weeks of exercise in the exercise program. In the population, the mean adult body weight for men is $\mu = 195.5$ pounds, $\sigma = 42.0$. For all of the following, assume that $\alpha = .05$, two-tailed.
    a. State the null and alternative hypotheses using symbols.
    b. Compute the power of the hypothesis test for a sample $n = 2,500$.
    c. Imagine that the researcher observes a sample mean of $M = 192.1$ pounds in a sample of $n = 2,500$ participants. Test the hypothesis that the exercise program reduced body weight. Compute Cohen's d.
    d. Repeat part c with a sample of $n = 25$.
    e. Compute Cohen's d for the results of parts c and d. Describe the distinction between effect size and statistical significance in these hypothetical studies.

Introduction to the *t* Statistic

clivewa/Shutterstock.com

Tools You Will Need

The following items are considered essential background material for this chapter. If you doubt your knowledge of any of these items, you should review the appropriate chapter or section before proceeding.

- Sample standard deviation (Chapter 4)
- Degrees of freedom (Chapter 4)
- Standard error (Chapter 7)
- Hypothesis testing (Chapter 8)

PREVIEW

PREVIEW

What information is conveyed in a dog's growl? Dogs respond appropriately to a variety of social signals, and an aggressive growl by another dog is no exception. They often respond by acting submissive or avoiding an encounter by backing off. This response is appropriate because it avoids a confrontation that could result in injury. What else does a dog "know" about a growl?

Taylor, Reby, and McComb (2011) examined whether dogs are able to use the sound of a growl to gain information about the size of the dog that growled. Each dog was tested by presenting a recorded sound of a small dog growl or a large dog growl. Two realistic model dogs were in front of the test dog during the presentation of the recording. The small model was a Jack Russell terrier and the large model was a German shepherd. The models were approximately 10 feet apart and separated by a partial screen. The researchers measured how long the test dogs viewed each model when the growl recording was presented. Can dogs infer the size of the growler from the sound of the growl? The researchers found that dogs spent significantly more time viewing the model (small or large dog) that matched the sound of the growl (small or large dog growl).

Suppose a group of students in an experimental psychology class decides to replicate a portion of the study as a project. They download audio files of dog growls from the Internet and select one large dog and one small dog growl to use as stimuli. They use two photographs (a small and a large dog) for dog models in the matching task. The photographs are presented simultaneously on separate computer monitors that are separated by a screen. For each test dog, the students play one of the growl recordings for 30 seconds. During this time, they record how much time the test dog looks at each dog photograph.

The students reason that if the test dogs do not get information about size from the growl, then they should show no preference in viewing time for either photograph. There should be no matching and the dogs would spend, on average, half their time, or 15 seconds, viewing the correct match. Alternatively, if the growl is conveying information about the size of the dog, viewing time of the photograph for the correct match should be greater than 15 seconds.

The students used a sample of $n = 16$ test subjects. Suppose the dogs showed an average viewing time of the correct dog size (a correct match of growl to photograph) of $M = 25$ seconds. Does this preference in viewing time provide evidence that dogs correctly match the sound of a growl to the size of a dog? Notice that for these data the population standard deviation, σ, is not known. Therefore, standard error, σ_M, cannot be computed. Instead the researcher must use the sample data to *estimate* standard error. A test statistic can be computed with the estimated standard error. Because it uses an estimate of error, the test statistic is not a *z*-score. In this chapter we will introduce the use of the *t* statistic for hypothesis tests when the value of the population standard deviation, σ, is not known. It has the same structure as the *z*-score, except that the value of standard error is estimated from the sample data.

9-1 The *t* Statistic: An Alternative to *z*

LEARNING OBJECTIVES

1. Describe and identify in a research example the circumstances in which a *t* statistic is used for hypothesis testing instead of a *z*-score and explain the fundamental difference between a *t* statistic and a *z*-score for a sample mean.

2. Calculate the estimated standard error of *M* for a specific sample size and sample variance and explain what it measures.

3. Explain the relationship between the *t* distribution and the normal distribution.

In the previous chapter, we presented the statistical procedures that permit researchers to use a sample mean to test hypotheses about an unknown population mean. These statistical procedures were based on a few basic concepts, which we summarize as follows:

1. A sample mean (*M*) is expected to approximate its population mean (μ). This permits us to use the sample mean to test a hypothesis about the population mean.

2. The standard error provides a measure of how much difference is reasonable to expect between a sample mean (*M*) and the population mean (μ).

$$\sigma_M = \frac{\sigma}{\sqrt{n}} \quad \text{or} \quad \sigma_M = \sqrt{\frac{\sigma^2}{n}}$$

3. To test the hypothesis, we compare the obtained sample mean (*M*) with the hypothesized population mean (μ) by computing a *z*-score test statistic.

$$z = \frac{M - \mu}{\sigma_M} = \frac{\text{sample mean} - \text{hypothesized population mean}}{\text{standard error between } M \text{ and } \mu}$$

$$z = \frac{\text{actual difference between sample } (M) \text{ and the hypothesis } (\mu)}{\text{expected difference between } M \text{ and } \mu \text{ with no treatment effect}}$$

The goal of the hypothesis test is to determine whether the obtained difference between the data and the hypothesis is significantly greater than would be expected by chance if no treatment effect exists. When the *z*-scores form a normal distribution, we are able to use the unit normal table (Appendix B) to find the critical region for the hypothesis test.

■ The Problem with *z*-Scores

The shortcoming of using a *z*-score for hypothesis testing is that the *z*-score formula requires more information than is usually available. Specifically, a *z*-score requires that we know the value of the population standard deviation (or variance), which is needed to compute the standard error. In most situations, however, the standard deviation for the population is not known. In fact, the whole reason for conducting a hypothesis test is to gain knowledge about an *unknown*, treated population. This situation appears to create a paradox: You want to use a *z*-score to find out about an unknown population, but you must know about the population before you can compute a *z*-score. Fortunately, there is a relatively simple solution to this problem. When the variance (or standard deviation) for the population is not known, we use the corresponding sample value in its place.

■ Introducing the *t* Statistic

In Chapter 4, the sample variance was developed specifically to provide an unbiased estimate of the corresponding population variance. Recall that the formulas for sample variance and sample standard deviation are as follows:

The concept of degrees of freedom, *df* = *n* − 1, was introduced in Chapter 4 (page 128) and is discussed later in this chapter (page 295).

$$\text{sample variance} = s^2 = \frac{SS}{n-1} = \frac{SS}{df}$$

$$\text{sample standard deviation} = s = \sqrt{\frac{SS}{n-1}} = \sqrt{\frac{SS}{df}}$$

Using the sample values, we can now *estimate* the standard error. Recall from Chapter 7 (page 224) that the value of the standard error can be computed using either standard deviation or variance:

$$\text{standard error} = \sigma_M = \frac{\sigma}{\sqrt{n}} \quad \text{or} \quad \sigma_M = \sqrt{\frac{\sigma^2}{n}}$$

Now we estimate the standard error by simply substituting the sample variance or standard deviation in place of the unknown population value:

$$\text{estimated standard error} = s_M = \frac{s}{\sqrt{n}} \quad \text{or} \quad s_M = \sqrt{\frac{s^2}{n}} \qquad \text{(9.1)}$$

Notice that the symbol for the *estimated standard error of* M is s_M instead of σ_M, indicating that the estimated value is computed from sample data rather than from the actual population parameter.

> The **estimated standard error** (s_M) is used as an estimate of the actual standard error, σ_M, when the value of σ is unknown. It is computed from the sample variance or sample standard deviation and provides an estimate of the standard distance between a sample mean M and the population mean μ.

Finally, you should recognize that we have shown formulas for standard error (actual or estimated) using both the standard deviation and the variance. In the past (Chapters 7 and 8), we concentrated on the formula using the standard deviation. At this point, however, we shift our focus to the formula based on variance. Thus, throughout the remainder of this chapter, and in the following chapters, the estimated standard error of M typically is presented and computed using

$$s_M = \sqrt{\frac{s^2}{n}}$$

There are two reasons for making this shift from standard deviation to variance:

1. In Chapter 4 (pages 130–132) we saw that the sample variance is an *unbiased* statistic; on average, the sample variance (s^2) provides an accurate and unbiased estimate of the population variance (σ^2).

2. In future chapters we will encounter other versions of the *t* statistic that require variance (instead of standard deviation) in the formulas for estimated standard error. To maximize the similarity from one version to another, we will use variance in the formula for *all* of the different *t* statistics. Thus, whenever we present a *t* statistic, the estimated standard error will be computed as

$$\text{estimated standard error} = \sqrt{\frac{\text{sample variance}}{\text{sample size}}}$$

Now we can substitute the estimated standard error in the denominator of the *z*-score formula. The result is a new test statistic called a *t* statistic:

$$t = \frac{M - \mu}{s_M} \tag{9.2}$$

> The *t* **statistic** is used to test hypotheses about an unknown population mean, μ, when the value of σ is unknown. The formula for the *t* statistic has the same structure as the *z*-score formula, except that the *t* statistic uses the estimated standard error in the denominator.

The only difference between the *t* formula and the *z*-score formula is that the *z*-score uses the actual population variance, σ^2 (or the standard deviation), and the *t* formula uses the corresponding sample variance (or standard deviation) when the population value is not known.

$$z = \frac{M - \mu}{\sigma_M} = \frac{M - \mu}{\sqrt{\sigma^2/n}} \qquad t = \frac{M - \mu}{s_M} = \frac{M - \mu}{\sqrt{s^2/n}}$$

The following example is an opportunity for you to test your understanding of the estimated standard error for a *t* statistic.

EXAMPLE 9.1

For a sample of $n = 9$ scores with $SS = 288$, compute the sample variance and the estimated standard error for the sample mean. You should obtain $s^2 = 36$ and $s_M = 2$. Good luck ∎

■ Degrees of Freedom and the *t* Statistic

In this chapter, we have introduced the *t* statistic as a substitute for a *z*-score. The basic difference between these two is that the *t* statistic uses sample variance (s^2) and the *z*-score uses the population variance (σ^2). To determine how well a *t* statistic approximates a *z*-score, we must determine how well the sample variance approximates the population variance.

According to the law of large numbers (Chapter 7, page 222), the larger the sample size (n), the more likely it is that the sample mean is close to the population mean. The same principle holds true for sample variance and the *t* statistic: For *t* statistics, however, this relationship is typically expressed in terms of the *degrees of freedom*, or the *df* value ($n - 1$) for the sample variance instead of sample size (n): As sample size increases, so does the value for degrees of freedom, and s^2 will be a better estimate of σ^2. Thus, the value for degrees of freedom associated with s^2 also describes how well *t* estimates *z*.

> The concept of degrees of freedom for sample variance was introduced in Chapter 4 (page 128).

$$\text{degrees of freedom} = df = n - 1 \tag{9.3}$$

> **Degrees of freedom** describe the number of scores in a sample that are independent and free to vary. Because the sample mean places a restriction on the value of one score in the sample, there are $n - 1$ degrees of freedom for a sample with n scores (see Chapter 4).

■ The *t* Distribution

Every sample from a population can be used to compute a *z*-score or a *t* statistic. If you select all the possible samples of a particular size (n), and compute the *z*-score for each sample mean, then the entire set of *z*-scores will form a *z*-score distribution. Note that if the distribution of sample means is normal, the distribution of *z*-scores is defined by the unit normal table (Chapter 7, page 228). In the same way, you can compute the *t* statistic for every sample and the entire set of *t* values will form a *t* *distribution*. As we saw in Chapter 7, the distribution of *z*-scores for sample means tends to be a normal distribution. Specifically, if the sample size is large (around $n = 30$ or more) then the distribution of sample means is a nearly perfect normal distribution, and if the sample is selected from a normally distributed population, then the distribution of sample means is guaranteed to be a perfect normal distribution. In these same situations, the *t* distribution approximates a normal distribution, just as a *t* statistic approximates a *z*-score. How well a *t* distribution approximates a normal distribution is determined by degrees of freedom. In general, the greater the sample size (n) is, the larger the degrees of freedom ($n - 1$) are, and the better the *t* distribution approximates the normal distribution. This fact is demonstrated in Figure 9.1, which shows a normal distribution and two *t* distributions with $df = 5$ and $df = 20$.

> A *t* **distribution** is the complete set of *t* values computed for every possible random sample for a specific sample size (n) or a specific degrees of freedom (df). The *t* distribution approximates the shape of a normal distribution.

FIGURE 9.1

Distributions of the *t* statistic for different values of degrees of freedom are compared to a normal *z*-score distribution. Like the normal distribution, *t* distributions are bell-shaped and symmetrical and have a mean of zero. However, *t* distributions are more variable than the normal distribution as indicated by the flatter and more spread-out shape. The larger the value of *df* is, the more closely the *t* distribution approximates a normal distribution.

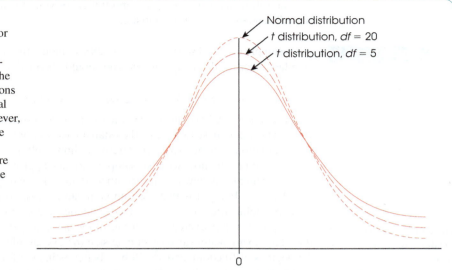

Normal distribution
t distribution, *df* = 20
t distribution, *df* = 5

0

■ The Shape of the *t* Distribution

The exact shape of a *t* distribution changes with degrees of freedom. In fact, statisticians speak of a "family" of *t* distributions. That is, there is a different sampling distribution of *t* (a distribution of all possible sample *t* values) for each possible number of degrees of freedom. As *df* gets very large, the *t* distribution gets closer in shape to a normal *z*-score distribution. A quick glance at Figure 9.1 reveals that distributions of *t* are bell-shaped and symmetrical and have a mean of zero. However, the *t* distribution has more variability than a normal *z* distribution, especially when *df* values are small (see Figure 9.1). The *t* distribution tends to be flatter and more spread out, whereas the normal *z* distribution has more of a central peak.

The reason that the *t* distribution is flatter and more variable than the normal *z*-score distribution becomes clear if you look at the structure of the formulas for *z* and *t*. For both *z* and *t*, the top of the formula, $M - \mu$, can take on different values because the sample mean (*M*) varies from one sample to another. For *z*-scores, however, the bottom of the formula does not vary, provided that all of the samples are the same size and are selected from the same population. Specifically, all the *z*-scores have the same standard error in the denominator, $\sigma_M = \sqrt{\sigma^2/n}$, because the population variance and the sample size are the same for every sample. For *t* statistics, on the other hand, the bottom of the formula varies from one sample to another. Specifically, the sample variance (s^2) changes from one sample to the next, so the estimated standard error also varies, $s_M = \sqrt{s^2/n}$. Thus, only the numerator of the *z*-score formula varies, but both the numerator and the denominator of the *t* statistic vary. As a result, *t* statistics are more variable than are *z*-scores, and the *t* distribution is flatter and more spread out. As sample size and *df* increase, however, the variability in the *t* distribution decreases, and it more closely resembles a normal distribution.

■ Determining Proportions and Probabilities for *t* Distributions

Just as we used the unit normal table to locate proportions associated with *z*-scores, we use a *t* distribution table to find proportions for *t* statistics. You should notice that the arrangement of the *t* table is different from the unit normal table because the *t* distribution changes as the value of *df* increases. The complete *t* distribution table is presented in Appendix B, page 595, and a portion of this table is reproduced in Table 9.1. The two rows at the

TABLE 9.1

A portion of the *t*-distribution table. The numbers in the table are the values of *t* that separate the tail from the main body of the distribution. Proportions for one or two tails are listed at the top of the table, and *df* values for *t* are listed in the first column.

	Proportion in One Tail					
	0.25	0.10	0.05	0.025	0.01	0.005
	Proportion in Two Tails Combined					
df	0.50	0.20	0.10	0.05	0.02	0.01
1	1.000	3.078	6.314	12.706	31.821	63.657
2	0.816	1.886	2.920	4.303	6.965	9.925
3	0.765	1.638	2.353	3.182	4.541	5.841
4	0.741	1.533	2.132	2.776	3.747	4.604
5	0.727	1.476	2.015	2.571	3.365	4.032
6	0.718	1.440	1.943	2.447	3.143	3.707

top of the table show proportions of the *t* distribution contained in either one or two tails, depending on which row is used. The first column of the table lists degrees of freedom for the *t* statistic. Finally, the numbers in the body of the table are the *t* values that mark the boundary between the tails and the rest of the *t* distribution.

For example, with *df* = 3, exactly 5% of the *t* distribution is located in the tail beyond *t* = 2.353 (Figure 9.2). The process of finding this value is highlighted in Table 9.1. Begin by locating *df* = 3 in the first column of the table. Then locate a proportion of 0.05 (5%) in the one-tail proportion row. When you line up these two values in the table, you should find *t* = 2.353. Because the distribution is symmetrical, 5% of the *t* distribution is also located in the tail beyond *t* = −2.353 (see Figure 9.2). Finally, notice that a total of 10% (or 0.10) is contained in the two tails beyond *t* = ±2.353 (check the proportion value in the "two-tails combined" row at the top of the table).

A close inspection of the *t* distribution table in Appendix B will demonstrate a point we made earlier: As the value for *df* increases, the *t* distribution becomes more similar to a normal distribution. For example, examine the column containing *t* values for a 0.05 proportion in two tails. You will find that when *df* = 1, the *t* values that separate the extreme 5% (0.05) from the rest of the distribution are *t* = ±12.706. As you read down the column, however, you should find that the critical *t* values become smaller and smaller, ultimately reaching ±1.96. You

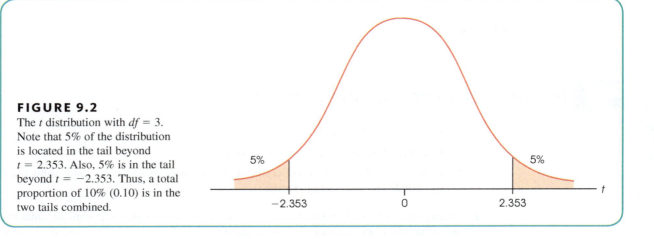

FIGURE 9.2

The *t* distribution with *df* = 3. Note that 5% of the distribution is located in the tail beyond *t* = 2.353. Also, 5% is in the tail beyond *t* = −2.353. Thus, a total proportion of 10% (0.10) is in the two tails combined.

should recognize ± 1.96 as the z-score values that separate the extreme 5% in a normal distribution. Thus, as df increases, the proportions in a t distribution become more like the proportions in a normal distribution. When the sample size (and degrees of freedom) is sufficiently large, the difference between a t distribution and the normal distribution becomes negligible.

Caution: The t distribution table printed in this book has been abridged and does not include entries for every possible df value. For example, the table lists t values for $df = 40$ and for $df = 60$, but does not list any entries for df values between 40 and 60. Occasionally, you will encounter a situation in which your t statistic has a df value that is not listed in the table. In these situations, you should look up the critical t and use the next-smallest value for degrees of freedom. If, for example, you have $df = 53$ (not listed), look up the critical t value for $df = 40$. If your sample t statistic is greater than the value listed, you can be certain that the data are in the critical region, and you can confidently reject the null hypothesis.

LEARNING CHECK   **LO1** **1.** In what circumstances is the t statistic used instead of a z-score for a hypothesis test?

    a. The t statistic is used when the sample size is $n = 30$ or larger.

    b. The t statistic is used when the population mean is known.

    c. The t statistic is used when the population variance (or standard deviation) is unknown.

    d. The t statistic is used if you are not sure that the population distribution is normal.

LO2 **2.** A sample of $n = 9$ scores has $SS = 72$. What is the estimated standard error for the sample mean?

    a. 9

    b. 3

    c. 1

    d. 2

LO3 **3.** On average, what value is expected for the t statistic when the null hypothesis is true?

    a. 0

    b. 1

    c. 1.96

    d. $t > 1.96$

ANSWERS   **1. c   2. c   3. a**

9-2   Hypothesis Tests with the *t* Statistic

LEARNING OBJECTIVES

4. Conduct a hypothesis test using the t statistic.

5. Explain how the likelihood of rejecting the null hypothesis for a t test is influenced by sample size and sample variance.

In the hypothesis-testing situation, we begin with a treated population with an unknown mean and an unknown variance. In this situation we are often considering a population that

FIGURE 9.3

The basic research situation for the *t* statistic hypothesis test. It is assumed that the parameter μ is known for the population before treatment. The purpose of the research study is to determine whether the treatment has an effect. Note that the population after treatment has unknown values for the mean and the variance. We will use a sample to test a hypothesis about the population mean.

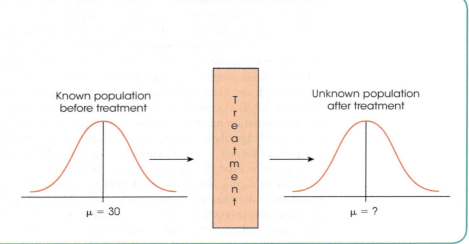

Known population before treatment

$\mu = 30$

Treatment

Unknown population after treatment

$\mu = ?$

has received some treatment (Figure 9.3). It might also involve a nonexperimental study using a nonequivalent group from a population (Chapter 1, page 26). The goal is to use a sample from the treated population (a treated sample) as the basis for determining whether the treatment has any effect.

■ Using the *t* Statistic for Hypothesis Testing

As always, the null hypothesis states that the treatment has no effect; specifically, H_0 states that the population mean is unchanged. Thus, the null hypothesis provides a specific value for the unknown population mean. The sample data provide a value for the sample mean. Finally, the variance and estimated standard error are computed from the sample data. When these values are used in the *t* formula, the result becomes

$$t = \frac{\begin{array}{c}\text{sample mean} \\ \text{(from the data)}\end{array} - \begin{array}{c}\text{population mean} \\ \text{(hypothesized from } H_0\text{)}\end{array}}{\begin{array}{c}\text{estimated standard error} \\ \text{(computed from the sample data)}\end{array}}$$

As with the *z*-score formula, the *t* statistic forms a ratio. The numerator measures the actual difference between the sample data (*M*) and the population hypothesis (μ). The estimated standard error in the denominator measures how much difference is reasonable to expect between a sample mean and the population mean. When the obtained difference between the data and the hypothesis (numerator) is much greater than expected (denominator), we obtain a large value for *t* (either large positive or large negative). In this case, we conclude that the data are not consistent with the hypothesis, and our decision is to "reject H_0." On the other hand, when the difference between the data and the hypothesis is small relative to the standard error, we obtain a *t* statistic near zero, and our decision is "fail to reject H_0." Notice how *z* and *t* are ratios that have the same basic structure.

$$z \text{ or } t = \frac{\text{actual difference between sample (}M\text{) and the hypothesis (}\mu\text{)}}{\text{expected difference between }M \text{ and } \mu \text{ with no treatment effect}}$$

As a ratio, they both compare the difference between *M* and the hypothesized μ to the expected difference between *M* and μ if H_0 is true.

The Unknown Population   As mentioned earlier, the hypothesis test often concerns a population that has received a treatment. This situation is shown in Figure 9.3. Note that the value of the mean is known for the population before treatment. The question is whether the treatment influences the scores and causes the mean to change. In this case, the unknown population is the one that exists after the treatment is administered, and the null hypothesis simply states that the value of the mean is not changed by the treatment.

Although the *t* statistic can be used in the "before and after" type of research shown in Figure 9.3, it also permits hypothesis testing in situations for which you do not have a known population mean to serve as a standard. Specifically, the *t* test does not require any prior knowledge about the population mean or the population variance. All you need to compute a *t* statistic is a null hypothesis and a sample from the unknown population. Thus, a *t* test can be used in situations for which the null hypothesis is obtained from a theory, a logical prediction, or just wishful thinking. For example, many studies use rating-scale questions to measure perceptions or attitudes. Participants are presented with a statement and asked to express their opinion on a scale from 1 to 7, with 1 indicating "strongly negative" and 7 indicating "strongly positive." A score of 4 indicates a neutral position, with no strong opinion one way or the other. In this situation, the null hypothesis would state that there is no preference, or no strong opinion, in the population, and use a null hypothesis of H_0: $\mu = 4$. The data from a sample is then used to evaluate the hypothesis. Note that the researcher has no prior knowledge about the population mean and states a hypothesis that is based on logic.

■ Hypothesis Testing Example

The following research situation demonstrates the procedures of hypothesis testing with the *t* statistic.

EXAMPLE 9.2   Chang, Aeschbach, Duffy, and Czeisler (2015) report that reading from a light-emitting eReader before bedtime can significantly affect sleep and lower alertness the next morning. To test this finding, a researcher obtains a sample of $n = 9$ volunteers who agree to spend at least 15 minutes using an eReader during the hour before sleeping and then take a standardized cognitive alertness test the next morning. For the general population, scores on the test average $\mu = 50$ and form a normal distribution. The sample of research participants had an average score of $M = 46$ with $SS = 162$.

STEP 1   **State the hypotheses and select an alpha level.**   In this case, the null hypothesis states that late-night reading from a light-emitting screen has no effect on alertness the following morning. In symbols, the null hypothesis states

H_0: $\mu_{\text{screen reading}} = 50$ (Even with the eReader, the mean alertness score is still 50.)

The alternative hypothesis states that reading from a screen at bedtime does affect alertness the next morning. A directional, one-tailed test would specify whether alertness is increased or decreased, but the nondirectional alternative hypothesis is expressed as follows:

H_1: $\mu_{\text{screen reading}} \neq 50$ (With the eReader, the mean alertness score is not 50.)

We will set the level of significance at $\alpha = .05$ for two tails.

STEP 2   **Locate the critical region.**   The test statistic is a *t* because the population variance is not known. Therefore, the value for degrees of freedom must be determined before the critical region can be located. For this sample

$$df = n - 1 = 9 - 1 = 8$$

For a two-tailed test at the .05 level of significance and with 8 degrees of freedom, the critical region consists of *t* values greater than $+2.306$ or less than -2.306. Figure 9.4 depicts the critical region in this *t* distribution.

STEP 3 **Calculate the test statistic.** The *t* statistic typically requires more computation than is necessary for a *z*-score. Therefore, we recommend that you divide the calculations into a three-stage process as follows:

a. First, calculate the sample variance. Remember that the population variance is unknown, and you must use the sample value in its place. (This is why we are using a *t* statistic instead of a *z*-score.)

$$s^2 = \frac{SS}{n-1} = \frac{SS}{df}$$

$$= \frac{162}{8} = 20.25$$

b. Next, use the sample variance (s^2) and the sample size (n) to compute the estimated standard error. This value is the denominator of the *t* statistic and measures how much difference is reasonable to expect by chance between a sample mean and the corresponding population mean.

$$s_M = \sqrt{\frac{s^2}{n}}$$

$$= \sqrt{\frac{20.25}{9}} = \sqrt{2.25} = 1.50$$

c. Finally, compute the *t* statistic for the sample data.

$$t = \frac{M - \mu}{s_M} = \frac{46 - 50}{1.50} = -2.67$$

STEP 4 **Make a decision regarding H_0.** The obtained *t* statistic of -2.67 falls into the critical region on the left-hand side of the *t* distribution (see Figure 9.4). Our statistical decision is to reject H_0, which is evidence that reading from a light-emitting screen at bedtime does affect alertness the following morning. As indicated by the sample mean, there is a tendency for the level of alertness to be reduced after reading a screen before bedtime. ■

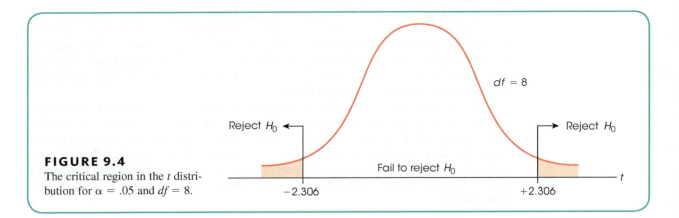

FIGURE 9.4
The critical region in the *t* distribution for $\alpha = .05$ and $df = 8$.

■ Assumptions of the *t* Test

Two basic assumptions are necessary for hypothesis tests with the *t* statistic.

1. The values in the sample must consist of *independent* observations.

   In everyday terms, two observations are independent if there is no consistent, predictable relationship between the first observation and the
   second. More precisely, two events (or observations) are independent if
   the occurrence of the first event has no effect on the probability of the
   second event. We examined specific examples of independence and non-
   independence in Box 8.1 (page 264).

2. The population sampled must be normal.

   This assumption is a necessary part of the mathematics underlying the
   development of the *t* statistic and the *t* distribution table. However, violating this assumption has little practical effect on the results obtained for a
   t statistic, especially when the sample size is relatively large. With very
   small samples, a normal population distribution is important. With larger
   samples, this assumption can be violated without affecting the validity
   of the hypothesis test. If you have reason to suspect that the population
   distribution is not normal, use a large sample to be safe.

■ The Influence of Sample Size and Sample Variance

As we noted in Chapter 8 (page 262), a variety of factors can influence the outcome of
a hypothesis test. In particular, the number of scores in the sample and the magnitude
of the sample variance both have a large effect on the *t* statistic and thereby influence
the statistical decision. The structure of the *t* formula makes these factors easier to
understand:

$$t = \frac{M - \mu}{s_M} \qquad \text{where } s_M = \sqrt{\frac{s^2}{n}}$$

Because the estimated standard error, s_M, appears in the denominator of the formula, a
larger value for s_M produces a smaller value (closer to zero) for *t*. Thus, any factor that
influences the standard error also affects the likelihood of rejecting the null hypothesis and
finding a significant treatment effect. The two factors that determine the size of the standard error are the sample variance, s^2, and the sample size, *n*.

The estimated standard error is directly related to the sample variance so that the larger
the variance, the larger the error. Thus, large variance means that you are less likely to
obtain a significant treatment effect. In general, large variance is bad for inferential statistics. Large variance means that the scores are widely scattered, which makes it difficult
to see any consistent patterns or trends in the data. In general, high variance reduces the
likelihood of rejecting the null hypothesis.

On the other hand, the estimated standard error is inversely related to the number of scores in the sample. The larger the sample is, the smaller the error is. If all
other factors are held constant, large samples tend to produce bigger *t* statistics and
therefore are more likely to produce significant results. For example, a 2-point mean
difference with a sample of $n = 4$ may not be convincing evidence of a treatment
effect. However, the same 2-point difference with a sample of $n = 100$ is much more
compelling.

LEARNING CHECK   **LO4 1.** A sample of $n = 25$ scores is selected from a population with a mean of $\mu = 73$, and a treatment is administered to the sample. After treatment, the sample has $M = 70$ and $s^2 = 100$. If a hypothesis test with a *t* statistic is used to evaluate the treatment effect, then what value will be obtained for the *t* statistic?

    a. $t = -0.75$

    b. $t = -3.00$

    c. $t = -1.50$

    d. $t = +1.50$

LO4 2. A hypothesis test produces a *t* statistic of $t = 2.30$. If the researcher is using a two-tailed test with $\alpha = .05$, how large does the sample have to be in order to reject the null hypothesis?

    a. At least $n = 8$

    b. At least $n = 9$

    c. At least $n = 10$

    d. At least $n = 11$

LO5 3. A sample is selected from a population and a treatment is administered to the sample. For a hypothesis test with a *t* statistic, if there is a 5-point difference between the sample mean and the original population mean, which set of sample characteristics is most likely to lead to a decision that there is a significant treatment effect?

    a. Small variance for a large sample.

    b. Small variance for a small sample.

    c. Large variance for a large sample.

    d. Large variance for a small sample.

ANSWERS   **1. c   2. c   3. a**

9-3 Measuring Effect Size for the *t* Statistic

LEARNING OBJECTIVES

6. Calculate Cohen's *d* or the percentage of variance accounted for (r^2) to measure effect size for a hypothesis test with a *t* statistic.

7. Explain how measures of effect size for a *t* test are influenced by sample size and sample variance.

8. Explain how a confidence interval can be used to describe the size of a treatment effect for a test and describe the factors that affect the width of a confidence interval.

9. Describe how the results from a hypothesis test using a *t* statistic are reported in the literature.

In Chapter 8 we noted that one criticism of a hypothesis test is that it does not really evaluate the size of the treatment effect. Instead, a hypothesis test simply determines whether the treatment effect is greater than chance, where "chance" is measured by the standard error. In particular, it is possible for a very small treatment effect to be "statistically significant," especially when the sample size is very large. To correct for this problem, it is recommended that the results from a hypothesis test be accompanied by a report of effect size such as Cohen's *d*.

■ Estimated Cohen's *d*

When Cohen's *d* was originally introduced (page 271), the formula was presented as

$$\text{Cohen's } d = \frac{\text{mean difference}}{\text{standard deviation}} = \frac{\mu_{\text{treatment}} - \mu_{\text{no treatment}}}{\sigma}$$

Cohen defined this measure of effect size in terms of the population mean difference and the population standard deviation. However, in most situations the population values are not known and you must substitute the corresponding sample values in their place. When this is done, many researchers prefer to identify the calculated value as an "*estimated* d" or name the value after one of the statisticians who first substituted sample statistics into Cohen's formula (e.g., Glass's *g* or Hedges's *g*). For hypothesis tests using the *t* statistic, the population mean with no treatment is the value specified by the null hypothesis. However, the population mean with treatment and the standard deviation are both unknown. Therefore, we use the mean for the treated sample and the standard deviation for the sample after treatment as estimates of the unknown parameters. With these substitutions, the formula for estimating Cohen's *d* becomes

$$\text{estimated } d = \frac{\text{mean difference}}{\text{sample standard deviation}} = \frac{M - \mu}{s} \tag{9.4}$$

The numerator measures that magnitude of the treatment effect by finding the difference between the mean for the treated sample and the mean for the untreated population (μ from H_0). The sample standard deviation in the denominator standardizes the mean difference into standard deviation units. Thus, an estimated *d* of 1.00 indicates that the size of the treatment effect is equivalent to one standard deviation. The following example demonstrates how the estimated *d* is used to measure effect size for a hypothesis test using a *t* statistic.

EXAMPLE 9.3   For the bedtime-reading study in Example 9.2, the participants averaged $M = 46$ on the alertness test. If the light-emitting eReader has no effect (as stated by the null hypothesis), the population mean would be $\mu = 50$. Thus, the results show a 4-point difference between the mean with bedtime reading ($M = 46$) and the mean for the general population ($\mu = 50$). Also, for this study the sample standard deviation is simply the square root of the sample variance, which was found to be $s^2 = 20.25$.

$$s = \sqrt{s^2} = \sqrt{20.25} = 4.50$$

Thus, Cohen's *d* for this example is estimated to be

$$\text{Cohen's } d = \frac{M - \mu}{s} = \frac{46 - 50}{4.50} = 0.89$$

According to the standards suggested by Cohen (Table 8.2, page 273), this is a large treatment effect. Remember, Cohen's *d* is always reported as a positive value (page 272).   ■

To help you visualize what is measured by Cohen's *d*, we have constructed a set of $n = 9$ scores with a mean of $M = 46$ and a standard deviation of $s = 4.5$ (the same values as in Examples 9.2 and 9.3). The set of scores is shown in Figure 9.5. Notice that the figure also includes an arrow that locates $\mu = 50$. Recall that $\mu = 50$ is the value specified by the null hypothesis and identifies what the mean ought to be if the treatment has no effect.

FIGURE 9.5

The sample distribution for the scores that were used in Examples 9.2 and 9.3. The population mean, $\mu = 50$, is the value that would be expected if eReader use before bedtime has no effect on alertness the next morning. Note that the sample mean is displaced from $\mu = 50$ by a distance close to one standard deviation.

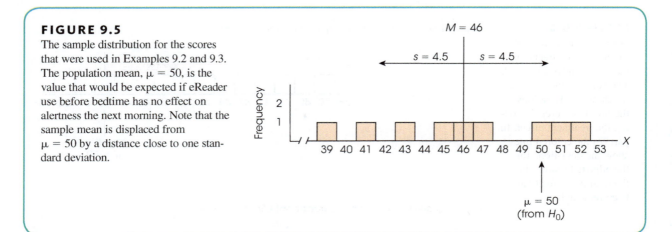

Clearly, our sample is not centered at $\mu = 50$. Instead, the scores have been shifted to the left so that the sample mean is $M = 46$. This shift, from 50 to 46, is the 4-point mean difference that was caused by the treatment effect. Also notice that the 4-point mean difference is almost equal to the standard deviation. Thus, the size of the treatment effect is close to one standard deviation. In other words, Cohen's $d = 0.89$ is an accurate description of the treatment effect.

The following example is an opportunity for you to test your understanding of hypothesis testing and effect size with the t statistic.

EXAMPLE 9.4

A sample of $n = 16$ individuals is selected from a population with a mean of $\mu = 40$. A treatment is administered to the individuals in the sample and, after treatment, the sample has a mean of $M = 44$ and a variance of $s^2 = 16$. Use a two-tailed test with $\alpha = .05$ to determine whether the treatment effect is significant and compute Cohen's d to measure the size of the treatment effect. You should obtain $t = 4.00$ with $df = 15$, which is large enough to reject H_0 with Cohen's $d = 1.00$.   ■

■ Measuring the Percentage of Variance Explained, r^2

An alternative method for measuring effect size is to determine how much of the variability in the scores is explained by the treatment effect. The concept behind this measure is that the treatment causes the scores to increase (or decrease), which means that the treatment is causing the scores to vary. If we can measure how much of the variability is explained by the treatment, we will obtain a measure of the size of the treatment effect.

To demonstrate this concept, we will use the data from the hypothesis test in Example 9.2. Recall that the null hypothesis stated that the treatment (bedtime reading from a light-emitting screen) has no effect on alertness the following morning. According to the null hypothesis, individuals who read from a light-emitting screen at bedtime should have the same alertness level as the general population, and therefore should average $\mu = 50$ on the standardized test.

However, if you look at the data in Figure 9.5, the scores are not centered at $\mu = 50$. Instead, the scores are shifted to the left so that they are centered around the sample mean, $M = 46$. This shift is the treatment effect. To measure the size of the treatment effect we calculate deviations from the mean and the sum of squared deviations, SS, two different ways.

FIGURE 9.6

Deviations from $\mu = 50$ (no treatment effect) for the scores in Example 9.2. The colored lines in part (a) show the deviations for the original scores, including the treatment effect. In part (b) the colored lines show the deviations for the adjusted scores after the treatment effect has been removed.

(a) Original scores, including the treatment effect

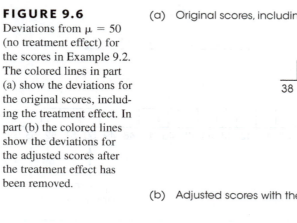
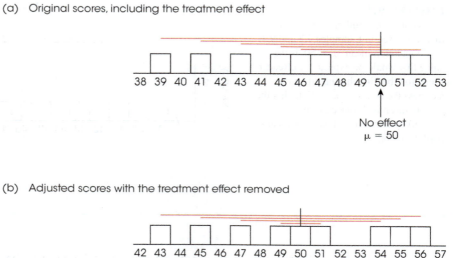

(b) Adjusted scores with the treatment effect removed

Figure 9.6(a) shows the original set of scores. For each score, the deviation from $\mu = 50$ is shown as a colored line. Recall that $\mu = 50$ comes from the null hypothesis and represents the population mean if the treatment has no effect. Note that almost all of the scores are located on the left-hand side of $\mu = 50$. This shift to the left is the treatment effect. Specifically, the late-night reading has caused a reduced level of alertness the following morning, which means that the participants' scores are generally lower than 50. Thus, the treatment has pushed the scores away from $\mu = 50$ (the null hypothesis) and has increased the size of the deviations.

Next, we will see what happens if the treatment effect is removed. In this example, the treatment has a 4-point effect (the average decreases from $\mu = 50$ to $M = 46$). To remove the treatment effect, we simply add 4 points to each score. The adjusted scores are shown in Figure 9.6(b) and, once again, the deviations from $\mu = 50$ are shown as colored lines. First, notice that the adjusted scores are centered at $\mu = 50$, indicating that there is no treatment effect. Also notice that the deviations—the colored lines—are noticeably smaller when the treatment effect is removed.

To measure how much the variability is reduced when the treatment effect is removed, we compute the sum of squared deviations, *SS*, for each set of scores. The left-hand columns of Table 9.2 show the calculations for the original scores [Figure 9.6(a)], and the right-hand columns show the calculations for the adjusted scores [Figure 9.6(b)]. Note that the total variability, including the treatment effect, is $SS = 306$. However, when the treatment effect is removed, the variability is reduced to $SS = 162$, which represents the variability that is not explained by the treatment effect. Total variability minus variability not explained by the treatment effect is equal to the amount of variability accounted for by the treatment. The difference between these two values, $306 - 162 = 144$ points, is the amount of variability that is accounted for by the treatment effect. This value is usually reported as a proportion or percentage of the total variability:

$$r^2 = \frac{\text{Variability accounted for by the treatment effect}}{\text{total variability}} = \frac{144}{306} = .47 \quad (\text{or } 47\%)$$

TABLE 9.2

Calculation of *SS*, the sum of squared deviations, for the data in Figure 9.6. The first three columns show the calculations for the original scores, including the treatment effect. The last three columns show the calculations for the adjusted scores after the treatment effect has been removed.

	Calculation of SS including the treatment effect			Calculation of SS after the treatment effect is removed	
Score	Deviation from $\mu = 50$	Squared Deviation	Adjusted Score	Deviation of adjusted score from $\mu = 50$	Squared Deviation
39	−11	121	39 + 4 = 43	−7	49
41	−9	81	41 + 4 = 45	−5	25
43	−7	49	43 + 4 = 47	−3	9
45	−5	25	45 + 4 = 49	−1	1
46	−4	16	46 + 4 = 50	0	0
47	−3	9	47 + 4 = 51	1	1
50	0	0	50 + 4 = 54	4	16
51	1	1	51 + 4 = 55	5	25
52	2	4	52 + 4 = 56	6	36
		$SS = 306$			$SS = 162$

Thus, removing the treatment effect reduces the variability by 47%. This value is called the *percentage of variance accounted for by the treatment* and is identified as r^2.

Rather than computing r^2 directly by comparing two different calculations for *SS*, the value can be found from a single equation based on the value of the *t* statistic:

$$r^2 = \frac{t^2}{t^2 + df} \tag{9.5}$$

The letter *r* is the traditional symbol used for a correlation, and the concept of r^2 is discussed again when we consider correlations in Chapter 14. Also, in the context of *t* statistics, the percentage of variance that we are calling r^2 is often identified by the Greek letter omega squared (ω^2).

For the hypothesis test in Example 9.2, we obtained $t = -2.67$ with $df = 8$. These values produce

$$\frac{(-2.67)^2}{(-2.67)^2 + 8} = \frac{7.13}{15.13} = 0.47$$

Note that this is the same value we obtained with the direct calculation of the percentage of variability accounted for by the treatment.

Interpreting r^2 In addition to developing the Cohen's *d* measure of effect size, Cohen (1988) also proposed criteria for evaluating the size of a treatment effect that is measured by r^2. The criteria were actually suggested for evaluating the size of a correlation, *r*, but are easily extended to apply to r^2. Cohen's standards for interpreting r^2 are shown in Table 9.3.

According to these standards, the data we constructed for Examples 9.1 and 9.2 show a very large effect size with $r^2 = 0.47$.

As a final note, we should remind you that, although sample size affects the hypothesis test, this factor has little or no effect on measures of effect size. In particular, estimates of Cohen's *d* are not influenced at all by sample size, and measures of r^2 are only slightly

TABLE 9.3

Criteria for interpreting the value of r^2 as proposed by Cohen (1988).

Percentage of Variance Explained, r^2	
$r^2 = 0.01$	Small effect
$r^2 = 0.09$	Medium effect
$r^2 = 0.25$	Large effect

affected by changes in the size of the sample. The sample variance, on the other hand, influences hypothesis tests and measures of effect size. Specifically, high variance reduces the likelihood of rejecting the null hypothesis and it reduces measures of effect size.

■ Confidence Intervals for Estimating μ

An alternative technique for describing the size of a treatment effect is to compute an estimate of the population mean after treatment. For example, if the mean before treatment is known to be $\mu = 80$ and the mean after treatment is estimated to be $\mu = 86$, then we can conclude that the size of the treatment effect is around 6 points.

Estimating an unknown population mean involves constructing a *confidence interval*. A confidence interval is based on the observation that a sample mean tends to provide a reasonably accurate estimate of the population mean. The fact that a sample mean tends to be near to the population mean implies that the population mean should be near to the sample mean. Thus, if we obtain a sample mean of $M = 86$, we can be reasonably confident that the population mean is around 86. Thus, a confidence interval consists of an interval of values around a sample mean, and we can be reasonably confident that the unknown population mean is located somewhere in the interval.

> A **confidence interval** is an interval, or range of values centered around a sample statistic. The logic behind a confidence interval is that a sample statistic, such as a sample mean, should be relatively near to the corresponding population parameter. Therefore, we can confidently estimate that the value of the parameter should be located in the interval near to the statistic.

■ Constructing a Confidence Interval

The construction of a confidence interval begins with the observation that every sample mean has a corresponding *t* value defined by the equation

$$t = \frac{M - \mu}{s_M}$$

Although the values for M and s_M are available from the sample data, we do not know the values for t or for μ. However, we can *estimate* the t value. For example, if the sample has $n = 9$ scores, then the t statistic has $df = 8$, and the distribution of all possible t values can be pictured as seen in Figure 9.7. Notice that the t values pile up around $t = 0$, so we can estimate that the t value for our sample should have a value around 0. Furthermore, the t distribution table lists a variety of different t values that correspond to specific proportions of the t distribution. With $df = 8$, for example, 95% of the t values are located between $t = +2.306$ and $t = -2.306$. To obtain these values, simply look up a two-tailed proportion of 0.05 (5%) for $df = 8$. Because 95% of all the possible t values are located between ± 2.306, we can be 95% confident that our sample mean corresponds to a t value in this interval. Similarly, we can be 80% confident that the mean for a sample of $n = 9$ scores corresponds to a t value between ± 1.397. Notice that we are able to estimate the value of t with a specific level of confidence. To construct a confidence interval for μ, we plug the estimated t value into the t equation, and then we can calculate the value of μ.

Before we demonstrate the process of constructing a confidence interval for an unknown population mean, we simplify the calculations by regrouping the terms in the t equation. Because the goal is to compute the value of μ, we use simple algebra to solve the equation for μ. The result is $\mu = M - t(s_M)$. However, we estimate that the t value is in an interval

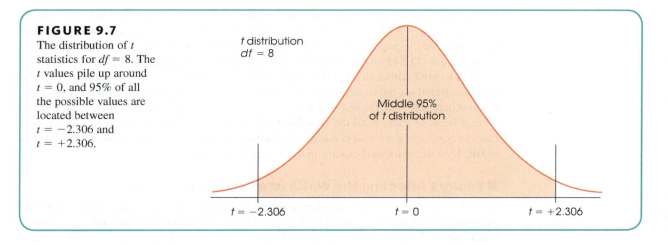

FIGURE 9.7

The distribution of *t* statistics for *df* = 8. The *t* values pile up around *t* = 0, and 95% of all the possible values are located between *t* = −2.306 and *t* = +2.306.

t distribution
df = 8

Middle 95% of *t* distribution

$t = -2.306$ $t = 0$ $t = +2.306$

around 0, with one end at +*t* and the other end at −*t*. The ±*t* can be incorporated in the equation to produce

$$\mu = M \pm t(s_M) \tag{9.6}$$

This is the basic equation for a confidence interval. Notice that the equation produces an interval around the sample mean. One end of the interval is located at $M + t(s_M)$ and the other end is at $M - t(s_M)$. The process of using this equation to construct a confidence interval is demonstrated in the following example.

EXAMPLE 9.5

Example 9.2 describes a study in which reading from a light-emitting screen just before bedtime resulted in lower alertness the following morning. Specifically, a sample of $n = 9$ participants who read from an eReader for at least 15 minutes during the hour before bedtime had an average score of $M = 46$ the next morning on an alertness test for which the general population mean is $\mu = 50$. The data produced an estimated standard error of $s_M = 1.50$. We will use this sample to construct a confidence interval to estimate the mean alertness score for the population of individuals who read from light-emitting screens at bedtime. That is, we will construct an interval of values that is likely to contain the unknown population mean.

Again, the estimation formula is

$$\mu = M \pm t(s_M)$$

In the equation, the value of $M = 46$ and $s_M = 1.50$ are obtained from the sample data. The next step is to select a level of confidence that will determine the value of *t* in the equation. The most commonly used confidence level is probably 95%, but values of 90% and 99% also have been used. For this example, we will use a confidence level of 95%, which means that we will construct the confidence interval so that we are 95% confident that the population mean is actually contained in the interval. Because we are using a confidence level of 95%, the resulting interval is called the *95% confidence interval for* μ.

To have 95% in the middle there must be 5% (or .05) in the tails. To find the *t* values, look under two tails, .05 in the *t* table.

To obtain the value for *t* in the equation, we simply estimate that the *t* statistic for our sample is located somewhere in the middle 95% of the *t* distribution. With $df = n - 1 = 8$, the middle 95% of the distribution is bounded by *t* values of +2.306 and −2.306 (see Figure 9.7). Using the sample data and the estimated range of *t* values, we obtain

$$\mu = M \pm t(s_M) = 46 \pm 2.306(1.50) = 46 \pm 3.459$$

At one end of the interval, we obtain $\mu = 46 + 3.459 = 49.459$, and at the other end we obtain $\mu = 46 - 3.459 = 42.541$. Our conclusion is that the average next-morning alertness score for the population of individuals who read from an eReader before bedtime is between $\mu = 42.541$ and $\mu = 49.459$, and we are 95% confident that the true population mean is located within this interval. The confidence comes from the fact that the calculation was based on only one assumption. Specifically, we assumed that the *t* statistic was located between $+2.306$ and -2.306, and we are 95% confident that this assumption is correct because 95% of all the possible *t* values are located in this interval. Finally, note that the confidence interval is constructed around the sample mean. As a result, the sample mean, $M = 46$, is located exactly in the center of the interval. ■

■ Factors Affecting the Width of a Confidence Interval

Two characteristics of the confidence interval should be noted. First, notice what happens to the width of the interval when you change the level of confidence (the percent confidence). To gain more confidence in your estimate, you must increase the width of the interval. Conversely, to have a smaller, more precise interval, you must give up confidence. In the estimation formula, the percentage of confidence influences the value of *t*. A larger level of confidence (the percentage), produces a larger *t* value and a wider interval. This relationship can be seen in Figure 9.7. In the figure, we identified the middle 95% of the *t* distribution in order to find an 95% confidence interval. It should be obvious that if we were to increase the confidence level to 99%, it would be necessary to increase the range of *t* values and thereby increase the width of the interval. Thus, there is a trade-off between precision (the width of the interval) and the confidence one has that the interval contains the population mean. If you set your confidence high, say 99%, that μ is in the interval, then the interval will have a large width.

Second, note what happens to the width of the interval if you change the sample size. This time the basic rule is as follows: the bigger the sample (*n*), the smaller the interval. This relationship is straightforward if you consider the sample size as a measure of the amount of information. A bigger sample gives you more information about the population and allows you to make a more precise estimate (a narrower interval). The sample size controls the magnitude of the standard error in the estimation formula. As the sample size increases, the standard error decreases, and the interval gets smaller. Notice that a researcher has the ability to control the width of a confidence interval by adjusting either the sample size or the level of confidence. For example, if a researcher feels that an interval is too broad (producing an imprecise estimate of the mean), the interval can be narrowed by either increasing the sample size or lowering the level of confidence. You also should note that because confidence intervals are influenced by sample size, they do not provide an unqualified measure of absolute effect size and are not an adequate substitute for Cohen's *d* or r^2. Nonetheless, they can be used in a research report to provide a description of the size of the treatment effect.

IN THE LITERATURE

Reporting the Results of a *t* Test

In Chapter 8, we noted the conventional style for reporting the results of a hypothesis test, according to APA format. First, recall that a scientific report typically uses the term *significant* to indicate that the null hypothesis has been rejected and the term *not significant* to indicate failure to reject H_0. Additionally, there is a prescribed format for reporting the calculated value of the test statistic, degrees of freedom, alpha level, and effect size for a *t* test. This format parallels the style introduced in Chapter 8 (page 261).

In Example 9.2 we calculated a *t* statistic of -2.67 with $df = 8$, and we decided to reject H_0 with alpha set at .05. Using the same data, we obtained $r^2 = 0.47$ (47%) for the percentage of variance explained by the treatment effect. In a scientific report, this information is conveyed in a concise statement, as follows:

> The participants had an average of $M = 46$ with $SD = 4.50$ on a standardized alertness test the morning following bedtime reading from a light-emitting screen. Statistical analysis indicates that the mean level of alertness was significantly lower than scores for the general population, $t(8) = -2.67$, $p < .05$, $r^2 = 0.47$.

The first statement reports the descriptive statistics, the mean ($M = 46$), and the standard deviation ($SD = 4.50$), as previously described (Chapter 4, page 135). The next statement provides the results of the inferential statistical analysis. Note that the degrees of freedom are reported in parentheses immediately after the symbol *t*. The value for the obtained *t* statistic follows (-2.67), and next is the probability of committing a Type I error (less than 5%). Finally, the effect size is reported, $r^2 = 47\%$. If the 95% confidence interval from Example 9.5 were included in the report as a description of effect size, it would be added after the results of the hypothesis test as follows:

$$t(8) = -2.67, p < .05, 95\% \text{ CI } [42.54, 49.46].$$

Often, researchers use a computer to perform a hypothesis test like the one in Example 9.2. In addition to calculating the mean, standard deviation, and the *t* statistic for the data, the computer usually calculates and reports the *exact probability* associated with the computed *t* value. In Example 9.2 we determined that any *t* value beyond ± 2.306 has a probability of less than .05 (see Figure 9.4). Thus, the obtained *t* value, $t = -2.67$, is reported as being very unlikely, $p < .05$. A computer printout, however, would have included an exact probability for our specific *t* value.

Whenever a specific probability value is available, you are encouraged to use it in a research report. For example, the computer analysis of these data reports an exact *p* value of $p = .029$, and the research report would state "$t(8) = -2.67$, $p = .029$" instead of using the less specific "$p < .05$." As one final caution, we note that occasionally a *t* value is so extreme that the computer reports $p = 0.000$. The zero value does not mean that the probability is literally zero; instead, it means that the computer has rounded off the probability value to three decimal places and obtained a result of 0.000. In this situation, you do not know the exact probability value, but you can report $p < .001$. ∎

LEARNING CHECK   **LO6** **1.** A sample of $n = 25$ is selected from a population with $\mu = 40$, and a treatment is administered to each individual in the sample. After treatment, the sample mean is $M = 44$ with a sample variance of $s^2 = 100$. Based on this information, what is the size of the treatment effect as measured by Cohen's *d*?

    a. $d = 0.04$
    b. $d = 0.40$
    c. $d = 1.00$
    d. $d = 2.00$

LO7 **2.** A sample is selected from a population with a mean of $\mu = 75$, and a treatment is administered to the individuals in the sample. The researcher intends to use a *t* statistic to evaluate the effect of the treatment. If the sample mean is $M = 79$, then which of the following outcomes would produce the largest value for Cohen's *d*?

    a. $n = 4$ and $s^2 = 30$
    b. $n = 16$ and $s^2 = 30$

   c. $n = 25$ and $s^2 = 30$

   d. All three samples would produce the same value for Cohen's *d*.

LO8 3. A sample of $n = 4$ scores is selected from a population with an unknown mean. The sample has a mean of $M = 40$ and a variance of $s^2 = 16$. Which of the following is the correct 90% confidence interval for μ?

   a. $\mu = 40 \pm 2.353(4)$

   b. $\mu = 40 \pm 1.638(4)$

   c. $\mu = 40 \pm 2.353(2)$

   d. $\mu = 40 \pm 1.638(2)$

LO9 4. A researcher uses a sample of $n = 25$ individuals to evaluate the effect of a treatment. The hypothesis test uses $\alpha = .05$ and produces a significant result with $t = 2.15$. How would this result be reported in the literature?

   a. $t(25) = 2.15, p < .05$

   b. $t(24) = 2.15, p < .05$

   c. $t(25) = 2.15, p > .05$

   d. $t(24) = 2.15, p > .05$

ANSWERS   1. b   2. d   3. c   4. b

9-4 ## Directional Hypotheses and One-Tailed Tests

LEARNING OBJECTIVE

10. Conduct a directional (one-tailed) hypothesis test using the *t* statistic.

As noted in Chapter 8, the nondirectional (two-tailed) test is more commonly used than the directional (one-tailed) alternative. On the other hand, a directional test may be used in some research situations, such as exploratory investigations or pilot studies, or when there is *a priori* justification (for example, a theory or previous findings). The following example demonstrates a directional hypothesis test with a *t* statistic, using the same experimental situation presented in Example 9.2.

EXAMPLE 9.6   The research question is whether reading from a light-emitting screen before bedtime affects alertness the following morning. Based on previous studies, the researcher is expecting the level of alertness to be reduced on the morning after late-night reading. Therefore, the researcher predicts that the participants will have an average alertness score that is lower than the mean for the general population, which is $\mu = 50$. For this example we will use the same sample data that were used in the original hypothesis test in Example 9.2. Specifically, the researcher tested a sample of $n = 9$ participants and obtained a mean score of $M = 46$ with $SS = 162$.

STEP 1   **State the hypotheses, and select an alpha level.**   With most directional tests, it is usually easier to state the hypothesis in words, including the directional prediction, and then convert the words into symbols. For this example, the researcher is predicting that reading from a light-emitting screen at bedtime will lower alertness scores the next morning. In general, the null hypothesis states that the predicted effect will not happen. For this study, the null hypothesis states that alertness scores will not be lowered by reading from a screen late at night. In symbols,

$$H_0: \mu_{\text{screen reading}} \geq 50 \qquad \text{(Scores will not be lower than the general population average.)}$$

Similarly, the alternative hypothesis states that the treatment will work. In this case, H_1 states that alertness scores will be lowered by reading from a light-emitting screen before bedtime. In symbols,

$$H_1: \mu_{\text{screen reading}} < 50 \qquad \text{(Alertness after late-night reading will be lower than the general population average.)}$$

We will set the level of significance at $\alpha = .05$.

STEP 2   **Locate the critical region.**   In this example, the researcher is predicting that the sample mean (M) will be less than 50. Thus, if the participants' average score is less than 50, the data will provide support for the researcher's prediction and will tend to refute the null hypothesis. Also note that a sample mean less than 50 will produce a negative value for the t statistic. Thus, the critical region for the one-tailed test will consist of negative t values located in the left-hand tail of the distribution. However, we must still determine exactly how large the t value must be to justify rejecting the null hypothesis. To find the critical value, you must look in the t distribution table using the one-tail proportions. With a sample of $n = 9$, the t statistic will have $df = 8$; using $\alpha = .05$, you should find a critical value of $t = 1.860$. Therefore, if we obtain a sample mean of less than 50 and the t statistic is beyond the -1.860 critical boundary on the left-hand side, we will reject the null hypothesis and conclude that reading from a light-emitting screen before bedtime significantly lowers alertness the next morning. Figure 9.8 shows the one-tailed critical region for this test.

STEP 3   **Calculate the test statistic.**   The computation of the t statistic is the same for either a one-tailed or a two-tailed test. Earlier (in Example 9.2), we found that the data for this experiment produce a test statistic of $t = -2.67$.

STEP 4   **Make a decision.**   The test statistic is in the critical region, so we reject H_0. In terms of the experimental variables, we have decided that reading from a light-emitting screen at

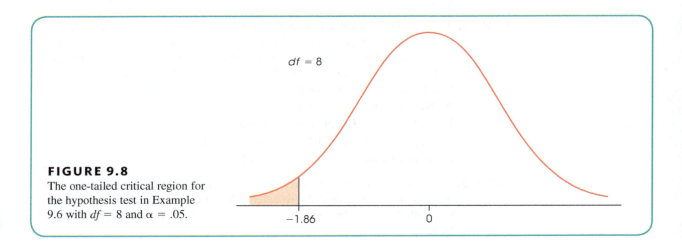

FIGURE 9.8
The one-tailed critical region for the hypothesis test in Example 9.6 with $df = 8$ and $\alpha = .05$.

bedtime reduces alertness the following morning. In a research report the results would be presented as follows:

> After reading from a light-emitting screen at bedtime, alertness scores the next morning were significantly lower than would be expected if there were no effect, $t(8) = -2.67$, $p < .05$, one tailed.

Note that the report clearly acknowledges that a one-tailed test was used. ■

■ The Critical Region for a One-Tailed Test

In Step 2 of Example 9.6, we determined that the critical region is in the left-hand tail of the distribution. However, it is possible to divide this step into two stages that eliminate the need to determine which tail (right or left) should contain the critical region. The first stage in this process is simply to determine whether the sample mean is in the direction predicted by the original research question. For this example, the researcher predicted that the alertness scores would be lowered. Specifically, the researcher expects the participants to have scores lower than the general population average of $\mu = 50$. The obtained sample mean, $M = 46$, is in the correct direction. This first stage eliminates the need to determine whether the critical region is in the left- or right-hand tail. Because we already have determined that the effect is in the correct direction, the sign of the t statistic ($+$ or $-$) no longer matters. The second stage of the process is to determine whether the effect is large enough to be significant. For this example, the requirement is that the sample produces a t statistic greater than 1.860. If the magnitude of the t statistic, independent of its sign, is greater than 1.860, the result is significant and H_0 is rejected.

LEARNING CHECK   **LO10**   **1.** A sample is selected from a population with a mean of $\mu = 50$, and a treatment is administered to the sample. If the treatment is expected to increase scores and a t statistic is used for a one-tailed hypothesis test, then which of the following is the correct null hypothesis?

    a. $\mu \leq 50$

    b. $\mu < 50$

    c. $\mu \geq 50$

    d. $\mu > 50$

LO10   **2.** A researcher predicts that a treatment will increase scores. To test the treatment effect, a sample of $n = 16$ is selected from a population with $\mu = 80$ and a treatment is administered to the individuals in the sample. After treatment, the sample mean is $M = 78$ with $s^2 = 16$. If the researcher uses a one-tailed test with $\alpha = .05$, then what decision should be made?

    a. Reject H_0 with $\alpha = .05$ or with $\alpha = .01$

    b. Fail to reject H_0 with $\alpha = .05$ or with $\alpha = .01$

    c. Reject H_0 with $\alpha = .05$ but not with $\alpha = .01$

    d. Reject H_0 with $\alpha = .01$ but not with $\alpha = .05$

LO10   **3.** A researcher fails to reject the null hypothesis with a regular two-tailed test using $\alpha = .05$. If instead the researcher had used a directional (one-tailed) test with the same data and the same alpha level, then what decision would be made?

a. Definitely reject the null hypothesis.

b. Definitely reject the null hypothesis if the treatment effect is in the predicted direction.

c. Definitely fail to reject the null hypothesis.

d. Possibly reject the null hypothesis if the treatment effect is in the predicted direction.

ANSWERS   1. a   2. b   3. d

SUMMARY

1. The t statistic is used instead of a z-score for hypothesis testing when the population standard deviation (or variance) is unknown.

2. To compute the t statistic, you must first calculate the sample variance (or standard deviation) as a substitute for the unknown population value.

$$\text{sample variance} = s^2 = \frac{SS}{df}$$

Next, the standard error is *estimated* by substituting s^2 in the formula for standard error. The estimated standard error is calculated in the following manner:

$$\text{estimated standard error} = s_M = \sqrt{\frac{s^2}{n}}$$

Finally, a t statistic is computed using the estimated standard error. The t statistic is used as a substitute for a z-score, which cannot be computed when the population variance or standard deviation is unknown.

$$t = \frac{M - \mu}{s_M}$$

3. The structure of the t formula is similar to that of the z-score.

$$z \text{ or } t = \frac{\text{sample mean} - \text{population mean}}{\text{(estimated) standard error}}$$

For a hypothesis test, you hypothesize a value for the unknown population mean and plug the hypothesized value into the equation along with the sample mean and the estimated standard error, which are computed from the sample data. If the hypothesized mean produces an extreme value for t, you conclude that the hypothesis was wrong.

4. The t distribution is symmetrical with a mean of zero. To evaluate a t statistic for a sample mean, the critical region must be located in a t distribution. There is a family of t distributions, with the exact shape of a particular distribution of t values depending on degrees of freedom ($n - 1$). Therefore, the critical t values depend on the value for df associated with the t test. As df increases, the shape of the t distribution approaches a normal distribution.

5. When a t statistic is used for a hypothesis test, Cohen's d can be computed to measure effect size. In this situation, the sample standard deviation is used in the formula to obtain an estimated value for d:

$$\text{estimated } d = \frac{\text{mean difference}}{\text{standard deviation}} = \frac{M - \mu}{s}$$

6. A second measure of effect size is r^2, which measures the percentage of the variability that is accounted for by the treatment effect. This value is computed as follows:

$$r^2 = \frac{t^2}{t^2 + df}$$

7. An alternative method for describing the size of a treatment effect is to use a confidence interval for μ. A confidence interval is a range of values that estimates the unknown population mean. The confidence interval uses the t equation, solved for the unknown mean:

$$\mu = M \pm t(s_M)$$

First, select a level of confidence and then look up the corresponding t values to use in the equation. For example, for 95% confidence, use the range of t values that determines the middle 95% of the distribution.

KEY TERMS

estimated standard error (294)

t statistic (294)

degrees of freedom or *df* (295)

t distribution (295)

estimated *d* (304)

percentage of variance accounted for
by the treatment (r^2) (307)

confidence interval (308)

FOCUS ON PROBLEM SOLVING

1. The first problem we confront in analyzing data is determining the appropriate statistical test. Remember that you can use a *z*-score for the test statistic only when the value for σ is known. If the value for σ is not provided, then you must use the *t* statistic.

2. For the *t* test, the sample variance is used to find the value for estimated standard error. Remember that when computing the sample variance, use $n - 1$ in the denominator (see Chapter 4). When computing estimated standard error, use n in the denominator.

DEMONSTRATION 9.1

A HYPOTHESIS TEST WITH THE *t* STATISTIC

A psychologist has prepared an "Optimism Test" that is administered yearly to graduating college seniors. The test measures how each graduating class feels about its future—the higher the score, the more optimistic the class. Last year's class had a mean score of $\mu = 15$. A sample of $n = 9$ seniors from this year's class was selected and tested. The scores for these seniors are 7, 12, 11, 15, 7, 8, 15, 9, and 6, which produce a sample mean of $M = 10$ with $SS = 94$.

On the basis of this sample, can the psychologist conclude that this year's class has a different level of optimism than last year's class?

Note that this hypothesis test will use a *t* statistic because the population variance (σ^2) is not known.

STEP 1   **State the hypotheses, and select an alpha level.**   The statements for the null hypothesis and the alternative hypothesis follow the same form for the *t* statistic and the *z*-score test.

$$H_0: \mu = 15 \qquad \text{(There is no change.)}$$
$$H_1: \mu \neq 15 \qquad \text{(This year's mean is different.)}$$

For this demonstration, we will use $\alpha = .05$, two tails.

STEP 2   **Locate the critical region.**   With a sample of $n = 9$ students, the *t* statistic has $df = n - 1 = 8$. For a two-tailed test with $\alpha = .05$ and $df = 8$, the critical *t* values are $t = \pm 2.306$. These critical *t* values define the boundaries of the critical region. The obtained *t* value must be more extreme than either of these critical values to reject H_0.

STEP 3   **Compute the test statistic.**   As we have noted, it is easier to separate the calculation of the *t* statistic into three stages.

Sample variance:

$$s^2 = \frac{SS}{n - 1} = \frac{94}{8} = 11.75$$

Estimated standard error. The estimated standard error for these data is

$$s_M = \sqrt{\frac{s^2}{n}} = \sqrt{\frac{11.75}{9}} = 1.14$$

The t statistic. Now that we have the estimated standard error and the sample mean, we can compute the *t* statistic. For this demonstration,

$$t = \frac{M - \mu}{s_M} = \frac{10 - 15}{1.14} = \frac{-5}{1.14} = -4.39$$

STEP 4 **Make a decision about H_0, and state a conclusion.**   The *t* statistic we obtained ($t = -4.39$) is in the critical region. Thus, our sample data are unusual enough to reject the null hypothesis at the .05 level of significance. We can conclude that there is a significant difference in the level of optimism between this year's and last year's graduating classes, $t(8) = -4.39$, $p < .05$, two-tailed.

DEMONSTRATION 9.2

EFFECT SIZE: ESTIMATING COHEN'S *d* AND COMPUTING r^2

We will estimate Cohen's *d* for the same data used for the hypothesis test in Demonstration 9.1. The mean optimism score for the sample from this year's class was 5 points lower than the mean from last year ($M = 10$ versus $\mu = 15$). In Demonstration 9.1 we computed a sample variance of $s^2 = 11.75$, so the standard deviation is $\sqrt{11.75} = 3.43$. With these values,

$$\text{estimated } d = \frac{\text{mean difference}}{\text{standard deviation}} = \frac{5}{3.43} = 1.46$$

To calculate the percentage of variance explained by the treatment effect, r^2, we need the value of *t* and the *df* value from the hypothesis test. In Demonstration 9.1 we obtained $t = -4.39$ with $df = 8$. Using these values in Equation 9.5, we obtain

$$r^2 = \frac{t^2}{t^2 + df} = \frac{(-4.39)^2}{(-4.39)^2 + 8} = \frac{19.27}{27.27} = 0.71$$

SPSS®

General instructions for using SPSS are presented in Appendix D. Following are detailed instructions for using SPSS to perform **the One-Sample *t* Test** presented in this chapter.

Demonstration Example

A teacher is interested in whether their students are generally satisfied with a new reading assignment. The teacher administers a survey that asks students to rate their satisfaction with the reading on a scale of -10 (very dissatisfied) to 0 (neither dissatisfied nor satisfied) to $+10$ (very satisfied). The teacher observes the following $n = 24$ ratings:

0	−5	4	8	0	5	4	3	−3	5	7	5
3	3	−1	−10	8	7	0	6	−5	5	10	9

The steps below will show you how to perform a one-sample *t* test to test the hypothesis that student ratings of satisfaction were different from zero.

Data Entry

1. Click the **Variable View** tab to enter information about the variables.
2. In the first row, enter "rating" (for rating score) in the Name field. Add a descriptive label for the variable (e.g., "Satisfaction with Reading") in the Label field. Fill in the remaining information about your variable where necessary. Be sure that **Type** = "Numeric", **Width** = "8", **Decimals** = "0", **Values** = "None", **Missing** = "None", **Columns** = "8", **Align** = "Right", and **Measure** = "Scale".
3. Enter all of the scores from the sample in the "rating" column.

Data Analysis

1. Click **Analyze** on the tool bar, select **Compare Means**, and click on **One-Sample T-Test**.
2. Highlight the column label for the set of scores (Satisfaction with Reading) in the left box and click the arrow to move it into the **Test Variable(s)** box.
3. In the **Test Value** box at the bottom of the One-Sample *t* Test window, enter the hypothesized value for the population mean from the null hypothesis. *Note:* The value is automatically set at zero until you type in a new value. In this example, the null hypothesis is that students are neither satisfied nor dissatisfied with the reading; thus, the value should remain zero.
4. In addition to performing the hypothesis test, the program will compute a confidence interval for the population mean difference. The confidence level is automatically set at 95% but you can select **Options** and change the percentage.
5. Click **OK**.

SPSS Output

The output shown in the figure below includes a table of sample statistics with the mean, standard deviation, and standard error for the sample mean. A second table shows the results of the hypothesis test, including the values for *t*, *df*, and the level of significance (the *p* value for the test), as well as the mean difference from the hypothesized value of $\mu = 0$ and a 95% confidence interval for the mean difference. To obtain a 95% confidence interval for the mean, simply add $\mu = 0$ points to the values in the table.

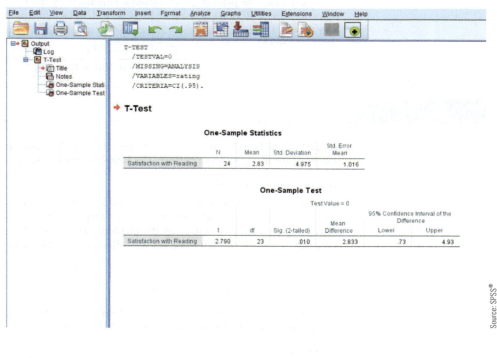

Source: SPSS®

Try It Yourself

Use the steps above to analyze the scores below:

100	94	98	102	123	92	107	127
104	103	120	117	103	127	125	90

The null hypothesis is that $\mu = 100$ points. Notice that your output should report that $t = 2.56$.

PROBLEMS

1. Under what circumstances is a t statistic used instead of a z-score for a hypothesis test?

2. Suppose that a researcher is interested in whether an exercise improves intelligence. The researcher randomly selects 100 participants, assigns them to an exercise program, and measures intelligence at the end of the exercise program. The measurement of intelligence has a known or assumed average score of $\mu = 100$ and a known or assumed standard deviation of $\sigma = 15$ in the population.
   a. What hypothesis test should the researcher use to evaluate whether the treatment affected intelligence?
   b. If σ and σ^2 were unknown, what hypothesis test should be used?

3. A sample of $n = 25$ scores has a mean of $M = 200$ and a variance of $s^2 = 100$.
   a. Explain what is measured by the sample variance.
   b. Compute the estimated standard error for the sample mean and explain what is measured by the standard error.

4. Find the estimated standard error for the sample mean for each of the following samples.
   a. $n = 9$ with $SS = 1,152$
   b. $n = 16$ with $SS = 540$
   c. $n = 25$ with $SS = 600$

5. The following sample of $n = 5$ scores was obtained from a population with unknown parameters. Scores: 20, 25, 30, 20, 30
   a. Compute the sample mean and variance. (*Note*: These are descriptive values that summarize the sample data.)
   b. Compute the estimated standard error for M. (*Note*: This is an inferential value that describes how accurately the sample mean represents the unknown population mean.)

6. The following sample of $n = 7$ scores was obtained from a population with unknown parameters. Scores: 2, 18, 15, 5, 15, 8, 7.
   a. Compute the sample mean and variance. (*Note*: These are descriptive values that summarize the sample data.)
   b. Compute the estimated standard error for M. (*Note*: This is an inferential value that describes how accurately the sample mean represents the unknown population mean.)

7. Explain why t distributions tend to be flatter and more spread out than the normal distribution.

8. Find the t values that form the boundaries of the critical region for a two-tailed test with $\alpha = .05$ for each of the following sample sizes:

   a. $n = 6$
   b. $n = 12$
   c. $n = 48$
   d. Repeat parts a–c assuming a one-tailed test, $\alpha = .05$.
   e. Repeat parts a–c assuming a two-tailed test, $\alpha = .01$.

9. Find the t value that forms the boundary of the critical region in the right-hand tail for a two-tailed test with $\alpha = .05$ for each of the following sample sizes.
   a. $n = 9$
   b. $n = 16$
   c. $n = 36$
   d. Repeat parts a–c assuming a one-tailed test, $\alpha = .05$.
   e. Repeat parts a–c assuming a two-tailed test, $\alpha = .01$.

10. A random sample of $n = 9$ individuals is selected from a population with $\mu = 20$, and a treatment is administered to each individual in the sample. After treatment, the following scores are observed:

    | 43 | 15 | 37 | 17 | 29 | 21 | 25 | 29 | 27 |

    a. Compute the sample mean and variance.
    b. How much difference is there between the mean for the treated sample and the mean for the original population? (*Note*: In a hypothesis test, this value forms the numerator of the t statistic.)
    c. If there is no treatment effect, what is the typical difference between the sample mean and its population mean? That is, find the standard error for M. (*Note*: In a hypothesis test, this value is the denominator of the t statistic.)
    d. Based on the sample data, does the treatment have a significant effect? Use a two-tailed test with $\alpha = .05$.

11. A random sample of $n = 7$ individuals is selected from a population with $\mu = 50$, and a treatment is administered to each individual in the sample. After treatment, the following scores are observed:

    | 37 | 49 | 47 | 47 | 47 | 43 | 45 |

    a. Compute the sample mean and variance.
    b. How much difference is there between the mean for the treated sample and the mean for the original population? (*Note*: In a hypothesis test, this value forms the numerator of the t statistic.)
    c. If there is no treatment effect, what is the typical difference between the sample mean and its population mean? That is, find the standard error for M. (*Note*: In a hypothesis test, this value is the denominator of the t statistic.)

d. Based on the sample data, does the treatment have a significant effect? Use a two-tailed test with $\alpha = .05$.

e. Repeat part d assuming that $\alpha = .01$.

12. A random sample of $n = 4$ individuals is selected from a population with $\mu = 35$, and a treatment is administered to each individual in the sample. After treatment, the sample mean is found to be $M = 40.1$ with $SS = 48$.

    a. How much difference is there between the mean for the treated sample and the mean for the original population? (*Note*: In a hypothesis test, this value forms the numerator of the *t* statistic.)

    b. If there is no treatment effect, how much difference is expected between the sample mean and its population mean? That is, find the standard error for *M*. (*Note*: In a hypothesis test, this value is the denominator of the *t* statistic.)

    c. Based on the sample data, does the treatment have a significant effect? Use a two-tailed test with $\alpha = .05$.

13. To evaluate the effect of a treatment, a sample is obtained from a population with a mean of $\mu = 40$, and the treatment is administered to the individuals in the sample. After treatment, the sample mean is found to be $M = 44.5$ with a variance of $s^2 = 36$.

    a. If the sample consists of $n = 4$ individuals, are the data sufficient to conclude that the treatment has a significant effect using a two-tailed test with $\alpha = .05$?

    b. If the sample consists of $n = 16$ individuals, are the data sufficient to conclude that the treatment has a significant effect using a two-tailed test with $\alpha = .05$?

    c. Comparing your answers for parts a and b, how does the size of the sample influence the outcome of a hypothesis test?

14. To evaluate the effect of a treatment, a sample of $n = 6$ is obtained from a population with a mean of $\mu = 80$, and the treatment is administered to the individuals in the sample. After treatment, the sample mean is found to be $M = 72$.

    a. If the sample variance is $s^2 = 54$, are the data sufficient to conclude that the treatment has a significant effect using a two-tailed test with $\alpha = .05$?

    b. If the sample variance is $s^2 = 150$, are the data sufficient to conclude that the treatment has a significant effect using a two-tailed test with $\alpha = .05$?

    c. Comparing your answers for parts a and b, how does the variability of the scores in the sample influence the outcome of a hypothesis test?

15. Weinstein, McDermott, and Roediger (2010) report that students who were given questions to be answered while studying new material had better scores when

tested on the material compared to students who were simply given an opportunity to reread the material. In a similar study, a group of students from a large psychology class received questions to be answered while studying for the final exam. The overall average for the exam was $\mu = 73.4$, but the $n = 16$ students who answered questions had a mean of $M = 78.3$ with a standard deviation of $s = 8.4$.

    a. Use a two-tailed test with $\alpha = .05$ to determine whether answering questions while studying produced significantly higher exam scores.

    b. Compute two different measurements of effect size.

16. People are poor at making judgments about probability. One source of error in judgments of probability is the base rate fallacy in which people ignore the *base rates* of low probability events. In a study of the base rate fallacy by Bar-Hillel (1980), participants were exposed to a vignette about a traffic accident. In the scenario, a taxicab was observed in a hit-and-run accident. In the city where the accident occurred, 85% of cabs are blue and 15% of cabs are green. Later, a witness testified that the cab in the accident was green and the witness was shown to be 80% accurate in identifying blue and green cabs (i.e., 20% of the time, the witness confused the cabs). What do you think is the probability that a green cab was in the hit-and-run? Most participants who encounter this problem report that the probability of the cab being green is much higher than the actual probability of 41%. That is, most participants ignore the fact that green cabs are relatively rare. Suppose that a researcher replicates the Bar-Hillel experiment with a sample of $n = 16$ participants. The researcher observes an average rated probability of $M = 60.06\%$ with $SS = 656.66$.

    a. Use a two-tailed test ($\alpha = .05$) of the hypothesis that participants showed a base rate fallacy. Assume that $\mu = 41$ if there is no base rate fallacy.

    b. Compute two different measurements of effect size.

17. To evaluate the effect of a treatment, a sample is obtained from a population with a mean of $\mu = 20$, and the treatment is administered to the individuals in the sample. After treatment, the sample mean is found to be $M = 22$ with a variance of $s^2 = 9$.

    a. Assuming that the sample consists of $n = 9$ individuals, use a two-tailed hypothesis test with $\alpha = .05$ to determine whether the treatment effect is significant and compute Cohen's *d* to measure effect size. Are the data sufficient to conclude that the treatment has a significant effect using a two-tailed test with $\alpha = .05$?

    b. Assuming that the sample consists of $n = 36$ individuals, repeat the test and compute Cohen's *d*.

    c. Comparing your answers for parts a and b, how does the size of the sample influence the outcome of a hypothesis test and Cohen's *d*?

18. To evaluate the effect of a treatment, a sample of $n = 8$ is obtained from a population with a mean of $\mu = 50$, and the treatment is administered to the individuals in the sample. After treatment, the sample mean is found to be $M = 55$.
   a. Assuming that the sample variance is $s^2 = 32$, use a two-tailed hypothesis test with $\alpha = .05$ to determine whether the treatment effect is significant and compute both Cohen's d and r^2 to measure effect size.
   b. Assuming that the sample variance is $s^2 = 72$, repeat the test and compute both measures of effect size.
   c. Comparing your answers for parts a and b, how does the variability of the scores in the sample influence the outcome of a hypothesis test and measures of effect size?

19. Your subjective experience of time is not fixed. You experience time "flying" during some activities and "dragging" during others. Researchers have shown that your experience of time can be altered by drugs that interact with the brain regions that are responsible for timing. Cheng, MacDonald, and Meck (2006) demonstrated that intervals are perceived as longer when under the influence of cocaine. In their experiment rats were trained to press Lever 1 after exposure to a short (two-second) sound and Lever 2 after exposure to a long (eight-second) sound. In a later test session, rats under the influence of cocaine were exposed to sounds of different duration and researchers measured lever pressing on Lever 1 and Lever 2. For each rat in the study, the researchers measured the duration of sound in seconds that was equally likely to be judged as long or short. Assume that the untreated population mean is $\mu = 4$ seconds and that the sample variance is $s^2 = 0.16$, the sample mean is $M = 3.78$, and the sample size is $n = 16$.
   a. Use a two-tailed test with $\alpha = .05$ to test whether cocaine influenced time perception.
   b. Construct the 95% confidence interval to estimate the μ for a population of rats under the influence of cocaine.
   c. Compute two different measures of effect size.

20. Oishi and Schimmack (2010) report that people who move from home to home frequently as children tend to have lower than average levels of well-being as adults. To further examine this relationship, a psychologist obtains a sample of $n = 12$ young adults who each experienced five or more different homes before they were 16 years old. These participants were given a standardized well-being questionnaire for which the general population has an average score of $\mu = 40$. The sample of well-being scores had an average of $M = 37$ and a variance of $s^2 = 10.73$.
   a. On the basis of this sample, is well-being for frequent movers significantly different from well-being in the general population? Use a two-tailed test with $\alpha = .05$.

   b. Compute the estimated Cohen's d to measure the size of the difference.
   c. Write a sentence showing how the outcome of the hypothesis test and the measure of effect size would appear in a research report.

21. In a classic study of procrastination by Lay (1986), in an introductory psychology class, students received a survey that measured procrastination. Participants were instructed to complete the survey and return it to the researcher by mail. High-procrastinating participants were identified by their degree of agreement with survey items like, "I often find myself performing tasks that I had intended to do days before." Interestingly, the researcher also measured the amount of time that it took for participants to return the survey. The following represents the number of days that some of the high procrastinators waited to return the survey:

   1   4   15   5   2   1   2   19   6   18

   a. Use a one-tailed test with $\alpha = .05$ to test the hypothesis that high procrastinators waited more than one day to return the survey.
   b. Compute the 95% confidence interval for the mean amount of time required to return the survey.
   c. Write a sentence showing the outcome of the hypothesis test and the confidence interval as it would appear in a research report.

22. The Muller-Lyer illusion is shown in the figure. Although the two horizontal lines are the same length, the line on the left appears to be much longer. To examine the strength of this illusion, Gillam and Chambers (1985) recruited 10 participants who reproduced the length of the horizontal line in the left panel of the figure. The strength of the illusion was measured by how much longer the reproduced line was than the actual length of the line in the figure. Below are data like those observed by the researchers. Each value represents how much longer (in millimeters) the reproduced line was than the line in the figure.

2.08   2.7   3.42   1.59   2.04   2.87   3.36   0.49   3.82   3.91

   a. Use a one-tailed hypothesis test with $\alpha = .01$ to demonstrate that the individuals in the sample significantly overestimate the true length of the line. (Note: Accurate estimation would produce a mean of $\mu = 0$ millimeters.)
   b. Calculate the estimated d and r^2, the percentage of variance accounted for, to measure the size of this effect.
   c. Construct a 95% confidence interval for the population mean estimated length of the vertical line.

The *t* Test for Two Independent Samples

CHAPTER 10

Tools You Will Need

The following items are considered essential background material for this chapter. If you doubt your knowledge of any of these items, you should review the appropriate chapter or section before proceeding.

- Sample variance (Chapter 4)
- Standard error formulas (Chapter 7)
- The *t* statistic (Chapter 9)
  - Distribution of *t* values
  - *df* for the *t* statistic
  - Estimated standard error

clivewa/Shutterstock.com

PREVIEW

The second research strategy, in which the two sets of data are obtained from the same group of participants, is called a *repeated-measures research design* or a *within-subjects design*. The statistics for evaluating the results from a repeated-measures design are introduced in Chapter 11. Also, at the end of Chapter 11, we discuss some of the advantages and disadvantages of independent-measures and repeated-measures designs.

LEARNING CHECK   LO1   1. Which of the following is most likely to be an independent-measures design?

a. A study comparing vocabulary size of 3-year-old children from lower socioeconomic status homes and 3-year-old children from higher socioeconomic status homes.

b. A study comparing classroom learning with and without background music.

c. A study comparing blood pressure before and after a workout.

d. A study evaluating jet lag by comparing cognitive performance at the beginning and end of a cross-country flight.

LO1   2. Which of the following is most likely to be a repeated-measures design?

a. A study comparing artistic skills performance for left-handed adolescents and right-handed adolescents.

b. A study comparing cholesterol levels before and after a diet featuring oatmeal.

c. A study comparing self-esteem for 6-year-old boys and 6-year-old girls.

d. A study comparing Facebook use for adolescents and over-30 adults.

LO1   3. An independent-measures study comparing two treatment conditions uses _____ groups of participants and obtains _____ score(s) for each participant.

a. 1, 1

b. 1, 2

c. 2, 1

d. 2, 2

ANSWERS   1. a   2. b   3. c

10-2   The Hypotheses and the Independent-Measures *t* Statistic

LEARNING OBJECTIVES

2. Describe the hypotheses for an independent-measures *t* test.

3. Describe the structure of the independent-measures *t* statistic and explain how it is related to the single-sample *t*.

4. Calculate the pooled variance for two samples and the estimated standard error for the sample mean difference, and explain what each one measures.

5. Calculate the complete independent-measures *t* statistic and its degrees of freedom.

Because an independent-measures study involves two separate samples, we need some special notation to help specify which data go with which sample. This notation involves the use of subscripts, which are small numbers written beside a sample statistic. For example, the number of scores in the first sample is identified by n_1; for the second sample, the number of scores is n_2. The sample means are identified by M_1 and M_2. The sums of squares are SS_1 and SS_2.

■ The Hypotheses for an Independent-Measures Test

The goal of an independent-measures research study is to evaluate the mean difference between two populations (or between two treatment conditions). Using subscripts to differentiate the two populations, the mean for the first population is μ_1, and the second population mean is μ_2. The difference between means is simply $\mu_1 - \mu_2$. As always, the null hypothesis states that there is no change, no effect, or, in this case, no difference. Thus, in symbols, the null hypothesis for the independent-measures test is

$$H_0: \mu_1 - \mu_2 = 0 \qquad \text{(No difference between the population means.)}$$

You should notice that the null hypothesis could also be stated as $\mu_1 = \mu_2$. However, the first version of H_0 produces a specific numerical value (zero) that is used in the calculation of the t statistic. Therefore, we prefer to phrase the null hypothesis in terms of the difference between the two population means.

The alternative hypothesis states that there is a mean difference between the two populations,

$$H_1: \mu_1 - \mu_2 \neq 0 \qquad \text{(There is a mean difference.)}$$

Equivalently, the alternative hypothesis can simply state that the two population means are not equal: $\mu_1 \neq \mu_2$.

■ The Formulas for an Independent-Measures Hypothesis Test

The independent-measures hypothesis test uses another version of the t statistic. The formula for this new t statistic has the same general structure as the t statistic formula that was introduced in Chapter 9. To help distinguish between the two t formulas, we refer to the original formula (Chapter 9) as the *single-sample* t *statistic* and the new formula as the *independent-measures* t *statistic*. Because the new independent-measures t includes data from two separate samples and hypotheses about two populations, the formulas may appear to be a bit overpowering. However, the new formulas are easier to understand if you view them in relation to the single-sample t formulas from Chapter 9. In particular, there are two points to remember:

1. The basic structure of the t statistic is the same for both the independent-measures and the single-sample hypothesis tests. In both cases,

$$t = \frac{\text{actual difference between sample data and the hypothesis}}{\text{expected difference between sample data and hypothesis with no treatment effect}}$$

2. The independent-measures t is basically a two-sample t that doubles all the elements of the single-sample t formulas.

To demonstrate the second point, we examine the two t formulas piece by piece.

The Overall *t* Formula   The single-sample t uses one sample mean to test a hypothesis about one population mean. The sample mean and the population mean appear in the numerator of the t formula, which measures how much difference there is between the sample data and the population hypothesis.

$$t = \frac{\text{sample mean} - \text{population mean}}{\text{estimated standard error}} = \frac{M - \mu}{s_M}$$

The independent-measures *t* uses the difference between *two* sample means to evaluate a hypothesis about the difference between *two* population means. Thus, the independent-measures *t* formula is

$$t = \frac{\text{sample mean difference} - \text{population mean difference}}{\text{estimated standard error of sample mean difference}} = \frac{(M_1 - M_2) - (\mu_1 - \mu_2)}{s_{(M_1 - M_2)}}$$

In this formula, the value of $M_1 - M_2$ is obtained from the sample data and the value for $\mu_1 - \mu_2$ comes from the null hypothesis. In a hypothesis test, the null hypothesis sets the population mean difference equal to zero, so the independent measures *t* formula can be simplified further,

$$t = \frac{\text{sample mean difference}}{\text{estimated standard error}}$$

In this form, the *t* statistic is a simple ratio comparing the actual mean difference (numerator) with the difference that is expected by chance (denominator).

The Estimated Standard Error   In each of the *t*-score formulas, the standard error in the denominator measures how much error is expected between the sample statistic and the population parameter. In the single-sample *t* formula, the standard error measures the amount of error expected for a sample mean and is represented by the symbol s_M. For the independent-measures *t* formula, the standard error measures the amount of error that is expected between a sample mean difference $(M_1 - M_2)$ and the population mean difference $(\mu_1 - \mu_2)$. The standard error for the sample mean difference is represented by the symbol $s_{(M_1 - M_2)}$.

Caution: Do not let the notation for standard error confuse you. In general, the symbol for standard error takes the form $s_{\text{statistic}}$. When the statistic is a single sample mean, *M*, the symbol for standard error is s_M. For the independent-measures test, the statistic is a sample mean difference $(M_1 - M_2)$, and the symbol for standard error is $s_{(M_1 - M_2)}$. In each case, the standard error tells how much discrepancy is reasonable to expect between the sample statistic and the corresponding population parameter.

Interpreting the Estimated Standard Error   The *estimated standard error of* $M_1 - M_2$ that appears in the bottom of the independent-measures *t* statistic can be interpreted in two ways. First, the standard error is defined as a measure of the standard or average distance between a sample statistic $(M_1 - M_2)$ and the corresponding population parameter $(\mu_1 - \mu_2)$. As always, samples are not expected to be perfectly accurate and the standard error measures how much difference is reasonable to expect between a sample statistic and the population parameter.

When the null hypothesis is true, however, the population mean difference is zero. In this case, the standard error is measuring how far, on average, the sample mean difference is from zero. However, measuring how far it is from zero is the same as measuring how big it is. Thus, there are two ways to interpret the estimated standard error of $(M_1 - M_2)$:

1. It measures the standard distance between $(M_1 - M_2)$ and $(\mu_1 - \mu_2)$.

2. When the null hypothesis is true, it measures the standard, or average size of $(M_1 - M_2)$. That is, it measures how much difference is reasonable to expect between the two sample means.

■ Calculating the Estimated Standard Error

To develop the formula for $s_{(M_1 - M_2)}$ we consider the following two points:

1. Each of the two sample means represents its own population mean, but in each case there is some error.

M_1 approximates μ_1 with some error.

M_2 approximates μ_2 with some error.

Thus, there are two sources of error. Each sample mean is not necessarily exactly equal to the population mean from which the sample was selected because of error. The amount of error associated with each sample mean is measured by the estimated standard error of *M*. Using Equation 9.1 (page 293), the estimated standard error for each sample mean is computed as follows:

$$\text{For } M_1, s_M = \sqrt{\frac{s_1^2}{n_1}} \qquad \text{For } M_2, s_M = \sqrt{\frac{s_2^2}{n_2}}$$

2. For the independent-measures *t* statistic, we want to know the total amount of error involved in using *two* sample means to approximate *two* population means. To do this, we will find the error from each sample separately and then add the two errors together. The resulting formula for standard error is

$$s_{(M_1 - M_2)} = \sqrt{\frac{s_1^2}{n_1} + \frac{s_2^2}{n_2}} \qquad (10.1)$$

Because the independent-measures *t* statistic uses two sample means, the formula for the estimated standard error simply combines the error for the first sample mean and the error for the second sample mean (Box 10.1).

■ Pooled Variance

Although Equation 10.1 accurately presents the concept of standard error for the independent-measures *t* statistic, this formula is limited to situations in which the two samples are exactly the same size (that is, $n_1 = n_2$). For situations in which the two sample sizes

---

BOX 10.1 The Variability of Difference Scores

It may seem odd that the independent-measures *t* statistic *adds* together the two sample errors when it *subtracts* to find the difference between the two sample means. The logic behind this apparently unusual procedure is demonstrated here.

We begin with two populations, I and II (see Figure 10.2). The scores in Population I range from a high of 70 to a low of 50. The scores in Population II range from 30 to 20. We will use the range as a measure of how spread out (variable) each population is:

For Population I, the scores cover a range of 20 points.

For Population II, the scores cover a range of 10 points.

If we randomly select one score from Population I and one score from Population II and compute the difference between these two scores ($X_1 - X_2$), what range of values is possible for these differences? To answer this question, we need to find the biggest possible difference and the smallest possible difference. Look at Figure 10.2; the biggest difference occurs when $X_1 = 70$ and $X_2 = 20$. This is a difference of $X_1 - X_2 = 50$ points. The smallest difference occurs when $X_1 = 50$ and $X_2 = 30$. This is a difference of $X_1 - X_2 = 20$ points. Notice that the differences go from a high of 50 to a low of 20. This is a range of 30 points:

Range for Population I (X_1 scores) = 20 points

Range for Population II (X_2 scores) = 10 points

Range for the differences ($X_1 - X_2$) = 30 points

Note that the variability for the difference in scores is found by *adding* together the variability for each of the two populations.

FIGURE 10.2

Two population distributions. The scores in Population I range from 50 to 70 (a 20-point spread) and the scores in Population II range from 20 to 30 (a 10-point spread). If you select one score from each of these two populations, the closest two values are $X_1 = 50$ and $X_2 = 30$. The two values that are farthest apart are $X_1 = 70$ and $X_2 = 20$.

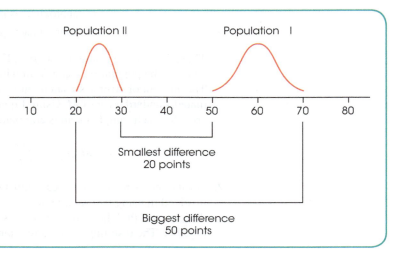

are different, the formula is *biased* and, therefore, inappropriate. The bias comes from the fact that Equation 10.1 treats the two sample variances equally. However, when the sample sizes are different, the two sample variances are not equally good estimates for error and should not be treated equally. In Chapter 7, we introduced the law of large numbers, which states that statistics obtained from large samples tend to be better (more accurate) estimates of population parameters than statistics obtained from small samples. This same fact holds for sample variances: The variance obtained from a large sample is a more accurate estimate of σ^2 than the variance obtained from a small sample.

One method for correcting the bias in the standard error is to combine the two sample variances into a single value called the *pooled variance*. The pooled variance is obtained by averaging or "pooling" the two sample variances using a procedure that allows the bigger sample to carry more weight in determining the final value.

You should recall that when there is only one sample, the sample variance is computed as

$$s^2 = \frac{SS}{df}$$

For the independent-measures *t* statistic, there are two *SS* values and two *df* values (one from each sample). The values from the two samples are combined to compute what is called the *pooled variance*. The pooled variance is identified by the symbol s_p^2 and is computed as

$$\text{pooled variance} = s_p^2 = \frac{SS_1 + SS_2}{df_1 + df_2} \tag{10.2}$$

With one sample, the variance is computed as *SS* divided by *df*. With two samples, the pooled variance is computed by combining the two *SS* values and then dividing by the combination of the two *df* values.

As we mentioned earlier, the pooled variance is actually an average of the two sample variances, but the average is computed so that the larger sample carries more weight in determining the final value. The following examples demonstrate this point.

Equal Samples Sizes   We begin by computing the pooled variance for two samples that are exactly the same size. The first sample has $n = 6$ scores with $SS = 50$, and the second sample has $n = 6$ scores with $SS = 30$. Individually, the two sample variances are

$$\text{Variance for sample 1:} \quad s^2 = \frac{SS}{df} = \frac{50}{5} = 10$$

$$\text{Variance for sample 2:} \quad s^2 = \frac{SS}{df} = \frac{30}{5} = 6$$

The pooled variance for these two samples is

$$s_p^2 = \frac{SS_1 + SS_2}{df_1 + df_2} = \frac{50 + 30}{5 + 5} = \frac{80}{10} = 8.00$$

Note that the pooled variance is exactly halfway between the two sample variances. Because the two samples are exactly the same size, the pooled variance is simply the average of the two sample variances.

Unequal Samples Sizes   Now consider what happens when the samples are not the same size. This time the first sample has $n = 3$ scores with $SS = 20$, and the second sample has $n = 9$ scores with $SS = 48$. Individually, the two sample variances are

$$\text{Variance for sample 1:} \quad s^2 = \frac{SS}{df} = \frac{20}{2} = 10$$

$$\text{Variance for sample 2:} \quad s^2 = \frac{SS}{df} = \frac{48}{8} = 6$$

The pooled variance for these two samples is

$$s_p^2 = \frac{SS_1 + SS_2}{df_1 + df_2} = \frac{20 + 48}{2 + 8} = \frac{68}{10} = 6.80$$

This time the pooled variance is not located halfway between the two sample variances. Instead, the pooled value is closer to the variance for the larger sample ($n = 9$ and $s^2 = 6$) than to the variance for the smaller sample ($n = 3$ and $s^2 = 10$). The larger sample carries more weight when the pooled variance is computed.

When computing the pooled variance, the weight for each of the individual sample variances is determined by its degrees of freedom. Because the larger sample has a larger df value, it carries more weight when averaging the two variances. This produces an alternative formula for computing pooled variance:

$$\text{pooled variance} = s_p^2 = \frac{df_1 s_1^2 + df_2 s_2^2}{df_1 + df_2} \tag{10.3}$$

For example, if the first sample has $df_1 = 2$ and the second sample has $df_2 = 8$, then the formula instructs you to take 2 of the first sample variance and 8 of the second sample variance for a total of 10 variances. You then divide by 10 to obtain the average. The alternative formula is especially useful if the sample data are summarized as means and variances. Finally, you should note that because the pooled variance is an average of the two sample variances, the value obtained for the pooled variance is always located between the two sample variances. When sample sizes are unequal, pooled variance will have a value closer to the variance of the larger sample.

■ Estimated Standard Error

Using the pooled variance in place of the individual sample variances, we can now obtain an unbiased measure of the standard error for a sample mean difference. The resulting formula for the independent-measures estimated standard error is

$$\text{estimated standard error of } M_1 - M_2 = s_{(M_1-M_2)} = \sqrt{\frac{s_p^2}{n_1} + \frac{s_p^2}{n_2}} \qquad (10.4)$$

Conceptually, this standard error measures how accurately the difference between two sample means represents the difference between the two population means. In a hypothesis test, H_0 specifies that $\mu_1 - \mu_2 = 0$, and the standard error also measures how much difference is expected on average between the two sample means. In either case, the formula combines the error for the first sample mean with the error for the second sample mean. Also note that the pooled variance from the two samples is used to compute the standard error for the two samples.

The following example is an opportunity to test your understanding of the pooled variance and the estimated standard error.

EXAMPLE 10.1   One sample from an independent-measures study has $n = 4$ with $SS = 72$. The other sample has $n = 8$ and $SS = 168$. For these data, compute the pooled variance and the estimated standard error for the mean difference. You should find that the pooled variance is $240/10 = 24$ and the estimated standard error is 3. Good luck.   ■

■ The Final Formula and Degrees of Freedom

The complete formula for the independent-measures *t* statistic is as follows:

$$t = \frac{(M_1 - M_2) - (\mu_1 - \mu_2)}{s_{(M_1-M_2)}}$$

$$= \frac{\text{sample mean difference} - \text{population mean difference}}{\text{estimated standard error}} \qquad (10.5)$$

In the formula, the estimated standard error in the denominator is calculated using Equation 10.4, and requires calculation of the pooled variance using either Equation 10.2 or 10.3.

The degrees of freedom for the independent-measures *t* statistic are determined by the *df* values for the two separate samples:

$$df \text{ for the } t \text{ statistic} = df \text{ for the first sample} + df \text{ for the second sample}$$
$$= df_1 + df_2$$
$$= (n_1 - 1) + (n_2 - 1) \qquad (10.6)$$

Equivalently, the *df* value for the independent-measures *t* statistic can be expressed as

$$df = n_1 + n_2 - 2 \qquad (10.7)$$

Note that the *df* formula subtracts 2 points from the total number of scores; 1 point for the first sample and 1 for the second.

The independent-measures *t* statistic is used for hypothesis testing. Specifically, we use the difference between two sample means $(M_1 - M_2)$ as the basis for testing hypotheses about the difference between two population means $(\mu_1 - \mu_2)$. In this context, the overall structure of the *t* statistic can be reduced to the following:

$$t = \frac{\text{data} - \text{hypothesis}}{\text{error}}$$

This same structure is used for both the single-sample *t* from Chapter 9 and the new independent-measures *t* that was introduced in the preceding pages. Table 10.1 identifies

TABLE 10.1

The basic elements of a *t* statistic for the single-sample *t* and the independent-measures *t*.

	Sample Data	Hypothesized Population Parameter	Estimated Standard Error	Sample Variance
Single-sample *t* statistic	M	μ	$\sqrt{\dfrac{s^2}{n}}$	$s^2 = \dfrac{SS}{df}$
Independent-measures	$(M_1 - M_2)$	$(\mu_1 - \mu_2)$	$\sqrt{\dfrac{s_p^2}{n_1} + \dfrac{s_p^2}{n_2}}$	$s_p^2 = \dfrac{SS_1 + SS_2}{df_1 + df_2}$

each component of these two *t* statistics and should help reinforce the point that we made earlier in the chapter; that is, the independent-measures *t* statistic simply doubles each aspect of the single-sample *t* statistic.

LEARNING CHECK

LO2 1. Which of the following is the correct null hypothesis for an independent-measures *t* test?

   a. There is no difference between the two sample means.

   b. There is no difference between the two population means.

   c. The difference between the two sample means is identical to the difference between the two population means.

   d. None of the other three choices is correct.

LO3 2. Which of the following does not accurately describe the relationship between the formulas for the single-sample *t* and the independent-measures *t*?

   a. The single-sample *t* has one sample mean and the independent-measures *t* has two.

   b. The single-sample *t* has one population mean and the independent-measures *t* has two.

   c. The single-sample *t* uses one sample variance to compute the standard error and the independent-measures *t* uses two.

   d. All of the above accurately describe the relationship.

LO4 3. One sample has $n = 21$ and a second sample has $n = 35$. If the pooled variance for the two samples is 210, then what is the estimated standard error for the sample mean difference?

   a. 9

   b. 4

   c. 3

   d. 2

LO5 4. A researcher obtains $M = 34$ with $SS = 190$ for a sample of $n = 10$ girls, and $M = 29$ with $SS = 170$ for a sample of $n = 10$ boys. If the two samples are used to evaluate the mean difference between the two populations, what value will be obtained for the *t* statistic?

   a. $\frac{5}{4} = 1.25$

   b. $\frac{5}{2} = 2.50$

   c. $\frac{5}{\sqrt{2}} = 3.54$

   d. $\frac{5}{1} = 5.00$

ANSWERS 1. b 2. d 3. b 4. b

10-3 Hypothesis Tests with the Independent-Measures *t* Statistic

LEARNING OBJECTIVES

6. Use the data from two samples to conduct an independent-measures *t* test evaluating the significance of the difference between two population means.

7. Conduct a directional (one-tailed) hypothesis test using the independent-measures *t* statistic.

8. Describe the basic assumptions underlying the independent-measures *t* hypothesis test, especially the homogeneity of variance assumption, and explain how the homogeneity assumption can be tested.

The independent-measures *t* statistic uses the data from two separate samples to help decide whether there is a significant mean difference between two populations or between two treatment conditions. A complete example of a hypothesis test with two independent samples follows.

EXAMPLE 10.2   Research has shown that people are more likely to show dishonest and self-interested behaviors in darkness than in a well-lit environment (Zhong, Bohns, & Gino, 2010). In one experiment, participants were given a set of 20 puzzles and were paid 50 cents for each one solved in a 5-minute period. However, the participants reported their own performance and there was no obvious method for checking their honesty. Thus, the task provided a clear opportunity to cheat and receive undeserved money. One group of participants was tested in a room with dimmed lighting and a second group was tested in a well-lit room. The reported number of solved puzzles was recorded for each individual. The following data represent results similar to those obtained in the study.

Number of Solved Puzzles			
Well-Lit Room		Dimly Lit Room	
11	6	7	9
9	7	13	11
4	12	14	15
5	10	16	11
$n_1 = 8$		$n_2 = 8$	
$M_1 = 8$		$M_2 = 12$	
$SS_1 = 60$		$SS_2 = 66$	

STEP 1   **State the hypotheses and select the alpha level.**   The null hypothesis says that for the general population, the brightness of the lighting in the room has no effect on the number of solved problems reported by the participants.

$$H_0: \mu_1 - \mu_2 = 0 \quad \text{(No difference.)}$$
$$H_1: \mu_1 - \mu_2 \neq 0 \quad \text{(There is a difference.)}$$

We will set $\alpha = .05$, two-tailed.

Directional hypotheses could be used and would specify whether the students who were tested in a dimly lit room should have higher or lower scores.

STEP 2   **Locate the critical region.**   This is an independent-measures design. The *t* statistic for these data has degrees of freedom determined by

$$df = df_1 = df_2$$
$$= (n_1 - 1) + (n_2 - 1)$$
$$= 7 + 7$$
$$= 14$$

With $df = 14$ and $\alpha = .05$, the *t* distribution has critical boundaries of $t = +2.145$ and $t = -2.145$ (see Figure 10.3).

STEP 3   **Obtain the data and compute the test statistic.**   As with the single-sample *t* test in Chapter 9, we recommend that the calculations be divided into three parts.

Caution: The pooled variance combines the two samples to obtain a single estimate of variance. In the formula, the two samples are combined in a single fraction.

First, find the pooled variance for the two samples:

$$s_p^2 = \frac{SS_1 + SS_2}{df_1 + df_2}$$
$$= \frac{60 + 66}{7 + 7} = \frac{126}{14} = 9$$

Caution: The standard error adds the errors from two separate samples. In the formula, these two errors are added as two separate fractions. In this case, the two errors are equal because the sample sizes are the same.

Second, use the pooled variance to compute the estimated standard error:

$$s_{(M_1 - M_2)} = \sqrt{\frac{s_p^2}{n_1} + \frac{s_p^2}{n_2}} = \sqrt{\frac{9}{8} + \frac{9}{8}}$$
$$= \sqrt{2.25}$$
$$= 1.50$$

Third, compute the *t* statistic:

$$t = \frac{(M_1 - M_2) - (\mu_1 - \mu_2)}{s_{(M_1 - M_2)}} = \frac{(8 - 12) - 0}{1.5}$$
$$= \frac{-4}{1.5} = -2.67$$

STEP 4   **Make a decision.**   The obtained value ($t = -2.67$) is in the critical region. In this example, the obtained sample mean difference is 2.67 times greater than would be expected if

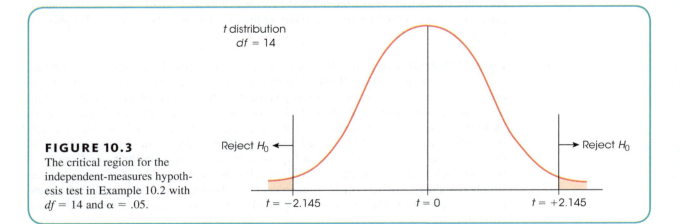

FIGURE 10.3

The critical region for the independent-measures hypothesis test in Example 10.2 with $df = 14$ and $\alpha = .05$.

there were no difference between the two populations. In other words, this result is very unlikely if H_0 is true. Therefore, we reject H_0 and conclude that there is a significant difference between the reported scores in the dimly lit room and the scores in the well-lit room. Specifically, the students in the dimly lit room reported significantly higher scores than those in the well-lit room. ■

■ Directional Hypotheses and One-Tailed Tests

When planning an independent-measures study, a researcher usually has some expectation or specific prediction for the outcome. For the cheating study in Example 10.2, the researchers expect the students in the dimly lit room to claim higher scores than the students in the well-lit room. This kind of directional prediction can be incorporated into the statement of the hypotheses, resulting in a directional, or one-tailed, test. Recall from Chapter 8 that one-tailed tests can lead to rejecting H_0 when the mean difference is relatively small compared to the magnitude required by a two-tailed test. As a result, one-tailed tests should be used when clearly justified by theory or previous findings. The following example demonstrates the procedure for stating hypotheses and locating the critical region for a one-tailed test using the independent-measures *t* statistic.

EXAMPLE 10.3   We will use the same research situation that was described in Example 10.2. The researcher is using an independent-measures design to examine the relationship between lighting and dishonest behavior. The prediction is that students in a well-lit room are less likely to cheat than students in a dimly lit room.

STEP 1   **State the hypotheses and select the alpha level.**   As always, the null hypothesis states that there is no effect, and the alternative hypothesis states that there is an effect. For this example, the predicted effect is that the students in the dimly lit room will claim to have higher scores. Thus, the two hypotheses are as follows.

$$H_0: \mu_{\text{Well-Lit}} \geq \mu_{\text{Dimly Lit}} \qquad \text{(Not less cheating in the well-lit vs. dimly lit room)}$$
$$H_1: \mu_{\text{Well-Lit}} < \mu_{\text{Dimly Lit}} \qquad \text{(Less cheating in the well-lit vs. dimly lit room)}$$

Note that it is usually easier to state the hypotheses in words before you try to write them in symbols. Also, it usually is easier to begin with the alternative hypothesis (H_1), which states that the treatment works as predicted. The alternative hypothesis predicts less cheating in the well-lit room; thus, the less-than sign is used. The greater-than-or-equal-to sign goes into the null hypothesis, indicating that the well-lit treatment does not decrease cheating. The idea of zero difference is the essence of the null hypothesis, and the numerical value of zero is used for ($\mu_1 - \mu_2$) during the calculation of the *t* statistic. For this test we will use $\alpha = .01$.

STEP 2   **Locate the critical region.**   For a directional test, the critical region is located entirely in one tail of the distribution. Rather than trying to determine which tail, positive or negative, is the correct location, we suggest you identify the criteria for the critical region in a two-step process as follows. First, look at the data and determine whether the sample mean difference is in the direction that was predicted. If the answer is no, then the data obviously do not support the predicted treatment effect, and you can stop the analysis. On the other hand, if the difference is in the predicted direction, then the second step is to determine whether the difference is large enough to be significant. To test for significance, simply find the one-tailed critical value in the *t* distribution table. If the calculated *t* statistic is more extreme (either positive or negative) than the critical value, then the difference is significant.

For this example, the students in the well lit room reported lower scores, as predicted. With $df = 14$, the one-tailed critical value for $\alpha = .01$ is $t = -2.624$.

STEP 3 **Collect the data and calculate the test statistic.** The details of the calculations were shown in Example 10.2. The data produce a t statistic of $t = -2.67$.

STEP 4 **Make a decision.** The t statistic of $t = -2.67$ is more extreme than the critical value of $t = -2.624$. Therefore, we reject the null hypothesis and conclude that the reported scores for students in the well-lit room are significantly lower than the scores for students in the dimly lit room. In a research report, the one-tailed test would be clearly noted:

Reported scores were significantly lower for students in the well-lit room, $t(14) = -2.67$, $p < .01$, one-tailed.   ◼

◼ Assumptions Underlying the Independent-Measures t Formula

There are three assumptions that should be satisfied before you use the independent-measures t formula for hypothesis testing:

1. The observations within each sample must be independent (see page 264).
2. The two populations from which the samples are selected must be normal.
3. The two populations from which the samples are selected must have equal variances.

The first two assumptions should be familiar from the single-sample t hypothesis test presented in Chapter 9. As before, the normality assumption is the less important of the two, especially with large samples. When there is reason to suspect that the populations are far from normal, you should compensate by ensuring that the samples are relatively large.

The third assumption is referred to as *homogeneity of variance* and states that the two populations being compared must have the same variance. You may recall a similar assumption for the z-score hypothesis test in Chapter 8. For that test, we assumed that the effect of the treatment was to add a constant amount to (or subtract a constant amount from) each individual score. As a result, the population standard deviation after treatment was the same as it had been before treatment. We now are making essentially the same assumption but phrasing it in terms of variances.

Recall that the pooled variance in the t-statistic formula is obtained by averaging together the two sample variances. It makes sense to average these two values only if they both are estimating the same population variance—that is, if the homogeneity of variance assumption is satisfied. If the two sample variances are estimating different population variances, then the average is meaningless. (*Note*: If two people are asked to estimate the same thing—for example, your weight—it is reasonable to average the two estimates. However, it is not meaningful to average estimates of two different things. If one person estimates your weight and another estimates the number of beans in a pound of whole-bean coffee, it is meaningless to average the two numbers.)

Homogeneity of variance is most important when there is a large discrepancy between the sample sizes. With equal (or nearly equal) sample sizes, this assumption is less critical, but still important. Violating the homogeneity of variance assumption can negate any meaningful interpretation of the data from an independent-measures experiment. Specifically, when you compute the t statistic in a hypothesis test, all the numbers in the formula come from the data except for the population mean difference, which you get from H_0. Thus, you are sure of all the numbers in the formula except one. If you obtain an extreme result for the t statistic (a value in the critical region), you conclude that the hypothesized value was wrong. But consider what happens when you violate the homogeneity of variance assumption. In this case, you have two questionable values in the formula (the hypothesized population value and the meaningless average of the two variances). Now if you obtain an extreme *t* statistic, you do not know which of these two values is responsible. Specifically, you cannot reject the hypothesis because it may have been the pooled variance that produced the extreme *t* statistic. Without satisfying the homogeneity of variance requirement, you cannot accurately interpret a *t* statistic, and the hypothesis test becomes meaningless.

■ Hartley's *F*-Max Test

How do you know whether the homogeneity of variance assumption is satisfied? One simple test involves just looking at the two sample variances. Logically, if the two population variances are equal, then the two sample variances should be very similar. When the two sample variances are close, you can be reasonably confident that the homogeneity assumption has been satisfied and proceed with the test. However, if one sample variance is more than three or four times larger than the other, then there is reason for concern. A more objective procedure involves a statistical test to evaluate the homogeneity assumption. Although there are many different statistical methods for determining whether the homogeneity of variance assumption has been satisfied, Hartley's *F*-max test is one of the simplest to compute and to understand. An additional advantage is that this test can also be used to check homogeneity of variance with more than two independent samples. Later, in Chapter 12, we examine statistical methods for comparing several different samples, and Hartley's test will be useful again. The following example demonstrates the *F*-max test for two independent samples.

EXAMPLE 10.4

The *F*-max test is based on the principle that a sample variance provides an unbiased estimate of the population variance. The null hypothesis for this test states that the population variances are equal; therefore, the sample variances should be very similar. The procedure for using the *F*-max test is as follows:

1. Compute the sample variance, $s^2 = \frac{SS}{df}$, for each of the separate samples.
2. Select the largest and the smallest of these sample variances and compute

$$F\text{-max} = \frac{s^2(\text{largest})}{s^2(\text{smallest})}$$

A relatively large value for *F*-max indicates a large difference between the sample variances. In this case, the data suggest that the population variances are different and that the homogeneity assumption has been violated. On the other hand, a small value of *F*-max (near 1.00) indicates that the sample variances are similar and that the homogeneity assumption is reasonable.

3. The *F*-max value computed for the sample data is compared with the critical value found in Table B.3 (Appendix B). If the sample value is larger than the table value, you conclude that the variances are different and that the homogeneity assumption is not valid.

To locate the critical value in the table, you need to know

a. *k* = number of separate samples. (For the independent-measures *t* test, *k* = 2.)
b. *df* = *n* − 1 for each sample variance. The Hartley test assumes that all samples are the same size.
c. the alpha level. The table provides critical values for α = .05 and α = .01. Generally, a test for homogeneity would use the larger alpha level.

Suppose, for example, that two independent samples each have $n = 10$ with sample variances of 12.34 and 9.15. For these data,

$$F\text{-max} = \frac{s^2(\text{largest})}{s^2(\text{smallest})} = \frac{12.34}{9.15} = 1.35$$

With $\alpha = .05$, $k = 2$, and $df = n - 1 = 9$, the critical value from the table is 4.03. Because the obtained *F*-max is smaller than this critical value, you conclude that the data do not provide evidence that the homogeneity of variance assumption has been violated. ■

The goal for most hypothesis tests is to reject the null hypothesis to demonstrate a significant difference or a significant treatment effect. However, when testing for homogeneity of variance, the preferred outcome is to fail to reject H_0. Failing to reject H_0 with the *F*-max test means that there is no significant difference between the two population variances and the homogeneity assumption is satisfied. In this case, you may proceed with the independent-measures *t* test using pooled variance.

If the *F*-max test rejects the hypothesis of equal variances, or if you simply suspect that the homogeneity of variance assumption is not justified, you should not compute an independent-measures *t* statistic using pooled variance. However, there is an alternative formula for the *t* statistic that is used by many computer applications for statistics. This alternative formula is presented in the SPSS section at the end of the chapter.

LEARNING CHECK   **LO6** **1.** What is the value of the independent-measures *t* statistic for a study with $n = 10$ participants in each treatment if the data produce $M = 38$ and $SS = 200$ for the first treatment, and $M = 33$ and $SS = 160$ for the second treatment?

    a. $t = 1.25$

    b. $t = 2.50$

    c. $t = 0.25$

    d. $t = \frac{5}{\sqrt{20}} = 1.12$

LO7 **2.** A researcher uses two samples, each with $n = 15$ participants, to evaluate the mean difference in performance scores between 8-year-old and 10-year-old children. The prediction is that the older children will have higher scores. The sample mean for the older children is five points higher than the mean for the younger children and the pooled variance for the two samples is 30. For a one-tailed test, what decision should be made?

    a. Reject the null hypothesis with $\alpha = .05$ but not with $\alpha = .01$.

    b. Reject the null hypothesis with either $\alpha = .05$ or $\alpha = .01$.

    c. Fail to reject the null hypothesis with $\alpha = .05$ but not with $\alpha = .01$.

    d. Fail to reject the null hypothesis with either $\alpha = .05$ or $\alpha = .01$.

LO8 **3.** Hartley's *F*-max test is used to evaluate the homogeneity of variance assumption. What is the null hypothesis for this test?

    a. The two sample variances are equal.

    b. The two sample variances are not equal.

    c. The two population variances are equal.

    d. The two population variances are not equal.

ANSWERS   **1. b   2. b   3. c**

The data from the study produced a mean score of $M = 12$ for the group in the dimly lit room and a mean of $M = 8$ for the group in the well-lit room. The estimated standard error for the mean difference was $s_{(M_1 - M_2)} = 1.5$. With $n = 8$ scores in each sample, the independent-measures *t* statistic has $df = 14$. To have 95% confidence, we simply estimate that the *t* statistic for the sample mean difference is located somewhere in the middle 95% of all the possible *t* values. According to the *t* distribution table, with $df = 14$, 95% of the *t* values are located between $t = +2.145$ and $t = -2.145$. Using these values in the estimation equation, we obtain

$$\mu_1 - \mu_2 = M_1 - M_2 \pm ts_{(M_1 - M_2)}$$
$$= 12 - 8 \pm 2.145(1.5)$$
$$= 4 \pm 3.218$$

This produces an interval of values ranging from $4 - 3.218 = 0.782$ to $4 + 3.218 = 7.218$. Thus, our conclusion is that students who were tested in the dimly lit room had higher scores than those who were tested in a well-lit room, and the mean difference between the two populations is somewhere between 0.782 points and 7.218 points. Furthermore, we are 95% confident that the true mean difference is in this interval because the only value estimated during the calculations was the *t* statistic, and we are 95% confident that the *t* value is located in the middle 95% of the distribution. Finally, note that the confidence interval is constructed around the sample mean difference. As a result, the sample mean difference, $M_1 - M_2 = 12 - 8 = 4$ points, is located exactly in the center of the interval. ■

As with the confidence interval for the single-sample *t* (page 310), the confidence interval for an independent-measures *t* is influenced by a variety of factors other than the actual size of the treatment effect. In particular, the width of the interval depends on the percentage of confidence used so that a larger percentage produces a wider interval. Also, the width of the interval depends on the sample size, so that a larger sample produces a narrower interval. Because the interval width is related to sample size, the confidence interval is not a pure measure of effect size like Cohen's *d* or r^2.

■ Confidence Intervals and Hypothesis Tests

In addition to describing the size of a treatment effect, estimation can be used to get an indication of the *significance* of the effect. Example 10.6 presented an independent-measures research study examining the effect of room lighting on performance scores (cheating). Based on the results of the study, the 95% confidence interval estimated that the population mean difference for the two groups of students was between 0.782 and 7.218 points. The confidence interval estimate is shown in Figure 10.4. In addition to the confidence interval for $\mu_1 - \mu_2$, we have marked the spot where the mean difference is equal to zero. You should recognize that a mean difference of zero is exactly what would be predicted by the null hypothesis if we were doing a hypothesis test. You also should realize that a zero difference ($\mu_1 - \mu_2 = 0$) is *outside* the 95% confidence interval. In other words, $\mu_1 - \mu_2 = 0$ is not an acceptable value if we want 95% confidence in our estimate. To conclude that a value of zero is *not acceptable* with 95% confidence is equivalent to concluding that a value of zero is *rejected* with 95% confidence. This conclusion is equivalent to rejecting H_0 with $\alpha = .05$. On the other hand, if a mean difference of zero were included within the 95% confidence interval, then we would have to conclude that $\mu_1 - \mu_2 = 0$ is an acceptable value, which is the same as failing to reject H_0.

The hypothesis test for these data was conducted in Example 10.2 (page 334) and the decision was to reject H_0 with $\alpha = .05$.

FIGURE 10.4

The 95% confidence interval for the population mean difference ($\mu_1 - \mu_2$) from Example 10.6. Note that $\mu_1 - \mu_2 = 0$ is excluded from the confidence interval, indicating that a zero difference is not an acceptable value (H_0 would be rejected in a hypothesis test with $\alpha = .05$).

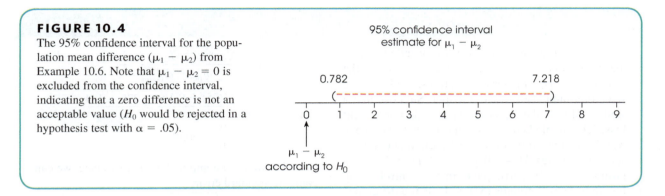

95% confidence interval estimate for $\mu_1 - \mu_2$

IN THE LITERATURE

Reporting the Results of an Independent-Measures *t* Test

Because the direction of the mean difference is described in the sentence, the t statistic can be reported as a positive value.

A research report typically presents the descriptive statistics followed by the results of the hypothesis test and measures of effect size (inferential statistics). In Chapter 4 (page 135), we demonstrated how the mean and the standard deviation are reported in APA format. In Chapter 9 (page 310), we illustrated the APA style for reporting the results of a *t* test. Now we use the APA format to report the results of Example 10.2, an independent-measures *t* test. A concise statement might read as follows:

The students who were tested in a dimly lit room reported higher performance scores ($M = 12$, $SD = 2.93$) than the students who were tested in the well-lit room ($M = 8$, $SD = 3.07$). The mean difference was significant, $t(14) = 2.67$, $p < .05$, $d = 1.33$.

You should note that standard deviation is not a step in the computations for the independent-measures *t* test presented in this chapter, yet it is useful when providing descriptive statistics for each treatment group. It is easily computed when doing the *t* test because you need *SS* and *df* for both groups to determine the pooled variance. Note that the format for reporting *t* is exactly the same as that described in Chapter 9 (page 311) and that the measure of effect size is reported immediately after the results of the hypothesis test. Box 10.2 describes how you can use the means, standard deviations, and size of samples that are reported in a published research article to compute *t*.

Also, as we noted in Chapter 9, if an exact probability is available from a computer analysis, it should be reported. For the data in Example 10.2, the computer analysis reports a probability value of $p = .018$ for $t = 2.67$ with $df = 14$. In the research report, this value would be included as follows:

The difference was significant, $t(14) = 2.67$, $p = .018$, $d = 1.33$.

Finally, if a confidence interval is reported to describe effect size, it appears immediately after the results from the hypothesis test. For the cheating behavior examples (Examples 10.2 and 10.6) the report would be as follows:

The difference was significant, $t(14) = 2.67$, $p = .018$, 95% CI [0.782, 7.218].

BOX 10.2 Computing *t* from Published Summary Statistics

Notice that you can compute *t* for an independent-measures study if you have access to only the sample means, standard deviations, and sample sizes in a published research paper. Suppose that a study evaluates the effect of a special diet on weight loss. The published study reports that the mean weight for the group of $n = 12$ participants on the special diet was $M = 10$ pounds ($SD = 5$) and a group of $n = 10$ participants in the control group lost only $M = 2$ pounds ($SD = 4$). Is the difference significant? The first step to answering this question is to compute the variance for each group. The standard deviation is the square root of variance, so to obtain variance you must square the standard deviation. Therefore,

$$\text{standard deviation} = \sqrt{\text{variance}}$$

and,

$$(\text{standard deviation})^2 = \text{variance}$$

Thus, variance, s^2, for the dieting group is 25 and s^2 for the control group is 16 pounds. Next, we can use Equation 10.3 to compute pooled variance as follows:

$$s_p^2 = \frac{df_{Diet}\, s_{Diet}^2 + df_{Control}\, s_{Control}^2}{df_{Diet} + df_{Control}}$$

$$= \frac{11(25) + 9(16)}{11 + 9} = \frac{275 + 144}{11 + 9}$$

$$= \frac{419}{20} = 20.95$$

Now that we have computed pooled variance, we can compute standard error:

$$s_{(M_{Diet} - M_{Control})} = \sqrt{\frac{s_p^2}{n_{Diet}} + \frac{s_p^2}{n_{Control}}}$$

$$= \sqrt{\frac{20.95}{12} + \frac{20.95}{10}}$$

$$= \sqrt{1.746 + 2.095} = \sqrt{3.84} = 1.96$$

and *t*:

$$t = \frac{(M_{Diet} - M_{Control}) - (\mu_{Diet} - \mu_{Contrrol})}{s_{(M_{Diet} - M_{Control})}}$$

$$= \frac{(10 - 2) - 0}{1.96} = \frac{8}{1.96} = 4.08$$

LEARNING CHECK   **LO9** **1.** A researcher obtains a mean of $M = 26$ for a sample in one treatment condition and $M = 28$ for a sample in another treatment. The pooled variance for the two samples is 16. What value would be obtained if Cohen's *d* were used to measure the effect size?

  a. $\frac{2}{16}$

  b. $\frac{4}{16}$

  c. $\frac{2}{4}$

  d. There is not enough information to determine Cohen's *d*.

LO10 **2.** Which of the following is not an accurate description of a confidence interval for a mean difference using the independent-measures *t* statistic?

  a. The sample mean difference, $M_1 - M_2$, will be located in the center of the interval.

  b. If other factors are held constant, the width of the interval will decrease if the sample size is increased.

  c. If other factors are held constant, the width of the interval will increase if the percentage of confidence is increased.

  d. If other factors are held constant, the width of the interval will increase if the difference between the two sample means is increased.

LO11 **3.** Which of the following accurately describes the 95% confidence interval for an independent-measures study for which a hypothesis test concludes that there is no significant mean difference with $\alpha = .05$?

    a. The confidence interval will include the value 0.

    b. The confidence interval will not include the value 0.

    c. The confidence interval will not include the value $M_1 - M_2$.

    d. None of the other options is accurate.

LO12 **4.** The results of a hypothesis test with an independent-measures *t* statistic are reported as follows: $t(22) = 2.48$, $p < .05$, $d = 0.27$. Which of the following is an accurate description of the study and the result?

    a. The study used a total of 24 participants and the null hypothesis was rejected.

    b. The study used a total of 22 participants and the null hypothesis was rejected.

    c. The study used a total of 24 participants and the null hypothesis was not rejected.

    d. The study used a total of 22 participants and the null hypothesis was not rejected.

ANSWERS   **1.** c **2.** d **3.** a **4.** a

10-5 | The Role of Sample Variance and Sample Size in the Independent-Measures *t* Test

LEARNING OBJECTIVE

13. Describe how sample size and sample variance influence the outcome of a hypothesis test and measures of effect size for the independent-measures *t* statistic.

In Chapter 9 (page 302) we identified several factors that can influence the outcome of a hypothesis test. Two factors that play important roles are the variability of the scores and the size of the samples. Both factors influence the magnitude of the estimated standard error in the denominator of the *t* statistic. The standard error is directly related to sample variance so that larger variance leads to larger error. As a result, larger variance produces a smaller value for the *t* statistic (closer to zero) and reduces the likelihood of finding a significant result. By contrast, the standard error is inversely related to sample size (larger size leads to smaller error). Thus, a larger sample produces a larger value for the *t* statistic (farther from zero) and increases the likelihood of rejecting H_0.

Although variance and sample size both influence the hypothesis test, only variance has a large influence on measures of effect size such as Cohen's *d* and r^2; larger variance produces smaller measures of effect size. Sample size, on the other hand, has no effect on the value of Cohen's *d* and only a small influence on r^2.

The following example provides a visual demonstration of how large sample variance can obscure a mean difference between samples and lower the likelihood of rejecting H_0 for an independent-measures study.

EXAMPLE 10.7   We will use the data in Figure 10.5 to demonstrate the influence of sample variance. The figure shows the results from a research study comparing two treatments. Notice that the

FIGURE 10.5

Two sample distributions representing two different treatments. These data show a significant difference between treatments, $t(16) = 8.62, p < .01$, and both measures of effect size indicate a very large treatment effect, $d = 4.10$ and $r^2 = 0.82$.

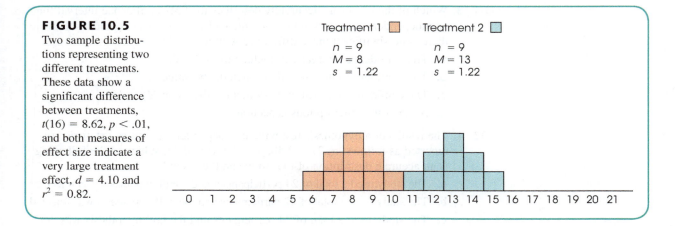

Treatment 1 ☐   Treatment 2 ☐

Treatment 1	Treatment 2
$n = 9$	$n = 9$
$M = 8$	$M = 13$
$s = 1.22$	$s = 1.22$

study uses two separate samples, each with $n = 9$, and there is a 5-point mean difference between the two samples: $M = 8$ for Treatment 1 and $M = 13$ for Treatment 2. Also notice that there is a clear difference between the two distributions; the scores for Treatment 2 are clearly higher than the scores for Treatment 1.

For the hypothesis test, the data produce a pooled variance of 1.50 and an estimated standard error of 0.58. The *t* statistic is

$$t = \frac{\text{mean difference}}{\text{estimated standard error}} = \frac{5}{0.58} = 8.62$$

With $df = 16$, this value is far into the critical region (for $\alpha = .05$ or $\alpha = .01$), so we reject the null hypothesis and conclude that there is a significant difference between the two treatments.

Now consider the effect of increasing sample variance. Figure 10.6 shows the results from a second research study comparing two treatments. Notice that there are still $n = 9$ scores in each sample, and the two sample means are still $M = 8$ and $M = 13$. However, the sample variances have been greatly increased: Each sample now has $s^2 = 44.25$ as compared with $s^2 = 1.5$ for the data in Figure 10.5. Notice that the increased variance means

Treatment 1 ☐   Treatment 2 ☐

Treatment 1	Treatment 2
$n = 9$	$n = 9$
$M = 8$	$M = 13$
$s = 6.65$	$s = 6.65$

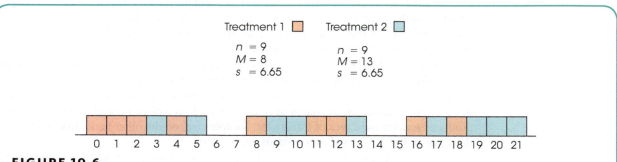

FIGURE 10.6

Two sample distributions representing two different treatments. These data show exactly the same mean difference as the scores in Figure 10.5; however the variance has been greatly increased. With the increased variance, there is no longer a significant difference between treatments, $t(16) = 1.59, p > .05$, and both measures of effect size are substantially reduced, $d = 0.75$ and $r^2 = 0.14$.

that the scores are now spread out over a wider range, with the result that the two samples are mixed together without any clear distinction between them.

The absence of a clear difference between the two samples is supported by the hypothesis test. The pooled variance is 44.25, the estimated standard error is 3.14, and the independent-measures *t* statistic is

$$t = \frac{\text{mean difference}}{\text{estimated standard error}} = \frac{5}{3.14} = 1.59$$

With $df = 16$ and $\alpha = .05$, this value is not in the critical region. Therefore, we fail to reject the null hypothesis and conclude that there is no significant difference between the two treatments. Although there is still a 5-point difference between sample means (as in Figure 10.6), the 5-point difference is not significant with the increased variance. In general, large sample variance can obscure any mean difference that exists in the data and reduces the likelihood of obtaining a significant difference in a hypothesis test.   ■

■ Error and the Role of Individual Differences

We have seen that one factor that contributes to size of standard error is the amount of variance in the sample data. (Another is sample size, *n*.) Individual differences are one source of this variability. In Chapter 1 we introduced participant variables as characteristics that can vary from person to person. These might consist of individual differences in, for example, personality, attitudes, past experiences, abilities, and motivation, just to name a few. The differences that exist among a sample of participants within each treatment group will have an influence on the variance (and standard deviation) of that treatment group. Thus, individual differences are one factor that have an effect on standard error.

In an independent-measures *t* test there are two separate groups of participants, one group of participants for Treatment 1 and another for Treatment 2. Ideally a random sample is selected for the study, and then the participants are randomly assigned to the treatment groups. Random assignment helps ensure that there is no bias between the groups in terms of individual differences. For example, it would help prevent one group from having, on average, more motivation than another group. It is possible, however, that assignment of participants to groups will be unintentionally biased and that the groups differ at the beginning of the experiment prior to administration of the treatments. Pretesting participants prior to treatment for differences in certain variables (such as attitudes or motivation) is one way to ensure the groups are equivalent at the start. Another method is using a matched sample study, which will be covered in Chapter 11.

LEARNING CHECK   **LO13**   **1.** Which of the following accurately describes how the outcome of a hypothesis test and measures of effect size with the independent-measures *t* statistic are affected when sample size is increased?

    a. The likelihood of rejecting the null hypothesis and measures of effect size both increase.

    b. The likelihood of rejecting the null hypothesis and measures of effect size both decrease.

    c. The likelihood of rejecting the null hypothesis increases and there is little or no effect on measures of effect size.

    d. The likelihood of rejecting the null hypothesis decreases and there is little or no effect on measures of effect size.

LO13   2. Which of the following accurately describes how the outcome of a hypothesis test and measures of effect size with the independent-measures *t* statistic are affected when sample variance increases?

    a. The likelihood of rejecting the null hypothesis and measures of effect size both increase.

    b. The likelihood of rejecting the null hypothesis and measures of effect size both decrease.

    c. The likelihood of rejecting the null hypothesis increases and there is little or no effect on measures of effect size.

    d. The likelihood of rejecting the null hypothesis decreases and there is little or no effect on measures of effect size.

LO13   3. Which of the following sets of data would produce the largest value for an independent-measures *t* statistic?

    a. Two sample means of 10 and 12 with sample variances of 20 and 25.

    b. Two sample means of 10 and 12 with variances of 120 and 125.

    c. Two sample means of 10 and 20 with variances of 20 and 25.

    d. Two sample means of 10 and 20 with variances of 120 and 125.

ANSWERS   **1. c   2. b   3. c**

SUMMARY

1. The independent-measures *t* statistic uses the data from two separate samples to draw inferences about the mean difference between two populations or between two different treatment conditions.

2. The formula for the independent-measures *t* statistic has the same structure as the original *z*-score or the single-sample *t*:

$$z \text{ or } t = \frac{\text{sample statistic} - \text{hypothesized population parameter}}{\text{a measure of standard error}}$$

For the independent-measures *t*, the sample statistic is the sample mean difference $(M_1 - M_2)$. The population parameter is the population mean difference, $(\mu_1 - \mu_2)$. The estimated standard error for the sample mean difference is computed by combining the errors for the two sample means. The resulting formula is

$$t = \frac{(M_1 - M_2) - (\mu_1 - \mu_2)}{s_{(M_1 - M_2)}}$$

where the estimated standard error is

$$s_{(M_1 - M_2)} = \sqrt{\frac{s_p^2}{n_1} + \frac{s_p^2}{n_2}}$$

The pooled variance in the formula, s_p^2, is the weighted mean of the two sample variances:

$$s_p^2 = \frac{SS_1 + SS_2}{df_1 + df_2}$$

This *t* statistic has degrees of freedom determined by the sum of the *df* values for the two samples:

$$df = df_1 + df_2$$
$$= (n_1 - 1) + (n_2 - 1)$$

3. For hypothesis testing, the null hypothesis states that there is no difference between the two population means:

$$H_0: \mu_1 = \mu_2 \text{ or } \mu_1 - \mu_2 = 0$$

4. When a hypothesis test with an independent-measures *t* statistic indicates a significant difference, it is recommended that you also compute a measure of the effect size. One measure of effect size is Cohen's *d*, which is a standardized measure of the mean difference. For the independent-measures *t* statistic, Cohen's *d* is estimated as follows:

$$\text{estimated } d = \frac{M_1 - M_2}{\sqrt{s_p^2}}$$

A second common measure of effect size is the percentage of variance accounted for by the treatment effect. This measure is identified by r^2 and is computed as

$$r^2 = \frac{t^2}{t^2 + df}$$

5. An alternative method for describing the size of the treatment effect is to construct a confidence interval for the population mean difference, $\mu_1 - \mu_2$. The confidence interval uses the independent-measures t equation, solved for the unknown mean difference:

$$\mu_1 - \mu_2 = M_1 - M_2 \pm ts_{(M_1 - M_2)}$$

First, select a level of confidence and then look up the corresponding t values. For example, for 95%

confidence, use the range of t values that determine the middle 95% of the distribution. The t values are then used in the equation along with the values for the sample mean difference and the standard error, which are computed from the sample data.

6. Appropriate use and interpretation of the t statistic using pooled variance require that the data satisfy the homogeneity of variance assumption. This assumption stipulates that the two populations have equal variances. An informal test of the assumption can be made by verifying that the two sample variances are approximately equal. Hartley's F-max test provides a statistical technique for determining whether the data satisfy the homogeneity assumption.

KEY TERMS

independent-measures research design *or* between-subjects design (325)

repeated-measures research design *or* within-subjects design (326)

independent-measures t statistic (327)

estimated standard error of $M_1 - M_2$ (328)

pooled variance (330)

homogeneity of variance (337)

FOCUS ON PROBLEM SOLVING

1. As you learn more about different statistical methods, one basic problem will be deciding which method is appropriate for a particular set of data. Fortunately, it is easy to identify situations in which the independent-measures t statistic is used. First, the data will always consist of two separate samples (two ns, two Ms, two SSs, and so on). Second, this t statistic is always used to answer questions about a mean difference: On the average, is one group different (better, faster, smarter) than the other group? If you examine the data and identify the type of question that a researcher is asking, you should be able to decide whether an independent-measures t is appropriate.

2. When computing an independent-measures t statistic from sample data, we suggest that you routinely divide the formula into separate stages rather than trying to do all the calculations at once. First, find the pooled variance. Second, compute the standard error. Third, compute the t statistic.

3. One of the most common errors for students involves confusing the formulas for pooled variance and standard error. When computing pooled variance, you are "pooling" the two samples together into a single variance. This variance is computed as a *single fraction*, with two SS values in the numerator and two df values in the denominator. When computing the standard error, you are adding the error from the first sample and the error from the second sample. These two separate errors are added as *two separate fractions* under the square root symbol.

DEMONSTRATION 10.1

THE INDEPENDENT-MEASURES t TEST

In a study of jury behavior, two samples of participants were provided details about a trial in which the defendant was obviously guilty. Although Group 2 received the same details as Group 1, the second group was also told that some evidence had been withheld from the jury by the judge. Later the participants were asked to recommend a jail sentence. The length of term suggested by each participant is presented here. Is there a significant difference between the two groups in their responses?

Group 1	Group 2	
4	3	
4	7	
3	8	for Group 1: $M = 3$ and $SS = 16$
2	5	
5	4	for Group 2: $M = 6$ and $SS = 24$
1	7	
1	6	
4	8	

There are two separate samples in this study. Therefore, the analysis will use the independent-measures *t* test.

STEP 1   **State the hypotheses, and select an alpha level.**

H_0: $\mu_1 - \mu_2 = 0$ (For the population, knowledge of withheld evidence has no effect on the suggested sentence.)

H_1: $\mu_1 - \mu_2 \neq 0$ (For the population, knowledge of withheld evidence has an effect on the jury's response.)

We will set the level of significance to $\alpha = .05$, two tails.

STEP 2   **Identify the critical region.**   For the independent-measures *t* statistic, degrees of freedom are determined by

$$df = df_1 + df_2$$
$$= 7 + 7$$
$$= 14$$

The *t* distribution table is consulted for a two-tailed test with $\alpha - .05$ and $df = 14$. The critical *t* values are $+2.145$ and -2.145.

STEP 3   **Compute the test statistic.**   As usual, we recommended that the calculation of the *t* statistic be separated into three stages.

Pooled variance   For these data, the pooled variance equals

$$s_p^2 = \frac{SS_1 + SS_2}{df_1 + df_2} = \frac{16 + 24}{7 + 7} = \frac{40}{14} = 2.86$$

Estimated standard error   Now we can calculate the estimated standard error for mean differences.

$$s_{(M_1 - M_2)} = \sqrt{\frac{s_p^2}{n_1} + \frac{s_p^2}{n_2}} = \sqrt{\frac{2.86}{8} + \frac{2.86}{8}} = \sqrt{0.358 + 0.358} = \sqrt{0.716} = 0.85$$

The *t* statistic   Finally, the *t* statistic can be computed.

$$t = \frac{(M_1 - M_2) - (\mu_1 - \mu_2)}{s_{(M_1 - M_2)}} = \frac{(3 - 6) - 0}{0.85} = \frac{-3}{0.85} = -3.53$$

STEP 4   **Make a decision about H_0, and state a conclusion.** The obtained t value of -3.53 falls in the critical region of the left tail (critical $t = \pm 2.145$). Therefore, the null hypothesis is rejected. The participants who were informed about the withheld evidence gave significantly longer sentences, $t(14) = -3.53$, $p > .05$, two tails.

DEMONSTRATION 10.2

EFFECT SIZE FOR THE INDEPENDENT-MEASURES t

Remember, Cohen's d is always reported as a positive number, regardless of the sign of the numerator (see Chapter 9, page 304).

We will estimate Cohen's d and compute r^2 for the jury decision data in Demonstration 10.1. For these data, the two sample means are $M_1 = 3$ and $M_2 = 6$, and the pooled variance is 2.86. Therefore, our estimate of Cohen's d is

$$\text{estimated } d = \frac{M_1 - M_2}{\sqrt{s_p^2}} = \frac{3 - 6}{\sqrt{2.86}} = \frac{-3}{1.69} = 1.78$$

With a t value of $t = 3.53$ and $df = 14$, the percentage of variance accounted for is

$$r^2 = \frac{t^2}{t^2 + df} = \frac{(3.53)^2}{(3.53)^2 + 14} = \frac{12.46}{26.46} = 0.47 \quad \text{(or 47\%)}$$

SPSS®

Are you just wasting your time while playing video games? Maybe not. Recent research suggests that playing action video games might improve your hand-eye coordination (Li, Chen, & Chen, 2016; Experiment 2). Researchers recruited participants who reported frequently playing action video games and participants who reported that they rarely or never played action video games. All participants then completed a computerized hand-eye coordination task in which the researchers measured participants' error in using a joystick to hold a moving dot in the center of a computer screen. Data like those observed by the researchers are listed below (lower numbers indicate less error and better hand-eye coordination).

Action Gamers		Non-Action Gamers	
13.0	9.5	16.0	10.5
15.0	11.0	20.5	18.0
9.0	14.5	13.0	10.5
13.0	13.0	15.0	15.0
6.0	11.0	15.0	17.5
18.0	15.0	16.5	15.0
11.0	11.0	15.0	15.0
11.0	11.0	15.0	20.5

The following steps will demonstrate how to use SPSS to perform an independent-samples t test to test the hypothesis that Action Gamers show lower error in a hand-eye coordination task.

Data Entry

1. Use the **Variable View** of the data editor to create two new variables for the data above. Enter "error" in the **Name** field of the first variable. Select **Numeric** in the **Type** field and **Scale** in the **Measure** field. Enter a brief, descriptive title for the variable in the **Label** field (here, "Error in holding dot position" was used). For the second variable, enter "group". Select **String** in the **Type** field and **Nominal** in the **Measure** field. Use "self-reported gaming" in the **Label** field.

2. Use the **Data View** of the data editor to enter the scores. The scores are entered in what is called *stacked format*, which means that all the scores from *both samples* are entered in one column of the data editor (named "error"). Enter the error scores for the Non-Action Gamers sample directly beneath the scores from the Action Gamers sample with no gaps or extra spaces.

3. Values are then entered into a second column ("group") to identify the sample or treatment condition corresponding to each of the scores. For example, enter "gamer" beside each score from the sample of Action Gamers and enter a "nongamer" beside each score from the sample of Non-Action Gamers. After you have successfully entered your data, your SPSS Data View should appear as below.

*Untitled3 [DataSet2] - IBM SPSS Statistics Data Editor

File   Edit   View   Data   Transform   Analyze

29 :

	error	group	var
1	13.00	gamer	
2	15.00	gamer	
3	9.00	gamer	
4	13.00	gamer	
5	6.00	gamer	
6	18.00	gamer	
7	11.00	gamer	
8	11.00	gamer	
9	9.50	gamer	
10	11.00	gamer	
11	14.50	gamer	
12	13.00	gamer	
13	11.00	gamer	
14	15.00	gamer	
15	11.00	gamer	
16	11.00	gamer	
17	16.00	nongamer	
18	20.50	nongamer	
19	13.00	nongamer	
20	15.00	nongamer	
21	15.00	nongamer	
22	16.50	nongamer	
23	15.00	nongamer	
24	15.00	nongamer	
25	10.50	nongamer	
26	18.00	nongamer	
27	10.50	nongamer	
28	15.00	nongamer	
29	17.50	nongamer	
30	15.00	nongamer	
31	15.00	nongamer	
32	20.50	nongamer	
33			

Data Analysis

1. Click **Analyze** on the tool bar, select **Compare Means**, and click on **Independent-Samples T Test**.
2. Highlight the column label for the set of scores ("Error in holding dot position") in the left box and click the arrow to move it into the **Test Variable(s)** box.
3. Highlight the label from the column containing the group labels ("Self-reported gaming") in the left box and click the arrow to move it into the **Group Variable** box.
4. Click **Define Groups**.
5. Assuming that you used "gamer" and "nongamer" to identify the two sets of scores, enter those labels into the appropriate group boxes.
6. Click **Continue**. After you have correctly identified the variables, the Independent-Samples T Test window should be like the figure below.

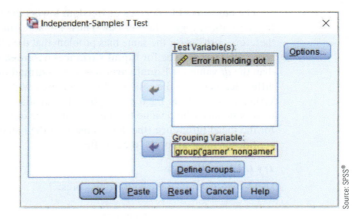

7. In addition to performing the hypothesis test, the program will compute a confidence interval for the population mean difference. The confidence level is automatically set at 95% but you can select **Options** and change the percentage.
8. Click **OK**.

SPSS Output

The output includes a table of sample statistics with the mean, standard deviation, and standard error of the mean for each group. A second table, which is split into two sections, begins with the results of Levene's test for homogeneity of variance. This test should *not* be significant. You do not want the two variances to be different because the independent-samples *t* test assumes that the sample variances are equal (i.e., homogeneity of variance). In this example, Levene's test is not significant ($p = .773$). Next, the results of the independent-measures *t* test are presented using two different assumptions. The top row shows the outcome assuming equal variances, using the pooled variance to compute *t*. If the Levene's test is not significant, use the top row outcomes to report computed *t*. The second row does not assume equal variances and computes the *t* statistic using an alternative method. The alternative procedure uses two steps to avoid the homogeneity of variance assumption:

1. The standard error is computed using the two separate sample variances as in Equation 10.1.
2. The value of degrees of freedom for the *t* statistic is adjusted using the following equation:

$$df = \frac{(V_1 + V_2)^2}{\dfrac{V_1^2}{n_1 - 1} + \dfrac{V_2^2}{n_2 - 1}} \text{ where } V_1 = \frac{s_1^2}{n_1} \text{ and } V_2 = \frac{s_2^2}{n_2}$$

T-Test

[DataSet1]

Group Statistics

	self-reported gaming status	N	Mean	Std. Deviation	Std. Error Mean
Error in holding dot position	gamer	16	12.0000	2.83431	.70858
	nongamer	16	15.5000	2.82253	.70563

Independent Samples Test

		Levene's Test for Equality of Variances		t-test for Equality of Means					95% Confidence Interval of the Difference	
		F	Sig.	t	df	Sig. (2-tailed)	Mean Difference	Std. Error Difference	Lower	Upper
Error in holding dot position	Equal variances assumed	.085	.773	-3.500	30	.001	-3.50000	1.00000	-5.54227	-1.45773
	Equal variances not assumed			-3.500	29.999	.001	-3.50000	1.00000	-5.54227	-1.45773

Source: SPSS®

The adjustment to degrees of freedom lowers the value of *df*, which pushes the boundaries for the critical region farther out. Thus, the adjustment makes the test more demanding and therefore corrects for the same bias problem that the pooled variance attempts to avoid. You will notice that the *df* value for **Equal variances assumed** (*df* = 30) is greater and less conservative than the *df* value for **Equal variances not assumed** (*df* = 29.999). Importantly, the size of the difference between these two values increases with greater violations of the homogeneity of variance assumption. Each row reports the calculated *t* value, the degrees of freedom, the level of significance (the *p* value for the test), the size of the mean difference, and the standard error for the mean difference (the denominator of the *t* statistic). Finally, the output includes a 95% confidence interval for the mean difference.

Try It Yourself

Use SPSS to analyze the following scores.

Action Gamers		Non-Action Gamers	
20.0	15.0	26.0	23.5
9.0	15.5	25.0	14.0
10.0	16.0	27.5	20.5
13.0	15.0	12.5	15.0
14.0	16.0	19.5	20.0
14.0	14.5	12.0	28.0
15.0	21.0	17.0	16.5
15.0	17.0	23.0	20.0

You should find that SPSS reports a significant difference between the Action Gamers and Non-Action Gamers, *t*(30) = −3.33, *p* = .002. You should also notice that the two samples differ with respect to their variances (i.e., Levene's test reveals a *p* of .019) and that the **Equal variances not assumed** row uses a smaller value for degrees of freedom (*df* = 23.96) than the **Equal variances assumed** row (*df* = 30). It is more conservative to report the ***Equal variances not assumed*** when reporting the results of the *t*-test.

PROBLEMS

1. Describe the basic characteristics that define an independent-measures, or a between-subjects, research study.

2. Describe what is measured by the estimated standard error in the bottom of the independent-measures *t* statistic.

3. One sample has *SS* = 72 and a second sample has *SS* = 24.
   a. If *n* = 7 for both samples, find each of the sample variances and compute the pooled variance. Because the samples are the same size, you should find that the pooled variance is exactly halfway between the two sample variances.

b. Now assume that $n = 7$ for the first sample and $n = 11$ for the second. Again, calculate the two sample variances and the pooled variance. You should find that the pooled variance is closer to the variance for the larger sample.

4. One sample has $SS = 80$ and a second sample has $SS = 48$.

a. If $n = 17$ for both samples, find each of the sample variances, and calculate the pooled variance. Because the samples are the same size, you should find that the pooled variance is exactly halfway between the two sample variances.

b. Now assume that $n = 17$ for the first sample and $n = 5$ for the second. Again, calculate the two sample variances and the pooled variance. You should find that the pooled variance is closer to the variance for the larger sample.

5. Two separate samples, each with $n = 9$ individuals, receive different treatments. After treatment, the first sample has $SS = 546$ and the second has $SS = 606$.

a. Find the pooled variance for the two samples.

b. Compute the estimated standard error for the sample mean difference.

c. If the sample mean difference is 8 points, is this enough to reject the null hypothesis and conclude that there is a significant difference for a two-tailed test at the .05 level?

6. Two separate samples receive different treatments. After treatment, the first sample has $n = 5$ with $SS = 60$, and the second has $n = 9$ with $SS = 84$.

a. Compute the pooled variance for the two samples.

b. Calculate the estimated standard error for the sample mean difference.

c. If the sample mean difference is 7 points, is this enough to reject the null hypothesis using a two-tailed test with $\alpha = .05$?

7. Research results suggest a relationship between the TV viewing habits of 5-year-old children and their future performance in high school. For example, Anderson, Huston, Wright, and Collins (1998) report that high school students who regularly watched *Sesame Street* as children had better grades in high school than their peers who did not watch *Sesame Street*. Suppose that a researcher intends to examine this phenomenon using a sample of 20 high school students.

The researcher first surveys the students' parents to obtain information on the family's TV viewing habits during the time that the students were 5 years old. Based on the survey results, the researcher selects a sample of $n = 10$ students with a history of watching *Sesame Street* and a sample of $n = 10$ students who did not watch the program. The average high school

grade is recorded for each student and the data are as follows:

Average High School Grade			
Watched *Sesame Street*		Did Not Watch *Sesame Street*	
86	99	90	79
87	97	89	83
91	94	82	86
97	89	83	81
98	92	85	92
$n = 10$		$n = 10$	
$M = 93$		$M = 85$	
$SS = 200$		$SS = 160$	

Use an independent-measures t test with $\alpha = .01$, two-tailed, to determine whether there is a significant difference between the two types of high school student.

8. Does posting calorie content for menu items affect people's choices in fast-food restaurants? According to results obtained by Elbel, Gyamfi, and Kersh (2011), the answer is no. The researchers monitored the calorie content of food purchases for children and adolescents in four large fast-food chains before and after mandatory labeling began in New York City. Although most of the adolescents reported noticing the calorie labels, apparently the labels had no effect on their choices. Data similar to the results obtained show an average of $M = 786$ calories per meal with $s = 85$ for $n = 100$ children and adolescents before the labeling, compared to an average of $M = 772$ calories with $s = 91$ for a similar sample of $n = 100$ children after the mandatory posting.

a. Use a two-tailed test with $\alpha = .05$ to determine whether the mean number of calories after the posting is significantly different than before calorie content was posted.

b. Calculate r^2 to measure effect size for the mean difference.

9. A long history of psychology research has demonstrated that memory is usually improved by studying material on multiple occasions rather than one time only. This effect is commonly known as distributed practice, or spacing effects. In a recent paper examining this effect, Cepeda et al. (2008) looked at the influence of different delays or gaps between study sessions. The results suggest that optimal long-term memory occurs when the study periods are spaced one to three weeks apart. In one part of the study, a group of participants studied a set of obscure trivia facts one day, returned the next day for a second study period, and then was tested five weeks later. A second group went through the same procedure but had a one-week gap between the two study sessions. The following data are similar to the results obtained in the study. Do the data indicate a significant difference between the two study conditions? Test with $\alpha = .05$, two-tailed.

One-Day Gap between Study Sessions	One-Week Gap between Study Sessions
$n = 20$	$n = 20$
$M = 26.4$	$M = 29.6$
$SS = 395$	$SS = 460$

10. Recent research has shown that creative people are more likely to cheat than their less-creative counterparts (Gino & Ariely, 2012). Participants in the study first completed creativity assessment questionnaires and then returned to the lab several days later for a series of tasks. One task was a multiple-choice general knowledge test for which the participants circled their answers on the test sheet. Afterward, they were asked to transfer their answers to a bubble sheet for computer scoring. However, the experimenter admitted that the wrong bubble sheet had been copied so that the correct answers were still faintly visible. Thus, the participants had an opportunity to cheat and inflate their test scores. Higher scores were valuable because participants were paid based on the number of correct answers. However, the researchers had secretly coded the original tests and the bubble sheets so that they could measure the degree of cheating for each participant. Assuming that the participants were divided into two groups based on their creativity scores, the following data are similar to the cheating scores obtained in the study.

High-Creativity Participants	Low-Creativity Participants
$n = 27$	$n = 27$
$M = 7.41$	$M = 4.78$
$SS = 749.5$	$SS = 830$

a. Use a one-tailed test with $\alpha = .05$ to determine whether these data are sufficient to conclude that high-creativity people are more likely to cheat than people with lower levels of creativity.
b. Compute Cohen's *d* to measure the size of the effect.
c. Write a sentence demonstrating how the results from the hypothesis test and the measure of effect size would appear in a research report.

11. Anxiety affects our ability to make decisions. Remmers and Zander (2018) demonstrated that anxiety also prevents us from intuiting about our environments. In their experiment, 111 participants were randomly assigned to receive either an anxiety-inducing statement (e.g., "Safety is guaranteed neither in our neighborhoods nor in our own homes") accompanied by a photograph of a dangerous situation, or an emotionally neutral statement (e.g., "A rolling pin is a kitchen tool that helps to extend dough") accompanied by an innocuous image.

Researchers then measured participants' intuition index, which assesses ability to identify a word (e.g., "sea") that was semantically related to a list of three words (e.g., "foam," "deep," and "salt"). Researchers observed reduced intuition among subjects who received anxiety-inducing statements and imagery than among those who received innocuous stimuli. Data like those observed by the researchers are listed below:

Neutral	Anxiety-Inducing
19	0
12	10
11	1
11	9
12	6
7	4

a. Are the test scores significantly lower for the participants who received anxiety-inducing statements? Use a two-tailed test with $\alpha = .05$.
b. Compute the value of r^2 (percentage of variance accounted for) for these data.

12. Positive events are great, but recent research suggests that unexpected positive outcomes (e.g., an unseasonably sunny day) predict greater-than-normal amounts of risk-taking and gambling (Otto, Fleming, & Glimcher, 2016). Researchers demonstrated this by comparing lottery sales—indicative of risk-taking—on normal days with lottery sales on days when some unexpected positive event occurred in the city. They observed increased sales after unexpected positive outcomes. Suppose that a researcher extends this observation to the laboratory and randomly assigns participants to two groups. Group 1 receives an unexpectedly large payment for participating and Group 2 receives the expected amount of compensation. The researcher then measures how much money the participants are willing to gamble in a game of chance.

Unexpected Positive Outcome	Expected Outcome
$n = 16$	$n = 16$
$M = 5.75$	$M = 5.00$
$SS = 6.5$	$SS = 10.0$

Test the one-tailed hypothesis that an unexpected positive outcome increased the amount of money that participants were willing to gamble. Use $\alpha = .01$.

13. Binge-watching a television show might not be the best way to enjoy a television series (Horvath, Horton, Lodge, & Hattie, 2017). Participants in an experiment watched an entire television series in the laboratory during either daily one-hour sessions or a single binge session. Participants were asked to rate their enjoyment

of the television series on a scale of 0–100. Data like those observed by the authors are listed below.

Binge-watched	Daily-watched
87	84
71	100
73	87
86	97
78	92

a. Test the hypothesis that binge-watching the television series resulted in less enjoyment of the show. Use $\alpha = .05$, two-tailed.
b. Compute Cohen's d to measure the size of the effect.
c. Write the results as they would appear in a scientific journal article.

14. What causes us to overeat? One surprising factor might be the material of the plate on which our food is served. Williamson, Block, and Keller (2016) gave $n = 68$ participants two donuts each and measured the amount of food that was wasted by each participant. In an independent-samples design, participants received their donuts either on a disposable paper plate or on a reusable plastic plate. Data like those observed by the authors are listed below.

Paper Plate (Grams of Wasted Food)	Plastic Plate (Grams of Wasted Food)
37	34
35	31
34	36
36	30
40	34
34	33
33	37
39	29

a. Test the hypothesis that participants who received donuts on a paper plate wasted more food than participants who were served donuts on a plastic, reusable plate. Use $\alpha = .05$, two-tailed.
b. Construct a 95% confidence interval to estimate the size of the mean difference.
c. Write the results as they would appear in a scientific journal article.

15. In a classic study in the area of problem solving, Katona (1940) compared the effectiveness of two methods of instruction. One group of participants was shown the exact, step-by-step procedure for solving a problem and was required to memorize the solution. Participants in a second group were encouraged to study the problem and find the solution on their own. They were given helpful hints and clues, but the exact solution was never explained. The study included the problem in the following figure showing a pattern of five squares made

of matchsticks. The problem is to change the pattern into exactly four squares by moving only three matches. (All matches must be used, none can be removed, and all the squares must be the same size.) After three weeks, both groups returned to be tested again. The two groups did equally well on the matchstick problem they had learned earlier. But when they were given new problems (similar to the matchstick problem), the memorization group had much lower scores than the group who explored and found the solution on their own. The following data demonstrate this result.

Find a Solution on Your Own	Memorization of the Solution
$n = 8$	$n = 8$
$M = 10.50$	$M = 6.16$
$SS = 108$	$SS = 116$

a. Is there a significant difference in performance on new problems for these two groups? Use a two-tailed test with $\alpha = .05$.
b. Construct a 90% confidence interval to estimate the size of the mean difference.

Incidentally, if you still have not discovered the solution to the matchstick problem, keep trying. According to Katona's results, it would be very poor teaching strategy for us to give you the answer. If you still have not discovered the solution, however, check Appendix C, page 627.

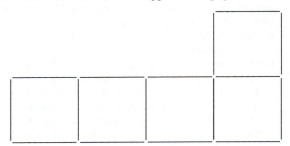

16. A researcher conducts an independent-measures study comparing two treatments and reports the t statistic as $t(20) = 2.09$.
   a. How many individuals participated in the entire study?
   b. Using a two-tailed test with $\alpha = .05$, is there a significant difference between the two treatments?
   c. Using a two-tailed test with $\alpha = .01$, is there a significant difference between the two treatments?
   d. Compute r^2 to measure the percentage of variance accounted for by the treatment effect.

17. In a recent study, Piff, Kraus, Côté, Cheng, and Keltner (2010) found that people from lower socioeconomic classes tend to display greater prosocial behavior than their higher-class counterparts. In one part of the study, participants played a game with an anonymous

partner. Part of the game involved sharing points with the partner. The lower economic class participants were significantly more generous with their points compared with the upper-class individuals. Results similar to those found in the study show that $n = 12$ lower-class participants shared an average of $M = 5.2$ points with $SS = 11.91$, compared to an average of $M = 4.3$ with $SS = 9.21$ for the $n = 12$ upper-class participants.

    a. Are the data sufficient to conclude that there is a significant mean difference between the two economic populations? Use a two-tailed test with $\alpha = .05$.

    b. Construct a 90% confidence interval to estimate the size of the population mean difference.

18. Describe the homogeneity of variance assumption and explain why it is important for the independent-measures *t* test.

19. If other factors are held constant, explain how each of the following influences the value of the independent-measures *t* statistic, the likelihood of rejecting the null hypothesis, and the magnitude of measures of effect size:

    a. Increasing the number of scores in each sample.

    b. Increasing the variance for each sample.

20. As noted on page 332, when the two population means are equal, the estimated standard error for the independent-measures *t* test provides a measure of how much difference to expect between two sample means. For each of the following situations, assume that $\mu_1 = \mu_2$ and calculate how much difference should be expected between the two sample means.

    a. One sample has $n = 6$ scores with $SS = 500$ and the second sample has $n = 12$ scores with $SS = 524$.

    b. One sample has $n = 6$ scores with $SS = 600$ and the second sample has $n = 12$ scores with $SS = 696$.

    c. In Part b, the samples have larger variability (bigger SS values) than in Part a, but the sample sizes are unchanged. How does larger variability affect the magnitude of the standard error for the sample mean difference?

21. Two samples are selected from the same population. For each of the following, calculate how much difference is expected, on average, between the two sample means.

    a. One sample has $n = 6$, the second has $n = 10$, and the pooled variance is 135.

    b. One sample has $n = 12$, the second has $n = 15$, and the pooled variance is 135.

    c. In Part b, the sample sizes are larger but the pooled variance is unchanged. How does larger sample size affect the magnitude of the standard error for the sample mean difference?

22. For each of the following, assume that the two samples are obtained from populations with the same mean, and calculate how much difference should be expected, on average, between the two sample means.

    a. Each sample has $n = 7$ scores with $s^2 = 142$ for the first sample and $s^2 = 110$ for the second. (*Note*: Because the two samples are the same size, the pooled variance is equal to the average of the two sample variances.)

    b. Each sample has $n = 28$ scores with $s^2 = 142$ for the first sample and $s^2 = 110$ for the second.

    c. In Part b, the two samples are bigger than in Part a, but the variances are unchanged. How does sample size affect the size of the standard error for the sample mean difference?

23. For each of the following, calculate the pooled variance and the estimated standard error for the sample mean difference

    a. The first sample has $n = 4$ scores and a variance of $s^2 = 17$, and the second sample has $n = 8$ scores and a variance of $s^2 = 27$.

    b. Now the sample variances are increased so that the first sample has $n = 4$ scores and a variance of $s^2 = 68$, and the second sample has $n = 8$ scores and a variance of $s^2 = 108$.

    c. Comparing your answers for Parts a and b, how does increased variance influence the size of the estimated standard error?

24. In 1974, Loftus and Palmer conducted a classic study demonstrating how the language used to ask a question can influence eyewitness memory. In the study, college students watched a film of an automobile accident and then were asked questions about what they saw. One group was asked, "About how fast were the cars going when they smashed into each other?" Another group was asked the same question, except the verb was changed to "hit" instead of "smashed into." The "smashed into" group reported significantly higher estimates of speed than the "hit" group. Suppose a researcher repeats this study with a sample of today's college students and obtains the following results:

Estimated Speed	
Smashed into	Hit
$n = 15$	$n = 15$
$M = 40.8$	$M = 34.9$
$SS = 510$	$SS = 414$

    a. Use an independent-measures *t* test with $\alpha = .05$ to determine whether there is a significant difference between the two conditions and compute r^2 to measure effect size.

    b. Now, increase the variability by doubling the two SS values to $SS_1 = 1,020$ and $SS_2 = 828$. Repeat the hypothesis test and the measure of effect size.

    c. Comparing your answers for Parts a and b, describe how sample variability influences the outcome of the hypothesis test and the measure of effect size.

The *t* Test for Two Related Samples

Tools You Will Need

The following items are considered essential background material for this chapter. If you doubt your knowledge of any of these items, you should review the appropriate chapter or section before proceeding.

- Introduction to the *t* statistic (Chapter 9)
  - Estimated standard error
  - Degrees of freedom
  - *t* distribution
  - Hypothesis tests with the *t* statistic
- Independent-measures design (Chapter 10)

clivewa/Shutterstock.com

PREVIEW

Dogs of all breeds end up in shelters for many reasons. Some dogs are lost and the goal is to reunite the pet with its owner. On the other hand, there are dogs that find their way to shelters because they are strays, abandoned, or victims of abuse or neglect. Unfortunately, some types of dogs are more likely to fall into these latter categories, and those that are identified as a pit bull type are among them. Pit bulls are perhaps the most misunderstood and maligned breed of dog (see Gorant, 2010; Dickey, 2016; and the ASPCA Policy and Position Statements). The goal of shelters and rescue groups has been to address public misperceptions and find homes for those dogs that are suitable for adoption. One approach to avoiding the euthanasia of shelter dogs is to present them to the public in a way that increases the adoption rate.

Many studies have identified appearance as the most important factor people use in selecting a dog (Ramirez, 2006; Weiss, Miller, Mohan-Gibbons, & Vela, 2012). So it is not surprising that many shelters post professionally taken photographs of dogs available for adoption on their websites and social media for people to review. Many shelters even include videos of dogs playing with shelter volunteers. Gunter, Barber, and Wynne (2016) conducted an experiment using a repeated-measures design that examined whether a photograph of a pit bull with a human handler in the photo would increase people's adoptability ratings of the dogs. As one part of a comprehensive study, participants rated their level of agreement on a six-point scale (from "strongly disagree" to "strongly agree") with the following statement: "If circumstances allowed, I'd consider adopting this dog." Each participant responded to this question after viewing a photograph of a pit bull with no handler and again after viewing a photograph of a pit bull with a male

child. The researchers found that participants reported higher adoptability ratings when viewing photographs of the pit bull with a male child than they did for photographs of the pit bull alone. Participants' perception of a pit bull was influenced by the context in which the dog was presented in a photograph.

Notice that the researchers recorded adoptability ratings for each participant in two conditions: after viewing a photograph of a pit bull with no handler and after viewing a photograph of a pit bull with the child. The goal of the analysis is to compare these two sets of scores to determine if perceptions of a dog's adoptability is influenced by the absence or presence of the child in the photograph.

In the previous chapter, we introduced a statistical procedure for evaluating the mean difference between two sets of data (the independent-measures *t* statistic). However, the independent-measures *t* statistic is intended for research situations involving two separate and independent samples of participants. You should realize that the two sets of rating scores in the present example are not from independent samples. In this instance the same group of individuals participated in both of the treatment conditions. That is, one sample of participants provided the two adoptability ratings, one for each type of photograph. What is needed is a new statistical analysis for comparing two means that are both obtained from the same group of participants.

In this chapter, we introduce the repeated-measures *t* statistic, which is used for hypothesis tests evaluating the mean difference between two sets of scores obtained from the same group of individuals. As you will see, however, this new *t* statistic is very similar to the original *t* statistic that was introduced in Chapter 9.

11-1 Introduction to Repeated-Measures Designs

LEARNING OBJECTIVE

1. Define a repeated-measures design, explain how it differs from an independent-measures design, and identify examples of each.

In the previous chapter, we introduced the independent-measures research design as one strategy for comparing two treatment conditions or two populations. The independent-measures design is characterized by the fact that two separate samples are used to obtain the two sets of scores that are to be compared. In this chapter, we examine an alternative

TABLE 11.1 An example of the data from a repeated-measures study using $n = 5$ participants to evaluate the difference between two treatments.

Participant	Treatment #1: First Score	Treatment #2: Second Score	
1	12	15	← the two scores
2	10	14	for one participant
3	15	17	
4	17	17	
5	12	18	

strategy known as a *repeated-measures design*, or a *within-subjects design*. With a repeated-measures design, one group of participants is measured in two different treatment conditions, so there are two separate scores for each individual in the sample. For example, a group of patients could be measured before therapy and then measured again after therapy. Or, response time could be measured in a driving simulation task for a group of individuals who are first tested when they are sober and then tested again after two alcoholic drinks. In each case, the same variable is being measured twice for the same set of individuals; that is, we are literally repeating measurements on the same sample.

> A research design that uses the same group of individuals in all of the different treatment conditions is called a **repeated-measures design** or a **within-subjects design**.

In a repeated-measures design comparing two treatments, each participant is measured twice, once in Treatment #1 and once in Treatment #2, to produce the two sets of scores that will be used to compare the treatments. An example of the data from a repeated-measures design is shown in Table 11.1. Notice that the scores are organized as pairs corresponding to the first and second scores for each participant. As a result, the first score for each individual is compared with the second score for that same individual to evaluate the difference between the two treatments.

The main advantage of a repeated-measures study is that it uses exactly the same individuals in all treatment conditions. Thus, there is no risk that the participants in one treatment are different from the participants in another. With an independent-measures design, on the other hand, there is always a risk that the results are biased because the individuals in one sample are systematically different (more skilled, more motivated, more extroverted, and so on) than the individuals in the other sample. At the end of this chapter, we present a more detailed comparison of repeated-measures studies and independent-measures studies, considering the advantages and disadvantages of both types of research.

Now we will examine the statistical techniques that allow a researcher to use the sample data from a repeated-measures study to draw inferences about the general population.

LEARNING CHECK   **LO1** **1.** In a repeated-measures study, the same group of individuals participates in all of the treatment conditions. Which of the following situations is not an example of a repeated-measures design?

    a. A researcher would like to study the effect of practice on performance in the same sample of participants.

    b. A researcher would like to compare individuals from two different populations.

   c. The effect of a treatment is studied in a small group of individuals with a rare disease by measuring their symptoms before and after treatment.

   d. A developmental psychologist examines how behavior unfolds by observing the same group of children at different ages.

LO1 **2.** A researcher conducts a research study comparing two treatment conditions and obtains 10 scores in each treatment. If the researcher used a repeated-measures design, then how many subjects participated in the research study?

   a. 10

   b. 20

   c. 21

   d. 40

LO1 **3.** For an experiment comparing two treatment conditions, an independent-measures design would obtain _____ score(s) for each subject and a repeated-measures design would obtain _____ score(s) for each subject.

   a. 1, 1

   b. 1, 2

   c. 2, 1

   d. 2, 2

ANSWERS   **1. b   2. a   3. b**

11-2 The *t* Statistic for a Repeated-Measures Research Design

LEARNING OBJECTIVES

2. Describe the data (difference scores) that are used for the repeated-measures *t* statistic.

3. Determine the hypotheses for a repeated-measures *t* test.

4. Describe the structure of the repeated-measures *t* statistic, including the estimated standard error and the degrees of freedom, and explain how the formula is related to the single-sample *t*.

5. Calculate the estimated standard error for the mean of the difference scores and explain what it measures.

The *t* statistic for a repeated-measures design is structurally similar to the other *t* statistics we have examined. As we shall see, it is essentially the same as the single-sample *t* statistic covered in Chapter 9. The major distinction of the repeated-measures *t* is that it is based on difference scores rather than raw scores (*X* values). In this section, we examine difference scores and develop the *t* statistic for repeated-measures designs.

■ Difference Scores: The Data for a Repeated-Measures Study

Many over-the-counter cold medications include the warning "may cause drowsiness." Table 11.2 shows an example of data from a study that examines this phenomenon. Note that there is one sample of $n = 4$ participants, and that each individual is measured twice. The first score for each person (X_1) is a measurement of reaction time before the medication was administered. The second score (X_2) measures reaction time one hour after taking

TABLE 11.2

Reaction time measurements taken before and after administering an over-the-counter cold medication.

Person	Before Medication (X_1)	After Medication (X_2)	Difference D
A	215	210	−5
B	221	242	21
C	196	219	23
D	203	228	25

$$\Sigma D = 64$$

Note that M_D is the mean for the sample of D scores.

$$M_D = \frac{\Sigma D}{n} = \frac{64}{4} = 16$$

the medication. Because we are interested in how the medication affects reaction time, we have computed the difference between the first score and the second score for each individual. The *difference scores*, or D values, are shown in the last column of the table. Notice that the difference scores measure the amount of change in reaction time for each person. Typically, the difference scores are obtained by subtracting the first score (before treatment) from the second score (after treatment) for each person:

$$\text{difference score} = D = X_2 - X_1 \tag{11.1}$$

Note that the sign of each D score tells you the direction of the change. Person A, for example, shows a decrease in reaction time after taking the medication (a negative change), but person B shows an increase (a positive change).

The sample of difference scores (D values) serves as the sample data for the hypothesis test and all calculations are done using the D scores. To compute the t statistic, for example, we use the number of D scores (n) as well as the mean for the sample of D scores (M_D) and the value of SS for the sample of D scores.

Because this new hypothesis test compares two sets of scores that are related to each other (they come from the same group of individuals), it often is called the *t* test for two related samples, in contrast to the *t* test for two independent samples described in Chapter 10.

■ The Hypotheses for a Repeated-Measures *t* Test

The researcher's goal is to use the sample of difference scores to answer questions about the general population. In particular, the researcher would like to know whether there is any difference between the two treatment conditions for the general population. Note that we are interested in a population of *difference scores*. That is, we would like to know what would happen if every individual in the population were measured in two treatment conditions (X_1 and X_2) and a difference score (D) were computed for everyone. Specifically, we are interested in the mean for the population of difference scores. We identify this population mean difference with the symbol μ_D (using the subscript letter D to indicate that we are dealing with D values rather than X scores).

As always, the null hypothesis states that for the general population there is no effect, no change, or no difference. For a repeated-measures study, the null hypothesis states that the mean difference for the general population is zero. In symbols,

$$H_0: \mu_D = 0$$

Again, this hypothesis refers to the mean for the entire population of difference scores. Although the population mean is zero, the individual scores in the population are not all equal to zero. Thus, even when the null hypothesis is true, we still expect some individuals to have positive difference scores and some to have negative difference scores. However, the positives and negatives are unsystematic and, in the long run, balance out to $\mu_D = 0$. Also note that a sample selected from this population will probably not have a mean exactly equal to zero. As always, there will be some error between a sample mean and the population mean, so even if $\mu_D = 0$ (H_0 is true), we do not expect M_D to be exactly equal to zero.

The alternative hypothesis states that there is a treatment effect that causes the scores in one treatment condition to be systematically higher (or lower) than the scores in the other condition. In symbols,

$$H_1: \mu_D \neq 0$$

According to H_1, the difference scores for the individuals in the population tend to be systematically positive (or negative), indicating a consistent, predictable difference between the two treatments.

■ The Repeated-Measures *t* Statistic

Figure 11.1 shows the general situation that exists for a repeated-measures hypothesis test. You may recognize that we are facing essentially the same situation that we encountered in Chapter 9. In particular, we have a population for which the mean and the standard deviation are unknown, and we have a sample that will be used to test a hypothesis about the unknown population. In Chapter 9, we introduced the single-sample *t* statistic, which allowed us to use a sample mean as a basis for testing hypotheses about an unknown population mean. This *t*-statistic formula will be used again here to develop the repeated-measures *t* test. To refresh your memory, the single-sample *t* statistic (Chapter 9) is defined by the formula

$$t = \frac{M - \mu}{s_M}$$

In this formula, the sample mean, M, is calculated from the data, and the value for the population mean, μ, is obtained from the null hypothesis. The estimated standard error, s_M, is also calculated from the data and provides a measure of how much difference is reasonable to expect between a sample mean and the population mean.

For the repeated-measures design, the sample data are difference scores and are identified by the letter D, rather than X. Accordingly, we will use the letter D in the formula to

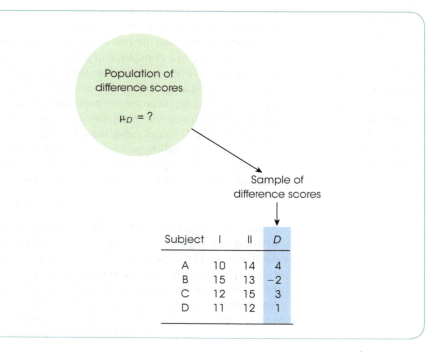

FIGURE 11.1
A sample of $n = 4$ people is selected from the population. Each individual is measured twice, once in treatment I and once in treatment II, and a difference score, D, is computed for each individual. This sample of difference scores is intended to represent the population. Note that we are using a sample of difference scores to represent a population of difference scores. Note that the mean for the population of difference scores is unknown. The null hypothesis states that there is no consistent or systematic difference between the two treatment conditions, so the population mean difference is $\mu_D = 0$.

Population of difference scores

$\mu_D = ?$

Sample of difference scores

Subject	I	II	D
A	10	14	4
B	15	13	−2
C	12	15	3
D	11	12	1

emphasize that we are dealing with difference scores instead of X values. Also, the population mean that is of interest to us is the population mean difference (the mean amount of change for the entire population), and is identified by the symbol μ_D. With these simple changes, the t formula for the repeated-measures design becomes

$$t = \frac{M_D - \mu_D}{s_{M_D}}$$ (11.2)

In this formula, the estimated standard error, s_{M_D}, is computed in exactly the same way as it is computed for the single-sample t statistic. To calculate the estimated standard error, the first step is to compute the variance (or the standard deviation) for the sample of D scores.

$$s^2 = \frac{SS}{n-1} = \frac{SS}{df} \quad \text{or} \quad s = \sqrt{\frac{SS}{df}}$$

The *estimated standard error for* M_D is then computed using the sample variance (or sample standard deviation) and the sample size, n.

$$s_{M_D} = \sqrt{\frac{s^2}{n}} \quad \text{or} \quad s_{M_D} = \frac{s}{\sqrt{n}}$$ (11.3)

The following example is an opportunity to test your understanding of variance and estimated standard error for the repeated-measures t statistic.

EXAMPLE 11.1   A repeated-measures study with a sample of $n = 10$ participants produces a mean difference of $M_D = 5.5$ points with $SS = 360$ for the difference scores. For these data, find the variance for the difference scores and the estimated standard error for the sample mean. You should obtain a variance of 40 and an estimated standard error of 2. Good luck.   ■

Notice that all of the calculations are done using the difference scores (the D scores) and that there is only one D score for each subject. With a sample of n participants, the number of D scores is n, and the t statistic has $df = n - 1$. Remember that n refers to the number of D scores, not the number of X scores in the original data.

You should also note that the *repeated-measures* t *statistic* is conceptually similar to the t statistics we have previously examined:

$$t = \frac{\text{sample statistic} - \text{population parameter}}{\text{estimated standard error}}$$

or

$$t = \frac{\text{actual difference between sample } (M_D) \text{ and the hypothesis } (\mu_D)}{\text{expected difference between } M_D \text{ and } \mu_D \text{ with no treatment effect}}$$

In this case, the sample data are represented by the mean for the sample of difference scores (M_D), the population parameter is the value predicted by H_0 ($\mu_D = 0$), and the amount of sampling error is measured by the standard error for the sample mean difference (s_{M_D}).

LEARNING CHECK   **LO2** **1.** What is the mean for the difference scores for the following data from a repeated-measures study?

	I	II
a. 16	5	13
b. 6	2	10
c. 8	6	6
d. 44	7	15

LO3 2. Which of the following is the correct statement of the null hypothesis for a repeated-measures hypothesis test?

   a. $M_D = 0$

   b. $\mu_D = 0$

   c. $\mu_1 = \mu_2$

   d. $M_1 = M_2$

LO4 3. Which of the following accurately describes the relationship between the repeated-measures *t* statistic and the single-sample *t* statistic?

   a. Each uses one sample mean.

   b. Each uses one population mean.

   c. Each uses one sample variance to compute the standard error.

   d. All of the above.

LO5 4. What is the value for the estimated standard error for a set of $n = 11$ difference scores with $SS = 990$?

   a. 40

   b. 3

   c. 2

   d. 1

ANSWERS **1. b** **2. b** **3. d** **4. b**

11-3 Hypothesis Tests for the Repeated-Measures Design

LEARNING OBJECTIVES

6. Conduct a repeated-measures *t* test to evaluate the significance of the population mean difference using the data from a repeated-measures study comparing two treatment conditions.

7. Conduct a directional (one-tailed) hypothesis test using the repeated-measures *t* statistic.

In a repeated-measures study, each individual is measured in two different treatment conditions and we are interested in whether there is a systematic difference between the scores in the first treatment condition and the scores in the second treatment condition. A difference score (*D* value) is computed for each person and the hypothesis test uses the difference scores from the sample to evaluate the overall mean difference, μ_D, for the entire population. The hypothesis test with the repeated-measures *t* statistic follows the same four-step process that we have used for other tests. The complete hypothesis-testing procedure is demonstrated in Example 11.2.

EXAMPLE 11.2   It's the night before an exam. It is getting late and you are trying to decide whether to study or sleep. There are obvious advantages to studying, especially if you feel that you do not have a good grasp of the material to be tested. On the other hand, a good night's sleep will leave you better prepared to deal with the stress of taking an exam. "To study or sleep?" was the question addressed by Gillen-O'Neel, Huynh, and Fuligni (2013). The researchers started with a sample of 535 ninth-grade students and followed up when the students were in the tenth and twelfth grades. Each year the students completed a diary every day for two weeks, recording how much time they spent studying outside of school and how much time they slept the night before. The students also reported the occurrence of academic problems each day such as "did not understand something taught in class" and "did poorly on a test, quiz, or homework." The primary result from the study is that the students reported more academic problems following nights with less-than-average sleep than they did after nights with more-than-average sleep, especially for the older students.

Recently, a researcher attempted to replicate the study using a sample of $n = 8$ college freshmen and obtained the data shown in Table 11.3.

STEP 1 **State the hypotheses, and select the alpha level.**

$$H_0: \mu_D = 0 \quad \text{(There is no difference between the two conditions.)}$$

$$H_1: \mu_D \neq 0 \quad \text{(There is a difference.)}$$

For this test, we use $\alpha = .05$.

STEP 2 **Locate the critical region.** For this example, $n = 8$, so the t statistic has $df = n - 1 = 7$. For $\alpha = .05$, the critical value listed in the t distribution table is ± 2.365.

STEP 3 **Calculate the t statistic.** Table 11.3 shows the sample data and the calculations of $M_D = 4$ and $SS = 112$. Note that all calculations are done with the difference scores. As we have done with the other t statistics, we present the calculation of the t statistic as a three-step process.

First, compute the sample variance for the D scores.

$$s^2 = \frac{SS}{n-1} = \frac{112}{7} = 16$$

Next, use the sample variance to compute the estimated standard error.

$$s_{M_D} = \sqrt{\frac{s^2}{n}} = \sqrt{\frac{16}{8}} = 1.41$$

TABLE 11.3
Academic problems for students after a night of below-average or above-average sleep.

Participant	Above-Average Sleep	Below-Average Sleep	D	D²
A	7	10	3	9
B	8	7	−1	1
C	4	14	10	100
D	6	13	7	49
E	3	11	8	64
F	9	10	1	1
G	4	4	0	0
H	7	11	4	16
			$\Sigma D = 32$	$\Sigma D^2 = 240$

$$M_D = \frac{32}{8} = 4 \qquad SS = \Sigma D^2 - \frac{(\Sigma D)^2}{n} = 240 - \frac{(32)^2}{8} = 112, \text{ and}$$

$$s^2 = \frac{SS}{df} = \frac{112}{7} = 16$$

Finally, use the sample mean (M_D) and the hypothesized population mean (μ_D) along with the estimated standard error to compute the value for the *t* statistic.

$$t = \frac{M_D - \mu_D}{s_{M_D}} = \frac{4 - 0}{1.41} = 2.84$$

STEP 4   **Make a decision.**   The *t* value we obtained is beyond the critical value of $+2.365$. The researcher rejects the null hypothesis and concludes that the amount of sleep at night has a statistically significant effect on academic problems the following day.   ■

■ Directional Hypotheses and One-Tailed Tests

In some repeated-measures studies, the researcher has a specific prediction concerning the direction of the treatment effect. For example, in the study described in Example 11.2, the researcher could predict that academic problems will be greater when a student has less than average sleep the previous night. This kind of directional prediction can be incorporated into the statement of the hypotheses, resulting in a directional, or one-tailed, hypothesis test. The following example demonstrates how the hypotheses and critical region are determined for a directional test.

EXAMPLE 11.3   We will reexamine the experiment presented in Example 11.2. The researcher is using a repeated-measures design to investigate how academic problems are influenced by the amount of sleep the night before. The researcher predicts that academic problems will increase when the participants have less-than-average sleep the previous night.

STEP 1   **State the hypotheses and select the alpha level.**   For this example, the researcher predicts that academic problems increase after participants get less-than-average sleep. On the other hand, the null hypothesis states that academic problems will not increase but rather will be unchanged or even decreased after nights with less-than-average sleep. In symbols,

H_0: $\mu_D \leq 0$ (There is no increase with less sleep.)

The alternative hypothesis says that the treatment does work. For this example, H_1 says that having less sleep will increase academic problems.

H_1: $\mu_D > 0$ (Academic problems are increased.)

We use $\alpha = .05$.

STEP 2   **Locate the critical region.**   As we demonstrated with the independent-measures *t* statistic (page 336), the critical region for a one-tailed test can be located using a two-stage process. Rather than trying to determine which tail of the distribution contains the critical region, you first look at the sample mean difference to verify that it is in the predicted direction. If not, then the treatment clearly did not work as expected and you can stop the test. If the change is in the correct direction, then the question is whether it is large enough to be significant. For this example, change is in the predicted direction (the researcher predicted increased problems and the sample mean shows an increase). With $n = 8$, we obtain $df = 7$ and a critical value of $t = 1.895$ for a one-tailed test with $\alpha = .05$. Thus, any *t* statistic beyond $+1.895$ is sufficient to reject the null hypothesis.

STEP 3   **Compute the *t* statistic.**   We calculated the *t* statistic in Example 11.2, and obtained $t = 2.84$.

STEP 4   **Make a decision.**   The obtained t statistic is beyond the critical boundary. Therefore, we reject the null hypothesis and conclude that less-than-average sleep resulted in a statistically significant increase in academic problems the following day.   ■

■ Assumptions of the Related-Samples t Test

The repeated-measures t statistic requires two basic assumptions:

1. The observations within each treatment condition must be independent (see page 264). Notice that the assumption of independence refers to the scores *within* each treatment. Inside each treatment, the scores are obtained from different individuals and should be independent of one another.

2. The population distribution of difference scores (D values) must be normal.

As before, the normality assumption is not a cause for concern unless the sample size is relatively small. In the case of severe departures from normality, the validity of the t test may be compromised with small samples. However, with relatively large samples ($n > 30$), this assumption can be ignored.

LEARNING CHECK   **LO6**   **1.** A researcher conducts a repeated-measures study comparing two treatment conditions with a sample of $n = 8$ participants and obtains a t statistic of $t = 2.381$. Which of the following is the correct decision for a two-tailed test?

   a. Reject the null hypothesis with $\alpha = .05$ but fail to reject with $\alpha = .01$

   b. Reject the null hypothesis with either $\alpha = .05$ or $\alpha = .01$

   c. Fail to reject the null hypothesis with either $\alpha = .05$ or $\alpha = .01$

   d. Cannot determine the correct decision without more information.

LO7   **2.** A researcher is using a one-tailed hypothesis test to evaluate the significance of a mean difference between two treatments in a repeated-measures study. If the treatment is expected to increase scores, then which of the following is the correct statement of the alternative hypothesis (H_1)?

   a. $u_D \geq 0$

   b. $u_D \leq 0$

   c. $u_D > 0$

   d. $u_D < 0$

ANSWERS   **1. a   2. c**

11-4   Effect Size, Confidence Intervals, and the Role of Sample Size and Sample Variance for the Repeated-Measures t

LEARNING OBJECTIVES

8. Measure effect size for a repeated-measures t test using either Cohen's d or r^2, the percentage of variance accounted for.

9. Use the data from a repeated-measures study to compute a confidence interval describing the size of the population mean difference.

10. Describe how the results of a repeated-measures *t* test and measures of effect size are reported in the scientific literature.

11. Describe how the outcome of a hypothesis test and measures of effect size using the repeated-measures *t* statistic are influenced by sample size and sample variance.

12. Describe how the consistency of the treatment effect is reflected in the variability of the difference scores and explain how this influences the outcome of a hypothesis test.

■ Effect Size for the Repeated-Measures *t*

As we noted with other hypothesis tests, whenever a treatment effect is found to be statistically significant, it is recommended that you also report a measure of the absolute magnitude of the effect. The most commonly used measures of effect size are Cohen's *d* and r^2, the percentage of variance accounted for. The size of the treatment effect also can be described with a confidence interval estimating the population mean difference, μ_D. Using the data from Example 11.2, we will demonstrate how these values are calculated to measure and describe effect size.

Cohen's *d* In Chapters 8 and 9 we introduced Cohen's *d* as a standardized measure of the mean difference between treatments. The standardization simply divides the population mean difference by the standard deviation. For a repeated-measures study, Cohen's *d* is defined as

$$d = \frac{\text{population mean difference}}{\text{standard deviation}} = \frac{\mu_D}{\sigma_D}$$

Because the population mean and standard deviation are unknown, we use the sample values instead. The sample mean, M_D, is the best estimate of the actual mean difference, and the sample standard deviation (square root of sample variance) provides the best estimate of the actual standard deviation. Thus, we are able to estimate the value of *d* as follows:

$$\text{estimated } d = \frac{\text{sample mean difference}}{\text{sample standard deviation}} = \frac{M_D}{s} \tag{11.4}$$

Because we are measuring the size of the effect and not the direction, it is customary to ignore a minus sign and report Cohen's *d* as a positive value.

For the repeated-measures study in Example 11.2, the sample mean difference is $M_D = 4$ and the sample variance is $s^2 = 16.00$, so the data produce

$$\text{estimated } d = \frac{M_D}{s} = \frac{4}{\sqrt{16}} = \frac{4}{4} = 1.00$$

Any value greater than 0.80 is considered to be a large effect, and these data are clearly in that category (see Table 8.2 on page 273).

The Percentage of Variance Accounted for, r^2 Percentage of variance is computed using the obtained *t* value and the *df* value from the hypothesis test, exactly as was done for the single-sample *t* (see page 307) and for the independent-measures *t* (see page 341). For the data in Example 11.2, we obtained $t = 2.84$ with $df = 7$, which produces

$$r^2 = \frac{t^2}{t^2 + df} = \frac{(2.84)^2}{(2.84)^2 + 7} = \frac{8.07}{15.07} = 0.536$$

For these data, 53.6% of the variance in the scores is explained by the amount of sleep. More specifically, the difference between below-average and above-average sleep produced consistently positive difference scores rather than differences near zero as predicted by the null hypothesis. Thus, the deviations from zero are largely explained by the difference between the two conditions.

The following example is an opportunity to test your understanding of Cohen's d and r^2 to measure effect size for the repeated-measures t statistic.

EXAMPLE 11.4

A repeated-measures study with $n = 16$ participants produces a mean difference of $M_D = 6$ points, $SS = 960$ for the difference scores, and $t = 3.00$. Calculate Cohen's d and r^2 to measure the effect size for this study. You should obtain $d = \frac{6}{8} = 0.75$ and $r^2 = \frac{9}{24} = 0.375$. Good luck. ∎

■ Confidence Intervals for Estimating μ_D

As noted in the previous two chapters, it is possible to compute a confidence interval as an alternative method for measuring and describing the size of the treatment effect. For the repeated-measures t, we use a sample mean difference, M_D, to estimate the population mean difference, μ_D. In this case, the confidence interval literally estimates the size of the treatment effect by estimating the population mean difference between the two treatment conditions.

As with the other t statistics, the first step is to solve the t equation for the unknown parameter. For the repeated-measures t statistic, we obtain

$$\mu_D = M_D \pm t s_{M_D} \tag{11.5}$$

In the equation, the values for M_D and for s_{M_D} are obtained from the sample data. Although the value for the t statistic is unknown, we can use the degrees of freedom for the t statistic and the t distribution table to estimate the t value. Using the estimated t and the known values from the sample, we can then compute the value of μ_D. The following example demonstrates the process of constructing a confidence interval for a population mean difference.

EXAMPLE 11.5

In Example 11.2 we presented a research study demonstrating how the amount of sleep influenced academic problems the next day. In the study, a sample of $n = 8$ college freshmen experienced significantly more academic problems following nights of less-than-average sleep compared to nights of above-average sleep. The mean difference between the two conditions was $M_D = 4$ points and the estimated standard error for the mean difference was $s_{M_D} = 1.41$. Now, we construct a 95% confidence interval to estimate the size of the population mean difference.

With a sample of $n = 8$ participants, the repeated-measures t statistic has $df = 7$. To have 95% confidence, we simply estimate that the t statistic for the sample mean difference is located somewhere in the middle 95% of all the possible t values. According to the t distribution table, with $df = 7$, 95% of the t values are located between $t = +2.365$ and $t = -2.365$. Using these values in the estimation equation, together with the values for the sample mean and the standard error, we obtain

$$\mu_D = M_D \pm t s_{M_D}$$
$$= 4 \pm 2.365(1.41)$$
$$= 4 \pm 3.33$$

This produces an interval of values ranging from $4 - 3.33 = 0.67$ to $4 + 3.33 = 7.33$. Our conclusion is that for the general population, a night of below-average sleep instead of above-average sleep increases academic problems between 0.67 and 7.33 points. We are 95% confident that the true mean difference is in this interval because the only value estimated during the calculations was the t statistic, and we are 95% confident that the t value is located in the middle 95% of the distribution. Finally, note that the confidence interval is constructed around the sample mean difference. As a result, the sample mean difference, $M_D = 4$ points, is located exactly in the center of the interval. ∎

As with the other confidence intervals presented in Chapters 9 and 10, the confidence interval for a repeated-measures *t* is influenced by a variety of factors other than the actual size of the treatment effect. In particular, the width of the interval depends on the percentage of confidence used so that a larger percentage produces a wider interval. Also, the width of the interval depends on the sample size, so that a larger sample produces a narrower interval. Because the interval width is related to sample size, the confidence interval is not a pure measure of effect size like Cohen's *d* or r^2.

Finally, we should note that the 95% confidence interval computed in Example 11.5 does not include the value $\mu_D = 0$. In other words, we are 95% confident that the population mean difference is not $\mu_D = 0$. This is equivalent to concluding that a null hypothesis specifying that $\mu_D = 0$ would be rejected with a test using $\alpha = .05$. If $\mu_D = 0$ were included in the 95% confidence interval, it would indicate that a hypothesis test would fail to reject H_0 with $\alpha = .05$.

IN THE LITERATURE

Reporting the Results of a Repeated-Measures *t* Test

As we have seen in Chapters 9 and 10, the APA format for reporting the results of *t* tests consists of a concise statement that incorporates the *t* value, degrees of freedom, alpha level, and effect size. One typically includes values for means and standard deviations, either in a statement or table (Chapter 4). For Example 11.2, we observed a mean difference of $M_D = 4.00$ with $s = 4.00$. Also, we obtained a *t* statistic of $t = 2.84$ with $df = 7$, and our decision was to reject the null hypothesis at the .05 level of significance. Finally, we measured effect size by computing the percentage of variance explained and obtained $r^2 = 0.536$. A published report of this study might summarize the results as follows:

> Experiencing a night of below-average sleep increased academic problems the following day by an average of $M = 4.00$ points with $SD = 4.00$. The treatment effect was statistically significant, $t(7) = 2.84, p < .05, r^2 = 0.536$.

When the hypothesis test is conducted with a computer program, the printout typically includes an exact probability for the level of significance. The *p*-value from the printout is then stated as the level of significance in the research report. For example, the data from Example 11.2 produced a significance level of $p = .025$, and the results would be reported as "statistically significant, $t(7) = 2.84, p = .017, r^2 = 0.536$." Occasionally, a probability is so small that the computer rounds it off to 3 decimal points and produces a value of zero. In this situation you do not know the exact probability value and should report $p < .001$.

If the confidence interval from Example 11.5 is reported as a description of effect size together with the results from the hypothesis test, it would appear as follows:

> A night of below-average sleep compared to above-average sleep significantly increased academic problems the next day, $t(7) = 2.84, p < .05$, 95% CI [0.67, 7.33].

■ Descriptive Statistics and the Hypothesis Test

Often, a close look at the sample data from a research study makes it easier to see the size of the treatment effect and to understand the outcome of the hypothesis test. In Example 11.2, we obtained a sample of $n = 8$ participants who produce a mean difference of $M_D = 4.00$ points with a standard deviation of $s = 4$ points. The sample mean and standard deviation describe a set of scores centered at $M_D = 4.00$ with most of the scores located within

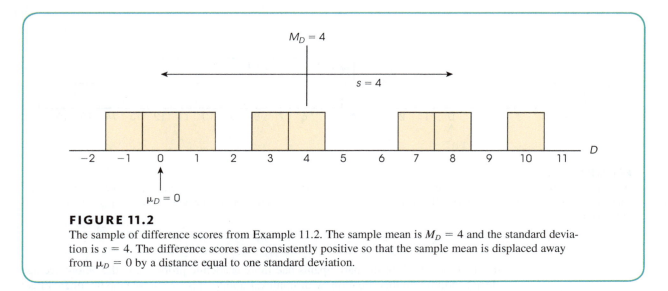

FIGURE 11.2

The sample of difference scores from Example 11.2. The sample mean is $M_D = 4$ and the standard deviation is $s = 4$. The difference scores are consistently positive so that the sample mean is displaced away from $\mu_D = 0$ by a distance equal to one standard deviation.

4 points of the mean. Figure 11.2 shows the actual set of difference scores that were obtained in Example 11.2. In addition to showing the scores in the sample, we have highlighted the position of $\mu_D = 0$; that is, the value specified in the null hypothesis. Notice that the scores in the sample are displaced away from zero. Specifically, the data are not consistent with a population mean of $\mu_D = 0$, which is why we rejected the null hypothesis. In addition, note that the sample mean is located one standard deviation above zero. This distance corresponds to the effect size measured by Cohen's $d = 1.00$. For these data, the picture of the sample distribution (see Figure 11.2) should help you to understand the measure of effect size and the outcome of the hypothesis test.

■ Sample Variance and Sample Size in the Repeated-Measures *t* Test

In previous chapters we identified sample variability and sample size as two factors that can influence the outcome of a hypothesis test. Both of these factors affect the magnitude of the estimated standard error in the denominator of the *t* statistic. The standard error is inversely related to sample size (larger size leads to smaller error) and is directly related to sample variance (larger variance leads to larger error). As a result, a bigger sample produces a larger value for the *t* statistic (farther from zero) and increases the likelihood of rejecting H_0. Larger variance, on the other hand, produces a smaller value for the *t* statistic (closer to zero) and reduces the likelihood of finding a significant result.

Although variance and sample size both influence the hypothesis test, only variance has a large influence on measures of effect size such as Cohen's d and r^2; larger variance produces smaller measures of effect size. Sample size, on the other hand, has no effect on the value of Cohen's d and only a small influence on r^2.

Variability as a Measure of Consistency for the Treatment Effect   In a repeated-measures study, the variability of the difference scores becomes a relatively concrete and easy-to-understand concept. In particular, the sample variability describes the *consistency* of the treatment effect. For example, if a treatment consistently adds a few points to each individual's score, then the set of difference scores will be clustered together with relatively small variability. This is the situation that we observed in Example 11.2 (see Figure 11.2)

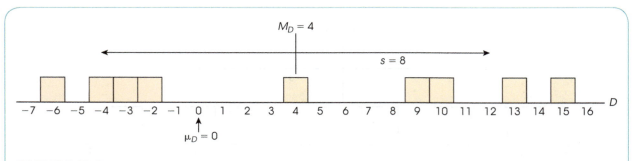

FIGURE 11.3
A sample of difference scores with a mean of $M_D = 4$ and a standard deviation of $s = 8$. The difference scores do not show a consistent increase or decrease. Because there is no consistent treatment effect, the null hypothesis $\mu_D = 0$ is not rejected.

in which nearly all of the participants had more academic problems in the below-average sleep condition. In this situation, with small variability, it is easy to see the treatment effect and it is likely to be significant.

Now consider what happens when the variability is large. Suppose that the sleep and academic problems study in Example 11.2 produced a sample of $n = 9$ difference scores consisting of $+13$, -6, $+10$, -2, -4, $+9$, -3, $+15$, and $+4$. These difference scores also have a mean of $M_D = 4.00$, but now the variability is substantially increased so that $SS = 512$ and the standard deviation is $s = 8$. Figure 11.3 shows the new set of difference scores. Again, we have highlighted the position of $\mu_D = 0$, which is the value specified in the null hypothesis. Notice that the high variability means that there is no consistent treatment effect. Some participants have more academic problems with below-average sleep (the positive differences) and some less (the negative differences). In the hypothesis test, the high variability increases the size of the estimated standard error and results in a hypothesis test that produces $t = 1.50$, which is not in the critical region. With these data, we would fail to reject the null hypothesis and conclude that the amount of sleep has no effect on academic problems the following day.

With small variability (see Figure 11.2), the 4-point treatment effect is easy to see and is statistically significant. With large variability (see Figure 11.3), the 4-point effect is not easy to see and is not significant. As we have noted several times in the past, large variability can obscure patterns in the data and reduces the likelihood of finding a significant treatment effect.

LEARNING CHECK   **LO8** **1.** The results of a repeated-measures study with $n = 5$ participants produce a mean difference of $M_D = 20$ points with $SS = 500$ for the difference scores and a t statistic of $t = 4.00$. If the percentage of variance, r^2, is used to measure effect size, then what is the value of r^2?

   a. $\frac{16}{20} = 0.8$

   b. $\frac{4}{20} = 0.2$

   c. $\frac{4}{8} = 0.5$

   d. $\frac{16}{8} = 2.0$

   LO9 **2.** For a repeated-measures study with $n = 16$ scores in each treatment, a researcher constructs an 95% confidence interval to describe the mean difference between treatments. What value is at the center of the interval and what t values are used to construct the interval?

a. The sample mean difference is at the center and $t = \pm 2.131$.

b. The sample mean difference is at the center and $t = \pm 1.753$.

c. Zero is at the center and $t = \pm 2.131$.

d. Zero is at the center and $t = \pm 1.753$.

LO10 **3.** A research report describing the results from a repeated-measures study states, "The data showed a significant difference between treatments, $t(22) = 4.71$, $p < .01$." From this report, what can you conclude about the outcome of the hypothesis test?

a. The test rejected the null hypothesis.

b. The test failed to reject the null hypothesis.

c. The test resulted in a Type I error.

d. The test resulted in a Type II error.

LO11 **4.** A repeated-measures study finds a mean difference of $M_D = 5$ points between two treatment conditions. Which of the following sample characteristics is most likely to produce a significant t statistic for the hypothesis test?

a. A large sample size (n) and a large variance.

b. A large sample size (n) and a small variance.

c. A small sample size (n) and a large variance.

d. A small sample size (n) and a small variance.

LO12 **5.** If the results of a repeated-measures study show that nearly all of the participants score around 5 points higher in Treatment A than in Treatment B, then which of the following accurately describes the data?

a. The variance of the difference scores is small and the likelihood of a significant result is low.

b. The variance of the difference scores is small and the likelihood of a significant result is high.

c. The variance of the difference scores is large and the likelihood of a significant result is low.

d. The variance of the difference scores is large and the likelihood of a significant result is high.

ANSWERS   1. a   2. a   3. a   4. b   5. b

11-5 Comparing Repeated- and Independent-Measures Designs

LEARNING OBJECTIVES

13. Describe the advantages and disadvantages of choosing a repeated-measures design instead of an independent-measures design to compare two treatment conditions, and with that information in mind, evaluate the situations in which each design would be more appropriate.

14. Define a matched-subjects design and explain how it differs from repeated-measures and independent-measures designs.

◼ Repeated-Measures versus Independent-Measures Designs

In many research situations, it is possible to use either a repeated-measures design or an independent-measures design to compare two treatment conditions. The independent-measures design would use two separate samples (one in each treatment condition) and the repeated-measures design would use only one sample with the same individuals participating in both treatments. The decision about which design to use is often made by considering the advantages and disadvantages of the two designs.

Number of Subjects   A repeated-measures design typically requires fewer subjects than an independent-measures design. The repeated-measures design uses subjects (or participants) more efficiently because each individual is measured in both of the treatment conditions. This can be especially important when there are relatively few subjects or participants available (for example, when you are studying a rare species or individuals with a rare disease).

Study Changes over Time   The repeated-measures design is especially well suited for studying learning, development, or other changes that take place over time. Remember that this design often involves measuring individuals at one time and then returning to measure the same individuals at a later time. In this way, a researcher can observe behaviors that change or develop over time.

Individual Differences   The primary advantage of a repeated-measures design is that it reduces or eliminates problems caused by individual differences (see Chapter 10, page 347). *Individual differences* are characteristics such as age, IQ, gender, and personality that vary from one individual to another (see Chapter 1, page 24). These individual differences can influence the scores obtained in a research study, and they can affect the outcome of a hypothesis test. Consider the data in Table 11.4. The first set of data represents the results from a typical independent-measures study and the second set represents a repeated-measures study. Note that we have identified each participant by name to help demonstrate the effects of individual differences.

For the independent-measures data, note that every score represents a different person. For the repeated-measures study, on the other hand, the same participants are measured in both of the treatment conditions. This difference between the two designs has some important consequences.

1. We have constructed the data so that both research studies have exactly the same scores and they both show the same 5-point mean difference between treatments.

TABLE 11.4

Hypothetical data showing the results from an independent-measures study and a repeated-measures study. The two sets of data use exactly the same numerical scores and they both show the same 5-point mean difference between treatments.

Independent-Measures Study (two separate samples)		Repeated-Measures Study (same sample in both treatments)		
Treatment 1	Treatment 2	Treatment 1	Treatment 2	D
(John) $X = 18$	(Sue) $X = 15$	(John) $X = 18$	(John) $X = 15$	-3
(Mary) $X = 27$	(Tom) $X = 20$	(Mary) $X = 27$	(Mary) $X = 20$	-7
(Bill) $X = 33$	(Dave) $X = 28$	(Bill) $X = 33$	(Bill) $X = 28$	-5
$M = 26$	$M = 21$			$M_D = -5$
$SS = 114$	$SS = 86$			$SS = 8$
$s^2 = 57$	$s^2 = 43$			$s^2 = 4$
$s_M = 5.77$				$s_{M_D} = 1.15$

In each case, the researcher would like to conclude that the 5-point difference was caused by the treatments. However, with the independent-measures design, there is always the possibility that the participants in Treatment 1 have different characteristics than those in Treatment 2, especially if there has been a problem with random assignment to treatment groups. For example, the three participants in Treatment 1 may be more skilled than those in Treatment 2 and their difference in skill level caused them to have higher scores. Note that this problem disappears with the repeated-measures design. Specifically, with repeated measures there is no possibility that the participants in one treatment are different from those in another treatment because the same participants are used in all the treatments.

2. Although the two sets of data contain exactly the same scores and have exactly the same 5-point mean difference, you should realize that they are very different in terms of the variance used to compute standard error. For the independent-measures study, you calculate the *SS* or variance for the scores in each of the two separate samples. Note that in each sample there are big differences between participants. In Treatment 1, for example, Bill has a score of 33 and John's score is only 18. These individual differences produce a relatively large sample variance and a large standard error. For the independent-measures study, the standard error is 5.77, which produces a *t* statistic of $t = 0.87$. For these data, the hypothesis test concludes that there is no significant difference between treatments.

In the repeated-measures study, the *SS* and variance are computed for the difference scores. If you examine the repeated-measures data in Table 11.4, you will see that the big differences between John and Bill that exist in Treatment 1 and in Treatment 2 are eliminated when you get to the difference scores. Because the individual differences are eliminated, the variance and standard error are dramatically reduced. For the repeated-measures study, the standard error is 1.15 and the *t* statistic is $t = -4.35$. With the repeated-measures *t*, the data show a significant difference between treatments. Thus, one big advantage of a repeated-measures study is that it reduces variance by removing individual differences, which increases the chances of finding a significant result.

Power   The repeated-measures design tends to have more power than an independent-measures design. As described in Chapter 8, page 275, power is the likelihood of detecting a real treatment effect. One factor that contributes to the size of standard error is the amount of variability in the sample data. The repeated-measures design eliminates the role of individual differences from the difference between the treatment conditions because the same participants provide scores for both treatments. This results in less error variability—that is, a lower value for standard error of mean difference. In general, lower values for standard error result in greater statistical power. (see Example 8.7, page 278).

Returning to Table 11.4, compare the standard error computed for the independent measures design, $s_M = 5.77$, to the value for the repeated-measures design, $s_{M_D} = 1.15$. The repeated measures design is more sensitive to the 5-point difference between sample means (i.e., power is greater) than the independent-samples design because standard error is lower. Recall that the standard error in an independent samples *t* test is based on the variability of *two* samples. In contrast, the standard error for a repeated-measures *t* statistic is influenced by the variability of just *one* sample.

■ Time-Related Factors and Order Effects

The primary disadvantage of a repeated-measures design is that the structure of the design allows for factors other than the treatment effect to cause a participant's score to change from one treatment to the next. Specifically, in a repeated-measures design, each individual is measured in two different treatment conditions, often *at two different times*. In this situation, outside factors that change over time may be responsible for changes in the participants' scores because they are confounded with the change in treatment. For example, a participant's health or mood may change over time and cause a difference in the participant's scores. Outside factors such as the weather can also change and may have an influence on participants' scores. Because a repeated-measures study often takes place over time, it is possible that time-related factors (other than the two treatments) are responsible for causing changes in the participants' scores.

Also, it is possible that participation in the first treatment influences the individual's score in the second treatment. If the researcher is measuring individual performance, for example, the participants may gain experience during the first treatment condition, and this extra practice helps their performance in the second condition. In this situation, the researcher would find a mean difference between the two conditions; however, the difference would not be caused by the treatments, instead it would be caused by practice effects. Changes in scores that are caused by participation in an earlier treatment are called *order effects* and can distort the mean differences found in repeated-measures research studies.

Counterbalancing One way to deal with time-related factors and order effects is to counterbalance the order of presentation of treatments. That is, the participants are randomly divided into two groups, with one group receiving Treatment 1 followed by Treatment 2, and the other group receiving Treatment 2 followed by Treatment 1. The goal of counterbalancing is to distribute any outside effects evenly over the two treatments. For example, if practice effects are a problem, then half of the participants will gain experience in Treatment 1, which then helps their performance in Treatment 2. However, the other half will gain experience in Treatment 2, which helps their performance in Treatment 1. Thus, prior experience helps the two treatments equally.

Finally, if there is reason to expect strong time-related effects or strong order effects, your best strategy is not to use a repeated-measures design. Instead, use independent-measures so that each individual participates in only one treatment and is measured only one time.

■ The Matched-Subjects Design

Occasionally, researchers try to approximate the advantages of independent-measures and repeated-measures designs by using a technique known as *matched subjects*. A *matched-subjects design* involves two separate samples, but each individual in one sample is matched one-to-one with an individual in the other sample. Typically, the individuals are matched on one or more variables that are considered to be especially important for the study. For example, a researcher studying verbal learning might want to be certain that the two samples are matched in terms of IQ. In this case, a participant with an IQ of 120 in one sample would be matched with another participant with an IQ of 120 in the other sample. Although the participants in one sample are not *identical* to the participants in the other sample, the matched-subjects design at least ensures that the two samples are equivalent (or matched) with respect to a specific variable.

> In a **matched-subjects design**, each individual in one sample is matched with an individual in the other sample. The matching is done so that the two individuals are equivalent (or nearly equivalent) with respect to a specific variable (or variables) that the researcher would like to control.

Notice that a matched-subjects design has characteristics of both an independent-measures design and a repeated-measures design. First, it uses a separate sample of participants in each of the two treatment conditions, which means that it literally is an independent-measures design. As a result, each participant is measured only one time in only one treatment condition and there is no risk of order effects or carry-over effects. At the same time, the matching process simulates a repeated-measures design because each individual in the first treatment is matched with an individual in the second treatment, and a difference score is computed for each matched pair. In a repeated-measures design, the matching is perfect because the same individual is used in both conditions. In a matched-subjects design the matching is based on the specific variable(s) that are matched. In each case, however, the data are used to compute difference scores and the hypothesis test for the matched-subjects design is the same as the *t* test used for the repeated-measures design. As a result, both designs are able to measure the individual differences and remove them from the variance in the data.

Thus, matched-subjects designs have the advantages of both an independent- and a repeated-measures design without the disadvantages of either one. We should note, however, that a matched-subjects design is not the same as a repeated-measures design. The matched pairs of participants in a matched-subjects design are not really the same people. Instead, they are merely "similar" individuals with the degree of similarity limited to the variable(s) that are used for the matching process.

LEARNING CHECK

LO13 1. Which of the following possibilities is a concern with a repeated-measures study?
   a. Negative values for the difference scores.
   b. Carryover effects.
   c. Obtaining a mean difference that is due to individual differences rather than treatment differences.
   d. All of the other options are major concerns.

LO13 2. For which of the following situations would an independent-measures design have the maximum advantage over a repeated-measures design?
   a. When individual differences are small and participating in one treatment is likely to produce a permanent change in the participant's performance.
   b. When individual differences are small and participating in one treatment is not likely to produce a permanent change in the participant's performance.
   c. When individual differences are large and participating in one treatment is likely to produce a permanent change in the participant's performance.
   d. When individual differences are large and participating in one treatment is not likely to produce a permanent change in the participant's performance.

LO14 3. A matched-subjects study comparing two treatments with 10 scores in each treatment requires a total of ___ participants and measures ___ score(s) for each individual.
   a. 10, 1
   b. 10, 2
   c. 20, 1
   d. 20, 2

ANSWERS 1. b   2. a   3. c

SUMMARY

1. In a repeated-measures research study, the same sample of individuals is tested in all the treatment conditions. This design literally repeats measurements on the same subjects.

2. The repeated-measures *t* test begins by computing a difference between the first and second measurements for each subject (or the difference for each matched pair). The difference scores, or *D* scores, are obtained by

$$D = X_2 - X_1$$

The sample mean, M_D, and sample variance, s^2, are used to summarize and describe the set of difference scores.

3. The formula for the repeated-measures *t* statistic is

$$t = \frac{M_D - \mu_D}{s_{M_D}}$$

In the formula, the null hypothesis specifies $\mu_D = 0$, and the estimated standard error is computed by

$$s_{M_D} = \sqrt{\frac{s^2}{n}}$$

4. For a repeated-measures design, effect size can be measured using either r^2 (the percentage of variance accounted for) or Cohen's *d* (the standardized mean difference). The value of r^2 is computed the same for both independent and repeated-measures designs:

$$r^2 = \frac{t^2}{t^2 + df}$$

Cohen's *d* is defined as the sample mean difference divided by standard deviation for both repeated- and independent-measures designs. For repeated-measures studies, Cohen's *d* is estimated as

$$\text{estimated } d = \frac{M_D}{s}$$

5. An alternative method for describing the size of the treatment effect is to construct a confidence interval for the population mean difference, μ_D. The confidence interval uses the repeated-measures *t* equation, solved for the unknown mean difference:

$$\mu_D = M_D \pm t s_{M_D}$$

First, select a level of confidence and then look up the corresponding *t* values. For example, for 95% confidence, use the range of *t* values that determine the middle 95% of the distribution. The *t* values are then used in the equation along with the values for the sample mean difference and the standard error, which are computed from the sample data.

6. A repeated-measures design may be preferred to an independent-measures study when one wants to observe changes in behavior in the same subjects, as in learning or developmental studies. An important advantage of the repeated-measures design is that it removes or reduces individual differences, which in turn lowers sample variability and tends to increase the chances for obtaining a significant result.

7. In a matched-subjects design the individuals in one sample are matched one-to-one with individuals in another sample. The matching is based on a variable (or variables) relevant to the study. The matched-subjects design has elements of an independent-measures study and a repeated-measures study, and is intended to produce the advantages of both designs without the disadvantages. However, the quality of a matched-subjects study is limited by the quality of the matching process.

KEY TERMS

repeated-measures design
  or within-subjects design (361)

difference scores (363)

estimated standard error for
  M_D (365)

repeated-measures *t* statistic (365)

individual differences (376)

order effects (378)

matched-subjects design (378)

FOCUS ON PROBLEM SOLVING

1. Once data have been collected, we must then select the appropriate statistical analysis. How can you tell whether the data call for a repeated-measures *t* test? Look at the experiment carefully. Is there only one sample of subjects? Are the same subjects tested

a second time? If your answers are yes to both of these questions, then a repeated-measures t test should be done. There is only one situation in which the repeated-measures t can be used for data from two samples, and that is for a *matched-subjects* study (page 378).

2. The repeated-measures t test is based on difference scores. In finding difference scores, be sure you are consistent with your method. That is, you may use either $X_2 - X_1$ or $X_1 - X_2$ to find D scores, but you must use the same method for all subjects.

DEMONSTRATION 11.1

A REPEATED-MEASURES *t* TEST

A major oil company would like to improve its tarnished image following a large oil spill. Its marketing department develops a short television commercial and tests it on a sample of $n = 7$ participants. People's attitudes about the company are measured with a short questionnaire, both before and after viewing the commercial. The data are as follows:

Person	X_1 (Before)	X_2 (After)	D (Difference)	
A	15	15	0	
B	11	13	+2	$\Sigma D = 21$
C	10	18	+8	
D	11	12	+1	$M_D = \frac{21}{7} = 3.00$
E	14	16	+2	
F	10	10	0	$SS = 74$
G	11	19	+8	

Was there a significant change? Note that participants are being tested twice—once before and once after viewing the commercial. Therefore, we have a repeated-measures design.

STEP 1   **State the hypotheses, and select an alpha level.**   The null hypothesis states that the commercial has no effect on people's attitude, or in symbols,

$$H_0: \mu_D = 0 \qquad \text{(The mean difference is zero.)}$$

The alternative hypothesis states that the commercial does alter attitudes about the company, or

$$H_1: \mu_D \neq 0 \qquad \text{(There is a mean change in attitudes.)}$$

For this demonstration, we will use an alpha level of .05 for a two-tailed test.

STEP 2   **Locate the critical region.**   Degrees of freedom for the repeated-measures t test are obtained by the formula

$$df = n - 1$$

For these data, degrees of freedom equal

$$df = 7 - 1 = 6$$

The t distribution table is consulted for a two-tailed test with $\alpha = .05$ for $df = 6$. The critical t values for the critical region are $t = \pm 2.447$.

STEP 3 **Compute the test statistic.** Once again, we suggest that the calculation of the *t* statistic be divided into a three-part process.

Variance for the D scores The variance for the sample of *D* scores is:

$$s^2 = \frac{SS}{n-1} = \frac{74}{6} = 12.33$$

Estimated standard error for M_D The estimated standard error for the sample mean difference is computed as follows:

$$s_{M_D} = \sqrt{\frac{s^2}{n}} = \sqrt{\frac{12.33}{7}} = \sqrt{1.76} = 1.33$$

The repeated-measures t statistic Now we have the information required to calculate the *t* statistic:

$$t = \frac{M_D - \mu_D}{s_{M_D}} = \frac{3-0}{1.33} = 2.26$$

STEP 4 **Make a decision about H_o, and state the conclusion.** The obtained *t* value is not extreme enough to fall in the critical region. Therefore, we fail to reject the null hypothesis. We conclude that the commercial did not produce a significant change in people's attitudes, $t(6) = 2.26$, $p > .05$, two-tailed. (Note that we state that *p* is *greater than* .05 because we failed to reject H_0.)

DEMONSTRATION 11.2

EFFECT SIZE FOR THE REPEATED-MEASURES *t*

We will estimate Cohen's *d* and calculate r^2 for the data in Demonstration 11.1. The data produced a sample mean difference of $M_D = 3.00$ with a sample variance of $s^2 = 12.33$. Based on these values, Cohen's *d* is

$$\text{estimated } d = \frac{\text{mean difference}}{\text{standard deviation}} = \frac{M_D}{s} = \frac{3.00}{\sqrt{12.33}} = \frac{3.00}{3.51} = 0.86$$

The hypothesis test produced $t = 2.26$ with $df = 6$. Based on these values,

$$r^2 = \frac{t^2}{t^2 + df} = \frac{(2.26)^2}{(2.26)^2 + 6} = \frac{5.11}{11.11} = 0.46 \quad (\text{or } 46\%)$$

SPSS®

While speeding tickets are unpleasant, speed enforcement is necessary to reduce the number of traffic fatalities and injuries. For safety's sake, you might think that perfect speed enforcement would be better than the imperfect speed enforcement that usually allows us to travel a few miles per hour over the speed limit without penalty. This idea was recently tested in a repeated-measures research design (Bowden, Loft, Tatsciore, & Visser, 2017). Participants in a video-game-like simulated driving task were asked to press a button on a steering wheel when they noticed a red dot positioned near street signs and pedestrians in the video game. They measured participants' delay in detecting the red dot. Each participant completed the simulated driving task under both normal speed enforcement (where participants were penalized for exceeding the speed limit by 4 miles per hour), and conservative speed enforcement (where participants were penalized for exceeding the speed limit by less than 1 mile per hour). Unsurprisingly, participants slowed down under conservative speed enforcement relative to the normal speed enforcement. However, participants were significantly slower to respond to red dots under conservative speed enforcement, which suggests that participants were distracted by the strict penalties for speeding. The data shown below are similar to those observed by the authors.

Participant	Normal Speed Enforcement	Conservative Speed Enforcement
A	7	10
B	8	7
C	4	14
D	6	13
E	3	11
F	9	10
G	4	4
H	7	11

Below are detailed instructions for using SPSS to perform the **Repeated-Measures *t* Test** that would be used to compare normal speed enforcement to conservative speed enforcement.

Data Entry

1. Use the **Variable View** of the data editor to create two new variables for the data above. Enter "Normal" in the **Name** field of the first variable. Select **Numeric** in the **Type** field and **Scale** in the **Measure** field. Enter a brief, descriptive title for the variable in the **Label** field (here, "Delay under normal speed enforcement" was used). Create the second variable in the same way, using "Conservative" in the **Name** field and "Delay under conservative speed enforcement" under the **Label** field.
2. Use the **Data View** of the data editor to enter the scores. Enter the data into two columns (Normal and Conservative) in the data editor with the first score for each participant in the first column and the second score in the second column. The two scores for each participant must be in the same row.

Data Analysis

1. Click **Analyze** on the tool bar, select **Compare Means**, and click on **Paired-Samples T Test**.
2. One at a time, highlight the column labels for the two data columns and click the arrow to move them into the **Paired Variables** box.
3. In addition to performing the hypothesis test, the program will compute a confidence interval for the population mean difference. The confidence level is automatically set at 95% but you can select **Options** and change the percentage.
4. Click **OK**.

SPSS Output

The output includes a table of sample statistics with the mean and standard deviation for each condition. A second table shows the correlation between the two sets of scores (correlations are presented in Chapter 15). The final table, which is split into two sections in the figure below, shows the results of the hypothesis test, including the mean and standard deviation for the difference scores, the standard error for the mean, a 95% confidence interval for the mean difference, and the values for *t*, *df*, and the level of significance (the *p* value for the test).

→ **T-Test**

Paired Samples Statistics

		Mean	N	Std. Deviation	Std. Error Mean
Pair 1	Delay under normal speed enforcement	6.0000	8	2.13809	.75593
	Delay under conservative speed enforcement	10.0000	8	3.20713	1.13389

Paired Samples Correlations

		N	Correlation	Sig.
Pair 1	Delay under normal speed enforcement & Delay under conservative speed enforcement	8	-.083	.844

Paired Samples Test

		Paired Differences							
					95% Confidence Interval of the Difference				
		Mean	Std. Deviation	Std. Error Mean	Lower	Upper	t	df	Sig. (2-tailed)
Pair 1	Delay under normal speed enforcement - Delay under conservative speed enforcement	-4.00000	4.00000	1.41421	-7.34408	-.65592	-2.828	7	.025

Source: SPSS®

Try It Yourself

Use SPSS to analyze the following scores.

Participant	Normal Speed Enforcement	Conservative Speed Enforcement
A	30.00	24.00
B	45.00	63.00
C	32.00	46.00
D	96.00	96.00
E	65.00	77.00
F	48.00	51.00
G	37.00	41.00
H	39.00	44.00
I	41.00	46.00
J	29.00	34.00
K	35.00	41.00

You should find that SPSS reports a significant difference between the participants for normal speed enforcement and conservative speed enforcement, $t(10) = -3.00$, $p = .013$.

PROBLEMS

1. For the each of the following studies determine whether a repeated-measures *t* test is the appropriate analysis. Explain your answers.

   a. A researcher is examining the effect of violent video games on behavior by comparing aggressive behaviors for one group who just finished playing a violent game with another group who played a neutral game.

   b. A researcher is examining the effect of humor on memory by presenting a group of participants with a series of humorous and not humorous sentences and then recording how many of each type of sentence is recalled by each participant.

   c. A researcher is evaluating the effectiveness of a new cholesterol medication by recording the cholesterol level for each individual in a sample before they start taking the medication and again after eight weeks with the medication.

2. What is the defining characteristic of a repeated-measures or within-subjects research design?

3. A researcher conducts an experiment comparing two treatment conditions with 22 scores in each treatment condition.
   a. If an independent-measures design is used, how many subjects are needed for the experiment?
   b. If a repeated-measures design is used, how many subjects are needed for the experiment?

4. A repeated-measures and an independent-measures study both produce a t statistic with $df = 15$. How many subjects participated in each experiment?

5. A sample of $n = 12$ individuals participates in a repeated-measures study that produces a sample mean difference of $M_D = 7.25$ with $SS = 396$ for the difference scores.
   a. Calculate the standard deviation for the sample of difference scores. Briefly explain what is measured by the standard deviation.
   b. Calculate the estimated standard error for the sample mean difference. Briefly explain what is measured by the estimated standard error.

6. How does the numerator of the repeated-measures t-statistic compare to the numerator of the single-sample t-statistic?

7. The following data are from a repeated-measures study examining the effect of a treatment by measuring a group of $n = 11$ participants before and after they receive the treatment.
   a. Calculate the difference scores and M_D.
   b. Compute SS, sample variance, and estimated standard error.
   c. Is there a significant treatment effect? Use $\alpha = .05$, two tails.

Participant	Before Treatment	After Treatment
A	66	84
B	50	44
C	38	52
D	58	56
E	50	52
F	34	42
G	44	51
H	42	49
I	62	67
J	50	57
K	56	62

8. The following data are from a repeated-measures study examining the effect of a treatment by measuring a group of $n = 9$ participants before and after they receive the treatment.

a. Calculate the difference scores and M_D.
b. Compute SS, sample variance, and estimated standard error.
c. Is there a significant treatment effect? Use $\alpha = .05$, two tails.

Participant	Before Treatment	After Treatment
A	82	89
B	64	67
C	76	79
D	6	8
E	38	40
F	150	147
G	10	14
H	4	11
I	16	18

9. When you get a surprisingly low price on a product do you assume that you got a really good deal or that you bought a low-quality product? Research indicates that you are more likely to associate low price and low quality if someone else makes the purchase rather than yourself (Yan & Sengupta, 2011). In a similar study, $n = 16$ participants were asked to rate the quality of low-priced items under two scenarios: purchased by a friend or purchased yourself. The results produced a mean difference of $M_D = 2.6$ and $SS = 135$, with self-purchases rated higher.
   a. Is the judged quality of objects significantly different for self-purchases than for purchases made by others? Use a two-tailed test with $\alpha = .05$.
   b. Compute Cohen's d to measure the size of the treatment effect.
   c. Write a sentence describing the outcome of the hypothesis test and the measure of effect size as it would appear in a research report.

10. The stimulant Ritalin has been shown to increase attention span and improve academic performance in children with ADHD (Evans et al., 2001). To demonstrate the effectiveness of the drug, a researcher selects a sample of $n = 20$ children diagnosed with the disorder and measures each child's attention span before and after taking the drug. The data show an average increase of attention span of $M_D = 4.8$ minutes with a variance of $s^2 = 125$ for the sample of difference scores.
   a. Is this result sufficient to conclude that Ritalin significantly improves attention span? Use a one-tailed test with $\alpha = .05$.
   b. Compute the 80% confidence interval for the mean change in attention span for the population.
   c. Write the results of the t test and the confidence interval as they would appear in a scientific journal article.

11. Callahan (2009) demonstrated that Tai Chi can significantly reduce symptoms for individuals with arthritis. Participants were 18 years old or older with doctor-diagnosed arthritis. Self-reports of pain and stiffness were measured at the beginning of an eight-week Tai Chi course and again at the end. Suppose that the data produced an average decrease in pain and stiffness of $M_D = 8.5$ points with a standard deviation of 21.5 for a sample of $n = 40$ participants.
    a. Use a two-tailed test with $\alpha = .05$ to determine whether the Tai Chi had a significant effect on pain and stiffness.
    b. Compute Cohen's *d* to measure the size of the treatment effect.

12. There is some evidence suggesting that you are likely to improve your test score if you rethink and change answers on a multiple-choice exam (Johnston, 1975). To examine this phenomenon, a teacher gave the same final exam to two sections of a psychology course. The students in one section were told to turn in their exams immediately after finishing, without changing any of their answers. In the other section, students were encouraged to reconsider each question and to change answers whenever they felt it was appropriate. Before the final exam, the teacher had matched nine students in the first section with nine students in the second section based on their midterm grades. For example, a student in the no-change section with an 89 on the midterm exam was matched with a student in the change section who also had an 89 on the midterm. The difference between the two final exam grades for each matched pair was computed, and the data showed that the students who were allowed to change answers scored higher by an average of $M_D = 7$ points with $SS = 288$.
    a. Do the data indicate a significant difference between the two conditions? Use a two-tailed test with $\alpha = .05$.
    b. Construct a 95% confidence interval to estimate the size of the population mean difference.
    c. Write a sentence demonstrating how the results of the hypothesis test and the confidence interval would appear in a research report.

13. Solve the following problems.
    a. A repeated-measures study with a sample of $n = 6$ participants produces a mean difference of $M_D = 4$ with $SS = 30$. Use a two-tailed hypothesis test with $\alpha = .05$ to determine whether this sample provides evidence of a significant treatment effect.
    b. Now assume that $SS = 480$ and repeat the hypothesis test.
    c. Explain how sample variability influences the likelihood of finding a significant mean difference.

14. Solve the following problems.
    a. A repeated-measures study with a sample of $n = 8$ participants produces a mean difference of $M_D = 3$ with a variance of $s^2 = 72$. Use a two-tailed hypothesis test with $\alpha = .05$ to determine whether it is likely that this sample came from a population with $\mu_D = 0$.
    b. Now assume that the sample mean difference is $M_D = 9$, and once again use a two-tailed hypothesis test with $\alpha = .05$ to determine whether it is likely that this sample came from a population with $\mu_D = 0$.
    c. Explain how the size of the sample mean difference influences the likelihood of finding a significant mean difference.

15. A sample of difference scores from a repeated-measures experiment has a mean of $M_D = 4$ with a standard deviation of $s = 6$.
    a. If $n = 9$, is this sample sufficient to reject the null hypothesis using a two-tailed test with $\alpha = .05$?
    b. Would you reject H_0 if $n = 36$? Again, assume a two-tailed test with $\alpha = .05$.
    c. Explain how the size of the sample influences the likelihood of finding a significant mean difference.

16. Participants enter a research study with unique characteristics that produce different scores from one person to another. For an independent-measures study, these individual differences can cause problems. Identify the problems and briefly explain how they are eliminated or reduced with a repeated-measures study.

17. Swearing is a common, almost reflexive, response to pain. Whether you knock your shin into the edge of a coffee table or smash your thumb with a hammer, most of us respond with a streak of obscenities. One question, however, is whether swearing has any effect on the amount of pain you feel. To address this issue, Stephens, Atkins, and Kingston (2009) conducted an experiment comparing swearing with other responses to pain. In the study, participants were asked to place one hand in icy cold water for as long as they could bear the pain. Half of the participants were told to repeat their favorite swear word over and over for as long as their hands were in the water. The other half repeated a neutral word. The researchers recorded how long each participant was able to tolerate the ice water. After a brief rest, the two groups switched words and repeated the ice water plunge. Thus, all the participants experienced both conditions (swearing and neutral) with half swearing on their first plunge and half on their second. The data in the following table are representative of the results obtained in the study and represented the reports of pain level of $n = 9$ participants.

Participant	Neutral Word	Swearing
A	9	7
B	9	8
C	9	5
D	4	5
E	10	8
F	9	4
G	6	5
H	10	10
I	6	2

a. Treat the data as if the scores are from an independent-measures study using two separate samples, each with $n = 9$ participants. Compute the pooled variance, the estimated standard error for the mean difference, and the independent-measures t statistic. Using $\alpha = .05$, two-tailed, is there a significant difference between the two sets of scores?

b. Now assume that the data are from a repeated-measures study using the same sample of $n = 9$ participants in both treatment conditions. Compute the variance for the sample of difference scores, the estimated standard error for the mean difference, and the repeated-measures t statistic. Using $\alpha = .05$, is there a significant difference between the two sets of scores? (You should find that the repeated-measures design substantially reduces the variance and increases the likelihood of rejecting H_0.)

18. Playing three-dimensional video games can improve cognitive function in older adults. In a recent experiment (West et al., 2017), a sample of $n = 15$ older adults were instructed to play Super Mario 64 and similar 3-D games for six months. Participants' scores on a cognitive assessment improved after the six-month treatment, relative to their scores on the cognitive assessment administered prior to the treatment. The authors observed a mean difference score of $M_D = 1.40$, with a standard deviation of the difference scores of $s = 2.59$.
   a. Test the hypothesis that the treatment significantly affected cognitive performance. Use a one-tailed test with $\alpha = .05$.
   b. Compute r^2 as a measurement of effect size.
   c. Write a sentence demonstrating how the results of the hypothesis test and the effect size would appear in a research report.

19. Problem 17 demonstrates that removing individual differences can substantially reduce variance and lower the standard error. However, this benefit only occurs if the individual differences are consistent across treatment conditions. In Problem 17, for example, the participants with the highest scores in the neutral-word condition also had the highest scores in the swear-word condition. Similarly, participants with the lowest scores in the first condition also had the lowest scores in the second condition. To construct the following data, we started with the scores in Problem 17 and scrambled the scores in Treatment 2 to eliminate the consistency of the individual differences.

Participant	Neutral Word	Swearing
A	9	5
B	9	2
C	9	5
D	4	10
E	10	8
F	9	4
G	6	7
H	10	5
I	6	8

a. If the data were from an independent-measures study using two separate samples, each with $n = 9$ participants, what value would be obtained for the independent-measures t statistic? Note: The scores in each treatment, the sample means, and the SS values are the same as in Problem 17. Nothing has changed. With $\alpha = .05$, is there a significant difference between the two treatment conditions?

b. Now assume that the data are from a repeated-measures study using the same sample of $n = 9$ participants in both treatment conditions. Compute the variance for the sample of difference scores, the estimated standard error for the mean difference, and the repeated-measures t statistic. Using $\alpha = .05$, is there a significant difference between the two sets of scores? (Because there no longer are consistent individual differences you should find that the repeated-measures t no longer reduces the variance.)

20. Exercise is known to produce positive psychological effects. Interestingly, not all exercise is equally effective. It turns out that exercising in a natural environment (e.g., jogging in the woods) produces better psychological outcomes than exercising in urban environments or in homes (Mackay & Neill, 2010). Suppose that a sports psychologist is interested in testing whether there is a difference between exercise in nature and exercise in the lab with respect to post-exercise anxiety levels. The researcher recruits $n = 7$ participants who exercise in the lab and exercise on a nature trail. The data below represent the anxiety scores that were measured after each exercise session.

Participant	Anxiety after Exercising in Lab	Anxiety after Exercising in Nature
A	32	8
B	66	68
C	52	48
D	48	37
E	52	44
F	48	38
G	52	44

a. Treat the data as if the scores are from an independent-measures study using two separate samples, each with $n = 7$ participants. Compute the pooled variance, the estimated standard error for the mean difference, and the independent-measures t statistic. Using $\alpha = .05$, is there a significant difference between the two sets of scores?

b. Now assume that the data are from a repeated-measures study using the same sample of $n = 7$ participants in both treatment conditions. Compute the variance for the sample of difference scores, the estimated standard error for the mean difference, and the repeated-measures t statistic. Using $\alpha = .05$, is there a significant difference between the two sets of scores?

21. Gamification refers to the application of game design and development to social, industrial, and educational settings. For example, a gamification program might award points or achievements to people for reaching specific goals. A recent experiment on gamification in the workplace revealed that machinists' motivation to work improved when they were given feedback about their job performance through a game-like smartphone app (Liu, Huang, & Zhang, 2017). Data like those observed by the authors are listed below.

Rated Motivation to Work before Gamification	Rated Motivation to Work after Gamification
2	7
2	1
3	4
9	10
6	8
9	10
6	9
5	6
7	12

a. Test the hypothesis that gamification affected participants' motivation to work. Use a two-tailed test with $\alpha = .05$.

b. Compute Cohen's d to measure the size of the treatment effect.

22. To construct the following data, we started with the scores in Problem 20 and scrambled the scores in Treatment 2 to eliminate the consistency of the individual differences.

Participant	Anxiety after Exercising in Lab	Anxiety after Exercising in Nature
A	32	37
B	66	68
C	52	44
D	48	8
E	52	44
F	48	48
G	52	38

a. If the data were from an independent-measures study using two separate samples, each with $n = 7$ participants, what value would be obtained for the independent-measures t statistic? With $\alpha = .05$, is there a significant difference between the two treatment conditions?

b. Now assume that the data are from a repeated-measures study using the same sample of $n = 7$ participants in both treatment conditions. Compute the variance for the sample of difference scores, the estimated standard error for the mean difference, and the repeated-measures t statistic. Using $\alpha = .05$, is there a significant difference between the two sets of scores?

23. Explain the difference between a matched-subjects design and a repeated-measures design.

24. A researcher conducts an experiment comparing two treatment conditions with 20 scores in each condition.
a. If an independent-measures design is used, how many participants are needed for the study?
b. If a repeated-measures design is used, how many participants are needed for the study?
c. If a matched-subjects design is used, how many participants are needed for the study?

25. A repeated-measures, a matched-subjects, and an independent-measures study all produce a t statistic with $df = 10$. How many participants were used in each study?

26. Traumatic brain injury (TBI) is a significant health problem. TBI is caused by impacts to the head that might occur during contact sports, motor vehicle accidents, and similar events. TBI is known to produce cognitive impairments and reductions in brain volume. In a recent, repeated-measures study on TBI, Zagorchev et al. (2016) observed that the size of the

amygdala among mild TBI patients was reduced at 12 months after injury, relative to two months after injury. Suppose that a researcher is interested in replicating and extending this observation. She recruits $n = 8$ participants with mild TBI and records the volume of a brain region at 2 months and again at 12 months. Her data are listed below.

Participant	Volume in One-Tenth of a Cubic Millimeter at 2 Months	Volume in One-Tenth of a Cubic Millimeter at 12 Months
A	15.6	15.7
B	21.6	16.9
C	22.6	18.7
D	17.5	16.8
E	15.1	11.2
F	21.7	20.2
G	20.1	18.2
H	24.0	22.1

a. Test the hypothesis that volume of the brain region changed between 2 and 12 months. Use $\alpha = .05$, two-tailed.
b. Compute the 80% confidence interval for the mean change in attention span for the population.

27. If you are using coffee to compensate for sleep loss, you might want to consider drinking your coffee under blue light. Beaven and Ekstrom (2013) recruited $n = 21$ participants who completed a series of cognitive alertness and reaction time tasks under conditions of caffeine consumption or exposure to blue light (or both). They discovered that one hour of exposure to blue light decreased reaction time among participants who had consumed caffeine, relative to a treatment that received one hour of exposure to white light. Suppose that a researcher replicates this observation with a sample of $n = 21$ participants. She measured the amount of delay in responding to a stimulus after exposure to white light and measured the amount of delay in responding to a stimulus after exposure to blue light.

a. The researcher observes a mean reaction time of $M = 432$ milliseconds ($SS = 1,280$) after exposure to white light and $M = 400$ milliseconds ($SS = 1,000$) after exposure to blue light. Do you have enough information to conduct a related-samples t-test? Explain your answer.
b. Now assume that $M_D = -32$ milliseconds and that the variance of the difference scores is $s^2 = 5,376$. Test the hypothesis that blue light decreased the delay to respond. Use $\alpha = .05$, one-tailed.

Introduction to Analysis of Variance

clivewa/Shutterstock.com

PREVIEW

Some of us have difficulty learning routes from maps. Fortunately, many automobiles have navigation systems to help those of us that get easily lost. What makes learning directions from maps difficult? Memory tasks involve encoding information in a form that makes it more readily recalled. Learning to navigate with maps requires encoding spatial information into memory storage. In a study of encoding method, researchers have examined the role of hand gestures in learning and recalling routes from maps (So, Ching, Lim, Cheng, & Ip, 2014).

Participants were given maps to study and asked to learn a route from a starting point to a destination. Then the maps were taken away and the participants were assigned to one of four rehearsal groups. In Group 1, participants were told to rehearse the route by visualizing a mental image of the map and moving a hand in the air through the route. For Group 2, participants rehearsed the route with hand movements by drawing it on a blank sheet of paper. In Group 3, participants were instructed to visualize the route, but their hand gestures were prevented because they held a softball with both hands during the rehearsal period. Group 4 was a control group that did no rehearsal. They were given a sheet of paper that contained alphabet letters, which they read aloud to prevent rehearsal. After the rehearsal period the participants were tested for how well they remembered the route.

The results revealed a statistically significant difference between groups. Rehearsal by hand gestures in the air resulted in the best recall of the map route, followed by the group that did rehearsal with a hand drawing of the route. There was no difference between the group that did rehearsal with restricted hand movements and the control group. The researchers concluded that gesturing with hands facilitates encoding the spatial information into memory, and thus, recall of the route was better. If we generalize these results to everyday life, then when you are studying directions to get someplace you have never visited, it should help if you use a finger to trace the route before departing. One could conclude from the results of this study that "it's okay to point."

Notice two things about this study. First, it is an independent-measures design. There is a different sample of participants for each treatment condition. Second, there are four treatment conditions. We have previously introduced the independent-measures design in Chapter 10, in which there were two independent samples—one for each of the two treatment conditions. In those circumstances, the independent-measures t test was used for the hypothesis test. When there are more than two treatment conditions for an independent-measures study, the t statistic cannot be used. In this chapter we will introduce analysis of variance, a hypothesis test that can be used for independent-measures studies in situations when there are more than two treatment conditions.

12-1 | Introduction: An Overview of Analysis of Variance

LEARNING OBJECTIVES

1. Describe the terminology that is used for ANOVA, especially the terms *factor* and *level*, identify the hypotheses for this test, and identify each in the context of a research example.
2. Identify the circumstances in which you should use ANOVA instead of t tests to evaluate mean differences, and explain why.
3. Describe the F-ratio that is used in ANOVA and explain how it is related to the t statistic.

Analysis of variance (ANOVA) is a hypothesis-testing procedure that is used to evaluate mean differences between two or more treatments (or populations). As with all inferential procedures, ANOVA uses sample data as the basis for drawing general conclusions about populations. It may appear that ANOVA and t tests are simply two different ways of doing exactly the same job: testing for mean differences. In some respects, this is true—both tests use sample data to test hypotheses about population means. However, ANOVA has a tremendous advantage over *t* tests. Specifically, *t* tests are limited to situations in which there are only two treatments to compare. The major advantage of ANOVA is that it can be used to compare *two or more treatments*. Thus, ANOVA provides researchers with much greater flexibility in designing experiments and interpreting results.

Figure 12.1 shows a typical research situation for which ANOVA would be used. Note that the study involves three samples representing three populations. The goal of the analysis is to determine whether the mean differences observed among the samples provide enough evidence to conclude that there are mean differences among the three populations. Specifically, we must decide between two interpretations:

1. There really are no differences between the populations (or treatments). The observed differences between the sample means are caused by random, unsystematic factors (sampling error) that differentiate one sample from another.

2. The populations (or treatments) really do have different means, and these population mean differences are responsible for causing systematic differences between the sample means.

You should recognize that these two interpretations correspond to the two hypotheses (null and alternative) that are part of the general hypothesis-testing procedure.

■ Terminology in Analysis of Variance

Before we continue, it is necessary to introduce some of the terminology that is used to describe the research situation shown in Figure 12.1. Recall (from Chapter 1) that when a researcher manipulates a variable to create the treatment conditions in an experiment, the variable is called an independent variable. For example, Figure 12.1 could represent a study examining driving performance under three different phone conditions: driving with no phone, talking on a hands-free phone, and talking on a hand-held phone. Note that the three conditions are created by the researcher. On the other hand, when a researcher uses a nonmanipulated variable to designate groups, the variable is called a *quasi-independent variable* (Chapter 1, page 27). For example, the three groups in Figure 12.1 could represent six-year-old, eight-year-old, and ten-year-old children. In the context of ANOVA, an independent variable or a quasi-independent variable is called a *factor*. Thus, Figure 12.1 could represent

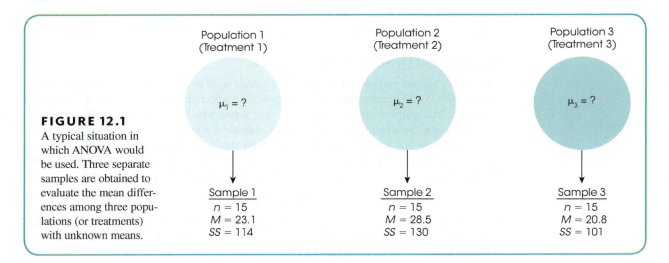

FIGURE 12.1
A typical situation in which ANOVA would be used. Three separate samples are obtained to evaluate the mean differences among three populations (or treatments) with unknown means.

Population 1 (Treatment 1)   Population 2 (Treatment 2)   Population 3 (Treatment 3)

$\mu_1 = ?$ $\mu_2 = ?$ $\mu_3 = ?$

Sample 1
$n = 15$
$M = 23.1$
$SS = 114$

Sample 2
$n = 15$
$M = 28.5$
$SS = 130$

Sample 3
$n = 15$
$M = 20.8$
$SS = 101$

an experimental study in which the telephone condition is the factor being evaluated, or it could represent a nonexperimental study in which age is the factor being examined.

> In analysis of variance, the variable (independent or quasi-independent) that designates the groups being compared is called a **factor**.

In addition, the individual groups or treatment conditions that are used to make up a factor are called the *levels* of the factor. For example, a study that examined performance under three different telephone conditions would have three levels of the factor.

> The individual conditions or values that make up a factor are called the **levels** of the factor.

Like the *t* tests presented in Chapters 10 and 11, ANOVA can be used with either an independent-measures or a repeated-measures design. Recall that an independent-measures design means that there is a separate group of participants for each of the treatments (or populations) being compared. In a repeated-measures design, on the other hand, the same group is tested in all of the different treatment conditions. In addition, ANOVA can be used to evaluate the results from a research study that involves more than one factor. For example, a researcher may want to compare two different therapy techniques, examining their immediate effectiveness as well as the persistence of their effectiveness over time. In this case, the ANOVA would evaluate mean differences between the two therapies as well as mean differences between the scores obtained at different times. A study that combines two factors is called a *two-factor design* or a *factorial design*. The ability to combine different factors and to mix different designs within one study provides researchers with the flexibility to develop studies that address scientific questions that could not be answered by a single design using a single factor.

Although ANOVA can be used in a wide variety of research situations, this chapter introduces ANOVA in its simplest form. Specifically, we consider only *single-factor designs*. That is, we examine studies that have only one independent variable (or only one quasi-independent variable). Second, we consider only *independent-measures designs*; that is, studies that use a separate group of participants for each treatment condition. The basic logic and procedures that are presented in this chapter form the foundation for more complex applications of ANOVA. For example, in Chapter 13 we extend the analysis to two-factor designs. But for now, in this chapter, we limit our discussion of ANOVA to *single-factor, independent-measures designs*.

■ Statistical Hypotheses for ANOVA

The research situation shown in Figure 12.1 can be used to introduce the statistical hypotheses for ANOVA. Three samples of participants are selected, one sample for each treatment condition. The purpose of the study is to determine whether there are significant differences between the treatment conditions. In statistical terms, we want to decide between two hypotheses: the null hypothesis (H_0), which states that the treatment conditions have no effect on the participant's scores; and the alternative hypothesis (H_1), which states that the treatment conditions do affect scores. In symbols, the null hypothesis states

$$H_0: \mu_1 = \mu_2 = \mu_3$$

In words, the null hypothesis states that the treatment conditions have no effect on performance. That is, the population means for the three conditions are all the same. In general, H_0 states that there is no treatment effect.

The alternative hypothesis states that the population means are not all the same:

H_1: There is at least one mean difference among the populations.

In general, H_1 states that the treatment conditions are not all the same; that is, there is a real *treatment effect*. As always, the hypotheses are stated in terms of population parameters, even though we use sample data to test them.

Notice that we are not stating a specific alternative hypothesis. This is because many different alternatives are possible, and it would be tedious to list them all. One alternative, for example, would be that the first two populations are identical, but that the third is different. Another alternative states that the last two means are the same, but that the first is different. Other alternatives might be

H_1: $\mu_1 \neq \mu_2 \neq \mu_3$ (All three means are different.)

H_1: $\mu_1 = \mu_3$, but μ_2 is different.

We should point out that a researcher typically entertains only one (or at most a few) of these alternative hypotheses. Usually a theory or the outcomes of previous studies will dictate a specific prediction concerning the treatment effect. For the sake of simplicity, we will state a general alternative hypothesis rather than try to list all possible specific alternatives.

■ Type I Errors and Multiple-Hypothesis Tests

If we already have t tests for comparing mean differences, you might wonder why ANOVA is necessary. Why create a whole new hypothesis-testing procedure that simply duplicates what the t tests can already do? The answer to this question is based on a concern about Type I errors.

Remember that each time you do a hypothesis test, you select an alpha level that determines the risk of a Type I error (see Chapter 8 page 257). With $\alpha = .05$, for example, there is a 5% (or a 1-in-20) risk of a Type I error, whenever your decision is to reject the null hypothesis. Often, a single experiment requires several hypothesis tests to evaluate all the mean differences. However, each test has a risk of a Type I error, and the more tests you do, the more risk there is.

For this reason, researchers often make a distinction between the *testwise alpha level* and the *experimentwise alpha level*. The testwise alpha level is simply the alpha level you select for each individual hypothesis test. The experimentwise alpha level is the total probability of a Type I error accumulated from all of the separate tests in the experiment. As the number of separate tests increases, so does the experimentwise alpha level.

> The **testwise alpha level** is the risk of a Type I error, or alpha level, for an individual hypothesis test.
>
> When an experiment involves several different hypothesis tests, the **experimentwise alpha level** is the total probability of a Type I error that is accumulated from all of the individual tests in the experiment. Typically, the experimentwise alpha level is substantially greater than the value of alpha used for any one of the individual tests.

For example, an experiment involving three treatments would require three separate t tests to compare all of the mean differences:

Test 1 compares Treatment I versus Treatment II.

Test 2 compares Treatment I versus Treatment III.

Test 3 compares Treatment II versus Treatment III.

If all tests use $\alpha = .05$, then there is a 5% risk of a Type I error for the first test, a 5% risk for the second test, and another 5% risk for the third test. The three separate tests accumulate to produce a relatively large experimentwise alpha level. The advantage of ANOVA is that it performs all three comparisons simultaneously in one hypothesis test. Thus, no matter how many different means are being compared, ANOVA uses one test with one alpha level to evaluate the mean differences and thereby avoids the problem of an inflated experiment-wise alpha level.

■ The Test Statistic for ANOVA

The test statistic for ANOVA is very similar to the t statistics used in earlier chapters. For the t statistic, we first computed the standard error, which measures the difference between two sample means that is reasonable to expect if there is no treatment effect (that is, if H_0 is true). Then we computed the t statistic with the following structure:

$$t = \frac{\text{obtained difference between two sample means}}{\text{standard error (the difference with no treatment effect)}}$$

For ANOVA, however, we want to compare differences among two *or more* sample means. With more than two samples, the concept of "difference between sample means" becomes difficult to define or measure. For example, if there are only two samples and they have means of $M = 20$ and $M = 30$, then there is a 10-point difference between the sample means. Suppose, however, that we add a third sample with a mean of $M = 35$. Now how much difference is there between the sample means? It should be clear that we have a problem. The solution to this problem is to use variance to define and measure the size of the differences among the sample means. Consider the following two sets of sample means:

Set 1	Set 2
$M_1 = 20$	$M_1 = 28$
$M_2 = 30$	$M_2 = 30$
$M_3 = 35$	$M_3 = 31$

If you compute the variance for the three numbers in each set, then the variance is $s^2 = 58.33$ for Set 1 and $s^2 = 2.33$ for Set 2. Notice that the two variances provide an accurate representation of the size of the differences. In Set 1 there are relatively large differences between sample means, and the variance is relatively large. In Set 2 the mean differences are small, and the variance is small.

Thus, we can use variance to measure sample mean differences when there are two or more samples. The test statistic for ANOVA uses this fact to compute an F-*ratio* with the following structure:

$$F = \frac{\text{variance (differences) between sample means}}{\text{variance (difference) expected with no treatment effect}}$$

Note that the F-ratio has the same basic structure as the t statistic but is based on *variance* instead of sample mean *difference*. The variance in the numerator of the F-ratio provides a single number that measures the differences among all of the sample means. The variance in the denominator of the F-ratio, like the standard error in the denominator of the t statistic, measures the mean differences that would be expected if there is no treatment effect. Thus, the t statistic and the F-ratio provide the same basic information. In each case, a large value for the test statistic provides evidence that the sample mean differences (numerator) are larger than would be expected if there were no treatment effects (denominator).

12-2 The Logic of Analysis of Variance

LEARNING OBJECTIVE

4. Identify the sources that contribute to the variance between-treatments and the variance within-treatments, and describe how these two variances are compared in the F-ratio to evaluate the null hypothesis.

The formulas and calculations required in ANOVA are somewhat complicated, but the logic that underlies the whole procedure is fairly straightforward. Therefore, this section gives a general picture of ANOVA before we start looking at the details. We will introduce the logic of ANOVA with the help of the data in Table 12.1. These data represent the results of an independent-measures experiment using three separate samples, each with $n = 5$ participants, to compare performance in three treatment conditions.

TABLE 12.1

Data from an experiment examining performance in three treatment conditions.

Treatment 1 (Sample 1)	Treatment 2 (Sample 2)	Treatment 3 (Sample 3)
4	0	1
3	1	2
6	3	2
3	1	0
4	0	0
$M = 4$	$M = 1$	$M = 1$

One obvious characteristic of the data in Table 12.1 is that the scores are not all the same. In everyday language, the scores are different; in statistical terms, the scores are variable. Our goal is to measure the amount of variability (the size of the differences) and to explain why the scores are different.

The first step is to determine the total variability for the entire set of data. To compute the total variability, we combine all the scores from all the separate samples to obtain one general measure of variability for the complete experiment. Once we have measured the total variability, we can begin to break it apart into separate components. The word *analysis* means dividing into smaller parts. Because we are going to analyze variability, the process is called *analysis of variance*. This analysis process divides the total variability in the entire data set into two basic components.

1. **Between-Treatments Variance**. Looking at the data in Table 12.1, we clearly see that much of the variability in the scores results from general differences between treatment conditions. For example, the scores in Treatment 1 tend to be much higher ($M = 4$) than the scores in Treatment 2 ($M = 1$). We will calculate the variance between treatments to provide a measure of the overall differences between treatment conditions. Notice that the variance between treatments is really measuring the differences between sample means.

2. **Within-Treatments Variance**. In addition to the general differences between treatment conditions, there is variability within each sample. Looking again at Table 12.1, we see that the scores in Treatment 1 are not all the same; they are variable. The within-treatments variance provides a measure of the variability inside each treatment condition.

Analyzing the total variability into these two components is the heart of ANOVA. We will now examine each of the components in more detail.

■ Between-Treatments Variance

Remember that calculating variance is simply a method for measuring how big the differences are for a set of numbers. When you see the term *variance*, you can automatically translate it into the term *differences*. Thus, the *between-treatments variance* simply measures how much difference exists between the treatment conditions. There are two possible explanations for these between-treatment differences:

1. The differences between treatments are not caused by any treatment effect but are simply the naturally occurring, random and unsystematic differences that exist between one sample and another. That is, the differences are the result of sampling error.

2. The differences between treatments have been caused by the *treatment effects*. For example, if treatments really do affect performance, then scores in one treatment should be systematically different from scores in another condition.

Thus, when we compute the between-treatments variance, we are measuring differences that could be caused by a systematic treatment effect or could simply be random and unsystematic mean differences caused by sampling error. To demonstrate that there really is a treatment effect, we must establish that the differences between treatments are bigger than would be expected by sampling error alone. To accomplish this goal, we determine how big the differences are when there is no systematic treatment effect; that is, we measure how much difference (or variance) can be explained by random and unsystematic factors. To measure these differences, we compute the variance within treatments.

■ Within-Treatments Variance

Inside each treatment condition, we have a set of individuals who all receive exactly the same treatment; that is, the researcher does not do anything that would cause these individuals to have different scores. In Table 12.1, for example, the data show that five individuals were tested in Treatment 2 (Sample 2). Although these five individuals all received exactly the same treatment, their scores are different. Why are the scores different? The answer is that there is no specific cause for the differences. Instead, the differences that exist within a treatment represent random and unsystematic differences that occur when there are no treatment effects causing the scores to be different. Thus, the *within-treatments variance* provides a measure of how big the differences are when H_0 is true.

Figure 12.2 shows the overall ANOVA and identifies the sources of variability that are measured by each of the two basic components.

■ The *F*-Ratio: The Test Statistic for ANOVA

Once we have analyzed the total variability into two basic components (between treatments and within treatments), we simply compare them. The comparison is made by computing an F-*ratio*. For the independent-measures ANOVA, the *F*-ratio has the following structure:

$$F = \frac{\text{variance between treatments}}{\text{variance within treatments}} = \frac{\text{differences including any treatment effects}}{\text{differences with no treatment effects}} \quad \text{(12.1)}$$

When we express each component of variability in terms of its sources (see Figure 12.2), the structure of the *F*-ratio is

$$F = \frac{\text{systematic treatment effects} + \text{random and unsystematic differences}}{\text{random and unsystematic differences}} \quad \text{(12.2)}$$

The value obtained for the *F*-ratio helps determine whether any treatment effects exist. Consider the following two possibilities:

1. When there are no systematic treatment effects, the differences between treatments (numerator) are entirely caused by random, unsystematic factors. In this case, the numerator and the denominator of the *F*-ratio are both measuring random differences and should be roughly the same size. With the numerator and denominator roughly

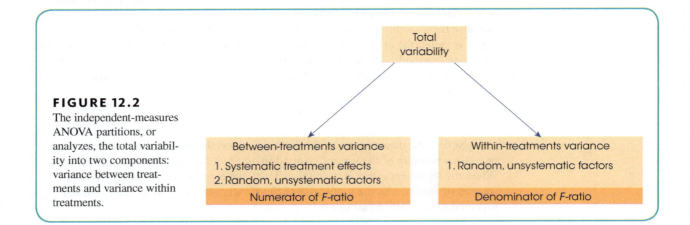

FIGURE 12.2

The independent-measures ANOVA partitions, or analyzes, the total variability into two components: variance between treatments and variance within treatments.

equal, the F-ratio should have a value around 1.00. In terms of the formula, when the treatment effect is zero, we obtain

$$F = \frac{0 + \text{random and unsystematic differences}}{\text{random and unsystematic differences}}$$

Thus, an F-ratio near 1.00 indicates that the differences between treatments (numerator) are random and unsystematic, just like the differences in the denominator. With an F-ratio near 1.00, we conclude that there is no evidence to suggest that the treatment has any effect.

2. When the treatment does have an effect, causing systematic differences between samples, then the combination of systematic and random differences in the numerator should be larger than the random differences alone in the denominator. In this case, the numerator of the F-ratio should be noticeably larger than the denominator, and we should obtain an F-ratio noticeably larger than 1.00. Thus, a large F-ratio is evidence for the existence of systematic treatment effects; that is, there are significant differences between treatments.

The random and unsystematic variability in the data can come from many sources. For example, we have already introduced participant variables as differences that exist between people before they receive any treatments. Participant variables may include individual differences in motivation, skills, attitudes, past experiences, and so on. The differences observed between *and* within treatment groups can be due to individual differences because samples of different participants are used. This variability is random and unsystematic and a result of sampling error.

It also is possible that there might be more variability both between and within treatment groups because the participants are unintentionally treated in a different way. Suppose the experiment requires that the researcher read a set of instructions to participants before they work on a problem-solving task. If there is variation in how the researcher reads the instructions from one participant to another in the study, even in tone of voice, it might introduce unsystematic variability.

Another possible example of random and unsystematic variability is *error of measurement*. Anytime a measurement is made, the possibility of error is introduced. This type of error is inherent in the measurement tool. For example, if you were given a small ruler to measure the dimensions of a room to the nearest millimeter, each time you measured the room you will likely come up with a different answer. Similarly, the tools and instruments we use to measure behavior may introduce some error, whether it is a stopwatch, a test of memory recall, or a personality inventory.

Because the denominator of the F-ratio measures only random and unsystematic variability, it is called the *error term*. The numerator of the F-ratio always includes the same unsystematic variability as in the error term, but it also includes any systematic differences caused by the treatment effect. The goal of ANOVA is to find out whether a treatment effect exists.

> For ANOVA, the denominator of the F-ratio is called the **error term**. The error term provides a measure of the variance caused by *random and unsystematic differences*. When the treatment effect is zero (H_0 is true), the error term measures the same sources of variance as the numerator of the F-ratio, so the value of the F-ratio is expected to be nearly equal to 1.00.

LEARNING CHECK **LO4** **1.** For an analysis of variance, the systematic treatment effects in a study contribute to the _____ and appears in the _____ of the F-ratio.

    a. variance between treatments, numerator

    b. variance between treatments, denominator

    c. variance within treatments, numerator

    d. variance within treatments, denominator

LO4 **2.** What is suggested by a value of 1 for the F-ratio in an ANOVA?

    a. There is a treatment effect and the null hypothesis should be rejected.

    b. There is no treatment effect and the null hypothesis should be rejected.

    c. There is a treatment effect and you should fail to reject the null hypothesis.

    d. There is no treatment effect and you should fail to reject the null hypothesis.

ANSWERS **1. a** **2. d**

12-3   ANOVA Notation and Formulas

LEARNING OBJECTIVE

5. Calculate the three SS values, the three df values, and the two mean squares (MS values) that are needed for the F-ratio and describe the relationships among them.

Because ANOVA typically is used to examine data from more than two treatment conditions (and more than two samples), we need a notational system to keep track of all the individual scores and totals. To help introduce this notational system, we use the data from the following example.

EXAMPLE 12.1   Over the years, students and teachers have developed a variety of strategies to help prepare for an upcoming test. But how do you know which strategy is best? A partial answer to this question comes from a research study comparing three different strategies (Weinstein, McDermott, & Roediger, 2010). In the study, students read a passage knowing that they would be tested on the material. In one condition, participants simply reread the material to be tested. In a second condition, the students answered prepared comprehension questions about the material, and in a third condition, the students generated and answered their own questions.

The data in Table 12.2 show the pattern of results obtained in the Weinstein et al. (2010) study. The data show the notation and statistics that will be described. ■

    1. The letter k is used to identify the number of treatment conditions—that is, the number of levels of the factor. For an independent-measures study, k also specifies the number of separate samples. For the data in Table 12.2, there are three treatments, so $k = 3$.

    2. The number of scores in each treatment is identified by a lowercase letter n. For the example in Table 12.2, $n = 6$ for all the treatments. If the samples are of different sizes, you can identify a specific sample by using a subscript. For example, n_2 is the number of scores in Treatment 2.

    3. The total number of scores in the entire study is specified by a capital letter N. When all the samples are the same size (n is constant), $N = kn$. For the data in

TABLE 12.2

Test scores for students using three different study strategies.

Read and Reread	Read, Then Answer Prepared Questions	Read, Then Create and Answer Questions	
2	5	8	
3	9	6	
8	10	12	
6	13	11	
5	8	11	
6	9	12	
$n_1 = 6$	$n_2 = 6$	$n_3 = 6$	$N = 18$
$T_1 = \Sigma X = 30$	$T_2 = 54$	$T_3 = 60$	$G = 30 + 54 + 60 = 144$
$M_1 = 5$	$M_2 = 9$	$M_3 = 10$	$k = 3$
$\Sigma X_1^2 = 174$	$\Sigma X_2^2 = 520$	$\Sigma X_3^2 = 630$	$\Sigma X^2 = 174 + 520 + 630 = 1324$
$SS_1 = 24$	$SS_2 = 34$	$SS_3 = 30$	

Table 12.2, there are $n = 6$ scores in each of the $k = 3$ treatments, so we have a total of $N = 3(6) = 18$ scores in the entire study.

Because ANOVA formulas require ΣX for each treatment and ΣX for the entire set of scores, we have introduced new notation (T and G) to help identify which ΣX is being used. Remember: *T* stands for *treatment total*, and *G* stands for *grand total*.

4. The sum of the scores (ΣX) for each treatment condition is identified by the capital letter T (for treatment total). The total for a specific treatment can be identified by adding a numerical subscript to the T. For example, the total for the second treatment in Table 12.2 is $T_2 = 54$.

5. The sum of all the scores in the research study (the grand total) is identified by G. You can compute G by adding up all N scores or by adding up the treatment totals: $G = \Sigma T$.

6. Although there is no new notation involved, we also have computed SS and M for each sample, and we have calculated ΣX^2 for the entire set of $N = 18$ scores in the study. These values are given in Table 12.2 and are important in the formulas and calculations for ANOVA.

Finally, we should note that there is no universally accepted notation for ANOVA. Although we are using Gs and Ts, for example, you may find that other sources use other symbols.

■ ANOVA Formulas

Because ANOVA requires extensive calculations and many formulas, one common problem for students is simply keeping track of the different formulas and numbers. Therefore, we will examine the general structure of the procedure and look at the organization of the calculations before we introduce the individual formulas.

1. The final calculation for ANOVA is the F-ratio, which is composed of two variances:

$$F = \frac{\text{variance between treatments}}{\text{variance within treatments}}$$

2. Each of the two variances in the F-ratio is calculated using the basic formula for sample variance:

$$\text{sample variance} = s^2 = \frac{SS}{df}$$

FIGURE 12.3

The structure and sequence of calculations for the ANOVA.

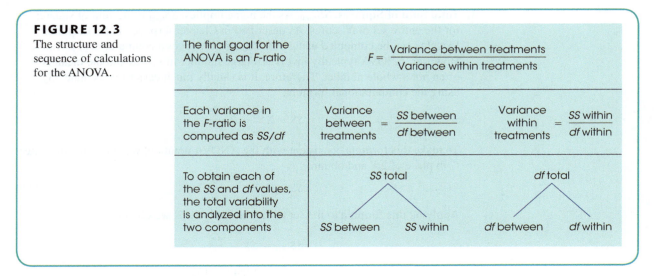

The final goal for the ANOVA is an F-ratio	$F = \dfrac{\text{Variance between treatments}}{\text{Variance within treatments}}$		
Each variance in the F-ratio is computed as SS/df	$\text{Variance between treatments} = \dfrac{SS \text{ between}}{df \text{ between}}$	$\text{Variance within treatments} = \dfrac{SS \text{ within}}{df \text{ within}}$	
To obtain each of the SS and df values, the total variability is analyzed into the two components	SS total → SS between, SS within	df total → df between, df within	

Therefore, we need to compute an *SS* and a *df* for the variance between treatments (numerator of *F*), and we need another *SS* and *df* for the variance within treatments (denominator of *F*). To obtain these *SS* and *df* values, we must go through two separate analyses: First, compute *SS* for the total study, and analyze it into two components (between and within). Then compute *df* for the total study, and analyze it into two components (between and within).

Thus, the entire process of ANOVA requires nine calculations: three values for *SS*, three values for *df*, two variances (between and within), and a final *F*-ratio. However, these nine calculations are all logically related and directed toward finding the final *F*-ratio. Figure 12.3 shows the logical structure of ANOVA calculations.

■ Analysis of Sum of Squares (SS)

The ANOVA requires that we first compute a total sum of squares and then partition this value into two components: between treatments and within treatments. This analysis is outlined in Figure 12.4. We will examine each of the three components separately.

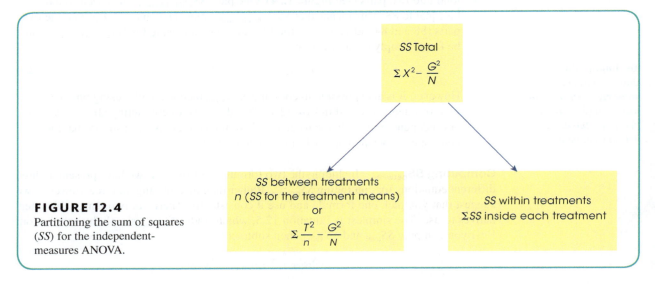

FIGURE 12.4

Partitioning the sum of squares (*SS*) for the independent-measures ANOVA.

SS Total

$$\Sigma X^2 - \dfrac{G^2}{N}$$

SS between treatments
n (SS for the treatment means)
or
$$\Sigma \dfrac{T^2}{n} - \dfrac{G^2}{N}$$

SS within treatments
ΣSS inside each treatment

1. **Total Sum of Squares, SS_{total}.** As the name implies, SS_{total} is the sum of squares for the entire set of N scores. As described in Chapter 4 (pages 121–122), this SS value can be computed using either a definitional or a computational formula. However, ANOVA typically involves a large number of scores and the mean is often not a whole number. Therefore, it is usually much easier to calculate SS_{total} using the computational formula:

$$SS = \Sigma X^2 - \frac{(\Sigma X)^2}{N}$$

To make this formula consistent with the ANOVA notation, we substitute the letter G in place of ΣX and obtain

$$SS_{total} = \Sigma X^2 - \frac{G^2}{N} \tag{12.3}$$

Applying this formula to the set of data in Table 12.2, we obtain

$$SS_{total} = \Sigma X^2 - \frac{G^2}{N}$$
$$= 1324 - 1152$$
$$= 172$$

2. **Within-Treatments Sum of Squares, $SS_{within\ treatments}$.** Now we are looking at the variability inside each of the treatment conditions. We already have computed the SS within each of the three treatment conditions (Table 12.2): $SS_1 = 24$, $SS_2 = 34$, and $SS_3 = 30$. To find the overall within-treatment sum of squares, we simply add these values together:

$$SS_{within\ treatments} = \Sigma SS_{inside\ each\ treatment} \tag{12.4}$$

For the data in Table 12.2, this formula gives

$$SS_{within\ treatments} = \Sigma SS_{inside\ each\ treatment} = 24 + 34 + 30 = 88$$

3. **Between-Treatments Sum of Squares, $SS_{between\ treatments}$.** Before we introduce any equations for $SS_{between\ treatments}$, consider what we have found so far. The total variability for the data in Table 12.2 is $SS_{total} = 172$. We intend to partition this total into two parts (see Figure 12.4). One part, $SS_{within\ treatments}$, has been found to be equal to 88. This means that $SS_{between\ treatments}$ must be equal to 84 so that the two parts (88 and 84) add up to the total (172). Thus, the value for $SS_{between\ treatments}$ can be found simply by subtraction:

> To simplify the notation we will use the subscripts *between* and *within* in place of *between treatments* and *within treatments*.

$$SS_{between} = SS_{total} - SS_{within} \tag{12.5}$$

However, it is also possible to compute $SS_{between}$ independently, using one of the two formulas presented in Box 12.1. The advantage of computing all three SS values independently is that you can check your calculations by ensuring that the two components, between and within, add up to the total.

Computing $SS_{between}$ Including the two formulas in Box 12.1, we have presented three different equations for computing $SS_{between}$. Rather than memorizing all three, however, we suggest that you pick one formula and use it consistently. There are two reasonable alternatives to use. The simplest is Equation 12.5, which finds $SS_{between}$ simply by subtraction: First you compute SS_{total} and SS_{within}, then subtract

$$SS_{between} = SS_{total} - SS_{within}$$

BOX 12.1 Alternative Formulas for $SS_{between}$

Recall that the variability between treatments is measuring the differences between treatment means. Conceptually, the most direct way of measuring the amount of variability among the treatment means is to compute the sum of squares for the set of sample means, SS_{means}. For the data in Table 12.2, the sample means are 5, 9, and 10. Computing SS for these three values produces $SS_{means} = 14$. However, each of the three means represents a group of $n = 6$ scores. Therefore, the final value for $SS_{between}$ is obtained by multiplying SS_{means} by n.

$$SS_{between} = n(SS_{means}) \qquad (12.6)$$

For the data in Table 12.2, we obtain

$$SS_{between} = n(SS_{means}) = 6(14) = 84$$

Unfortunately, Equation 12.6 can only be used when all of the samples are exactly the same size (equal n's), and the equation can be very awkward, especially when the treatment means are not whole numbers. Therefore, we also present a computational formula for $SS_{between}$ that uses the treatment totals (T) instead of the treatment means.

$$SS_{between} = \Sigma \frac{T^2}{n} - \frac{G^2}{N} \qquad (12.7)$$

For the data in Table 12.2 this formula produces:

$$SS_{between} = \frac{30^2}{6} + \frac{54^2}{6} + \frac{60^2}{6} - \frac{144^2}{18}$$

$$= 150 + 486 + 600 - 1152$$

$$= 1236 - 1152$$

$$= 84$$

Note that all three techniques (Equations 12.5, 12.6, and 12.7) produce the same result, $SS_{between} = 84$.

The second alternative is to use Equation 12.7, which computes $SS_{between}$ using the treatment totals (the T values). The advantage of this alternative is that it provides a way to check your arithmetic: Calculate SS_{total}, $SS_{between}$, and SS_{within} separately, and then check to be sure that the two components add up to equal SS_{total}.

Using Equation 12.6, which computes SS for the set of sample means, is usually not a good choice. Unless the sample means are all whole numbers, this equation can produce very tedious calculations. In most situations, one of the other two equations is a better alternative.

The following example is an opportunity for you to test your understanding of the analysis of SS in ANOVA.

EXAMPLE 12.2   Three samples, each with $n = 5$ participants, are used to evaluate the mean differences among three treatment conditions. The three sample totals and SS values are $T_1 = 10$ with $SS_1 = 16$, $T_2 = 25$ with $SS_2 = 20$, and $T_3 = 40$ with $SS_3 = 24$. If $SS_{total} = 150$, then what are the values for $SS_{between}$ and SS_{within}? You should find that $SS_{between} = 90$ and $SS_{within} = 60$. Good luck. ∎

■ The Analysis of Degrees of Freedom (*df*)

The analysis of degrees of freedom (*df*) follows the same pattern as the analysis of SS. First, we find *df* for the total set of N scores, and then we partition this value into two components: degrees of freedom between treatments and degrees of freedom within treatments. In computing degrees of freedom, there are two important considerations to keep in mind:

1. Each *df* value is associated with a specific SS value.

2. Normally, the value of *df* is obtained by counting the number of items that were used to calculate SS and then subtracting 1. For example, if you compute SS for a set of n scores, then $df = n - 1$.

With this in mind, we will examine the degrees of freedom for each part of the analysis.

1. **Total Degrees of Freedom, df_{total}.** To find the df associated with SS_{total}, you must first recall that this SS value measures variability for the entire set of N scores. Therefore, the df value is

$$df_{total} = N - 1 \qquad\qquad (12.8)$$

For the data in Table 12.2, the total number of scores is $N = 18$, so the total degrees of freedom are

$$df_{total} = 18 - 1$$
$$= 17$$

2. **Within-Treatments Degrees of Freedom, df_{within}.** To find the df associated with SS_{within}, we must look at how this SS value is computed. Remember, we first find SS inside of each of the treatments and then add these values together. Each of the treatment SS values measures variability for the n scores in the treatment, so each SS has $df = n - 1$. When all these individual treatment values are added together, we obtain

$$df_{within} = \Sigma(n - 1) = \Sigma df_{\text{in each treatment}} \qquad\qquad (12.9)$$

For the experiment we have been considering, each treatment has $n = 6$ scores. This means there are $n - 1 = 5$ degrees of freedom inside each treatment. Because there are three different treatment conditions, this gives a total of 15 for the within-treatments degrees of freedom. Notice that this formula for df simply adds up the number of scores in each treatment (the n values) and subtracts 1 for each treatment. If these two stages are done separately, you obtain

$$df_{within} = N - k \qquad\qquad (12.10)$$

(Adding up all the n values gives N. If you subtract 1 for each treatment, then altogether you have subtracted k because there are k treatments.) For the data in Table 12.2, $N = 18$ and $k = 3$, so

$$df_{within} = 18 - 3$$
$$= 15$$

3. **Between-Treatments Degrees of Freedom, $df_{between}$.** The df associated with $SS_{between}$ can be found by considering how the SS value is obtained. These SS formulas measure the variability for the set of treatments (totals or means). To find $df_{between}$, simply count the number of treatments and subtract 1. Because the number of treatments is specified by the letter k, the formula for df is

$$df_{between} = k - 1 \qquad\qquad (12.11)$$

For the data in Table 12.2, there are three different treatment conditions (three T values or three sample means), so the between-treatments degrees of freedom are computed as follows:

$$df_{between} = 3 - 1$$
$$= 2$$

FIGURE 12.5
Partitioning the degrees of freedom (*df*)
for the independent-measures ANOVA.

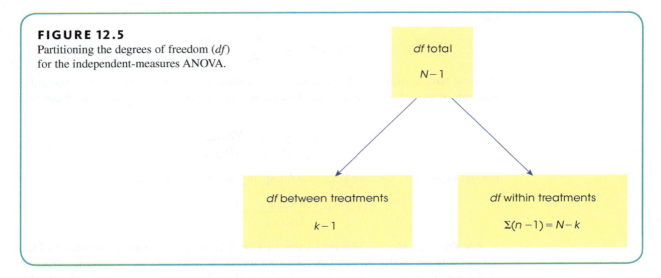

Notice that the two parts we obtained from this analysis of degrees of freedom add
up to equal the total degrees of freedom:

$$df_{total} = df_{within} + df_{between}$$
$$17 = 15 + 2$$

The complete analysis of degrees of freedom is shown in Figure 12.5.

As you are computing the *SS* and *df* values for ANOVA, keep in mind that the labels that
are used for each value can help you understand the formulas. Specifically:

1. The term *total* refers to the entire set of scores. We compute *SS* for the whole set of
   N scores, and the *df* value is simply $N - 1$.

2. The term *within treatments* refers to differences that exist inside the individual treatment conditions. Thus, we compute *SS* and *df* inside each of the separate treatments.

3. The term *between treatments* refers to differences from one treatment to another.
   With three treatments, for example, we are comparing three different means (or
   totals) and have $df = 3 - 1 = 2$.

■ Calculation of Variances (*MS*) and the *F*-Ratio

After computing the three *SS* and three *df* values, the next step in the ANOVA procedure is
to compute the variance between treatments and the variance within treatments, which are
used to calculate the *F*-ratio (see Figure 12.3).

In ANOVA, it is customary to use the term *mean square*, or simply *MS*, in place of the
term *variance*. Recall (from Chapter 4) that variance is defined as the mean of the squared
deviations. In the same way that we use *SS* to stand for the sum of the squared deviations,
we now will use *MS* to stand for the mean of the squared deviations. For the final *F*-ratio
we will need an *MS* (variance) between treatments for the numerator and an *MS* (variance)
within treatments for the denominator. In each case

$$MS = s^2 = \frac{SS}{df} \tag{12.12}$$

For the data we have been considering,

$$MS_{between} = s^2_{between} = \frac{SS_{between}}{df_{between}} = \frac{84}{2} = 42$$

and

$$MS_{within} = s^2_{within} = \frac{SS_{within}}{df_{within}} = \frac{88}{15} = 5.87$$

We now have a measure of the variance (or differences) between the treatments and a measure of the variance within the treatments. The F-ratio simply compares these two variances:

$$F = \frac{s^2_{between}}{s^2_{within}} = \frac{MS_{between}}{MS_{within}} \qquad (12.13)$$

For the experiment we have been examining, the data give an F-ratio of

$$F = \frac{MS_{between}}{MS_{within}} = \frac{42}{5.87} = 7.16$$

For this example, the obtained value of $F = 7.16$ indicates that the numerator of the F-ratio is substantially bigger than the denominator. If you recall the conceptual structure of the F-ratio as presented in Equations 12.1 and 12.2, the F value we obtained indicates that the differences between treatments are more than seven times bigger than what would be expected if there is no treatment effect. Stated in terms of the experimental variables, the strategy used for studying does appear to have an effect on test performance. However, to properly evaluate the F-ratio, we must select an α level and consult the F-distribution table that is discussed in the next section.

LEARNING CHECK   **LO5**   **1.** An analysis of variances produces $df_{between\ treatments} = 3$ and $df_{within\ treatments} = 26$. For this analysis, what is df_{total}?

    a. 27

    b. 28

    c. 29

    d. Cannot be determined without additional information.

LO5   **2.** An analysis of variance is used to evaluate the mean differences among five treatment conditions. The analysis produces $SS_{within\ treatments} = 20$, $SS_{between\ treatments} = 40$, and $SS_{total} = 60$. For this analysis, what is $MS_{between\ treatments}$?

    a. $\frac{20}{5}$

    b. $\frac{20}{4}$

    c. $\frac{40}{5}$

    d. $\frac{40}{4}$

LO5   **3.** A research study compares three treatments with $n = 5$ in each treatment. If the SS values for the three treatments are 25, 20, and 15, then the analysis of variance would produce SS_{within} equal to _____.

    a. 4

    b. 12

    c. 60

    d. Cannot be determined from the information given.

ANSWERS   **1. c   2. d   3. c**

12-4 Examples of Hypothesis Testing and Effect Size with ANOVA

LEARNING OBJECTIVES

6. Define the *df* values for an *F*-ratio and use the *df* values, together with an alpha level, to locate the critical region in the distribution of *F*-ratios.

7. Conduct a complete ANOVA to evaluate the differences among a set of means and compute a measure of effect size to describe the mean differences.

8. Explain how the results from an ANOVA and measures of effect size are reported in the scientific literature.

■ The Distribution of *F*-Ratios

In analysis of variance, the *F*-ratio is constructed so that the numerator and denominator of the ratio are measuring exactly the same variance when the null hypothesis is true (see Equation 12.2). In this situation, we expect the value of *F* to be around 1.00.

If the null hypothesis is false, the *F*-ratio should be much greater than 1.00. The problem now is to define precisely which values are "around 1.00" and which are "much greater than 1.00." To answer this question, we need to look at all the possible *F* values that can be obtained when the null hypothesis is true—that is, the *distribution of* F-*ratios*.

Before we examine this distribution in detail, you should note two obvious characteristics:

1. Because *F*-ratios are computed from two variances (the numerator and denominator of the ratio), *F* values always are positive numbers. Remember that variance is always positive.

2. When H_0 is true, the numerator and denominator of the *F*-ratio are measuring the same variance. In this case, the two sample variances should be about the same size, so the ratio should be near 1. In other words, the distribution of *F*-ratios should pile up around 1.00.

With these two factors in mind, we can sketch the distribution of *F*-ratios. The distribution is cut off at zero (all positive values), piles up around 1.00, and then tapers off to the right (Figure 12.6). The exact shape of the *F* distribution depends on the degrees of freedom for the two variances in the *F*-ratio. You should recall that the precision of a sample variance depends on the number of scores or the degrees of freedom. In general, the variance for a large sample (large *df*) provides a more accurate estimate of the population variance. Because the precision of the *MS* values depends on *df*, the shape of the *F* distribution also depends on the *df* values for the numerator and denominator of the *F*-ratio. With very large *df* values, nearly all the *F*-ratios are clustered very near to 1.00. With the smaller *df* values, the *F* distribution is more spread out.

■ The *F* Distribution Table

For ANOVA, we expect *F* near 1.00 if H_0 is true. An *F*-ratio that is much larger than 1.00 is an indication that H_0 is not true. In the *F* distribution, we need to separate those values that are reasonably near 1.00 from the values that are significantly greater than 1.00. These critical values are presented in an *F* distribution table in Appendix B, page 597. A portion of the *F* distribution table is shown in Table 12.3. To use the table, you must know the *df* values for the *F*-ratio (numerator and denominator), and you must know the alpha level for the hypothesis test. It is customary for an *F* table to have the *df* values for the numerator of

FIGURE 12.6

The distribution of *F*-ratios with $df = 2, 15$. Of all the values in the distribution, only 5% are larger than $F = 3.68$ and only 1% are larger than $F = 6.36$.

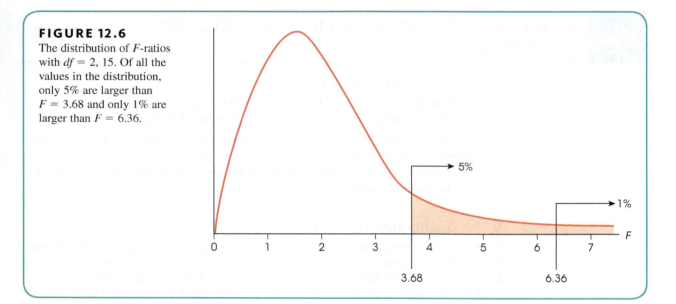

the *F*-ratio printed across the top of the table. The *df* values for the denominator of *F* are printed in a column on the left-hand side. For the experiment we have been considering, the numerator of the *F*-ratio (between treatments) has $df = 2$, and the denominator of the *F*-ratio (within treatments) has $df = 15$. This *F*-ratio is said to have "degrees of freedom equal to 2 and 15." The degrees of freedom would be written as $df = 2, 15$. To use the table, you would first find $df = 2$ across the top of the table and $df = 15$ in the first column. When you line up these two values, they point to a pair of numbers in the middle of the table. These numbers give the critical cutoffs for $\alpha = .05$ and $\alpha = .01$. With $df = 2, 15$, for example, the numbers in the table are 3.68 and 6.36. Thus, only 5% of the distribution ($\alpha = .05$) corresponds to values greater than 3.68 and only 1% of the distribution ($\alpha = .01$) corresponds to values greater than 6.36 (see Figure 12.6).

TABLE 12.3

A portion of the *F* distribution table. Entries in roman type are critical values for the .05 level of significance, and values in bold type are for the .01 level of significance. The critical values for $df = 2$, 15 have been highlighted (see text).

Degrees of Freedom: Denominator	Degrees of Freedom: Numerator					
	1	2	3	4	5	6
11	4.84	3.98	3.59	3.36	3.20	3.09
	9.65	**7.20**	**6.22**	**5.67**	**5.32**	**5.07**
12	4.75	3.88	3.49	3.26	3.11	3.00
	9.33	**6.93**	**5.95**	**5.41**	**5.06**	**4.82**
13	4.67	3.80	3.41	3.18	3.02	2.92
	9.07	**6.70**	**5.74**	**5.20**	**4.86**	**4.62**
14	4.60	3.74	3.34	3.11	2.96	2.85
	8.86	**6.51**	**5.56**	**5.03**	**4.69**	**4.46**
15	4.54	3.68	3.29	3.06	2.90	2.79
	8.68	**6.36**	**5.42**	**4.89**	**4.56**	**4.32**
16	4.49	3.63	3.24	3.01	2.85	2.74
	8.53	**6.23**	**5.29**	**4.77**	**4.44**	**4.20**

TABLE 12.4
An ANOVA summary
showing the results of the
ANOVA calculations for
the data in Example 12.1.

Source	SS	df	MS	
Between treatments	84	2	42.00	$F = 7.16$
Within treatments	88	15	5.87	
Total	172	17		

■ An Example of Hypothesis Testing and Effect Size with Anova

The Hypothesis Test   Although we have now seen all the individual components of ANOVA, we now demonstrate the complete ANOVA process using the research examining different strategies for studying presented in Example 12.1. All of the calculations for the ANOVA were completed in the previous section and are summarized in Table 12.4, which is called an *ANOVA summary table*. The table shows the source of variability (between treatments, within treatments, and total variability): *SS*, *df*, *MS*, and the final *F*-ratio.

The current APA format for reporting the results from an ANOVA is presented on page 412.

Although these tables are no longer used in published reports, they are a common part of computer printouts, and they do provide a concise method for presenting the results of an analysis. (Note that you can conveniently check your work: Adding the first two entries in the *SS* column, $84 + 88$, produces SS_{total}. The same applies to the *df* column.) When using ANOVA, you might start with a blank ANOVA summary table and then fill in the values as they are calculated. With this method, you are less likely to "get lost" in the analysis, wondering what to do next.

Using the results in Table 12.4, we can now present the complete ANOVA using the standard four-step procedure for hypothesis testing.

STEP 1   **State the hypotheses and select an alpha level.**

$$H_0: \mu_1 = \mu_2 = \mu_3 \quad \text{(There is no treatment effect.)}$$

H_1: At least one of the treatment means is different.

We will use $\alpha = .05$.

STEP 2   **Locate the critical region.**   We have found that $df_{between} = 2$ and $df_{within} = 15$. Thus, the *F*-ratio has $df = 2, 15$ and with $\alpha = .05$ the critical region consists of *F*-ratios greater than 3.68.

STEP 3   **Compute the *F*-ratio.**   The calculations were completed in the previous section and are summarized in Table 12.4. The data produce $F = 7.16$.

STEP 4   **Make a decision.**   The *F* value we obtained, $F = 7.16$, is in the critical region. It is very unlikely ($p < .05$) that we would obtain a value this large if H_0 is true. Therefore, we reject H_0 and conclude that there are significant differences among the three strategies for studying.

This completes the step-by-step demonstration of the ANOVA procedure. However, there is one additional point that can be made using this example.

The research study compared the effectiveness of three different strategies for studying: simply rereading the material, answering prepared questions about the material, or creating and answering your own questions. The statistical decision is to reject H_0, which means that the three strategies are not all the same. However, we have not determined which ones are different. Is answering prepared questions different from making up and answering your own questions? Is answering prepared questions different from simply rereading? Unfortunately, these questions remain unanswered. We do know that at least one difference exists (we rejected H_0), but additional analysis is necessary to find out exactly where this difference is. We address this problem in Section 12.5.

■ Measuring Effect Size for ANOVA

As we noted previously, a *significant* mean difference simply indicates that the difference observed in the sample data is very unlikely to have occurred just by chance. Thus, the term significant does not necessarily mean *large*, it simply means larger than expected by chance. To provide an indication of how large the effect actually is, it is recommended that researchers report a measure of effect size in addition to the measure of significance.

For ANOVA, the simplest and most direct way to measure effect size is to compute the percentage of variance accounted for by the treatment conditions. Like the r^2 value used to measure effect size for the *t* tests in Chapters 9, 10, and 11, this percentage measures how much of the variability in the scores is accounted for by the differences between treatments. For ANOVA, the calculation and the concept of the percentage of variance is extremely straightforward. Specifically, we determine how much of the total *SS* is accounted for by the $SS_{between\ treatments}$.

$$\text{The percentage of variance accounted for} = \frac{SS_{between\ treatments}}{SS_{total}} \tag{12.14}$$

For the data in Example 12.1, we obtain:

$$\text{The percentage of variance accounted for} = \frac{84}{172} = 0.488 \text{ (or 48.8\%)}$$

In published reports of ANOVA results, the percentage of variance accounted for by the treatment effect is usually called η^2 (the Greek letter *eta squared*) instead of using r^2. Thus, for the study in Example 12.1, $\eta^2 = 0.488$.

The following example is an opportunity for you to test your understanding of computing η^2 to measure effect size in ANOVA.

EXAMPLE 12.3   The ANOVA from an independent-measures study is summarized in the following table.

Source	SS	df	MS	
Between treatments	84	2	42	$F(2, 12) = 7.00$
Within treatments	72	12	6	
Total	156	14		

Compute η^2 to measure effect size for this study. You should find that $\eta^2 = 0.538$. Good luck. ■

IN THE LITERATURE

Reporting the Results of Analysis of Variance

The APA format for reporting the results of ANOVA begins with a presentation of the treatment means and standard deviations in the narrative of an article, a table, or a graph. These descriptive statistics are not needed in the calculations of the actual ANOVA, but you can easily determine the treatment means from *n* and *T* $(M = T/n)$ and the standard deviations from the *SS* values for each treatment $[s = \sqrt{SS/(n - 1)}]$. Next, report the results of the ANOVA. For the study described in Example 12.1, the report might state the following:

> The means and standard deviations are presented in Table 12.5. The analysis of variance indicates that there are significant differences among the three strategies for studying, $F(2, 15) = 7.16, p < .05, \eta^2 = 0.488$.

TABLE 12.5

Quiz scores following three different strategies for studying.

	Simply Reread	Answer Prepared Questions	Create and Answer Your Own Questions
M	5.00	9.00	10.00
SD	2.19	2.61	2.45

Note how the F-ratio is reported. In this example, degrees of freedom for between and within treatments are $df = 2, 15$, respectively. These values are placed in parentheses immediately following the symbol F. Next, the calculated value for F is reported, followed by the probability of committing a Type I error (the alpha level) and the measure of effect size.

When an ANOVA is done using a computer program, the F-ratio is usually accompanied by an exact value for p. The data from Example 12.1 were analyzed using the SPSS program (see the SPSS section at the end of this chapter) and the computer output included a significance level of $p = .007$. Using the exact p value from the computer output, the research report would conclude, "The analysis of variance revealed significant differences among the three strategies for studying, $F(2, 15) = 7.16$, $p = .007$, $\eta^2 = 0.488$."

■ An Example with Unequal Sample Sizes

In the previous examples, all the samples were exactly the same size (equal n's). However, the formulas for ANOVA can be used when the sample size varies within an experiment. You also should note, however, that the general ANOVA procedure is most accurate when used to examine experimental data with equal sample sizes. Therefore, researchers generally try to plan experiments with equal n's. However, there are circumstances in which it is impossible or impractical to have an equal number of participants in every treatment condition. In these situations, ANOVA still provides a valid test, especially when the samples are relatively large and when the discrepancy between sample sizes is not extreme.

The following example demonstrates an ANOVA with samples of different sizes.

EXAMPLE 12.4   A researcher is interested in the amount of homework required by different academic majors. Students were recruited from Biology, English, and Psychology to participate in the study. The researcher randomly selects one course that each student is currently taking and asks the student to record the amount of out-of-class work required each week for the course. The researcher used all of the volunteer participants, which resulted in unequal sample sizes. The data are summarized in Table 12.6.

STEP 1   **State the hypotheses, and select the alpha level.**

$$H_0: \mu_1 = \mu_2 = \mu_3$$
$$H_1: \text{At least one population is different.}$$
$$\alpha = .05$$

TABLE 12.6

Average hours of homework per week for one course for students in three academic majors.

Biology	English	Psychology	
$n = 4$	$n = 10$	$n = 6$	$N = 20$
$M = 9$	$M = 13$	$M = 14$	$G = 250$
$T = 36$	$T = 130$	$T = 84$	$\Sigma X^2 = 3{,}377$
$SS = 37$	$SS = 90$	$SS = 60$	

STEP 2 **Locate the critical region.** To find the critical region, we first must determine the df values for the F-ratio:

$$df_{total} = N - 1 = 20 - 1 = 19$$
$$df_{between} = k - 1 = 3 - 1 = 2$$
$$df_{within} = N - k = 20 - 3 = 17$$

The F-ratio for these data has $df = 2, 17$. With $\alpha = .05$, the critical value for the F-ratio is 3.59.

STEP 3 **Compute the F-ratio.** First, compute the three SS values. As usual, SS_{total} is the SS for the total set of $N = 20$ scores, and SS_{within} combines the SS values from inside each of the treatment conditions.

$$SS_{total} = \Sigma X^2 - \frac{G^2}{N} \qquad SS_{within} = \Sigma SS_{inside\ each\ treatment}$$

$$= 3377 - \frac{250^2}{20} \qquad\qquad = 37 + 90 + 60$$

$$= 3377 - 3125$$

$$= 252 \qquad\qquad\qquad\qquad = 187$$

$SS_{between}$ can be found by subtraction (Equation 12.5).

$$SS_{between} = SS_{total} - SS_{within}$$
$$= 252 - 187$$
$$= 65$$

Or, $SS_{between}$ can be calculated using the computation formula (Equation 12.7). If you use the computational formula, be careful to match each treatment total (T) with the appropriate sample size (n) as follows:

$$SS_{between} = \Sigma \frac{T^2}{n} - \frac{G^2}{N}$$

$$= \frac{36^2}{4} + \frac{130^2}{10} + \frac{84^2}{6} - \frac{250}{20}$$

$$= 324 + 1690 + 1176 - 3125$$

$$= 65$$

Finally, compute the MS values and the F-ratio:

$$MS_{between} = \frac{SS}{df} = \frac{65}{2} = 32.5$$

$$MS_{within} = \frac{SS}{df} = \frac{187}{17} = 11$$

$$F = \frac{MS_{between}}{MS_{within}} = \frac{32.5}{11} = 2.95$$

STEP 4   **Make a decision.**   Because the obtained *F*-ratio is not in the critical region, we fail to reject the null hypothesis and conclude that there are no significant differences among the three populations of students in terms of the average amount of homework each week.   ■

■ Assumptions for the Independent-Measures ANOVA

The independent-measures ANOVA requires the same three assumptions that were necessary for the independent-measures *t* hypothesis test:

1. The observations within each sample must be independent (see page 337).
2. The populations from which the samples are selected must be normal.
3. The populations from which the samples are selected must have equal variances (homogeneity of variance).

Ordinarily, researchers are not overly concerned with the assumption of normality, especially when large samples are used, unless there are strong reasons to suspect the assumption has not been satisfied. The assumption of homogeneity of variance is an important one. If a researcher suspects it has been violated, it can be tested by Hartley's *F*-max test for homogeneity of variance (Chapter 10, page 338).

LEARNING CHECK   **LO6** **1.** A researcher uses analysis of variance to test for mean differences among three treatments with a sample of $n = 10$ in each treatment. The *F*-ratio for this analysis would have what *df* values?
   a. $df = 3, 10$
   b. $df = 3, 30$
   c. $df = 3, 27$
   d. $df = 2, 27$

LO7 **2.** The following table shows the results of an analysis of variance comparing three treatment conditions with a sample of $n = 11$ participants in each treatment. Note that several values are missing in the table. What is the missing value for the *F*-ratio?
   a. 3.33
   b. 4.2
   c. 14
   d. 28

Source	SS	df	MS	
Between	xx	xx	14	F = xx
Within	xx	xx	xx	
Total	154	xx		

LO8 **3.** A research report concludes that there are significant differences among treatments, with "$F(2, 27) = 8.62, p < .01, \eta^2 = 0.46$." How many treatment conditions were compared in this study?
   a. 2
   b. 3
   c. 29
   d. 30

ANSWERS   **1. d   2. a   3. b**

12-5 Post Hoc Tests

LEARNING OBJECTIVE

9. Describe the circumstances in which post hoc tests are necessary and explain what the tests accomplish.

As noted earlier, the primary advantage of ANOVA (compared to t tests) is it allows researchers to test for significant mean differences when there are *more than two* treatment conditions. ANOVA accomplishes this feat by comparing all the individual mean differences simultaneously within a single test. Unfortunately, the process of combining several mean differences into a single test statistic creates some difficulty when it is time to interpret the outcome of the test. Specifically, when you obtain a significant F-ratio (reject H_0), it simply indicates that somewhere among the entire set of mean differences there is at least one that is statistically significant. In other words, the overall F-ratio only tells you that a significant difference exists; it does not tell exactly which means are significantly different and which are not.

In Example 12.1 we presented an independent-measures study using three samples to compare three strategies for studying in preparation for a quiz: rereading the material to be tested, answering prepared questions on the material, creating and answering your own questions. The three sample means were $M_1 = 5$, $M_2 = 9$, and $M_3 = 10$. In this study there are three mean differences:

1. There is a 4-point difference between M_1 and M_2.

2. There is a 1-point difference between M_2 and M_3.

3. There is a 5-point difference between M_1 and M_3.

The ANOVA used to evaluate these data produced a significant F-ratio indicating that at least one of the sample mean differences is large enough to satisfy the criterion of statistical significance. In this example, the 5-point difference is the biggest of the three and, therefore, it must indicate a significant difference between the first treatment and the third treatment ($\mu_1 \neq \mu_3$). But what about the 4-point difference? Is it also large enough to be significant? And what about the 1-point difference between M_2 and M_3? Is it also significant? The purpose of *post hoc tests* is to answer these questions.

> **Post hoc tests** (or **posttests**) are additional hypothesis tests that are done after an ANOVA to determine exactly which mean differences are significant and which are not.

As the name implies, post hoc tests are done after an ANOVA. More specifically, these tests are done after ANOVA when

1. you reject H_0 and

2. there are three or more treatments ($k \geq 3$).

Rejecting H_0 indicates that at least one difference exists among the treatments. If there are only two treatments, then there is no question about which means are different and, therefore, there is no need for posttests. However, with three or more treatments ($k \geq 3$), the problem is to determine exactly which means are significantly different.

■ Posttests and Type I Errors

In general, a post hoc test enables you to go back through the data and compare the individual treatments two at a time. In statistical terms, this is called making *pairwise comparisons*.

For example, with $k = 3$, we would compare μ_1 versus μ_2, then μ_2 versus μ_3, and then μ_1 versus μ_3. In each case, we are looking for a significant mean difference. The process of conducting pairwise comparisons involves performing a series of separate hypothesis tests, and each of these tests includes the risk of a Type I error. As you do more and more separate tests, the risk of a Type I error accumulates and is called the *experimentwise alpha level*.

We have seen, for example, that a research study with three treatment conditions produces three separate mean differences, each of which could be evaluated using a post hoc test. If each test uses $\alpha = .05$, then there is a 5% risk of a Type I error for the first posttest, another 5% risk for the second test, and one more 5% risk for the third test. Although the probability of error is not simply the sum across the three tests, it should be clear that increasing the number of separate tests definitely increases the total, experimentwise probability of a Type I error.

Whenever you are conducting posttests, you must be concerned about the experimentwise alpha level. Statisticians have worked with this problem and have developed several methods for trying to control Type I errors in the context of post hoc tests. We will consider two alternatives.

■ Tukey's Honestly Significant Difference (HSD) Test

The first post hoc test we consider is *Tukey's HSD test*. We selected Tukey's HSD test because it is a commonly used test in psychological research. Tukey's test allows you to compute a single value that determines the minimum difference between treatment means that is necessary for significance. This value, called the *honestly significant difference*, or HSD, is then used to compare any two treatment conditions. If the mean difference exceeds Tukey's HSD, you conclude that there is a significant difference between the treatments. Otherwise, you cannot conclude that the treatments are significantly different. The formula for Tukey's HSD is

$$HSD = q\sqrt{\frac{MS_{\text{within}}}{n}} \qquad (12.15)$$

The q value used in Tukey's HSD test is called a Studentized range statistic.

where the value of q is found in Table B.5 (Appendix B, page 600), $MS_{\text{within treatments}}$ is the within-treatments variance from the ANOVA, and n is the number of scores in each treatment. Tukey's test requires that the sample size, n, be the same for all treatments. To locate the appropriate value of q, you must know the number of treatments in the overall experiment (k), the degrees of freedom for $MS_{\text{within treatments}}$ (the error term in the F-ratio), and you must select an alpha level (generally the same α used for the ANOVA).

EXAMPLE 12.5

To demonstrate the procedure for conducting post hoc tests with Tukey's HSD, we use the data from Example 12.1, which are summarized in Table 12.7. Note that the table displays summary statistics for each sample and the results from the overall ANOVA. With $k = 3$ treatments, $n = 6$, and $\alpha = .05$, you should find that the value of q for the test is $q = 3.67$ (see Table B.5). Therefore, Tukey's HSD is

$$HSD = q\sqrt{\frac{MS_{\text{within}}}{n}} = 3.67\sqrt{\frac{5.87}{6}} = 3.63$$

Thus, the mean difference between any two samples must be at least 3.63 to be significant. Using this value, we can make the following conclusions:

1. Treatment A is significantly different from Treatment B ($M_A - M_B = 4.00$).
2. Treatment A is also significantly different from Treatment C ($M_A - M_C = 5.00$).
3. Treatment B is not significantly different from Treatment C ($M_B - M_C = 1.00$).

TABLE 12.7

Results from the research study in Example 12.1. Summary statistics are presented for each treatment along with the outcome from the ANOVA.

	Treatment A: Reread	Treatment B: Prepared Questions	Treatment C: Create Questions
	$n = 6$	$n = 6$	$n = 6$
	$T = 30$	$T = 54$	$T = 60$
	$M = 5.00$	$M = 9.00$	$M = 10.00$

Source	SS	df	MS
Between	84	2	42
Within	88	15	5.87
Total	172	17	
Overall $F(2, 15) = 7.16$			

■ The Scheffé Test

Because it uses an extremely cautious method for reducing the risk of a Type I error, the *Scheffé test* has the distinction of being one of the safest of all possible post hoc tests (smallest risk of a Type I error). The Scheffé test uses an *F*-ratio to evaluate the significance of the difference between any two treatment conditions. The numerator of the *F*-ratio is an *MS* between treatments that is calculated using *only the two treatments you want to compare*. The denominator is the same MS_{within} that was used for the overall ANOVA. The "safety factor" for the Scheffé test comes from the following two considerations:

1. Although you are comparing only two treatments, the Scheffé test uses the value of *k* from the original experiment to compute *df* between treatments. Thus, *df* for the numerator of the *F*-ratio is $k - 1$.

2. The critical value for the Scheffé *F*-ratio is the same as was used to evaluate the *F*-ratio from the overall ANOVA. Thus, Scheffé requires that every posttest satisfy the same criterion that was used for the complete ANOVA. The following example uses the data from Example 12.1 (see Table 12.6) to demonstrate the Scheffé posttest procedure.

EXAMPLE 12.6

Remember that the Scheffé procedure requires a separate $SS_{between}$, $MS_{between}$, and *F*-ratio for each comparison being made. Although Scheffé computes $SS_{between}$ using the regular computational formula (Equation 12.7), you must remember that all the numbers in the formula are entirely determined by the two treatment conditions being compared. We begin with the smallest mean difference, which involves comparing Treatment B (with $T = 54$ and $n = 6$) and Treatment C (with $T = 60$ and $n = 6$). The first step is to compute $SS_{between}$ for these two groups. In the formula for *SS*, notice that the grand total for the two groups is $G = 54 + 60 = 114$, and the total number of scores for the two groups is $N = 6 + 6 = 12$.

$$SS_{between} = \Sigma \frac{T^2}{n} - \frac{G^2}{N}$$
$$= \frac{(54)^2}{6} + \frac{(60)^2}{6} - \frac{(114)^2}{12}$$
$$= 486 + 600 - 1083$$
$$= 3$$

Although we are comparing only two groups, these two were selected from a study consisting of $k = 3$ samples. The Scheffé test uses the overall study to determine the degrees of freedom between treatments. Therefore, $df_{between} = 3 - 1 = 2$, and the *MS* between treatments is

$$MS_{between} = \frac{SS_{between}}{df_{between}} = \frac{3}{2} = 1.5$$

Finally, the Scheffé procedure uses the error term from the overall ANOVA to compute the F-ratio. In this case, $MS_{within} = 5.87$ with $df_{within} = 15$. Thus, the Scheffé test produces an F-ratio of

$$F = \frac{MS_{between}}{MS_{within}} = \frac{1.5}{5.87} = 0.26$$

With $df = 2$, 15 and $\alpha = .05$, the critical value for F is 3.68 (see Table B.4). Therefore, our obtained F-ratio is not in the critical region, and we conclude that these data show no significant difference between Treatment B and Treatment C.

The second-largest mean difference involves Treatment A ($T = 30$) versus Treatment B ($T = 54$). This time the data produce $SS_{between} = 48$, $MS_{between} = 24$, and $F(2, 15) = 4.09$ (check the calculations for yourself). Once again, the critical value for F is 3.68, so we conclude that there is a significant difference between Treatment A and Treatment B.

The final comparison is Treatment A ($M = 5$) versus Treatment C ($M = 10$). We have already found that the 4-point mean difference between A and B is significant, so the 5-point difference between A and C also must be significant. Thus, the Scheffé posttest indicates that both B and C (answering prepared questions and creating and answering your own questions) are significantly different from Treatment A (simply rereading), but there is no significant difference between B and C.   ■

In this case, the two post-test procedures, Tukey's HSD and Scheffé, produce exactly the same results. You should be aware, however, that there are situations in which Tukey's test will find a significant difference but Scheffé will not. Again, the Scheffé test is one of the safest of the posttest techniques because it provides the greatest protection from Type I errors. To provide this protection, the Scheffé test simply requires a larger difference between sample means before you may conclude that the difference is significant.

LEARNING CHECK   **LO9** **1.** Under what circumstances are post hoc tests necessary after an ANOVA?
  a. When H_0 is rejected.
  b. When there are more than two treatments.
  c. When H_0 is rejected and there are more than two treatments.
  d. You *always* should do post hoc tests after an ANOVA.

LO9 **2.** An ANOVA finds significant treatment effects for a study comparing three treatments with means of $M_1 = 10$, $M_2 = 5$, $M_3 = 2$. If Tukey's HSD is computed to be HSD $= 2.50$, then which of the treatments are significantly different?
  a. 1 vs. 2 and 2 vs. 3
  b. 1 vs. 2 and 1 vs. 3
  c. 1 vs. 3 and 2 vs. 3
  d. 1 vs. 2 and 1 vs. 3 and 2 vs. 3

ANSWERS   **1. c   2. d**

12-6 More about ANOVA

LEARNING OBJECTIVES

10. Explain how the outcome of an ANOVA and measures of effect size are influenced by sample size, sample variance, and sample mean differences.

11. Explain the relationship between the independent-measures t test and an ANOVA when evaluating the difference between two means from an independent-measures study.

■ A Conceptual View of ANOVA

Because analysis of variance requires relatively complex calculations, students encountering this statistical technique for the first time often tend to be overwhelmed by the formulas and arithmetic and lose sight of the general purpose for the analysis. The following two examples are intended to minimize the role of the formulas and shift attention back to the conceptual goal of the ANOVA process.

EXAMPLE 12.7   The following data represent the outcome of an experiment using two separate samples to evaluate the mean difference between two treatment conditions. Take a minute to look at the data and, without doing any calculations, try to predict the outcome of an ANOVA for these values. Specifically, predict what values should be obtained for the between-treatments variance (MS) and the F-ratio. If you do not "see" the answer after 20 or 30 seconds, try reading the hints that follow the data.

Treatment I	Treatment II	
4	2	$N = 8$
0	1	$G = 16$
1	0	$\Sigma X^2 = 56$
3	5	
$T = 8$	$T = 8$	
$SS = 10$	$SS = 14$	

Again, if you are having trouble predicting the outcome of the ANOVA, read the following hints, and then go back and look at the data.

Hint 1:   Remember that $SS_{between}$ and $MS_{between}$ provide a measure of how much difference there is between treatment conditions.

Hint 2:   Find the mean or total (T) for each treatment, and determine how much difference there is between the two treatments.

You should realize by now that the data have been constructed so that there is zero difference between treatments. The two sample means (and totals) are identical, so $SS_{between} = 0$, $MS_{between} = 0$, and the F-ratio is zero.   ■

Conceptually, the numerator of the F-ratio always measures how much difference exists between treatments. In Example 12.7, we constructed an extreme set of scores with zero difference. However, you should be able to look at any set of data and quickly compare the means (or totals) to determine whether there are big differences between treatments or small differences between treatments.

Being able to estimate the magnitude of between-treatment differences is a good first step in understanding ANOVA and should help you to predict the outcome of an ANOVA. However, the *between-treatment* differences are only one part of the analysis. You must also understand the *within-treatment* differences that form the denominator of the *F*-ratio. The following example is intended to demonstrate the concepts underlying SS_{within} and MS_{within}. In addition, the example should give you a better understanding of how the between-treatment differences and the within-treatment differences act together within the ANOVA.

EXAMPLE 12.8

The purpose of this example is to present a visual image for the concepts of between-treatments variability and within-treatments variability. In this example, we compare two hypothetical outcomes for the same experiment. In each case, the experiment uses two separate samples to evaluate the mean difference between two treatments. The following data represent the two outcomes, which we call Experiment A and Experiment B.

Experiment A		Experiment B	
Treatment		Treatment	
I	II	I	II
8	12	4	12
8	13	11	9
7	12	2	20
9	11	17	6
8	13	0	16
9	12	8	18
7	11	14	3
$M = 8$	$M = 12$	$M = 8$	$M = 12$
$s = 0.82$	$s = 0.82$	$s = 6.35$	$s = 6.35$

The data from Experiment A are displayed in a frequency distribution graph in Figure 12.7(a). Notice that there is a 4-point difference between the treatment means ($M_1 = 8$ and $M_2 = 12$). This is the *between-treatments* difference that contributes to the numerator of the *F*-ratio. Also notice that the scores in each treatment are clustered close around the mean, indicating that the variance inside each treatment is relatively small. This is the *within-treatments* variance that contributes to the denominator of the *F*-ratio. Finally, you should realize that it is easy to see the mean difference between the two samples. The fact that there is a clear mean difference between the two treatments is confirmed by computing the *F*-ratio for Experiment A.

$$F = \frac{\text{between-treatments difference}}{\text{within-treatments differences}} = \frac{MS_{between}}{MS_{within}} = \frac{56}{0.667} = 83.96$$

An *F*-ratio of $F = 83.96$ is sufficient to reject the null hypothesis, so we conclude that there is a significant difference between the two treatments.

Now consider the data from Experiment B, which are shown in Figure 12.7(b) and present a very different picture. This experiment has the same 4-point difference between treatment means that we found in Experiment A ($M_1 = 8$ and $M_2 = 12$). However, for these data the scores in each treatment are scattered across the entire scale, indicating relatively large variance inside each treatment. In this case, the large variance within treatments overwhelms the relatively small mean difference between treatments. In the figure it is almost impossible to see the mean difference between treatments. The within-treatments variance

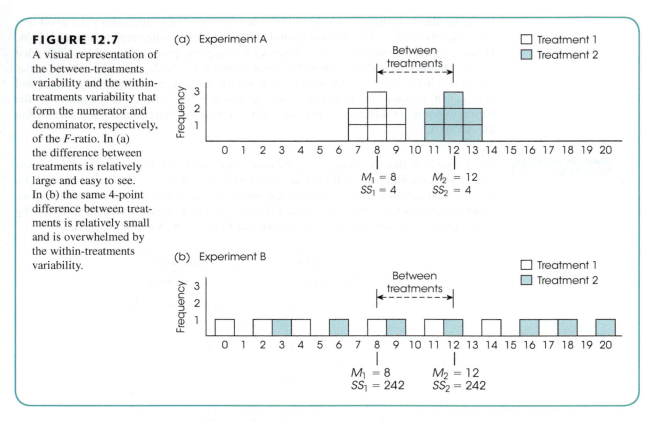

FIGURE 12.7

A visual representation of the between-treatments variability and the within-treatments variability that form the numerator and denominator, respectively, of the F-ratio. In (a) the difference between treatments is relatively large and easy to see. In (b) the same 4-point difference between treatments is relatively small and is overwhelmed by the within-treatments variability.

appears in the bottom of the F-ratio and, for these data, the F-ratio confirms that there is no clear mean difference between treatments.

$$F = \frac{\text{between-treatments difference}}{\text{within-treatments differences}} = \frac{MS_{\text{between}}}{MS_{\text{within}}} = \frac{56}{40.33} = 1.39$$

For Experiment B, the F-ratio is not large enough to reject the null hypothesis, so we conclude that there is no significant difference between the two treatments. Once again, the statistical conclusion is consistent with the appearance of the data in Figure 12.7(b). Looking at the figure, we see that the scores from the two samples appear to be intermixed randomly with no clear distinction between treatments.

As a final point, note that the denominator of the F-ratio, MS_{within}, is a measure of the variability (or variance) within each of the separate samples. As we have noted in previous chapters, high variability makes it difficult to see any patterns in the data. In Figure 12.7(a), the 4-point mean difference between treatments is easy to see because the sample variability is small. In Figure 12.7(b), the 4-point difference gets lost because the sample variability is large. In general, you can think of variance as measuring the amount of "noise" or "confusion" in the data. With large variance there is a lot of noise and confusion, and it is difficult to see any clear patterns. ■

Although Examples 12.7 and 12.8 present somewhat simplified demonstrations with exaggerated data, the general point of the examples is to help you *see* what happens when you perform an ANOVA. Specifically:

1. The numerator of the F-ratio (MS_{between}) measures how much difference exists between the treatment means. The bigger the mean differences, the bigger the F-ratio.

2. The denominator of the F-ratio (MS_{within}) measures the variance of the scores inside each treatment; that is, the variance for each of the separate samples. In general, larger sample variance produces a smaller F-ratio.

We should note that the number of scores in the samples also influences the outcome of an ANOVA. As with most other hypothesis tests, if other factors are held constant, increasing the sample size tends to increase the likelihood of rejecting the null hypothesis. However, changes in sample size have little or no effect on measures of effect size such as η^2.

■ The Relationship Between ANOVA and *t* Tests

When you are evaluating the mean difference from an independent-measures study comparing only two treatments (two separate samples), you can use either an independent-measures t test (Chapter 10) or the ANOVA presented in this chapter. In practical terms, it makes no difference which you choose. These two statistical techniques always result in the same statistical decision. In fact, the two methods use many of the same calculations and are very closely related in several other respects. The basic relationship between t statistics and F-ratios can be stated in an equation:

$$F = t^2$$

This relationship can be explained by first looking at the structure of the formulas for F and t. The t statistic compares *distances*: the distance between two sample means (numerator) and the distance computed for the standard error (denominator). The F-ratio, on the other hand, compares *variances*. You should recall that variance is a measure of squared distance. Hence, the relationship:

$$F = t^2$$

There are several other points to consider in comparing the t statistic to the F-ratio.

1. It should be obvious that you will be testing the same hypotheses whether you choose a t test or an ANOVA. With only two treatments, the hypotheses for either test are

$$H_0: \mu_1 = \mu_2$$
$$H_1: \mu_1 \neq \mu_2$$

2. The degrees of freedom for the t statistic and the df for the denominator of the F-ratio (df_{within}) are identical. For example, if you have two samples, each with six scores, the independent-measures t statistic will have $df = 10$, and the F-ratio will have $df = 1, 10$. In each case, you are adding the df from the first sample ($n - 1$) and the df from the second sample ($n - 1$).

3. The distribution of t and the distribution of F-ratios match perfectly if you take into consideration the relationship $F = t^2$. Consider the t distribution with $df = 18$ and the corresponding F distribution with $df = 1, 18$ that are presented in Figure 12.8. Notice the following relationships:
   a. If each of the t values is squared, then all of the negative values become positive. As a result, the whole left-hand side of the t distribution (below zero) will be flipped over to the positive side. This creates an asymmetrical, positively skewed distribution—that is, the F distribution.
   b. For $\alpha = .05$, the critical region for t is determined by values greater than $+2.101$ or less than -2.101. When these boundaries are squared, you get $\pm2.101^2 = 4.41$.

Notice that 4.41 is the critical value for $\alpha = .05$ in the F distribution. Any value that is in the critical region for t will end up in the critical region for F-ratios after it is squared.

FIGURE 12.8

The distribution of t statistics with $df = 18$ and the corresponding distribution of F-ratios with $df = 1, 18$. Notice that the critical values for $\alpha = .05$ are $t = \pm 2.101$ and $F = 2.101^2 = 4.41$.

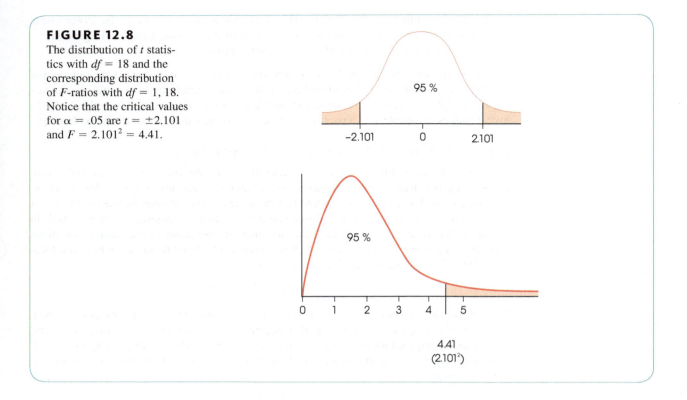

LO10   **1.** Which combination of factors is most likely to produce a large value for the F-ratio and a large value for η^2?

    a. Large mean differences and large sample variances.

    b. Large mean differences and small sample variances.

    c. Small mean differences and large sample variances.

    d. Small mean differences and small sample variances.

LO10   **2.** If an analysis of variance is used for the following data, what would be the effect of changing the value of SS_2 to 100?

    a. Increase SS_{within} and increase the size of the F-ratio.

    b. Increase SS_{within} and decrease the size of the F-ratio.

    c. Decrease SS_{within} and increase the size of the F-ratio.

    d. Decrease SS_{within} and decrease the size of the F-ratio.

Sample Data	
$M_1 = 15$	$M_2 = 25$
$SS_1 = 90$	$SS_2 = 70$

LO11   **3.** A researcher uses an ANOVA to evaluate the mean difference between two treatment conditions and obtains $F = 9.00$ with $df = 1, 17$. If an independent-measures t statistic had been used instead of the ANOVA, then what t value would be obtained and what is the df value for t?

    a. $t = 3.00$ with $df = 16$

    b. $t = 3.00$ with $df = 17$

    c. $t = 16$ with $df = 16$

    d. $t = 16$ with $df = 17$

ANSWERS   **1. b   2. b   3. b**

SUMMARY

1. Analysis of variance (ANOVA) is a statistical technique that is used to test for mean differences among two or more treatment conditions. The null hypothesis for this test states that in the general population there are no mean differences among the treatments. The alternative states that at least one mean is different from another.

2. The test statistic for ANOVA is a ratio of two variances called an F-ratio. The variances in the F-ratio are called mean squares, or MS values. Each MS is computed by

$$MS = \frac{SS}{df}$$

3. For the independent-measures ANOVA, the F-ratio is

$$F = \frac{MS_{between}}{MS_{within}}$$

The $MS_{between}$ measures differences between the treatments by computing the variability of the treatment means or totals. These differences are assumed to be produced by

   a. treatment effects (if they exist).
   b. differences resulting from chance.

The MS_{within} measures variability inside each of the treatment conditions. Because individuals inside a treatment condition are all treated exactly the same, any differences within treatments cannot be caused by treatment effects. Thus, the within-treatments MS is produced only by differences caused by chance. With these factors in mind, the F-ratio has the following structure:

$$F = \frac{\text{treatment effect} + \text{differences due to chance}}{\text{differences due to chance}}$$

When there is no treatment effect (H_0 is true), the numerator and the denominator of the F-ratio are measuring the same variance, and the obtained ratio should be near 1.00. If there is a significant treatment effect, the numerator of the ratio should be larger than the denominator, and the obtained F value should be much greater than 1.00.

4. The formulas for computing each SS, df, and MS value are presented in Figure 12.9, which also shows the general structure for the ANOVA.

5. The F-ratio has two values for degrees of freedom, one associated with the MS in the numerator and one associated with the MS in the denominator. These df values are used to find the critical value for the F-ratio in the F distribution table.

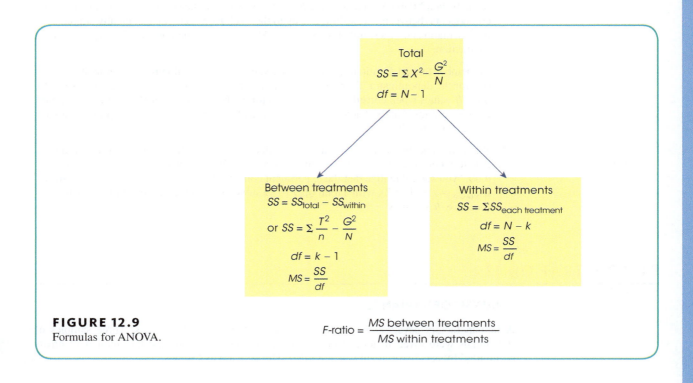

FIGURE 12.9
Formulas for ANOVA.

6. Effect size for the independent-measures ANOVA is measured by computing eta squared (η^2), the percentage of variance accounted for by the treatment effect.

$$\eta^2 = \frac{SS_{between}}{SS_{between} + SS_{within}}$$
$$= \frac{SS_{between}}{SS_{total}}$$

7. When the decision from an ANOVA is to reject the null hypothesis and when the experiment has more than two treatment conditions, it is necessary to continue the analysis with a post hoc test, such as Tukey's HSD test or the Scheffé test. The purpose of these tests is to determine exactly which treatments are significantly different and which are not.

KEY TERMS

analysis of variance (ANOVA) (392)

factor (394)

levels (394)

two-factor design or factorial design (394)

single-factor design (394)

single-factor, independent-measures design (394)

testwise alpha level (395)

experimentwise alpha level (395)

between-treatments variance (398)

treatment effect (398)

within-treatments variance (399)

F-ratio (399)

error term (400)

mean square (MS) (407)

distribution of F-ratios (409)

ANOVA summary table (411)

eta squared (η^2) (412)

post hoc tests or posttests (416)

pairwise comparisons (416)

Tukey's HSD test (417)

Scheffé test (418)

FOCUS ON PROBLEM SOLVING

1. It can be helpful to compute all three SS values separately, then check to verify that the two components (between and within) add up to the total. However, you can greatly simplify the calculations if you simply find SS_{total} and $SS_{within\ treatments}$, then obtain $SS_{between\ treatments}$ by subtraction.

2. Remember that an F-ratio has two separate values for df: a value for the numerator and one for the denominator. Properly reported, the $df_{between}$ value is stated first. You will need both df values when consulting the F distribution table for the critical F value. You should recognize immediately that an error has been made if you see an F-ratio reported with a single value for df.

3. When you encounter an F-ratio and its df values reported in the literature, you should be able to reconstruct much of the original experiment. For example, if you see "$F(2, 36) = 4.80$," you should realize that the experiment compared $k = 3$ treatment groups (because $df_{between} = k - 1 = 2$), with a total of $N = 39$ subjects participating in the experiment (because $df_{within} = N - k = 36$).

DEMONSTRATION 12.1

ANALYSIS OF VARIANCE

A human factors psychologist studied three computer keyboard designs. Three samples of individuals were given material to type on a particular keyboard, and the number of errors committed by each participant was recorded. The data are as follows:

Keyboard A	Keyboard B	Keyboard C	
0	6	6	$N = 15$
4	8	5	$G = 60$
0	5	9	$\Sigma X^2 = 356$
1	4	4	
0	2	6	
$T = 5$	$T = 25$	$T = 30$	
$SS = 12$	$SS = 20$	$SS = 14$	

Are these data sufficient to conclude that there are significant differences in typing performance among the three keyboard designs?

STEP 1 **State the hypotheses, and specify the alpha level.** The null hypothesis states that there is no difference among the keyboards in terms of number of errors committed. In symbols,

$$H_0: \mu_1 = \mu_2 = \mu_3 \quad \text{(Type of keyboard used has no effect.)}$$

As noted previously in this chapter, there are a number of possible statements for the alternative hypothesis. Here we state the general alternative hypothesis:

$$H_1: \text{At least one of the treatment means is different.}$$

We will set alpha at $\alpha = .05$.

STEP 2 **Locate the critical region.** To locate the critical region, we must obtain the values for $df_{between}$ and df_{within}.

$$df_{between} = k - 1 = 3 - 1 = 2$$
$$df_{within} = N - k = 15 - 3 = 12$$

The F-ratio for this problem has $df = 2, 12$. Consult the F-distribution table for $df = 2$ in the numerator and $df = 12$ in the denominator. The critical F value for $\alpha = .05$ is $F = 3.88$. The obtained F-ratio must exceed this value to reject H_0.

STEP 3 **Perform the analysis.** The analysis involves the following steps, which produce the nine values needed to fill an ANOVA summary table:

Source	SS	df	MS	
Between treatments	—	—	—	$F = $ __
Within treatments	—	—	—	
Total	—	—		

The analysis of SS. We will compute SS_{total} followed by its two components.

$$SS_{total} = \Sigma X^2 - \frac{G^2}{N} = 356 - \frac{60^2}{15} = 356 - \frac{3600}{15}$$
$$= 356 - 240 = 116$$

$$SS_{within} = \Sigma SS_{inside} \text{ each treatment}$$
$$= 12 + 20 + 14$$
$$= 46$$

$$SS_{between} = SS_{total} - SS_{within}$$
$$= 116 - 46$$
$$= 70$$

Analyze degrees of freedom.   We will compute df_{total}. Its components, $df_{between}$ and df_{within}, were previously calculated (Step 2).

$$df_{total} = N - 1 = 15 - 1 = 14$$
$$df_{between} = 2$$
$$df_{within} = 12$$

Calculate the MS values.   The values for $MS_{between}$ and MS_{within} are determined.

$$MS_{between} = \frac{SS_{between}}{df_{between}} = \frac{70}{2} = 35$$

$$MS_{within} = \frac{SS_{within}}{df_{within}} = \frac{46}{12} = 3.83$$

Compute the F-ratio.   Finally, we can compute F.

$$F = \frac{MS_{between}}{MS_{within}} = \frac{35}{3.83} = 9.14$$

STEP 4   **Make a decision about H_o, and state a conclusion.**   The obtained F of 9.14 exceeds the critical value of 3.88. Therefore, we can reject the null hypothesis. The type of keyboard used has a significant effect on the number of errors committed, $F(2, 12) = 9.14$, $p < .05$.

DEMONSTRATION 12.2

COMPUTING EFFECT SIZE FOR ANALYSIS OF VARIANCE

We will compute eta squared (η^2), the percentage of variance explained, for the data that were analyzed in Demonstration 12.1. The data produced a between-treatments SS of 70 and a total SS of 116. Thus,

$$\eta^2 = \frac{SS_{between}}{SS_{total}} = \frac{70}{116} = 0.60 \text{ (or 60\%)}$$

SPSS®

General instructions for using SPSS are presented in Appendix D. Following are detailed instructions for using SPSS to perform the **Single-Factor, Independent-Measures Analysis of Variance (ANOVA)** presented in this chapter.

In a study of perceptions of dog behavior, researchers surveyed dog-owners about the behavior (Starling, Branson, Thomson, & McGreevy, 2013). Each participant rated their dog's boldness by indicating their agreement with a series of statements like, "Approaches unfamiliar children away from home in a friendly manner." The researchers observed that some dog breeds are bolder than others. For example, Staffordshire bull terriers were rated as bolder than Australian cattle dogs.

Suppose that a researcher conducts a similar study by asking participants to rate the boldness of Labrador retrievers, boxers, and greyhounds. The hypothetical data are listed below. Each value represents the boldness score on a series of questions in the study.

Labrador Retriever	Boxer	Greyhound
28	5	11
20	13	11
32	17	16
19	2	6
16	7	24
19	13	7
25	11	0
19	18	11
8	11	10
14	23	4

The following steps will demonstrate how to use SPSS to perform a one-way analysis of variance to test the hypothesis that different dog breeds have different levels of boldness.

Data Entry

1. Use the **Variable View** of the data editor to create two new variables for the data above. Enter "boldness" in the **Name** field of the first variable. Select **Numeric** in the **Type** field and **Scale** in the **Measure** field. Enter a brief, descriptive title for the variable in the **Label** field (here, "Boldness Score" was used).

2. For the second variable, enter "breed" in the **Name** field. Select **Numeric** in the **Type** field and **Nominal** in the **Measure** field. Use "Dog Breed" in the **Label** field. In the **Values** field, click the "…" button to assign labels to group numbers. In the dialog box that follows, enter a "1" for value and "Labrador" for label. Repeat this process until the dialog box is like the box below.

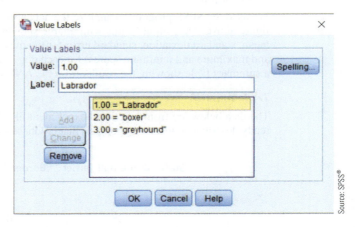

3. Return to the Data View. The scores are entered in a *stacked format* in the data editor, which means that all the scores from all of the different treatments are entered in a single column ("boldness"). Enter the scores for boxers directly beneath the scores for Labradors with no gaps or extra spaces. Continue in the same column with the scores for greyhounds.

4. In the second column ("breed"), enter a number to identify the treatment condition for each score. Enter a 1 beside each score from the first treatment, a 2 beside each score from the second treatment, and so on.

Data Analysis

1. Click **Analyze** on the tool bar, select **Compare Means**, and click on **One-Way ANOVA**.

2. Highlight the column label for the set of scores ("Boldness Score") in the left box and click the arrow to move it into the **Dependent List** box.

3. Highlight the label for the column containing the treatment numbers ("Dog Breed") in the left box and click the arrow to move it into the **Factor** box.

4. To select descriptive statistics for each treatment, click on the **Options** box, select **Descriptives**, and click **Continue**.

5. Click **OK**.

➡ **Oneway**

Descriptives

Boldness Score

	N	Mean	Std. Deviation	Std. Error	95% Confidence Interval for Mean Lower Bound	Upper Bound	Minimum	Maximum
Labrador	10	20.0000	6.92820	2.19089	15.0439	24.9561	8.00	32.00
boxer	10	12.0000	6.32456	2.00000	7.4757	16.5243	2.00	23.00
greyhound	10	10.0000	6.63325	2.09762	5.2549	14.7451	.00	24.00
Total	30	14.0000	7.76375	1.41746	11.1010	16.8990	.00	32.00

ANOVA

Boldness Score

	Sum of Squares	df	Mean Square	F	Sig.
Between Groups	560.000	2	280.000	6.364	.005
Within Groups	1188.000	27	44.000		
Total	1748.000	29			

SPSS Output

We used the SPSS program to analyze the data above, and the program output is shown in the figure above. The output begins with a table showing descriptive statistics (number of scores, mean, standard deviation, standard error for the mean, a 95% confidence interval for the mean, and maximum and minimum scores) for each dog breed. The second part of the output presents a summary table showing the results from the ANOVA.

Try It Yourself

The data below are from a fictional experiment like that described above. Follow the steps already described to test the hypothesis that boldness scores are different for different dog breeds.

Staffordshire Bull Terrier	Border Collie	Jack Russell Terrier
35	25	22
5	0	0
28	0	20
12	24	0
28	22	4
12	2	20
25	17	4
15	7	10
20	18	6
20	5	14

After you have successfully analyzed the above data, you should find that the ANOVA is not significant, $F(2, 27) = 3.231$. You should also notice that the group means are identical to those for Labradors, boxers, and greyhounds above. Thus, $MS_{between}$ in the ANOVA comparing Staffordshire bull terrier, Border collie, and Jack Russell terrier dog breeds is identical to

$MS_{between}$ in the ANOVA comparing Labrador, boxer, and greyhound dog breeds. The important difference between the two sets of scores is the variation within groups. In the first ANOVA, MS_{within} was equal to 44 and in the second ANOVA, MS_{within} was equal to 86.67.

PROBLEMS

1. Why should you use ANOVA instead of several t tests to evaluate mean differences when an experiment consists of three or more treatment conditions?

2. Suppose that you conduct three different t-tests to analyze the results of an experiment with three independent samples. Each individual t-test uses $\alpha = .05$. What is the probability of a Type 1 error occurring in any of the hypothesis tests? Is it greater than, equal to, or less than .05?

3. What value is expected for the F-ratio, on average, if the null hypothesis is true in an ANOVA? Explain why.

4. Describe some of the reasons that group means might be different from each other in an analysis of variance. Describe some of the reasons that individual scores might be different from each other.

5. Describe the similarities between an F-ratio and a t statistic.

6. Calculate SS_{total}, $SS_{between}$, and SS_{within} for the following set of data:

Treatment 1	Treatment 2	Treatment 3	
$n = 12$	$n = 12$	$n = 12$	$N = 36$
$T = 60$	$T = 72$	$T = 24$	$G = 156$
$SS = 30$	$SS = 46$	$SS = 40$	$\Sigma X^2 = 896$

7. Calculate SS_{total}, $SS_{between}$, and SS_{within} for the following set of data:

Treatment 1	Treatment 2	Treatment 3	
$n = 10$	$n = 10$	$n = 10$	$N = 30$
$T = 105$	$T = 180$	$T = 110$	$G = 395$
$SS = 350.5$	$SS = 190.0$	$SS = 424$	$\Sigma X^2 = 6,517$

8. A researcher uses an ANOVA to compare six treatment conditions with a sample of $n = 12$ in each treatment. For this analysis, find df_{total}, $df_{between}$, and df_{within}.

9. A researcher reports an F-ratio with $df_{between} = 3$ and $df_{within} = 40$ for an independent-measures ANOVA.
   a. How many treatment conditions were compared in the experiment?
   b. How many subjects participated in the experiment?
   c. Use Appendix B to find the critical value for F. Use $\alpha = .05$.
   d. What is the critical region for $\alpha = .01$?

10. A researcher reports an F-ratio with $df = 4, 62$ from an independent-measures research study.
    a. How many treatment conditions were compared in the study?
    b. What was the total number of participants in the study?
    c. Use Appendix B to find the critical value for F. Use $\alpha = .05$.
    d. What is the critical region for $\alpha = .01$?

11. The following values are from an independent-measures study comparing three treatment conditions.

	Treatment	
I	II	III
$n = 12$	$n = 12$	$n = 12$
$SS = 220$	$SS = 242$	$SS = 198$

   a. Compute the variance for each sample.
   b. Compute MS_{within}, which would be the denominator of the F-ratio for an ANOVA. Because the samples are all the same size, you should find that MS_{within} is equal to the average of the three sample variances.

12. The following values are from an independent-measures study comparing three treatment conditions.

	Treatment	
I	II	III
$n = 7$	$n = 7$	$n = 7$
$SS = 48$	$SS = 60$	$SS = 36$

   a. Compute the variance for each sample.
   b. Compute MS_{within}, which would be the denominator of the F-ratio for an ANOVA.

13. A researcher conducts an experiment comparing three treatment conditions with a separate sample of $n = 6$ in each treatment. An ANOVA is used to evaluate the data, and the results of the ANOVA are presented in the following table. Complete all missing values in the table. *Hint*: Begin with the values in the df column.

Source	SS	df	MS	
Between treatments	____	____	____	$F = $ ____
Within treatments	____	____	4	
Total	92	____		

14. The following summary table presents the results from an ANOVA comparing five treatment conditions with

$n = 5$ participants in each condition. Complete all missing values. (*Hint*: Start with the *df* column.)

Source	SS	df	MS	
Between treatments	____	____	125	F = ____
Within treatments	____	____	____	
Total	1500	____		

15. A developmental psychologist is examining the development of language skills from age 2 to age 4. Three different groups of children are obtained, one for each age, with $n = 16$ children in each group. Each child is given a language-skills assessment test. The resulting data were analyzed with an ANOVA to test for mean differences between age groups. The results of the ANOVA are presented in the following table. Fill in all missing values.

Source	SS	df	MS	
Between treatments	48	____	____	F = ____
Within treatments	____	____	____	
Total	252	____		

16. The following data were obtained from an independent-measures research study comparing three treatment conditions. Use an ANOVA with $\alpha = .05$ to determine whether there are any significant mean differences among the treatments.

	Treatment	
I	II	III
5	2	7
1	6	3
2	2	2
3	3	4
0	5	5
1	3	2
2	0	4
2	3	5

17. Many know the feeling of being too groggy to take an 8 a.m. exam—we need to be more alert and aroused to do our best. Some of us also know the feeling of being too aroused by the prospect of an exam to do our best. Thus, the best level of performance occurs when we are in the Goldilocks zone—not too aroused and not too groggy. Simon and Moghaddam (2016) recently replicated this observation in rats. Subjects were randomly assigned to one of three groups. Each group received a different dose (none, a medium dose, or a high dose) of Ritalin, which is a stimulant. Researchers measured the rats' learning and cognitive performance and observed the highest level of performance in the group that received a medium dose of Ritalin.

Suppose that the following data were obtained from a similar independent-measures research study comparing three treatment conditions.

	Treatment	
I – No Drug	II – Medium Dose	III – High Dose
17	29	16
14	21	24
22	27	20
19	23	21
18	25	19

a. Use an ANOVA with $\alpha = .05$ to determine whether there are any significant mean differences among the treatments.

b. What test would a researcher use to compare the No Drug group to the High Dose group?

c. Write the results of the ANOVA as they would appear in a research article.

18. Open positions in the highest-paying jobs, the best internships, and the most prestigious graduate programs can attract many applicants. Initial evaluations of applications are usually based on the applicant's résumé or curriculum vitae. In a recent study, researchers demonstrated that two different kinds of fancy "graphical résumés," which include graphic descriptions of things like the timeline of an applicant's education, are no more effective than traditional text-based résumés (Popham, Lee, Sublette, Kent, & Carswell, 2017). In many ways, they actually might be worse. Suppose that a researcher is interested in studying the effect of résumé type on participants' ratings of the competence of the job applicant. She randomly assigns participants to one of three groups. Group 1 receives a traditional text-based résumé. Group 2 receives a graphic résumé with photos of the college that was listed under the applicant's education. Group 3 receives a graphic résumé with charts summarizing the amount of time the hypothetical applicant spent at each of their previous jobs. All participants rate the perceived competence of the job applicant. The hypothetical data are listed below.

	Treatment	
Text	Graphic Type 1	Graphic Type 2
9	4	6
8	4	6
10	6	6
8	8	7
10	8	5

a. Use an ANOVA with $\alpha = .05$ to determine whether there are any significant mean differences among the treatments.

b. Suppose that a researcher wanted to compare text résumés to graphic type 1 résumés only. What test statistic would the researcher use?

c. Write the results of the ANOVA as they would appear in a research article.

19. The following data were obtained from an independent-measures research study comparing three treatment conditions. Use an ANOVA with $\alpha = .05$ to determine whether there are any significant mean differences among the treatments.

Treatment			
I	II	III	
$n = 7$	$n = 5$	$n = 9$	$N = 21$
$T = 168$	$T = 140$	$T = 279$	$G = 587$
$SS = 186$	$SS = 80$	$SS = 168$	$\Sigma X^2 = 17{,}035$

20. A research study comparing three treatment conditions produces $T = 20$ with $n = 4$ for the first treatment, $T = 10$ with $n = 5$ for the second treatment, and $T = 30$ with $n = 6$ for the third treatment. Calculate $SS_{\text{between treatments}}$ for these data.

21. A research study comparing three treatment conditions produces $T = 28$ with $n = 7$ for the first treatment, $T = 32$ with $n = 8$ for the second treatment, and $T = 108$ with $n = 9$ for the third treatment. Calculate $SS_{\text{between treatments}}$ for these data.

22. The following values are from an independent-measures study comparing three treatment conditions.

Treatments			
I	II	III	IV
$n = 12$	$n = 12$	$n = 12$	$n = 12$
$SS = 77$	$SS = 110$	$SS = 66$	$SS = 99$

a. Compute the variance for each sample.

b. Compute MS_{within}, which would be the denominator of the F-ratio for an ANOVA. Because the samples are all the same size, you should find that MS_{within} is equal to the average of the three sample variances.

23. A research report from an independent-measures study states that there are significant differences between treatments, $F(4, 40) = 3.45, p < .05$.

a. How many treatment conditions were compared in the study?

b. What was the total number of participants in the study?

c. Would the result be significant if $\alpha = .01$?

24. Several factors influence the size of the F-ratio. For each of the following, indicate whether it would influence the numerator or the denominator of the F-ratio, and indicate whether the size of the F-ratio would increase or decrease.

a. Increase the differences between the sample means.

b. Increase the size of the sample variances within each group.

25. A researcher used ANOVA and computed $F = 4.25$ for the following data.

Treatments		
I	II	III
$n = 10$	$n = 10$	$n = 10$
$M = 20$	$M = 28$	$M = 35$
$SS = 1{,}005$	$SS = 1{,}391$	$SS = 1{,}180$

a. If the mean for Treatment III were changed to $M = 25$, what would happen to the size of the F-ratio (increase or decrease)? Explain your answer.

b. If the SS for Treatment I were changed to $SS = 1{,}400$, what would happen to the size of the F-ratio (increase or decrease)? Explain your answer.

26. The following data were obtained from an independent-measures study comparing three treatment conditions.

Treatment			
I	II	III	
$n = 6$	$n = 6$	$n = 6$	$N = 18$
$M = 1$	$M = 2$	$M = 6$	$G = 54$
$SS = 60$	$SS = 65$	$SS = 40$	$\Sigma X^2 = 411$

a. Calculate the sample variance for each of the three samples.

b. Use an ANOVA with $\alpha = .05$ to determine whether there are any significant differences among the three treatment means. (*Note:* In the ANOVA you should find that MS_{within} is equal to the average of the three sample variances.)

27. For the preceding problem you should find that there are significant differences among the three treatments. One reason for the significance is that the sample variances are relatively small. To create the following data, we kept the same sample means that appeared in problem 26 but doubled the SS values within each sample.

Treatment			
I	II	III	
$n = 6$	$n = 6$	$n = 6$	$N = 18$
$M = 1$	$M = 2$	$M = 6$	$G = 54$
$SS = 120$	$SS = 130$	$SS = 80$	$\Sigma X^2 = 576$

a. Calculate the sample variance for each of the three samples. Describe how these sample variances compare with those from problem 26.

b. Predict how the increase in sample variance should influence the outcome of the analysis. That is, how will the F-ratio for these data compare with the value obtained in problem 19?

c. Use an ANOVA with $\alpha = .05$ to determine whether there are any significant differences among the three treatment means. (Does your answer agree with your prediction in part b?)

28. The following data were observed in an independent-measures study comparing three treatment conditions.

	Treatment	
I	II	III
4	10	11
0	14	17
0	8	16
6	7	6
10	8	11
4	7	10
4	2	13

a. Use an ANOVA with $\alpha = .05$ to determine whether there are any significant differences among the three treatment means. *Note:* Because the samples are all the same size, MS_{within} is the average of the three sample variances.

b. Calculate η^2 to measure the effect size for this study.

29. The following data summarize the results from an independent-measures study comparing three treatment conditions.

	Treatment		
I	II	III	
$n = 5$	$n = 5$	$n = 5$	
$M = 1$	$M = 5$	$M = 6$	$N = 15$
$T = 5$	$T = 25$	$T = 30$	$G = 60$
$s^2 = 9.00$	$s^2 = 10.00$	$s^2 = 11.00$	$\Sigma X^2 = 430$
$SS = 36$	$SS = 40$	$SS = 44$	

a. Use an ANOVA with $\alpha = .05$ to determine whether there are any significant differences among the three treatment means. *Note:* Because the samples are all the same size, MS_{within} is the average of the three sample variances.

b. Calculate η^2 to measure the effect size for this study.

30. An ANOVA produces an F-ratio with $df = 1, 34$. Could the data have been analyzed with a t test? What would be the degrees of freedom for the t statistic?

31. To create the following data we started with the same sample means and variances that appeared in problem 29, but doubled the sample size to $n = 10$.

	Treatment		
I	II	III	
$n = 10$	$n = 10$	$n = 10$	
$M = 1$	$M = 5$	$M = 6$	$N = 30$
$T = 10$	$T = 50$	$T = 60$	$G = 120$
$s^2 = 9.00$	$s^2 = 10.00$	$s^2 = 11.00$	$\Sigma X^2 = 890$
$SS = 81$	$SS = 90$	$SS = 99$	

a. Predict how the increase in sample size should affect the F-ratio for these data compared to the values obtained in problem 29.

b. Use an ANOVA with $\alpha = .05$ to check your prediction. *Note:* Because the samples are all the same size, MS_{within} is the average of the three sample variances.

c. Predict how the increase in sample size should affect the value of η^2 for these data compared to the η^2 in problem 29. Calculate η^2 to check your prediction.

32. The following scores are from an independent-measures study comparing two treatment conditions.

Treatment I	Treatment II	
10	7	
8	4	
7	9	$N = 16$
9	3	$G = 120$
13	7	$\Sigma X^2 = 1036$
7	6	
6	10	
12	2	

a. Use an independent-measures t test with $\alpha = .05$ to determine whether there is a significant mean difference between the two treatments.

b. Use an ANOVA with $\alpha = .05$ to determine whether there is a significant mean difference between the two treatments. You should find that $F = t^2$.

Two-Factor Analysis of Variance

Tools You Will Need

The following items are considered essential background material for this chapter. If you doubt your knowledge of any of these items, you should review the appropriate chapter or section before proceeding.

- Independent-measures analysis of variance (Chapter 12)
- Individual differences (page 347)

clivewa/Shutterstock.com

PREVIEW

Does browsing social media have an effect on people's self-esteem? Wilcox and Stephen (2013) conducted an experiment to see if self-esteem would be enhanced when people with close relationships use Facebook. The researchers randomly assigned adults to one of two conditions. The participants either browsed Facebook or browsed a popular news website. The researchers then divided these two browsing groups of participants in half based on the strength of their relationships with Facebook friends. These groups reflected either weak or strong relationship connections (called "tie strength" by the authors of the study) to their friends based on the participants' self-reports. The design of the experiment is shown in Table 13.1. Notice that there are two independent variables (or factors): type of browsing and strength of relationships with Facebook friends. Combining these two factors resulted in four separate groups of participants.

After browsing Facebook or a news website, each participant was asked to complete a self-esteem questionnaire. The results of the study showed that participants who browsed Facebook had higher self-esteem scores only when they had strong relationships with their friends. On the other hand, when participants browsed the news site, there was no difference in self-esteem between the strong and weak relationship groups (Figure 13.1).

The Wilcox and Stephen study is an example of research that involves two independent (or quasi-independent) variables in the same study. These variables are:

1. Type of media browsing (Facebook or the news website)

2. Strength of relationships with Facebook friends (weak or strong)

TABLE 13.1

The structure of a two-factor experiment for the Wilcox and Stephen (2013) study. The two factors are type of media browsing and strength of relationships with Facebook friends. There are two levels for each factor.

Strength of Relationship

		Weak	Strong
Type of Browsing	Facebook	Browse Facebook and weak relationship	Browse Facebook and strong relationship
	News Website	Browse news website and weak relationship	Browse news website and strong relationship

The results of the study indicate that the effect of one variable (strength of friendship relationships) on self-esteem *depends on* another variable (type of web browsing).

You should realize that it is quite common to have two variables that interact in this way. For example, changing the amount of a medication may have profoundly different effects on elderly patients than on younger ones. In this instance there is an interaction between the dose of a drug and the age of the patients because the effect of changing the dose depends on the age of the patient. To observe how one variable interacts with another, it is necessary to study both variables simultaneously in one study. However, the analysis of variance (ANOVA) procedures introduced in Chapter 12 are limited to evaluating mean differences produced by one independent

FIGURE 13.1

Results similar to Wilcox and Stephen (2013, Study 3) are shown. There was no effect of strength of relationship on self-esteem for the groups that browsed the news website. For participants who browsed Facebook, self-esteem was higher for those participants with strong relationships with Facebook friends than for those with weak relationships.

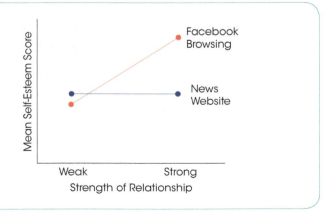

variable and are not appropriate for mean differences involving two (or more) independent variables.

Fortunately, ANOVA is a very flexible hypothesis testing procedure that can be modified to evaluate the mean differences produced in a research study with two (or more) independent variables. These studies use *factorial designs*. In this chapter we introduce the analysis of a basic factorial design, called the *two-factor* ANOVA, which tests the significance of each independent variable acting alone as well as the interaction between variables.

You should note that the results of the Wilcox and Stephen study are an example of an interaction. The effect of strength of relationships with friends on self-esteem depended on the type of browsing the participants were using. The results show that under certain conditions (having close friends on Facebook), spending time browsing Facebook bolsters self-esteem.

■ Chapter Overview

In Chapter 12 we introduced analysis of variance (ANOVA) as a hypothesis-testing procedure for evaluating differences among two or more sample means. The specific advantage of ANOVA, especially in contrast to *t* tests, is that ANOVA can be used to evaluate the significance of mean differences in situations in which there are more than two sample means being compared. However, the presentation of ANOVA in Chapter 12 was limited to single-factor, independent-measures research designs. Recall that *single factor* indicates that the research study involves only one independent variable (or only one quasi-independent variable), and the term *independent-measures* indicates that the study uses a separate sample for each of the different treatment conditions being compared.

In this chapter we extend the ANOVA procedure to some more sophisticated research situations in which ANOVA is used. Specifically, we introduce the two-factor ANOVA. As an example of a research situation in which a two-factor ANOVA is used, consider a researcher who wants to examine how weight loss is related to different combinations of diet and exercise. In this situation, two variables are manipulated (diet and exercise) while a third variable is observed (weight loss). In statistical terminology, the research study has two independent variables, or two factors. In this chapter, we show how the general ANOVA procedure from Chapter 12 can be used to test for mean differences in a two-factor research study.

13-1 An Overview of the Two-Factor, Independent-Measures ANOVA

> **LEARNING OBJECTIVES**
>
> **1.** Describe the structure of a factorial research design—especially a two-factor independent-measures design—using the terms factor and level and identify the *factors* and *levels* for a specific example of a two-factor design.
>
> **2.** Define a main effect and an interaction and identify the patterns of data that produce main effects and interactions.
>
> **3.** Identify the three *F*-ratios for a two-factor ANOVA and explain how they are related to each other.

In most research situations, the goal is to examine the relationship between two variables. Typically, the research study attempts to isolate the two variables to eliminate or reduce the influence of any outside variables that may distort the relationship being studied. A typical experiment, for example, focuses on one independent variable (which is expected to influence behavior) and one dependent variable (which is a measure of the behavior).

In real life, however, variables rarely exist in isolation. That is, behavior usually is influenced by a variety of different variables acting and interacting simultaneously. To examine these more complex, real-life situations, researchers often design research studies

An independent variable is a manipulated variable in an experiment. A quasi-independent variable is not manipulated but defines the groups of scores in a nonexperimental study.

that include more than one independent (or quasi-independent) variable. Thus, researchers systematically change two (or more) variables and then observe how the changes influence another (dependent) variable.

In Chapter 12, we examined ANOVA for *single-factor* research designs—that is, designs that included only one independent variable or only one quasi-independent variable. When a research study involves more than one factor, it is called a *factorial design*. In this chapter, we consider the simplest version of a factorial design. Specifically, we examine ANOVA as it applies to research studies with exactly two factors. In addition, we limit our discussion to studies that use a separate sample for each treatment condition—that is, independent-measures designs. Finally, we consider only *two-factor designs* for which the sample size (n) is the same for all treatment conditions. In the terminology of ANOVA, this chapter examines *two-factor, independent-measures, equal* n *designs*.

We will use the Wilcox and Stephen (2013) study described in the Chapter Preview to introduce the two-factor research design. Wilcox and Stephen (2013) conducted a research study to determine whether type of media browsing has an effect on the self-esteem of participants. The researchers randomly assigned participants to browse either Facebook or a popular news website. The researchers also divided the two browsing groups of participants into two groups based on the strength of their relationships with Facebook friends based on their self-reports. For the dependent variable, participants completed the brief version of Rosenberg's (1989) self-esteem questionnaire after five minutes of online browsing.

Table 13.2 shows the structure of the study. Note that the study involves two separate factors: One factor is manipulated by the researcher, assigning participants to either Facebook browsing or news website browsing. The second factor is the strength of relationships to Facebook friends. This was a quasi-independent variable (not manipulated) because participants were divided into weak versus strong relationships based on a pre-existing participant variable. The two factors are used to create a *matrix* with the two Facebook conditions defining the rows and the different levels of relationship ties with friends defining the columns. The resulting two-by-two matrix shows four different combinations of the variables, producing four different conditions. Thus, the research study would require four separate samples, one for each *cell*, or box, in the matrix. The dependent variable for the study is the level of self-esteem for the participants in each of the four conditions.

Notice that this two-factor, independent measures design has one independent variable (Facebook browsing) and one quasi-independent variable (strength of relationship with Facebook friends).

TABLE 13.2

The structure of a two-factor experiment presented as a matrix. The two factors are type of media browsing and strength of Facebook relationships. There are two levels for each factor.

| | | Factor *B*: Strength of Facebook Relationships | |
		Weak	Strong
Factor *A*: Type of Browsing	Facebook	Scores for a group of participants who have weak Facebook relationships and are assigned to browse Facebook	Scores for a group of participants who have strong Facebook relationships and are assigned to browse Facebook
	News Website	Scores for a group of participants who have weak Facebook relationships and are assigned to browse the news website	Scores for a group of participants who have strong Facebook relationships and are assigned to browse the news website

The two-factor ANOVA tests for mean differences in research studies that are structured like the Wilcox and Stephen study in Table 13.2. For this example, the two-factor ANOVA evaluates three separate sets of mean differences:

1. Is there a difference in self-esteem between browsing either Facebook or the news website?

2. Is there a difference in self-esteem for participants who have weak versus strong relationships with Facebook friends?

3. If there is a difference in self-esteem, is it related to specific combinations of the two factors; that is, a certain combination of the type of browsing and the strength of relationship with friends? For example, perhaps self-esteem depends on levels of relationship strength for those participants browsing Facebook, but not for those who are browsing a news website.

Thus, the two-factor ANOVA allows us to examine three types of mean differences within one analysis. In particular, we conduct three separate hypothesis tests for the same data, with a separate F-ratio for each test. The three F-ratios have the same basic structure:

$$F = \frac{\text{variance (differences) between treatments}}{\text{variance (differences) expected if there is no treatment effect}}$$

In each case, the numerator of the F-ratio measures the actual mean differences in the data, and the denominator measures the differences that would be expected if there is no treatment effect. As always, a large value for the F-ratio indicates that the sample mean differences are greater than would be expected by chance alone, and therefore provides evidence of a treatment effect. To determine whether the obtained F-ratios are *significant*, we need to compare each F-ratio with the critical values found in the F-distribution table in Appendix B.

■ Main Effects and Interactions

As noted in the previous section, a two-factor ANOVA actually involves three distinct hypothesis tests. In this section, we examine these three tests in more detail.

For the purpose of a two-factor ANOVA, it is arbitrary which independent variable is identified as factor A and which as factor B.

Traditionally, the two independent variables in a two-factor experiment are identified as factor A and factor B. For the study presented in Table 13.2, we have identified type of browsing as factor A, and the level of strength of Facebook relationships as factor B. The goal of the study is to evaluate the mean differences that may be produced by either of these factors acting independently or by the two factors acting together.

■ Main Effects

One purpose of the study is to determine whether differences in browsing type (factor A) result in differences in behavior. To answer this question, we compare the mean score for all participants in the Facebook condition with the mean for all the participants in the News Website condition. Note that this process evaluates the mean difference between the top row and the bottom row in Table 13.2.

To make this process more concrete, we present a set of hypothetical data in Table 13.3. The table shows the mean score for each of the treatment conditions (cells) as well as the overall mean for each column (each level of strength of relationships and the overall mean for each row [each type of browsing]). Note that these hypothetical data were developed with the full version of the Rosenberg (1989) self-esteem scale. These data indicate that participants who browsed Facebook (the top row) had an overall mean of $M = 19$. This overall mean is the average of all scores for participants who browsed Facebook. In contrast, the participants who browsed the news website had an overall mean of $M = 15$ (the

TABLE 13.3

Hypothetical data for an experiment examining the effect of type of web browsing on self-esteem for participants with weak versus strong relationships with friends. The data assumes that the sample size, n, is the same for every group. The Wilcox and Stephen study used the brief 5-point version of the Rosenberg Self-Esteem Scale. For these hypothetical data we are assuming the 30-point full Rosenberg Self-Esteem Scale (1965, 1989) is used.

	Weak	Strong	
Facebook	$M = 16$	$M = 22$	$M = 19$
News Website	$M = 12$	$M = 18$	$M = 15$
	$M = 14$	$M = 20$	

mean for the bottom row). The difference between these means constitutes what is called the *main effect* for type of browsing , or the *main effect for factor* A.

Similarly, the main effect for factor B (relationship strength with friends) is defined by the mean difference between the columns of the matrix. For the data in Table 13.3, all participants with weak relationships with friends had an overall mean score of $M = 14$. Participants who had strong relationships with friends had an overall average score of $M = 20$. The difference between these means constitutes the *main effect* for the levels of relationship strength, or the *main effect for factor* B.

> The mean differences among the levels of one factor are referred to as the **main effect** of that factor. When the design of the research study is represented as a matrix with one factor determining the rows and the second factor determining the columns, then the mean differences among the rows describe the main effect of one factor, and the mean differences among the columns describe the main effect for the second factor.

The mean differences between columns or rows simply *describe* the main effects for a two-factor study. As we have observed in earlier chapters, the existence of sample mean differences does not necessarily imply that the differences are *statistically significant*. In general, two samples are not expected to have exactly the same means. There will always be small differences from one sample to another, and you should not automatically assume that these differences are an indication of a systematic treatment effect. Small differences may reflect error variability due to chance. In the case of a two-factor study, any main effects that are observed in the data must be evaluated with a hypothesis test to determine whether they are statistically significant effects. Unless the hypothesis test demonstrates that the main effects are significant, you must conclude that the observed mean differences are simply the result of sampling error.

The evaluation of main effects accounts for two of the three hypothesis tests in a two-factor ANOVA. We state hypotheses concerning the main effect of factor A and the main effect of factor B and then calculate two separate F-ratios to evaluate the hypotheses.

For the example we are considering, factor A involves the comparison of two different web browsing conditions. The null hypothesis would state that there is no difference between the two conditions; that is, type of browsing has no effect on self-esteem. In symbols,

$$H_0: \mu_{A_1} = \mu_{A_2}$$

The alternative hypothesis is that the two different types of browsing do result in different self-esteem means:

$$H_1: \mu_{A_1} \neq \mu_{A_2}$$

To evaluate these hypotheses, we compute an F-ratio that compares the actual mean differences between the two types of browsing conditions versus the amount of difference that would be expected without any treatment effect.

$$F = \frac{\text{variance (differences) between the means for factor } A}{\text{variance (differences) expected if there is no treatment effect}}$$

$$F = \frac{\text{variance (differences) between the row means}}{\text{variance (differences) expected if there is no treatment effect}}$$

Similarly, factor B involves the comparison of the two levels of strength of relationship. The null hypothesis states that there is no difference in the mean amount of self-esteem between the two groups. In symbols,

$$H_0: \mu_{B_1} = \mu_{B_2}$$

As always, the alternative hypothesis states that the means are different:

$$H_0: \mu_{B_1} \neq \mu_{B_2}$$

Again, the F-ratio compares the obtained mean difference between the two levels of strength of relationship versus the amount of difference that would be expected if there is no systematic treatment effect.

$$F = \frac{\text{variance (differences) between the means for factor } B}{\text{variance (differences) expected if there is no treatment effect}}$$

$$F = \frac{\text{variance (differences) between the column means}}{\text{variance (differences) expected if there is no treatment effect}}$$

■ Interactions

In addition to evaluating the main effect of each factor individually, the two-factor ANOVA allows you to evaluate other mean differences that may result from unique combinations of the two factors. For example, specific combinations of type of browsing and strength of relationship acting together may have effects that are different from the main effects of either factor by themselves. Any "extra" mean differences that are not explained by the main effects are called an *interaction,* or an *interaction between factors.* The real advantage of combining two factors within the same study is the ability to examine the unique effects caused by an interaction.

> An **interaction** between two factors occurs whenever the mean differences between individual treatment conditions, or cells, are different from what would be predicted from the overall main effects of the factors.

To make the concept of an interaction more concrete, we reexamine the data shown in Table 13.3. For these data, there is no interaction; that is, there are no extra mean differences that are not explained by the main effects. For example, within each relationship strength condition (each column of the matrix) the average level of self-esteem for participants browsing Facebook is 4 points higher than the average for those browsing the news website. This 4-point mean difference is exactly what is predicted by the overall main effect for type of browsing. Now consider the data in Table 13.4 showing a different set of hypothetical data and illustrate an interaction.

TABLE 13.4

Hypothetical data for an experiment examining the effect of type of web browsing on self-esteem for participants who have weak versus strong relationships with friends. The data show the same main effects as the values in Table 13.3, but the individual treatment means have been modified to create an interaction. The data assumes that the sample size, n, is the same for every group. They also are based on the 30-point version of the Rosenberg Self-Esteem Scale.

	Weak	Strong	
Facebook	$M = 13$	$M = 25$	$M = 19$
News Website	$M = 15$	$M = 15$	$M = 15$
	$M = 14$	$M = 20$	

These new data show exactly the same main effects that existed in Table 13.3 (the column means and the row means have not been changed). There is still a 4-point mean difference between the two rows (the main effect for type of browsing) and a 6-point mean difference between the two columns (the main effect for strength of relationship). But now there is an interaction between the two factors. For example, among the participants who browsed Facebook (top row), there is a 12-point difference in self-esteem between participants with strong relationships ($M = 25$) and those with weak ties ($M = 13$). This 12-point difference cannot be explained by the main effect for the relationship strength because this main effect was only 6 points. Also, for the participants who browsed the news (bottom row of Table 13.4), the data show no difference between the two relationship strength groups. Again, the zero difference is not what would be expected based on the 6-point main effect for the relationship strength factor. Mean differences that are not explained by the main effects are an indication of an interaction between the two factors.

To evaluate the interaction, the two-factor ANOVA first identifies mean differences that are not explained by the main effects. The extra mean differences are then evaluated by an F-ratio with the following structure:

$$F = \frac{\text{Variance (mean differences) not explained by the main effects}}{\text{Variance (mean differences) expected if there are no treatment effects}}$$

The null hypothesis for this F-ratio simply states that there is no interaction:

H_0: There is no interaction between factors A and B. The mean differences between treatment conditions are explained by the main effects of the two factors.

The alternative hypothesis is that there is an interaction between the two factors:

H_1: There is an interaction between factors. The mean differences between treatment conditions are not what would be predicted from the overall main effects of the two factors.

■ More about Interactions

In the previous section, we introduced the concept of an interaction as the unique effect produced by two factors working together. This section presents two alternative definitions of an interaction. These alternatives are intended to help you understand the concept of an interaction and to help you identify an interaction when you encounter one in a set of data. You should realize that the new definitions are equivalent to the original and simply present slightly different perspectives on the same concept.

The first new perspective on the concept of an interaction focuses on the notion of independence for the two factors. More specifically, if the two factors are independent, so that one factor does not influence the effect of the other, then there is no interaction. On the other hand, when the two factors are not independent, so that the effect of one factor *depends on* the other, then there is an interaction. The notion of dependence between factors is consistent with our earlier discussion of interactions. If one factor influences the effect of the other, then unique combinations of the factors produce unique effects.

> When the effect of one factor depends on the different levels of a second factor, then there is an **interaction** between the factors.

This definition of an interaction should be familiar in the context of a "drug interaction." Your doctor and pharmacist are always concerned that the effect of one medication may be altered or distorted by a second medication that is being taken at the same time. Thus, if the effect of one drug (factor *A*) depends on a second drug (factor *B*), then you have an interaction between the two drugs.

Returning to Table 13.3, you will notice that the type of browsing effect (top row versus bottom row) *does not depend* on relationship strength (shown in the two columns). For these data, both Facebook and news website browsing result in the same 4-point increase in self-esteem regardless of the relationship strength group. Thus, the effect of type of browsing does not depend on relationship tie strength and there is no interaction. Now consider the data in Table 13.4. This time, the effect of type of web browsing *depends on* the relationship strength of participants. Facebook browsing results in a 12-point increase in self-esteem with strong relationships, but there is no mean difference in relationship strength when browsing the news website. Thus, the effect of relationship strength on self-esteem depends on which type of browsing the participants use. This result indicates that there is an interaction between the two factors.

The second alternative definition of an interaction is obtained when the results of a two-factor study are presented in a graph. In this case, the concept of an interaction can be defined in terms of the pattern displayed in the graph. Figure 13.2 shows the two sets of data we have been considering. The original data from Table 13.3, where there is no interaction, are presented in Figure 13.2(a). To construct this figure, we selected one of the factors to be displayed on the horizontal axis; in this case, the different levels of type of browsing are displayed. The dependent variable, the level of self-esteem, is shown on the vertical axis. Note that the figure actually contains two separate graphs. The top line shows the relationship between strength of friendships and self-esteem for those participants who browsed Facebook. The bottom line shows the relationship between strength of friendship

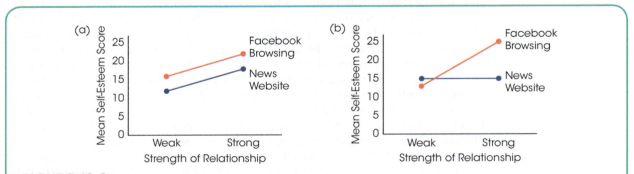

FIGURE 13.2
(a) Graph showing the treatment means for Table 13.3, for which there is no interaction. (b) Graph for Table 13.4, for which there is an interaction.

and self-esteem for those that browsed the news website. In general, the picture in the graph matches the structure of the data matrix; the columns of the matrix appear as values along the X axis, and the rows of the matrix appear as separate lines in the graph.

For the original set of means displayed in Figure 13.2(a), note that the two lines are parallel; that is, the distance between lines is constant. In this case, the distance between lines reflects the 4-point difference in the mean self-esteem scores between Facebook and news website browsing, and this same difference exists across both levels of relationship strength.

Now look at a graph that is obtained when there is an interaction in the data. Figure 13.2(b) shows the data from Table 13.4. This time, note that the lines in the graph are not parallel. The distance between the lines changes as you scan from left to right. The lines are diverging, which indicates that there is an interaction between type of browsing and relationship strength. The effect of type of browsing on self-esteem depends on the level of relationship strength.

> When the results of a two-factor study are presented in a graph, the existence of nonparallel lines (lines that cross, converge, or diverge) indicates an **interaction** between the two factors.

For many students, the concept of an interaction is easiest to understand using the perspective of interdependency; that is, an interaction exists when the effects of one factor *depend* on the levels of another factor. However, the easiest way to identify an interaction within a set of data is to draw a graph showing the treatment means. The presence of non-parallel lines is an easy way to spot an interaction.

■ Independence of Main Effects and Interactions

The $A \times B$ interaction typically is called "the A by B" interaction. For example, if there is an interaction between type of browsing and relationship strength, it may be called the "browsing by relationship strength" interaction.

The two-factor ANOVA consists of three hypothesis tests, each evaluating specific mean differences: the A effect, the B effect, and the $A \times B$ interaction. As we have noted, these are three *separate* tests, but you should also realize that the three tests are *independent*. That is, the outcome for any one of the three tests is totally unrelated to the outcome for either of the other two. Thus, it is possible for data from a two-factor study to display any possible combination of significant and/or not significant main effects and interactions. The data sets in Table 13.5 show several possibilities.

Table 13.5(a) shows data with mean differences between levels of factor A (an A effect) but no mean differences for factor B and no interaction. To identify the A effect, notice that the overall mean for A_1 (the top row) is 10 points higher than the overall mean for A_2 (the bottom row). This 10-point difference is the main effect for factor A. To evaluate the B effect, notice that both columns have exactly the same overall mean, indicating no difference between levels of factor B; hence, there is no B effect. Finally, the absence of an interaction is indicated by the fact that the overall A effect (the 10-point difference) is constant within each column; that is, the A effect *does not depend* on the levels of factor B. (Alternatively, the data indicate that the overall B effect is constant within each row.)

Table 13.5(b) shows data with an A effect and a B effect, but no interaction. For these data, the A effect is indicated by the 10-point mean difference between rows, and the B effect is indicated by the 20-point mean difference between columns. The fact that the 10-point A effect is constant within each column indicates no interaction.

Finally, Table 13.5(c) shows data that display an interaction but no main effect for factor A or for factor B. For these data, there is no mean difference between rows (no A effect) and no mean difference between columns (no B effect). However, within each row (or within each column), there are mean differences. The "extra" mean differences within the rows and columns cannot be explained by the overall main effects and therefore indicate an interaction.

TABLE 13.5

Three sets of data showing different combinations of main effects and an interaction for a two-factor study. (The numerical value in each cell of the matrices represents the mean value obtained for the sample in that treatment condition.)

(a) Data showing a main effect for factor A but no B effect and no interaction.

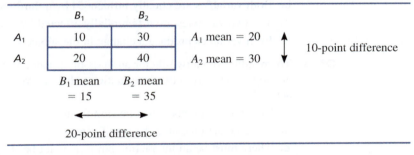

(b) Data showing main effects for both factor A and factor B but no interaction.

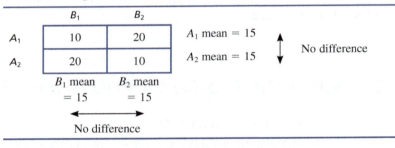

(c) Data showing no main effect for either factor but an interaction.

	B_1	B_2	
A_1	10	20	A_1 mean = 15
A_2	20	10	A_2 mean = 15

B_1 mean = 15 B_2 mean = 15

No difference

No difference

The following example is an opportunity to test your understanding of main effects and interactions.

EXAMPLE 13.1

The following matrix represents the outcome of a two-factor experiment. Describe the main effect for factor A and the main effect for factor B. Does there appear to be an interaction between the two factors?

Experiment I

	B_1	B_2
A_1	$M = 10$	$M = 20$
A_2	$M = 30$	$M = 40$

You should conclude that there is a main effect for factor A (the scores in A_2 average 20 points higher than in A_1) and there is a main effect for factor B (the scores in B_2 average 10 points higher than in B_1) but there is no interaction; there is a constant 20-point difference between A_1 and A_2 that does not depend on the levels of factor B.

LO1 **1.** A two-factor study with two levels of factor A and three levels of factor B uses a separate sample of $n = 6$ participants in each treatment condition. How many participants are needed for the entire study?

    a. 6

    b. 12

    c. 30

    d. 36

LO2 **2.** Which of the following accurately describes an interaction between two variables?

    a. The effect of one variable depends on the levels of the second variable.

    b. Both variables are equally influenced by a third factor.

    c. The two variables are differentially affected by a third variable.

    d. Both variables produce a change in the subjects' scores.

LO3 **3.** The results from a two-factor analysis of variance show that both main effects are significant. From this information, what can you conclude about the interaction?

    a. The interaction also must be significant.

    b. The interaction cannot be significant.

    c. There must be an interaction but it may not be statistically significant.

    d. You can make no conclusions about the significance of the interaction.

ANSWERS   **1. d   2. a   3. d**

13-2 An Example of the Two-Factor ANOVA and Effect Size

LEARNING OBJECTIVES

4. Describe the two-stage structure of a two-factor ANOVA and explain what happens in each stage.

5. Compute the SS, df, and MS values needed for a two-factor ANOVA, explain the relationships among them, and use the df values from a specific ANOVA to describe the structure of the study and the number of participants.

6. Conduct a two-factor ANOVA including measures of effect size for both main effects and the interaction.

The two-factor ANOVA is composed of three distinct hypothesis tests:

1. The main effect of factor A (often called the A-effect). Assuming that factor A is used to define the rows of the matrix, the main effect of factor A evaluates the mean differences between rows.

2. The main effect of factor B (called the B-effect). Assuming that factor B is used to define the columns of the matrix, the main effect of factor B evaluates the mean differences between columns.

3. The interaction (called the $A \times B$ interaction). The interaction evaluates mean differences between treatment conditions that are not predicted from the overall main effects from factor A or factor B.

For each of these three tests, we are looking for mean differences between treatments that are larger than would be expected if there are no treatment effects. In each case, the significance of the treatment effect is evaluated by an F-ratio. All three F-ratios have the same basic structure:

$$F = \frac{\text{variance (mean differences) between treatments}}{\text{variance (mean differences) expected if there are no treatment effects}}$$ (13.1)

The general structure of the two-factor ANOVA is shown in Figure 13.3. Note that the overall analysis is divided into two stages. In the first stage, the total variability is separated into two components: between-treatments variability and within-treatments variability. This first stage is identical to the single-factor ANOVA introduced in Chapter 12, with each cell in the two-factor matrix viewed as a separate treatment condition. The within-treatments variability that is obtained in Stage 1 of the analysis is used as the denominator for the F-ratios. As we noted in Chapter 12, within each treatment, all of the participants are treated exactly the same. Thus, any differences that exist within the treatments cannot be caused by treatment effects. As a result, the within-treatments variability provides a measure of the differences that exist when there are no systematic treatment effects influencing the scores (see pages 339 to 400).

The between-treatments variability obtained in Stage 1 of the analysis combines all the mean differences produced by factor A, factor B, and the interaction. The purpose of the second stage is to partition the differences into three separate components: differences attributed to factor A, differences attributed to factor B, and any remaining mean differences that define the interaction. These three components form the numerators for the three F-ratios in the analysis.

The goal of this analysis is to compute the variance values needed for the three F-ratios. We need three between-treatments variances (one for factor A, one for factor B, and one for

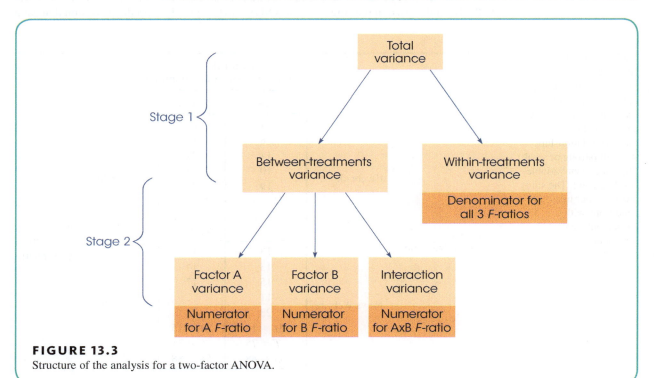

FIGURE 13.3
Structure of the analysis for a two-factor ANOVA.

Remember that in ANOVA a variance is called a mean square, or *MS*.

the interaction), and we need a within-treatments variance. Each of these variances (or mean squares) is determined by a sum of squares value (*SS*) and a degrees of freedom value (*df*):

$$\text{mean square} = MS = \frac{SS}{df}$$

EXAMPLE 13.2

To demonstrate the two-factor ANOVA, we return to the work of Wilcox and Stephen (2013). Table 13.6 shows another set of hypothetical data from the two-factor study, this time including individual participant scores. The two factors are, as in Tables 13.3 and 13.4, type of browsing and strength of relationship with Facebook friends. In this hypothetical example, a separate group of $n = 5$ students was tested in each of the four conditions. The dependent variable is self-esteem on the Rosenberg Scale, again using the 30-point full Rosenberg Scale.

The data are displayed in a matrix, with the two levels of browsing type (factor *A*) making up the rows and the two levels of relationship strength (factor *B*) making up the columns. Note that the data matrix has a total of four *cells* or treatment conditions with a separate sample of $n = 5$ participants in each condition. Most of the notation should be familiar from the single-factor ANOVA presented in Chapter 12. Specifically, the treatment totals are identified by *T* values, the total number of scores in the entire study is $N = 20$, and the grand total (sum) of all 20 scores is $G = 340$. In addition to these familiar values, we have included the totals for each row and for each column in the matrix. The goal of the ANOVA is to determine whether the mean differences observed in the data are significantly greater than would be expected if there were no treatment effects.

■ Stage 1 of the Two-Factor Analysis

The first stage of the two-factor analysis separates the total variability into two components: between-treatments and within-treatments. The formulas for this stage are identical to the formulas used in the single-factor ANOVA in Chapter 12 with the provision that each

TABLE 13.6

Hypothetical data for a two-factor study comparing two levels of type of browsing (Facebook or news website) and two levels of relationship strength (strong or weak). The dependent variable is self-esteem. The study involves four treatment conditions with $n = 5$ participants in each treatment.

		Factor *B*: Strength of Facebook Relationships		
		Weak	Strong	
	Facebook	14	29	
		8	19	
		14	29	
		10	23	
		19	25	$T_{row} = 190$
		$M = 13$	$M = 25$	
		$T = 65$	$T = 125$	
Factor *A*: Type of Browsing		$SS = 72$	$SS = 72$	
	News Website	13	10	
		13	14	
		15	17	
		11	12	
		23	22	$T_{row} = 150$
		$M = 15$	$M = 15$	
		$T = 75$	$T = 75$	
		$SS = 88$	$SS = 88$	
		$T_{col} = 140$	$T_{col} = 200$	

$N = 20$
$G = 340$
$\Sigma X^2 = 6540$

cell in the two-factor matrix is treated as a separate treatment condition. The formulas and the calculations for the data in Table 13.6 are as follows:

Total Variability

$$SS_{total} = \Sigma X^2 - \frac{G^2}{N} \tag{13.2}$$

For these data,

$$SS_{total} = 6540 - \frac{340^2}{20} = 760$$

This SS value measures the variability for all $N = 20$ scores and has degrees of freedom given by

$$df_{total} = N - 1 \tag{13.3}$$

For the data in Table 13.6, $df_{total} = 19$.

Within-Treatments Variability   To compute the variance within treatments, we first compute SS and $df = n - 1$ for each of the individual treatment conditions. Then the within-treatments SS is defined as

$$SS_{within\ treatments} = \Sigma SS_{each\ treatment} \tag{13.4}$$

And the within-treatments df is defined as

$$df_{within\ treatments} = \Sigma df_{each\ treatment} \tag{13.5}$$

For the four treatment conditions in Table 13.6,

$$SS_{within\ treatments} = 72 + 72 + 88 + 88$$
$$= 320$$
$$df_{within\ treatments} = 4 + 4 + 4 + 4$$
$$= 16$$

Between-Treatments Variability   Because the two components in Stage 1 must add up to the total, the easiest way to find $SS_{between\ treatments}$ is by subtraction.

$$SS_{between\ treatments} = SS_{total} - SS_{within} \tag{13.6}$$

For the data in Table 13.6, we obtain

$$SS_{between\ treatments} = 760 - 320 = 440$$

However, you can also use the computational formula to calculate $SS_{between\ treatments}$ directly.

$$SS_{between\ treatments} = \Sigma \frac{T^2}{n} - \frac{G^2}{N} \tag{13.7}$$

For the data in Table 13.6, there are four treatments (four T values), each with $n = 5$ scores, and the between-treatments SS is

$$SS_{between\ treatments} = \frac{65^2}{5} + \frac{125^2}{5} + \frac{75^2}{5} + \frac{75^2}{5} - \frac{340^2}{20}$$
$$= 845 + 3125 + 1125 + 1125 - 5780$$
$$= 6220 - 5780 = 440$$

The between-treatments df value is determined by the number of treatments (or the number of T values) minus one. For a two-factor study, the number of treatments is equal to the number of cells in the matrix. Thus,

$$df_{\text{between treatments}} = \text{number of cells} - 1 \qquad (13.8)$$

For these data,

$$df_{\text{between treatments}} = 4 - 1 = 3$$

This completes the first stage of the analysis. Note that the two components add to equal the total for both SS values and df values.

$$SS_{\text{between treatments}} + SS_{\text{within treatments}} = SS_{\text{total}}$$
$$440 + 320 = 760$$
$$df_{\text{between treatments}} + df_{\text{within treatments}} = df_{\text{total}}$$
$$3 + 16 = 19$$

■ Stage 2 of the Two-Factor Analysis

The second stage of the analysis determines the numerators for the three F-ratios. Specifically, this stage determines the between-treatments variance for factor A, factor B, and the interaction.

1. **Factor A.** The main effect for factor A evaluates the mean differences between the levels of factor A. For this example, factor A defines the rows of the matrix, so we are evaluating the mean differences between rows. To compute the SS for factor A, we calculate a between-treatment SS using the row totals exactly the same as we computed $SS_{\text{between treatments}}$ using the treatment totals (T values) earlier. For factor A, the row totals are 190 and 150, and each total was obtained by adding 10 scores.

   Therefore,

   $$SS_A = \sum \frac{T_{ROW}^2}{n_{ROW}} - \frac{G^2}{N} \qquad (13.9)$$

   For our data,

   $$SS_A = \frac{190^2}{10} + \frac{150^2}{10} - \frac{340^2}{20}$$
   $$= 3610 + 2250 - 5780$$
   $$= 80$$

   Factor A involves two treatments (or two rows), easy and difficult, so the df value is

   $$df_A = \text{number of rows} - 1 \qquad (13.10)$$
   $$= 2 - 1$$
   $$= 1$$

2. **Factor B.** The calculations for factor B follow exactly the same pattern that was used for factor A, except for substituting columns in place of rows. The main effect for factor B evaluates the mean differences between the levels of factor B, which define the columns of the matrix.

   $$SS_B = \sum \frac{T_{COL}^2}{n_{COL}} - \frac{G^2}{N} \qquad (13.11)$$

For our data, the column totals are 140 and 200, and each total was obtained by adding 10 scores. Thus,

$$SS_B = \frac{140^2}{10} + \frac{200^2}{10} - \frac{340^2}{20}$$
$$= 1960 + 4000 - 5780$$
$$= 180$$

$$df_B = \text{number of columns} - 1 \tag{13.12}$$
$$= 2 - 1$$
$$= 1$$

3. **The $A \times B$ Interaction.** The $A \times B$ interaction is defined as the "extra" mean differences not accounted for by the main effects of the two factors. We use this definition to find the SS and df values for the interaction by simple subtraction. Specifically, the between-treatments variability is partitioned into three parts: the A effect, the B effect, and the interaction (see Figure 13.3). We have already computed the SS and df values for A and B, so we can find the interaction values by subtracting to find out how much is left. Thus,

$$SS_{A \times B} = SS_{\text{between treatments}} - SS_A - SS_B \tag{13.13}$$

For our data,

$$SS_{A \times B} = 440 - 80 - 180$$
$$= 180$$

Similarly,

$$df_{A \times B} = df_{\text{between treatments}} - df_A - df_B \tag{13.14}$$
$$= 3 - 1 - 1$$
$$= 1$$

An easy-to-remember alternative formula for $df_{A \times B}$ is

$$df_{A \times B} = df_A \times df_B \tag{13.15}$$
$$= 1 \times 1 = 1$$

■ Mean Squares and *F*-Ratios for the Two-Factor Anova

The two-factor ANOVA consists of three separate hypothesis tests with three separate F-ratios. The denominator for each F-ratio is intended to measure the variance (differences) that would be expected if there are no treatment effects. As we saw in Chapter 12, the within-treatments variance is the appropriate denominator for an independent-measures design (see page 399). The within-treatments variance is called a *mean square*, or *MS*, and is computed as follows:

$$MS_{\text{within treatments}} = \frac{SS_{\text{within treatments}}}{df_{\text{within treatments}}}$$

For the data in Table 13.6,

$$MS_{\text{within tretments}} = \frac{320}{16} = 20$$

This value forms the denominator for all three F-ratios.

The numerators of the three F-ratios all measured variance or differences between treatments: differences between levels of factor A, differences between levels of factor B, and extra differences that are attributed to the $A \times B$ interaction. These three variances are computed as follows:

$$MS_A = \frac{SS_A}{df_A} \quad MS_B = \frac{SS_B}{df_B} \quad MS_{A \times B} = \frac{SS_{A \times B}}{df_{A \times B}}$$

For the data in Table 13.6, the three MS values are

$$MS_A = \frac{80}{1} = 80 \quad MS_B = \frac{180}{1} = 180 \quad MS_{A \times B} = \frac{180}{1} = 180$$

Finally, the three F-ratios are

$$F_A = \frac{MS_A}{MS_{\text{within treatments}}} = \frac{80}{20} = 4.00$$

$$F_B = \frac{MS_B}{MS_{\text{within treatments}}} = \frac{180}{20} = 9.00$$

$$F_{A \times B} = \frac{MS_{A \times B}}{MS_{\text{within treatments}}} = \frac{180}{20} = 9.00$$

To determine whether each F-ratio is statistically significant, we find critical values for F in the F distribution table in Appendix B. For this example, all F-ratios have $df = 1$ for the numerator (MS value for main effects and interactions) and $df = 16$ for the denominator for all F-ratios. Checking the table with $df = 1, 16$, we find a critical value of 4.49 for $\alpha = .05$ and a critical value of 8.53 for $\alpha = .01$. For the main effect of A (type of browsing), we obtained $F = 4.00$. The main effect of A is not significant because the F-ratio of $F = 4.00$ is less than the critical value of $F = 4.49$. This means that there was no significant effect of type of browsing. Note that this does *not* mean that type of browsing is unimportant because the analysis of the main effect of type of browsing doesn't consider that factor's potential interaction with relationship strength. The main effect of factor B (relationship strength) has an F-ratio of $F = 9.00$, which is greater than the critical F-ratio. Thus, relationship strength had a significant effect on self-esteem. Similarly, the interaction between factor A and factor B is significant because $F = 9.00$ is greater than the critical F-ratio. This means that the effect of browsing Facebook on self-esteem depends on the strength of a person's relationship with their Facebook friends. ∎

Table 13.7 is a summary table for the complete two-factor ANOVA from Example 13.2. Although these tables are no longer commonly used in research reports, they provide a concise format for displaying all of the elements of the analysis. Moreover, tables like these are routinely reported by statistical software like SPSS.

The following example is an opportunity to test your understanding of the calculations required for a two-factor ANOVA.

TABLE 13.7

A summary table for the two-factor ANOVA for the data from Example 13.2.

Source	SS	df	MS	F
Between treatments	440	3		
Factor A (browsing type)	80	1	80	4.00
Factor B (relationship strength)	180	1	180	9.00
$A \times B$	180	1	180	9.00
Within treatments	320	16	20	
Total	760	19		

EXAMPLE 13.3   The following data summarize the results from a two-factor independent-measures experiment:

	Factor B		
	B_1	B_2	B_3
Factor A A_1	$n = 10$ $T = 0$ $SS = 30$	$n = 10$ $T = 10$ $SS = 40$	$n = 10$ $T = 20$ $SS = 50$
A_2	$n = 10$ $T = 40$ $SS = 60$	$n = 10$ $T = 30$ $SS = 50$	$n = 10$ $T = 20$ $SS = 40$

Calculate the total for each level of factor A and compute SS for factor A, then calculate the totals for factor B, and compute SS for this factor. You should find that the totals for factor A are 30 and 90, and $SS_A = 60$. All three totals for factor B are equal to 40, thus the means for all levels of factor B are the same. Because the means are equal, there is no variability, and $SS_B = 0$.   ∎

■ An Example of Hypothesis Testing with a Two-Factor ANOVA

The Hypothesis Test   We have seen all the individual components of a two-factor ANOVA and their calculations. We can now use the four-step procedure for hypothesis testing with the information in Table 13.7.

STEP 1   **State the hypotheses and select an alpha level.**

H_0 for main effect of type of browsing: $\mu_{Facebook} = \mu_{NewsWebsite}$.

H_0 for main effect of strength of relationship: $\mu_{Weak} = \mu_{Strong}$.

H_0 for interaction: The effect of factor type of browsing does not depend on the levels of strength of relationships.

We will use $\alpha = .05$.

STEP 2   **Locate the critical region.**   We have found that $df_{within} = 16$, $df_{browsing} = 1$, $df_{strength} = 1$, and $df_{browsing \times strength} = 1$. Thus, each of the F-ratios has $df = 1, 16$ and with $\alpha = .05$ the critical region consists of F-ratios greater than 4.49 for all three tests.

STEP 3   **Compute the F-ratios.**   The calculations were completed in the previous section and are summarized in Table 13.7. For the main effect of A (type of browsing), we obtained $F = 4.00$. The main effect of factor B (relationship strength) has an F-ratio of $F = 9.00$. The interaction between factor A and factor B has an F-ratio of $F = 9.00$.

STEP 4   **Make a decision.**   Because each of the F-ratios has $df = 1, 16$ and with $\alpha = .05$ the critical region consists of F-ratios greater than 4.49 for all three tests. The main effect of type of browsing is not significant, therefore we fail to reject the null hypothesis. Both the strength of the relationship main effect and the interaction between browsing type and relationship strength are significant.

■ Measuring Effect Size for the Two-Factor ANOVA

The general technique for measuring effect size with an ANOVA is to compute a value for η^2 (eta squared), the percentage of variance that is explained by the treatment effects.

For a two-factor ANOVA, we compute three separate values for η^2: one measuring how much of the variance is explained by the main effect for factor A, one for factor B, and a third for the interaction. We remove any variability that can be explained by other sources before we calculate the percentage for each of the three specific treatment effects. Thus, for example, before we compute the η^2 for factor A, we remove the variability that is explained by factor B and the variability explained by the interaction. The resulting equation is

$$\text{for factor } A, \ \eta^2 = \frac{SS_A}{SS_{\text{total}} - SS_B - SS_{A \times B}} \tag{13.16}$$

Note that the denominator of Equation 13.16 consists of the variability that is explained by factor A and the other *unexplained* variability. Thus, an equivalent version of the equation is

$$\text{for factor } A, \ \eta^2 = \frac{SS_A}{SS_A + SS_{\text{within treatments}}} \tag{13.17}$$

Similarly, the η^2 formulas for factor B and for the interaction are as follows:

$$\text{for factor } B, \ \eta^2 = \frac{SS_B}{SS_{\text{total}} - SS_A - SS_{A \times B}} = \frac{SS_B}{SS_B + SS_{\text{within treatments}}} \tag{13.18}$$

$$\text{for } A \times B, \ \eta^2 = \frac{SS_{A \times B}}{SS_{\text{total}} - SS_A - SS_B} = \frac{SS_{A \times B}}{SS_{A \times B} + SS_{\text{within treatments}}} \tag{13.19}$$

Because each of the η^2 equations computes a percentage that is not based on the total variability of the scores, the results are often called *partial* eta squares. For the data in Example 13.2, the equations produce the following values:

$$\eta^2 \text{ for factor } A \text{ (browsing type)} = \frac{80}{80 + 320} = 0.20$$

$$\eta^2 \text{ for factor } B \text{ (relationship strength)} = \frac{180}{180 + 320} = 0.36$$

$$\eta^2 \text{ for factor } A \times B \text{ interaction} = \frac{180}{180 + 320} = 0.36$$

IN THE LITERATURE

Reporting the Results of a Two-Factor ANOVA

The APA format for reporting the results of a two-factor ANOVA follows the same basic guidelines as the single-factor report. First, the means and standard deviations are reported. Because a two-factor design typically involves several treatment conditions, these descriptive statistics often are presented in a table or a graph. Next, the results of all three hypothesis tests (*F*-ratios) are reported. The results for the study in Example 13.2 could be reported as follows:

The means and standard deviations for all treatment conditions are shown in Table 13.8. The two-factor analysis of variance showed no significant main effect for type of browsing, $F(1, 16) = 4.00$, $p > .05$, $\eta^2 = 0.20$. However, the analysis revealed a significant main effect of relationship strength, $F(1, 16) = 9.00$, $p < .05$, $\eta^2 = 0.36$, and a significant interaction between type of browsing and relationship strength, $F(1, 16) = 9.00$, $p < .05$, $\eta^2 = 0.36$.

TABLE 13.8

Mean self-esteem score for each treatment condition.

Type of Browsing		Relationship Strength	
		Weak	Strong
	Facebook	$M = 13.00$ $SD = 4.24$	$M = 25.00$ $SD = 4.24$
	News Website	$M = 15.00$ $SD = 4.69$	$M = 15.00$ $SD = 4.69$

■ Interpreting the Results from a Two-Factor ANOVA

Because the two-factor ANOVA involves three separate tests, you must consider the overall pattern of results rather than focusing on the individual main effects or the interaction. In particular, whenever there is a significant interaction, you should be cautious about accepting the main effects at face value (whether they are significant or not). Remember, an interaction means that the effect of one factor *depends on* the level of the second factor. Because the effect changes from one level to the next, there is no consistent "main effect."

Figure 13.4 shows the sample means obtained from the Facebook versus news browsing study. Recall that the analysis showed that the main effect of browsing type (Facebook vs. the news website) was not significant. Although the main effect was too small to be significant, it would be incorrect to conclude that type of browsing did not influence self-esteem. For this example, the difference between browsing Facebook and browsing a news website depends on the strength of the participants' relationships with their Facebook friends. Specifically, there is little or no difference between Facebook and news when the relationships are weak. However, browsing Facebook produces much higher self-esteem scores when participants had strong ties to their Facebook friends. Thus, the difference between browsing Facebook and browsing a news website *depends on* the strength of the relationship. This interdependence between factors is the source of the significant interaction.

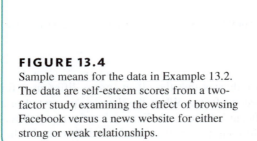

FIGURE 13.4

Sample means for the data in Example 13.2. The data are self-esteem scores from a two-factor study examining the effect of browsing Facebook versus a news website for either strong or weak relationships.

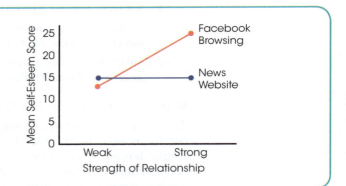

LO4 **1.** Which of the following accurately describes the two stages of a two-factor ANOVA?

    a. The first stage partitions the total variability and the second stage partitions the within-treatment variability.

    b. The first stage partitions the total variability and the second stage partitions the between-treatment variability.

    c. The first stage partitions the between-treatment variability and the second stage partitions the within-treatment variability.

    d. None of the other options is accurate.

LO5 **2.** In a two-factor analysis of variance, the F-ratio for factor A has $df = 2, 60$ and the F-ratio for factor B has $df = 3, 60$. Based on this information, what are the df values for the F-ratio for the interaction?

    a. 3, 60

    b. 5, 60

    c. 6, 60

    d. Cannot be determined without additional information.

LO6 **3.** In a two-factor ANOVA with three levels of factor A and three levels of factor B, $SSA = 50$ and SS within treatments $= 150$. With $n = 11$ scores for each of the nine groups in the analysis, which of the following is the correct value for η^2 for factor A?

    a. $\eta^2 = \dfrac{50}{150 + 50} = .25$

    b. $\eta^2 = \dfrac{50}{150} = .33$

    c. $\eta^2 = \dfrac{25}{2.5} = 10.00$

    d. $\eta^2 = \dfrac{50}{2.5} = 20.00$

ANSWERS **1. b** **2. c** **3. a**

13-3 More about the Two-Factor ANOVA

LEARNING OBJECTIVES

7. Explain how simple main effects can be used to analyze and describe the details of main effects and interactions.

8. Explain how adding a participant variable as a second factor can reduce variability caused by individual differences.

■ Testing Simple Main Effects

The existence of a significant interaction indicates that the effect (mean difference) for one factor depends on the levels of the second factor. When the data are presented in a matrix showing treatment means, a significant interaction indicates that the mean differences within one column (or row) show a different pattern than the mean differences within another column (or row). In this case, a researcher may want to perform a separate analysis for each of the individual columns (or rows). In effect, the researcher is separating the two-factor experiment into a series of separate single-factor experiments. The process of testing the significance of mean differences within one column (or one row) of a two-factor design is called testing *simple main effects*.

To demonstrate this process, we once again use the data from the Browsing Type versus Relationship Strength study (Example 13.2), which is summarized in Table 13.6.

EXAMPLE 13.4   For this demonstration we test for significant mean differences within each column of the two-factor data matrix. That is, we test for significant mean differences between Facebook versus news browsing for the strong relationship condition, and then repeat the test for the weak relationship condition. In terms of the two-factor notational system, we test the simple main effect of factor A for each level of factor B.

For the Strong Relationship Condition   Because we are restricting the data to the first row of the data matrix, the data effectively have been reduced to a single-factor study comparing only two treatment conditions. Therefore, the analysis is essentially a single-factor ANOVA duplicating the procedure presented in Chapter 12. To facilitate the change from a two-factor to a single-factor analysis, the data for the strong relationship condition (second column of the matrix) are reproduced as follows using the notation for a single-factor study.

	Browsing Type	
Facebook	News	
$n = 5$	$n = 5$	$N = 10$
$M = 25$	$M = 15$	$G = 200$
$T = 125$	$T = 75$	

State the hypothesis and select alpha level.   For this restricted set of the data, the null hypothesis would state that there is no difference between the mean for the Facebook condition and the mean for the news condition. In symbols,

$$H_0: \mu_{Facebook} = \mu_{news} \text{ for strong relationships}$$

Alpha is set at .05.

Compute the *F*-ratio for the simple main effect.   To evaluate this hypothesis, we use an F-ratio for which the numerator, $MS_{between\ treatments}$, is determined by the mean differences between these two groups and the denominator consists of $MS_{within\ treatments}$ from the original ANOVA. Thus, the F-ratio has the structure

$$F = \frac{\text{variance (differences) for the means in column 2}}{\text{variance (differences) expected by chance}}$$
$$= \frac{MS_{between} \text{ for the two treatments in column 2}}{MS_{within} \text{ from the original ANOVA}}$$

To compute the $MS_{between\ treatments}$, we begin with the two treatment totals $T = 125$ and $T = 75$. Each of these totals is based on $n = 5$ scores, and the two totals add up to a grand total of $G = 200$. The $SS_{between\ treatments}$ for the two treatments is

$$SS_{between} = \Sigma \frac{T^2}{n} - \frac{G^2}{N}$$
$$= \frac{125^2}{5} + \frac{75^2}{5} - \frac{200^2}{10}$$
$$= 3125 + 1125 - 4000$$
$$= 250$$

TABLE 13.9

A single-factor study comparing two treatments (a) can be transformed into a two-factor study (b) by using a participant characteristic (gender) as a second factor. This process creates smaller, more homogeneous groups, which reduces the variance within groups.

(a)

Treatment I	Treatment II
1 (F)	6 (F)
10 (M)	10 (M)
7 (F)	10 (F)
10 (M)	12 (M)
0 (F)	15 (M)
10 (M)	6 (F)
4 (F)	15 (M)
6 (M)	6 (F)
$M = 6$	$M = 10$
$SS = 114$	$SS = 102$

(b)

	Treatment I	Treatment II
Males	10	10
	10	12
	10	15
	6	15
	$M = 9$	$M = 13$
	$SS = 12$	$SS = 18$
Females	1	6
	7	10
	0	6
	4	6
	$M = 3$	$M = 7$
	$SS = 30$	$SS = 12$

EXAMPLE 13.5 We will use the data in Table 13.9 to demonstrate how the variance caused by individual differences can be reduced by adding a participant characteristic, such as age or gender, as a second factor. For the single-factor study in Table 13.9(a), the two treatments produce

$$SS_{\text{within treatments}} = 114 + 102 = 216$$

With $n = 8$ in each treatment, we obtain

$$df_{\text{within treatments}} = 7 + 7 = 14$$

These values produce

$$MS_{\text{within treatments}} = \frac{SS_{\text{within treatments}}}{df_{\text{within treatments}}} = \frac{216}{14} = 15.43$$

which will be the denominator of the F-ratio evaluating the mean difference between treatments. For the two-factor study in Table 13.9(b), the four treatments produce

$$SS_{\text{within teatments}} = 12 + 18 + 30 + 12 = 72$$

With $n = 4$ in each treatment, we obtain

$$df_{\text{within treatments}} = 3 + 3 + 3 + 3 = 12$$

These values produce

$$MS_{\text{within treatments}} = \frac{72}{12} = 6.00$$

which will be the denominator of the F-ratio evaluating the main effect for the treatments. Notice that the error term for the single-factor F is much larger than the error term for the two-factor F. Reducing the individual differences within each group has greatly reduced the within-treatment variance that forms the denominator of the F-ratio.

Both designs, single-factor and two-factor, will evaluate the difference between the two treatment means, $M = 6$ and $M = 10$, with $n = 8$ in each treatment. These values produce $SS_{\text{between treatments}} = 64$ and, with $k = 2$ treatments, we obtain $df_{\text{between treatments}} = 1$. Thus,

$$MS_{\text{between treatments}} = \frac{64}{1} = 64$$

For the two-factor design, this is the MS for the main effect of the treatment factor. With different denominators, however, the two designs produce very different F-ratios. For the single-factor design, we obtain

$$F = \frac{MS_{\text{between treatments}}}{MS_{\text{within treatments}}} = \frac{64}{15.43} = 4.15$$

With $df = 1, 14$, the critical value for $\alpha = .05$ is $F = 4.60$. Our F-ratio is not in the critical region so we fail to reject the null hypothesis and must conclude that there is no significant difference between the two treatments.

For the two-factor design, however, we obtain

$$F = \frac{MS_{\text{between treatments}}}{MS_{\text{within treatments}}} = \frac{64.00}{6.00} = 10.67$$

With $df = 1, 12$, the critical value for $\alpha = .05$ is $F = 4.75$. Our F-ratio is well beyond this value so we reject the null hypothesis and conclude that there is a significant difference between the two treatments. ■

For the single-factor study in Example 13.5, the individual differences caused by gender were part of the variance within each treatment condition. This increased variance reduced the F-ratio and resulted in a conclusion of no significant difference between treatments. In the two-factor analysis, the individual differences caused by gender are measured by the main effect for gender, which is a between-groups factor. Because the gender differences are now between-groups rather than within-groups, they no longer contribute to the variance.

The two-factor analysis has other advantages beyond reducing the variance. Specifically, it allows you to evaluate mean differences between genders as well as differences between treatments, and it reveals any interaction between treatment and gender.

■ Assumptions for the Two-Factor Anova

The validity of the ANOVA presented in this chapter depends on the same three assumptions we have encountered with other hypothesis tests for independent-measures designs (the t test in Chapter 10 and the single-factor ANOVA in Chapter 12):

1. The observations within each sample must be independent (see page 337).

2. The populations from which the samples are selected must be normal.

3. The populations from which the samples are selected must have equal variances (homogeneity of variance).

As before, the assumption of normality generally is not a cause for concern, especially when the sample size is relatively large. The homogeneity of variance assumption is more important, and if it appears that your data fail to satisfy this requirement, you should conduct a test for homogeneity before you attempt the ANOVA. Hartley's F-max test (see page 338) allows you to use the sample variances from your data to determine whether there is evidence for any differences among the population variances. Remember, for the two-factor ANOVA, there is a separate sample for each cell in the data matrix. The test for homogeneity applies to all these samples and the populations they represent.

LEARNING CHECK   **LO7**   **1.** After performing a factorial ANOVA with three levels of factor A and two levels of factor B, you analyze the simple main effect of factor A at one level of factor B. Assuming that each n equals 6, what are the degrees of freedom for the simple main effect?

   a. $df = 1, 10$
   b. $df = 2, 10$
   c. $df = 1, 30$
   d. $df = 2, 30$

LO8   **2.** A researcher is interested in the effect of caffeine on students' test scores in an introductory statistics class. What is the consequence of adding major as a factor in the ANOVA?

   a. The F-ratio for the caffeine factor will decrease.
   b. The $MS_{\text{within treatments}}$ value will increase.
   c. The $MS_{\text{within treatments}}$ value will decrease.
   d. The $MS_{\text{within treatments}}$ value and the F-ratio for the caffeine factor will both decrease.

ANSWERS   **1. d   2. c**

SUMMARY

1. A research study with two independent variables is called a two-factor design. Such a design can be diagrammed as a matrix, with the levels of one factor defining the rows and the levels of the other factor defining the columns. Each cell in the matrix corresponds to a specific combination of the two factors.

2. Traditionally, the two factors are identified as factor A and factor B. The purpose of the ANOVA is to determine whether there are any significant mean differences among the treatment conditions or cells in the experimental matrix. These treatment effects are classified as follows:

   a. The A effect: overall mean differences among the levels of factor A.

   b. The B effect: overall mean differences among the levels of factor B.

   c. The $A \times B$ interaction: extra mean differences that are not accounted for by the main effects.

3. The two-factor ANOVA produces three F-ratios: one for factor A, one for factor B, and one for the $A \times B$ interaction. Each F-ratio has the same basic structure:

$$F = \frac{MS_{\text{treatment effect}} \text{ (either } A \text{ or } B \text{ or } A \times B)}{MS_{\text{within treatments}}}$$

The formulas for the SS, df, and MS values for the two-factor ANOVA are presented in Figure 13.5.

FIGURE 13.5
The ANOVA for an independent-measures two-factor design.

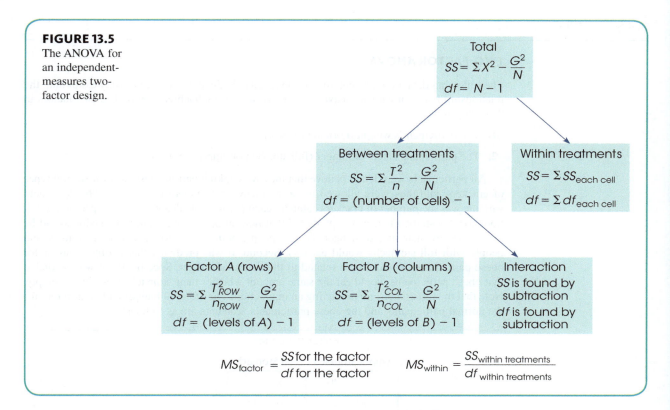

Total

$$SS = \Sigma X^2 - \frac{G^2}{N}$$

$$df = N - 1$$

Between treatments

$$SS = \Sigma \frac{T^2}{n} - \frac{G^2}{N}$$

$$df = \text{(number of cells)} - 1$$

Within treatments

$$SS = \Sigma SS_{\text{each cell}}$$

$$df = \Sigma df_{\text{each cell}}$$

Factor A (rows)

$$SS = \Sigma \frac{T^2_{ROW}}{n_{ROW}} - \frac{G^2}{N}$$

$$df = \text{(levels of A)} - 1$$

Factor B (columns)

$$SS = \Sigma \frac{T^2_{COL}}{n_{COL}} - \frac{G^2}{N}$$

$$df = \text{(levels of B)} - 1$$

Interaction

SS is found by subtraction

df is found by subtraction

$$MS_{\text{factor}} = \frac{SS \text{ for the factor}}{df \text{ for the factor}} \qquad MS_{\text{within}} = \frac{SS_{\text{within treatments}}}{df_{\text{within treatments}}}$$

KEY TERMS

factorial design (438)

two-factor designs (438)

two-factor, independent-measures, equal n designs (438)

matrix (438)

cell (438)

main effect (440)

interaction (441, 443)

simple main effects (457)

FOCUS ON PROBLEM SOLVING

1. Before you begin a two-factor ANOVA, take time to organize and summarize the data. It is best if you summarize the data in a matrix with rows corresponding to the levels of one factor and columns corresponding to the levels of the other factor. In each cell of the matrix, show the number of scores (n), the total and mean for the cell, and the SS within the cell. Also compute the row totals and column totals that are needed to calculate main effects.

2. For a two-factor ANOVA, there are three separate F-ratios. These three F-ratios use the same error term in the denominator (MS_{within}). On the other hand, these F-ratios have different numerators and may have different df values associated with each of these numerators.

DEMONSTRATION 13.1

TWO-FACTOR ANOVA

The following data are representative of the results obtained in a research study examining the relationship between eating behavior and body weight (Schachter, 1968). The two factors in this study were:

1. The participant's weight (normal or obese)

2. The participant's state of hunger (full stomach or empty stomach)

All participants were led to believe that they were taking part in a taste test for several types of crackers, and they were allowed to eat as many crackers as they wanted. The dependent variable was the number of crackers eaten by each participant. There were two specific predictions for this study. First, it was predicted that normal participants' eating behavior would be determined by their state of hunger. That is, people with empty stomachs would eat more and people with full stomachs would eat less. Second, it was predicted that eating behavior for obese participants would not be related to their state of hunger. Specifically, it was predicted that obese participants would eat the same amount whether their stomachs were full or empty. Note that the researchers are predicting an interaction: The effect of hunger will be different for the normal participants and the obese participants. The data are as follows:

		Factor B: Hunger			
		Empty Stomach	Full Stomach		
Factor A: Weight	Normal	$n = 20$ $M = 22$ $T = 440$ $SS = 1540$	$n = 20$ $M = 15$ $T = 300$ $SS = 1270$	$T_{normal} = 740$	$G = 1440$ $N = 80$ $\Sigma X^2 = 31{,}836$
	Obese	$n = 20$ $M = 17$ $T = 340$ $SS = 1320$	$n = 20$ $M = 18$ $T = 360$ $SS = 1266$	$T_{obese} = 700$	
		$T_{empty} = 780$	$T_{full} = 660$		

STEP 1 **State the hypotheses and select alpha.** For a two-factor study, there are three separate hypotheses: the two main effects and the interaction.

For factor A, the null hypothesis states that there is no difference in the amount eaten for normal participants versus obese participants. In symbols,

$$H_0: \mu_{normal} = \mu_{obese}$$

For factor B, the null hypothesis states that there is no difference in the amount eaten for full-stomach versus empty-stomach conditions. In symbols,

$$H_0: \mu_{full} = \mu_{empty}$$

For the $A \times B$ interaction, the null hypothesis can be stated two different ways. First, the difference in eating between the full-stomach and empty-stomach conditions will be the same for normal and obese participants. Second, the difference in eating between the normal and obese participants will be the same for the full-stomach and empty-stomach conditions. In more general terms,

$$H_0: \text{The effect of factor } A \text{ does not depend on the levels}$$
$$\text{of factor } B \text{ (and } B \text{ does not depend on } A\text{)}.$$

We will use $\alpha = .05$ for all tests.

STEP 2 **Find critical region.**

$$df_{\text{within treatments}} = \Sigma df = 19 + 19 + 19 + 19 = 76$$
$$df_A = \text{number of rows} - 1 = 1$$
$$df_B = \text{number of columns} - 1 = 1$$
$$df_{A \times B} = df_{\text{between treatments}} - df_A - df_B$$
$$= 3 - 1 - 1 = 1$$

All three F-ratios have $df = 1, 76$. With $\alpha = .05$, the critical F value is 3.98 for all three tests.

STEP 3 **The two-factor analysis.** Rather than compute the df values and look up critical values for F at this time, we will proceed directly to the ANOVA.

Stage 1

The first stage of the analysis is identical to the independent-measures ANOVA presented in Chapter 12, where each cell in the data matrix is considered a separate treatment condition.

$$SS_{\text{total}} = \Sigma X^2 - \frac{G^2}{N}$$
$$= 31{,}836 - \frac{1440^2}{80} = 5916$$

$$SS_{\text{within treatments}} = \Sigma SS_{\text{inside each treatment}} = 1540 + 1270 + 1320 + 1266 = 5396$$

$$SS_{\text{between treatments}} = \Sigma \frac{T^2}{n} - \frac{G^2}{N}$$
$$= \frac{440^2}{20} + \frac{300^2}{20} + \frac{340^2}{20} + \frac{360^2}{20} - \frac{1440^2}{80}$$
$$= 520$$

The corresponding degrees of freedom are

$$df_{\text{total}} = N - 1 = 79$$
$$df_{\text{within treatments}} = \Sigma df = 19 + 19 + 19 + 19 = 76$$
$$df_{\text{between treatments}} = \text{number of treatments} - 1 = 3$$

Stage 2

The second stage of the analysis partitions the between-treatments variability into three components: the main effect for factor A, the main effect for factor B, and the $A \times B$ interaction.

For factor A (normal/obese),

$$SS_A = \Sigma \frac{T_{ROWS}^2}{n_{ROWS}} - \frac{G^2}{N}$$
$$= \frac{740^2}{40} + \frac{700^2}{40} - \frac{1440^2}{80}$$
$$= 20$$

For factor B (full/empty),

$$SS_B = \Sigma \frac{T_{COLS}^2}{n_{COLS}} - \frac{G^2}{N}$$
$$= \frac{780^2}{40} + \frac{660^2}{40} - \frac{1440^2}{80}$$
$$= 180$$

For the $A \times B$ interaction,

$$SS_{A \times B} = SS_{\text{between treatments}} - SS_A - SS_B$$
$$= 520 - 20 - 180$$
$$= 320$$

The corresponding degrees of freedom are

$$df_A = \text{number of rows} - 1 = 1$$

$$df_B = \text{number of columns} - 1 = 1$$

$$df_{A \times B} = df_{\text{between treatments}} - df_A - df_B$$
$$= 3 - 1 - 1 = 1$$

The MS values needed for the F-ratios are

$$MS_A = \frac{SS_A}{df_A} = \frac{20}{1} = 20$$

$$MS_B = \frac{SS_B}{df_B} = \frac{180}{1} = 180$$

$$MS_{A \times B} = \frac{SS_{A \times B}}{df_{A \times B}} = \frac{320}{1} = 320$$

$$MS_{\text{within treatments}} = \frac{SS_{\text{within treatments}}}{df_{\text{within treatments}}} = \frac{5396}{76} = 71$$

Finally, the F-ratios are

$$F_A = \frac{MS_A}{MS_{\text{within treatments}}} = \frac{20}{71} = 0.28$$

$$F_B = \frac{MS_B}{MS_{\text{within treatments}}} = \frac{180}{71} = 2.54$$

$$F_{A \times B} = \frac{MS_{A \times B}}{MS_{\text{within treatments}}} = \frac{320}{71} = 4.51$$

STEP 4   **Make a decision and state a conclusion.**   All three F-ratios have $df = 1, 76$. With $\alpha = .05$, the critical F value is 3.98 for all three tests.

For these data, factor A (weight) has no significant effect; $F(1, 76) = 0.28$. Statistically, there is no difference in the number of crackers eaten by normal versus obese participants.

Similarly, factor B (fullness) has no significant effect; $F(1, 76) = 2.54$. Statistically, the number of crackers eaten by full participants is no different from the number eaten by hungry participants. (*Note*: This conclusion concerns the combined group of normal and obese participants. The interaction concerns these two groups separately.)

These data produce a significant interaction; $F(1, 76) = 4.51, p < .05$. This means that the effect of fullness does depend on weight. A closer look at the original data shows that the degree of fullness did affect the normal participants, but it had no effect on the obese participants.

SPSS®

General instructions for using SPSS are presented in Appendix D. Following are detailed instructions for using SPSS to perform the **Two-Factor, Independent-Measures Analysis of Variance (ANOVA)** presented in this chapter.

You might be surprised that psychologists disagree about the benefits of rewards in education. Some psychologists claim that rewarding a student for strong performance should result in lasting improvement in the student's performance. Other psychologists claim that rewards reduce motivation, creativity, and performance. It's also possible that the effect of a reward on students' behavior *depends on* the difficulty of the task. Cameron, Pierce, and So (2004) conducted a two-factor, independent-measures experiment to test this idea. All participants in this experiment were trained to identify the differences between two cartoons. Factor A was the difficulty of the training task. Participants in the low-difficulty condition were trained to find only two differences between the cartoons. Participants in the high-difficulty condition were trained to find four differences between the cartoons. Factor B was whether participants were rewarded for finding the differences between the cartoons during training. After training, the researchers measured the number of differences detected by participants in a new set of five pairs of cartoons. The authors observed an interaction of the crossover type: Rewarding participants for success during training improved test performance when the task was difficult. However, rewarding participants during training decreased performance when the task was easy. Thus, it seems that rewarding students for hard work helps students to learn, but rewarding students for easy tasks reduces performance on a later test. Scores like those observed by the researchers are listed below.

		Factor B: Reward during Training	
		No Reward	Reward
Factor A: Task Difficulty during Training	Easy	19	15
		26	15
		17	12
		18	10
		19	11
		15	21
	Difficult	13	19
		19	23
		12	25
		11	16
		20	15
		15	22

Data Entry

1. Use the **Variable View** of the data editor to create three new variables for the data above. Enter "test" in the **Name** field of the first variable. Select **Numeric** in the **Type** field and **Scale** in the **Measure** field. Enter a brief, descriptive title for the variable in the **Label** field (here, "Number of differences detected at test" was used).

2. For the second variable, enter "difficulty" in the **Name** field. Select **Numeric** in the **Type** field and **Nominal** in the **Measure** field. Use "Difficulty of task during training" in the **Label** field. In the **Values** field, click the "…" button to assign labels to group numbers. In the dialog box that follows, enter a "1" for value and "Easy" for label and enter a "2" for value and "Difficult" for label.

3. For the third variable, enter "reward" in the **Name** field. Select **Numeric** in the **Type** field and **Nominal** in the **Measure** field. Use "Reward during training" in the **Label** field. In the **Values** field, click the "…" button to assign labels to group numbers. In the dialog box that follows, enter a "1" for value and "No Reward" for label and enter a "2" for value and "Reward" for label. When you have successfully created your variables, your Variable View should look like the figure below.

	Name	Type	Width	Decimals	Label	Values	Missing	Columns	Align	Measure	Role
1	test	Numeric	8	2	Number of differences detected at test	None	None	8	≣ Right	⬤ Scale	⬎ Input
2	difficulty	Numeric	8	2	Difficulty of task during training	(1.00, Easy)..	None	8	≣ Right	⬤ Nominal	⬎ Input
3	reward	Numeric	8	2	Reward during training	(1.00, No reward)...	None	8	≣ Right	⬤ Nominal	⬎ Input

Source: SPSS®

4. The scores are entered into the SPSS data editor in a *stacked format*, which means that all the scores from all the different treatment conditions are entered in a single column ("test").

5. In the second column ("difficulty"), enter a code number to identify the level of factor *A* for each score. Enter a 1 for scores that were collected in the Easy level of factor *A* and a 2 for scores from the Difficult level of factor *A*.

6. In the third column ("reward"), enter a 1 if the score comes from the No Reward level of factor *B* and a 2 if the score comes from the Reward level of factor *B*. When you have successfully entered your data, the Data View should be like the figure below.

	📏 test	⬤ difficulty	⬤ reward
1	19.00	1.00	1.00
2	26.00	1.00	1.00
3	17.00	1.00	1.00
4	18.00	1.00	1.00
5	19.00	1.00	1.00
6	15.00	1.00	1.00
7	13.00	2.00	1.00
8	19.00	2.00	1.00
9	12.00	2.00	1.00
10	11.00	2.00	1.00
11	20.00	2.00	1.00
12	15.00	2.00	1.00
13	15.00	1.00	2.00
14	15.00	1.00	2.00
15	12.00	1.00	2.00
16	10.00	1.00	2.00
17	11.00	1.00	2.00
18	21.00	1.00	2.00
19	19.00	2.00	2.00
20	23.00	2.00	2.00
21	25.00	2.00	2.00
22	16.00	2.00	2.00
23	15.00	2.00	2.00
24	22.00	2.00	2.00

Source: SPSS®

Data Analysis

1. Click **Analyze** on the tool bar, select **General Linear Model**, and click on **Univariate**.

2. Highlight the column label for the set of scores ("test") in the left box and click the arrow to move it into the **Dependent Variable** box.

3. One by one, highlight the column labels for the two factor codes ("difficulty" and "reward") and click the arrow to move them into the **Fixed Factors** box.

4. If you want descriptive statistics for each treatment, click on the **Options** box, select **Descriptives**, and click **Continue**.

5. If you want to visualize the results with a graph, click **Plots** and highlight the "reward" factor and click the arrow to move it to the **Horizontal Axis** field. Click the "difficulty" factor and click the arrow to move it to the **Separate Lines** field. Click the **Add** button. When you have successfully added your plot, the **Profile Plots** window should be like the figure below. Click **Continue**.

6. To conduct the factorial ANOVA, Click **OK** in the **Univariate** window.

SPSS Output

The output begins with a table listing the factors, followed by a table showing descriptive statistics including the mean and standard deviation for each cell or treatment condition. The results of the ANOVA are shown in the table labeled **Tests of Between-Subjects Effects**. The top row (*Corrected Model*) presents the between-treatments SS and *df* values. The second row (*Intercept*) is not relevant for our purposes. The next three rows present the two main effects and the interaction (the SS, *df*, and MS values, as well as the *F*-ratio and the level of significance), with each factor identified by its column number from the SPSS data editor. The next row (*Error*) describes the error term (denominator of the *F*-ratio), and the final row (*Corrected Total*) describes the total variability for the entire set of scores (ignore the row labeled *Total*).

coffe
answ
... of the six treatment con-
... Note that one of the treatment means is missing.

Factor B

		B_1	B_2	B_3
Factor A	A_1	$M = 8$	$M = 12$	$M = 28$
	A_2	$M = 4$	$M = 8$	$M = ?$

a. What value for the missing mean would result in no main effect for factor A?

b. What value for the missing mean would result in no interaction?

7. A researcher conducts an independent-measures, two-factor study with two levels of factor A and two levels of factor B, using a sample of $n = 8$ participants in each treatment condition.

a. What are the df values for the F-ratio evaluating the main effect of factor A?

b. What are the df values for the F-ratio evaluating the main effect of factor B?

c. What are the df values for the F-ratio evaluating the interaction?

8. A researcher conducts an independent-measures, two-factor study with two levels of factor A and three levels of factor B, using a sample of $n = 10$ participants in each treatment condition.

a. What are the df values for the F-ratio evaluating the main effect of factor A?

b. What are the df values for the F-ratio evaluating the main effect of factor B?

c. What are the df values for the F-ratio evaluating the interaction?

9. The following results are from an independent-measures, two-factor study with $n = 10$ participants in each treatment condition.

Factor B

		B_1	B_2
Factor A	A_1	$T = 40$ $M = 4$ $SS = 50$	$T = 10$ $M = 1$ $SS = 30$
	A_2	$T = 50$ $M = 5$ $SS = 60$	$T = 20$ $M = 2$ $SS = 40$

$$N = 40$$
$$G = 120$$
$$\Sigma X^2 = 640$$

a. Use a two-factor ANOVA with $\alpha = .05$ to evaluate the main effects and the interaction.

b. Compute η^2 to measure the effect size for each of the main effects and the interaction.

10. The following results are from an independent-measures, two-factor study with $n = 4$ participants in each treatment condition. Use a two-factor ANOVA with $\alpha = .05$ to evaluate the main effects and the interaction.

Factor B

		B_1	B_2	B_3
Factor A	A_1	17 9 9 9 $T = 44$ $M = 11$ $SS = 48$	21 18 17 12 $T = 68$ $M = 17$ $SS = 42$	20 21 21 30 $T = 92$ $M = 23$ $SS = 66$
	A_2	17 7 13 7 $T = 44$ $M = 11$ $SS = 72$	1 10 7 2 $T = 20$ $M = 5$ $SS = 54$	13 16 22 17 $T = 68$ $M = 17$ $SS = 42$

$$N = 24$$
$$G = 336$$
$$SX^2 = 5820$$

11. Most sports injuries are immediate and obvious, like a broken leg. However, some can be more subtle, like the neurological damage that may occur when soccer players repeatedly head a soccer ball. To examine effects of repeated heading, McAllister et al. (2013) examined a group of football and ice hockey players and a group of athletes in noncontact sports before and shortly after the season. The dependent variable was performance on a conceptual thinking task. Following are hypothetical data from an independent-measures study similar to the one by McAllister et al. The researchers measured conceptual thinking for contact and noncontact athletes at the beginning of their first season and for separate groups of athletes at the end of their second season.

a. Use a two-factor ANOVA with $\alpha = .05$ to evaluate the main effects and interaction.

b. Calculate the effects size (η^2) for the main effects and the interaction.

c. Briefly describe the outcome of the study.

	Factor B: Time	
	Before the First Season	After the Second Season
Contact Sport	$n = 20$ $M = 9$ $T = 180$ $SS = 380$	$n = 20$ $M = 4$ $T = 80$ $SS = 390$
Noncontact Sport	$n = 20$ $M = 9$ $T = 180$ $SS = 350$	$n = 20$ $M = 8$ $T = 160$ $SS = 400$

Factor A: Sport

$$\Sigma X^2 = 6,360$$

12. The following table summarizes the results from a two-factor study with two levels of factor A and three levels of factor B using a separate sample of $n = 5$ participants in each treatment condition. Fill in the missing values. (*Hint:* Start with the df values.)

Source	SS	df	MS	
Between treatments	___	___		
Factor A	48	___	___	$F = $ ___
Factor B	___	___	24	$F = $ ___
$A \times B$ Interaction	___	___	12	$F = $ ___
Within treatments	144	___	___	
Total	___	___		

13. The following table summarizes the results from a two-factor study with two levels of factor A and three levels of factor B using a separate sample of $n = 8$ participants in each treatment condition. Fill in the missing values. (*Hint:* Start with the df values.)

Source	SS	df	MS	
Between treatments	72	___		
Factor A	___	___	___	$F = $ ___
Factor B	___	___	___	$F = 6.0$
$A \times B$ Interaction	___	___	12	$F = $ ___
Within treatments	126	___	___	
Total	___	___		

14. Emoticons, like ☺ and ☹, are helpful for expressing emotion in communications that otherwise have limited emotional content (e.g., emails, text messages, and social media posts). Derks, Bos, and von Grumbkow (2007) conducted an independent sample experiment to study the effect of social context and emotion on the use of emoticons. Four groups of participants were instructed to use an internet chat program and the researchers measured the number of emoticons produced by participants. The first group of participants was instructed to communicate about an emotionally positive *task* (e.g., organizing a group project). The second group of participants was instructed to communicate about an emotionally positive *socio-emotional event* (e.g., brainstorming about a friend's birthday). The third group was treated like the first group, except that task was emotionally negative and the fourth group was treated like the second group, except that the socio-emotional event was emotionally negative. They observed data like those listed below.

	Factor B: Emotion	
	Positive	Negative
Task	14 5 13 9 9	1 5 1 9 4
Socioemotional	7 13 11 13 16	13 9 15 14 19

Factor A: Context

a. Use a two-factor ANOVA with $\alpha = .05$ to evaluate the main effects and interaction.
b. How do the results change if $\alpha = .01$?
c. Calculate the effects size (η^2) for the main effects and the interaction.
d. Briefly describe the outcome of the study.

15. You might have heard the claim that students have specific "learning styles" and that each student learns best when the method of instruction matches their specific learning style. This claim has not held up to experimental scrutiny (Pashler, McDaniel, Rohrer, & Bjork, 2008). For example, Massa and Mayer (2006) divided participants into two groups—visual learning preference and verbal learning preference—based on participants' responses to questionnaires about learning style. In addition, all participants received either text-based verbal instruction or visual instruction. The researchers measured participants' performance in a learning test. They observed no evidence of an interaction between learning preference and instructional method. Data like those observed by the authors are listed below. Use a two-factor ANOVA with $\alpha = .05$ to evaluate the data. Describe the effect of the instructional method on test scores for visual and verbal learning styles.

	Factor B: Instructional Method	
	Visual	Verbal
Visual	26	14
	12	14
	20	12
	24	16
	18	2
	8	2
Verbal	28	20
	8	2
	20	13
	22	4
	10	15
	8	6

Factor A: Learning Style

16. The diathesis stress approach to mental illness proposes that neither environmental stress alone nor genetic factors alone are enough to produce mental illness. Instead, both environmental stress and genetic predisposition to mental illness are required for mental illnesses like schizophrenia and depression to be expressed. In a recent test of this idea (Sachs, Ni, & Caron, 2015), either normal rats or rats that were genetically modified to have low levels of the neurotransmitter serotonin received either no social stress or a social stress treatment. Researchers measured the number of social interactions produced by subjects after the stress treatment. Data like those observed by the author are listed below. Use a two-factor ANOVA with $\alpha = .05$ to evaluate the data.

	Factor B: Gene	
	Low serotonin	High serotonin
Social Stress	0	15
	8	17
	11	23
	5	14
	8	16
	10	23
None	19	10
	12	16
	22	17
	20	17
	15	23
	20	19

Factor A: Stress

17. Eyewitnesses in jury trials are influenced by memory processes like forgetting. Jurors seem to also be influenced by instructions that encourage skepticism and the language used in eyewitness testimony. In a recent study of jury decision making (Kurinec & Weaver, 2018), participants were asked to play the role of juror by rating defendant culpability after reading eyewitness testimony and juror instructions. Participants read eyewitness testimony that was written in abstract language (e.g., "a shady character committed the crime") or concrete language (e.g., "the defendant was observed wearing a dark-colored mask"). Before giving their ratings, participant jurors read either jury instructions that increased skepticism or an equivalent amount of unrelated text. The pattern of results below is similar to those observed by the researchers. Each fictitious score represents a participant's rating of the suspect's guilt on a scale of 0 ("least likely to be guilty") to 10 ("most likely to be guilty"). Use a two-factor ANOVA with $\alpha = .05$ to evaluate the data.

	Factor B: Instructions	
	Juror instructions	Irrelevant text
Concrete	3	4
	7	10
	3	8
	1	8
	6	10
Abstract	4	4
	2	8
	3	6
	6	1
	0	6

Factor A: Language of Testimony

18. In a classic study of the effect of memory on caffeine, Loke (1988) studied the effect of caffeine on the serial position effect. In memory tests, the serial position effect refers to the observation that memory for items at the beginning and end of a list are remembered better than items in the middle of a list. Loke observed an interaction between caffeine dose (low, moderate, and high) and position (1st, 2nd, 3rd, and 4th block) in a list of recalled items. A pattern of results like those reported by the researcher is described in the table below.

		Factor B: Serial Position			
		1st	2nd	3rd	4th
Factor A: Dose of Caffeine	High	$M = 11$	$M = 8$	$M = 11$	$M = 14$
	Low	$M = 5$	$M = 4$	$M = 12$	$M = 15$

Complete the following ANOVA table. Use an ANOVA with $\alpha = .05$ to evaluate the data. Describe the outcome of the study.

Source	SS	df	MS	
Between treatments	224	___	___	
Factor A (*dose of caffeine*)	16	___	___	$F =$ ___
Factor B (*serial position*)	170	___	___	$F =$ ___
$A \times B$ Interaction	38	___	___	$F =$ ___
Within treatments	144	72	___	
Total	___	___		

19. The following results are from an independent-measures, two-factor study with $n = 5$ participants in each treatment condition.

		Factor B		
		B_1	B_2	B_3
Factor A	A_1	9 9 13 5 9 $SS = 32$ $M = 9$ $T = 45$	18 15 13 13 11 $SS = 28$ $M = 14$ $T = 70$	9 13 9 6 13 $SS = 36$ $M = 10$ $T = 50$
	A_2	8 14 9 10 14 $SS = 32$ $M = 11$ $T = 55$	0 6 6 0 3 $SS = 36$ $M = 3$ $T = 15$	8 10 3 6 8 $SS = 28$ $M = 7$ $T = 35$

a. Use a two-factor ANOVA with $\alpha = .05$ to evaluate the main effects and the interaction.
b. Test the simple effect of factor A at level B_2.

20. The following results are from an independent-measures, two-factor study with $n = 4$ participants in each treatment condition.

		Factor B	
		B_1	B_2
Factor A	A_1	10 7 7 0 $SS = 54$ $M = 6$ $T = 24$	22 12 16 22 $SS = 72$ $M = 18$ $T = 72$
	A_2	21 13 21 21 $SS = 48$ $M = 19$ $T = 76$	14 12 5 5 $SS = 66$ $M = 9$ $T = 36$

a. Use a two-factor ANOVA with $\alpha = .05$ to evaluate the main effects and the interaction.
b. Test the simple effect of factor A at level B_1.

21. Suppose that a researcher conducts an independent samples experiment comparing three treatments. Participants serve in the experiment either online or by visiting the researcher's lab. Scores for this hypothetical experiment are listed below.

Treatment I	Treatment II	Treatment III
13 (online)	10 (online)	1 (online)
1 (online)	10 (online)	7 (online)
7 (online)	19 (online)	6 (online)
9 (online)	8 (online)	13 (online)
5 (online)	8 (online)	3 (online)
13 (in-lab)	13 (in-lab)	6 (in-lab)
17 (in-lab)	17 (in-lab)	8 (in-lab)
15 (in-lab)	15 (in-lab)	15 (in-lab)
6 (in-lab)	23 (in-lab)	10 (in-lab)
9 (in-lab)	12 (in-lab)	16 (in-lab)

a. Use a one-way ANOVA with $\alpha = .05$ to evaluate the effect of treatment on the scores.
b. Use a two-way ANOVA with factor A as treatment and factor B as online versus in-lab to evaluate the effect of the treatment on the scores.
c. Compare the results from part a with your results from part b and explain any differences.

Correlation and Regression

CHAPTER 14

Tools You Will Need

The following items are considered essential background material for this chapter. If you doubt your knowledge of any of these items, you should review the appropriate chapter or section before proceeding.

- Sum of squares (*SS*) (Chapter 4)
  - Computational formula
  - Definitional formula
- *z*-scores (Chapter 5)
  - Hypothesis testing (Chapter 8)
  - Analysis of variance (ANOVA) (Chapter 12)

PREVIEW

LO1 **1.** Which of the following is a justified conclusion if a correlation is negative?

 a. Increases in X tend to be accompanied by increases in Y.

 b. Increases in X tend to be accompanied by decreases in Y.

 c. Increases in X are always accompanied by increases in Y.

 d. Increases in X are always accompanied by decreases in Y.

LO1 **2.** Which of the following correlations indicates the most consistent relationship between X and Y?

 a. 0.80

 b. 0.40

 c. −0.10

 d. −0.90

ANSWERS **1. b** **2. d**

14-2 The Pearson Correlation

LEARNING OBJECTIVES

2. Calculate the sum of products of deviations (SP) for a set of scores using the definitional and computational formulas.

3. Calculate the Pearson correlation for a set of scores and explain what it measures.

4. Explain how the value of the Pearson correlation is affected when a constant is added to each of the X scores and/or the Y scores, and when the X and/or Y scores are all multiplied by a constant.

By far the most common correlation is the *Pearson correlation* (or the Pearson product–moment correlation), which measures the degree of *linear relationship*; that is, how well the data points fit a straight line.

> The **Pearson correlation** measures the degree and the direction of the linear relationship between two variables.

The Pearson correlation for a sample is identified by the letter r. The corresponding correlation for the entire population is identified by the Greek letter rho (ρ), which is the Greek equivalent of the letter r. Conceptually, this correlation is computed by

$$r = \frac{\text{degree to which } X \text{ and } Y \text{ vary together}}{\text{degree to which } X \text{ and } Y \text{ vary separately}}$$

$$= \frac{\text{covariability of } X \text{ and } Y}{\text{variability of } X \text{ and } Y \text{ separately}}$$

When there is a perfect linear relationship, every change in the X variable is accompanied by a corresponding change in the Y variable. In Figure 14.4(a), for example, every time the value of X increases, there is a perfectly predictable decrease in the value of Y. The result is a perfect linear relationship, with X and Y always varying together. In this case, the covariability (X and Y together) is identical to the variability of X and Y separately, and the formula produces a correlation with a magnitude of 1.00 or −1.00. At the other extreme, when there is no linear relationship, a change in the X variable does not correspond to any predictable change in the Y variable. In this case, there is no covariability, and the resulting correlation is zero.

■ The Sum of Products of Deviations

The calculation of the Pearson correlation requires one new concept: the *sum of products of deviations,* or *SP.* This new value is similar to *SS* (the sum of squared deviations), which is used to measure variability for a single variable. Now, we use *SP* to measure the amount of covariability between two variables. The value for *SP* can be calculated with either a definitional formula or a computational formula.

The *definitional formula* for the sum of products is

$$SP = \Sigma(X - M_X)(Y - M_Y) \tag{14.1}$$

where M_X is the mean for the X scores and M_Y is the mean for the Y scores.

The definitional formula instructs you to perform the following sequence of operations:

1. Find the X deviation and the Y deviation for each individual.

2. Find the product of the deviations for each individual.

3. Add the products.

Notice that this process literally *defines* the value being calculated; that is, the formula actually computes the sum of the products of the deviations.

The *computational formula* for the sum of products of deviations is

Caution: **The n in this formula refers to the number of pairs of scores.**

$$SP = \Sigma XY - \frac{\Sigma X \Sigma Y}{n} \tag{14.2}$$

Because the computational formula uses the original scores (X and Y values), it usually results in easier calculations than those required with the definitional formula, especially if M_X or M_Y is not a whole number. However, both formulas will always produce the same value for *SP*.

You may have noticed that the formulas for *SP* are similar to the formulas you have learned for *SS* (sum of squares). Specifically, the two sets of formulas have exactly the same structure, but the *SS* formulas use squared values (X times X) and the *SP* formulas use products (X times Y).

Definitional Formulas **Computational Formulas**

$$SS = \Sigma(X - M_X)^2 \qquad\qquad SS = \Sigma X^2 - \frac{(\Sigma X)^2}{n}$$

or,

$$SS = \Sigma(X - M_X)(X - M_X) \qquad SS = \Sigma XX - \frac{(\Sigma X)(\Sigma X)}{n}$$

therefore,

$$SP = \Sigma(X - M_X)(Y - M_Y) \qquad SP = \Sigma XY - \frac{(\Sigma X)(\Sigma Y)}{n}$$

The following example demonstrates the calculation of *SP* with both formulas.

EXAMPLE 14.1

The same set of $n = 4$ pairs of scores are used to calculate SP, first using the definitional formula and then using the computational formula.

For the definitional formula, you need deviation scores for each of the X values and each of the Y values. Note that the mean for the Xs is $M_X = 2.5$ and the mean for the Ys is $M_Y = 5$. The deviations and the products of deviations are shown in the following table:

Caution: The signs (+ and −) are critical in determining the sum of products, *SP*.

Scores		Deviations		Products
X	Y	$X - M_X$	$Y - M_Y$	$(X - M_X)(Y - M_Y)$
1	3	−1.5	−2	+3
2	6	−0.5	+1	−0.5
4	4	+1.5	−1	−1.5
3	7	+0.5	+2	+1
				$SP = +2$

For these scores, the sum of the products of the deviations is $SP = +2$.

For the computational formula, you need the X value, the Y value, and the XY product for each individual. Then you find the sum of the Xs, the sum of the Ys, and the sum of the XY products. These values are as follows:

X	Y	XY	
1	3	3	
2	6	12	
4	4	16	
3	7	21	
10	20	52	Totals

Substituting the totals in the formula gives

$$SP = \Sigma XY - \frac{\Sigma X \Sigma Y}{n}$$
$$= 52 - \frac{10(20)}{4}$$
$$= 52 - 50$$
$$= 2$$

Both formulas produce the same result, $SP = 2$. ∎

The following example is an opportunity to test your understanding of the calculation of *SP* (the sum of products of deviations).

EXAMPLE 14.2

Calculate the sum of products of deviations (*SP*) for the following set of scores. Use the definitional formula and then the computational formula. You should obtain $SP = 5$ with both formulas. Good luck.

X	Y
0	1
3	3
2	3
5	2
0	1

FIGURE 14.5
Scatter plot for the data
from Example 14.3.

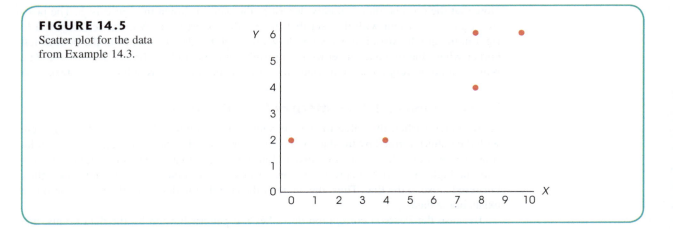

■ Calculation of the Pearson Correlation

As noted earlier, the Pearson correlation consists of a ratio comparing the covariability of X and Y (the numerator) with the variability of X and Y separately (the denominator). In the formula for the Pearson r, we use SP to measure the covariability of X and Y. The variability of X is measured by computing SS for the X scores and the variability of Y is measured by SS for the Y scores. With these definitions, the formula for the Pearson correlation becomes

$$r = \frac{SP}{\sqrt{SS_X SS_Y}} \qquad (14.3)$$

Note that you *multiply* SS for X by SS for Y in the denominator of the Pearson formula.

The following example demonstrates the use of this formula with a simple set of scores.

EXAMPLE 14.3

X	Y
0	2
10	6
4	2
8	4
8	6

The Pearson correlation is computed for the set of $n = 5$ pairs of scores shown in the margin.

Before starting any calculations, it is useful to put the data in a scatter plot and make a preliminary estimate of the correlation. These data have been graphed in Figure 14.5. Looking at the scatter plot, it appears that there is a very good (but not perfect) positive correlation. You should expect an approximate value of $r = +0.8$ or $+0.9$. To find the Pearson correlation, we need SP, SS for X, and SS for Y. The calculations for each of these values, using the definitional formulas, are presented in Table 14.1. (Note that the mean for the X values is $M_X = 6$ and the mean for the Y scores is $M_Y = 4$.)

Using the values from Table 14.1, the Pearson correlation is

$$r = \frac{SP}{\sqrt{(SS_X)(SS_Y)}} = \frac{28}{\sqrt{(64)(16)}} = \frac{28}{32} = +0.875$$

TABLE 14.1
Calculation of SS_X, SS_Y, and SP for a sample of $n = 5$ pairs of scores.

Scores		Deviations		Squared Deviations		Products
X	Y	$X - M_X$	$Y - M_Y$	$(X - M_X)^2$	$(Y - M_Y)^2$	$(X - M_X)(Y - M_Y)$
0	2	−6	−2	36	4	+12
10	6	+4	+2	16	4	+8
4	2	−2	−2	4	4	+4
8	4	+2	0	4	0	0
8	6	+2	+2	4	4	+4
				$SS_X = 64$	$SS_Y = 16$	$SP = +28$

Note that the correlation accurately describes the pattern shown in Figure 14.5. The positive sign is consistent with the fact that the Y values increase as X increases from left to right in the graph. Also, if a line were drawn through the data points, from the bottom-left corner where the two axes meet to the top-right data point ($X = 10$ and $Y = 6$), the data points would be very close to the line, indicating a very good correlation (near 1.00). ■

■ Correlation and the Pattern of Data Points

As we noted earlier, the value for the correlation in Example 14.3 is perfectly consistent with the pattern formed by the data points in Figure 14.5. The positive sign for the correlation indicates that the points are clustered around a line that slopes up to the right. Second, the high value for the correlation (near 1.00) indicates that the points are very tightly clustered close to the line. Thus, the value of the correlation describes the relationship that exists in the data.

Because the Pearson correlation describes the pattern formed by the data points, any factor that does not change the pattern also does not change the correlation. For example, if 5 points were added to each of the X values in Figure 14.5, then each data point would move to the right. However, because all of the data points shift to the right by the same amount, the overall pattern is not changed—it is simply moved to a new location. Similarly, if 5 points were subtracted from each X value, the pattern would shift to the left. In either case, the overall pattern stays the same and the correlation is not changed. In the same way, adding a constant to (or subtracting a constant from) each Y value simply shifts the pattern up (or down) but does not change the pattern and, therefore, does not change the correlation. Similarly, multiplying each X and/or Y value by a positive constant also does not change the pattern formed by the data points and does not change the correlation. For example, if each of the X values in Figure 14.5 were multiplied by 2, the same scatter plot could be used to display either the original scores or the new scores. The current figure shows the original scores, but if the values on the X-axis (0, 1, 2, 3, and so on) were doubled (0, 2, 4, 6, and so on), then the same figure would show the pattern formed by the new scores. Multiplying either the X or the Y values by a negative number, however, does not change the numerical value of the correlation but it does change the sign. For example, if each X value in Figure 14.5 were multiplied by -1, then the current data points would be moved to the left-hand side of the Y-axis, forming a mirror image of the current pattern. Instead of the positive correlation in the current figure, the new pattern would produce a negative correlation with exactly the same numerical value.

In summary, adding a constant to (or subtracting a constant from) each X and/or Y value does not change the pattern of data points and does not change the correlation. Also, multiplying (or dividing) each X or each Y value by a positive constant does not change the pattern and does not change the value of the correlation. Multiplying by a negative constant, however, produces a mirror image of the pattern and, therefore, changes the sign of the correlation.

■ The Pearson Correlation and *z*-Scores

The Pearson correlation measures the relationship between an individual's location in the X distribution and his or her location in the Y distribution. For example, a positive correlation means that individuals who score high on X also tend to score high on Y. Similarly, a negative correlation indicates that individuals with high X scores tend to have low Y scores.

Recall from Chapter 5 that *z*-scores identify the exact location of each individual score within a distribution. With this in mind, each X value can be transformed into a *z*-score, z_X, using the mean and standard deviation for the set of Xs. Similarly, each Y score can be transformed into z_Y. If the X and Y values are viewed as a sample, the transformation is completed using the sample formula for z (Equation 5.3, page 155). If the X and Y values form a complete population, the z-scores are computed using Equation 5.1 (page 153). After the transformation, the formula for the Pearson correlation can be expressed entirely in terms of z-scores.

$$\text{For a sample, } r = \frac{\Sigma z_X z_Y}{(n-1)} \tag{14.4}$$

$$\text{For a population, } \rho = \frac{\Sigma z_X z_Y}{N} \tag{14.5}$$

Note that the population value is identified with a Greek letter rho (ρ).

LEARNING CHECK **LO2 1.** What is the value of SP for a set of $n = 5$ pairs of X and Y values with $\Sigma X = 10$, $\Sigma Y = 15$, and $\Sigma XY = 75$?

    a. -20

    b. -28

    c. 45

    d. 60

LO3 2. A set of $n = 50$ pairs of X and Y scores has $SS_X = 180$, $SS_Y = 80$, $\Sigma X = 50$, $\Sigma Y = 150$, and $\Sigma XY = 180$. What is the Pearson correlation for these scores?

    a. $\frac{180}{120} = 1.50$

    b. $\frac{180}{1440} = 0.125$

    c. $\frac{30}{120} = 0.25$

    d. $\frac{30}{1440} = 0.21$

LO4 3. A set of $n = 15$ pairs of X and Y values has a Pearson correlation of $r = 0.40$. If 2 points were added to each of the X values, then what is the correlation for the resulting data?

    a. 0.40

    b. -0.40

    c. 0.60

    d. -0.60

ANSWERS **1. c   2. c   3. a**

14-3 Using and Interpreting the Pearson Correlation

LEARNING OBJECTIVES

5. Explain why a cause-and-effect explanation is not justified by a correlation between two variables.

6. Explain how a correlation can be influenced by a restricted range of scores or by outliers.

7. Define the coefficient of determination and explain what it measures.

■ Where and Why Correlations Are Used

Although correlations have a number of different applications, a few specific examples are presented next to give an indication of the value of this statistical measure.

1. **Prediction.** If two variables are known to be related in some systematic way, it is possible to use one of the variables to make accurate predictions about the other. For example, when you applied for admission to college, you were required to submit a great deal of personal information, including your SAT scores. College officials want this information so they can predict your chances of success in college. It has been demonstrated that SAT scores and college grade point averages are correlated. Students who do well on the SAT tend to do well in college; students who have difficulty with the SAT tend to have difficulty in college. Based on this relationship, college admissions officers can make a prediction about the potential success of each applicant. You should note that this prediction is not perfectly accurate. Not everyone who does poorly on the SAT will have trouble in college. That is why you also submit letters of recommendation, high school grades, and other information with your application. The process of using relationships to make predictions is called *regression* and is discussed at the end of this chapter.

2. **Validity.** Suppose a psychologist develops a new test for measuring intelligence. How could you show that this test truly measures what it claims; that is, how could you demonstrate the validity of the test? One common technique for demonstrating validity is to use a correlation. If the test actually measures intelligence, then the scores on the test should be related to other measures of intelligence—for example, standardized IQ tests, performance on learning tasks, problem-solving ability, and so on. The psychologist could measure the correlation between the new test and each of these other measures of intelligence to demonstrate that the new test is valid.

3. **Reliability.** In addition to evaluating the validity of a measurement procedure, correlations are used to determine reliability. A measurement procedure is considered reliable to the extent that it produces stable, consistent measurements. That is, a reliable measurement procedure will produce the same (or nearly the same) scores when the same individuals are measured twice under the same conditions. For example, if your IQ were measured as 113 last week, you would expect to obtain nearly the same score if your IQ were measured again this week. One way to evaluate reliability is to use correlations to determine the relationship between two sets of measurements. When reliability is high, the correlation between two measurements should be strong and positive.

4. **Theory Verification.** Many psychological theories make specific predictions about the relationship between two variables. For example, a developmental theory may predict a relationship between the parents' IQs and the child's IQ, or a social psychologist may have a theory predicting a relationship between early father/daughter relationships and the daughter's future success in romantic relationships. In each case, the prediction of the theory could be tested by determining the correlation between the two variables.

■ Interpreting Correlations

When you encounter correlations, there are four additional considerations that you should bear in mind:

1. Correlation simply describes a relationship between two variables. It does not explain why the two variables are related. Specifically, a correlation should not

and cannot be interpreted as proof of a cause-and-effect relationship between the two variables.

2. The value of a correlation can be affected greatly by the range of scores represented in the data.

3. One or two extreme data points, often called *outliers*, can have a dramatic effect on the value of a correlation.

4. When judging how "good" a relationship is, it is tempting to focus on the numerical value of the correlation. For example, a correlation of $+0.50$ is halfway between 0 and 1.00 and therefore appears to represent a moderate degree of relationship. However, a correlation should not be interpreted as a proportion. Although a correlation of 1.00 does mean that there is a 100% perfectly predictable relationship between X and Y, a correlation of 0.50 does not mean that you can make predictions with 50% accuracy. To describe how accurately one variable predicts the other, you must square the correlation. Thus, a correlation of $r = 0.50$ means that one variable *partially* predicts the other, but the predictable portion is only $r^2 = 0.50^2 = 0.25$ (or 25%) of the total variability.

We now discuss each of these four points in detail.

■ Correlation and Causation

One of the most common errors in interpreting correlations is to assume that a correlation necessarily implies a cause-and-effect relationship between the two variables. (Even Pearson blundered by asserting causation from correlational data [Blum, 1978].) We are constantly bombarded with reports of relationships: Cigarette smoking is related to heart disease; alcohol consumption is related to birth defects; carrot consumption is related to good eyesight. Do these relationships mean that cigarettes cause heart disease or carrots cause good eyesight? The answer is *no*. Although there may be a causal relationship, the simple existence of a correlation does not prove it. Earlier, for example, we discussed a study showing a relationship between high school grades and family income. However, this result does not mean that having a higher family income *causes* students to get better grades. For example, if mom gets an unexpected bonus at work, it is unlikely that her child's grades will also show a sudden increase. To establish a cause-and-effect relationship, it is necessary to conduct a true experiment (see page 23) in which one variable is manipulated by a researcher and other variables are rigorously controlled. The fact that a correlation does not establish causation is demonstrated in the following example.

EXAMPLE 14.4   Suppose we select a variety of different cities and towns throughout the United States and measure the number of churches (X variable) and the number of serious crimes (Y variable) for each. A scatter plot showing hypothetical data for this study is presented in Figure 14.6. Notice that this scatter plot shows a strong, positive correlation between churches and crime. You also should note that these are realistic data. It is reasonable that the small towns would have less crime and fewer churches and that the large cities would have large values for both variables. Does this relationship mean that churches cause crime? Does it mean that crime causes churches? It should be clear that both answers are *no*. Although a strong correlation exists between churches and crime, the real cause of the relationship is the size of the population.

FIGURE 14.6

Hypothetical data showing the relationship between the number of churches and the number of serious crimes for a sample of U.S. cities.

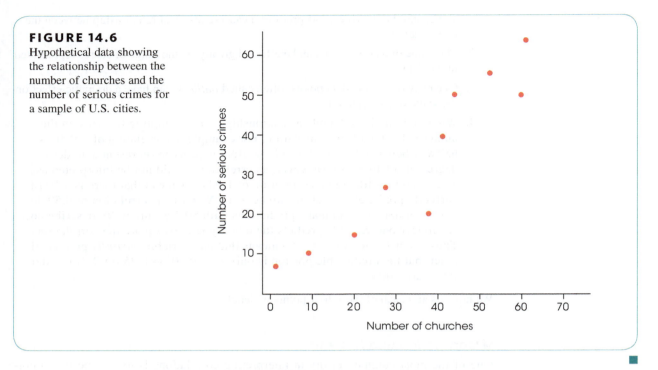

■ Correlation and Restricted Range

Whenever a correlation is computed from scores that do not represent the full range of possible values, you should be cautious in interpreting the correlation. Suppose, for example, you are interested in the relationship between IQ and creativity. If you select a sample of your fellow college students, your data probably will represent only a limited range of IQ scores (most likely from 110 to 130). The correlation within this *restricted range* could be completely different from the correlation that would be obtained from a full range of IQ scores. For example, Figure 14.7 shows a strong positive relationship between X and Y when the entire range of scores is considered. However, this relationship is obscured when the data are limited to a restricted range.

To be safe, you should not generalize any correlation beyond the range of data represented in the sample. For a correlation to provide an accurate description for the general population, there should be a wide range of X and Y values in the data.

FIGURE 14.7

In this example, the full range of X and Y values shows a strong, positive correlation, but the restricted range of scores produces a correlation near zero.

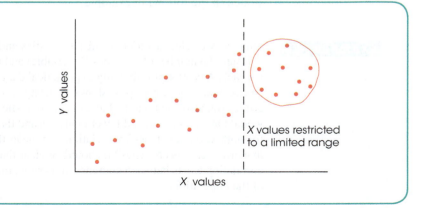

■ Outliers

An outlier is an individual with X and/or Y values that are substantially different (larger or smaller) from the values obtained for the other individuals in the data set. The data point of a single outlier can have a dramatic influence on the value obtained for the correlation. This effect is illustrated in Figure 14.8. Figure 14.8(a) shows a set of $n = 5$ data points for which the correlation between the X and Y variables is nearly zero (actually, $r = -0.08$). In Figure 14.8(b), one extreme data point (14, 12) has been added to the original data set. When this outlier is included in the analysis, a strong, positive correlation emerges (now, $r = +0.85$). Note that the single outlier drastically alters the value for the correlation and thereby can affect one's interpretation of the relationship between variables X and Y. Without the outlier, one would conclude there is no relationship between the two variables. With the extreme data point, $r = +0.85$ implies a strong relationship with Y increasing consistently as X increases. The problem of outliers is a good reason for looking at a scatter plot instead of simply basing your interpretation on the numerical value of the correlation. If you only "go by the numbers," you might overlook the fact that one extreme data point inflated the size of the correlation.

■ Correlation and the Strength of the Relationship

A correlation measures the degree of relationship between two variables on a scale from 0 to 1.00. Although this number provides a measure of the degree of relationship, the squared correlation provides a better measure of the strength of the relationship.

One of the common uses of correlation is for prediction. For example, college admissions officers do not just guess which applicants are likely to do well; they use variables

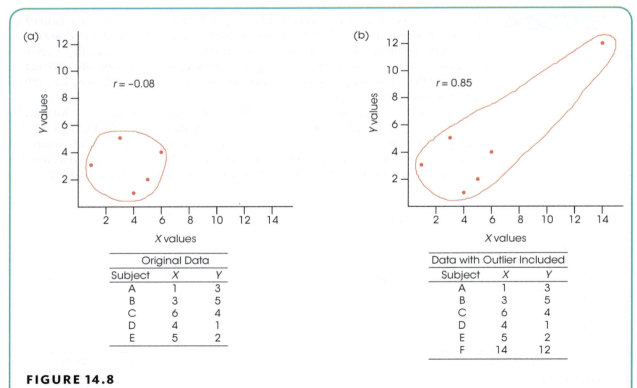

FIGURE 14.8

A demonstration of how one extreme data point (an outlier) can influence the value of a correlation.

such as SAT scores and high school grades to predict which students are most likely to be successful. These predictions are based on correlations. By using correlations, the admissions officers expect to make more accurate predictions than would be obtained by chance. In general, the squared correlation (r^2) measures the gain in accuracy that is obtained from using the correlation for prediction. The squared correlation measures the proportion of variability in the data that is explained by the relationship between X and Y. It is sometimes called the *coefficient of determination*.

> The value r^2 is called the **coefficient of determination** because it measures the proportion of variability in one variable that can be determined from the relationship with the other variable. A correlation of $r = 0.80$ (or -0.80), for example, means that $r^2 = 0.64$ (or 64%) of the variability in the Y scores can be predicted from the relationship with X.

In earlier chapters (see pages 305, 341, and 370) we introduced r^2 as a method for measuring effect size for research studies where mean differences were used to compare treatments. Specifically, we measured how much of the variance in the scores was accounted for by the differences between treatments. In experimental terminology, r^2 measures how much of the variance in the dependent variable is accounted for by the independent variable. Now we are doing the same thing, except that there is no independent or dependent variable. Instead, we simply have two variables, X and Y, and we use r^2 to measure how much of the variance in one variable can be determined from its relationship with the other variable. The following example demonstrates this concept.

EXAMPLE 14.5   Figure 14.9 shows three sets of data representing different degrees of linear relationship. The first set of data [Figure 14.9(a)] shows the relationship between IQ and shoe size. In this case, the correlation is $r = 0$ (and $r^2 = 0$), and you have no ability to predict a person's IQ based on his or her shoe size. Knowing a person's shoe size provides no information (0%) about the person's IQ. In this case, shoe size provides no help explaining why different people have different IQs.

Now consider the data in Figure 14.9(b). These data show a moderate, positive correlation, $r = +0.60$, between IQ scores and college grade point averages (GPA). Students with high IQs tend to have higher grades than students with low IQs. From this relationship, it is possible to predict a student's GPA based on his or her IQ. However, you should realize that the prediction is not perfect. Although students with high IQs *tend* to have high GPAs, this

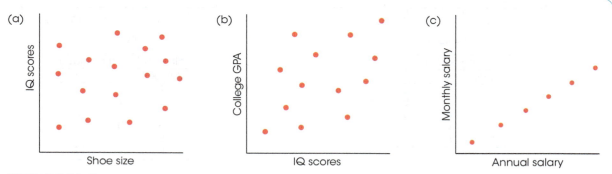

FIGURE 14.9
Three sets of data showing three different degrees of linear relationship.

is not always true. Thus, knowing a student's IQ provides some information about the student's grades, or knowing a student's grades provides some information about the student's IQ. In this case, IQ scores help explain the fact that different students have different GPAs. Specifically, you can say that *part* of the differences in GPA are accounted for by IQ. With a correlation of $r = +0.60$, we obtain $r^2 = 0.36$, which means that 36% of the variance in GPA can be explained by IQ.

Finally, consider the data in Figure 14.9(c). This time we show a perfect linear relationship ($r = +1.00$) between monthly salary and yearly salary for a group of college employees. With $r = 1.00$ and $r^2 = 1.00$, there is 100% predictability. If you know a person's monthly salary, you can predict the person's annual salary with perfect accuracy. If two people have different annual salaries, the difference can be completely explained (100%) by the difference in their monthly salaries. ■

Just as r^2 was used to evaluate effect size for mean differences in Chapters 9, 10, and 11, r^2 can now be used to evaluate the size or strength of the correlation. The same standards that were introduced in Table 9.3 (page 307) apply to both uses of the r^2 measure. Specifically, an r^2 value of 0.01 indicates a small effect or a small correlation, an r^2 value of 0.09 indicates a medium correlation, and r^2 of 0.25 or larger indicates a large correlation.

More information about the coefficient of determination (r^2) is presented in Section 14-5. For now, you should realize that whenever two variables are consistently related, it is possible to use one variable to predict values for the second variable.

LEARNING CHECK   **LO5** **1.** A researcher obtains a strong positive correlation between aggressive behavior for six-year-old children and the amount of violence they watch on television. Based on this correlation, which of the following conclusions is justified?

   a. Decreasing the amount of violence that the children see on TV will reduce their aggressive behavior.

   b. Increasing the amount of violence that the children see on TV will increase their aggressive behavior.

   c. Children who watch more TV violence tend to exhibit more aggressive behavior.

   d. All of the above.

LO6 **2.** A set of $n = 5$ pairs of X and Y scores produces a Pearson correlation of $r = 0.10$. The X values vary from 40 to 50 and the Y values vary from 30 to 60. If one new individual with $X = 4$ and $Y = 4$ is added to the sample, then what is the most likely value for the new correlation?

   a. -0.60

   b. 0.10

   c. 0.20

   d. 0.60

LO7 **3.** A set of $n = 12$ pairs of X and Y values produces a Pearson correlation of $r = -0.70$. How much of the variability in the Y scores can be predicted from the relationship with X?

   a. 16%

   b. 49%

   c. 0.16%

   d. 0.49%

ANSWERS   **1. c   2. d   3. b**

Hypothesis Tests with the Pearson Correlation

LEARNING OBJECTIVE

8. Conduct a hypothesis test evaluating the significance of a correlation.

The Pearson correlation is generally computed for sample data. As with most sample statistics, however, a sample correlation is often used to answer questions about the corresponding population correlation. For example, a psychologist would like to know whether there is a relationship between IQ and creativity. This is a general question concerning a population. To answer the question, a sample would be selected, and the sample data would be used to compute the correlation value. You should recognize this process as an example of inferential statistics: using samples to draw inferences about populations. In the past, we have been concerned primarily with using sample means as the basis for answering questions about population means. In this section, we examine the procedures for using a sample correlation as the basis for testing hypotheses about the corresponding population correlation.

■ The Hypotheses

The basic question for this hypothesis test is whether a correlation exists in the population. The null hypothesis is "No. There is no correlation in the population," or "The population correlation is zero." The alternative hypothesis is "Yes. There is a real, nonzero correlation in the population." Because the population correlation is traditionally represented by ρ (the Greek letter rho), these hypotheses would be stated in symbols as

$$H_0: \rho = 0 \qquad \text{(There is no population correlation.)}$$
$$H_1: \rho \neq 0 \qquad \text{(There is a real correlation.)}$$

When there is a specific prediction about the direction of the correlation, it is possible to do a directional, or one-tailed test. For example, if a researcher is predicting a positive relationship, the hypotheses would be

$$H_0: \rho \leq 0 \qquad \text{(The population correlation is not positive.)}$$
$$H_1: \rho > 0 \qquad \text{(The population correlation is positive.)}$$

The correlation from the sample data is used to evaluate the hypotheses. For the regular, nondirectional test, a sample correlation near zero provides support for H_0 and a sample value far from zero tends to refute H_0. For a directional test, a positive value for the sample correlation would tend to refute a null hypothesis stating that the population correlation is not positive.

Although sample correlations are used to test hypotheses about population correlations, you should keep in mind that samples are not expected to be identical to the populations from which they come; there will be some discrepancy (sampling error) between a sample statistic and the corresponding population parameter. Specifically, you should always expect some error between a sample correlation and the population correlation it represents. One implication of this fact is that even when there is no correlation in the population ($\rho = 0$), you are still likely to obtain a nonzero value for the sample correlation. This is particularly true for small samples. Figure 14.10 illustrates how a small sample from a population with a near-zero correlation could result in a correlation that deviates from zero. The colored dots in the figure represent the entire population and the three circled dots represent a random sample. Note that the three sample points show a relatively good, positive correlation even though there is no linear trend ($\rho = 0$) for the population.

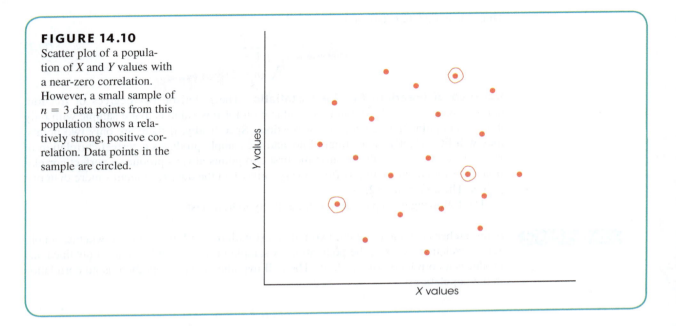

FIGURE 14.10

Scatter plot of a population of X and Y values with a near-zero correlation. However, a small sample of $n = 3$ data points from this population shows a relatively strong, positive correlation. Data points in the sample are circled.

When you obtain a nonzero correlation for a sample, the purpose of the hypothesis test is to decide between the following two interpretations:

1. There is no correlation in the population ($\rho = 0$) and the sample value is the result of sampling error. Remember, a sample is not expected to be identical to the population. There always is some error between a sample statistic and the corresponding population parameter. This is the situation specified by H_0.

2. The nonzero sample correlation accurately represents a real, nonzero correlation in the population. This is the alternative stated in H_1.

The correlation from the sample will help to determine which of these two interpretations is more likely. A sample correlation near zero supports the conclusion that the population correlation is also zero. A sample correlation that is substantially different from zero supports the conclusion that there is a real, nonzero correlation in the population.

■ The Hypothesis Test

The hypothesis test evaluating the significance of a correlation can be conducted using either a t statistic or an F-ratio. The F-ratio is discussed later (pages 517–519) and we focus on the t statistic here. The t statistic for a correlation has the same general structure as t statistics introduced in Chapters 9, 10, and 11.

$$t = \frac{\text{sample statistic} - \text{population parameter}}{\text{standard error}}$$

In this case, the sample statistic is the sample correlation (r) and the corresponding parameter is the population correlation (ρ). The null hypothesis specifies that the population correlation is $\rho = 0$. The final part of the equation is the standard error, which is determined by

$$\text{standard error for } r = s_r = \sqrt{\frac{1 - r^2}{n - 2}} \tag{14.6}$$

Thus, the complete t statistic is

$$t = \frac{r - \rho}{\sqrt{\dfrac{(1 - r^2)}{(n - 2)}}} \qquad (14.7)$$

Degrees of Freedom for the t Statistic The t statistic has degrees of freedom defined by $df = n - 2$. An intuitive explanation for this value is that a sample with only $n = 2$ data points has no degrees of freedom. Specifically, if there are only two points, they will fit perfectly on a straight line, and the sample produces a perfect correlation of $r = +1.00$ or $r = -1.00$. Because the first two points always produce a perfect correlation, the sample correlation is free to vary only when the data set contains more than two points. Thus, $df = n - 2$.

The following examples demonstrate the hypothesis test.

EXAMPLE 14.6

A researcher is using a regular, two-tailed test with $\alpha = .05$ to determine whether a non-zero correlation exists in the population. A sample of $n = 30$ individuals is obtained and produces a correlation of $r = 0.35$. The null hypothesis states that there is no correlation in the population:

$$H_0: \rho = 0$$

For this example, $df = 28$ and the critical values are $t = \pm 2.048$. With $r^2 = 0.35^2 = 0.1225$, the data produce

$$t = \frac{0.35 - 0}{\sqrt{(1 - 0.1225)/28}} = \frac{0.35}{0.177} = 1.98$$

The t value is not in the critical region, so we fail to reject the null hypothesis. The sample correlation is not large enough to reject the null hypothesis. ∎

EXAMPLE 14.7

Once again we begin with a sample of $n = 30$ and a correlation of $r = 0.35$. This time we use a directional, one-tailed test to determine whether there is a positive correlation in the population.

$$H_0: \rho \leq 0 \qquad \text{(There is not a positive correlation.)}$$
$$H_1: \rho > 0 \qquad \text{(There is a positive correlation.)}$$

The sample correlation is positive, as predicted, so we simply need to determine whether it is large enough to be significant. For a one-tailed test with $df = 28$ and $\alpha = .05$, the critical value is $t = 1.701$. In the previous example, we found that this sample produces $t = 1.97$, which is beyond the critical boundary. For the one-tailed test, we reject the null hypothesis and conclude that there is a significant positive correlation in the population. ∎

Instead of computing a t statistic for the hypothesis test, you can simply compare the sample correlation with the list of critical values in Table B.6 in Appendix B. To use the table, you need to know the sample size (n) and the alpha level. In Examples 14.6 and 14.7, we used a sample of $n = 30$, a correlation of $r = 0.35$, and an alpha level of .05. In the table, you locate $df = n - 2 = 28$ in the left-hand column and the value .05 for either one tail or two tails across the top of the table. For $df = 28$ and $\alpha = .05$ for a two-tailed test, the table shows a critical value of 0.361. Because our sample correlation is not greater than this critical value, we fail to reject the null hypothesis (as in Example 14.6). For a one-tailed test, the table lists a critical value of 0.306. This time, our sample correlation is greater than the critical value so we reject the null hypothesis and conclude that the correlation is significantly greater than zero (as in Example 14.7).

As with most hypothesis tests, if other factors are held constant, the likelihood of finding a significant correlation increases as the sample size increases. For example, a sample correlation of $r = 0.50$ produces a nonsignificant $t(8) = 1.63$ for a sample of $n = 10$, but the same correlation produces a significant $t(18) = 2.45$ if the sample size is increased to $n = 20$.

The following example is an opportunity to test your understanding of the hypothesis test for the significance of a correlation.

EXAMPLE 14.8   A researcher obtains a correlation of $r = -0.39$ for a sample of $n = 25$ individuals. For a two-tailed test with $\alpha = .05$, does this sample provide sufficient evidence to conclude that there is a significant, nonzero correlation in the population? Calculate the t statistic and then check your conclusion using the critical value in Table B6. You should obtain $t(23) = 2.03$. With a critical value of $t = 2.069$, the correlation is not significant. From Table B6, the critical value is 0.396. Again, the correlation is not significant. ∎

IN THE LITERATURE

Reporting Correlations

There is not a standard APA format for reporting correlations. However, it is useful for the report to include information such as the sample size, the calculated value for the correlation, whether it is a statistically significant relationship, the probability level, and the type of test used (one- or two-tailed). For example, a correlation might be reported as follows:

APA format does not use a zero before the decimal when reporting a correlation.

> A correlation for the data revealed a significant relationship between amount of education and annual income, $r = +.65$, $n = 30$, $p < .01$, two tails.

Sometimes a study might look at several variables, and correlations between all possible variable pairings are computed. Suppose, for example, that a study measured people's annual income, amount of education, age, and intelligence. With four variables, there are six possible pairings leading to six different correlations. The results from multiple correlations are most easily reported in a table called a *correlation matrix*, using footnotes to indicate which correlations are significant. For example, the report might state:

> The analysis examined the relationships among income, amount of education, age, and intelligence for $n = 30$ participants. The correlations between pairs of variables are reported in Table 14.2. Significant correlations are noted in the table.

TABLE 14.2

Correlation matrix for income, amount of education, age, and intelligence.

	Education	Age	IQ
Income	+.65**	+.41*	+.27
Education	–	+.11	+.38*
Age		–	+.02

Note: $n = 30$, *$p < .05$, two tails, and
**$p < .01$, two tails

LEARNING CHECK   **LO8** **1.** A researcher selects a sample of $n = 25$ high school students and measures the grade point average and the amount of time spent using their smartphone for each student. The researcher plans to use a hypothesis test to determine whether there is a significant relationship between the two variables. Which of the following is the correct null hypothesis for the test?

    a. $\rho = 0$

    b. $\rho \neq 0$

    c. $\rho = 1.00$

    d. $\rho \neq 1.00$

    LO8 **2.** The Pearson correlation is calculated for a sample of $n = 26$ individuals. What value of df should be used to test the significance of the correlation?

    a. 24

    b. 25

    c. 26

    d. Cannot be determined without additional information.

ANSWERS   **1. a   2. a**

14-5 Alternatives to the Pearson Correlation

LEARNING OBJECTIVES

  9. Explain how ranks are assigned to a set of scores, especially tied scores.

 10. Compute the Spearman correlation for a set of data and explain what it measures.

 11. Describe the circumstances in which the point-biserial correlation is used and explain what it measures.

 12. Describe the circumstances in which the phi-coefficient is used and explain what it measures.

The Pearson correlation measures the degree of linear relationship between two variables when the data (X and Y values) consist of numerical scores from an interval or ratio scale of measurement. However, other correlations have been developed for nonlinear relationships and for other types of data. In this section we examine three additional correlations: the Spearman correlation, the point-biserial correlation, and the phi-coefficient. As you will see, all three can be viewed as special applications of the Pearson correlation.

■ The Spearman Correlation

When the Pearson correlation formula is used with data from an ordinal scale (ranks), the result is called the *Spearman correlation*. The Spearman correlation is used in two situations.

First, the Spearman correlation is used to measure the relationship between X and Y when both variables are measured on ordinal scales. Recall from Chapter 1 that an ordinal scale typically involves ranking individuals rather than obtaining numerical scores. Rank-order data are fairly common because they are often easier to obtain than interval or ratio scale data. For example, a teacher may feel confident about rank-ordering students' leadership abilities but would find it difficult to measure leadership on some other scale.

In addition to measuring relationships for ordinal data, the Spearman correlation can be used as a valuable alternative to the Pearson correlation, even when the original raw scores

FIGURE 14.11
The relationship between practice and performance. Although this relationship is not linear, there is a consistent positive relationship: an increase in performance tends to accompany an increase in practice.

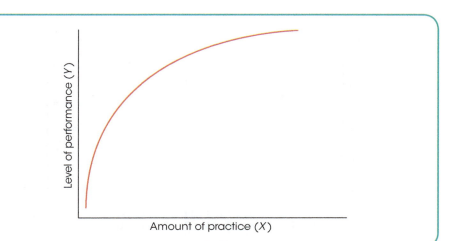

are on an interval or a ratio scale. As we have noted, the Pearson correlation measures the degree of *linear relationship* between two variables—that is, how well the data points fit on a straight line. However, a researcher often expects the data to show a consistently one-directional relationship but not necessarily a linear relationship. For example, Figure 14.11 shows the typical relationship between practice and performance. For nearly any skill, increasing amounts of practice tend to be associated with improvements in performance (the more you practice, the better you get). However, it is not a straight-line relationship. When you are first learning a new skill, practice produces large improvements in performance. After you have been performing a skill for several years, however, additional practice produces only minor changes in performance. Although there is a consistent relationship between the amount of practice and the quality of performance, it clearly is not linear. If the Pearson correlation were computed for these data, it would not produce a correlation of 1.00 because the data do not fit perfectly on a straight line. In a situation like this, the Spearman correlation can be used to measure the degree to which a relationship is consistently one directional, independent of its form. Incidentally, when there is a consistently one-directional relationship between two variables, the relationship is said to be *monotonic*. Thus, the Spearman correlation measures the degree of monotonic relationship between two variables.

The word *monotonic* describes a sequence that is consistently increasing (or decreasing). Like the word *monotonous*, it means constant and unchanging.

The reason that the Spearman correlation measures consistency, rather than form, comes from a simple observation: When two variables are consistently related, their ranks are linearly related. For example, a perfectly consistent positive relationship means that every time the X variable increases, the Y variable also increases. Thus, the smallest value of X is paired with the smallest value of Y, the second-smallest value of X is paired with the second-smallest value of Y, and so on. Every time the rank for X goes up by 1 point, the rank for Y also goes up by 1 point. As a result, the ranks fit perfectly on a straight line. This phenomenon is demonstrated in the following example.

EXAMPLE 14.9   Table 14.3 presents X and Y scores for a sample of $n = 4$ people. Note that the data show a perfectly consistent relationship. Each increase in X is accompanied by an increase in Y. However, the relationship is not linear—as can be seen in the graph of the data in Figure 14.12(a).

TABLE 14.3
Scores and ranks for Example 14.9.

Person	X	Y	X-Rank	Y-Rank
A	2	2	1	1
B	3	8	2	2
C	4	9	3	3
D	10	10	4	4

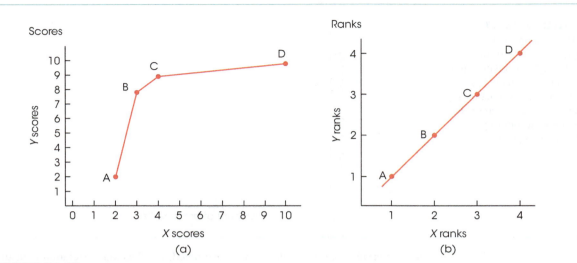

FIGURE 14.12

Scatter plots showing (a) the scores and (b) the ranks for the data in Example 14.9. Notice that there is a consistent, positive relationship between the X and Y scores, although it is not a linear relationship. Also notice that the scatter plot for the ranks shows a perfect linear relationship.

Next, we convert the scores to ranks. The lowest X is assigned a rank of 1, the next lowest a rank of 2, and so on. The Y scores are then ranked in the same way. The ranks are listed in Table 14.3 and shown in Figure 14.12(b). Note that the perfect consistency for the scores produces a perfect linear relationship for the ranks. ■

The preceding example demonstrates that a consistent relationship among scores produces a linear relationship when the scores are converted to ranks. Thus, if you want to measure the consistency of a relationship for a set of scores, you can simply convert the scores to ranks and then use the Pearson correlation formula to measure the linear relationship for the ranked data. The degree of linear relationship for the ranks provides a measure of the degree of consistency for the original scores.

To summarize, the Spearman correlation measures the relationship between two variables when both are measured on ordinal scales (ranks). There are two general situations in which the Spearman correlation is used:

1. Spearman is used when the original data are ordinal; that is, when the X and Y values are ranks. In this case, you simply apply the Pearson correlation formula to the set of ranks.

2. The Spearman correlation is used when a researcher wants to measure the degree to which the relationship between X and Y is consistently one directional, independent of the specific form of the relationship. In this case, the original scores are first converted to ranks; then the Pearson correlation formula is used with the ranks. Because the Pearson formula measures the degree to which the ranks fit on a straight line, it also measures the degree of consistency in the relationship for the original scores.

In either case, the Spearman correlation is identified by the symbol r_S to differentiate it from the Pearson correlation. The complete process of computing the Spearman correlation, including ranking scores, is demonstrated in Example 14.10.

EXAMPLE 14.10

The following data show a nearly perfect monotonic relationship between X and Y. When X increases, Y tends to decrease, and there is only one reversal in this general trend. To compute the Spearman correlation, we first rank the X and Y values, and we then compute the Pearson correlation for the ranks.

We have listed the X values in order so that the trend is easier to recognize.

Original Data		Ranks		
X	Y	X	Y	XY
3	12	1	5	5
4	10	2	3	6
10	11	3	4	12
11	9	4	2	8
12	2	5	1	5
				$36 = \Sigma XY$

To compute the correlation, we need SS for X, SS for Y, and SP. Remember that all these values are computed with the ranks, not the original scores. The X ranks are simply the integers 1, 2, 3, 4, and 5. These values have $\Sigma X = 15$ and $\Sigma X^2 = 55$. The SS for the X ranks is

$$SS_X = \Sigma X^2 - \frac{(\Sigma X)^2}{n} = 55 - \frac{(15)^2}{5} = 10$$

Note that the ranks for Y are identical to the ranks for X; that is, they are the integers 1, 2, 3, 4, and 5. Therefore, the SS for Y is identical to the SS for X:

$$SS_Y = 10$$

To compute the SP value, we need ΣX, ΣY, and ΣXY for the ranks. The XY values are listed in the table with the ranks, and we already have found that both the Xs and the Ys have a sum of 15. Using these values, we obtain

$$SP = \Sigma XY - \frac{(\Sigma X)(\Sigma Y)}{n} = 36 - \frac{(15)(15)}{5} = -9$$

Finally, the Spearman correlation simply uses the Pearson formula for the ranks.

$$r_S = \frac{SP}{\sqrt{(SS_X)(SS_Y)}} = \frac{-9}{\sqrt{10(10)}} = -0.9$$

The Spearman correlation indicates that the data show a consistent (nearly perfect) negative trend. ∎

■ Ranking Tied Scores

When you are converting scores into ranks for the Spearman correlation, you may encounter two (or more) identical scores. Whenever two scores have exactly the same value, their ranks should also be the same. This is accomplished by the following procedure:

1. List the scores in order from smallest to largest. Include tied values in the list.

2. Assign a rank (first, second, and so on) to each position in the ordered list.

3. When two (or more) scores are tied, compute the mean of their ranked positions, and assign this mean value as the final rank for each score.

The process of finding ranks for tied scores is demonstrated here. These scores have been listed in order from smallest to largest.

Scores	Rank Position	Final Rank	
3	1	1.5	Mean of 1 and 2
3	2	1.5	
5	3	3	
6	4	5	Mean of 4, 5, and 6
6	5	5	
6	6	5	
12	7	7	

Note that this example has seven scores and uses all seven ranks. For $X = 12$, the largest score, the appropriate rank is 7. It cannot be given a rank of 6 because that rank has been used for the tied scores.

■ Special Formula for the Spearman Correlation

When the X values and Y values are ranks, the calculations necessary for SS and SP can be greatly simplified. First, you should note that the X ranks and the Y ranks are simply integers: 1, 2, 3, 4, . . . , n. To compute the mean for these integers, you can locate the midpoint of the series by $M = (n + 1)/2$. Similarly, the SS for this series of integers can be computed by

$$SS = \frac{n(n^2 - 1)}{12} \text{ (Try it out.)}$$

Also, because the X ranks and the Y ranks are the same values, the SS for X is identical to the SS for Y.

Because calculations with ranks can be simplified and because the Spearman correlation uses ranked data, these simplifications can be incorporated into the final calculations for the Spearman correlation. Instead of using the Pearson formula after ranking the data, you can put the ranks directly into a simplified formula,

Caution: **In this formula, you compute the value of the fraction and then subtract from 1. The 1 is not part of the fraction.**

$$r_S = 1 - \frac{6\Sigma D^2}{n(n^2 - 1)} \tag{14.8}$$

where D is the difference between the X rank and the Y rank for each individual. This special formula produces the same result that would be obtained from the Pearson formula. However, note that this special formula should be used only after the scores have been converted to ranks and when there are no ties among the ranks. If there are relatively few tied ranks, the formula still may be used, but it loses accuracy as the number of ties increases. The application of this formula is demonstrated in the following example.

EXAMPLE 14.11   To demonstrate the special formula for the Spearman correlation, we use the same data that were presented in Example 14.10. The ranks for these data are shown again here:

Ranks		Difference	
X	Y	D	D²
1	5	4	16
2	3	1	1
3	4	1	1
4	2	−2	4
5	1	−4	16
			38 = ΣD^2

Using the special formula for the Spearman correlation, we obtain

$$r_S = 1 - \frac{6\Sigma D^2}{n(n^2 - 1)}$$

$$= 1 - \frac{6(38)^2}{5(25 - 1)}$$

$$= 1 - \frac{228}{120}$$

$$= 1 - 1.90$$

$$= -0.90$$

This is exactly the same answer that we obtained in Example 14.10, using the Pearson formula on the ranks. ■

The following example is an opportunity to test your understanding of the Spearman correlation.

EXAMPLE 14.12   Compute the Spearman correlation for the following set of scores:

X	Y
2	7
12	38
9	6
10	19

You should obtain $r_S = 0.80$. Good luck. ■

■ The Point-Biserial Correlation and Measuring Effect Size with r^2

In Chapters 9, 10, and 11 we introduced r^2 as a measure of effect size that often accompanies a hypothesis test using the t statistic. This measure of effect size is related to correlation, r, and we now have an opportunity to demonstrate the relationship. Specifically, we compare the independent-measures t test (Chapter 10) and a special version of the Pearson correlation known as the *point-biserial correlation*.

The point-biserial correlation is used to measure the relationship between two variables in situations in which one variable consists of regular, numerical scores, but the second variable has only two values. A variable with only two values is called a *dichotomous variable* or a *binomial variable*. Here are some examples of dichotomous variables:

1. College graduate versus not a college graduate
2. First-born child versus later-born child
3. Success versus failure on a particular task
4. Older than 30 years old versus younger than 30 years old

It is customary to use the numerical values 0 and 1, but any two different numbers would work equally well and would not affect the value of the correlation.

To compute the point-biserial correlation, the dichotomous variable is first converted to numerical values by assigning a value of zero (0) to one category and a value of one (1) to the other category. Then the regular Pearson correlation formula is used with the converted data.

To demonstrate the point-biserial correlation and its association with the r^2 measure of effect size, we use the data from Example 10.2 (page 334). The original example compared cheating behavior in a dimly lit room compared to a well-lit room. The results showed that participants in the dimly lit room claimed to have solved significantly more puzzles than the participants in the well-lit room. The data from the independent-measures study are presented on the left side of Table 14.4. Notice that the data consist of two separate samples, and the independent-measures t was used to determine whether there was a significant mean difference between the two populations represented by the samples.

On the right-hand side of Table 14.4 we have reorganized the data into a form that is suitable for a point-biserial correlation. Specifically, we used each participant's puzzle-solving score as the X value and we have created a new variable, Y, to represent the group or condition for each individual. In this case, we have used $Y = 0$ for individuals in the well-lit room and $Y = 1$ for participants in the dimly lit room.

When the data in Table 14.4 were originally presented in Chapter 10, we conducted an independent-measures t hypothesis test and obtained $t = -2.67$ with $df = 14$. We measured the size of the treatment effect by calculating r^2, the percentage of variance accounted for, and obtained $r^2 = 0.337$.

Calculating the point-biserial correlation for these data also produces a value for r. Specifically, the X scores produce $SS = 190$; the Y values produce $SS = 4.00$, and the sum of the products of the X and Y deviations produces $SP = 16$. The point-biserial correlation is

$$r = \frac{SP}{\sqrt{SS_X SS_Y}}$$

$$= \frac{16}{\sqrt{(190)(4)}}$$

$$= \frac{16}{27.57}$$

$$= 0.5803$$

TABLE 14.4

The same data are organized in two different formats. On the left-hand side, the data appear as two separate samples appropriate for an independent-measures t hypothesis test. On the right-hand side, the same data are shown as a single sample, with two scores for each individual: the number of puzzles solved (X) and a dichotomous score (Y) that identifies the group in which the participant is located (Well-lit = 0 and Dimly lit = 1). The data on the right are appropriate for a point-biserial correlation.

Number of Solved Puzzles				Data for the Point-Biserial Correlation. Two Scores, X and Y, for each of the n = 16 participants		
Well-Lit Room		Dimly Lit Room		Participant	Puzzles Solved (X)	Group (Y)
11	6	7	9	A	11	0
9	7	13	11	B	9	0
4	12	14	15	C	4	0
5	10	16	11	D	5	0
$n = 8$		$n = 8$		E	6	0
$M = 8$		$M = 12$		F	7	0
$SS = 60$		$SS = 66$		G	12	0
				H	10	0
				I	7	1
				J	13	1
				K	14	1
				L	16	1
				M	9	1
				N	11	1
				O	15	1
				P	11	1

Notice that squaring the value of the point-biserial correlation produces $r^2 = (0.5803)^2 = 0.337$, which is the same as the value of r^2 we obtained measuring effect size.

In some respects, the point-biserial correlation and the independent-measures hypothesis test are evaluating the same thing. Specifically, both are examining the relationship between room lighting and cheating behavior.

1. The correlation is measuring the *strength* of the relationship between the two variables. A large correlation (near 1.00 or -1.00) would indicate that there is a consistent, predictable relationship between cheating and the amount of light in the room. In particular, the value of r^2 measures how much of the variability in cheating can be predicted by knowing whether the participants were tested in a well-lit or dimly lit room.

2. The t test evaluates the *significance* of the relationship. The hypothesis test determines whether the mean difference in grades between the two groups is greater than can be reasonably explained by chance alone.

As we noted in Chapter 10 (pages 340–344), the outcome of the hypothesis test and the value of r^2 are often reported together. The t value measures statistical significance and r^2 measures the effect size. Also, as we noted in Chapter 10, the values for t and r^2 are directly related. In fact, either can be calculated from the other by the equations

$$r^2 = \frac{t^2}{t^2 + df} \quad \text{and} \quad t^2 = \frac{r^2}{(1 - r^2)/df}$$

where *df* is the degrees of freedom for the t statistic.

However, you should note that r^2 is determined entirely by the size of the correlation, whereas t is influenced by the size of the correlation and the size of the sample. For example, a correlation of $r = 0.30$ produces $r^2 = 0.09$ (9%) no matter how large the sample may be. Using Equation 14.7, a point-biserial correlation of $r = 0.30$ for a total sample of 10 people ($n = 5$ in each group) produces a nonsignificant value of $t = 0.890$. If the sample is increased to 50 people ($n = 25$ in each group), the same correlation produces a significant t value of $t = 2.17$. Although t and r are related, they are measuring different things.

■ The Phi-Coefficient

When both variables (X and Y) measured for each individual are dichotomous, the correlation between the two variables is called the *phi-coefficient*. To compute phi (ϕ), you follow a two-step procedure:

1. Convert each of the dichotomous variables to numerical values by assigning a 0 to one category and a 1 to the other category for each of the variables.

2. Use the regular Pearson formula with the converted scores.

This process is demonstrated in the following example.

EXAMPLE 14.13   A researcher is interested in examining the relationship between birth-order position and personality. A random sample of $n = 8$ individuals is obtained, and each individual is classified in terms of birth-order position as first-born or only child versus later-born. Then each individual's personality is classified as either introvert or extrovert.

The original measurements are then converted to numerical values by the following assignments:

Birth Order	Personality
1st or only child = 0	Introvert = 0
Later-born child = 1	Extrovert = 1

The original data and the converted scores are as follows:

Original Data		Converted Scores	
Birth Order X	Personality Y	Birth Order X	Personality Y
1st	Introvert	0	0
3rd	Extrovert	1	1
Only	Extrovert	0	1
2nd	Extrovert	1	1
4th	Extrovert	1	1
2nd	Introvert	1	0
Only	Introvert	0	0
3rd	Extrovert	1	1

The Pearson correlation formula is then used with the converted data to compute the phi-coefficient.

Because the assignment of numerical values is arbitrary (either category could be designated 0 or 1), the sign of the resulting correlation is meaningless. As with most correlations, the *strength* of the relationship is best described by the value of r^2, the coefficient of determination, which measures how much of the variability in one variable is predicted or determined by the association with the second variable.

We also should note that although the phi-coefficient can be used to assess the relationship between two dichotomous variables, the more common statistical procedure is a chi-square statistic, which is examined in Chapter 15. ∎

LEARNING CHECK   **LO9**   **1.** If the following scores are converted to ranks (1 = smallest), then what rank is assigned to the score X = 6? Scores: 4, 5, 5, 6, 6, 6, 7, 9, 10

    a. 4

    b. 5

    c. 6

    d. 7

LO10   **2.** What is the Spearman correlation for the following set of ranked data?

    a. 0.9

    b. −0.9

    c. 0.375

    d. −0.375

X	Y
1	5
2	4
3	2
4	3
5	1

LO11   **3.** Which of the following correlations can be computed for data that are also suitable for an independent-measures *t* test?

    a. Pearson

    b. Spearman

    c. Point-biserial

    d. Phi-coefficient

LO12 4. A researcher would like to measure the relationship between success in a class (pass/fail) and voter registration (yes/no). Which of the following correlations would be appropriate?

   a. Pearson

   b. Spearman

   c. Point-biserial

   d. Phi-coefficient

ANSWERS   1. b   2. b   3. c   4. d

14-6   Introduction to Linear Equations and Regression

LEARNING OBJECTIVES

13. Define the equation that describes a linear relationship between two variables.

14. Compute the regression equation (slope and Y-intercept) for a set of X and Y scores.

15. Compute the standard error of estimate for a regression equation and explain what it measures.

16. Conduct an analysis of regression to evaluate the significance of a regression equation.

Earlier in this chapter, we introduced the Pearson correlation as a technique for describing and measuring the linear relationship between two variables. Figure 14.13 presents hypothetical data showing the relationship between Math SAT scores and college grade point average (GPA) for engineering students. Note that the figure shows a strong, but not perfect, positive relationship. Also note that we have drawn a line through the middle of the data points. This line serves several purposes:

1. The line makes the relationship between Math SAT scores and GPA easier to see.

2. The line identifies the center, or *central tendency*, of the relationship, just as the mean describes central tendency for a set of scores. Thus, the line provides a simplified description of the relationship. For example, if the data points were removed, the straight line would still give a general picture of the relationship between Math SAT scores and GPA.

3. Finally, the line can be used for prediction. The line establishes a precise, one-to-one relationship between each X value (SAT score) and a corresponding Y value (GPA). For example, a Math SAT score of 620 corresponds to a GPA of 3.25 (see Figure 14.13). Thus, the college admissions officers could use the straight-line relationship to predict that an engineering student entering college with a Math SAT score of 620 should achieve a college GPA of approximately 3.25.

Our goal in this section is to develop a procedure that identifies and defines the straight line that provides the best fit for any specific set of data. This straight line does not have to be drawn on a graph; it can be presented in a simple equation. Thus, our goal is to find the equation for the line that best describes the relationship for a set of X and Y data.

FIGURE 14.15

The distance between the actual data point (Y) and the predicted point on the line ($\hat{Y}$) is defined as $Y - \hat{Y}$. The goal of regression is to find the equation for the line that minimizes these distances.

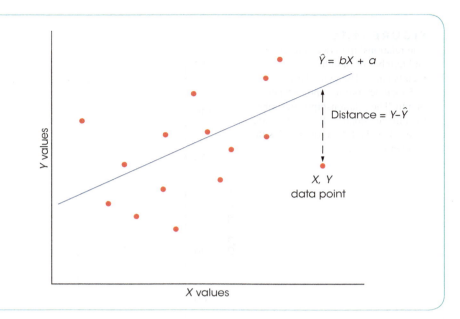

values for b and a. The problem is to find the specific line that provides the best fit to the actual data points.

The Least-Squares Solution To determine how well a line fits the data points, the first step is to define mathematically the distance between the line and each data point. For every X value in the data, the linear equation determines a Y value on the line. This value is the predicted Y and is called $\hat{Y}$ ("Y hat"). The distance between this predicted value and the actual Y value in the data is determined by

$$\text{distance} = Y - \hat{Y}$$

Note that we simply are measuring the vertical distance between the actual data point (Y) and the predicted point on the line. This distance measures the error between the predicted value of Y on the line and the actual value in the data (Figure 14.15).

Because some of these distances will be positive and some will be negative, the next step is to square each distance to obtain a uniformly positive measure of error. Finally, to determine the total error between the line and the data, we add the squared errors for all of the data points. The result is a measure of overall squared error between the line and the data:

$$\text{total squared error} = \Sigma(Y - \hat{Y})^2$$

Now we can define the *best-fitting* line as the one that has the smallest total squared error. For obvious reasons, the resulting line is commonly called the *least-squared-error solution*. In symbols, we are looking for a linear equation of the form

$$\hat{Y} = bX + a$$

For each value of X in the data, this equation determines the point on the line ($\hat{Y}$) that gives the best prediction of Y. The problem is to find the specific values for a and b that make this the best-fitting line.

The calculations that are needed to find this equation require calculus and some sophisticated algebra, so we will not present the details of the solution. The results, however, are relatively straightforward, and the solutions for b and a are as follows:

$$b = \frac{SP}{SS_X} \qquad (14.10)$$

where SP is the sum of products and SS_X is the sum of squares for the X scores.

A commonly used alternative formula for the slope is based on the standard deviations for X and Y. The alternative formula is

$$b = r\frac{s_Y}{s_X} \qquad (14.11)$$

where s_Y is the standard deviation for the Y scores, s_X is the standard deviation for the X scores, and r is the Pearson correlation for X and Y. After the value of b is computed, the value of the constant a in the equation is determined by

$$a = M_Y - bM_X \qquad (14.12)$$

Note that these formulas determine the linear equation that provides the best prediction of Y values. This equation is called the *regression equation for* Y.

> The **regression equation for Y** is the linear equation
>
> $$\hat{Y} = bX + a \qquad (14.13)$$
>
> where the constant b is determined by Equation 14.10 or 14.11, and the constant a is determined by Equation 14.12. This equation results in the least squared error between the data points and the line.

EXAMPLE 14.14   The scores in the following table are used to demonstrate the calculation and use of the regression equation for predicting Y.

X	Y	$X - M_X$	$Y - M_Y$	$(X - M_X)^2$	$(Y - M_Y)^2$	$(X - M_X)(Y - M_Y)$
5	10	1	3	1	9	3
1	4	-3	-3	9	9	9
4	5	0	-2	0	4	0
7	11	3	4	9	16	12
6	15	2	8	4	64	16
4	6	0	-1	0	1	0
3	5	-1	-2	1	4	2
2	0	-2	-7	4	49	14
				$SS_X = 28$	$SS_Y = 156$	$SP = 56$

For these data, $\Sigma X = 32$, so $M_X = 4$. Also, $\Sigma Y = 56$, so $M_Y = 7$. These values have been used to compute the deviation scores for each X and Y value. The final three columns show the squared deviations for X and for Y and the products of the deviation scores.

Our goal is to find the values for b and a in the regression equation. Using Equations 14.10 and 14.12, the solutions for b and a are

$$b = \frac{SP}{SS_X} = \frac{56}{28} = 2$$

$$a = M_Y - bM_X = 7 - 2(4) = -1$$

Because the process of transforming all of the original scores into z-scores can be tedious, researchers usually compute the raw-score version of the regression equation (Equation 14.13) instead of the standardized form. However, most computer programs report the value of beta as part of the output from linear regression, and you should understand what this value represents.

■ The Standard Error of Estimate

It is possible to determine a regression equation for any set of data by simply using the formulas already presented. The linear equation you obtain is then used to generate predicted Y values for any known value of X. However, it should be clear that the accuracy of this prediction depends on how well the points on the line correspond to the actual data points—that is, the amount of error between the predicted values, $\hat{Y}$, and the actual scores, Y values. Figure 14.17 shows two different sets of data that have exactly the same regression equation. In one case, there is a perfect correlation ($r = +1$) between X and Y, so the linear equation fits the data perfectly. For the second set of data, the predicted Y values on the line only approximate the real data points.

A regression equation by itself allows you to make predictions, but it does not provide any information about the accuracy of the predictions. To measure the precision of the regression, it is customary to compute a *standard error of estimate*.

> The **standard error of estimate** gives a measure of the standard distance between the predicted Y values on the regression line and the actual Y values in the data.

Conceptually, the standard error of estimate is very much like a standard deviation: both provide a measure of standard distance. Also, you will see that the calculation of the standard error of estimate is very similar to the calculation of standard deviation.

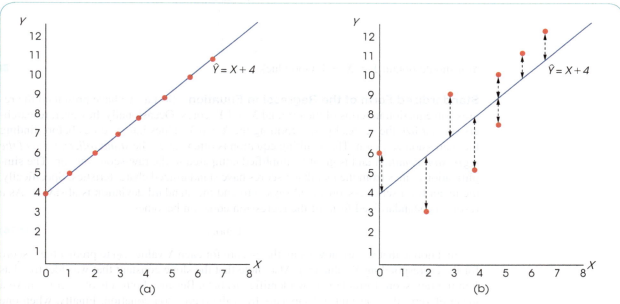

FIGURE 14.17
(a) A scatter plot showing data points that perfectly fit the regression line defined by $\hat{Y} = X + 4$. Note that the correlation is $r = +1.00$. (b) A scatter plot for another set of data with a regression equation of $\hat{Y} = X + 4$. Notice that there is error between the actual data points and the predicted Y values on the regression line.

To calculate the standard error of estimate, we first find a sum of squared deviations (SS). Each deviation measures the distance between the actual Y value (from the data) and the predicted Y value (from the regression line). This sum of squares is commonly called $SS_{residual}$ because it is based on the remaining distance between the actual Y scores and the predicted values.

$$SS_{residual} = \Sigma(Y - \hat{Y})^2 \tag{14.16}$$

The obtained SS value is then divided by its degrees of freedom to obtain a measure of variance. This procedure for computing variance should be very familiar (Chapter 4, page 118).

$$\text{Variance} = \frac{SS}{df}$$

The degrees of freedom for the standard error of estimate are $df = n - 2$. The reason for having $n - 2$ degrees of freedom, rather than the customary $n - 1$, is that we now are measuring deviations from a line rather than deviations from a mean. To find the equation for the regression line, you must know the means for both the X and the Y scores. Specifying these two means places two restrictions on the variability of the data, with the result that the scores have only $n - 2$ degrees of freedom. (*Note*: the $df = n - 2$ for $SS_{residual}$ is the same $df = n - 2$ that we encountered when testing the significance of the Pearson correlation on page 496.)

Recall that variance measures the average squared distance.

The final step in the calculation of the standard error of estimate is to take the square root of the variance to obtain a measure of standard distance. The final equation is

$$\text{Standard error of estimate } = \sqrt{\frac{SS_{residual}}{df}} = \sqrt{\frac{\Sigma(Y - \hat{Y})^2}{n - 2}} \tag{14.17}$$

The following example demonstrates the calculation of this standard error.

EXAMPLE 14.16 The same data that were used in Example 14.14 are used here to demonstrate the calculation of the standard error of estimate. These data have the regression equation

$$\hat{Y} = 2X - 1$$

Using this regression equation, we have computed the predicted Y value, the residual, and the squared residual for each individual.

Data		Predicted Y value	Residual	Squared Residual
X	Y	$\hat{Y} = 2X - 1$	$Y - \hat{Y}$	$(Y - \hat{Y})^2$
5	10	9	1	1
1	4	1	3	9
4	5	7	−2	4
7	11	13	−2	4
6	15	11	4	16
4	6	7	−1	1
3	5	5	0	0
2	0	3	−3	9
			0	$SS_{residual} = 44$

First note that the sum of the residuals is equal to zero. In other words, the sum of the distances above the line is equal to the sum of the distances below the line. This is true for any set of data and provides a way to check the accuracy of your calculations. The squared residuals are listed in the final column. For these data, the sum of the squared residuals is $SS_{residual} = 44$. With $n = 8$, the data have $df = n - 2 = 6$, so the standard error of estimate is

$$\text{standard error of estimate} = \sqrt{\frac{SS_{residual}}{df}} = \sqrt{\frac{44}{6}} = 2.708$$

Remember: The standard error of estimate provides a measure of how accurately the regression equation predicts the Y values. In this case, the standard distance between the actual data points and the regression line is measured by the standard error of estimate = 2.708. ∎

Relationship between the Standard Error and the Correlation It should be clear from Example 14.16 that the standard error of estimate is directly related to the magnitude of the correlation between X and Y. If the correlation is near 1.00 (or -1.00), the data points are clustered close to the line, and the standard error of estimate is small. As the correlation gets nearer to zero, the data points become more widely scattered, the line provides less accurate predictions, and the standard error of estimate grows larger.

Earlier (page 492), we observed that squaring the correlation provides a measure of the accuracy of prediction. The squared correlation, r^2, is called the coefficient of determination because it determines what proportion of the variability in Y is predicted by the relationship with X. Because r^2 measures the predicted portion of the variability in the Y scores, we can use the expression $(1 - r^2)$ to measure the unpredicted portion. Thus,

$$\text{Predicted variability} = SS_{regression} = r^2 SS_Y \tag{14.18}$$

$$\text{Unpredicted variability} = SS_{residual} = (1 - r^2)SS_Y \tag{14.19}$$

For example, if $r = 0.80$, then $r^2 = 0.64$ (or 64%) of the variability for the Y scores is predicted by the relationship with X, and the remaining 36% $(1 - r^2)$ is the unpredicted portion. Note that when $r = 1.00$, the prediction is perfect and there are no residuals. As the correlation approaches zero, the data points move farther off the line and the residuals grow larger. Using Equation 14.19 to compute $SS_{residual}$, the standard error of estimate can be computed as

$$\text{standard error of estimate} = \sqrt{\frac{SS_{residual}}{df}} = \sqrt{\frac{(1 - r^2)SS_Y}{n - 2}} \tag{14.20}$$

Because it is usually much easier to compute the Pearson correlation than to compute the individual $(Y - \hat{Y})^2$ values, Equation 14.19 is usually the easiest way to compute $SS_{residual}$, and Equation 14.20 is usually the easiest way to compute the standard error of estimate for a regression equation. The following example demonstrates this new formula.

EXAMPLE 14.17 We use the same data used in Examples 14.14 and 14.16, which produced $SS_X = 28$, $SS_Y = 156$, and $SP = 56$. For these data, the Pearson correlation is

$$r = \frac{56}{\sqrt{28(156)}} = \frac{56}{66.09} = 0.847$$

With $SS_Y = 156$ and a correlation of $r = 0.847$, the *predicted variability* from the regression equation is

$$SS_{regression} = r^2 SS_Y = (0.847^2)(156) = 0.718(156) = 112.01$$

Similarly, the *unpredicted variability* is

$$SS_{residual} = (1 - r^2)SS_Y = (1 - 0.847^2)(156) = 0.282(156) = 43.99$$

Notice that the new formula for $SS_{residual}$ produces the same value, within rounding error, that we obtained by adding the squared residuals in Example 14.16. Also note that this new formula is generally much easier to use because it requires only the correlation value (r) and the SS for Y. The primary point of this example, however, is that $SS_{residual}$ and the standard error of estimate are closely related to the value of the correlation. As correlations get larger (near $+1.00$ or -1.00), the data points move closer to the regression line, and the standard error of estimate gets smaller. ∎

Because it is possible to have the same regression line for sets of data that have different correlations, it is also important to examine r^2 and the standard error of estimate. The regression equation simply describes the best-fitting line and is used for making predictions. However, r^2 and the standard error of estimate indicate how accurate these predictions will be.

■ Analysis of Regression: The Significance of the Regression Equation

As we noted earlier in this chapter, a sample correlation is expected to be representative of its population correlation. For example, if the population correlation is zero, the sample correlation is expected to be near zero. Note that we do not expect the sample correlation to be exactly equal to zero. This is the general concept of *sampling error* that was introduced in Chapter 1 (page 7). The principle of sampling error is that there is typically some discrepancy or error between the value obtained for a sample statistic and the corresponding population parameter. Thus, when there is no relationship whatsoever in the population, a correlation of $\rho = 0$, you are still likely to obtain a nonzero value for the sample correlation. In this situation, however, the sample correlation is meaningless and a hypothesis test usually demonstrates that the correlation is not significant.

Whenever you obtain a nonzero value for a sample correlation, you will also obtain real, numerical values for the regression equation. However, if there is no real relationship in the population, both the sample correlation and the regression equation are meaningless—they are simply the result of sampling error and should not be viewed as an indication of any relationship between X and Y. In the same way that we tested the significance of a Pearson correlation, we can test the significance of the regression equation. In fact, when a single variable X is being used to predict a single variable Y, the two tests are equivalent. In each case, the purpose of the test is to determine whether the sample correlation represents a real relationship or is simply the result of sampling error. For both tests, the null hypothesis states that there is no relationship between the two variables in the population. For a correlation,

$$H_0: \text{the population correlation is } \rho = 0$$

For the regression equation,

$$H_0: \text{the slope of the regression equation } (b \text{ or beta}) \text{ is zero}$$

For regression, an equivalent version of H_0 states that the regression equation does not predict a significant portion of the variability in the Y scores.

The process of testing the significance of a regression equation is called *analysis of regression* and is very similar to the analysis of variance (ANOVA) presented in Chapter 12. As with ANOVA, the regression analysis uses an F-ratio to determine whether the variance predicted by the regression equation is significantly greater than would be expected if there were no relationship between X and Y. The F-ratio is a ratio of two variances, or mean square (MS) values, and each variance is obtained by dividing an SS value by its

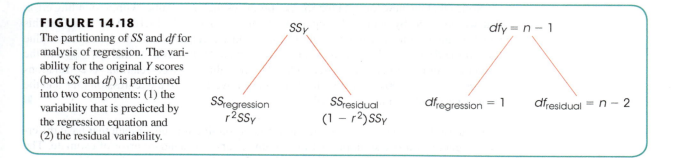

FIGURE 14.18
The partitioning of *SS* and *df* for analysis of regression. The variability for the original *Y* scores (both *SS* and *df*) is partitioned into two components: (1) the variability that is predicted by the regression equation and (2) the residual variability.

corresponding degrees of freedom. The numerator of the *F*-ratio is $MS_{\text{regression}}$, which is the variance in the *Y* scores that is predicted by the regression equation. This variance measures the systematic changes in *Y* that occur when the value of *X* increases or decreases. The denominator is MS_{residual}, which is the unpredicted variance in the *Y* scores. This variance measures the changes in *Y* that are independent of changes in *X*. The two *MS* value are defined as

$$MS_{\text{regression}} = \frac{SS_{\text{regression}}}{df_{\text{regression}}} \text{ with } df = 1 \quad \text{and} \quad MS_{\text{residuals}} = \frac{SS_{\text{residual}}}{df_{\text{residual}}} \text{ with } df = n - 2$$

The *F*-ratio is

$$F = \frac{MS_{\text{regression}}}{MS_{\text{residual}}} \quad \text{with } df = 1, n - 2 \tag{14.21}$$

The complete analysis of *SS* and degrees of freedom is diagrammed in Figure 14.18. The analysis of regression procedure is demonstrated in the following example, using the same data that we used in Examples 14.14, 14.16, and 14.17.

EXAMPLE 14.18   The data consist of $n = 8$ pairs of scores with a correlation of $r = 0.847$ and $SS_Y = 156$. The null hypothesis either states that there is no relationship between *X* and *Y* in the population or that the regression equation has $b = 0$ and does not account for a significant portion of the variance for the *Y* scores.

The *F*-ratio for the analysis of regression has $df = 1, n - 2$. For these data, $df = 1, 6$. With $\alpha = .05$, the critical value is 5.99.

As noted in the previous section, the *SS* for the *Y* scores can be separated into two components: the predicted portion corresponding to r^2 and the unpredicted, or residual, portion corresponding to $(1 - r^2)$. With $r = 0.847$, we obtain $r^2 = 0.718$ and

$$\text{predicted variability} = SS_{\text{regression}} = 0.718(156) = 112.01$$

$$\text{unpredicted variability} = SS_{\text{residual}} = (1 - 0.718)(156) = 0.282(156) = 43.99$$

Using these *SS* values and the corresponding *df* values, we calculate a variance or *MS* for each component. For these data the *MS* values are:

$$MS_{\text{regression}} = \frac{SS_{\text{regression}}}{df_{\text{regression}}} = \frac{112.01}{1} = 112.01$$

$$MS_{\text{residual}} = \frac{SS_{\text{residual}}}{df_{\text{residual}}} = \frac{43.99}{6} = 7.33$$

TABLE 14.5
A summary table showing the results of the analysis of regression in Example 14.18.

Source	SS	df	MS	F
Regression	112.01	1	112.01	15.28
Residual	43.99	6	7.33	
Total	156	7		

Finally, the F-ratio for evaluating the significance of the regression equation is

$$F = \frac{MS_{\text{regression}}}{MS_{\text{residual}}} = \frac{112.01}{7.33} = 15.28$$

The F-ratio is in the critical region, so we reject the null hypothesis and conclude that the regression equation does account for a significant portion of the variance for the Y scores. The complete analysis of regression is summarized in Table 14.5, which is a common format for computer printouts of regression analysis. ■

Significance of Regression and Significance of the Correlation   As noted earlier, in a situation with a single X variable and a single Y variable, testing the significance of the regression equation is equivalent to testing the significance of the Pearson correlation. Therefore, whenever the correlation between two variables is significant, you can conclude that the regression equation is also significant. Similarly, if a correlation is not significant, the regression equation is also not significant. For the data in Example 14.18, we concluded that the regression equation is significant.

To demonstrate the equivalence of the two tests, we will show that the t statistic used to test the significance of a correlation (Equation 14.7, page 496) is equivalent to the F-ratio used to test the significance of the regression equation (Equation 14.21). We begin with the t statistic

$$t = \frac{r - \rho}{\sqrt{\dfrac{(1 - r^2)}{(n - 2)}}}$$

First, note that the population correlation, ρ, is always zero, as specified by the null hypothesis, so we can simply remove it from the equation. Next, we square the t statistic to produce the corresponding F-ratio.

$$t^2 = F = \frac{r^2}{\dfrac{(1 - r^2)}{(n - 2)}}$$

Finally, multiply the numerator and the denominator by SS_Y to produce

$$t^2 = F = \frac{r^2(SS_Y)}{\dfrac{(1 - r^2)(SS_Y)}{(n - 2)}}$$

You should recognize the numerator as $SS_{\text{regression}}$, which is equivalent to $MS_{\text{regression}}$ because $df_{\text{regression}} = 1$. Also, the denominator is identical to MS_{residual}. Thus, the squared t statistic used to test the significance of a correlation is identical to the F-ratio used to test the significance of a regression equation.

LEARNING CHECK **LO13** **1.** In the general linear equation $Y = bX + a$, what is measured by the value of a?

   a. The point at which the line crosses the X-axis.

   b. The point at which the line crosses the Y-axis.

   c. The amount that X changes each time Y increases by 1 point.

   d. The amount that Y changes each time X increases by 1 point.

LO14 **2.** A set of $n = 25$ pairs of X and Y values has $M_X = 5$, $SS_X = 5$, $M_Y = 2$, $SS_Y = 20$, and $SP = 10$. What is the regression equation for predicting Y from X?

   a. $Y = 2X - 2$

   b. $Y = 2X - 8$

   c. $Y = 0.5X + 4$

   d. $Y = 0.5X + 1$

LO15 **3.** What is measured by the standard error of estimate for a regression equation?

   a. The standard distance between a predicted Y value and the mean for the Y scores.

   b. The standard distance between a predicted Y value and the center of the regression line.

   c. The standard distance between a predicted Y value and the actual Y value.

   d. The standard distance between an actual Y value and the center of the regression line.

LO16 **4.** A researcher computes the regression equation for predicting Y for a sample of $n = 26$ pairs of X and Y values. If the significance of the equation is evaluated with an analysis of regression, then what are the df values for the F-ratio?

   a. 1, 24

   b. 1, 23

   c. 2, 23

   d. 2, 22

ANSWERS **1. b** **2. b** **3. c** **4. a**

SUMMARY

1. A correlation measures the relationship between two variables, X and Y. The relationship is described by three characteristics:

   a. *Direction*. A relationship can be either positive or negative. A positive relationship means that X and Y vary in the same direction. A negative relationship means that X and Y vary in opposite directions. The sign of the correlation ($+$ or $-$) specifies the direction.

   b. *Form*. The most common form for a relationship is a straight line. However, special correlations exist for measuring other forms. The form is specified by the type of correlation used. For example, the Pearson correlation measures linear form.

   c. *Strength or consistency*. The numerical value of the correlation measures the strength or consistency of the relationship. A correlation of 1.00 indicates a perfectly consistent relationship and 0.00 indicates no relationship at all. For the Pearson correlation, $r = 1.00$ (or -1.00) means that the data points fit perfectly on a straight line.

2. The most commonly used correlation is the Pearson correlation, which measures the degree of linear

relationship. The Pearson correlation is identified by the letter r and is computed by

$$r = \frac{SP}{\sqrt{SS_X SS_Y}}$$

In this formula, SP is the sum of products of deviations and can be calculated with either a definitional formula or a computational formula:

definitional formula: $SP = \Sigma(X - M_X)(Y - M_Y)$

computational formula: $SP = \Sigma XY - \frac{\Sigma X \Sigma Y}{n}$

3. A correlation between two variables should not be interpreted as implying a causal relationship. Simply because X and Y are related does not mean that X causes Y or that Y causes X.

4. To evaluate the strength of a relationship, you square the value of the correlation. The resulting value, r^2, is called the *coefficient of determination* because it measures the portion of the variability in one variable that can be determined using the relationship with the second variable.

5. The Spearman correlation (r_S) measures the consistency of direction in the relationship between X and Y—that is, the degree to which the relationship is one-directional, or monotonic. The Spearman correlation is computed by a two-stage process:
   a. Rank the X scores and the Y scores separately.
   b. Compute the Pearson correlation using the ranks.

6. The point-biserial correlation is used to measure the strength of the relationship when one of the two variables is dichotomous. The dichotomous variable is coded using values of 0 and 1, and the regular Pearson formula is applied. Squaring the point-biserial correlation produces the same r^2 value that is obtained to measure effect size for the independent-measures t test. When both variables, X and Y, are dichotomous, the phi-coefficient can be used to measure the strength of the relationship. Both variables are coded 0 and 1, and the Pearson formula is used to compute the correlation.

7. When there is a general linear relationship between two variables, X and Y, it is possible to construct a linear equation that allows you to predict the Y value corresponding to any known value of X:

predicted Y value $= \hat{Y} = bX + a$

The technique for determining this equation is called regression. By using a *least-squares* method to minimize the error between the predicted Y values and the actual Y values, the best-fitting line is achieved when the linear equation has

$$b = \frac{SP}{SS_X} = r\frac{s_Y}{s_X} \quad \text{and} \quad a = M_Y - bM_X$$

8. The linear equation generated by regression (called the regression equation) can be used to compute a predicted Y value for any value of X. However, the prediction is not perfect, so for each Y value, there is a predicted portion and an unpredicted, or residual, portion. Overall, the predicted portion of the Y score variability is measured by r^2, and the residual portion is measured by $1 - r^2$.

Predicted variability $= SS_{regression} = r^2 SS_Y$

Unpredicted variability $= SS_{residual} = (1 - r^2)SS_Y$

9. The residual variability can be used to compute the standard error of estimate, which provides a measure of the standard distance (or error) between the predicted Y values on the line and the actual data points. The standard error of estimate is computed by

$$\text{standard error of estimate} = \sqrt{\frac{SS_{residual}}{n - 2}} = \sqrt{MS_{residual}}$$

10. It is also possible to compute an F-ratio to evaluate the significance of the regression equation. The process is called analysis of regression and determines whether the equation predicts a significant portion of the variance for the Y scores. First a variance, or MS, value is computed for the predicted variability and the residual variability,

$$MS_{regression} = \frac{SS_{regression}}{df_{regression}} \quad MS_{residual} = \frac{SS_{residual}}{df_{residual}}$$

where $df_{regression} = 1$ and $df_{residual} = n - 2$. Next, an F-ratio is computed to evaluate the significance of the regression equation.

$$F = \frac{MS_{regression}}{MS_{residual}} \quad \text{with } df = 1, n - 2$$

KEY TERMS

correlation (479)

scatter plot (479)

positive correlation (480)

negative correlation (480)

perfect correlation (481)

envelope (481)

Pearson correlation (482)

linear relationship (482)

sum of products (SP) (483)

SP definitional formula (483)

SP computational formula (483)

outliers (489, 491)

is simply the Pearson correlation between X and Y. Finally, the table uses a *t* statistic to evaluate the significance of the predictor variable. This is identical to the significance of the regression equation, and you should find that *t* is equal to the square root of the *F*-ratio from the analysis of regression.

Regression

Variables Entered/Removed[a]

Model	Variables Entered	Variables Removed	Method
1	Engine displacement[b]	.	Enter

a. Dependent Variable: Estimated annual fuel cost

b. All requested variables entered.

Model Summary[b]

Model	R	R Square	Adjusted R Square	Std. Error of the Estimate
1	.818[a]	.670	.629	597.17700

a. Predictors: (Constant), Engine displacement

b. Dependent Variable: Estimated annual fuel cost

ANOVA[a]

Model		Sum of Squares	df	Mean Square	F	Sig.
1	Regression	5787037.037	1	5787037.037	16.227	.004[b]
	Residual	2852962.963	8	356620.370		
	Total	8640000.000	9			

a. Dependent Variable: Estimated annual fuel cost

b. Predictors: (Constant), Engine displacement

Coefficients[a]

Model		Unstandardized Coefficients		Standardized Coefficients	t	Sig.
		B	Std. Error	Beta		
1	(Constant)	79.630	444.367		.179	.862
	Engine displacement	462.963	114.927	.818	4.028	.004

a. Dependent Variable: Estimated annual fuel cost

Source: SPSS®

Try It Yourself

Below are scores for a different set of vehicles from the EPA data. Use SPSS to compute the bivariate correlation between displacement and fuel cost and the linear equation for the relationship between displacement and fuel cost. Your results should reveal a significant correlation between displacement and fuel cost, $r(12) = 0.81$, $p = .001$, a significant regression analysis, $F(1, 12) = 23.60$, and an unstandardized coefficient for engine displacement of 255.437.

Engine Displacement	Estimated Annual Fuel Cost
2.0	2050
1.4	1300
2.0	1800
2.0	1650
3.6	1750
5.3	2250
1.6	1300
2.0	1150
5.5	3200
2.0	1750
2.4	1450
2.0	1600
5.7	2200
6.2	2500

OTHER CORRELATIONAL ANALYSES

Phi-Coefficient Data Entry

The **phi-coefficient** can also be computed by entering the complete string of 0s and 1s into two columns of the SPSS data editor, then following the same Data Analysis instructions that were presented for the Pearson correlation. However, this can be tedious, especially with a large set of scores. The following is an alternative procedure for computing the phi-coefficient with large data sets.

1. Enter the values 0, 0, 1, 1 (in order) into the first column of the SPSS data editor.
2. Enter the values 0, 1, 0, 1 (in order) into the second column.
3. Count the number of individuals in the sample who are classified with $X = 0$ and $Y = 0$. Enter this frequency in the top box of the third column of the data editor. Then, count how many have $X = 0$ and $Y = 1$ and enter the frequency in the second box of the third column. Continue with the number who have $X = 1$ and $Y = 0$, and finally the number who have $X = 1$ and $Y = 1$. You should end up with four values in Column 3.
4. Click **Data** on the Tool Bar at the top of the SPSS Data Editor page and select **Weight Cases** at the bottom of the list.
5. Click the circle labeled **weight cases by**, and then highlight the label for the column containing your frequencies (VAR00003) on the left and move it into the **Frequency Variable** box by clicking on the arrow.
6. Click **OK**.
7. Click **Analyze** on the tool bar, select **Correlate**, and click on **Bivariate**.
8. One by one move the labels for the two data columns containing the 0s and 1s (probably VAR00001 and VAR00002) into the **Variables** box. (Highlight each label and click the arrow to move it into the box.)
9. Verify that the **Pearson** box is checked.
10. Click **OK**.

Phi-Coefficient SPSS Output

The program produces the same correlation matrix that was described for the Pearson correlation. Again, you want the correlation between X and Y, which is in either the upper-right or lower-left corner. Remember, with the phi-coefficient the sign of the correlation is meaningless.

To compute the **Spearman** correlation, enter either the X and Y ranks or the X and Y scores into the first two columns. Then follow the same Data Analysis instructions that were presented for the Pearson correlation. At Step 3 in the instructions, click on the **Spearman** box before the final OK. (*Note*: If you enter X and Y scores into the data editor, SPSS converts the scores to ranks before computing the Spearman correlation.)

To compute the **point-biserial** correlation, enter the scores (X values) in the first column and enter the numerical values (usually 0 and 1) for the dichotomous variable in the second column. Then, follow the same Data Analysis instructions that were presented for the Pearson correlation.

PROBLEMS

1. Calculate SP (the sum of products of deviations) for the following scores. *Note*: Both means are whole numbers, so the definitional formula works well.

X	Y
4	8
3	11
9	8
0	1

2. Calculate SP (the sum of products of deviations) for the following scores. *Note*: Both means are decimal values, so the computational formula works well.

X	Y
0	4
1	1
0	5
4	1
2	1
1	3

3. For the following scores,

X	Y
2	5
5	6
4	0
6	3
5	12
8	4

   a. Sketch a scatter plot showing the six data points.
   b. Just looking at the scatter plot, estimate the value of the Pearson correlation.
   c. Compute the Pearson correlation.

4. For the following scores,

X	Y
3	4
0	1
6	6
3	1

   a. Sketch a scatter plot and estimate the Pearson correlation.
   b. Compute the Pearson correlation.

5. For the following scores,

X	Y
3	11
4	9
1	13
7	2
5	5

   a. Sketch a scatter plot and estimate the Pearson correlation.
   b. Compute the Pearson correlation.

6. For the following scores,

X	Y
11	1
3	15
5	7
6	8
5	9

   a. Compute SS for X and Y and SP.
   b. Compute the Pearson correlation.

7. The scores below are a modification of the scores in Problem 6:

X	Y
11	15
3	1
5	7
6	8
5	9

   a. Compute SS for X and Y and SP. Compare these values to your answer for part a of Problem 6.
   b. Compute the Pearson correlation. Compare your results to what you got for part b of Problem 6.

8. For the following scores,

X	Y
3	6
5	5
6	0
6	2
5	2

a. Sketch a scatter plot and estimate the value of the Pearson correlation.

b. Compute the Pearson correlation.

9. With a small sample, a single point can have a large effect on the magnitude of the correlation. To create the following data, we started with the scores from Problem 8 and changed the first X value from $X = 3$ to $X = 8$.

X	Y
8	6
5	5
6	0
6	2
5	2

a. Sketch a scatter plot and estimate the value of the Pearson correlation.

b. Compute the Pearson correlation.

10. For the following set of scores,

X	Y
4	5
6	5
3	2
9	4
6	5
2	3

a. Compute the Pearson correlation.

b. Add 2 points to each X value and compute the correlation for the modified scores. How does adding a constant to every score affect the value of the correlation?

c. Multiply each of the original X values by 2 and compute the correlation for the modified scores. How does multiplying each score by a constant affect the value of the correlation?

11. Judge and Cable (2010) demonstrated a positive relationship between weight and income for a group of men. The following are data similar to those obtained in the study. To simplify the weight variable, the men are classified into five categories that measure actual weight relative to height, from 1 = thinnest to 5 = heaviest. Income is recorded as thousands earned annually.

Weight (X)	Income (Y)
4	151
5	88
3	52
2	73
1	49
3	92
1	56
5	143

a. Calculate the Pearson correlation for these data.

b. Is the correlation statistically significant? Use a two-tailed test with $\alpha = .05$.

12. In recent years, researchers have differentiated between two types of Internet harassment: cyberbullying and Internet trolling. In a recent study of cyber harassment, a large sample of online participants answered survey questions related to personality, cyberbullying history, and Internet trolling. The authors observed a correlation between Internet trolling and cyberbullying (Zezulka & Seigried-Spellar, 2016). Below are scores that capture the relationship observed by the authors.

Participant	Cyberbullying Score	Internet Trolling Score
A	2	1
B	4	8
C	7	9
D	7	9
E	6	9
F	3	5
G	6	8

a. Calculate the Pearson correlation for these data.

b. Is the correlation statistically significant? Use a two-tailed test with $\alpha = .05$.

13. For a two-tailed test with $\alpha = .05$, use Table B.6 to determine how large a Pearson correlation is necessary to be statistically significant for each of the following samples:

a. A sample of $n = 6$

b. A sample of $n = 12$

c. A sample of $n = 24$

14. It appears that there is a significant relationship between cognitive ability and social status, at least for birds. Boogert, Reader, and Laland (2006) measured social status and individual learning ability for a group of starlings. The following data represent results similar to those obtained in the study. Because social status is an ordinal variable consisting of five ordered categories, the Spearman correlation is appropriate for these data. Convert the social status categories and the learning scores to ranks and compute the Spearman correlation.

Subject	Social Status	Learning Score
A	1	3
B	3	10
C	2	7
D	3	11
E	5	19
F	4	17
G	5	17
H	2	4
I	4	12
J	2	3

15. In Problem 14, do the data suggest that increased learning ability caused starlings to have greater social status? Explain.

16. Problem 11 presented data showing a positive relationship between weight and income for a sample of professional men. However, weight was coded in five categories that could be viewed as an ordinal scale rather than an interval or ratio scale. If so, a Spearman correlation is more appropriate than a Pearson correlation. Convert the weights and the incomes into ranks and compute the Spearman correlation for the scores in Problem 11.

17. Problem 13 in Chapter 10 presented data demonstrating that participants who binge-watched a television series enjoyed the show less than participants who watched the series in daily sessions. In the study, one group watched the complete television series in a single session and the other group watched the show in daily, one-hour sessions. After watching the series, each group rated their enjoyment of the series on a scale of 0–100.
   a. Convert the data from this problem into a form suitable for the point-biserial correlation (use 1 for the binge-watching participants and 0 for participants who watched the show in daily sessions), and then compute the correlation.
   b. Square the value of the point-biserial correlation to obtain r^2.
   c. The t test in Chapter 10 produced $t = -2.94$ with $df = 8$. Use the equation on page 505 to compute the value of r^2 directly from the t statistic and its df. Within rounding error, the value of r^2 from the equation should be equal to the value obtained from the point-biserial correlation.

18. Sketch a graph showing the line for the equation $Y = 2X - 1$. On the same graph, show the line for $Y = -X + 8$.

19. A set of $n = 18$ pairs of scores (X and Y values) has $SS_X = 16$, $SS_Y = 64$, and $SP = 20$. If the mean for the X values is $M_X = 6$ and the mean for the Y values is $M_Y = 8$:
   a. Calculate the Pearson correlation for the scores.
   b. Find the regression equation for predicting Y from the X values.

20. A set of $n = 15$ pairs of scores (X and Y values) produces a regression equation of $\hat{Y} = 3X + 8$. Find the predicted Y value for each of the following X scores: 1, 2, 3, and 6.

21. Briefly explain what is measured by the standard error of estimate.

22. In general, how is the magnitude of the standard error of estimate related to the value of the correlation?

23. For the following set of data, compute the Pearson correlation statistic and find the linear regression equation for predicting Y from X:

X	Y
1	5
2	10
0	9
3	12
2	11
4	13

24. The following set of X values is the same as those used in Problem 23. For the following set of data:

X	Y
1	0
2	10
0	8
3	14
2	12
4	16

   a. Compute the Pearson correlation statistic and compare your answer to the answer that you found for Problem 23.
   b. Find the linear regression equation for predicting Y from X. Compare your answer to Problem 23.
   c. Explain the difference between b in a linear regression equation and the Pearson correlation statistic.

25. For the following data:

X	Y
7	7
5	2
0	11
3	12
2	15
7	1

a. Find the regression equation for predicting Y from X.

b. Calculate the Pearson correlation for these data. Use r^2 and SS_Y to compute $SS_{residual}$ and the standard error of estimate for the equation.

26. For the following scores:

X	Y
3	8
5	8
2	6
2	3
4	6
1	4
4	7

a. Find the regression equation for predicting Y from X.

b. Calculate the predicted Y value for each X.

27. The regression equation is computed for a set of $n = 18$ pairs of X and Y values with a correlation of $r = +0.50$ and $SS_Y = 48$.

a. Find the standard error of estimate for the regression equation.

b. How big would the standard error be if the sample size were $n = 66$?

28. Solve the following problems.

a. One set of 10 pairs of scores, X and Y values, produces a correlation of $r = 0.60$. If $SS_Y = 200$, find the standard error of estimate for the regression line.

b. A second set of 10 pairs of X and Y values produces a correlation of $r = 0.40$. If $SS_Y = 200$, find the standard error of estimate for the regression line.

29. Does the regression equation from Problem 25 account for a significant portion of the variance in the Y scores? Use $\alpha = .05$ to evaluate the F-ratio.

30. Solve the following problems.

a. A researcher computes the linear regression equation for a sample of $n = 20$ pairs of scores, X and Y values. If an analysis of regression is used to test the significance of the equation, what are the df values for the F-ratio?

b. A researcher evaluating the significance of a regression equation obtains an F-ratio with $df = 1, 23$. How many pairs of scores, X and Y values, are in the sample?

The Chi-Square Statistic: Tests for Goodness of Fit and Independence

CHAPTER 15

Tools You Will Need

The following items are considered essential background material for this chapter. If you doubt your knowledge of any of these items, you should review the appropriate chapter or section before proceeding.

- Proportions (math review, Appendix A)
- Frequency distributions (Chapter 2)

clivewa/Shutterstock.com

PREVIEW

PREVIEW

We have all seen or heard of examples of prejudice and bias, if not experienced it directly in our daily lives. Expressions of prejudice can run the gamut from blatant and overt to subtle and implicit, and they may occur in any context. Boysen and Vogel (2009) examined instances of bias in a context that you are familiar with—the classroom on college campuses. The researchers had a large sample ($n = 333$) of professors respond to a questionnaire about the occurrence of prejudiced statements and behaviors in their classes. For example, one question required only a "yes" or "no" response: "In the last year has a student said or done something obviously prejudiced during class?" It was found that 27% of the professors reported observing overt bias in the classroom. The researchers also had the professors report the different types of prejudiced behavior that was noticed in the classroom.

Consider a hypothetical example inspired by the work of Boysen and Vogel. Suppose we asked a large number of college professors if they have observed something that was clearly prejudiced in their classes during the past year, and it was found that $n = 75$ professors from the initial group responded "yes." Next, we asked these 75 professors to tell us which of the following biased behaviors they observed most often: an offensive joke, a remark that was a stereotype, or a slur/insult. Each professor must choose only *one* type of behavior—the one that they observed the most. The following table shows the hypothetical data.

Biased behavior professors reported observing most often in the classroom.

	Offensive Joke	Stereotype	Slur/Insult
Observed Frequencies	25	45	5

Notice that the table consists of three categories of prejudiced behaviors observed most often. Furthermore, the values in the table are frequencies. They reflect the number of professors that selected each of the categories. Remember, it was a forced-choice question because they had to choose the biased behavior observed most often. Thus, the frequency for each category is made up of different professors. There were 75 professors responding to the questionnaire, so the frequencies also sum to 75.

Basically, the table depicts a frequency distribution and the frequency values for each category are called *observed frequencies*. We can ask some questions about these data. For example, are all types of bias observed in the classroom equally likely, or are any types more likely to be observed? Another question is, how would you expect all 75 frequencies to be distributed across the bias categories if no one type of bias occurred more than others?

In this chapter you will learn how to test hypotheses about frequency distributions using chi-square tests. You also will learn that the chi-square statistic can be used to test hypotheses about relationships between variables.

15-1 Introduction to Chi-Square: The Test for Goodness of Fit

LEARNING OBJECTIVES

1. Describe parametric and nonparametric hypothesis tests.
2. Describe the data (observed frequencies) for a chi-square test for goodness of fit.
3. Describe the hypotheses for a chi-square test for goodness of fit, explain how the expected frequencies are obtained, and find the expected frequencies for a specific research example.

■ Parametric and Nonparametric Statistical Tests

All the statistical tests we have examined thus far are designed to test hypotheses about specific population parameters. For example, we used t tests to assess hypotheses about a population mean (μ) or mean difference ($\mu_1 - \mu_2$). In addition, these tests typically make assumptions about other population parameters. Recall that, for analysis of variance (ANOVA), the population distributions are assumed to be normal and homogeneity of variance is required. Because these tests all concern parameters and require assumptions about parameters, they are called *parametric tests*.

Another general characteristic of parametric tests is that they require a numerical score for each individual in the sample. The scores then are added, squared, averaged, and otherwise manipulated using basic arithmetic. In terms of measurement scales, parametric tests require data from an interval or a ratio scale (see Chapter 1).

Often, researchers are confronted with experimental situations that do not conform to the requirements of parametric tests. In these situations, it may not be appropriate to use a parametric test. Remember that when the assumptions of a test are violated, the test may lead to an erroneous interpretation of the data. Fortunately, there are several hypothesis-testing techniques that provide alternatives to parametric tests. These alternatives are called *nonparametric tests*.

In this chapter, we introduce two commonly used examples of nonparametric tests. Both tests are based on a statistic known as chi-square, and both tests use sample data to evaluate hypotheses about the proportions or relationships that exist within populations. Note that the two chi-square tests, like most nonparametric tests, do not state hypotheses in terms of a specific parameter and they make few (if any) assumptions about the population distribution. For the latter reason, nonparametric tests sometimes are called *distribution-free tests*.

One of the most obvious differences between parametric and nonparametric tests is the type of data they use. All of the parametric tests that we have examined so far require numerical scores. For nonparametric tests, on the other hand, the participants are usually just classified into categories resulting in frequencies, such as the number of Democrats and Republicans in a town, or the number of small, medium, and large cups of coffee sold at the corner café. Note that these classifications involve measurement on nominal or ordinal scales, and they do not produce numerical values that can be used to calculate means and variances. Instead, the data for many nonparametric tests are simply frequencies—such as the number of students enrolled in elementary, middle, and high schools in a town.

■ The Chi-Square Test for Goodness of Fit

Parameters such as the mean and the standard deviation are the most common way to describe a population, but there are situations in which a researcher has questions about the proportions or relative frequencies for a distribution. For example:

> How does the number of teachers under the age of 40 compare with how many are 40 or older in the profession?

> Of the two leading brands of cola, which is preferred by most Americans?

> In the past 10 years, has there been a significant change in the proportion of 10-year-old children who have their own cell phone?

Note that each of the preceding examples asks a question about proportions in the population. In particular, we are not measuring a numerical score for each individual. Instead, the individuals are simply classified into categories and we want to know what proportion of the population is in each category. The *chi-square test for goodness of fit* is specifically designed to answer this type of question. In general terms, this chi-square test uses the frequencies obtained for sample data to test hypotheses about the corresponding proportions in the population.

The name of the test comes from the Greek letter χ (chi, pronounced "kye"), which is used to identify the test statistic.

> The **chi-square test for goodness of fit** uses sample data consisting of frequencies to test hypotheses about the proportions for a population distribution. The test determines how well the obtained sample frequencies fit the population proportions specified by the null hypothesis.

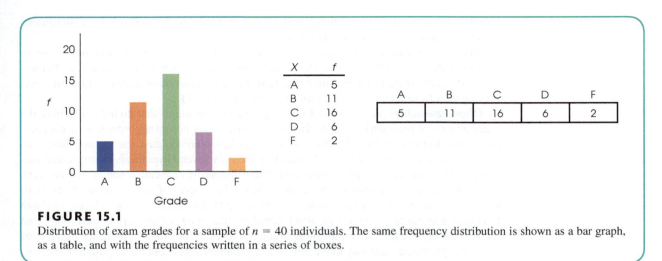

FIGURE 15.1

Distribution of exam grades for a sample of $n = 40$ individuals. The same frequency distribution is shown as a bar graph, as a table, and with the frequencies written in a series of boxes.

Recall from Chapter 2 that a frequency distribution is defined as a tabulation of the number of individuals located in each category of the scale of measurement. In a frequency distribution graph, the categories that make up the scale of measurement are listed on the X axis. In a frequency distribution table, the categories are listed in the first column. With chi-square tests, however, it is customary to present the scale of measurement as a series of boxes (often called cells), with each box corresponding to a separate category on the scale. The frequency corresponding to each category is simply presented as a number written inside the box. Figure 15.1 shows how a distribution of exam grades for a set of $n = 40$ students can be presented as a graph, a table, or a series of boxes. The scale of measurement for this example consists of five categories of grades (A, B, C, D, and F).

■ The Null Hypothesis for the Goodness-of-Fit Test

For the chi-square test for goodness of fit, the null hypothesis specifies the proportion (or percentage) of the population in each category. For example, a hypothesis might state that 50% of all college students graduating in 2020 are men and 50% are women. The simplest way of presenting this hypothesis is to put the hypothesized proportions in the series of boxes representing the scale of measurement:

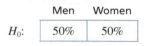

Although it is conceivable that a researcher could choose any proportions for the null hypothesis, there usually is some well-defined rationale for stating a null hypothesis. Generally, H_0 falls into one of the following categories.

1. **No Preference, Equal Proportions**. The null hypothesis often states that the population is divided equally among the categories or that there is no preference among the different categories. For example, a hypothesis stating that there is no preference among the three leading brands of soft drinks would specify a population distribution as follows:

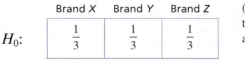

(Preferences in the population are equally divided among the three soft drinks.)

The no-preference hypothesis is used in situations in which a researcher wants to determine whether there are any preferences among the categories, or whether the proportions differ from one category to another.

Because the null hypothesis for the goodness-of-fit test specifies an exact distribution for the population, the alternative hypothesis (H_1) simply states that the population distribution has a different shape from that specified in H_0. If the null hypothesis states that the population is equally divided among three categories, the alternative hypothesis says that the population is not divided equally.

2. **No Difference from a Known Population**. The null hypothesis can state that the proportions for one population are not different from the proportions than are known to exist for another population. For example, suppose it is known that 28% of the licensed drivers in the state are younger than 30 years old and 72% are 30 or older. A researcher might wonder whether this same proportion holds for the distribution of speeding tickets. The null hypothesis would state that tickets are handed out across the population of drivers in the same proportion of their age representation. In other words, there is no difference between the age distribution for drivers in the population and the age distribution for speeding tickets. Specifically, the null hypothesis would be

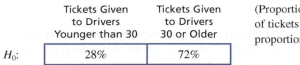

	Tickets Given to Drivers Younger than 30	Tickets Given to Drivers 30 or Older	(Proportions for the population of tickets are not different from proportions for drivers.)
H_0:	28%	72%	

The no-difference hypothesis is used when a specific population distribution is already known. For example, you may have a known distribution from an earlier time, and the question is whether there has been any change in the proportions. Or, you may have a known distribution for one population (drivers) and the question is whether a second population (speeding tickets) has the same proportions.

Again, the alternative hypothesis (H_1) simply states that the population proportions are not equal to the values specified by the null hypothesis. For this example, H_1 would state that the number of speeding tickets is disproportionately high for one age group and disproportionately low for the other.

■ The Data for the Goodness-of-Fit Test

The data for a chi-square test are remarkably simple. There is no need to calculate a sample mean or SS; you just select a sample of n individuals and count how many are in each category. The resulting values are called *observed frequencies*. The symbol for observed frequency is f_o. For example, the following data represent observed frequencies for a sample of 40 college students. The students were classified into three categories based on the number of times they reported exercising each week.

No Exercise	Once a Week	More than Once a Week	
15	19	6	$n = 40$

Notice that each individual in the sample is classified into one and only one of the categories. Thus, the frequencies in this example represent three completely separate groups of students: 15 who do not exercise regularly, 19 who average once a week, and 6 who exercise more than once a week. Also note that the observed frequencies add up to the total sample size: $\Sigma f_o = n$. Finally, you should realize that we are not assigning individuals to categories. Instead, we are simply measuring individuals to determine the category in which they belong.

> The **observed frequency** is the number of individuals from the sample who are classified in a particular category. Each individual is counted in one and only one category.

■ Expected Frequencies

The general goal of the chi-square test for goodness of fit is to compare the data (the observed frequencies) with the null hypothesis. The problem is to determine how well the data fit the distribution specified in H_0—hence the name *goodness of fit*.

The first step in the chi-square test is to construct a hypothetical sample that represents how the sample distribution would look if it were in perfect agreement with the proportions stated in the null hypothesis. Suppose, for example, the null hypothesis states that the population is distributed in three categories with the following proportions:

	Category A	Category B	Category C
H_0:	25%	50%	25%

(The population is distributed across the three categories with 25% in Category A, 50% in Category B, and 25% in Category C.)

If this hypothesis is correct, how would you expect a random sample of $n = 40$ individuals to be distributed among the three categories? It should be clear that your best strategy is to predict that 25% of the sample would be in Category A, 50% would be in Category B, and 25% would be in Category C. To find the exact frequency expected for each category, multiply the sample size (n) by the proportion (or percentage) from the null hypothesis. For this example, you would expect:

25% of 40 = 0.25(40) = 10 individuals in Category A

50% of 40 = 0.50(40) = 20 individuals in Category B

25% of 40 = 0.25(40) = 10 individuals in Category C

The frequency values predicted from the null hypothesis are called *expected frequencies*. The symbol for expected frequency is f_e, and the expected frequency for each category is computed by

$$\text{expected frequency} = f_e = pn \tag{15.1}$$

where p is the proportion stated in the null hypothesis and n is the sample size.

> The **expected frequency** for each category is the frequency value that is predicted from the proportions in the null hypothesis and the sample size (n). The expected frequencies define an ideal, *hypothetical* sample distribution that would be obtained if the sample proportions were in perfect agreement with the proportions specified in the null hypothesis.

Note that the no-preference null hypothesis will always produce equal expected frequencies (f_e values) for all categories because the proportions (p) are the same for all categories. On the other hand, the no-difference null hypothesis typically will not produce equal values for the expected frequencies because the hypothesized proportions typically vary from one category to another. You also should note that the expected frequencies are calculated, hypothetical values, and the numbers that you obtain may be decimals or fractions. The observed frequencies, on the other hand, always represent real individuals and always are whole numbers.

■ The Chi-Square Statistic

The general purpose of any hypothesis test is to determine whether the sample data support or refute a hypothesis about the population. In the chi-square test for goodness of fit, the sample is expressed as a set of observed frequencies (f_o values), and the null hypothesis is used to generate a set of expected frequencies (f_e values). The *chi-square statistic* simply measures how well the data (f_o) fit the hypothesis (f_e). The symbol for the chi-square statistic is χ^2. The formula for the chi-square statistic is

$$\text{chi-square} = \chi^2 = \Sigma\frac{(f_o - f_e)^2}{f_e} \tag{15.2}$$

As the formula indicates, the value of chi-square is computed by the following steps:

1. Find the difference between f_o (the data) and f_e (the hypothesis) for each category.
2. Square the difference. This ensures that all values are positive.
3. Next, divide the squared difference by f_e.
4. Finally, sum the values from all the categories.

The first two steps determine the numerator of the chi-square statistic and should be easy to understand. Specifically, the numerator measures how much difference there is between the data (the f_o values) and the hypothesis (represented by the f_e values). The final step is also reasonable: we add the values to obtain the total discrepancy between the data and the hypothesis. Thus, a large value for chi-square indicates that the data do not fit the hypothesis, and leads us to reject the null hypothesis.

However, the third step, which determines the denominator of the chi-square statistic, is not so obvious. Why must we divide by f_e before we add the category values? The answer to this question is that the obtained discrepancy between f_o and f_e is viewed as *relatively* large or *relatively* small depending on the size of the expected frequency. This point is demonstrated in the following analogy.

Suppose you were going to throw a party and you *expected* 1,000 people to show up. However, at the party you counted the number of guests and *observed* that 1,040 actually showed up. Forty more guests than expected are no major problem when all along you were planning for 1,000. There will still probably be enough cola, popcorn, and potato chips for everyone. On the other hand, suppose you had a party and you expected 10 people to attend but instead 50 actually showed up. Forty more guests in this case spell big trouble. How "significant" the discrepancy is depends in part on what you were originally expecting. With very large expected frequencies, allowances are made for more error between f_o and f_e. This is accomplished in the chi-square formula by dividing the squared discrepancy for each category, $(f_o - f_e)^2$, by its expected frequency.

LEARNING CHECK   **LO1**   **1.** Which of the following is a characteristic of nonparametric tests?
    a. They require a numerical score for each individual.
    b. They require assumptions about the population distribution(s).
    c. They evaluate hypotheses about population means or variances.
    d. None of the above is a characteristic of a nonparametric test.

LO2 **2.** Which of the following accurately describes the observed frequencies for a chi-square test for goodness of fit?

   a. They are always positive whole numbers.

   b. They are always positive but can include fractions or decimals.

   c. They can be positive or negative but are always whole numbers.

   d. They can be positive or negative and can include fractions or decimals.

LO3 **3.** A researcher uses a sample of $n = 90$ participants to test whether people have any preferences among three kinds of apples. Each person tastes all three types and then picks a favorite. What are the expected frequencies for the chi-square test for goodness of fit?

   a. $\frac{1}{3}, \frac{1}{3}, \frac{1}{3}$

   b. 10, 10, 10

   c. 30, 30, 30

   d. 60, 60, 60

ANSWERS **1.** d **2.** a **3.** c

15-2 An Example of the Chi-Square Test for Goodness of Fit

LEARNING OBJECTIVES

4. Define the degrees of freedom for the chi-square test for goodness of fit and locate the critical value for a specific alpha level in the chi-square distribution.

5. Conduct a chi-square test for goodness of fit and report the results as they would appear in the scientific literature.

■ The Chi-Square Distribution and Degrees of Freedom

It should be clear from the chi-square formula that the numerical value of chi-square is a measure of the discrepancy between the observed frequencies (data) and the expected frequencies (H_0). As usual, the sample data are not expected to provide a perfectly accurate representation of the population. In this case, the proportions or observed frequencies in the sample are not expected to be exactly equal to the proportions in the population. Thus, if there are small discrepancies between the f_o and f_e values, we obtain a small value for chi-square and we conclude that there is a good fit between the data and the hypothesis (fail to reject H_0). However, when there are large discrepancies between f_o and f_e, we obtain a large value for chi-square and conclude that the data do not fit the hypothesis (reject H_0). To decide whether a particular chi-square value is "large" or "small," we must refer to a *chi-square distribution*. This distribution is the set of chi-square values for all the possible random samples when H_0 is true. Much like other distributions we have examined (*t* distribution, *F* distribution), the chi-square distribution is a theoretical distribution with well-defined characteristics. Some of these characteristics are easy to infer from the chi-square formula.

1. The formula for chi-square involves adding squared values, so you can never obtain a negative value. Thus, all chi-square values are zero or larger.

2. When H_0 is true, you expect the data (f_o values) to be close to the hypothesis (f_e values). Thus, we expect chi-square values to be small when H_0 is true.

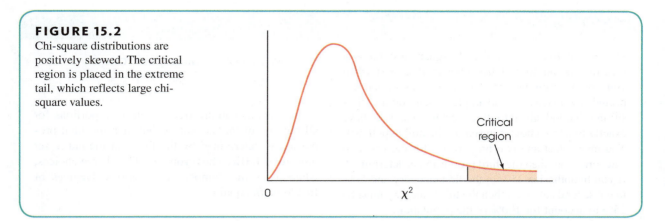

FIGURE 15.2
Chi-square distributions are positively skewed. The critical region is placed in the extreme tail, which reflects large chi-square values.

These two factors suggest that the typical chi-square distribution will be positively skewed (Figure 15.2). Note that small values near zero are expected when H_0 is true and large values (in the right-hand tail) are very unlikely. Thus, unusually large values of chi-square form the critical region for the hypothesis test.

Although the typical chi-square distribution is positively skewed, there is one other factor that plays a role in the exact shape of the chi-square distribution—the number of categories. Recall that the chi-square formula requires that you add values from every category. The more categories you have, the more likely it is that you will obtain a large sum for the chi-square value. On average, chi-square will be larger when you are adding values from 10 categories than when you are adding values from only three categories. As a result, there is a whole family of chi-square distributions, with the exact shape of each distribution determined by the number of categories used in the study. Technically, each specific chi-square distribution is identified by degrees of freedom (*df*) rather than the number of categories. For the goodness-of-fit test, the degrees of freedom are determined by

Caution: The *df* for a chi-square test is *not* related to sample size (*n*), as it is in most other tests.

$$df = C - 1 \tag{15.3}$$

where C is the number of categories. A brief discussion of this *df* formula is presented in Box 15.1. Figure 15.3 shows the general relationship between *df* and the shape of the

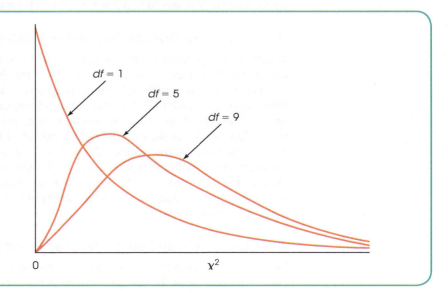

FIGURE 15.3
The shape of the chi-square distribution for different values of *df*. As the number of categories increases, the peak (mode) of the distribution has a larger chi-square value.

BOX 15.1 A Closer Look at Degrees of Freedom

Degrees of freedom for the chi-square test literally measure the number of free choices that exist when you are determining the null hypothesis or the expected frequencies. For example, when you are classifying individuals into three categories, you have exactly two free choices in stating the null hypothesis. You may select any two proportions for the first two categories, but then the third proportion is determined. If you hypothesize 25% in the first category and 50% in the second category, then the third category must be 25% to account for 100% of the population.

Category A	Category B	Category C
25%	50%	?

In general, you are free to select proportions for all but one of the categories, but then the final proportion is determined by the fact that the entire set must total 100%. Thus, you have $C - 1$ free choices, where C is the number of categories: Degrees of freedom, df, equal $C - 1$.

chi-square distribution. Note that the chi-square values tend to get larger (shift to the right) as the number of categories and the degrees of freedom increase.

The following example is an opportunity to test your understanding of the expected frequencies and the df value for the chi-square test for goodness of fit.

EXAMPLE 15.1

A researcher has developed three different designs for a computer keyboard. A sample of $n = 60$ participants is obtained, and each individual tests all three keyboards and identifies his or her favorite. The frequency distribution of preferences is as follows:

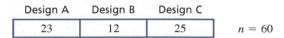

Design A	Design B	Design C	
23	12	25	$n = 60$

The values for each category are the observed frequencies, f_o, and note that the $\Sigma f_o = n$. Assume that the null hypothesis states that there are no preferences among the three designs. Find the expected frequencies for the chi-square test and determine the df value for the chi-square statistic. You should find that $f_e = 20$ for all three designs and $df = 2$. Good luck. ■

■ Locating the Critical Region for a Chi-Square Test

Recall that a large value for the chi-square statistic indicates a big discrepancy between the data and the hypothesis, and suggests that we reject H_0. To determine whether a particular chi-square value is significantly large, you must consult the table titled The Chi-Square Distribution (Appendix B). A portion of the chi-square table is shown in Table 15.1. The first column lists df values for the chi-square test, and the top row of the table lists proportions (alpha levels) in the extreme right-hand tail of the distribution. The numbers in the body of the table are the critical values of chi-square. The table shows, for example, that when the null hypothesis is true and $df = 3$, only 5% (.05) of the chi-square values are greater than 7.81, and only 1% (.01) are greater than 11.34. Thus, with $df = 3$, any chi-square value greater than 7.81 has a probability of $p < .05$, and any value greater than 11.34 has a probability of $p < .01$.

■ A Complete Chi-Square Test for Goodness of Fit

We use the same step-by-step process for testing hypotheses with chi-square as we used for other hypothesis tests. In general, the steps consist of stating the hypotheses, locating the

TABLE 15.1
A portion of the table of critical values for the chi-square distribution.

df	Proportion in Critical Region				
	0.10	0.05	0.025	0.01	0.005
1	2.71	3.84	5.02	6.63	7.88
2	4.61	5.99	7.38	9.21	10.60
3	6.25	7.81	9.35	11.34	12.84
4	7.78	9.49	11.14	13.28	14.86
5	9.24	11.07	12.83	15.09	16.75
6	10.64	12.59	14.45	16.81	18.55
7	12.02	14.07	16.01	18.48	20.28
8	13.36	15.51	17.53	20.09	21.96
9	14.68	16.92	19.02	21.67	23.59

critical region, computing the test statistic, and making a decision about H_0. The following example demonstrates the complete process of hypothesis testing with the goodness-of-fit test.

EXAMPLE 15.2   Humans tend to associate some colors, especially red and yellow, with increased hunger (Singh, 2006). Many fast food restaurants use this relationship when designing the signs and décor of their restaurants. To examine this phenomenon, a psychologist presents participants with a series of words describing moods/emotions (calm, happy, hungry, sleepy, anxious, and so on) and asks each person to choose the color that they associate with each. Each participant is given four color choices: red, yellow, green, and blue. The following data indicate how many people identified each color as associated with hunger.

Red	Yellow	Green	Blue
19	16	10	5

The question for the hypothesis test is whether there are any preferences among the four color choices. Are any of the colors associated with hunger more (or less) often than would be expected simply by chance?

STEP 1   **State the hypotheses and select an alpha level.**   The hypotheses can be stated as follows:

> H_0: In the general population, no specific color is associated with hunger more than any other. Thus, the four colors are selected equally often, and the population distribution has the following proportions:

Red	Yellow	Green	Blue
25%	25%	25%	25%

> H_1: In the general population, one or more of the colors is more likely to be associated with hunger than the others.

> We will use $\alpha = .05$.

STEP 2   **Locate the critical region.**   For this example, the value for degrees of freedom is

$$df = C - 1 = 4 - 1 = 3$$

For $df = 3$ and $\alpha = .05$, the table of critical values for chi-square indicates that the critical χ^2 has a value of 7.81. The critical region is sketched in Figure 15.4.

FIGURE 15.4

For Example 15.2, the critical region begins at a chi-square value of 7.81.

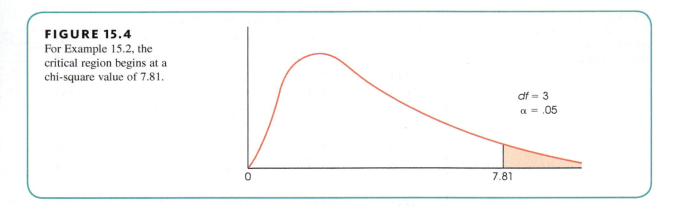

$df = 3$
$\alpha = .05$

0   7.81

STEP 3   Calculate the chi-square statistic. The calculation of chi-square is actually a two-stage process. First, you must compute the expected frequencies from H_0 and then calculate the value of the chi-square statistic. For this example, the null hypothesis specifies that one-quarter of the population ($p = 25\%$) will be in each of the four categories. According to this hypothesis, we should expect one-quarter of the sample to be in each category. With a sample of $n = 50$ individuals, the expected frequency for each category is

Expected frequencies are computed and may be decimal values. Observed frequencies are always whole numbers.

$$f_e = pn = 25\% \ of \ 50 = \frac{1}{4}(50) = 0.25(50) = 12.5$$

The observed frequencies and the expected frequencies are presented in Table 15.2. Using these values, the chi-square statistic may now be calculated.

$$\chi^2 = \Sigma\frac{(f_o - f_e)^2}{f_e}$$

$$= \frac{(19 - 12.5)^2}{12.5} + \frac{(16 - 12.5)^2}{12.5} + \frac{(10 - 12.5)^2}{12.5} + \frac{(5 - 12.5)^2}{12.5}$$

$$= \frac{42.25}{12.5} + \frac{12.25}{12.5} + \frac{6.25}{12.5} + \frac{56.25}{12.5}$$

$$= 3.38 + 0.98 + 0.50 + 4.50$$

$$= 9.36$$

STEP 4   State a decision and a conclusion. The obtained chi-square value is in the critical region. Therefore, H_0 is rejected, and the researcher may conclude that the four colors are not equally likely to be associated with hunger. Instead, there are significant differences among the four colors, with some selected more often and others less often than would be expected by chance. Looking at the data, it is clear that red and yellow are associated with hunger more than expected but green and blue are associated less than expected.

TABLE 15.2

The observed frequencies and the expected frequencies for the chi-square test in Example 15.2.

	Red	Yellow	Green	Blue
Observed Frequencies	19	16	10	5

	Red	Yellow	Green	Blue
Expected Frequencies	12.5	12.5	12.5	12.5

IN THE LITERATURE

Reporting the Results for Chi-Square

APA style specifies the format for reporting the chi-square statistic in scientific journals. For the results of Example 15.2, the report might state:

> The data showed that some of the four colors were significantly more likely to be associated with hunger than the others, $\chi^2(3, n = 50) = 9.36, p < .05$.

Note that the form of the report is similar to that of other statistical tests we have examined. Degrees of freedom are indicated in parentheses following the chi-square symbol. Also contained in the parentheses is the sample size (n). This additional information is important because the degrees of freedom value is based on the number of categories (C), not sample size. Next, the calculated value of chi-square is presented, followed by the probability that a Type I error has been committed. Because we obtained an extreme, very unlikely value for the chi-square statistic, the probability is reported as *less than* the alpha level. Additionally, the report may provide the observed frequencies (f_o) for each category. This information may be presented in a simple sentence or in a table. ∎

■ Goodness of Fit and the Single-Sample *t* Test

We began this chapter with a general discussion of the difference between parametric tests and nonparametric tests. In this context, the chi-square test for goodness of fit is an example of a nonparametric test; that is, it makes no assumptions about the parameters of the population distribution, and it does not require data from an interval or ratio scale. In contrast, the single-sample *t* test introduced in Chapter 9 is an example of a parametric test: It assumes a normal population, it tests hypotheses about the population mean (a parameter), and it requires numerical scores that can be added, squared, divided, and so on.

Although the chi-square test and the single-sample *t* are clearly distinct, they are also very similar. In particular, both tests are intended to use the data from a single sample to test hypotheses about a single population.

The primary factor that determines whether you should use the chi-square test or the *t* test is the type of measurement that is obtained for each participant. If the sample data consist of numerical scores (from an interval or ratio scale), it is appropriate to compute a sample mean and use a *t* test to evaluate a hypothesis about the population mean. For example, a researcher could measure the IQ for each individual in a sample of registered voters. A *t* test could then be used to evaluate a hypothesis about the mean IQ for the entire population of registered voters. On the other hand, if the individuals in the sample are classified into non-numerical categories (on a nominal or ordinal scale), you would use a chi-square test to evaluate a hypothesis about the population proportions. For example, a researcher could classify people according to gender by simply counting the number of males and females in a sample of registered voters. A chi-square test would then be appropriate to evaluate a hypothesis about the population proportions.

LEARNING CHECK   **LO4** **1.** A researcher is conducting a chi-square test for goodness of fit to evaluate preferences among different designs for a new automobile. With a sample of $n = 30$ the researcher obtains a chi-square statistic of $\chi^2 = 6.81$. What is the appropriate statistical decision for this outcome?

   a. Reject the null hypothesis with $\alpha = .05$, but not with $\alpha = .01$.

   b. Reject the null hypothesis with either $\alpha = .05$ or $\alpha = .01$.

    c. Fail to reject the null hypothesis with either $\alpha = .05$ or $\alpha = .01$.

    d. There is not enough information to determine the appropriate decision.

LO4 **2.** Which of the following is the correct equation to compute df for the chi-square test for goodness of fit?

    a. $n - 1$

    b. $C - 1$

    c. $n - C$ (where C is the number of categories)

    d. None of the above.

LO5 **3.** A researcher uses a sample of 20 college sophomores to determine whether they have any preference between two smartphones. Each student uses each phone for one day and then selects a favorite. If 14 students select the first phone and only 6 choose the second, then what is the value for χ^2?

    a. 0.80

    b. 1.60

    c. 3.20

    d. 11.0

ANSWERS **1. d** **2. b** **3. c**

15-3 The Chi-Square Test for Independence

LEARNING OBJECTIVES

6. Define the degrees of freedom for the chi-square test for independence and locate the critical value for a specific alpha level in the chi-square distribution.

7. Describe the hypotheses for a chi-square test for independence and explain how the expected frequencies are obtained.

8. Conduct a chi-square test for independence and report the results as they would appear in the scientific literature.

The chi-square statistic may also be used to test whether there is a relationship between two variables. In this situation, each individual in the sample is measured or classified on two separate variables. For example, a group of students could be classified in terms of personality (introvert, extrovert) and in terms of color preference (red, yellow, green, or blue). Usually, the data from this classification are presented in the form of a matrix, where the rows correspond to the categories of one variable and the columns correspond to the categories of the second variable. Table 15.3 presents hypothetical data for a sample of $n = 200$ students who have been classified by personality and color preference. The number in each box, or cell, of the matrix indicates the frequency, or number of individuals in that particular group. In Table 15.3, for example, there are 10 students who were classified as introverted and who selected red as their preferred color. To obtain these data, the researcher first selects a random sample of $n = 200$ students. Each student is then given a personality test and is asked to select a preferred color from among the four choices. Note that the classification is based on the measurements for each student; the researcher does not assign students to categories. Also, note that the data consist of frequencies, not scores, from a sample. The goal is to use the frequencies from the sample to test a hypothesis about the population frequency distribution. Specifically, are these data sufficient to

TABLE 15.3

Color preferences according to personality types.

	Red	Yellow	Green	Blue	
Introvert	10	3	15	22	50
Extrovert	90	17	25	18	150
	100	20	40	40	$n = 200$

conclude that there is a significant relationship between personality and color preference in the population of students?

You should realize that the color preference study shown in Table 15.3 is an example of nonexperimental research (pages 25–27). The researcher did not manipulate any variable and the participants were not randomly assigned to groups or treatment conditions. However, similar data are often obtained from true experiments. The following example is a demonstration of frequency data from an experimental study.

EXAMPLE 15.3   Guéguen, Jacob, and Lamy (2010) demonstrated that romantic background music increases the likelihood that a heterosexual woman will give her phone number to a new male acquaintance. The participants were recruited to take part in research on product evaluation. Each woman was taken to a waiting room with background music playing. For some of the women, the music was a popular love song and for the others, the music was a neutral song. After three minutes the participant was moved to another room in which a man was already waiting. The men were posing as participants, but were working with the experimenter. The participant and the confederate were instructed to eat two cookies, one organic and one without organic ingredients, and then talk about the differences between the two for a few minutes. After five minutes, the experimenter returned to end the study and asked the pair to wait alone for a few minutes. During this time, the man used a scripted line to ask the woman for her phone number. Data similar to the results obtained in the study are shown in Table 15.4. Note that the researchers manipulated the type of background music (independent variable) and recorded the number of yes and no responses (dependent variable) for each type of music. As with the color preference data, the researchers would like to use the frequencies from the sample to test a hypothesis about the corresponding frequency distribution in the population. In this case, the researchers would like to know whether the sample data provide enough evidence to conclude that there is a significant relationship between the type of music and a woman's response to a request for her phone number. ■

The procedure for using sample frequencies to evaluate hypotheses concerning relationships between variables involves another test using the chi-square statistic. In this situation, however, the test is called the *chi-square test for independence*.

> The **chi-square test for independence** uses the frequency data from a sample to evaluate the relationship between two variables in the population. Each individual in the sample is classified on both of the two variables, creating a two-dimensional frequency distribution matrix. The frequency distribution for the sample is then used to test hypotheses about the corresponding frequency distribution in the population.

TABLE 15.4

A frequency distribution table showing the number of participants who answered either yes or no when asked for their phone numbers. One group listened to romantic music while in a waiting room and the second group listened to neutral music.

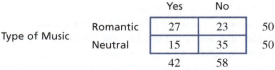

		Gave Phone Number?		
		Yes	No	
Type of Music	Romantic	27	23	50
	Neutral	15	35	50
		42	58	

■ The Null Hypothesis for the Test for Independence

The null hypothesis for the chi-square test for independence states that the two variables being measured are *independent*; that is, for each individual, the value obtained for one variable is not related to (or influenced by) the value for the second variable. This general hypothesis can be expressed in two different conceptual forms, each viewing the data and the test from slightly different perspectives. The study in Example 15.3 examining background music and the likelihood of giving a phone number is used to present both versions of the null hypothesis.

> Two variables are **independent** when there is no consistent, predictable relationship between them. In this case, the frequency distribution for one variable is not related to (or dependent on) the categories of the second variable. As a result, when two variables are independent, the frequency distribution for one variable will have the same shape (same proportions) for all categories of the second variable.

H_0 **Version 1** For this version of H_0, the data are viewed as a single sample with each individual measured on two variables. The goal of the chi-square test is to evaluate the relationship between the two variables. For the example we are considering, the goal is to determine whether there is a consistent, predictable relationship between the type of music and whether a woman gives her phone number. That is, if I know the type of background music, will it help me to predict whether she will give her number? The null hypothesis states that there is no relationship. The alternative hypothesis, H_1, states that there is a relationship between the two variables.

H_0: For the general population of students, there is no relationship between the type of background music and whether a woman will give her phone number.

This version of H_0 demonstrates the similarity between the chi-square test for independence and a correlation. In each case, the data consist of two measurements (X and Y) for each individual, and the goal is to evaluate the relationship between the two variables. The correlation, however, requires numerical scores for X and Y. The chi-square test, on the other hand, simply uses frequencies for individuals classified into categories.

H_0 **Version 2** For this version of H_0, the data are viewed as two (or more) separate samples representing two (or more) populations or treatment conditions. The goal of the chi-square test is to determine whether there are significant differences between the populations. For the example we are considering, the data in Table 15.4 would be viewed as a sample of $n = 50$ women who hear romantic music (top row) and a separate sample of $n = 50$ women who hear neutral music (bottom row). The chi-square test will determine whether the proportion of women giving phone numbers with romantic music is significantly different from the proportion with neutral music. From this perspective, the null hypothesis is stated as follows:

H_0: In the population of female undergraduates, the proportions of yes and no responses with romantic music are not different from the proportions with neutral music. The two distributions have the same shape (same proportions).

This version of H_0 demonstrates the similarity between the chi-square test and an independent-measures t test (or ANOVA). In each case, the data consist of two (or more) separate samples that are being used to test for differences between two (or more) populations. The t test (or ANOVA) requires numerical scores to compute means and mean differences. However, the chi-square test simply uses frequencies for individuals classified into categories. The null hypothesis for the chi-square test states that the populations have the same proportions (same shape). The alternative hypothesis, H_1, simply states that the populations have different proportions. For the example we are considering, H_1 states that the proportions of yes and no responses with romantic music are different from the proportions with neutral music.

Equivalence of H_0 Version 1 and H_0 Version 2   Although we have presented two different statements of the null hypothesis, these two versions are equivalent. The first version of H_0 states that the likelihood that a woman will give her phone number to a man she has just met is not related to the type of background music. If this hypothesis is correct, then the distribution of yes and no responses should not depend on the type of music. In other words, the distribution of yes and no responses should have the same proportions for romantic and neutral music, which is the second version of H_0.

For example, if we found that 60% of the women said no with neutral music, then H_0 would predict that we also should find that 60% say no with romantic music. In this case, knowing the type of background music does not help you predict whether she will say yes or no. Note that finding the *same proportions* indicates *no relationship*.

On the other hand, if the proportions were different, it would suggest that there is a relationship. For example, if 60% of the women say no with neutral music but only 30% say no with romantic music, then there is a clear, predictable relationship between the type of music and the woman's response. (If I know the type of music, I can predict the woman's response.) Thus, finding *different proportions* means that there is *a relationship* between the two variables.

Thus, stating that there is no relationship between two variables (version 1 of H_0) is equivalent to stating that the distributions have equal proportions (version 2 of H_0).

■ Observed and Expected Frequencies

The chi-square test for independence uses the same basic logic that was used for the goodness-of-fit test. First, a sample is selected, and each individual is classified or categorized. Because the test for independence considers two variables, every individual is classified on both variables, and the resulting frequency distribution is presented as a two-dimensional matrix (see Table 15.4). As before, the frequencies in the sample distribution are called *observed frequencies* and are identified by the symbol f_o.

The next step is to find the expected frequencies, or f_e values, for this chi-square test. As before, the *expected frequencies* define an ideal hypothetical distribution that is in perfect agreement with the null hypothesis. Once the expected frequencies are obtained, we compute a chi-square statistic to determine how well the data (observed frequencies) fit the null hypothesis (expected frequencies).

Although you can use either version of the null hypothesis to find the expected frequencies, the logic of the process is much easier when you use H_0 stated in terms of equal proportions. For the example we are considering, the null hypothesis states:

H_0: The frequency distribution of yes and no responses has the same shape (same proportions) for both categories of background music.

To find the expected frequencies, we first determine the overall distribution of yes and no responses and then apply this distribution to both categories of music. Table 15.5 shows an empty matrix corresponding to the data from Table 15.4. Notice that the empty matrix includes all of the row totals and column totals from the original sample data. The row totals and column totals are essential for computing the expected frequencies.

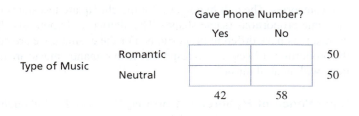

		Gave Phone Number?	
		Yes	No
Type of Music	Romantic			50
	Neutral			50
		42	58

The column totals for the matrix describe the overall distribution of yes/no responses. For these data, 42 women said yes. Because the total sample consists of 100 women, the proportion saying yes is 42 out of 100, or 42%. Similarly, 58 out of 100, or 58%, said no.

The row totals in the matrix define the two types of music. For example, the matrix in Table 15.5 shows a total of 50 women who heard romantic music (the top row) and a sample of 50 women who heard neutral music (the bottom row). According to the null hypothesis, both groups should have the same proportions of yes and no responses. To find the expected frequencies, we simply apply the overall distribution of yes and no responses to each group. Beginning with the sample of 50 women in the top row, we obtain expected frequencies of

42% say yes: $f_e = 42\%$ of $50 = 0.42(50) = 21$

58% say no: $f_e = 58\%$ of $50 = 0.58(50) = 29$

Using exactly the same proportions for the sample of $n = 50$ women who heard neutral music in the bottom row, we obtain expected frequencies of

42% say yes: $f_e = 42\%$ of $50 = 0.42(50) = 21$

58% say no: $f_e = 58\%$ of $50 = 0.58(50) = 29$

The complete set of expected frequencies is shown in Table 15.6. Notice that the row totals and the column totals for the expected frequencies are the same as those for the original data (the observed frequencies) in Table 15.4.

A Simple Formula for Determining Expected Frequencies Although expected frequencies are derived directly from the null hypothesis and the sample characteristics, it is not necessary to go through extensive calculations to find f_e values. In fact, there is a simple formula that determines f_e for any cell in the frequency distribution matrix:

$$f_e = \frac{f_c f_r}{n} \tag{15.4}$$

where f_c is the frequency total for the column (column total), f_r is the frequency total for the row (row total), and n is the number of individuals in the entire sample. To demonstrate this formula, we compute the expected frequency for romantic music and no phone number in Table 15.6. First, note that this cell is located in the top row and second column in the table.

		Gave Phone Number?	
		Yes	No
Type of Music	Romantic	21	29	50
	Neutral	21	29	50
		42	58

The column total is $f_c = 58$, the row total is $f_r = 50$, and the sample size is $n = 100$. Using these values in Formula 15.4, we obtain

$$f_e = \frac{58(50)}{100} = \frac{2900}{100} = 29$$

This is identical to the expected frequency we obtained using percentages from the overall distribution.

■ The Chi-Square Statistic and Degrees of Freedom

The chi-square test of independence uses exactly the same chi-square formula as the test for goodness of fit:

$$\chi^2 = \Sigma \frac{(f_o - f_e)^2}{f_e}$$

For the observed frequencies in Table 15.4 and the expected frequencies in Table 15.6, we obtain

$$\chi^2 = \frac{(27 - 21)^2}{21} + \frac{(23 - 29)^2}{29} + \frac{(15 - 21)^2}{21} + \frac{(35 - 29)^2}{29}$$
$$= 1.714 + 1.241 + 1.714 + 1.241$$
$$= 5.91$$

As before, the formula measures the discrepancy between the data (f_o values) and the hypothesis (f_e values). A large discrepancy produces a large value for chi-square and indicates that H_0 should be rejected. To determine whether a particular chi-square statistic is significantly large, you must first determine degrees of freedom (df) for the statistic and then consult the chi-square distribution in the appendix. For the chi-square test of independence, degrees of freedom are based on the number of cells for which you can freely choose expected frequencies. Recall that the f_e values are partially determined by the sample size (n) and by the row totals and column totals from the original data. These various totals restrict your freedom in selecting expected frequencies. This point is illustrated in Table 15.7. After one of the f_e values has been determined, all the other f_e values in the table are also determined. In general, the row totals and the column totals restrict the final choices in each row and column. As a result, we may freely choose all but one f_e in each row and all but one f_e in each column. If R is the number of rows and C is the number of columns, and you remove the last column and the bottom row from the matrix, you are left with a smaller matrix that has $C - 1$ columns and $R - 1$ rows. The number of cells in the smaller matrix determines the df value. Thus, the total number of f_e values that you can freely choose is $(R - 1)(C - 1)$, and the degrees of freedom for the chi-square test of independence are given by the formula

$$df = (R - 1)(C - 1) \tag{15.5}$$

Also note that once you calculate the expected frequencies to fill the smaller matrix, the rest of the f_e values can be found by subtraction.

TABLE 15.7

Degrees of freedom and expected frequencies. (After one value has been determined, all the remaining expected frequencies are determined by the row totals and the column totals. This example has only one free choice, so $df = 1$.)

		Gave Phone Number?		
		Yes	No	
Type of Music	Romantic	21	?	50
	Neutral	?	?	50
		42	58	

The following example is an opportunity to test your understanding of the expected frequencies and the *df* value for the chi-square test for independence.

EXAMPLE 15.4   A researcher would like to know which factors are most important to people who are buying a new car. Each individual in a sample of $n = 200$ customers is asked to identify the most important factor in the decision process: Performance, Reliability, or Style. The researcher would like to know whether there is a difference between the factors identified by younger adults (age 35 or younger) compared to those identified by older adults (age greater than 35). The data are as follows:

Observed Frequencies of Most Important Factor, According to Age

	Performance	Reliability	Style	Totals
Younger	21	33	26	80
Older	19	67	34	120
Totals	40	100	60	

Compute the expected frequencies and determine the value for *df* for the chi-square test. You should find expected frequencies of 16, 40, and 24 for the younger adults; 24, 60, and 36 for the older adults; and *df* = 2. Good luck. ∎

■ A Summary of the Chi-Square Test for Independence

At this point we have presented essentially all of the elements of a chi-square test for independence. Using the romantic music study in Example 15.3 (page 547), the test is summarized as follows.

STEP 1   **State the hypotheses and select an alpha level.**   For this example, the null hypothesis states that there is no relationship between the type of background music and the likelihood that a woman will give her phone number to a man she has just met; the two variables are independent. The alternative hypothesis states that there is a relationship between the two variables, or that the likelihood of giving a phone number depends on the type of background music.

STEP 2   **Locate the critical region.**   The chi-square test has degrees of freedom given by

$$df = (R - 1)(C - 1) = 1(1) = 1$$

With $df = 1$ and $\alpha = .05$, the critical value is $\chi^2 = 3.84$.

STEP 3   **Compute the test statistic.**   Earlier, we computed $\chi^2 = 5.91$ for the data from Example 15.3 using the observed frequencies in Table 15.4 and the expected frequencies in Table 15.6.

STEP 4   **Make a decision.**   The chi-square value that we obtained is beyond the critical boundary, so we reject H_0 and conclude that there is a statistically significant relationship between the likelihood that a woman will give her phone number to a man she has just met and the type of background music. Looking at the data, it is clear that the proportion of women who give their numbers after listening to romantic music is higher than the proportion of women who have been listening to neutral music.

LEARNING CHECK   **LO6** **1.** If a chi-square test for independence has $df = 2$, then how many cells are in the matrix of observed frequencies?

  a. 4

  b. 5

  c. 6

  d. 8

LO7 **2.** Which of the following can be evaluated with a chi-square test for independence?

  a. The relationship between two variables.

  b. Differences between two or more population frequency distributions.

  c. Either the relationship between two variables or the differences between distributions.

  d. Neither the relationship between two variables nor the differences between distributions.

LO8 **3.** A researcher classifies a group of people into three age groups and measures whether each person used Facebook during the previous week (yes/no). The researcher uses a chi-square test for independence to determine if there is a significant relationship between the two variables. If the researcher obtains $\chi^2 = 5.75$, then what is the correct decision for the test?

  a. Reject H_0 for $\alpha = .05$ but not for $\alpha = .01$.

  b. Reject H_0 for $\alpha = .01$ but not for $\alpha = .05$.

  c. Reject H_0 for either $\alpha = .05$ or $\alpha = .01$.

  d. Fail to reject H_0 for $\alpha = .05$ and $\alpha = .01$.

ANSWERS   **1. c**   **2. c**   **3. d**

15-4   Effect Size and Assumptions for the Chi-Square Tests

LEARNING OBJECTIVES

9. Compute Cohen's w to measure effect size for both chi-square tests.

10. Compute the phi-coefficient or Cramér's V to measure effect size for the chi-square test for independence.

11. Identify the basic assumptions and restrictions for chi-square tests.

■ Cohen's w

Hypothesis tests, like the chi-square test for goodness of fit or for independence, evaluate the statistical significance of the results from a research study. Specifically, the intent of the test is to determine whether it is likely that the patterns or relationships observed in the sample data could have occurred without any corresponding patterns or relationships in the population. Tests of significance are influenced not only by the size or strength of the treatment effects but also by the size of the samples. As a result, even a small effect can be statistically significant if it is observed in a very large sample. Because a significant effect does not necessarily mean a large effect, it is generally recommended that the outcome of a hypothesis test be accompanied by a measure of the effect size. This general recommendation also applies to the chi-square tests presented in this chapter.

Jacob Cohen (1992) introduced a statistic called w that provides a measure of effect size for either of the chi-square tests. The formula for *Cohen's w* is very similar to the chi-square formula but uses proportions instead of frequencies.

$$w = \sqrt{\sum \frac{(p_o - p_e)^2}{p_e}}$$
(15.6)

In the formula, the p_o values are the observed proportions in the data and are obtained by dividing each observed frequency by the total number of participants.

$$\text{observed proportion} = p_o = \frac{f_o}{n}$$

Similarly, the p_e values are the expected proportions that are specified in the null hypothesis. The formula instructs you to:

1. Compute the difference between the observed proportion and the expected proportion for each cell (category).

2. For each cell, square the difference and divide by the expected proportion.

3. Add the values from Step 2 and take the square root of the sum.

The following example demonstrates this process.

EXAMPLE 15.5   A researcher would like to determine whether students have any preferences among four pizza shops in town. A sample of $n = 40$ students is obtained and fresh pizza is ordered from each of the four shops. Each student tastes all four pizzas and then selects a favorite. The observed frequencies are as follows:

Shop A	Shop B	Shop C	Shop D	
6	12	8	14	40

The null hypothesis states that there are no preferences among the four shops, so the expected proportion is $p = 0.25$ for each. The observed proportions are $\frac{6}{40} = 0.15$ for Shop A, $\frac{12}{40} = 0.30$ for Shop B, $\frac{8}{40} = 0.20$ for Shop C, and $\frac{14}{40} = 0.35$ for Shop D. The calculations for w are summarized in the table below.

	P_o	P_e	$(P_o - P_e)$	$(P_o - P_e)^2$	$(P_o - P_e)^2/P_e$
Shop A	0.15	0.25	−0.10	0.01	0.04
Shop B	0.30	0.25	0.05	0.0025	0.01
Shop C	0.25	0.25	−0.05	0.0025	0.01
Shop D	0.35	0.25	0.10	0.01	0.04
					0.10

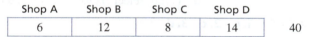

$$\sum \frac{(p_o - p_e)^2}{p_e} = 0.10 \text{ and } w = \sqrt{0.10} = 0.316$$

■

Cohen (1992) also suggested guidelines for interpreting the magnitude of w, with values near 0.10 indicating a small effect, 0.30 a medium effect, and 0.50 a large effect. By these standards, the value obtained in Example 15.5 is a medium effect.

The Role of Sample Size   You may have noticed that the formula for computing w does not contain any reference to the sample size. Instead, w is calculated using only the sample proportions and the proportions from the null hypothesis. As a result, the size of the sample has no influence on the magnitude of w. This is one of the basic characteristics of all measures of effect size. Specifically, the number of scores in the sample has little or no influence on effect size. On the other hand, sample size does have a large impact on the outcome of a hypothesis test. For example, the data in Example 15.5 produce $\chi^2 = 4.00$. With $df = 3$, the critical value for $\alpha = .05$ is 7.81 and we conclude that there are no significant preferences among the four pizza shops. However, if the number of individuals in each category is doubled, so that the observed frequencies become 12, 24, 16, and 28, then the new $\chi^2 = 8.00$. Now the statistic is in the critical region so we reject H_0 and conclude that there are significant preferences. Thus, increasing the size of the sample increases the likelihood of rejecting the null hypothesis. You should realize, however, that the proportions for the new sample are exactly the same as the proportions for the original sample, so the value of w does not change. For both sample sizes, $w = 0.316$.

Chi-Square and w   Although the chi-square statistic and effect size as measured by w are intended for different purposes and are affected by different factors, they are algebraically related. In particular, the portion of the formula for w that is under the square root can be obtained by dividing the formula for chi-square by n. Dividing by the sample size converts each of the frequencies (observed and expected) into a proportion, which produces the formula for w. As a result, you can determine the value of w directly from the chi-square value by the following equation:

$$w = \sqrt{\frac{\chi^2}{n}} \qquad (15.7)$$

For the data in Example 15.5, we obtained $\chi^2 = 4.00$ and $w = 0.316$. Substituting in the formula produces

$$w = \sqrt{\frac{\chi^2}{n}} = \sqrt{\frac{4.00}{40}} = \sqrt{0.10} = 0.316$$

Although Cohen's w statistic also can be used to measure effect size for the chi-square test for independence, two other measures have been developed specifically for this hypothesis test. These two measures, known as the phi-coefficient and Cramér's V, make allowances for the size of the data matrix and are considered to be superior to w, especially with very large data matrices.

■ The Phi-Coefficient and Cramér's V

In Chapter 14 (page 505), we introduced the *phi-coefficient* as a measure of correlation for data consisting of two dichotomous variables (both variables have exactly two values). This same situation exists when the data for a chi-square test for independence form a 2×2 matrix (again, each variable has exactly two values). In this case, it is possible to compute the correlation phi (ϕ) in addition to the chi-square hypothesis test for the same set of data. Because phi is a correlation, it measures the strength of the relationship, rather than the significance, and thus provides a measure of effect size. The value for the phi-coefficient can be computed directly from chi-square by the following formula:

Note that the value of χ^2 is already a squared value. Do not square it again.

$$\phi = \sqrt{\frac{\chi^2}{n}} \qquad (15.8)$$

Note that Cohen's w and ϕ (Equations 15.7 and 15.8) are the same for a 2×2 data matrix.

The value of the phi-coefficient is determined entirely by the *proportions* in the 2×2 data matrix and is completely independent of the absolute size of the frequencies. The chi-square value, however, is influenced by the proportions and by the size of the frequencies. This distinction is demonstrated in the following example.

EXAMPLE 15.6   The following data show a frequency distribution evaluating the relationship between self-assigned gender of male and female and preference between two candidates for student president.

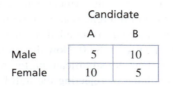

	Candidate A	Candidate B
Male	5	10
Female	10	5

Note that the data show that males prefer Candidate B by a 2-to-1 margin and females prefer Candidate A by 2 to 1. Also note that the sample includes a total of 15 males and 15 females. We will not perform all the arithmetic here, but these data produce a chi-square value equal to 3.33 (which is not significant) and a phi-coefficient of 0.333.

Next, we keep exactly the same proportions in the data, but double all of the frequencies. The resulting data are as follows:

	Candidate A	Candidate B
Male	10	20
Female	20	10

Once again, males prefer Candidate B by 2 to 1 and females prefer Candidate A by 2 to 1. However, the sample now contains 30 males and 30 females. For these new data, the value of chi-square is 6.67, twice as big as it was before (and now significant with $\alpha = .05$), but the value of the phi-coefficient is still 0.333.

Because the proportions are the same for the two samples, the value of the phi-coefficient is unchanged. However, the larger sample provides more convincing evidence than the smaller sample, so the larger sample is more likely to produce a significant result. ■

The interpretation of ϕ follows the same standards used to evaluate a correlation (Table 9.3, page 307, shows the standards for squared correlations): a correlation of 0.10 is a small effect, 0.30 is a medium effect, and 0.50 is a large effect. Occasionally, the value of ϕ is squared (ϕ^2) and is reported as a percentage of variance accounted for, exactly the same as r^2.

When the chi-square test involves a matrix larger than 2×2, a modification of the phi-coefficient, known as *Cramér's* V, can be used to measure effect size.

$$V = \sqrt{\frac{\chi^2}{n(df^*)}}$$ (15.9)

Note that the formula for Cramér's V (Equation 15.9) is identical to the formula for the phi-coefficient (Equation 15.8) except for the addition of df^* in the denominator. The df^* value is *not* the same as the degrees of freedom for the chi-square test, but it is related. Recall that the chi-square test for independence has $df = (R - 1)(C - 1)$, where R is the number of rows in the table and C is the number of columns. For Cramér's V, the value of df^* is the smaller of either $(R - 1)$ or $(C - 1)$.

Cohen (1988) has also suggested standards for interpreting Cramér's V that are shown in Table 15.8. Note that when $df^* = 1$, as in a 2×2 matrix, the criteria for interpreting V are exactly the same as the criteria for interpreting a regular correlation or a phi-coefficient.

In a research report, the measure of effect size appears immediately after the results of the hypothesis test. For the romantic music study in Example 15.3, for example, we

TABLE 15.8
Standards for interpreting Cramér's V as proposed by Cohen (1988).

	Small Effect	Medium Effect	Large Effect
For $df^* = 1$	0.10	0.30	0.50
For $df^* = 2$	0.07	0.21	0.35
For $df^* = 3$	0.06	0.17	0.29

obtained $\chi^2 = 5.91$ for a sample of $n = 100$ participants. Because the data form a 2×2 matrix, the phi-coefficient is the appropriate measure of effect size and the data produce

$$\phi = \sqrt{\frac{\chi^2}{n}} = \sqrt{\frac{5.91}{100}} = 0.243$$

For these data, the results from the hypothesis test and the measure of effect size would be reported as follows:

> The results showed a significant relationship between the type of background music and a woman's willingness to give her phone number, $\chi^2(1, n = 100) = 5.91, p < .05, \phi = 0.243$. Specifically, women who listened to romantic music were much more likely to give their phone numbers to men they had just met.

■ Assumptions and Restrictions for Chi-Square Tests

To use a chi-square test for goodness of fit or a test of independence, several conditions must be satisfied. For any statistical test, violation of assumptions and restrictions casts doubt on the results. For example, the probability of committing a Type I error may be distorted when assumptions of statistical tests are not satisfied. Some important assumptions and restrictions for using chi-square tests are the following:

1. **Independence of Observations**. This is *not* to be confused with the concept of independence between *variables*, as seen in the chi-square test for independence (Section 15.3). One consequence of independent observations is that each observed frequency is generated by a different individual. A chi-square test would be inappropriate if a person could produce responses that can be classified in more than one category or contribute more than one frequency count to a single category. (See page 264 for more information on independence.)

2. **Size of Expected Frequencies**. A chi-square test should not be performed when the expected frequency of any cell is less than 5. The chi-square statistic can be distorted when f_e is very small. Consider the chi-square computations for a single cell. Suppose that the cell has values of $f_e = 1$ and $f_o = 5$. Note the difference between the observed and expected frequencies is four. However, the total contribution of this cell to the total chi-square value is

$$\text{cell} = \frac{(f_o - f_e)^2}{f_e} = \frac{(5 - 1)^2}{1} = \frac{4^2}{1} = 16$$

Now consider another instance, in which $f_e = 10$ and $f_o = 14$. The difference between the observed and the expected frequencies is still 4, but the contribution of this cell to the total chi-square value differs from that of the first case:

$$\text{cell} = \frac{(f_o - f_e)^2}{f_e} = \frac{(14 - 10)^2}{10} = \frac{4^2}{10} = 1.6$$

It should be clear that a small f_e value can have a great influence on the chi-square value. This problem becomes serious when f_e values are less than 5. When f_e is very small, what would otherwise be a minor discrepancy between f_o and f_e results in large chi-square values. The test is too sensitive when f_e values are extremely small. One way to avoid small expected frequencies is to use large samples.

LEARNING CHECK   **LO9**   **1.** Which of the following is an appropriate measure of effect size for the chi-square test for goodness of fit?

    a. Cohen's w

    b. The phi-coefficient

    c. Cramér's V

    d. Either the phi-coefficient or Cramér's V

LO10   **2.** A researcher obtains $\chi^2 = 4.0$ for a test for independence using observed frequencies in a 3×3 matrix. If the sample contained a total of $n = 50$ people, then what is the value of Cramér's V?

    a. 0.04

    b. 0.16

    c. 0.20

    d. 0.40

LO11   **3.** Under what circumstances should the chi-square statistic not be used?

    a. When the expected frequency is greater than 5 for any cell.

    b. When the expected frequency is less than 5 for any cell.

    c. When the expected frequency equals the observed frequency for any cell.

    d. None of the above.

ANSWERS   **1. a   2. c   3. b**

SUMMARY

1. Chi-square tests are nonparametric techniques that test hypotheses about the form of the entire frequency distribution. Two types of chi-square tests are the test for goodness of fit and the test for independence. The data for these tests consist of the frequency or number of individuals who are located in each category.

2. The test for goodness of fit compares the frequency distribution for a sample to the population distribution that is predicted by H_0. The test determines how well the observed frequencies (sample data) fit the expected frequencies (data predicted by H_0).

3. The expected frequencies for the goodness-of-fit test are determined by

$$\text{expected frequency} = f_e = pn$$

where p is the hypothesized proportion (according to H_0) of observations falling into a category and n is the size of the sample.

4. The chi-square statistic is computed by

$$\text{chi-square} = \chi^2 = \Sigma \frac{(f_o - f_e)^2}{f_e}$$

where f_o is the observed frequency for a particular category and f_e is the expected frequency for that category. Large values for χ^2 indicate that there is a large discrepancy between the observed (f_o) and the expected (f_e) frequencies, which may warrant rejection of the null hypothesis.

5. Degrees of freedom for the test for goodness of fit are

$$df = C - 1$$

where C is the number of categories in the variable. Degrees of freedom measure the number of categories for which f_e values can be freely chosen. As can be seen from the formula, all but the last f_e value to be determined are free to vary.

6. The chi-square distribution is positively skewed and begins at the value of zero. Its exact shape is determined by degrees of freedom.

7. The chi-square test for independence is used to assess the relationship between two variables. The null hypothesis states that the two variables in question are independent of each other. That is, the frequency distribution for one variable does not depend on the categories of the second variable. On the other hand, if a relationship does exist, then the form of the distribution for one variable depends on the categories of the other variable.

8. For the test for independence, the expected frequencies for H_0 can be directly calculated from the marginal frequency totals,

$$f_e = \frac{f_c f_r}{n}$$

where f_c is the total column frequency and f_r is the total row frequency for the cell in question.

9. Degrees of freedom for the test for independence are computed by

$$df = (R - 1)(C - 1)$$

where R is the number of row categories and C is the number of column categories.

10. For the test of independence, a large chi-square value means there is a large discrepancy between the f_o and f_e values. Rejecting H_0 in this test provides support for a relationship between the two variables.

11. Both chi-square tests (for goodness of fit and independence) are based on the assumption that each observation is independent of the others. That is, each observed frequency reflects a different individual, and no individual can produce a response that would be classified in more than one category or more than one frequency in a single category.

12. The chi-square statistic is distorted when f_e values are small. Chi-square tests, therefore, should not be performed when the expected frequency of any cell is less than 5.

13. Cohen's w is a measure of effect size that can be used for both chi-square tests.

$$w = \sqrt{\Sigma \frac{(p_o - p_e)^2}{p_e}}$$

The effect size for a chi-square test for independence is measured by computing a phi-coefficient for data that form a 2×2 matrix or computing Cramér's V for a matrix that is larger than 2×2.

$$phi = \sqrt{\frac{\chi^2}{n}} \qquad Cramér's\ V = \sqrt{\frac{\chi^2}{n(df^*)}}$$

where df^* is the smaller of $(R - 1)$ and $(C - 1)$. Both phi and Cramér's V are evaluated using the criteria in Table 15.8.

KEY TERMS

parametric test (535)

nonparametric test (535)

chi-square test for goodness of fit (535)

observed frequency (538)

expected frequency (538)

chi-square statistic (539)

chi-square distribution (540)

chi-square test for independence (547)

independent (548)

Cohen's w (554)

phi-coefficient (555)

Cramér's V (556)

FOCUS ON PROBLEM SOLVING

1. The expected frequencies that you calculate must satisfy the constraints of the sample. For the goodness-of-fit test, $\Sigma f_e = \Sigma f_o = n$. For the test of independence, the row totals and column totals for the expected frequencies should be identical to the corresponding totals for the observed frequencies.

2. It is entirely possible to have fractional (decimal) values for expected frequencies. Observed frequencies, however, are always whole numbers.

DEMONSTRATION 15.2

EFFECT SIZE WITH CRAMÉR'S *V*

Because the data matrix is larger than 2×2, we will compute Cramér's V to measure effect size.

$$\text{Cramér's } V = \sqrt{\frac{\chi^2}{n(df^*)}} = \sqrt{\frac{38.09}{200(1)}} = \sqrt{0.19} = 0.436$$

SPSS®

General instructions for using SPSS are presented in Appendix D. Following are detailed instructions for using SPSS to perform **The Chi-Square Tests for Goodness of Fit and for Independence** that are presented in this chapter.

The Chi-Square Test for Independence

Remember that Pavlov's dogs learned to prepare for food by salivating in response to the sound of a bell. Classic research on substance dependence and drug overdose suggests that a similar effect occurs when addicts prepare to self-administer drugs. For example, Gutierrez et al. (1994) studied the circumstances surrounding overdoses in a sample of emergency room patients. Some of the patients in their study were admitted to the hospital because of a heroin overdose. Other patients in the study were admitted because of injuries or illnesses that were unrelated to substance abuse but were discovered to have, coincidentally, administered heroin shortly before admission to the hospital. Thus, all the patients in the study had recently administered heroin, but only some of the patients overdosed. The researchers observed that 100% of the non-overdose patients administered heroin in the usual place of administration—for example, in their own homes. In contrast, only 48% of overdose patients had administered heroin in the place where they typically administered heroin. A pattern of results like those observed by Gutierrez et al. are listed below.

	Place of Heroin Administration	
	Usual Environment	Unusual Environment
Overdose Patient	37	39
Non-Overdose Patient	22	1

Data Entry

1. In the **Variable View** create three new variables: one variable for the observed frequencies ("frequency"), one variable for the row ("patientType"), and one variable for the column ("place"). Enter descriptive labels in the **Label** fields for each of the variables. For the row variable, click the "..." in the **Values** field and assign value labels for each row in the analysis. Here, we assigned "Overdose patient" to the value of 1 and "Non-overdose patient" to the value of 2. Repeat this procedure for the column variable. When you have successfully entered your variables, your Variable View window should look like the figure below.

2. In the **Data View**, enter the complete set of observed frequencies in one column of the SPSS data editor ("frequency").

3. In the second column ("patientType"), enter the value that identifies the row corresponding to each observed frequency. For the current example, enter a 1 beside each observed frequency that came from the first row and a 2 beside each frequency that came from the second row.

4. In a third column ("place"), enter the value that identifies the column corresponding to each observed frequency. For this example, a 1 is entered for the value from the first column and a 2 is entered for the value from the second column. When you have successfully entered your data, the Data View should be as below.

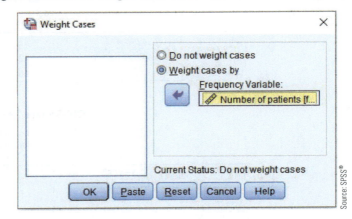

Source: SPSS®

Data Analysis

1. Click **Data** on the tool bar at the top of the page and select **weight cases** at the bottom of the list.

2. Click the **weight cases by** circle, then highlight the label for the column containing the observed frequencies ("Number of patients . . .") on the left and move it into the **Frequency Variable** box by clicking on the arrow. The **Weight Cases** window should look like the figure below.

Source: SPSS®

3. Click **OK**. An output window will appear informing you that the cases have been weighted. You can close the window.

4. Click **Analyze** on the tool bar at the top of the page, select **Descriptive Statistics**, and click on **Crosstabs**.

5. Highlight the label for the column containing the rows ("Patient Type . . .") and move it into the **Rows** box by clicking on the arrow.

6. Highlight the label for the column containing the columns ("Place of heroin self . . .") and move it into the **Columns** box by clicking on the arrow.

7. Click on **Statistics**, select **Chi-Square**, and click **Continue**.

8. Click **OK**.

SPSS Output

The output is shown in the figure below. The first table in the output simply lists the variables and is not shown in the figure. The **Crosstabulation** table simply shows the matrix of observed frequencies. The final table, labeled **Chi-Square Tests**, reports the results. Focus on the top row, the **Pearson Chi-Square**, which reports the calculated chi-square value, the degrees of freedom, and the level of significance (the *p* value or the alpha level for the test). For this example, the chi-square test was significant, $\chi^2(1) = 16.176, p < .001$. The note below the Chi-Square Tests table reports whether any expected frequencies are less than 5, which would violate the assumptions of the chi-square test.

→ **Crosstabs**

Case Processing Summary

	Cases					
	Valid		Missing		Total	
	N	Percent	N	Percent	N	Percent
Patient type * Place of heroin self-administration	99	100.0%	0	0.0%	99	100.0%

Patient type * Place of heroin self-administration Crosstabulation

Count

		Place of heroin self-administration		Total
		Usual environment	Unusual environment	
Patient type	Overdose patient	37	39	76
	Non-overdose patient	22	1	23
Total		59	40	99

Chi-Square Tests

	Value	df	Asymptotic Significance (2-sided)	Exact Sig. (2-sided)	Exact Sig. (1-sided)
Pearson Chi-Square	16.176[a]	1	.000		
Continuity Correction[b]	14.284	1	.000		
Likelihood Ratio	20.041	1	.000		
Fisher's Exact Test				.000	.000
N of Valid Cases	99				

a. 0 cells (0.0%) have expected count less than 5. The minimum expected count is 9.29.

b. Computed only for a 2x2 table

Try It Yourself

For the scores below, perform a chi-square test of independence. If you have done this correctly you should find that $\chi^2(1) = 59.490, p < .001$.

	Place of Heroin Administration	
	Usual Environment	Unusual Environment
Overdose Patient	32	34
Non-Overdose Patient	102	2

THE CHI-SQUARE TEST FOR GOODNESS OF FIT

Data Entry

1. Enter the set of observed frequencies in the first column of the SPSS data editor. If there are four categories, for example, enter the four observed frequencies.
2. In the second column, enter the numbers 1, 2, 3, and so on, so there is a number beside each of the observed frequencies in the first column.

Data Analysis

1. Click **Data** on the tool bar at the top of the page and select **weight cases** at the bottom of the list.
2. Click the **weight cases by** circle, then highlight the label for the column containing the observed frequencies (VAR00001) on the left and move it into the **Frequency Variable** box by clicking on the arrow.
3. Click **OK**.
4. Click **Analyze** on the tool bar, select **Nonparametric Tests**, select **Legacy Dialogs,** and click on **Chi-Square.**
5. Highlight the label for the column containing the digits 1, 2, and 3, and move it into the **Test Variables** box by clicking on the arrow.
6. To specify the expected frequencies, you can either use the **all categories equal** option, which automatically computes expected frequencies, or you can enter your own values. To enter your own expected frequencies, click on the **values** option, and one by one enter the expected frequencies into the small box and click **Add** to add each new value to the bottom of the list.
7. Click **OK**.

SPSS Output

The program produces a table showing the complete set of observed and expected frequencies. A second table provides the value for the chi-square statistic, the degrees of freedom, and the level of significance (the p value or alpha level for the test).

PROBLEMS

1. Parametric tests (such as t or ANOVA) differ from nonparametric tests (such as chi-square) primarily in terms of the assumptions they require and the data they use. Explain these differences.

2. The student population at the state college consists of 30% freshmen, 25% sophomores, 25% juniors, and 20% seniors. The college theater department recently staged a production of a modern musical. A researcher recorded the class status of each student entering the theater and found a total of 20 freshmen, 23 sophomores, 22 juniors, and 15 seniors. Is the distribution of class status for theatergoers significantly different from the distribution for the general college? Test at the .05 level of significance.

3. A developmental psychologist would like to determine whether infants display any color preferences. A stimulus consisting of four color patches (red, green, blue, and yellow) is projected onto the ceiling above a crib. Infants are placed in the crib, one at a time, and the psychologist records how much time each infant spends looking at each of the four colors. The color that receives the most attention during a 100-second test period is identified as the preferred color for that infant. The preferred colors for a sample of 80 infants are shown in the following table:

Red	Green	Blue	Yellow
30	13	23	14

a. Do the data indicate any significant preferences among the four colors? Test at the .05 level of significance.
b. Write a sentence demonstrating how the outcome of the hypothesis test would appear in a research report.

4. Data from the Department of Motor Vehicles indicate that 80% of all licensed drivers are older than age 25.
a. In a sample of $n = 50$ people who recently received speeding tickets, 33 were older than age 25 and the other 17 were age 25 or younger. Is the age distribution for this sample significantly different from the distribution for the population of licensed drivers? Use $\alpha = .05$.
b. In a sample of $n = 50$ people who recently received parking tickets, 36 were older than age 25 and the other 14 were age 25 or younger. Is the age distribution for this sample significantly different from the distribution for the population of licensed drivers? Use $\alpha = .05$.

5. A psychologist examining art appreciation selected an abstract painting that had no obvious top or bottom. Hangers were placed on the painting so that it could be hung with any one of the four sides at the top. The painting was shown to a sample of $n = 60$ participants, and each was asked to hang the painting in the orientation that looked correct. The following data indicate how many people chose each of the four sides to be placed at the top. Are any of the orientations selected more (or less) often than would be expected simply by chance? Test with $\alpha = .05$.

Top up (correct)	Bottom up	Left side up	Right side up
20	20	10	10

6. A professor in the psychology department would like to determine whether there has been a significant change in grading practices over the years. It is known that the overall grade distribution for the department in 1985 had 14% As, 26% Bs, 31% Cs, 19% Ds, and 10% Fs. A sample of $n = 200$ psychology students from last semester produced the following grade distribution:

A	B	C	D	F
30	62	64	32	12

Do the data indicate a significant change in the grade distribution? Test at the .05 level of significance.

7. Automobile insurance is much more expensive for teenage drivers than for older drivers. To justify this cost difference, insurance companies claim that the younger drivers are much more likely to be involved in costly accidents. To test this claim, a researcher obtains information about registered drivers from the department of motor vehicles and selects a sample of $n = 300$ accident reports from the police department.

The motor vehicle department reports the percentage of registered drivers in each age category as follows: 16% are younger than age 20; 28% are 20 to 29 years old; and 56% are age 30 or older. The number of accident reports for each age group is as follows:

Under Age 20	Age 20–29	Age 30 or Older
68	92	140

a. Do the data indicate that the distribution of accidents for the three age groups is significantly different from the distribution of drivers? Test with $\alpha = .05$.
b. Compute Cohen's w to measure the size of the effect.
c. Write a sentence demonstrating how the outcome of the hypothesis test and the measure of effect size would appear in a research report.

8. A communications company has developed three new designs for a smartphone. To evaluate consumer response, a sample of 240 college students is selected and each student is given all three phones to use for one week. At the end of the week, the students must identify which of the three designs they prefer. The distribution of preference is as follows:

Design 1	Design 2	Design 3
108	76	56

a. Do the results indicate any significant preferences among the three designs?
b. Compute Cohen's w to measure the size of the effect.

9. In Problem 8, a researcher asked college students to evaluate three new smartphone designs. However, the researcher suspects that college students may have criteria that are different from those used by older adults. To test this hypothesis, the researcher repeats the study using a sample of $n = 60$ older adults in addition to a sample of $n = 60$ students. The distribution of preference is as follows:

	Design 1	Design 2	Design 3	
Student	40	10	10	60
Older Adult	20	30	10	60
	60	40	20	

Do the data indicate that the distribution of preferences for older adults is significantly different from the distribution for college students? Test with $\alpha = .05$.

10. Earlier in the chapter, we introduced the chi-square test of independence with a study examining the relationship between personality and color preference. The following table shows the frequency distribution for a group of $n = 200$ students who were classified in terms of personality (introvert, extrovert) and in terms

of color preference (red, yellow, green, or blue). Do the data indicate a significant relationship between the two variables? Test with $\alpha = .05$.

	Red	Yellow	Green	Blue	
Introvert	10	3	15	22	50
Extrovert	90	17	25	18	150
	100	20	40	40	$n = 200$

11. Liu et al. (2015) recently reported the results of a study examining whether happy people live longer. The study followed a large sample of British women, aged 50 to 69 over a 10-year period. At the beginning of the study the women were asked several questions, including how often they felt happy. After 10 years, roughly 4% of the women had died. The following table shows a frequency distribution similar to the results obtained in the study.

	Lived	Died	
Happy Most of the Time	382	18	400
Unhappy Most of the Time	194	6	200
	576	24	

   a. Do the data indicate a significant relationship between living longer and being happy most of the time? Test with $\alpha = .05$.
   b. Compute the phi-coefficient to measure the size of the treatment effect.

12. Many businesses use some type of customer loyalty program to encourage repeat customers. A common example is the buy-ten-get-one-free punch card. Drèze and Nunes (2006) examined a simple variation of this program that appears to give customers a head start on completing their cards. One group of customers at a car wash was given a buy-eight-get-one-free card and a second group was given a buy-ten-get-one-free card that had already been punched twice. Although both groups needed eight punches to earn a free wash, the group with the two free punches appeared to be closer to reaching their goal. A few months later, the researchers recorded the number of customers who had completed their cards and earned their free car wash. The following data are similar to the results obtained in the study. Do the data indicate a significant difference between the two card programs? Test with $\alpha = .05$.

	Completed	Not Completed	
Buy-Eight-Get-One-Free	10	40	50
Buy-Ten (with Two Free Punches)	19	31	50
	29	71	

13. In a classic study, Loftus and Palmer (1974) investigated the relationship between memory for eyewitnesses and the questions they are asked. In the study, participants watched a film of an automobile accident and then were questioned about the accident. One group was asked how fast the cars were going when they "smashed into" each other. A second group was asked about the speed when the cars "hit" each other, and a third group was not asked any question about the speed of the cars. A week later, the participants returned to answer additional questions about the accident, including whether they recalled seeing any broken glass. Although there was no broken glass in the film, several students claimed to remember seeing it. The following table shows the frequency distribution of responses for each group.

		Response to the Question "Did You See Any Broken Glass?"	
		Yes	No
Verb Used to Ask About the Speed	"Smashed into"	16	34
	"Hit"	7	43
	Control (Not Asked)	6	44

   a. Does the proportion of participants who claim to remember broken glass differ significantly from group to group? Test with $\alpha = .05$.
   b. Compute Cramér's V to measure the size of the treatment effect.
   c. Describe how the phrasing of the question influenced the participants' memories.
   d. Write a sentence demonstrating how the outcome of the hypothesis test and the measure of effect size would be reported in a journal article.

14. The Internet is rapidly becoming an essential source of information about health, nutrition, finances, and current events. Neunschwander, Abbott, and Mobley (2012) were interested in inequality of access to the Internet as a function of the characteristics of the participant. They recruited a very large sample of participants from the Indiana Supplemental Nutrition Assistance Program and surveyed them about their access to the technology. They observed that racial minorities, older persons, and persons with lower educational attainment were less likely to have a functioning computer at home and, relatedly, were less likely to have access to the Internet. Below are frequencies similar to those observed by the researchers.

	Owns a Computer	Does Not Own a Computer
Age: 18–30	378	336
Age: 71 and Older	23	127

   a. Does the proportion of participants who own a computer differ significantly from group to group? Test with $\alpha = .05$.

b. Compute Cramér's V to measure the size of the treatment effect.

c. Write a sentence demonstrating how the outcome of the hypothesis test and the measure of effect size would be reported in a journal article.

15. Captive animals in laboratories or zoos benefit from environmental enrichment. In a recent experiment on the effects of enrichment on animal behavior, Robbins and Margulis (2014) compared the effects of different types of auditory enrichment on captive gorillas. In their experiment, three gorillas—Koga, Lily, and Sidney—were exposed to either natural sounds, classical music, or rock music. The researchers counted the number of times the gorillas oriented toward the sound source. Frequencies like those observed by the researchers are listed below.

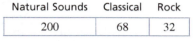

Natural Sounds	Classical	Rock
200	68	32

a. Do the results indicate any significant preferences among the three types of music?

b. Write a sentence demonstrating how the outcome of the hypothesis test would be reported in a journal article.

16. Many parents allow their underage children to drink alcohol in limited situations when an adult is present to supervise. The idea is that teens will learn responsible drinking habits if they first experience alcohol in a controlled environment. Other parents take a strict no-drinking approach with the idea that they are sending a clear message about what is right and what is wrong. Recent research, however, suggests that the more permissive approach may actually result in more negative consequences (McMorris et al., 2011). The researchers surveyed a sample of 200 students each year from ages 14 to 17. The students were asked about their alcohol use and about alcohol-related problems such as binge drinking, fights, and blackouts. The following table shows data similar to the results from the study.

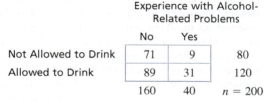

	Experience with Alcohol-Related Problems		
	No	Yes	
Not Allowed to Drink	71	9	80
Allowed to Drink	89	31	120
	160	40	$n = 200$

a. Do the data show a significant relationship between the parents' rules about alcohol and subsequent alcohol-related problems? Test with $\alpha = .05$.

b. Compute Cramér's V to measure the strength of the relationship.

17. A recent study indicates that people tend to select video game avatars with characteristics similar to those of their creators (Bélisle & Bodur, 2010). Participants who had created avatars for a virtual community game completed a questionnaire about their personalities. An independent group of viewers examined the avatars and recorded their impressions of the avatars. One personality characteristic considered was introverted/extroverted. The following table shows the frequency distribution of personalities for participants and the avatars they created.

	Participant Personality		
	Introverted	Extroverted	
Introverted Avatar	22	23	45
Extroverted Avatar	16	39	55
	38	62	

a. Is there a significant relationship between the personalities of the participants and the personalities of their avatars? Test with $\alpha = .05$.

b. Compute the phi-coefficient to measure the size of the effect.

18. Suppose that a researcher is interested in differences between young adults and older adults with respect to social media preferences. The researcher asked participants to indicate their preference for a specific social media application by checking all that apply among the following: Twitter, Facebook, and Snapchat™. The researcher observes the following:

	Twitter	Facebook	Snapchat
Younger Adults	10	12	3
Older Adults	6	15	4

Identify the assumptions of the chi-square test that would be violated if the researcher performed a chi-square test on the frequencies above.

19. Research indicates that people who volunteer to participate in research studies tend to have higher intelligence than nonvolunteers. To test this phenomenon, a researcher obtains a sample of 200 high school students. The students are given a description of a psychological research study and asked whether they would volunteer to participate. The researcher also obtains an IQ score for each student and classifies the students into high, medium, and low IQ groups.

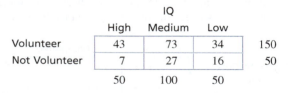

	IQ		
	High	Medium	Low
Volunteer	43	73	34	150
Not Volunteer	7	27	16	50
	50	100	50	

Do the preceding data indicate a significant relationship between IQ and volunteering? Test at the .05 level of significance.

Basic Mathematics Review

Preview

This appendix reviews some of the basic math skills that are necessary for the statistical calculations presented in this book. Many students already will know some or all of this material. Others will need to do extensive work and review. To help you assess your own skills, we include a skills assessment exam here. You should allow approximately 30 minutes to complete the test. When you finish, grade your test using the answer key on pages 588–589.

Notice that the test is divided into five sections. If you miss more than three questions in any section of the test, you probably need help in that area. Turn to the section of this appendix that corresponds to your problem area. In each section, you will find a general review, examples, and additional practice problems. After reviewing the appropriate section and doing the practice problems, turn to the end of the appendix. You will find another version of the skills assessment exam. If you still miss more than three questions in any section of the exam, continue studying. Get assistance from an instructor or a tutor if necessary. At the end of this appendix is a list of recommended books for individuals who need a more extensive review than can be provided here. We stress that mastering this material now will make the rest of the course much easier.

Skills Assessment Preview Exam

■ Section 1

(corresponding to Section A.1 of this appendix)

1. $3 + 2 \times 7 = ?$
2. $(3 + 2) \times 7 = ?$
3. $3 + 2^2 - 1 = ?$
4. $(3 + 2)^2 - 1 = ?$
5. $12/4 + 2 = ?$
6. $12/(4 + 2) = ?$
7. $12/(4 + 2)^2 = ?$
8. $2 \times (8 - 2^2) = ?$
9. $2 \times (8 - 2)^2 = ?$
10. $3 \times 2 + 8 - 1 \times 6 = ?$
11. $3 \times (2 + 8) - 1 \times 6 = ?$
12. $3 \times 2 + (8 - 1) \times 6 = ?$

■ Section 2

(corresponding to Section A.2 of this appendix)

1. The fraction $\frac{3}{4}$ corresponds to a percentage of _____.
2. Express 30% as a fraction.
3. Convert $\frac{12}{40}$ to a decimal.
4. $\frac{2}{13} + \frac{8}{13} = ?$
5. $1.375 + 0.25 = ?$
6. $\frac{2}{5} \times \frac{1}{4} = ?$
7. $\frac{1}{8} + \frac{2}{3} = ?$
8. $3.5 \times 0.4 = ?$
9. $\frac{1}{5} \div \frac{3}{4} = ?$
10. $3.75/0.5 = ?$
11. In a group of 80 students, 20% are psychology majors. How many psychology majors are in this group?
12. A company reports that two-fifths of its employees are women. If there are 90 employees, how many are women?

■ Section 3

(corresponding to Section A.3 of this appendix)

1. $3 + (-2) + (-1) + 4 = ?$
2. $6 - (-2) = ?$
3. $-2 - (-4) = ?$
4. $6 + (-1) - 3 - (-2) - (-5) = ?$
5. $4 \times (-3) = ?$

6. $-2 \times (-6) = ?$
7. $-3 \times 5 = ?$
8. $-2 \times (-4) \times (-3) = ?$
9. $12 \div (-3) = ?$
10. $-18 \div (-6) = ?$
11. $-16 \div 8 = ?$
12. $-100 \div (-4) = ?$

■ Section 4

(corresponding to Section A.4 of this appendix)
For each equation, find the value of X.

1. $X + 6 = 13$
2. $X - 14 = 15$
3. $5 = X - 4$
4. $3X = 12$
5. $72 = 3X$
6. $X/5 = 3$
7. $10 = X/8$
8. $3X + 5 = -4$
9. $24 = 2X + 2$
10. $(X + 3)/2 = 14$
11. $(X - 5)/3 = 2$
12. $17 = 4X - 11$

■ Section 5

(corresponding to Section A.5 of this appendix)

1. $4^3 = ?$
2. $\sqrt{25 - 9} = ?$
3. If $X = 2$ and $Y = 3$, then $XY^3 = ?$
4. If $X = 2$ and $Y = 3$, then $(X + Y)^2 = ?$
5. If $a = 3$ and $b = 2$, then $a^2 + b^2 = ?$
6. $(-3)^3 = ?$
7. $(-4)^4 = ?$
8. $\sqrt{4} \times 4 = ?$
9. $36/\sqrt{9} = ?$
10. $(9 + 2)^2 = ?$
11. $5^2 + 2^3 = ?$
12. If $a = 3$ and $b = -1$, then $a^2b^3 = ?$

The answers to the skills assessment exam are at the end of the appendix (pages 588–589).

A-1   Symbols and Notation

Table A.1 presents the basic mathematical symbols that you should know, along with examples of their use. Statistical symbols and notation are introduced and explained throughout this book as they are needed. Notation for exponents and square roots is covered separately at the end of this appendix.

Parentheses are a useful notation because they specify and control the order of computations. Everything inside the parentheses is calculated first. For example,

$$(5 + 3) \times 2 = 8 \times 2 = 16$$

Changing the placement of the parentheses also changes the order of calculations. For example,

$$5 + (3 \times 2) = 5 + 6 = 11$$

■ Order of Operations

Often a formula or a mathematical expression will involve several different arithmetic operations, such as adding, multiplying, squaring, and so on. When you encounter these situations, you must perform the different operations in the correct sequence. Following is a list of mathematical operations, showing the order in which they are to be performed.

1. Any calculation contained within parentheses is done first.
2. Squaring (or raising to other exponents) is done second.
3. Multiplying and/or dividing is done third. A series of multiplication and/or division operations should be done in order from left to right.
4. Adding and/or subtracting is done fourth.

The following examples demonstrate how this sequence of operations is applied in different situations.

To evaluate the expression

$$(3 + 1)^2 - 4 \times 7/2$$

first, perform the calculation within parentheses:

$$(4)^2 - 4 \times 7/2$$

Next, square the value as indicated:

$$16 - 4 \times 7/2$$

TABLE A.1

Symbol	Meaning	Example
+	Addition	$5 + 7 = 12$
−	Subtraction	$8 - 3 = 5$
×, ( )	Multiplication	$3 \times 9 = 27$, $3(9) = 27$
÷, /	Division	$15 \div 3 = 5$, $15/3 = 5$, $\frac{15}{3} = 5$
>	Greater than	$20 > 10$
<	Less than	$7 < 11$
≠	Not equal to	$5 \neq 6$

Then perform the multiplication and division:

$$16 - 14$$

Finally, do the subtraction:

$$16 - 14 = 2$$

A sequence of operations involving multiplication and division should be performed in order from left to right. For example, to compute $12/2 \times 3$, you divide 12 by 2 and then multiply the result by 3:

$$12/2 \times 3 = 6 \times 3 = 18$$

Notice that violating the left-to-right sequence can change the result. For this example, if you multiply before dividing, you will obtain

$$12/2 \times 3 = 12/6 = 2 \qquad \textbf{(This is wrong.)}$$

A sequence of operations involving only addition and subtraction can be performed in any order. For example, to compute $3 + 8 - 5$, you can add 3 and 8 and then subtract 5:

$$(3 + 8) - 5 = 11 - 5 = 6$$

or you can subtract 5 from 8 and then add the result to 3:

$$3 + (8 - 5) = 3 + 3 = 6$$

A mathematical expression or formula is simply a concise way to write a set of instructions. When you evaluate an expression by performing the calculation, you simply follow the instructions. For example, assume you are given these instructions:

1. First, add 3 and 8.
2. Next, square the result.
3. Next, multiply the resulting value by 6.
4. Finally, subtract 50 from the value you have obtained.

You can write these instructions as a mathematical expression.

1. The first step involves addition. Because addition is normally done last, use parentheses to give this operation priority in the sequence of calculations:

$$(3 + 8)$$

2. The instruction to square a value is noted by using the exponent 2 beside the value to be squared:

$$(3 + 8)^2$$

3. Because squaring has priority over multiplication, you can simply introduce the multiplication into the expression:

$$6 \times (3 + 8)^2$$

4. Addition and subtraction are done last, so simply write in the requested subtraction:

$$6 \times (3 + 8)^2 - 50$$

To calculate the value of the expression, you work through the sequence of operations in the proper order:

$$6 \times (3 + 8)^2 - 50 = 6 \times (11)^2 - 50$$
$$= 6 \times (121) - 50$$
$$= 726 - 50$$
$$= 676$$

As a final note, you should realize that the operation of squaring (or raising to any exponent) applies only to the value that immediately precedes the exponent. For example,

$$2 \times 3^2 = 2 \times 9 = 18 \qquad \text{(Only the 3 is squared.)}$$

If the instructions require multiplying values and then squaring the product, you must use parentheses to give the multiplication priority over squaring. For example, to multiply 2 times 3 and then square the product, you would write

$$(2 \times 3)^2 = (6)^2 = 36$$

LEARNING CHECK

1. Evaluate each of the following expressions:
   a. $4 \times 8/2^2$
   b. $4 \times (8/2)^2$
   c. $100 - 3 \times 12/(6 - 4)^2$
   d. $(4 + 6) \times (3 - 1)^2$
   e. $(8 - 2)/(9 - 8)^2$
   f. $6 + (4 - 1)^2 - 3 \times 4^2$
   g. $4 \times (8 - 3) + 8 - 3$

ANSWERS

1. a. 8   b. 64   c. 91   d. 40   e. 6   f. −33   g. 25

A-2 Proportions: Fractions, Decimals, and Percentages

A proportion is a part of a whole and can be expressed as a fraction, a decimal, or a percentage. For example, in a class of 40 students, only 3 failed the final exam.

The proportion of the class that failed can be expressed as a fraction

$$\text{fraction} = \frac{3}{40}$$

or as a decimal value

$$\text{decimal} = 0.075$$

or as a percentage

$$\text{percentage} = 7.5\%$$

Suppose you divided a pie into five equal pieces (fifths). If you first ate two-fifths of the pie and then another one-fifth, the total amount eaten would be three-fifths of the pie:

If the two fractions do not have the same denominator, you must first find equivalent fractions with a common denominator before you can add or subtract. The product of the two denominators will always work as a common denominator for equivalent fractions (although it may not be the lowest common denominator). For example,

$$\frac{2}{3} + \frac{1}{10} = ?$$

Because these two fractions have different denominators, it is necessary to convert each into an equivalent fraction and find a common denominator. We will use $3 \times 10 = 30$ as the common denominator. Thus, the equivalent fraction of each is

$$\frac{2}{3} = \frac{20}{30} \quad \text{and} \quad \frac{1}{10} = \frac{3}{30}$$

Now the two fractions can be added:

$$\frac{20}{30} + \frac{3}{30} = \frac{23}{30}$$

5. **Comparing the Size of Fractions**   When comparing the size of two fractions with the same denominator, the larger fraction will have the larger numerator. For example,

$$\frac{5}{8} > \frac{3}{8}$$

The denominators are the same, so the whole is partitioned into pieces of the same size. Five of these pieces are more than three of them:

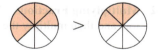

When two fractions have different denominators, you must first convert them to fractions with a common denominator to determine which is larger. Consider the following fractions:

$$\frac{3}{8} \quad \text{and} \quad \frac{7}{16}$$

If the numerator and denominator of $\frac{3}{8}$ are multiplied by 2, the resulting equivalent fraction will have a denominator of 16:

$$\frac{3}{8} = \frac{3 \times 2}{8 \times 2} = \frac{6}{16}$$

Now a comparison can be made between the two fractions:

$$\frac{6}{16} < \frac{7}{16}$$

Therefore,

$$\frac{3}{18} < \frac{7}{16}$$

■ Decimals

1. **Converting Decimals to Fractions**   Like a fraction, a decimal represents part of the whole. The first decimal place to the right of the decimal point indicates how many tenths are used. For example,

$$0.1 = \frac{1}{10} \quad 0.7 = \frac{7}{10}$$

The next decimal place represents $\frac{1}{100}$, the next $\frac{1}{1000}$, the next $\frac{1}{10,000}$, and so on. To change a decimal to a fraction, just use the number without the decimal point for the numerator. Use the denominator that the last (on the right) decimal place represents. For example,

$$0.32 = \frac{32}{100} \quad 0.5333 = \frac{5333}{10,000} \quad 0.05 = \frac{5}{100} \quad 0.001 = \frac{1}{1000}$$

2. **Adding and Subtracting Decimals**   To add and subtract decimals, the only rule is that you must keep the decimal points in a straight vertical line. For example,

$$
\begin{array}{r}
0.27 \\
+1.326 \\
\hline
1.596
\end{array}
\qquad
\begin{array}{r}
3.595 \\
-0.67 \\
\hline
2.925
\end{array}
$$

3. **Multiplying Decimals**   To multiply two decimal values, you first multiply the two numbers, ignoring the decimal points. Then you position the decimal point in the answer so that the number of digits to the right of the decimal point is equal to the total number of decimal places in the two numbers being multiplied. For example,

1.73	**(two decimal places)**
×0.251	**(three decimal places)**
173	
865	
346	
0.43423	**(five decimal places)**

0.25	**(two decimal places)**
×0.005	**(three decimal places)**
125	
00	
00	
0.00125	**(five decimal places)**

4. **Dividing Decimals**   The simplest procedure for dividing decimals is based on the fact that dividing two numbers is identical to expressing them as a fraction:

$$0.25 \div 1.6 \text{ is identical to } \frac{0.25}{1.6}$$

You now can multiply both the numerator and the denominator of the fraction by 10, 100, 1000, or whatever number is necessary to remove the decimal places. Remember that multiplying both the numerator and the denominator of a fraction by the *same* value will create an equivalent fraction. Therefore,

$$\frac{0.25}{1.6} = \frac{0.25 \times 100}{1.6 \times 100} = \frac{25}{160} = \frac{5}{32}$$

The result is a division problem without any decimal places in the two numbers.

■ Percentages

1. **Converting a Percentage to a Fraction or a Decimal**   To convert a percentage to a fraction, remove the percent sign, place the number in the numerator, and use 100 for the denominator. For example,

$$52\% = \frac{52}{100} \qquad 5\% = \frac{5}{100}$$

To convert a percentage to a decimal, remove the percent sign and divide by 100, or simply move the decimal point two places to the left. For example,

$$83\% = 83. = 0.83$$
$$14.5\% = 14.5 = 0.145$$
$$5\% = 5. = 0.05$$

2. **Performing Arithmetic Operations with Percentages**   There are situations in which it is best to express percent values as decimals in order to perform certain arithmetic operations. For example, what is 45% of 60? This question may be stated as

$$45\% \times 60 = ?$$

The 45% should be converted to decimal form to find the solution to this question. Therefore,

$$0.45 \times 60 = 27$$

LEARNING CHECK

1. Convert $\frac{3}{25}$ to a decimal.

2. Convert $\frac{3}{8}$ to a percentage.

3. Next to each set of fractions, write "True" if they are equivalent and "False" if they are not:

   a. $\frac{3}{8} = \frac{9}{24}$ _____   b. $\frac{7}{9} = \frac{17}{19}$ _____

   c. $\frac{2}{7} = \frac{4}{14}$ _____

4. Compute the following:

   a. $\frac{1}{6} \times \frac{7}{10}$ b. $\frac{7}{8} - \frac{1}{2}$ c. $\frac{9}{10} \div \frac{2}{3}$ d. $\frac{7}{22} + \frac{2}{3}$

5. Identify the larger fraction of each pair:

   a. $\frac{7}{10}, \frac{21}{100}$ b. $\frac{3}{4}, \frac{7}{12}$ c. $\frac{22}{3}, \frac{19}{3}$

6. Convert the following decimals into fractions:

   a. 0.012   b. 0.77   c. 0.005

7. $2.59 \times 0.015 = ?$

8. $1.8 \div 0.02 = ?$

9. What is 28% of 45?

ANSWERS   1. 0.12   2. 37.5%   3. a. True   b. False   c. True

4. a. $\frac{7}{60}$ b. $\frac{3}{8}$ c. $\frac{27}{20}$ d. $\frac{65}{66}$ 5. a. $\frac{7}{10}$ b. $\frac{3}{4}$ c. $\frac{22}{3}$

6. a. $\frac{12}{1000} = \frac{3}{250}$ b. $\frac{77}{100}$ c. $\frac{5}{1000} = \frac{1}{200}$ 7. 0.03885   8. 90   9. 12.6

A-3 Negative Numbers

Negative numbers are used to represent values less than zero. Negative numbers may occur when you are measuring the difference between two scores. For example, a researcher may want to evaluate the effectiveness of a propaganda film by measuring people's attitudes with a test both before and after viewing the film:

	Before	After	Amount of Change
Person A	23	27	+4
Person B	18	15	−3
Person C	21	16	−5

Notice that the negative sign provides information about the direction of the difference: a plus sign indicates an increase in value, and a minus sign indicates a decrease.

Because negative numbers are frequently encountered, you should be comfortable working with these values. This section reviews basic arithmetic operations using negative numbers. You should also note that any number without a sign (+ or −) is assumed to be positive.

1. **Adding Negative Numbers** When adding numbers that include negative values, simply interpret the negative sign as subtraction. For example,

$$3 + (-2) + 5 = 3 - 2 + 5 = 6$$

When adding a long string of numbers, it often is easier to add all the positive values to obtain the positive sum and then to add all of the negative values to obtain the negative sum. Finally, you subtract the negative sum from the positive sum. For example,

$$-1 + 3 + (-4) + 3 + (-6) + (-2)$$

$$\text{positive sum} = 6 \qquad \text{negative sum} = 13$$

$$\text{Answer: } 6 - 13 = -7$$

2. **Subtracting Negative Numbers** To subtract a negative number, change it to a positive number, and add. For example,

$$4 - (-3) = 4 + 3 = 7$$

This rule is easier to understand if you think of positive numbers as financial gains and negative numbers as financial losses. In this context, taking away a debt is equivalent to a financial gain. In mathematical terms, taking away a negative number is equivalent to adding a positive number. For example, suppose you are meeting a friend for lunch. You have $7, but you owe your friend $3. Thus, you really have only $4 to spend for lunch. But your friend forgives (takes away) the $3 debt. The result is that you now have $7 to spend. Expressed as an equation,

$$\$4 \text{ minus a } \$3 \text{ debt} = \$7$$

$$4 - (-3) = 4 + 3 = 7$$

A-5 Exponents and Square Roots

■ Exponential Notation

A simplified notation is used whenever a number is being multiplied by itself. The notation consists of placing a value, called an *exponent*, on the right-hand side of and raised above another number, called a *base*. For example,

$$7^{3} \leftarrow \text{exponent}$$
$$\uparrow$$
$$\text{base}$$

The exponent indicates how many times the base is used as a factor in multiplication. Following are some examples:

$7^3 = 7(7)(7)$ **(Read "7 cubed" or "7 raised to the third power")**

$5^2 = 5(5)$ **(Read "5 squared")**

$2^5 = 2(2)(2)(2)(2)$ **(Read "2 raised to the fifth power")**

There are a few basic rules about exponents that you will need to know for this course. They are outlined here.

1. **Numbers Raised to One or Zero**   Any number raised to the first power equals itself. For example,

$$6^1 = 6$$

Any number (except zero) raised to the zero power equals 1. For example,

$$9^0 = 1$$

2. **Exponents for Multiple Terms**   The exponent applies only to the base that is just in front of it. For example,

$$XY^2 = XYY$$
$$a^2b^3 = aabbb$$

3. **Negative Bases Raised to an Exponent**   If a negative number is raised to a power, then the result will be positive for exponents that are even and negative for exponents that are odd. For example,

$$(-4)^3 = -4(-4)(-4)$$
$$= 16(-4)$$
$$= -64$$

and

$$(-3)^4 = -3(-3)(-3)(-3)$$
$$= 9(-3)(-3)$$
$$= 9(9)$$
$$= 81$$

Note: The parentheses are used to ensure that the exponent applies to the entire negative number, including the sign. Without the parentheses there is some ambiguity as to how the exponent should be applied. For example, the expression -3^2 could have two interpretations:

$$-3^2 = (-3)(-3) = 9 \quad \text{or} \quad -3^2 = -(3)(3) = -9$$

4. **Exponents and Parentheses**   If an exponent is present outside of parentheses, then the computations within the parentheses are done first, and the exponential computation is done last:

$$(3 + 5)^2 = 8^2 = 64$$

Notice that the meaning of the expression is changed when each term in the parentheses is raised to the exponent individually:

$$3^2 + 5^2 = 9 + 25 = 34$$

Therefore,

$$X^2 + Y^2 \neq (X + Y)^2$$

5. **Fractions Raised to a Power**   If the numerator and denominator of a fraction are each raised to the same exponent, then the entire fraction can be raised to that exponent. That is,

$$\frac{a^2}{b^2} = \left(\frac{a}{b}\right)^2$$

For example,

$$\frac{3^2}{4^2} = \left(\frac{3}{4}\right)^2$$

$$\frac{9}{16} = \frac{3}{4}\left(\frac{3}{4}\right)$$

$$\frac{9}{16} = \frac{9}{16}$$

■ Square Roots

The square root of a value equals a number that when multiplied by itself yields the original value. For example, the square root of 16 equals 4 because 4 times 4 equals 16. The symbol for the square root is called a *radical,* $\sqrt{}$. The square root is taken for the number under the radical. For example,

$$\sqrt{16} = 4$$

Finding the square root is the inverse of raising a number to the second power (squaring). Thus,

$$\sqrt{a^2} = a$$

For example,

$$\sqrt{3^2} = \sqrt{9} = 3$$

Also,

$$(\sqrt{b})^2 = b$$

For example,

$$(\sqrt{64})^2 = 8^2 = 64$$

Computations under the same radical are performed *before* the square root is taken. For example,

$$\sqrt{9 + 16} = \sqrt{25} = 5$$

Note that with addition (or subtraction), separate radicals yield a different result:

$$\sqrt{9} + \sqrt{16} = 3 + 4 = 7$$

Therefore,

$$\sqrt{X} + \sqrt{Y} \neq \sqrt{X + Y}$$
$$\sqrt{X} - \sqrt{Y} \neq \sqrt{X - Y}$$

If the numerator and denominator of a fraction each have a radical, then the entire fraction can be placed under a single radical:

$$\frac{\sqrt{16}}{\sqrt{4}} = \sqrt{\frac{16}{4}}$$
$$\frac{4}{2} = \sqrt{4}$$
$$2 = 2$$

Therefore,

$$\frac{\sqrt{X}}{\sqrt{Y}} = \sqrt{\frac{X}{Y}}$$

Also, if the square root of one number is multiplied by the square root of another number, then the same result would be obtained by taking the square root of the product of both numbers. For example,

$$\sqrt{9} \times \sqrt{16} = \sqrt{9 \times 16}$$
$$3 \times 4 = \sqrt{144}$$
$$12 = 12$$

Therefore,

$$\sqrt{a} \times \sqrt{b} = \sqrt{ab}$$

LEARNING CHECK

1. Perform the following computations:
   a. $(-6)^3$
   b. $(3 + 7)^2$
   c. a^3b^2 when $a = 2$ and $b = -5$
   d. a^4b^3 when $a = 2$ and $b = 3$
   e. $(XY)^2$ when $X = 3$ and $Y = 5$
   f. $X^2 + Y^2$ when $X = 3$ and $Y = 5$
   g. $(X + Y)^2$ when $X = 3$ and $Y = 5$
   h. $\sqrt{5 + 4}$
   i. $(\sqrt{9})^2$
   j. $\dfrac{\sqrt{16}}{\sqrt{4}}$

1. a. −216   b. 100   c. 200   d. 432   e. 225
   f. 34   g. 64   h. 3   i. 9   j. 2

Problems for Appendix A   Basic Mathematics Review

1. $50/(10 - 8) = ?$
2. $(2 + 3)^2 = ?$
3. $20/10 \times 3 = ?$
4. $12 - 4 \times 2 + 6/3 = ?$
5. $24/(12 - 4) + 2 \times (6 + 3) = ?$
6. Convert $\frac{7}{20}$ to a decimal.
7. Express $\frac{9}{25}$ as a percentage.
8. Convert 0.91 to a fraction.
9. Express 0.0031 as a fraction.
10. Next to each set of fractions, write "True" if they are equivalent and "False" if they are not:
    a. $\dfrac{4}{1000} = \dfrac{2}{100}$ _____
    b. $\dfrac{5}{6} = \dfrac{52}{62}$ _____
    c. $\dfrac{1}{8} = \dfrac{7}{56}$ _____

11. Perform the following calculations:
    a. $\dfrac{4}{5} \times \dfrac{2}{3} = ?$ b. $\dfrac{7}{9} \div \dfrac{2}{3} = ?$
    c. $\dfrac{3}{8} + \dfrac{1}{5} = ?$ d. $\dfrac{5}{18} - \dfrac{1}{6} = ?$
12. $2.51 \times 0.017 = ?$
13. $3.88 \times 0.0002 = ?$
14. $3.17 + 17.0132 = ?$

15. $5.55 + 10.7 + 0.711 + 3.33 + 0.031 = ?$
16. $2.04 \div 0.2 = ?$
17. $0.36 \div 0.4 = ?$
18. $5 + 3 - 6 - 4 + 3 = ?$
19. $9 - (-1) - 17 + 3 - (-4) + 5 = ?$
20. $5 + 3 - (-8) - (-1) + (-3) - 4 + 10 = ?$
21. $8 \times (-3) = ?$
22. $-22 \div (-2) = ?$
23. $-2(-4) - (-3) = ?$
24. $84 \div (-4) = ?$

Solve the equations in problems 25−32 for X.

25. $X - 7 = -2$
26. $9 = X + 3$
27. $\dfrac{X}{4} = 11$
28. $-3 = \dfrac{X}{3}$
29. $\dfrac{X + 3}{5} = 2$
30. $\dfrac{X + 1}{3} = -8$
31. $6X - 1 = 11$
32. $2X + 3 = -11$
33. $(-5)^2 = ?$
34. $(-5)^3 = ?$
35. If $a = 4$ and $b = 3$, then $a^2 + b^4 = ?$
36. If $a = -1$ and $b = 4$, then $(a + b)^2 = ?$
37. If $a = -1$ and $b = 5$, then $ab^2 = ?$
38. $\dfrac{18}{\sqrt{4}} = ?$
39. $\sqrt{\dfrac{20}{5}} = ?$

Skills Assessment Final Exam

■ Section 1

1. $4 + 8/4 = ?$
2. $(4 + 8)/4 = ?$
3. $4 \times 3^2 = ?$
4. $(4 \times 3)^2 = ?$
5. $10/5 \times 2 = ?$
6. $10/(5 \times 2) = ?$
7. $40 - 10 \times 4/2 = ?$
8. $(5 - 1)^2/2 = ?$
9. $3 \times 6 - 3^2 = ?$
10. $2 \times (6 - 3)^2 = ?$
11. $4 \times 3 - 1 + 8 \times 2 = ?$
12. $4 \times (3 - 1 + 8) \times 2 = ?$

■ Section 2

1. Express $\frac{14}{80}$ as a decimal.
2. Convert $\frac{6}{25}$ to a percentage.
3. Convert 18% to a fraction.
4. $\frac{3}{5} \times \frac{2}{3} = ?$
5. $\frac{5}{24} + \frac{5}{6} = ?$
6. $\frac{7}{12} \div \frac{5}{6} = ?$
7. $\frac{5}{9} - \frac{1}{3} = ?$
8. $6.11 \times 0.22 = ?$
9. $0.18 \div 0.9 = ?$
10. $8.742 + 0.76 = ?$
11. In a statistics class of 72 students, three-eighths of the students received a B on the first test. How many Bs were earned?
12. What is 15% of 64?

■ Section 3

1. $3 - 1 - 3 + 5 - 2 + 6 = ?$
2. $-8 - (-6) = ?$
3. $2 - (-7) - 3 + (-11) - 20 = ?$
4. $-8 - 3 - (-1) - 2 - 1 = ?$

5. $8(-2) = ?$
6. $-7(-7) = ?$
7. $-3(-2)(-5) = ?$
8. $-3(5)(-3) = ?$
9. $-24 \div (-4) = ?$
10. $36 \div (-6) = ?$
11. $-56/7 = ?$
12. $-7/(-1) = ?$

■ Section 4

Solve for X.

1. $X + 5 = 12$
2. $X - 11 = 3$
3. $10 = X + 4$
4. $4X = 20$
5. $\frac{X}{2} = 15$
6. $18 = 9X$
7. $\frac{X}{5} = 35$
8. $2X + 8 = 4$
9. $\frac{X + 1}{3} = 6$
10. $4X + 3 = -13$
11. $\frac{X + 3}{3} = -7$
12. $23 = 2X - 5$

■ Section 5

1. $5^3 = ?$
2. $(-4)^3 = ?$
3. $(-2)^5 = ?$
4. $(-2)^6 = ?$
5. If $a = 4$ and $b = 2$, then $ab^2 = ?$
6. If $a = 4$ and $b = 2$, then $(a + b)^3 = ?$
7. If $a = 4$ and $b = 2$, then $a^2 + b^2 = ?$
8. $(11 + 4)^2 = ?$
9. $\sqrt{7^2} = ?$
10. If $a = 36$ and $b = 64$, then $\sqrt{a + b} = ?$
11. $\frac{25}{\sqrt{25}} = ?$
12. If $a = -1$ and $b = 2$, then $a^3b^4 = ?$

Answer Key   Skills Assessment Exams

PREVIEW EXAM
■ Section 1

1. 17
2. 35
3. 6
4. 24
5. 5
6. 2
7. $\frac{1}{3}$
8. 8
9. 72
10. 8
11. 24
12. 48

FINAL EXAM
■ Section 1

1. 6
2. 3
3. 36
4. 144
5. 4
6. 1
7. 20
8. 8
9. 9
10. 18
11. 27
12. 80

PREVIEW EXAM
■ Section 2

1. 75%
2. $\dfrac{30}{100}$, or $\dfrac{3}{10}$
3. 0.3
4. $\dfrac{10}{13}$
5. 1.625
6. $\dfrac{2}{20}$, or $\dfrac{1}{10}$
7. $\dfrac{19}{24}$
8. 1.4
9. $\dfrac{4}{15}$
10. 7.5
11. 16
12. 36

■ Section 3

1. 4
2. 8
3. 2
4. 9
5. −12
6. 12
7. −15
8. −24
9. −4
10. 3
11. −2
12. 25

■ Section 4

1. $X = 7$
2. $X = 29$
3. $X = 9$
4. $X = 4$
5. $X = 24$
6. $X = 15$
7. $X = 80$
8. $X = -3$
9. $X = 11$
10. $X = 25$
11. $X = 11$
12. $X = 7$

■ Section 5

1. 64
2. 4
3. 54
4. 25
5. 13
6. −27
7. 256
8. 8
9. 12
10. 121
11. 33
12. −9

FINAL EXAM
■ Section 2

1. 0.175
2. 24%
3. $\dfrac{18}{100}$, or $\dfrac{9}{50}$
4. $\dfrac{6}{15}$, or $\dfrac{2}{5}$
5. $\dfrac{25}{24}$
6. $\dfrac{42}{60}$, or $\dfrac{7}{10}$
7. $\dfrac{2}{9}$
8. 1.3442
9. 0.2
10. 9.502
11. 27
12. 9.6

■ Section 3

1. 8
2. −2
3. −25
4. −13
5. −16
6. 49
7. −30
8. 45
9. 6
10. −6
11. −8
12. 7

■ Section 4

1. $X = 7$
2. $X = 14$
3. $X = 6$
4. $X = 5$
5. $X = 30$
6. $X = 2$
7. $X = 175$
8. $X = -2$
9. $X = 17$
10. $X = -4$
11. $X = -24$
12. $X = 14$

■ Section 5

1. 125
2. −64
3. −32
4. 64
5. 16
6. 216
7. 20
8. 225
9. 7
10. 10
11. 5
12. −16

Solutions to Selected Problems for Appendix A
Basic Mathematics Review

1. 25
3. 6
5. 21
6. 0.35
7. 36%
9. $\dfrac{31}{10,000}$
10. **b.** False
11. **a.** $\dfrac{8}{15}$ **b.** $\dfrac{21}{18}$ **c.** $\dfrac{23}{40}$
12. 0.04267
14. 20.1832
17. 0.9
19. 5
21. −24
22. 11
25. $X = 5$
28. $X = -9$
30. $X = -25$
31. $X = 2$
34. −125
36. 9
37. −25
39. 2

Suggested Review Books

There are many basic mathematics books available if you need a more extensive review than this appendix can provide. Several are probably available in your library. The following books are but a few of the many that you may find helpful:

Karr, R., Massey, M., & Gustafson, R. D. (2013). *Beginning algebra: A guided approach* (10th ed.). Boston, MA: Cengage.

Lial, M. L., Salzman, S. A., & Hestwood, D. L. (2017). *Basic college mathematics* (10th ed.). New York, NY: Pearson.

McKeague, C. P. (2013). *Basic mathematics: A text/workbook* (8th ed.). Boston, MA: Cengage.

Statistical Tables

TABLE B.1   The Unit Normal Table*

---

*Column A lists *z*-score values. A vertical line drawn through a normal distribution at a *z*-score location divides the distribution into two sections.
Column B identifies the proportion in the larger section, called the *body*.
Column C identifies the proportion in the smaller section, called the *tail*.
Column D identifies the proportion between the mean and the *z*-score.
Note: Because the normal distribution is symmetrical, the proportions for negative *z*-scores are the same as those for positive *z*-scores.

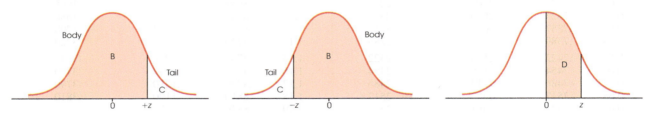

(A) z	(B) Proportion in Body	(C) Proportion in Tail	(D) Proportion Between Mean and z	(A) z	(B) Proportion in Body	(C) Proportion in Tail	(D) Proportion Between Mean and z
0.00	.5000	.5000	.0000	0.25	.5987	.4013	.0987
0.01	.5040	.4960	.0040	0.26	.6026	.3974	.1026
0.02	.5080	.4920	.0080	0.27	.6064	.3936	.1064
0.03	.5120	.4880	.0120	0.28	.6103	.3897	.1103
0.04	.5160	.4840	.0160	0.29	.6141	.3859	.1141
0.05	.5199	.4801	.0199	0.30	.6179	.3821	.1179
0.06	.5239	.4761	.0239	0.31	.6217	.3783	.1217
0.07	.5279	.4721	.0279	0.32	.6255	.3745	.1255
0.08	.5319	.4681	.0319	0.33	.6293	.3707	.1293
0.09	.5359	.4641	.0359	0.34	.6331	.3669	.1331
0.10	.5398	.4602	.0398	0.35	.6368	.3632	.1368
0.11	.5438	.4562	.0438	0.36	.6406	.3594	.1406
0.12	.5478	.4522	.0478	0.37	.6443	.3557	.1443
0.13	.5517	.4483	.0517	0.38	.6480	.3520	.1480
0.14	.5557	.4443	.0557	0.39	.6517	.3483	.1517
0.15	.5596	.4404	.0596	0.40	.6554	.3446	.1554
0.16	.5636	.4364	.0636	0.41	.6591	.3409	.1591
0.17	.5675	.4325	.0675	0.42	.6628	.3372	.1628
0.18	.5714	.4286	.0714	0.43	.6664	.3336	.1664
0.19	.5753	.4247	.0753	0.44	.6700	.3300	.1700
0.20	.5793	.4207	.0793	0.45	.6736	.3264	.1736
0.21	.5832	.4168	.0832	0.46	.6772	.3228	.1772
0.22	.5871	.4129	.0871	0.47	.6808	.3192	.1808
0.23	.5910	.4090	.0910	0.48	.6844	.3156	.1844
0.24	.5948	.4052	.0948	0.49	.6879	.3121	.1879

TABLE B.3   Critical Values for the *F*-Max Statistic*

*The critical values for α = .05 are in lightface type, and for α = .01, they are in boldface type.

n − 1	*k* = Number of Samples										
	2	3	4	5	6	7	8	9	10	11	12
4	9.60	15.5	20.6	25.2	29.5	33.6	37.5	41.4	44.6	48.0	51.4
	23.2	**37.**	**49.**	**59.**	**69.**	**79.**	**89.**	**97.**	**106.**	**113.**	**120.**
5	7.15	10.8	13.7	16.3	18.7	20.8	22.9	24.7	26.5	28.2	29.9
	14.9	**22.**	**28.**	**33.**	**38.**	**42.**	**46.**	**50.**	**54.**	**57.**	**60.**
6	5.82	8.38	10.4	12.1	13.7	15.0	16.3	17.5	18.6	19.7	20.7
	11.1	**15.5**	**19.1**	**22.**	**25.**	**27.**	**30.**	**32.**	**34.**	**36.**	**37.**
7	4.99	6.94	8.44	9.70	10.8	11.8	12.7	13.5	14.3	15.1	15.8
	8.89	**12.1**	**14.5**	**16.5**	**18.4**	**20.**	**22.**	**23.**	**24.**	**26.**	**27.**
8	4.43	6.00	7.18	8.12	9.03	9.78	10.5	11.1	11.7	12.2	12.7
	7.50	**9.9**	**11.7**	**13.2**	**14.5**	**15.8**	**16.9**	**17.9**	**18.9**	**19.8**	**21.**
9	4.03	5.34	6.31	7.11	7.80	8.41	8.95	9.45	9.91	10.3	10.7
	6.54	**8.5**	**9.9**	**11.1**	**12.1**	**13.1**	**13.9**	**14.7**	**15.3**	**16.0**	**16.6**
10	3.72	4.85	5.67	6.34	6.92	7.42	7.87	8.28	8.66	9.01	9.34
	5.85	**7.4**	**8.6**	**9.6**	**10.4**	**11.1**	**11.8**	**12.4**	**12.9**	**13.4**	**13.9**
12	3.28	4.16	4.79	5.30	5.72	6.09	6.42	6.72	7.00	7.25	7.48
	4.91	**6.1**	**6.9**	**7.6**	**8.2**	**8.7**	**9.1**	**9.5**	**9.9**	**10.2**	**10.6**
15	2.86	3.54	4.01	4.37	4.68	4.95	5.19	5.40	5.59	5.77	5.93
	4.07	**4.9**	**5.5**	**6.0**	**6.4**	**6.7**	**7.1**	**7.3**	**7.5**	**7.8**	**8.0**
20	2.46	2.95	3.29	3.54	3.76	3.94	4.10	4.24	4.37	4.49	4.59
	3.32	**3.8**	**4.3**	**4.6**	**4.9**	**5.1**	**5.3**	**5.5**	**5.6**	**5.8**	**5.9**
30	2.07	2.40	2.61	2.78	2.91	3.02	3.12	3.21	3.29	3.36	3.39
	2.63	**3.0**	**3.3**	**3.5**	**3.6**	**3.7**	**3.8**	**3.9**	**4.0**	**4.1**	**4.2**
60	1.67	1.85	1.96	2.04	2.11	2.17	2.22	2.26	2.30	2.33	2.36
	1.96	**2.2**	**2.3**	**2.4**	**2.4**	**2.5**	**2.5**	**2.6**	**2.6**	**2.7**	**2.7**

TABLE B.4   The *F* Distribution*

*Table entries in lightface type are critical values for the .05 level of significance. Boldface type values are for the .01 level of significance.

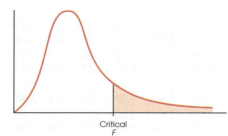

Critical
F

Degrees of Freedom: Denominator	Degrees of Freedom: Numerator														
	1	2	3	4	5	6	7	8	9	10	11	12	14	16	20
1	161	200	216	225	230	234	237	239	241	242	243	244	245	246	248
	4052	**4999**	**5403**	**5625**	**5764**	**5859**	**5928**	**5981**	**6022**	**6056**	**6082**	**6106**	**6142**	**6169**	**6208**
2	18.51	19.00	19.16	19.25	19.30	19.33	19.36	19.37	19.38	19.39	19.40	19.41	19.42	19.43	19.44
	98.49	**99.00**	**99.17**	**99.25**	**99.30**	**99.33**	**99.34**	**99.36**	**99.38**	**99.40**	**99.41**	**99.42**	**99.43**	**99.44**	**99.45**
3	10.13	9.55	9.28	9.12	9.01	8.94	8.88	8.84	8.81	8.78	8.76	8.74	8.71	8.69	8.66
	34.12	**30.92**	**29.46**	**28.71**	**28.24**	**27.91**	**27.67**	**27.49**	**27.34**	**27.23**	**27.13**	**27.05**	**26.92**	**26.83**	**26.69**
4	7.71	6.94	6.59	6.39	6.26	6.16	6.09	6.04	6.00	5.96	5.93	5.91	5.87	5.84	5.80
	21.20	**18.00**	**16.69**	**15.98**	**15.52**	**15.21**	**14.98**	**14.80**	**14.66**	**14.54**	**14.45**	**14.37**	**14.24**	**14.15**	**14.02**
5	6.61	5.79	5.41	5.19	5.05	4.95	4.88	4.82	4.78	4.74	4.70	4.68	4.64	4.60	4.56
	16.26	**13.27**	**12.06**	**11.39**	**10.97**	**10.67**	**10.45**	**10.27**	**10.15**	**10.05**	**9.96**	**9.89**	**9.77**	**9.68**	**9.55**
6	5.99	5.14	4.76	4.53	4.39	4.28	4.21	4.15	4.10	4.06	4.03	4.00	3.96	3.92	3.87
	13.74	**10.92**	**9.78**	**9.15**	**8.75**	**8.47**	**8.26**	**8.10**	**7.98**	**7.87**	**7.79**	**7.72**	**7.60**	**7.52**	**7.39**
7	5.59	4.74	4.35	4.12	3.97	3.87	3.79	3.73	3.68	3.63	3.60	3.57	3.52	3.49	3.44
	12.25	**9.55**	**8.45**	**7.85**	**7.46**	**7.19**	**7.00**	**6.84**	**6.71**	**6.62**	**6.54**	**6.47**	**6.35**	**6.27**	**6.15**
8	5.32	4.46	4.07	3.84	3.69	3.58	3.50	3.44	3.39	3.34	3.31	3.28	3.23	3.20	3.15
	11.26	**8.65**	**7.59**	**7.01**	**6.63**	**6.37**	**6.19**	**6.03**	**5.91**	**5.82**	**5.74**	**5.67**	**5.56**	**5.48**	**5.36**
9	5.12	4.26	3.86	3.63	3.48	3.37	3.29	3.23	3.18	3.13	3.10	3.07	3.02	2.98	2.93
	10.56	**8.02**	**6.99**	**6.42**	**6.06**	**5.80**	**5.62**	**5.47**	**5.35**	**5.26**	**5.18**	**5.11**	**5.00**	**4.92**	**4.80**
10	4.96	4.10	3.71	3.48	3.33	3.22	3.14	3.07	3.02	2.97	2.94	2.91	2.86	2.82	2.77
	10.04	**7.56**	**6.55**	**5.99**	**5.64**	**5.39**	**5.21**	**5.06**	**4.95**	**4.85**	**4.78**	**4.71**	**4.60**	**4.52**	**4.41**
11	4.84	3.98	3.59	3.36	3.20	3.09	3.01	2.95	2.90	2.86	2.82	2.79	2.74	2.70	2.65
	9.65	**7.20**	**6.22**	**5.67**	**5.32**	**5.07**	**4.88**	**4.74**	**4.63**	**4.54**	**4.46**	**4.40**	**4.29**	**4.21**	**4.10**
12	4.75	3.88	3.49	3.26	3.11	3.00	2.92	2.85	2.80	2.76	2.72	2.69	2.64	2.60	2.54
	9.33	**6.93**	**5.95**	**5.41**	**5.06**	**4.82**	**4.65**	**4.50**	**4.39**	**4.30**	**4.22**	**4.16**	**4.05**	**3.98**	**3.86**
13	4.67	3.80	3.41	3.18	3.02	2.92	2.84	2.77	2.72	2.67	2.63	2.60	2.55	2.51	2.46
	9.07	**6.70**	**5.74**	**5.20**	**4.86**	**4.62**	**4.44**	**4.30**	**4.19**	**4.10**	**4.02**	**3.96**	**3.85**	**3.78**	**3.67**
14	4.60	3.74	3.34	3.11	2.96	2.85	2.77	2.70	2.65	2.60	2.56	2.53	2.48	2.44	2.39
	8.86	**6.51**	**5.56**	**5.03**	**4.69**	**4.46**	**4.28**	**4.14**	**4.03**	**3.94**	**3.86**	**3.80**	**3.70**	**3.62**	**3.51**
15	4.54	3.68	3.29	3.06	2.90	2.79	2.70	2.64	2.59	2.55	2.51	2.48	2.43	2.39	2.33
	8.68	**6.36**	**5.42**	**4.89**	**4.56**	**4.32**	**4.14**	**4.00**	**3.89**	**3.80**	**3.73**	**3.67**	**3.56**	**3.48**	**3.36**
16	4.49	3.63	3.24	3.01	2.85	2.74	2.66	2.59	2.54	2.49	2.45	2.42	2.37	2.33	2.28
	8.53	**6.23**	**5.29**	**4.77**	**4.44**	**4.20**	**4.03**	**3.89**	**3.78**	**3.69**	**3.61**	**3.55**	**3.45**	**3.37**	**3.25**

TABLE B.4   The *F* Distribution*   *(continued)*

Degrees of Freedom: Denominator	Degrees of Freedom: Numerator														
	1	2	3	4	5	6	7	8	9	10	11	12	14	16	20
17	4.45	3.59	3.20	2.96	2.81	2.70	2.62	2.55	2.50	2.45	2.41	2.38	2.33	2.29	2.23
	8.40	**6.11**	**5.18**	**4.67**	**4.34**	**4.10**	**3.93**	**3.79**	**3.68**	**3.59**	**3.52**	**3.45**	**3.35**	**3.27**	**3.16**
18	4.41	3.55	3.16	2.93	2.77	2.66	2.58	2.51	2.46	2.41	2.37	2.34	2.29	2.25	2.19
	8.28	**6.01**	**5.09**	**4.58**	**4.25**	**4.01**	**3.85**	**3.71**	**3.60**	**3.51**	**3.44**	**3.37**	**3.27**	**3.19**	**3.07**
19	4.38	3.52	3.13	2.90	2.74	2.63	2.55	2.48	2.43	2.38	2.34	2.31	2.26	2.21	2.15
	8.18	**5.93**	**5.01**	**4.50**	**4.17**	**3.94**	**3.77**	**3.63**	**3.52**	**3.43**	**3.36**	**3.30**	**3.19**	**3.12**	**3.00**
20	4.35	3.49	3.10	2.87	2.71	2.60	2.52	2.45	2.40	2.35	2.31	2.28	2.23	2.18	2.12
	8.10	**5.85**	**4.94**	**4.43**	**4.10**	**3.87**	**3.71**	**3.56**	**3.45**	**3.37**	**3.30**	**3.23**	**3.13**	**3.05**	**2.94**
21	4.32	3.47	3.07	2.84	2.68	2.57	2.49	2.42	2.37	2.32	2.28	2.25	2.20	2.15	2.09
	8.02	**5.78**	**4.87**	**4.37**	**4.04**	**3.81**	**3.65**	**3.51**	**3.40**	**3.31**	**3.24**	**3.17**	**3.07**	**2.99**	**2.88**
22	4.30	3.44	3.05	2.82	2.66	2.55	2.47	2.40	2.35	2.30	2.26	2.23	2.18	2.13	2.07
	7.94	**5.72**	**4.82**	**4.31**	**3.99**	**3.76**	**3.59**	**3.45**	**3.35**	**3.26**	**3.18**	**3.12**	**3.02**	**2.94**	**2.83**
23	4.28	3.42	3.03	2.80	2.64	2.53	2.45	2.38	2.32	2.28	2.24	2.20	2.14	2.10	2.04
	7.88	**5.66**	**4.76**	**4.26**	**3.94**	**3.71**	**3.54**	**3.41**	**3.30**	**3.21**	**3.14**	**3.07**	**2.97**	**2.89**	**2.78**
24	4.26	3.40	3.01	2.78	2.62	2.51	2.43	2.36	2.30	2.26	2.22	2.18	2.13	2.09	2.02
	7.82	**5.61**	**4.72**	**4.22**	**3.90**	**3.67**	**3.50**	**3.36**	**3.25**	**3.17**	**3.09**	**3.03**	**2.93**	**2.85**	**2.74**
25	4.24	3.38	2.99	2.76	2.60	2.49	2.41	2.34	2.28	2.24	2.20	2.16	2.11	2.06	2.00
	7.77	**5.57**	**4.68**	**4.18**	**3.86**	**3.63**	**3.46**	**3.32**	**3.21**	**3.13**	**3.05**	**2.99**	**2.89**	**2.81**	**2.70**
26	4.22	3.37	2.98	2.74	2.59	2.47	2.39	2.32	2.27	2.22	2.18	2.15	2.10	2.05	1.99
	7.72	**5.53**	**4.64**	**4.14**	**3.82**	**3.59**	**3.42**	**3.29**	**3.17**	**3.09**	**3.02**	**2.96**	**2.86**	**2.77**	**2.66**
27	4.21	3.35	2.96	2.73	2.57	2.46	2.37	2.30	2.25	2.20	2.16	2.13	2.08	2.03	1.97
	7.68	**5.49**	**4.60**	**4.11**	**3.79**	**3.56**	**3.39**	**3.26**	**3.14**	**3.06**	**2.98**	**2.93**	**2.83**	**2.74**	**2.63**
28	4.20	3.34	2.95	2.71	2.56	2.44	2.36	2.29	2.24	2.19	2.15	2.12	2.06	2.02	1.96
	7.64	**5.45**	**4.57**	**4.07**	**3.76**	**3.53**	**3.36**	**3.23**	**3.11**	**3.03**	**2.95**	**2.90**	**2.80**	**2.71**	**2.60**
29	4.18	3.33	2.93	2.70	2.54	2.43	2.35	2.28	2.22	2.18	2.14	2.10	2.05	2.00	1.94
	7.60	**5.42**	**4.54**	**4.04**	**3.73**	**3.50**	**3.33**	**3.20**	**3.08**	**3.00**	**2.92**	**2.87**	**2.77**	**2.68**	**2.57**
30	4.17	3.32	2.92	2.69	2.53	2.42	2.34	2.27	2.21	2.16	2.12	2.09	2.04	1.99	1.93
	7.56	**5.39**	**4.51**	**4.02**	**3.70**	**3.47**	**3.30**	**3.17**	**3.06**	**2.98**	**2.90**	**2.84**	**2.74**	**2.66**	**2.55**
32	4.15	3.30	2.90	2.67	2.51	2.40	2.32	2.25	2.19	2.14	2.10	2.07	2.02	1.97	1.91
	7.50	**5.34**	**4.46**	**3.97**	**3.66**	**3.42**	**3.25**	**3.12**	**3.01**	**2.94**	**2.86**	**2.80**	**2.70**	**2.62**	**2.51**
34	4.13	3.28	2.88	2.65	2.49	2.38	2.30	2.23	2.17	2.12	2.08	2.05	2.00	1.95	1.89
	7.44	**5.29**	**4.42**	**3.93**	**3.61**	**3.38**	**3.21**	**3.08**	**2.97**	**2.89**	**2.82**	**2.76**	**2.66**	**2.58**	**2.47**
36	4.11	3.26	2.86	2.63	2.48	2.36	2.28	2.21	2.15	2.10	2.06	2.03	1.98	1.93	1.87
	7.39	**5.25**	**4.38**	**3.89**	**3.58**	**3.35**	**3.18**	**3.04**	**2.94**	**2.86**	**2.78**	**2.72**	**2.62**	**2.54**	**2.43**
38	4.10	3.25	2.85	2.62	2.46	2.35	2.26	2.19	2.14	2.09	2.05	2.02	1.96	1.92	1.85
	7.35	**5.21**	**4.34**	**3.86**	**3.54**	**3.32**	**3.15**	**3.02**	**2.91**	**2.82**	**2.75**	**2.69**	**2.59**	**2.51**	**2.40**
40	4.08	3.23	2.84	2.61	2.45	2.34	2.25	2.18	2.12	2.07	2.04	2.00	1.95	1.90	1.84
	7.31	**5.18**	**4.31**	**3.83**	**3.51**	**3.29**	**3.12**	**2.99**	**2.88**	**2.80**	**2.73**	**2.66**	**2.56**	**2.49**	**2.37**

TABLE B.4   The *F* Distribution* *(continued)*

Degrees of Freedom: Denominator	Degrees of Freedom: Numerator														
	1	2	3	4	5	6	7	8	9	10	11	12	14	16	20
42	4.07	3.22	2.83	2.59	2.44	2.32	2.24	2.17	2.11	2.06	2.02	1.99	1.94	1.89	1.82
	7.27	**5.15**	**4.29**	**3.80**	**3.49**	**3.26**	**3.10**	**2.96**	**2.86**	**2.77**	**2.70**	**2.64**	**2.54**	**2.46**	**2.35**
44	4.06	3.21	2.82	2.58	2.43	2.31	2.23	2.16	2.10	2.05	2.01	1.98	1.92	1.88	1.81
	7.24	**5.12**	**4.26**	**3.78**	**3.46**	**3.24**	**3.07**	**2.94**	**2.84**	**2.75**	**2.68**	**2.62**	**2.52**	**2.44**	**2.32**
46	4.05	3.20	2.81	2.57	2.42	2.30	2.22	2.14	2.09	2.04	2.00	1.97	1.91	1.87	1.80
	7.21	**5.10**	**4.24**	**3.76**	**3.44**	**3.22**	**3.05**	**2.92**	**2.82**	**2.73**	**2.66**	**2.60**	**2.50**	**2.42**	**2.30**
48	4.04	3.19	2.80	2.56	2.41	2.30	2.21	2.14	2.08	2.03	1.99	1.96	1.90	1.86	1.79
	7.19	**5.08**	**4.22**	**3.74**	**3.42**	**3.20**	**3.04**	**2.90**	**2.80**	**2.71**	**2.64**	**2.58**	**2.48**	**2.40**	**2.28**
50	4.03	3.18	2.79	2.56	2.40	2.29	2.20	2.13	2.07	2.02	1.98	1.95	1.90	1.85	1.78
	7.17	**5.06**	**4.20**	**3.72**	**3.41**	**3.18**	**3.02**	**2.88**	**2.78**	**2.70**	**2.62**	**2.56**	**2.46**	**2.39**	**2.26**
55	4.02	3.17	2.78	2.54	2.38	2.27	2.18	2.11	2.05	2.00	1.97	1.93	1.88	1.83	1.76
	7.12	**5.01**	**4.16**	**3.68**	**3.37**	**3.15**	**2.98**	**2.85**	**2.75**	**2.66**	**2.59**	**2.53**	**2.43**	**2.35**	**2.23**
60	4.00	3.15	2.76	2.52	2.37	2.25	2.17	2.10	2.04	1.99	1.95	1.92	1.86	1.81	1.75
	7.08	**4.98**	**4.13**	**3.65**	**3.34**	**3.12**	**2.95**	**2.82**	**2.72**	**2.63**	**2.56**	**2.50**	**2.40**	**2.32**	**2.20**
65	3.99	3.14	2.75	2.51	2.36	2.24	2.15	2.08	2.02	1.98	1.94	1.90	1.85	1.80	1.73
	7.04	**4.95**	**4.10**	**3.62**	**3.31**	**3.09**	**2.93**	**2.79**	**2.70**	**2.61**	**2.54**	**2.47**	**2.37**	**2.30**	**2.18**
70	3.98	3.13	2.74	2.50	2.35	2.23	2.14	2.07	2.01	1.97	1.93	1.89	1.84	1.79	1.72
	7.01	**4.92**	**4.08**	**3.60**	**3.29**	**3.07**	**2.91**	**2.77**	**2.67**	**2.59**	**2.51**	**2.45**	**2.35**	**2.28**	**2.15**
80	3.96	3.11	2.72	2.48	2.33	2.21	2.12	2.05	1.99	1.95	1.91	1.88	1.82	1.77	1.70
	6.96	**4.88**	**4.04**	**3.56**	**3.25**	**3.04**	**2.87**	**2.74**	**2.64**	**2.55**	**2.48**	**2.41**	**2.32**	**2.24**	**2.11**
100	3.94	3.09	2.70	2.46	2.30	2.19	2.10	2.03	1.97	1.92	1.88	1.85	1.79	1.75	1.68
	6.90	**4.82**	**3.98**	**3.51**	**3.20**	**2.99**	**2.82**	**2.69**	**2.59**	**2.51**	**2.43**	**2.36**	**2.26**	**2.19**	**2.06**
125	3.92	3.07	2.68	2.44	2.29	2.17	2.08	2.01	1.95	1.90	1.86	1.83	1.77	1.72	1.65
	6.84	**4.78**	**3.94**	**3.47**	**3.17**	**2.95**	**2.79**	**2.65**	**2.56**	**2.47**	**2.40**	**2.33**	**2.23**	**2.15**	**2.03**
150	3.91	3.06	2.67	2.43	2.27	2.16	2.07	2.00	1.94	1.89	1.85	1.82	1.76	1.71	1.64
	6.81	**4.75**	**3.91**	**3.44**	**3.14**	**2.92**	**2.76**	**2.62**	**2.53**	**2.44**	**2.37**	**2.30**	**2.20**	**2.12**	**2.00**
200	3.89	3.04	2.65	2.41	2.26	2.14	2.05	1.98	1.92	1.87	1.83	1.80	1.74	1.69	1.62
	6.76	**4.71**	**3.88**	**3.41**	**3.11**	**2.90**	**2.73**	**2.60**	**2.50**	**2.41**	**2.34**	**2.28**	**2.17**	**2.09**	**1.97**
400	3.86	3.02	2.62	2.39	2.23	2.12	2.03	1.96	1.90	1.85	1.81	1.78	1.72	1.67	1.60
	6.70	**4.66**	**3.83**	**3.36**	**3.06**	**2.85**	**2.69**	**2.55**	**2.46**	**2.37**	**2.29**	**2.23**	**2.12**	**2.04**	**1.92**
1000	3.85	3.00	2.61	2.38	2.22	2.10	2.02	1.95	1.89	1.84	1.80	1.76	1.70	1.65	1.58
	6.66	**4.62**	**3.80**	**3.34**	**3.04**	**2.82**	**2.66**	**2.53**	**2.43**	**2.34**	**2.26**	**2.20**	**2.09**	**2.01**	**1.89**
∞	3.84	2.99	2.60	2.37	2.21	2.09	2.01	1.94	1.88	1.83	1.79	1.75	1.69	1.64	1.57
	6.64	**4.60**	**3.78**	**3.32**	**3.02**	**2.80**	**2.64**	**2.51**	**2.41**	**2.32**	**2.24**	**2.18**	**2.07**	**1.99**	**1.87**

Source: Table A14 of Snedecor, G. W., and Cochran, W. G. (1980). *Statistical Methods* (7th ed.). Ames, Iowa: Iowa State University Press. Copyright © 1980 by the Iowa State University Press, 2121 South State Avenue, Ames, Iowa 50010. Reprinted with permission of the Iowa State University Press.

TABLE B.5   The Studentized Range Statistic (q)*

*The critical values for q corresponding to $\alpha = .05$ (lightface type) and $\alpha = .01$ (boldface type).

df for Error Term	\multicolumn{11}{c}{k = Number of Treatments}										
	2	3	4	5	6	7	8	9	10	11	12
5	3.64	4.60	5.22	5.67	6.03	6.33	6.58	6.80	6.99	7.17	7.32
	5.70	**6.98**	**7.80**	**8.42**	**8.91**	**9.32**	**9.67**	**9.97**	**10.24**	**10.48**	**10.70**
6	3.46	4.34	4.90	5.30	5.63	5.90	6.12	6.32	6.49	6.65	6.79
	5.24	**6.33**	**7.03**	**7.56**	**7.97**	**8.32**	**8.61**	**8.87**	**9.10**	**9.30**	**9.48**
7	3.34	4.16	4.68	5.06	5.36	5.61	5.82	6.00	6.16	6.30	6.43
	4.95	**5.92**	**6.54**	**7.01**	**7.37**	**7.68**	**7.94**	**8.17**	**8.37**	**8.55**	**8.71**
8	3.26	4.04	4.53	4.89	5.17	5.40	5.60	5.77	5.92	6.05	6.18
	4.75	**5.64**	**6.20**	**6.62**	**6.96**	**7.24**	**7.47**	**7.68**	**7.86**	**8.03**	**8.18**
9	3.20	3.95	4.41	4.76	5.02	5.24	5.43	5.59	5.74	5.87	5.98
	4.60	**5.43**	**5.96**	**6.35**	**6.66**	**6.91**	**7.13**	**7.33**	**7.49**	**7.65**	**7.78**
10	3.15	3.88	4.33	4.65	4.91	5.12	5.30	5.46	5.60	5.72	5.83
	4.48	**5.27**	**5.77**	**6.14**	**6.43**	**6.67**	**6.87**	**7.05**	**7.21**	**7.36**	**7.49**
11	3.11	3.82	4.26	4.57	4.82	5.03	5.20	5.35	5.49	5.61	5.71
	4.39	**5.15**	**5.62**	**5.97**	**6.25**	**6.48**	**6.67**	**6.84**	**6.99**	**7.13**	**7.25**
12	3.08	3.77	4.20	4.51	4.75	4.95	5.12	5.27	5.39	5.51	5.61
	4.32	**5.05**	**5.50**	**5.84**	**6.10**	**6.32**	**6.51**	**6.67**	**6.81**	**6.94**	**7.06**
13	3.06	3.73	4.15	4.45	4.69	4.88	5.05	5.19	5.32	5.43	5.53
	4.26	**4.96**	**5.40**	**5.73**	**5.98**	**6.19**	**6.37**	**6.53**	**6.67**	**6.79**	**6.90**
14	3.03	3.70	4.11	4.41	4.64	4.83	4.99	5.13	5.25	5.36	5.46
	4.21	**4.89**	**5.32**	**5.63**	**5.88**	**6.08**	**6.26**	**6.41**	**6.54**	**6.66**	**6.77**
15	3.01	3.67	4.08	4.37	4.59	4.78	4.94	5.08	5.20	5.31	5.40
	4.17	**4.84**	**5.25**	**5.56**	**5.80**	**5.99**	**6.16**	**6.31**	**6.44**	**6.55**	**6.66**
16	3.00	3.65	4.05	4.33	4.56	4.74	4.90	5.03	5.15	5.26	5.35
	4.13	**4.79**	**5.19**	**5.49**	**5.72**	**5.92**	**6.08**	**6.22**	**6.35**	**6.46**	**6.56**
17	2.98	3.63	4.02	4.30	4.52	4.70	4.86	4.99	5.11	5.21	5.31
	4.10	**4.74**	**5.14**	**5.43**	**5.66**	**5.85**	**6.01**	**6.15**	**6.27**	**6.38**	**6.48**
18	2.97	3.61	4.00	4.28	4.49	4.67	4.82	4.96	5.07	5.17	5.27
	4.07	**4.70**	**5.09**	**5.38**	**5.60**	**5.79**	**5.94**	**6.08**	**6.20**	**6.31**	**6.41**
19	2.96	3.59	3.98	4.25	4.47	4.65	4.79	4.92	5.04	5.14	5.23
	4.05	**4.67**	**5.05**	**5.33**	**5.55**	**5.73**	**5.89**	**6.02**	**6.14**	**6.25**	**6.34**
20	2.95	3.58	3.96	4.23	4.45	4.62	4.77	4.90	5.01	5.11	5.20
	4.02	**4.64**	**5.02**	**5.29**	**5.51**	**5.69**	**5.84**	**5.97**	**6.09**	**6.19**	**6.28**
24	2.92	3.53	3.90	4.17	4.37	4.54	4.68	4.81	4.92	5.01	5.10
	3.96	**4.55**	**4.91**	**5.17**	**5.37**	**5.54**	**5.69**	**5.81**	**5.92**	**6.02**	**6.11**
30	2.89	3.49	3.85	4.10	4.30	4.46	4.60	4.72	4.82	4.92	5.00
	3.89	**4.45**	**4.80**	**5.05**	**5.24**	**5.40**	**5.54**	**5.65**	**5.76**	**5.85**	**5.93**
40	2.86	3.44	3.79	4.04	4.23	4.39	4.52	4.63	4.73	4.82	4.90
	3.82	**4.37**	**4.70**	**4.93**	**5.11**	**5.26**	**5.39**	**5.50**	**5.60**	**5.69**	**5.76**
60	2.83	3.40	3.74	3.98	4.16	4.31	4.44	4.55	4.65	4.73	4.81
	3.76	**4.28**	**4.59**	**4.82**	**4.99**	**5.13**	**5.25**	**5.36**	**5.45**	**5.53**	**5.60**
120	2.80	3.36	3.68	3.92	4.10	4.24	4.36	4.47	4.56	4.64	4.71
	3.70	**4.20**	**4.50**	**4.71**	**4.87**	**5.01**	**5.12**	**5.21**	**5.30**	**5.37**	**5.44**
∞	2.77	3.31	3.63	3.86	4.03	4.17	4.28	4.39	4.47	4.55	4.62
	3.64	**4.12**	**4.40**	**4.60**	**4.76**	**4.88**	**4.99**	**5.08**	**5.16**	**5.23**	**5.29**

TABLE B.6   Critical Values for the Pearson Correlation*

*To be significant, the sample correlation, r, must be greater than or equal to the critical value in the table.

| | Level of Significance for One-Tailed Test | | | |
	.05	.025	.01	.005
	Level of Significance for Two-Tailed Test			
$df = n - 2$	.10	.05	.02	.01
1	.988	.997	.9995	.9999
2	.900	.950	.980	.990
3	.805	.878	.934	.959
4	.729	.811	.882	.917
5	.669	.754	.833	.874
6	.622	.707	.789	.834
7	.582	.666	.750	.798
8	.549	.632	.716	.765
9	.521	.602	.685	.735
10	.497	.576	.658	.708
11	.476	.553	.634	.684
12	.458	.532	.612	.661
13	.441	.514	.592	.641
14	.426	.497	.574	.623
15	.412	.482	.558	.606
16	.400	.468	.542	.590
17	.389	.456	.528	.575
18	.378	.444	.516	.561
19	.369	.433	.503	.549
20	.360	.423	.492	.537
21	.352	.413	.482	.526
22	.344	.404	.472	.515
23	.337	.396	.462	.505
24	.330	.388	.453	.496
25	.323	.381	.445	.487
26	.317	.374	.437	.479
27	.311	.367	.430	.471
28	.306	.361	.423	.463
29	.301	.355	.416	.456
30	.296	.349	.409	.449
35	.275	.325	.381	.418
40	.257	.304	.358	.393
45	.243	.288	.338	.372
50	.231	.273	.322	.354
60	.211	.250	.295	.325
70	.195	.232	.274	.302
80	.183	.217	.256	.283
90	.173	.205	.242	.267
100	.164	.195	.230	.254

Source: Table VI of Fisher, R. A., and Yates, F. (1974). *Statistical Tables for Biological, Agricultural and Medical Research* (6th ed.). London: Longman Group Ltd. (previously published by Oliver and Boyd Ltd., Edinburgh). Copyright ©1963 R. A. Fisher and F. Yates. Adapted and reprinted with permission of Pearson Education Limited.

TABLE B.7   The Chi-Square Distribution*

*The table entries are critical values of χ^2.

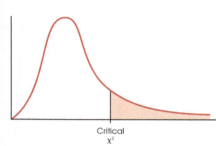

Critical
χ^2

df	Proportion in Critical Region				
	0.10	0.05	0.025	0.01	0.005
1	2.71	3.84	5.02	6.63	7.88
2	4.61	5.99	7.38	9.21	10.60
3	6.25	7.81	9.35	11.34	12.84
4	7.78	9.49	11.14	13.28	14.86
5	9.24	11.07	12.83	15.09	16.75
6	10.64	12.59	14.45	16.81	18.55
7	12.02	14.07	16.01	18.48	20.28
8	13.36	15.51	17.53	20.09	21.96
9	14.68	16.92	19.02	21.67	23.59
10	15.99	18.31	20.48	23.21	25.19
11	17.28	19.68	21.92	24.72	26.76
12	18.55	21.03	23.34	26.22	28.30
13	19.81	22.36	24.74	27.69	29.82
14	21.06	23.68	26.12	29.14	31.32
15	22.31	25.00	27.49	30.58	32.80
16	23.54	26.30	28.85	32.00	34.27
17	24.77	27.59	30.19	33.41	35.72
18	25.99	28.87	31.53	34.81	37.16
19	27.20	30.14	32.85	36.19	38.58
20	28.41	31.41	34.17	37.57	40.00
21	29.62	32.67	35.48	38.93	41.40
22	30.81	33.92	36.78	40.29	42.80
23	32.01	35.17	38.08	41.64	44.18
24	33.20	36.42	39.36	42.98	45.56
25	34.38	37.65	40.65	44.31	46.93
26	35.56	38.89	41.92	45.64	48.29
27	36.74	40.11	43.19	46.96	49.64
28	37.92	41.34	44.46	48.28	50.99
29	39.09	42.56	45.72	49.59	52.34
30	40.26	43.77	46.98	50.89	53.67
40	51.81	55.76	59.34	63.69	66.77
50	63.17	67.50	71.42	76.15	79.49
60	74.40	79.08	83.30	88.38	91.95
70	85.53	90.53	95.02	100.42	104.22
80	96.58	101.88	106.63	112.33	116.32
90	107.56	113.14	118.14	124.12	128.30
100	118.50	124.34	129.56	135.81	140.17

Source: Table 8 of Pearson, E., and Hartley, H. O. (1966). *Biometrika Tables for Statisticians* (3rd ed.). New York: Cambridge University Press. Adapted and reprinted with permission of the Biometrika trustees.

Solutions for Odd-Numbered Problems in the Text

CHAPTER 1 Introduction to Statistics

1. **a.** The population consists of all high school students in the United States.
   b. The sample is the group of 100 students who were measured in the study.
   c. The average number is a statistic. Notice that you might be more specific and say "descriptive" statistic. Inferential statistic or parameter would be incorrect because the calculated average describes only the data measured in the sample.

3. **a.** The population consists of all college students in the United States.
   b. The sample consists of the 100 students who participated in the study.
   c. The group that received decaffeinated coffee is in a control condition (that is, no caffeine).
   d. The group that received the caffeinated coffee is in an experimental condition.
   e. The sample contains 100 participants (50 in each group). The population is either infinitely large or too large for it to be practical to measure all members. If you said that the population consisted of 100 students, you might have mistakenly thought that the population consisted of everyone in the study.
   f. The average calculated after the memory test is a "statistic" or, more specifically, "descriptive statistic." "Inferential statistic" or "parameter" would be incorrect because the average describes only the data in the sample.

5. **a.** Statistic (or descriptive statistic)
   b. Parameter

7. **a.** The average score in the afternoon was 80 and the average score in the morning was 76, so you might be tempted to think that there is some real advantage for testing in the afternoon. However, the difference between means could be due to random chance alone—sampling error. Based on the descriptive statistics given in this sample, we just don't know whether an advantage exists or not.
   b. Inferential statistics

9. Age: ratio scale and continuous. Although people usually report whole-number years, the variable is the amount of time and time is infinitely divisible.

Income: ratio scale and discrete. Income is determined by units of currency. For U.S. dollars, the smallest unit is the penny and there are no intermediate values between 1 cent and 2 cents.

Dependents: ratio scale and discrete. Family size consists of whole-number categories with no intermediate values.

Social Security: nominal scale and discrete. Social security numbers are essentially names that are coded as 9-digit numbers. There are no intermediate values between two consecutive social security numbers.

11. **a.** An ordinal scale provides information about the direction of difference (greater or less) between two measurements.
    b. An interval scale provides information about the magnitude of the difference between two measurements.
    c. A ratio scale provides information about the ratio of two measurements, which allow comparisons such as "twice as much."

13. A correlational study has only one group of individuals and measures two (or more) different variables for each individual. Other research methods evaluating relationships between variables compare two (or more) different groups of scores.

15. **a.** This is not an experiment because no independent variable is manipulated and participants are not randomly assigned to groups that receive different amounts of milkfat.
    b. It is possible that participants in the reduced milkfat (skim or 1% milk) group (that is, children who regularly drank reduced-fat milk) also tended to be more sedentary.
    c. Possibility 1: A researcher could randomly assign participants to groups that receive different amounts of milkfat.

Possibility 2: A researcher could assign participants to two groups that receive different amounts of milkfat, holding constant characteristics like the amount of physical activity by participants in each group.

Possibility 3: A researcher could assign participants to two groups that receive different amounts of milkfat, matching the two groups in the amount of physical activity.

17. a. Loneliness is a continuous variable. If it is measured with ratings of 1 to 4, it may appear to be discrete but it could be measured with a 1 to 40 rating, which means that each category could be further divided. The UCLA Loneliness Scale is an interval scale of measurement because a value of zero does not represent a complete absence of loneliness.

b. $n = 86$

c. This is an experimental study because participants were randomly assigned to groups.

d. The group that was instructed to post more status updates is an experimental group.

19. a. The dependent variable is the number of correct answers on the test, which is a measure of knowledge of the material.

b. Knowledge is a continuous variable. If it is measured with a 10-question test, it may appear to be discrete but it could be measured with a 100-question test, which means that each category can be further divided.

c. Ratio scale. Zero is absolute, which means a complete absence of correct answers.

21. a. This study used the experimental method because participants were randomly assigned to groups that received different instructions.

b. The independent variable was the instructions received by participants (that is, being told that their group waited and the other didn't versus being told that their

group didn't wait and the other group waited). The dependent variable was whether or not children chose to wait for a larger reward.

23. a. $\Sigma X = 15$

b. $(\Sigma X)^2 = (15)^2 = 225$. Note that if you answered 65, you were incorrect because you squared the scores before summing them.

c. $\Sigma X - 3 = 15 - 3 = 12$. Note that if your answer was 3, you were incorrect because you subtracted 3 from each score before summing.

d. $\Sigma(X - 3) = (4 - 3) + (2 - 3) + (6 - 3) + (3 - 3) = (1) + (-1) + (3) + (0) = 3$. Note that if your answer was 12, you were incorrect because you summed the scores before subtracting 3.

25. a. $\Sigma(X - 4)^2 = 158$

b. $(\Sigma X)^2 = (-2)^2 = 4$

c. $\Sigma X^2 = 62$

d. $\Sigma(X + 3) = 13$

27. a. $\Sigma XY = 2$

b. $\Sigma X \Sigma Y = 56$

c. $\Sigma Y = 7$

d. $n = 4$

29. a. $(\Sigma X)^2$

b. ΣX^2

c. $\Sigma(X - 2)$

d. $\Sigma(X - 1)^2$

31. a. $n\Sigma X^2 = 195$

b. $(\Sigma Y)^2 = 361$

c. $\Sigma XY = 22$

d. $\Sigma X \Sigma Y = 209$

CHAPTER 2 Frequency Distributions

1. Distribution table:

a.

X	f	$p = \frac{f}{n}$	$\% = p(100)$
15	1	0.05	5%
14	2	0.10	10%
13	3	0.15	15%
12	3	0.15	15%
11	2	0.10	10%
10	4	0.20	20%
9	2	0.10	10%
8	1	0.05	5%
7	2	0.10	10%

b. $n = 20$

3. a. $n = 14$

b. $\Sigma X = 48$. If your answer was 21, you were incorrect because you did not list all instances of each score. ΣX in this problem is the sum of 6, 5, 5, 4, 4, 4, 4, 3, 3, 3, 2, 2, 2, 1.

c. $\Sigma X^2 = 190$. If your answer was 2304, you were incorrect because you summed all scores before squaring. Remember that squaring of each score and multiplying by its frequency is done before summing of scores, unless Σ is inside of parentheses, which looks like $(\Sigma X)^2$.

5. a. $n = 17$

b. $\Sigma X = 55$

c. $\Sigma X^2 = 197$

7. a.

X	f	cf	c%
20	1	20	100%
19	2	19	95%
18	2	17	85%
17	4	15	75%
16	4	11	55%
15	3	7	35%
14	2	4	20%
13	2	2	10%

9. a.

X	f	Lower real limit	Upper real limit
70−79	1	69.5	79.5
60−69	2	59.5	69.5
50−59	1	49.5	59.5
40−49	2	39.5	49.5
30−39	5	29.5	39.5
20−29	7	19.5	29.5
10−19	3	9.5	19.5

b. Positively skewed
c. See table. If you answered 70 and 79, 60 and 69, and so on, you were incorrect because you did not identify time as a continuous variable.

11. Adjacent bars touch in a histogram, but there is a gap between adjacent bars in a bar graph. Bar graphs are used to display nominal or ordinal data and histograms are used to display interval or ratio data. In a population, curves are often used to display distributions.

13. You can compute ΣX, ΣX^2, and the mean from a regular frequency distribution, but you cannot compute those statistics from a grouped frequency distribution. In a regular frequency distribution, you can read the exact values of scores. You can compute n from both types of frequency distributions.

15. a.

X	f
14	2
13	4
12	3
11	0
10	1
9	3
8	4
7	1

b. Bimodal

17. a. Age is a ratio scale so a histogram should be used.
b. Birth order is an ordinal scale so a bar graph should be used.
c. Academic major is a nominal scale so a bar graph should be used.
d. Voter registration status is a nominal scale so a bar graph should be used.

19. a. The size of the drink is an ordinal scale so a bar graph would be most appropriate.
b.

Regular prices

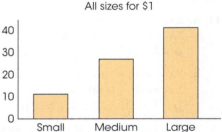
All sizes for $1

c. More drinks were sold during the sale and a larger proportion of large-sized drinks were sold during the sale.

21.

X	f Short Time Between Flash Cards	f Long Time Between Flash Cards
4	0	2
3	4	6
2	3	1
1	2	1
0	1	0

There are more high scores among participants who studied the flash cards with a long amount of time between flash cards than among participants who studied flash cards with a short amount of time between flash cards.

23. a.

2	5
3	78
4	—
5	25
6	02357789
7	268
8	37
9	46

b. Normal distribution

CHAPTER 3 Central Tendency

1. $M = \frac{\Sigma X}{n} = \frac{108}{9} = 12$

3. $\Sigma X = 2 + 7 + 9 + 4 + 5 + 3 + 0 + 6 = 36.$
$M = \frac{\Sigma X}{n} = \frac{36}{8} = 4.5$

5. $\Sigma X = 1 + 2 + 2 + 3 + 3 + 3 + 3 + 3 + 10 + 10 = 40.$
$M = \frac{\Sigma X}{n} = \frac{40}{10} = 4$. The mean is the balance point of the distribution.

7. The mean is the statistic for central tendency that is equivalent to diving the sum of scores equally across all members of a sample.

9. $\Sigma X = N\mu = 7(13) = 91$

11. a. $\Sigma X_1 = n_1 M_1 = 4(6) = 24$
$\Sigma X_2 = n_2 M_2 = 4(12) = 48$
$M = \frac{\Sigma X_1 + \Sigma X_2}{n_1 + n_2} = \frac{24 + 48}{4 + 4} = \frac{72}{8} = 9$

b. $\Sigma X_1 = n_1 M_1 = 3(6) = 18$
$\Sigma X_2 = n_2 M_2 = 6(12) = 72$
$M = \frac{\Sigma X_1 + \Sigma X_2}{n_1 + n_2} = \frac{18 + 72}{3 + 6} = \frac{90}{9} = 10$

If you answered that $M = 9$, your answer was incorrect because your calculation did not account for the fact that the groups had different numbers of scores. That is, you did not *weight* the mean by the size of the group contributing to the overall, combined mean.

c. $\Sigma X_1 = n_1 M_1 = 6(6) = 36$
$\Sigma X_2 = n_2 M_2 = 3(12) = 36.$
$M = \frac{\Sigma X_1 + \Sigma X_2}{n_1 + n_2} = \frac{36 + 36}{6 + 3} = \frac{72}{9} = 8$

13. The new score of $X = 11$ is 10 points lower than the old score of $X = 21$. Thus, we subtract 10 points from ΣX and divide the new value of ΣX by n as below.
$\Sigma X - 10 = nM - 10 = 10 (7) - 10 = 60.$
$M = \frac{\Sigma X}{n} = \frac{60}{10} = 6$

15. To calculate the effect of removing a score on the mean, we subtract the value of that score (12) from ΣX and divide by the new value for n (5) as below.
$\Sigma X - 12 = nM - 12 = 6 (10) - 12 = 48.$
$M = \frac{\Sigma X}{n} = \frac{48}{5} = 9.60$

17. To calculate the effect of removing a score on the mean, we subtract the value of that score (21) from ΣX and divide by the new value for N (9) as below.
$\Sigma X - 21 = N\mu - 21 = 10 (12) - 21 = 99.$
$M = \frac{\Sigma X}{n} = \frac{99}{9} = 11$

19. a. $\mu = 100$
b. $\mu = 0$
c. $\mu = 100$
d. $\mu = 1$

21. $Mdn = 5$

23. a. To find the median of scores from a discrete variable, we use the sorting method. The middle score in the sorted list is $X = 3$. Thus, $Mdn = 3$.
b. In this question, the median falls somewhere within the upper and lower real limits of the several tied $X = 3$ scores. We use the following equation to find the precise median for these scores:

$$\text{Median} = X_{LRL} + \left(\frac{0.5N - f_{\text{BELOW LRL}}}{f_{\text{TIED}}}\right)$$

$$\text{Median} = 2.5 + \left(\frac{0.5(10) - 4}{3}\right) = 2.5 + \left(\frac{5 - 4}{3}\right) =$$
$$2.5 + \left(\frac{1}{3}\right) = 2.83$$

25. $\Sigma X = \Sigma fX = 1(9) + 1(8) + 3(7) + 4(6) + 1(5) = 67.$
$M = 6.7$. If you answered 7.0, your answer was incorrect because you did not multiply each score by its corresponding frequency. $Mdn = 6.5$. $Mode = 6$.

27. The distribution is bimodal. The major mode corresponds to a score of $X = 3$. The minor mode corresponds to a score of $X = 8$.

29. a. $M = 7.5$, $Mdn = 8$, $Mode = 9$.
b. Based on these relative values, the distribution is negatively skewed.

31. a. $M = 3.0$, $Mdn = 3.5$, $Mode = 4$.
b. Based on these relative values, the distribution is negatively skewed.

CHAPTER 4 Variability

1. A measure of variability describes the degree to which the scores in a distribution are spread out or clustered together. Variability also measures the size of the distance between scores.

3. range = URL for X_{max} − LRL for X_{min} = 12.5 − 0.5 = 12. IQR = $Q3$ − $Q1$ = 6.5 − 4.5 = 2.0. The 75th percentile corresponds to 6.5 and the 25th percentile corresponds to 4.5. The IQR is a better measurement of variability than the simple range because most of the scores are clustered together within a range of two points.

5. Variance measures the average squared deviation between each score and the mean. Standard deviation measures the average deviation between each score and the mean.

7. $SS = 36$, $\sigma^2 = 9$, and $\sigma = 3$. If you answered $\sigma^2 = 12$, and $\sigma = 3.46$, your answer was incorrect because you used the formula for sample variance. If you answered $SS = 2.67$, $\sigma^2 = 0.67$, and $\sigma = 0.82$, your answer was incorrect because you divided $(\Sigma X)^2$ by $N − 1$ instead of dividing by N in computing SS.

9. In a sample with a standard deviation of zero, all scores have the same value. A standard deviation of zero occurs only when the set of scores has no variability.

11. a.

	Definitional	Computational
Set A	11.34	11.33
Set B	30	30

The difference between the definitional and computational formulas for Set A arises because of rounding error in computing the squared deviations between each score and the mean. For Set A, $M = 9.\overline{6}$ (note that the bar over the 6 indicates that the 6 repeats). Rounded to two decimal places, $M = 9.67$. For each squared deviation, rounding to two decimal places produces some error. When you sum those squared deviations, the rounding error is cumulative. Thus, the value produced by the computational formula is correct and the value produced by the definitional formula is incorrect.

13. a. $SS = 72$, $\sigma^2 = 9$, $\sigma = 3$
b. The computational formula should be used because $\mu = 2.5$.

15. a. If the scores are a population, $SS = 20$, $\sigma^2 = 4$, $\sigma = 2$.
b. If the scores are a sample, $SS = 20$, $s^2 = 5$, $s = 2.24$.

17. A value of 11 should be used in the formula for variance. A value of 12 should be used in the denominator of the formula for the mean. These two values are different because the sample mean statistic is an unbiased estimator of the population mean, and sample variance uses $n − 1$ to be an unbiased estimator of σ^2.

19. $SS = 128$, $s^2 = 16$, $s = 4$.

21. a. $\mu = 4$, $\sigma^2 = 6$
b.

Sample	Score 1	Score 2	$M = \frac{\Sigma X}{n}$	SS	$\frac{SS}{n-1}$	$\frac{SS}{n}$
a	1	1	1.00	0.00	0.00	0.00
b	1	4	2.50	4.50	4.50	2.25
c	1	7	4.00	18.00	18.00	9.00
d	4	1	2.50	4.50	4.50	2.25
e	4	4	4.00	0.00	0.00	0.00
f	4	7	5.50	4.50	4.50	2.25
g	7	1	4.00	18.00	18.00	9.00
h	7	4	5.50	4.50	4.50	2.25
i	7	7	7.00	0.00	0.00	0.00

c. Mean of M column = 4. Mean of $\frac{SS}{n-1}$ column = 6. Mean of column $\frac{SS}{n}$ = 3. The mean of M column matches μ. The mean of the $\frac{SS}{n-1}$ column matches σ^2. The sample mean is an unbiased estimate of μ. $\frac{SS}{n-1}$ is an unbiased estimate of σ^2. $\frac{SS}{n}$ is a biased estimate of σ^2.

23. $\Sigma X = 720$, $SS = df(s^2) = 11\,(3^2) = 11\,(9) = 99$. If you answered $SS = 108$, your answer was incorrect because you multiplied s^2 by n instead of $df = n − 1$.

25.

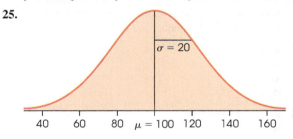

27. a. Original sample $M = 64$ and $s = 13$. If your answer was $s = 7$, your answer was incorrect because adding or subtracting a constant value from each score does not change the standard deviation. Adding or subtracting a constant value from each score does not change the distances between scores and the mean.
b. Original sample $M = 16$ and $s = 6$. If your answer was $s = 18$, you were incorrect because multiplying or dividing each score by a constant value multiplies or divides the standard deviation by the same value.

29. a. The transformed sample scores are 0, 2, 1, 0, 6, 1, 0, 2. For the new sample, $M = 1.5$ and $s = 2$.
   b. For the original sample, $M = 0.75$ and $s = 1$.

31. a. range $= X_{max} - X_{min} = 12 - 0 = 12$, $s^2 = 24$, and $s = 4.90$.
   b. range $= X_{max} - X_{min} = 12 - 0 = 12$, $s^2 = 48$, and $s = 6.93$.
   c. The range is unchanged by increasing the distance between scores in the center of the distribution. Standard deviation and variance increase when the distance between scores in the center of the distribution is increased.

33. a. 1991: $M = 8$, $SS = 200$, $s^2 = 25$, and $s = 5$. 2006: $M = 15$, $SS = 512$, $s^2 = 64$, and $s = 8$.

   b.

1991	2006
$M = 8.00$	$M = 15.00$
$SD = 5.00$	$SD = 8.00$

35. Pre-adaptation: $M = 12$, $SS = 294$, $s^2 = 49$, $s = 7$. Post-adaptation: $M = 5$, $SS = 96$, $s^2 = 16$, $s = 4$. The adaptation procedure decreased both the central tendency and the variability in the distance between participants' pointing and the target.

37. a.

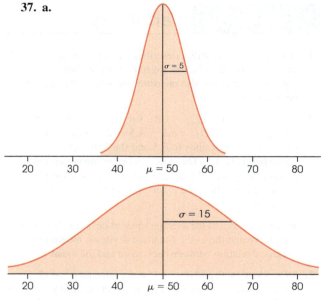

b. A score of $X = 65$ would be considered an extreme value in the distribution with a mean of $\mu = 50$ and a standard deviation of $\sigma = 5$ because this value is 3 standard deviations above the mean.

CHAPTER 5 z-Scores

1. The z-score represents the distance and direction of a score's location relative to the mean in either a sample or a population distribution.

3. a. d
   b. c
   c. a
   d. b

5. a. $z = +1.00$
   b. $z = +0.50$
   c. $z = -2.00$
   d. $z = -0.60$

7. a.

$X = 50$ $X = 62$ $X = 53$
$z = \frac{X - \mu}{\sigma} = \frac{50 - 50}{6} = 0.00$ $z = +2.00$ $z = +0.50$

$X = 44$ $X = 47$ $X = 38$
$z = -1.00*$ $z = -0.50$ $z = -2.00$

b. $z = +1.00$ $z = +2.50$ $z = +1.50$
   $X = \mu + z\sigma$ $X = 65$ $X = 59$
   $= 50 + (+1.00)6$
   $= 56$

   $z = -1.50$ $z = -3.00$ $z = -2.50$
   $X = 41**$ $X = 32$ $X = 35$

9. A sample has a mean of $M = 90$ and a standard deviation of $s = 20$.
   a. $X = 95$ $X = 98$ $X = 105$
   $z = +0.25$ $z = +0.40$ $z = +0.75$

   $X = 80$ $X = 88$ $X = 76$
   $z = -0.50$ $z = -0.10$ $z = -0.70$

   b. Find the X value for each of the following z-scores.
   $z = -1.00$ $z = +0.50$ $z = -1.50$
   $X = 70$ $X = 100$ $X = 60$
   $z = +0.75$ $z = -1.25$ $z = +2.60$
   $X = 105$ $X = 65$ $X = 142$

*If your answer was 1.00, it was incorrect because you ignored the sign of the z-score.

**If your answer was $X = 59$, it was incorrect because you ignored the sign of the z-score.

11. a. $z = 0.00$. The exam score was equal to the mean.
  b. $z = +1.00$. The exam score was above average.
  c. $z = -2.00$. The exam score was extremely low.
  d. $z = -0.50$. The exam score was below average.

13. a. $z = +0.42, X = 104.80$
  b. $z = +1.25, X = 101.60$
  c. $z = +1.79, X = 85.60$
  d. $z = +4.17, X = 82.40$

15. $\sigma = 40$

17. $\mu = X - z\sigma = 24 - (-1.5)4 = 24 - (-6.0) = 30.0$.
  If you answered 18, you were incorrect because your calculation did not consider that $z = -1.50$ is below the mean.

19. $\sigma = \frac{X - \mu}{z} = \frac{54 - 45}{+1.50} = \frac{9}{+1.50} = 6.00$

21. $s = \frac{X - M}{z} = \frac{54 - 63}{-0.75} = \frac{-9}{-0.75} = 12.00$

23. $X = 21$ is 9 points higher than $X = 12$. 9 points corresponds to 1.50 standard deviations ($z = -1.00$ is 1.50 standard deviations greater than $z = -2.50$).
  $s = 9 \div 1.50 = 6.00$. $X = 21$ is one standard deviation (6 points) below the mean. Thus, $M = 27$. You can check your work by recalculating the z-scores based on the values for s and M in your answer. That is $z = \frac{X - M}{s} = \frac{12 - 27}{6} = \frac{-15}{6} = -2.50$ and $z = \frac{X - M}{s} = \frac{21 - 27}{6} = \frac{-6}{6} = -1.00$.

25. a. $X = 70$: $z = \frac{X - \mu}{\sigma} = \frac{70 - 82}{8} = \frac{-12}{8} = -1.50$
  $X = 60$: $z = \frac{X - \mu}{\sigma} = \frac{60 - 72}{12} = \frac{-12}{12} = -1.00$

$X = 60$ will lead to the better grade because it is 1 standard deviation below the mean and $X = 70$ is 1.5 standard deviations below the mean.
  b. $X = 58$ corresponds to a z-score of $+1.50$. $X = 85$ corresponds to a z-score of $+1.50$. The two scores should lead to the same grade.
  c. $X = 32$ corresponds to a z-score of $+2.00$. $X = 26$ corresponds to a z-score of $+3.00$. $X = 26$ should lead to a better grade.

27. a. $X = 39$ corresponds to $z = -0.50$.
  $X_{\text{Transformed}} = \mu + z\sigma = 100 + (-0.50)20$
  $= 100 + (-10) = 90$
  b. $X = 36$ corresponds to $z = -1.25$. $X_{\text{Transformed}} = 75$.
  c. $X = 45$ corresponds to $z = +1.00$. $X_{\text{Transformed}} = 120$.
  d. $X = 50$ corresponds to $z = +2.25$. $X_{\text{Transformed}} = 145$.

29. a. $\mu = 5$ and $\sigma = 4$
  b. and **c.**

Original X	z-score	Transformed X
6	+0.25	55
1	-1.00	30
0	-1.25	25
7	+0.50	60
4	-0.25	45
13	+2.00	90
4	-0.25	45

31. No. $X = 220$ corresponds to a z-score of $z = +0.40$. That is, $X = 220$ is not an extreme or unusual score.

CHAPTER 6 Probability

1. The two requirements for a random sample are: (1) each individual has an equal chance of being selected and (2) if more than one individual is selected, the probabilities must stay constant for all selections.

3. a. $p(\text{freshman}) = \frac{\text{frequency of freshman}}{\text{total students in the class}} = \frac{32}{32 + 48} = \frac{32}{80} = 0.40$
  b. $p(\text{freshman}) = 40\%$. Because random sampling is used, the first five samples are returned to the population.
  c. $p(\text{freshman}) = \frac{\text{frequency of freshman}}{\text{total students in the class}} = \frac{32}{32 + 58} = \frac{32}{90} = 0.36$

5. a. Body to the left of z, $p = .9772$

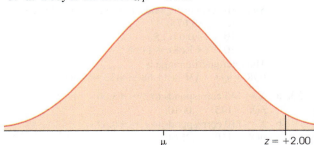

b. Body to the left. $p = .6915$
  c. Body to the right. $p = .9332$
  d. Body to the right. $p = .9525$

7. a. $p = .4452$
  b. $p = .3159$
  c. $p = .4332$
  d. $p = .1554$

9. a. $p(-1.64 < z < +1.64) = .4495 + .4495 = .8990$
  b. $p(-1.96 < z < +1.96) = .4750 + .4750 = .9500$
  c. $p(-1.00 < z < +1.00) = .3413 + .3413 = .6826$

11. a. $z = +1.65$
  b. $z = -0.84$
  c. $z = -1.28$
  d. $z = 0.00$

13. a. $-1.96 < z < +1.96$
  b. $-0.67 < z < +0.67$
  c. $-1.15 < z < +1.15$
  d. $-0.84 < z < +0.84$

15. a. Body is to the left of the distribution because the score $X = 74$ is greater than the mean

$$z = \frac{X - \mu}{\sigma} = \frac{74 - 70}{12} = \frac{+4}{12} = +0.33$$

$p\,(z < +0.33) = .6293$

b. Body is to the left of the distribution because the score is greater than the mean.

$$z = \frac{X - \mu}{\sigma} = \frac{84 - 70}{12} = \frac{+14}{12} = +1.17$$

$p\,(z < +1.17) = .8790$

c. Body is to the right of the distribution because the score is less than the mean.

$$z = \frac{X - \mu}{\sigma} = \frac{54 - 70}{12} = \frac{-16}{12} = -1.33$$

$p\,(z > -1.33) = .9082$

d. Body is to the right of the distribution because the score is less than the mean.

$$z = \frac{X - \mu}{\sigma} = \frac{58 - 70}{12} = \frac{-12}{12} = -1.00$$

$p\,(z > +1.00) = .8413$

17. a. $z = \frac{X - \mu}{\sigma} = \frac{140 - 100}{15} = \frac{+40}{15} = +2.67$

$p(z > +2.67) = .0038$

b. $z = \frac{X - \mu}{\sigma} = \frac{120 - 100}{15} = \frac{+20}{15} = +1.33$ and $z = +2.67$ from previous answer.

$p(+1.33 < z < +2.67) = .0918 - .0038 = .0880$

c. $z = \frac{X - \mu}{\sigma} = \frac{90 - 100}{15} = \frac{-10}{15} = -0.67$

$z = \frac{X - \mu}{\sigma} = \frac{109 - 100}{15} = \frac{+9}{15} = +0.60$

$p(-0.67 < z < +0.60) = .7468 - .2743 = .4743$

d. $p(z > +1.65) = .05$

$X = \mu + z\sigma$

$= 100 + (+1.65)15$

$= 100 + (+24.75)$

$= 124.75$

$p(X > 124.75) = .05$

e. $p(z < +0.67) = .75$

$X = \mu + z\sigma$

$= 100 + (+0.67)15$

$= 110.05$

$p(X < 110.05) = .75$

19. a. $Q1$ corresponds to a z-score of -0.67. Thus,

$Q1 = \mu + (-0.67)\sigma$

$= 35 + (-0.67)6$

$= 35 + (-4.02)$

$= 30.98$

Similarly, $Q3$ corresponds to a z-score of $+0.67$. Thus,

$Q3 = \mu + (+0.67)\sigma$

$= 35 + (+0.67)6$

$= 35 + (+4.02)$

$= 39.02$

The interquartile range is

$IQR = Q3 - Q1 = 39.02 - 30.98 = 8.04$.

b.

Actual Exam Score	z	percentile rank
33	−0.33	37.07
30	−0.83	20.33
36	+0.17	56.75
36	+0.17	56.75
26	−1.50	6.68
35	0.00	50.00
40	+0.83	79.67
38	+0.50	69.15
44	+1.50	93.32
42	+1.17	87.90
21	−2.33	0.99
35	0.00	50.00
41	+1.00	84.13
29	−1.00	15.87
36	+0.17	56.75

c. Bottom 25%

Actual Exam Score	Perceived Exam Score
30	35
26	34
21	35
29	32

Mean perceived exam score = 34.00

Top 25%

Actual Exam Score	Perceived Exam Score
40	37
44	40
42	41
41	43

Mean perceived exam score = 40.25. Notice that the difference in perceived exam score is small compared to the difference in actual exam score.

21. a. $X = 9$ corresponds to a z-score of -0.40.

$p(z > -0.40) = .6554$

b. $X = 8$ corresponds to a z-score of $z = -0.80$.

$X = 12$ corresponds to a z-score of $z = +0.80$.

$p(8 < X < 12) = .5762$.

c. $Q1$ corresponds to a z-score of -0.67. Thus,

$Q1 = \mu + (-0.67)\sigma$

$= 10 + (-0.67)2.5$

$= 10 + (-1.68) = 8.32$

Similarly, $Q3$ corresponds to a z-score of $+0.67$. Thus,

$Q3 = \mu + (-0.67)\sigma$

$= 10 + (+0.67)2.5$

$= 10 + (+1.68) = 11.68$

The interquartile range is

$IQR = Q3 - Q1 = 11.68 - 8.32 = 3.36$.

23. a. $X = 145$ corresponds to $z = +3.00$

$p(X > 145) = .0010$.

b. $X = 110$ corresponds to $z = +0.67$

$p(X > 110) = .2514$.

CHAPTER 7 Probability and Samples

1. **a.** The distribution of sample means consists of the sample means for all the possible random samples of a specific size (n) from a specific population.
   b. The central limit theorem specifies the basic characteristics of the distribution of sample means for any size samples from any population. Specifically, the shape will approach a normal distribution as the sample size increases, the mean is equal to the population mean, and the standard deviation of the distribution of sample means (standard error) equals the population standard deviation divided by the square root of the sample size (n).
   c. The expected value of M is the mean of the distribution of sample means (μ).
   d. The standard error of M is the standard deviation of the distribution of sample means $(\sigma_M = \frac{\sigma}{\sqrt{n}})$.

3. **a.** s is the sample standard deviation, σ is the population standard deviation, σ_M is the standard deviation of the distribution of sample means (i.e., it is the standard error).
   b. M is the mean of a set of sample scores, μ is the population mean, and μ_M is the mean of the distribution of sample means.

5. The distribution will be normal because $n > 30$, with an expected value of $\mu = 90$ and a standard error of
$$\sigma_M = \frac{\sigma}{\sqrt{n}} = \frac{32}{\sqrt{64}} = \frac{32}{8} = 4$$

7. **a.** $\sigma_M = \frac{\sigma}{\sqrt{n}} = \frac{18}{\sqrt{4}} = \frac{18}{2} = 9$
   b. $\sigma_M = \frac{\sigma}{\sqrt{n}} = \frac{18}{\sqrt{9}} = \frac{18}{3} = 6$
   c. $\sigma_M = \frac{\sigma}{\sqrt{n}} = \frac{18}{\sqrt{36}} = \frac{18}{6} = 3$

9. The expected value of the mean is equal to $\mu = 75$. The standard deviation of the distribution of study group means is $\sigma_M = \frac{\sigma}{\sqrt{n}} = \frac{10}{\sqrt{4}} = \frac{10}{2} = 5$

11. $\sigma_M = \frac{\sigma}{\sqrt{n}} = \frac{50}{\sqrt{25}} = \frac{50}{5} = 10$
$$z = \frac{M - \mu}{\sigma_M} = \frac{220 - 200}{10} = \frac{+20}{10} = +2.00$$

13. **a.** $\sigma_M = \frac{\sigma}{\sqrt{n}} = \frac{24}{\sqrt{4}} = \frac{24}{2} = 12$
$$z = \frac{M - \mu}{\sigma_M} = \frac{91 - 85}{12} = \frac{+6}{12} = +0.50$$
   b. $\sigma_M = \frac{\sigma}{\sqrt{n}} = \frac{24}{\sqrt{9}} = \frac{24}{3} = 8$
$$z = \frac{M - \mu}{\sigma_M} = \frac{91 - 85}{8} = \frac{+6}{8} = +0.75$$
   c. $\sigma_M = \frac{\sigma}{\sqrt{n}} = \frac{24}{\sqrt{16}} = \frac{24}{4} = 6$
$$z = \frac{M - \mu}{\sigma_M} = \frac{91 - 85}{6} = \frac{+6}{6} = +1.00$$
   d. $\sigma_M = \frac{\sigma}{\sqrt{n}} = \frac{24}{\sqrt{36}} = \frac{24}{6} = 4$
$$z = \frac{M - \mu}{\sigma_M} = \frac{91 - 85}{4} = \frac{+6}{4} = +1.50$$

15. **a.** $\sigma_M = \frac{\sigma}{\sqrt{n}} = \frac{10}{\sqrt{4}} = \frac{10}{2} = 5.0$
$$z = \frac{M - \mu}{\sigma_M} = \frac{53 - 50}{5.0} = \frac{+3}{5.0} = +0.60$$
$$p(M > 53) = .2743$$

b. $\sigma_M = \frac{\sigma}{\sqrt{n}} = \frac{10}{\sqrt{16}} = \frac{10}{4} = 2.5$
$$z = \frac{M - \mu}{\sigma_M} = \frac{53 - 50}{2.5} = \frac{+3}{2.5} = +1.20$$
$$p(M > 53) = .1151$$
c. $\sigma_M = \frac{\sigma}{\sqrt{n}} = \frac{10}{\sqrt{25}} = \frac{10}{5} = 2.0$
$$z = \frac{M - \mu}{\sigma_M} = \frac{53 - 50}{2.0} = \frac{+3}{2.0} = +1.50$$
$$p(M > 53) = .0668$$

17. **a.** $\sigma_M = \frac{\sigma}{\sqrt{n}} = \frac{8}{\sqrt{4}} = \frac{8}{2} = 4$
$$z = \frac{M - \mu}{\sigma_M} = \frac{32 - 30}{4} = \frac{+2}{4} = +0.50$$
$$p(M > 32) = .3085.$$
   b. Cannot answer because the distribution of sample means is not normal with $n = 4$.
   c. $\sigma_M = \frac{\sigma}{\sqrt{n}} = \frac{8}{\sqrt{64}} = \frac{8}{8} = 1$
$$z = \frac{M - \mu}{\sigma_M} = \frac{32 - 30}{1} = \frac{+2}{1} = +2.00$$
$$p(M > 32) = .0228$$
   d. With $n = 64$, the distribution of sample means is normal. $\sigma_M = 1$, $z = +2.00$, and $p = .0228$.

19. **a.** Cannot answer because the distribution of sample means is not normal with $n = 9$.
   b. $\sigma_M = \frac{\sigma}{\sqrt{n}} = \frac{12}{\sqrt{36}} = \frac{12}{6} = 2$
$$z = \frac{M - \mu}{\sigma_M} = \frac{75 - 71.5}{2} = \frac{+3.5}{2} = +1.75$$
$$p(M > 75) = .0401.$$
   c. For $M = 75$, $p(\mu < M < 75) = .4599$
   For $M = 70$, $z = \frac{M - \mu}{\sigma_M} = \frac{70 - 71.5}{2} = \frac{-1.5}{2} = -0.75$
   $p(70 < M < \mu) = .2734$. $p(70 < M < 75) = .4599 + .2734 = .7333$

21. **a.** $\sigma_M = \frac{\sigma}{\sqrt{n}} = \frac{10}{\sqrt{4}} = \frac{10}{2} = 5$
   b. $\sigma_M = \frac{\sigma}{\sqrt{n}} = \frac{10}{\sqrt{25}} = \frac{10}{5} = 2$

23. **a.** $n = 16$
   b. $n = 64$
   c. $n = 144$

25. **a.**

   b. Observing a human seemed to increase the amount of time to choose the end of the tube to inspect.

27. **a.** $\sigma_M = \frac{\sigma}{\sqrt{n}} = \frac{12}{\sqrt{36}} = \frac{12}{6} = 2$
$$z = \frac{M - \mu}{\sigma_M} = \frac{59 - 65}{2} = \frac{-6}{2} = -3.00$$
   The sample mean is an extreme value.
   b. $\sigma_M = \frac{\sigma}{\sqrt{n}} = \frac{30}{\sqrt{36}} = \frac{30}{6} = 5$
$$z = \frac{M - \mu}{\sigma_M} = \frac{59 - 65}{5} = \frac{-6}{5} = -1.20$$
   The sample mean is not an extreme value.

CHAPTER 8   Introduction to Hypothesis Testing

1. A hypothesis test does not allow a researcher to claim that an alternative hypothesis is true. A hypothesis test compares (1) the probability of obtaining the sample data if the null hypothesis were true to (2) α, which is the criterion for rejecting the null hypothesis.

3. a. Null: There is no effect of the college preparation course. Alternative: There is an effect of the college preparation course.
b. H_0: $\mu_{\text{PrepCourse}} = 20$; H_1: $\mu_{\text{PrepCourse}} \neq 20$

5. Both types of errors arise after decisions about the data are made. A Type I error occurs when the researcher rejects the null hypothesis but the null hypothesis is true (e.g., the treatment did not have an effect). A Type II error occurs when the researcher fails to reject the null hypothesis but the null hypothesis is false (e.g., the treatment really does have an effect). Type I errors are worse than Type II errors because Type I errors result in false reports in the literature but Type II errors usually do not.

7. a. Increasing the size of the treatment effect *increases* the value of z.
b. Increasing the population standard deviation *decreases* the value of z.
c. Increasing the number of scores in the sample *increases* the value of z.

9. a. The null hypothesis is that the program did not change hours spent studying. The alternative hypothesis is that the program affected the number of hours spent studying.
b. Step 1: H_0: $\mu_{\text{Program}} = 15$; H_1: $\mu_{\text{Program}} \neq 15$. $\alpha = .05$
Step 2: Critical region $z = +/-1.96$
Step 3: $\sigma_M = \frac{\sigma}{\sqrt{n}} = \frac{9}{\sqrt{36}} = \frac{9}{6} = 1.5$
$z = \frac{M - \mu}{\sigma_M} = \frac{18 - 15}{1.5} = \frac{+3}{1.5} = +2.00$
Step 4: z obtained is in the critical region. Reject the null hypothesis. There is evidence that the motivational program affected amount of time that students spent studying.

11. a. The null hypothesis is that using an electronic textbook has no effect on exam scores. The alternative hypothesis is that using an electronic textbook has an effect on exam scores.
b. Step 1: H_0: $\mu_{\text{Electronic}} = 77$; H_1: $\mu_{\text{Electronic}} \neq 77$. $\alpha = .05$
Step 2: Critical region $z = +/-1.96$
Step 3: $\sigma_M = \frac{\sigma}{\sqrt{n}} = \frac{8}{\sqrt{16}} = \frac{8}{4} = 2$
$z = \frac{M - \mu}{\sigma_M} = \frac{72.5 - 77.0}{2} = \frac{-4.5}{2} = -2.25$
Step 4: z obtained is in the critical region. Reject the null hypothesis. There is evidence that studying from a screen affects exam scores.

13. a. Step 1: H_0: $\mu_{\text{treatment}} = 20$; H_1: $\mu_{\text{treatment}} \neq 20$. $\alpha = .05$
Step 2: Critical region $z = +/-1.96$

Step 3: $\sigma_M = \frac{\sigma}{\sqrt{n}} = \frac{10}{\sqrt{25}} = \frac{10}{5} = 2$
$z = \frac{M - \mu}{\sigma_M} = \frac{25 - 20}{2} = \frac{+5}{2} = +2.50$
Step 4: z obtained is in the critical region. Reject the null hypothesis.
b. Step 1: H_0: $\mu_{\text{treatment}} = 20$; H_1: $\mu_{\text{treatment}} \neq 20$. $\alpha = .05$
Step 2: Critical region $z = +/-1.96$
Step 3: $\sigma_M = \frac{\sigma}{\sqrt{n}} = \frac{10}{\sqrt{4}} = \frac{10}{2} = 5$
$z = \frac{M - \mu}{\sigma_M} = \frac{25 - 20}{5} = \frac{+5}{5} = +1.00$
Step 4: z obtained is not in the critical region. Fail to reject the null hypothesis.
c. Increasing the sample size decreases the value of σ_M, increases the value of z, and increases the likelihood that the hypothesis test will reject the null hypothesis.

15. a. With a 6-point treatment effect, for the z-score to be greater than $+1.96$, the standard error must be smaller than 3.06 because *critical* $z = 1.96 = \frac{M - \mu}{\sigma_M} = \frac{6}{\sigma_M}$. Multiply both sides by σ_M and then divide both sides by 1.96. If $\sigma = 10$, then $3.06 = \frac{10}{\sqrt{n}}$. Multiply both sides by $\sqrt{n}$, divide both sides by 3.06, and square both sides to remove the square root sign over n. The sample size must be greater than 10.68; a sample of $n = 11$ or larger is needed.
b. With a 3-point treatment effect, for the z-score to be greater than 1.96, the standard error must be smaller than 1.53. The sample size must be greater than 42.72; a sample of $n = 43$ or larger is needed.

17. a. Step 1: H_0: $\mu_{\text{Course}} \leq 500$; H_1: $\mu_{\text{Course}} > 500$. $\alpha = .01$, one-tailed.
Step 2: Critical region for $z = +2.33$
Step 3: $\sigma_M = \frac{\sigma}{\sqrt{n}} = \frac{100}{\sqrt{20}} = \frac{100}{4.47} = 22.37$
$z = \frac{M - \mu}{\sigma_M} = \frac{562 - 500}{22.37} = \frac{+62}{22.37} = +2.77$
Step 4: z obtained is in the critical region. Reject the null hypothesis. There is evidence that the new course affected SAT scores.
b. $d = \frac{M - \mu}{\sigma} = \frac{562 - 500}{100} = \frac{62}{100} = 0.62$
c. The new course had a significant effect on SAT scores, $z = 2.77, p < .05, d = 0.62$.

19. The z-test cannot be conducted because the test assumes that the treatment affects the mean but not the standard deviation.

21. a. **STEP 1: Sketch the Distributions for the Null and Alternative Hypotheses.** $\sigma_M = \frac{\sigma}{\sqrt{n}} = \frac{10}{\sqrt{4}} = \frac{10}{2} = 5$. See figure at the top of page 613.
STEP 2: Locate the Critical Regions and Compute M_{critical}. The hypothesis test will be two-tailed, $\alpha = .05$. Thus, the critical boundaries are ± 1.96. The critical sample mean in the right tail of the distribution is
$M_{\text{critical}} = \mu_{\text{null}} + 1.96(\sigma_M) = 50 + 1.96(5) = 59.80$

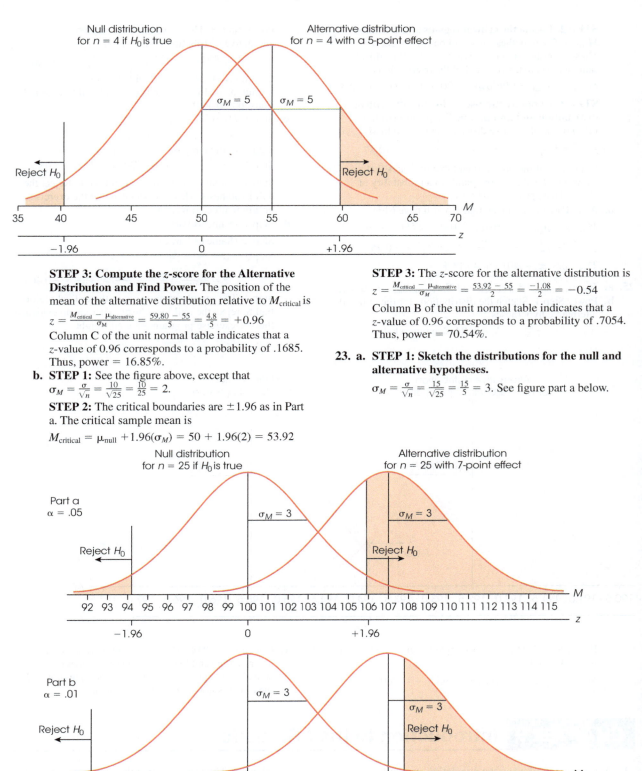

STEP 3: Compute the z-score for the Alternative Distribution and Find Power. The position of the mean of the alternative distribution relative to $M_{critical}$ is

$$z = \frac{M_{critical} - \mu_{alternative}}{\sigma_M} = \frac{59.80 - 55}{5} = \frac{4.8}{5} = +0.96$$

Column C of the unit normal table indicates that a z-value of 0.96 corresponds to a probability of .1685. Thus, power = 16.85%.

b. STEP 1: See the figure above, except that

$$\sigma_M = \frac{\sigma}{\sqrt{n}} = \frac{10}{\sqrt{25}} = \frac{10}{25} = 2.$$

STEP 2: The critical boundaries are ±1.96 as in Part a. The critical sample mean is

$$M_{critical} = \mu_{null} + 1.96(\sigma_M) = 50 + 1.96(2) = 53.92$$

STEP 3: The z-score for the alternative distribution is

$$z = \frac{M_{critical} - \mu_{alternative}}{\sigma_M} = \frac{53.92 - 55}{2} = \frac{-1.08}{2} = -0.54$$

Column B of the unit normal table indicates that a z-value of 0.96 corresponds to a probability of .7054. Thus, power = 70.54%.

23. a. STEP 1: Sketch the distributions for the null and alternative hypotheses.

$$\sigma_M = \frac{\sigma}{\sqrt{n}} = \frac{15}{\sqrt{25}} = \frac{15}{5} = 3. \text{ See figure part a below.}$$

STEP 2: Locate the critical regions and compute $M_{critical}$. The hypothesis test will be two-tailed, $\alpha = .05$. Thus, the critical boundaries are ± 1.96. The critical sample mean in the right tail of the distribution is

$$M_{critical} = \mu_{null} + 1.96(\sigma_M) = 100 + 1.96(3) = 105.88$$

STEP 3: Compute the z-score for the alternative distribution and find power. The position of the mean of the alternative distribution relative to $M_{critical}$ is

$$z = \frac{M_{critical} - \mu_{alternative}}{\sigma_M} = \frac{105.88 - 107}{3} = \frac{-1.12}{3} = -0.37$$

Column B of the unit normal table indicates that a z-value of -0.37 corresponds to a probability of .6443. Thus, power = 64.43%.

b. As in Part a except (see figure part b page 613)

$$M_{critical} = \mu_{null} + 2.58(\sigma_M) = 100 + 2.58(3) = 107.74$$

$$z = \frac{M_{critical} - \mu_{alternative}}{\sigma_M} = \frac{107.74 - 107}{3} = \frac{0.74}{3} = +0.25$$

Thus, power = .4013 or 40.13%.

25. a. $H_0: \mu_{NoTreatment} = \mu_{Excercise}$ and $H_1: \mu_{NoTreatment} \neq \mu_{Exercise}$

b. Power Step 1: Sketch the distributions for the null and alternative hypotheses.

$$\sigma_M = \frac{\sigma}{\sqrt{n}} = \frac{42}{\sqrt{2500}} = \frac{42}{50} = 0.84. \text{ See figure below.}$$

Power Step 3: The position of the mean of the alternative distribution relative to $M_{critical}$ is

$$z = \frac{M_{critical} - \mu_{alternative}}{\sigma_M} = \frac{193.85 - 192.5}{0.84} = \frac{+1.335}{0.84} = +1.61$$

Column B of the unit normal table indicates that a z-value of 1.61 corresponds to a probability of .9463. Thus, power = 94.63%.

c. Step 1: $H_0: \mu_{NoTreatment} = \mu_{Exercise}$
   $H_1: \mu_{NoTreatment} \neq \mu_{Exercise}$

Step 2: Critical region for $z = \pm 1.96$

Step 3: $z = \frac{M - \mu}{\sigma_M} = \frac{192.1 - 195.5}{0.84} = \frac{-3.40}{0.84} = -4.05$

Step 4: z obtained is in the critical region. Reject the null hypothesis. There is evidence that the exercise program affected weight.

d. Step 1: (Same as Part c)

Step 2: (Same as Part c)

Step 3: $\sigma_M = \frac{\sigma}{\sqrt{n}} = \frac{42}{\sqrt{25}} = \frac{42}{5} = 8.40$

$$z = \frac{M - \mu}{\sigma_M} = \frac{192.1 - 195.5}{8.40} = \frac{-3.40}{8.40} = -0.40$$

Step 4: The sample mean is not in the critical region. Fail to reject the null hypothesis. There is no evidence that the exercise program affected weight.

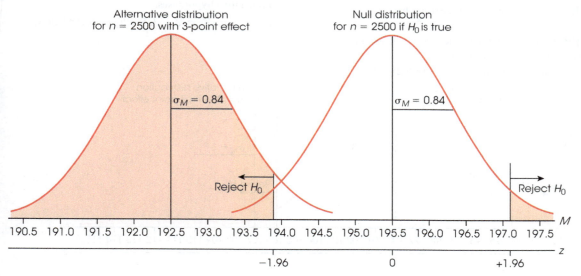

Alternative distribution for $n = 2500$ with 3-point effect

Null distribution for $n = 2500$ if H_0 is true

$\sigma_M = 0.84$

$\sigma_M = 0.84$

Reject H_0

Reject H_0

190.5 191.0 191.5 192.0 192.5 193.0 193.5 194.0 194.5 195.0 195.5 196.0 196.5 197.0 197.5   M

-1.96 0 $+1.96$ z

Power Step 2: The critical sample mean in the left tail of the distribution is

$$M_{critical} = \mu_{null} - 1.96(\sigma_M) = 195.5 - 1.96(0.84) = 193.85$$

e. Cohen's $d = 0.08$. The effect size is small. With a very large sample of $n = 2500$, the effect is statistically significant. With a sample size of $n = 25$, the effect is not statistically significant.

CHAPTER 9   Introduction to the *t* Statistic

1. A z-score is used when the population standard deviation (or variance) is known. The *t* statistic is used when the population variance and standard deviation is unknown. The *t* statistic uses the sample variance or standard deviation in place of the unknown population parameters.

3. a. Sample variance measures the variability in the sample.

b. $s_M = \sqrt{\frac{s^2}{n}} = \sqrt{\frac{100}{25}} = \sqrt{4} = 2$. The standard error ($s_M$) measures the typical distance between a sample mean and a population mean in the distribution of sample means.

5. a. $M = \frac{\Sigma X}{n} = \frac{125}{5} = 25$,

$SS = \Sigma(X - M)^2 = (20 - 25)^2 + (25 - 25)^2$
$\qquad + (30 - 25)^2 + (20 - 25)^2 + (30 - 25)^2$

$\qquad = (-5)^2 + (0)^2 + (5)^2 + (-5)^2 + (5)^2$

$\qquad = 25 + 0 + 25 + 25 + 25 = 100.$

$df = n - 1 = 5 - 1 = 4$

$s^2 = \frac{SS}{df} = \frac{100}{4} = 25.$

b. $s_M = \sqrt{\frac{s^2}{n}} = \sqrt{\frac{25}{5}} = \sqrt{5} = 2.24.$

7. The sample variance (s^2) or sample standard deviation (s) used to compute t changes from one sample to another and contributes to the variability of the t statistics. A z-score uses the population variance, which is constant from one sample to another.

9. a. $df = n - 1 = 9 - 1 = 8$ and critical $t = 2.306$

b. $df = 15$ and critical $t = 2.131$

c. $df = 35$ and critical $t = 2.042$

d. ts for a, b, and c equal 1.860, 1.753, and 1.697, respectively

e. ts for a, b, and c equal 3.355, 2.947, and 2.750, respectively

11. a. $df = n - 1 = 7 - 1 = 6$, $M = \frac{\Sigma X}{n} = \frac{315}{7} = 45$, and $s^2 = \frac{SS}{df} = \frac{96}{6} = 16$. If your value for SS was incorrect, the solution is:

$SS = \Sigma(X - M)^2$

$\qquad = (37 - 45)^2 + (49 - 45)^2 + (47 - 45)^2 + (47 - 45)^2 + (47 - 45)^2 + (43 - 45)^2 + (45 - 45)^2$

$\qquad = (-8)^2 + (4)^2 + (2)^2 + (2)^2 + (2)^2 + (-2)^2 + (0)^2$

$\qquad = 64 + 16 + 4 + 4 + 4 + 4 + 0 = 96$

b. $M - \mu = 45 - 50 = -5$

c. $s_M = \sqrt{\frac{s^2}{n}} = \sqrt{\frac{16}{7}} = \sqrt{2.29} = 1.51.$

d. Step 1: H_0: $\mu_{Treated} = 50$, H_1: $\mu_{Treated} \neq 50$, and $\alpha = .05$, two-tailed.

Step 2: Critical $t = 2.447$

Step 3: $t = \frac{M - \mu}{s_M} = \frac{45 - 50}{1.51} = \frac{-5}{1.51} = -3.31$

Step 4: t from Step 3 is more extreme than t critical from Step 2, so reject null hypothesis. There is evidence that the treatment affected the scores.

e. Step 1: H_0: $\mu_{Treated} = 50$, H_1: $\mu_{Treated} \neq 50$, and $\alpha = .01$, two-tailed.

Step 2: Critical $t = 3.707$

Step 3: $t = \frac{M - \mu}{s_M} = \frac{45 - 50}{1.51} = \frac{-5}{1.51} = -3.31$

Step 4: t from Step 3 is less extreme than t critical from Step 2 so fail to reject null hypothesis. There is no evidence that the treatment affected the scores.

13. a. Step 1: H_0: $\mu_{Treated} = 40$, H_1: $\mu_{Treated} \neq 40$, and $\alpha = .05$, two-tailed.

Step 2: Critical $t = 3.182$

Step 3: $s_M = \sqrt{\frac{s^2}{n}} = \sqrt{\frac{36}{4}} = \sqrt{9} = 3$

$t = \frac{M - \mu}{s_M} = \frac{44.5 - 40.0}{3} = \frac{4.5}{3} = 1.50$

Step 4: t from Step 3 is less extreme than t critical from Step 2 so fail to reject null hypothesis. There is no evidence that the treatment affected the scores.

b. Step 1: as in Part a.

Step 2: Critical $t = 2.131$

Step 3: $s_M = \sqrt{\frac{s^2}{n}} = \sqrt{\frac{36}{16}} = \sqrt{2.25} = 1.5$

$t = \frac{M - \mu}{s_M} = \frac{44.5 - 40.0}{1.5} = \frac{4.5}{1.5} = 3.00.$

Step 4: t from Step 3 is more extreme than t critical from Step 2 so reject null hypothesis. There is evidence that the treatment affected the scores.

c. Increasing the sample size increases the likelihood that the hypothesis test will reject the null hypothesis if the null hypothesis is false.

15. a. Step 1: H_0: $\mu_{Treated} = 73.4$, H_1: $\mu_{Treated} \neq 73.4$, and $\alpha = .05$, two-tailed.

Step 2: Critical $t = 2.131$

Step 3: $s_M = \frac{s}{\sqrt{n}} = \frac{8.4}{\sqrt{16}} = \frac{8.4}{4} = 2.1$

$t = \frac{M - \mu}{s_M} = \frac{78.3 - 73.4}{2.1} = \frac{4.9}{2.1} = 2.33$

Step 4: t from Step 3 is more extreme than t critical from Step 2 so reject null hypothesis. There is evidence that answering questions while studying affected the scores.

b. estimated $d = \frac{\text{mean difference}}{\text{standard deviation}} = \frac{78.3 - 73.4}{8.4} = \frac{4.9}{8.4} = 0.58$

$r^2 = \frac{t^2}{t^2 + df} = \frac{(2.33)^2}{(2.33)^2 + 15} = \frac{5.43}{5.43 + 15} = \frac{5.43}{20.43} = 0.27$

17. a. Step 1: H_0: $\mu_{Treated} = 20$, H_1: $\mu_{Treated} \neq 20$, and $\alpha = .05$, two-tailed.

Step 2: Critical $t = 2.306$

Step 3: $s_M = \sqrt{\frac{s^2}{n}} = \sqrt{\frac{9}{9}} = \sqrt{1} = 1.00$

$t = \frac{M - \mu}{s_M} = \frac{22 - 20}{1.00} = \frac{22 - 20}{1.00} = 2.00$

Step 4: t from Step 3 is less extreme than t critical from Step 2 so fail to reject null hypothesis. There is no evidence that the treatment affected the scores.

estimated $d = \frac{\text{mean difference}}{\text{standard deviation}} = \frac{22 - 20}{3} = \frac{2}{3} = 0.67$

b. Step 1: H_0: $\mu_{Treated} = 20$, H_1: $\mu_{Treated} \neq 20$, and $\alpha = .05$, two-tailed.

Step 2: Critical $t = 2.042$

Step 3: $s_M = \sqrt{\frac{s^2}{n}} = \sqrt{\frac{9}{36}} = \sqrt{0.25} = 0.50$

$t = \frac{M - \mu}{s_M} = \frac{22 - 20}{0.50} = \frac{22 - 20}{0.50} = 4.00$

Step 4: t from Step 3 is more extreme than t critical from Step 2 so reject the null hypothesis. There is evidence that the treatment affected the scores.

estimated $d = \frac{\text{mean difference}}{\text{standard deviation}} = \frac{22 - 20}{3} = \frac{2}{3} = 0.67$

c. Increasing the sample size increases the likelihood of rejecting the null hypothesis but has little or no effect on Cohen's d.

19. a. Step 1: H_0: $\mu_{\text{Treated}} = 4$, H_1: $\mu_{\text{Treated}} \neq 4$, and $\alpha = .05$, two-tailed.

Step 2: Critical $t = 2.131$ with 15 degrees of freedom

$$s_M = \sqrt{\frac{s^2}{n}} = \sqrt{\frac{0.16}{16}} = \sqrt{0.01} = 0.1$$

$$t = \frac{M - \mu}{s_M} = \frac{3.78 - 4.00}{.1} = \frac{-0.22}{.1} = -2.20$$

Step 4: t from Step 3 is more extreme than t critical from Step 2 so reject null hypothesis. There is evidence that the drug affected the scores.

b. For 95% confidence interval, use $t = \pm 2.131$. The interval is

$$\mu = M \pm t(s_M) = 3.78 \pm 2.131(.1) = 3.78 \pm .2131$$

The interval extends from 3.57 to 3.99.

c. estimated $d = \frac{\text{mean difference}}{\text{standard deviation}} = \frac{3.78 - 4.00}{0.4} = \frac{0.22}{0.4} = 0.55$

$$r^2 = \frac{t^2}{t^2 + df} = \frac{(-2.20)^2}{(-2.20)^2 + 15} = \frac{4.84}{4.84 + 15} = \frac{4.84}{19.84} = 0.24$$

21. a. Step 1: H_0: $\mu_{\text{procrastinators}} \leq$, 1 day H_1: $\mu_{\text{procrastinators}} >$ 1 day, and $\alpha = .05$, one-tailed.

Step 2: Critical $t = 1.833$ with 9 degrees of freedom

Step 3:

$$SS = \Sigma X^2 - \frac{(\Sigma X)^2}{n} = 997 - \frac{(73)^2}{10} = 997 - \frac{5329}{10}$$

$$= 997 - 532.9 = 464.1$$

$$s^2 = \frac{SS}{df} = \frac{464.1}{9} = 51.57.$$

$$M = \frac{\Sigma X}{n} = \frac{73}{10} = 7.3$$

$$s_M = \sqrt{\frac{s^2}{n}} = \sqrt{\frac{51.57}{10}} = \sqrt{5.157} = 2.27$$

$$t = \frac{M - \mu}{s_M} = \frac{7.3 - 1.0}{2.27} = \frac{6.3}{2.27} = 2.78$$

Step 4: t from Step 3 is more extreme than t critical from Step 2, so reject the null hypothesis. There is evidence that high procrastinators waited more than one day to return the survey.

b. For a 95% confidence interval, use $t = \pm 2.262$. The interval is

$$\mu = M \pm t(s_M) = 7.3 \pm 2.262(2.27) = 7.3 \pm 5.13$$

The interval extends from 2.17 to 12.43.

c. High procrastinators waited significantly longer than one day to return the survey, $t(9) = 2.78$, $p < .05$, one-tailed, 95%CI[2.17, 12.43].

CHAPTER 10 The t Test for Two Independent Samples

1. An independent-measures study uses a separate sample for each of the treatments or populations being compared.

3. a. $df_1 = n_1 - 1 = 7 - 1 = 6$

$df_2 = n_2 - 1 = 7 - 1 = 6$

$s_1^2 = \frac{SS_1}{df_1} = \frac{72}{6} = 12$ and $s_2^2 = \frac{SS_2}{df_2} = \frac{24}{6} = 4$

$s_P^2 = \frac{SS_1 + SS_2}{df_1 + df_2} = \frac{72 + 24}{6 + 6} = \frac{96}{12} = 8$

b. $df_1 = 6$ and $df_2 = n_2 - 1 = 11 - 1 = 10$

$s_1^2 = 12$ and $s_2^2 = \frac{SS_2}{df_2} = \frac{24}{10} = 2.4$

$s_P^2 = \frac{SS_1 + SS_2}{df_1 + df_2} = \frac{72 + 24}{6 + 10} = \frac{96}{16} = 6$

5. a. $df_1 = n_1 - 1 = 9 - 1 = 8$

$df_2 = n_2 - 1 = 9 - 1 = 8$

$s_P^2 = \frac{SS_1 + SS_2}{df_1 + df_2} = \frac{546 + 606}{8 + 8} = \frac{1152}{16} = 72$

b. $s_{M_1 - M_2} = \sqrt{\frac{s_P^2}{n_1} + \frac{s_P^2}{n_2}} = \sqrt{\frac{72}{9} + \frac{72}{9}} = \sqrt{8 + 8} =$

$\sqrt{16} = 4.00$

c. $t = \frac{(M_1 - M_2) - (\mu_1 - \mu_2)}{s_{M_1 - M_2}} = \frac{8}{4.00} = 2.00$. With $df_{\text{total}} = df_1 + df_2 = 16$, the critical t value is ± 2.120. We fail to reject the null hypothesis.

7. Step 1: H_0: $\mu_{\text{Watch}} - \mu_{\text{No Watch}} = 0$

H_1: $\mu_{\text{Watch}} - \mu_{\text{No Watch}} \neq 0$

Step 2: $df_1 = n_1 - 1 = 10 - 1 = 9$

$df_2 = n_2 - 1 = 10 - 1 = 9$

$df = df_1 + df_2 = 9 + 9 = 18$

The critical value is ± 2.878.

Step 3: $s_P^2 = \frac{SS_1 + SS_2}{df_1 + df_2} = \frac{200 + 160}{9 + 9} = \frac{360}{18} = 20$

$s_{M_1 - M_2} = \sqrt{\frac{s_P^2}{n_1} + \frac{s_P^2}{n_2}} = \sqrt{\frac{20}{10} + \frac{20}{10}} = \sqrt{2 + 2} = \sqrt{4} = 2.00$

$t = \frac{(M_1 - M_2) - (\mu_1 - \mu_2)}{s_{M_1 - M_2}} = \frac{(93 - 85) - 0}{2.00} = \frac{8}{2.00} = 4.00$

Step 4: Reject the null hypothesis because the t-value from Step 3 is more extreme than the critical region identified in Step 2. There is evidence that participants who watched *Sesame Street* had significantly higher grades than those participants who did not watch.

9. a. Step 1: H_0: $\mu_{\text{One Day}} - \mu_{\text{One Week}} = 0$

H_1: $\mu_{\text{One Day}} - \mu_{\text{One Week}} \neq 0$

Step 2: $df_1 = n_1 - 1 = 20 - 1 = 19$

$df_2 = n_2 - 1 = 20 - 1 = 19$

$df = df_1 + df_2 = 19 + 19 = 38$

The critical t-value is ± 2.042.

Step 3: $s_P^2 = \frac{SS_1 + SS_2}{df_1 + df_2} = \frac{395 + 460}{19 + 19} = \frac{855}{38} = 22.5$

$s_{M_1 - M_2} = \sqrt{\frac{s_P^2}{n_1} + \frac{s_P^2}{n_2}} = \sqrt{\frac{22.5}{20} + \frac{22.5}{20}}$

$= \sqrt{2.25} = 1.50$

$t = \frac{(M_1 - M_2) - (\mu_1 - \mu_2)}{s_{M_1 - M_2}} = \frac{(26.4 - 29.6) - 0}{1.50}$

$= \frac{-3.20}{1.50} = -2.13$

Step 4: Reject the null hypothesis because the t-value from Step 3 is more extreme than the critical region identified in Step 2. There is evidence that a one-week gap is better for memory than a one-day gap.

11. a. Step 1: H_0: $\mu_{\text{Neutral}} - \mu_{\text{Anxiety}} = 0$

H_1: $\mu_{\text{Neutral}} - \mu_{\text{Anxiety}} \neq 0$

Step 2: $df_1 = n_1 - 1 = 6 - 1 = 5$

$df_2 = n_2 - 1 = 6 - 1 = 5$

$df = df_1 + df_2 = 5 + 5 = 10$

The critical t-value is ± 2.228.

Step 3: $s_P^2 = \frac{SS_1 + SS_2}{df_1 + df_2} = \frac{76 + 84}{5 + 5} = \frac{160}{10} = 16$

$s_{M_1 - M_2} = \sqrt{\frac{s_P^2}{n_1} + \frac{s_P^2}{n_2}} = \sqrt{\frac{16}{6} + \frac{16}{6}} = \sqrt{5.33} = 2.31$

$t = \frac{(M_1 - M_2) - (\mu_1 - \mu_2)}{s_{M_1 - M_2}} = \frac{(12 - 5) - 0}{2.31} = \frac{7}{2.31} = 3.03$

Step 4: Reject the null hypothesis. There is a significant difference between the group that received anxiety-inducing statements and the group that received neutral statements.

b. $r^2 = \frac{t^2}{t^2 + df} = \frac{9.18}{9.18 + 10} = \frac{9.18}{19.18} = 0.48$

13. a. Step 1: H_0: $\mu_{\text{Binge}} - \mu_{\text{Daily}} = 0$, H_1: $\mu_{\text{Binge}} - \mu_{\text{Daily}} \neq 0$

Step 2: $df_1 = n_1 - 1 = 5 - 1 = 4$

$df_2 = n_2 - 1 = 5 - 1 = 4$

$df = df_1 + df_2 = 4 + 4 = 8$

The critical t-value is ± 2.306.

Step 3: $s_P^2 = \frac{SS_1 + SS_2}{df_1 + df_2} = \frac{214 + 178}{4 + 4} = \frac{392}{8} = 49$

$s_{M_1 - M_2} = \sqrt{\frac{s_P^2}{n_1} + \frac{s_P^2}{n_2}} = \sqrt{\frac{49}{5} + \frac{49}{5}} = \sqrt{19.6} = 4.43$

$t = \frac{(M_1 - M_2) - (\mu_1 - \mu_2)}{s_{M_1 - M_2}} = \frac{(79 - 92) - 0}{4.43} = \frac{-13}{4.43} = -2.93$

Step 4: Reject the null hypothesis. There is a significant difference between the group that watched the show in a single binge session and the group that watched the show in daily sessions.

b. estimated $d = \frac{M_1 - M_2}{\sqrt{s_P^2}} = \frac{79 - 92}{\sqrt{49}} = \frac{-13}{7} = 1.86$

c. The results indicate that binge-watching the television series resulted in significantly lower ratings of enjoyment than watching in daily sessions, $t(8) = -2.93, p < .05, d = 1.86$.

15. a. Step 1: H_0: $\mu_{\text{Solve}} - \mu_{\text{Memorize}} = 0$

H_1: $\mu_{\text{Solve}} - \mu_{\text{Memorize}} \neq 0$

Step 2: $df_1 = n_1 - 1 = 8 - 1 = 7$

$df_2 = n_2 - 1 = 8 - 1 = 7$

$df = df_1 + df_2 = 7 + 7 = 14$

The critical t-value is ± 2.145.

Step 3: $s_P^2 = \frac{SS_1 + SS_2}{df_1 + df_2} = \frac{108 + 116}{7 + 7} = \frac{224}{14} = 16$

$s_{M_1 - M_2} = \sqrt{\frac{s_P^2}{n_1} + \frac{s_P^2}{n_2}} = \sqrt{\frac{16}{8} + \frac{16}{8}} = \sqrt{4} = 2.00$

$t = \frac{(M_1 - M_2) - (\mu_1 - \mu_2)}{s_{M_1 - M_2}} = \frac{(10.5 - 6.16) - 0}{2.00} = \frac{4.34}{2.00} = 2.17$

Step 4: Reject the null hypothesis. There is a significant difference between the group that solved the problem independently and the group that memorized the solution.

b. $\mu_1 - \mu_2 = M_1 - M_2 \pm t(s_{M_1 - M_2}) =$
$10.5 - 6.16 \pm 1.761(2.00) = 4.34 \pm 3.52$

17. a. Step 1: H_0: $\mu_{\text{lower}} - \mu_{\text{upper}} = 0$, H_1: $\mu_{\text{lower}} - \mu_{\text{upper}} \neq 0$

Step 2: $df = df_1 + df_2 = 11 + 11 = 22$. With $df = 20$ and $\alpha = .05$, the critical region consists of t values beyond ± 2.074.

Step 3: $s_P^2 = \frac{SS_1 + SS_2}{df_1 + df_2} = \frac{11.91 + 9.21}{11 + 11} = \frac{21.12}{22} = 0.96$

$s_{M_1 - M_2} = \sqrt{\frac{s_P^2}{n_1} + \frac{s_P^2}{n_2}} = \sqrt{\frac{0.96}{12} + \frac{0.96}{12}} = \sqrt{0.08 + 0.08}$

$= \sqrt{0.16} = 0.40$

$t = \frac{(M_1 - M_2) - (\mu_1 - \mu_2)}{s_{M_1 - M_2}} = \frac{(5.2 - 4.3) - 0}{0.40} = \frac{0.90}{0.40} = 2.25$

Step 4: The t statistic is in the critical region. Reject the null hypothesis and conclude that there was a significant difference in point-sharing between lower socioeconomic status participants and upper-class participants.

b. $\mu_1 - \mu_2 = M_1 - M_2 \pm t(s_{M_1 - M_2}) =$
$5.2 - 4.3 \pm 1.717(0.40) = 0.90 \pm 0.6868$

19. a. The size of the two samples influences the magnitude of the estimated standard error in the denominator of the t statistic. As sample size increases, the value of t also increases (moves farther from zero), and the likelihood of rejecting H_0 also increases; however, sample size has little or no effect on measures of effect size.

b. The variability of the scores influences the estimated standard error in the denominator.
As the variability of the scores increases, the value of t decreases (becomes closer to zero), the likelihood of rejecting H_0 decreases, and measures of effect size decrease.

21. a. $s_{M_1 - M_2} = \sqrt{\frac{s_P^2}{n_1} + \frac{s_P^2}{n_2}} = \sqrt{\frac{135}{6} + \frac{135}{10}} = \sqrt{36} = 6.00$

b. $s_{M_1 - M_2} = \sqrt{\frac{s_P^2}{n_1} + \frac{s_P^2}{n_2}} = \sqrt{\frac{135}{12} + \frac{135}{15}} = \sqrt{20.25} = 4.50$

c. Larger samples produce smaller standard error.

23. a. $s_P^2 = \frac{df_1 s_1^2 + df_2 s_2^2}{df_1 + df_2} = \frac{3(17) + 7(27)}{3 + 7} = \frac{51 + 189}{3 + 7} = \frac{240}{10} = 24$

$s_{M_1 - M_2} = \sqrt{\frac{s_P^2}{n_1} + \frac{s_P^2}{n_2}} = \sqrt{\frac{24}{4} + \frac{24}{8}} = \sqrt{9} = 3.00$

b. $s_P^2 = \frac{df_1 s_1^2 + df_2 s_2^2}{df_1 + df_2} = \frac{3(68) + 7(108)}{3 + 7} = \frac{204 + 756}{3 + 7} = \frac{960}{10} = 96$

$s_{M_1 - M_2} = \sqrt{\frac{s_P^2}{n_1} + \frac{s_P^2}{n_2}} = \sqrt{\frac{96}{4} + \frac{96}{8}} = \sqrt{36} = 6.00$

CHAPTER 11 The *t* Test for Two Related Samples

1. a. Independent-measures: The researcher is comparing two separate groups.
 b. Repeated-measures: There are two scores (humorous and not humorous) for each individual.
 c. Repeated-measures: There are two scores (before and after) for each individual.

3. a. An independent-measures design would require two separate samples, each with 22 participants, for a total of 44 participants.
 b. A repeated-measures design would use the same sample of $n = 22$ participants in both treatment conditions.

5. a. $df = n - 1 = 12 - 1 = 11$

$$s^2 = \frac{SS}{df} = \frac{396}{11} = 36$$

$$s = \sqrt{s^2} = \sqrt{36} = 6$$

 b. $s_{M_D} = \frac{s}{\sqrt{n}} = \frac{6}{\sqrt{12}} = \frac{6}{3.46} = 1.73$

7. a.

Participant	Before Treatment	After Treatment	$D = X_2 - X_1$	$D^2 = (X_2 - X_1)^2$
A	66	84	18	324
B	50	44	−6	36
C	38	52	14	196
D	58	56	−2	4
E	50	52	2	4
F	34	42	8	64
G	44	51	7	49
H	42	49	7	49
I	62	67	5	25
J	50	57	7	49
K	56	62	6	36

$$M_D = \frac{\Sigma D}{n} = \frac{66}{11} = 6$$

 b. $SS = \Sigma D^2 - \frac{(\Sigma D)^2}{n}$

$$= 836 - \frac{(66)^2}{11}$$

$$= 836 - \frac{4356}{11} = 836 - 396 = 440$$

$$df = n - 1 = 11 - 1 = 10$$

$$s^2 = \frac{SS}{df} = \frac{440}{10} = 44$$

$$s_{M_D} = \sqrt{\frac{s^2}{n}} = \sqrt{\frac{44}{11}} = \sqrt{4} = 2$$

 c. Step 1: H_0: $\mu_D = 0$, H_0: $\mu_D \neq 0$

 Step 2: For $\alpha = .05$, two-tailed, and $df = 10$, the critical t value is ± 2.228.

 Step 3: $t = \frac{M_D - \mu_D}{s_{M_D}} = \frac{6 - 0}{2} = 3$

 Step 4: The t-value calculated in Step 3 is more extreme than the critical value obtained in Step 2 so reject the null hypothesis and conclude that the effect of the treatment was significant.

9. a. Step 1: H_0: $\mu_D = 0$, H_1: $\mu_D \neq 0$

 Step 2: For $\alpha = .05$, two-tailed, and $df = 15$, the critical t value is ± 2.131.

 Step 3: $s^2 = \frac{SS}{df} = \frac{135}{15} = 9$

$$s_{M_D} = \sqrt{\frac{s^2}{n}} = \sqrt{\frac{9}{16}} = \sqrt{0.5625} = 0.75$$

$$t = \frac{M_D - \mu_D}{s_{M_D}} = \frac{2.6 - 0}{0.75} = 3.47$$

 Step 4: The t-value calculated in Step 3 is more extreme than the critical value obtained in Step 2, so reject the null hypothesis and conclude that the judged quality of objects was significantly different for self-purchases than for purchases made by others.

 b. estimated $d = \frac{M_D}{s} = \frac{2.6}{3} = 0.87$

 c. Participants rated the quality of items purchased by others significantly lower than self-purchased items, $t(15) = 3.47$, $p < .05$, $d = 0.87$.

11. a. Step 1: H_0: $\mu_D = 0$, H_1: $\mu_D \neq 0$

 Step 2: For $\alpha = .05$, two-tailed, and $df = n - 1 = 40 - 1 = 39$, the critical t value is ± 2.042. Notice that we used the critical t value for $df = 30$ because $df = 39$ is not listed.

 Step 3: $s_{M_D} = \frac{s}{\sqrt{n}} = \frac{21.5}{\sqrt{40}} = \frac{21.5}{6.32} = 3.40$

$$t = \frac{M_D - \mu_D}{s_{M_D}} = \frac{8.5 - 0}{3.40} = 2.50$$

 Step 4: The t-value calculated in Step 3 is more extreme than the critical value obtained in Step 2, so reject the null hypothesis and conclude that Tai Chi significantly affected pain and stiffness.

 b. estimated $d = \frac{M_D}{s} = \frac{8.5}{21.5} = 0.395$

13. a. Step 1: H_0: $\mu_D = 0$, H_1: $\mu_D \neq 0$

 Step 2: With $df = 5$ and $\alpha = .05$, the critical values are $t = \pm 2.571$.

 Step 3: $s^2 = \frac{SS}{df} = \frac{30}{5} = 6$

$$s_{M_D} = \sqrt{\frac{s^2}{n}} = \sqrt{\frac{6}{6}} = \sqrt{1.00} = 1.00$$

$$t = \frac{M_D - \mu_D}{s_{M_D}} = \frac{4 - 0}{1.00} = 4.00$$

 Step 4: The t-value calculated in Step 3 is more extreme than the critical value, so reject the null hypothesis.

 b. As in Part a, except:

 Step 3: $s^2 = \frac{SS}{df} = \frac{480}{5} = 96$

$$s_{M_D} = \sqrt{\frac{s^2}{n}} = \sqrt{\frac{96}{6}} = \sqrt{16.00} = 4.00$$

$$t = \frac{M_D - \mu_D}{s_{M_D}} = \frac{4 - 0}{4.00} = 1.00$$

 Step 4: Fail to reject the null hypothesis. There is no evidence that the treatment produced an effect.

 c. Low sample variability increases the likelihood of rejecting the null hypothesis if the null hypothesis is false.

15. a. Step 1: H_0: $\mu_D = 0$, H_1: $\mu_D \neq 0$

Step 2: With $df = 8$ and $\alpha = .05$, the critical values are $t = \pm 2.306$.

Step 3: $s_{M_D} = \frac{s}{\sqrt{n}} = \frac{6}{\sqrt{9}} = \frac{6}{3.00} = 2.00$

$$t = \frac{M_D - \mu_D}{s_{M_D}} = \frac{4 - 0}{2.00} = 2.00$$

Step 4: Fail to reject the null hypothesis. There is no evidence that the treatment produced an effect.

b. As above, except:

Step 2: With $df = 35$ and $\alpha = .05$, the critical values are ± 2.042.

Step 3: $s_{M_D} = \frac{s}{\sqrt{n}} = \frac{6}{\sqrt{36}} = \frac{6}{6.00} = 1.00$

$$t = \frac{M_D - \mu_D}{s_{M_D}} = \frac{4 - 0}{1.00} = 4.00$$

Step 4: Reject the null hypothesis. There is evidence that the treatment had a significant effect.

c. If other factors are held constant, a larger sample increases the likelihood of finding a significant mean difference.

17. a. Step 1: H_0: $\mu_{\text{Swear}} - \mu_{\text{Neutral}} = 0$,
H_1: $\mu_{\text{Swear}} - \mu_{\text{Neutral}} \neq 0$

Step 2: With $df = 16$ and $\alpha = .05$, the critical t value is ± 2.120.

Step 3: $SS_{\text{Neutral}} = \Sigma X^2 - \frac{(\Sigma X)^2}{n}$

$$= 612 - \frac{(72)^2}{9}$$

$$= 612 - \frac{5184}{9} = 612 - 576 = 36$$

$M_{\text{Neutral}} = \frac{\Sigma X}{n} = \frac{72}{9} = 8$

$SS_{\text{Swear}} = \Sigma X^2 - \frac{(\Sigma X)^2}{n}$

$$= 372 - \frac{(54)^2}{9} = 372 - \frac{2916}{9}$$

$$= 372 - 324 = 48$$

$M_{\text{Swear}} = \frac{\Sigma X}{n} = \frac{54}{9} = 6$

$s_P^2 = \frac{SS_{\text{Neutral}} + SS_{\text{Swear}}}{df_{\text{Neutral}} + df_{\text{Swear}}} = \frac{36 + 48}{8 + 8} = \frac{84}{16} = 5.25$

$s_{M_1 - M_2} = \sqrt{\frac{s_P^2}{n_1} + \frac{s_P^2}{n_2}} = \sqrt{\frac{5.25}{9} + \frac{5.25}{9}} = \sqrt{1.166} = 1.08$

$t = \frac{(M_1 - M_2) - (\mu_1 - \mu_2)}{s_{M_1 - M_2}} = \frac{(8 - 6) - 0}{1.08} = \frac{2}{1.08} = 1.85$

Step 4: The t value calculated in Step 3 is less extreme than the critical t value from Step 2. Fail to reject the null hypothesis and conclude that there is no evidence that swearing affected pain level.

b. Step 1: H_0: $\mu_D = 0$, H_1: $\mu_D \neq 0$

Step 2: With $df = 8$ and $\alpha = .05$, the critical values are $t = \pm 2.306$.

Step 3: $M_D = \frac{\Sigma D}{n} = \frac{-18}{9} = -2$

$SS = \Sigma D^2 - \frac{(\Sigma D)^2}{n} = 68 - \frac{(-18)^2}{9}$

$$= 68 - \frac{324}{9} = 68 - 36 = 32$$

$s^2 = \frac{SS}{df} = \frac{32}{8} = 4$

$s_{M_D} = \sqrt{\frac{s^2}{n}} = \sqrt{\frac{4}{9}} = \sqrt{0.44} = 0.67$
$t = \frac{M_D - \mu_D}{s_{M_D}} = \frac{-2 - 0}{0.67} = -2.99$. You might have

noticed that standard error is equal to $\frac{2}{3}$ and that -2 divided by $\frac{2}{3}$ equals -3.00. Thus, you may have obtained a value of -3.00 for the t statistic.

Step 4: The t value calculated in Step 3 is more extreme than the critical t value from Step 2. Reject the null hypothesis and conclude that there was significantly less pain while swearing.

19. a. Because the scores in each sample are the same as in Problem 17, the results are also the same. The 2-point mean difference has an estimated standard error of 1.08 and $t(16) = 1.85$. Fail to reject the null hypothesis.

b. As in Problem 22, except:

Step 3: $SS = 140$

$s^2 = \frac{SS}{df} = \frac{140}{8} = 17.5$

$s_{M_D} = \sqrt{\frac{s^2}{n}} = \sqrt{\frac{17.5}{9}} = \sqrt{1.94} = 1.39$

$t = \frac{M_D - \mu_D}{s_{M_D}} = \frac{-2 - 0}{1.39} = -1.44$

Step 4: The t value calculated in Step 3 is less extreme than the critical t value from Step 2. Fail to reject the null hypothesis and conclude that there is no evidence that swearing affected pain level.

21. a. Step 1: H_0: $\mu_D = 0$, H_1: $\mu_D \neq 0$

Step 2: With $df = 8$ and $\alpha = .05$, the critical t value is ± 2.306.

Step 3: $M_D = 2$ and $SS = 32$

$s^2 = \frac{SS}{df} = \frac{32}{8} = 4$

$s_{M_D} = \sqrt{\frac{s^2}{n}} = \sqrt{\frac{4}{9}} = \sqrt{0.44} = 0.67$

$t = \frac{M_D - \mu_D}{s_{M_D}} = \frac{2 - 0}{0.67} = 2.99$

You might have noticed that standard error is equal to $\frac{2}{3}$ and that 2 divided by $\frac{2}{3}$ is equal to 3.00. Thus, you may have obtained a value of 3.00 for the t statistic.

Step 4: Reject the null hypothesis because t from Step 3 is more extreme than the t from Step 4. There is evidence that motivation to work was significantly affected by gamification.

b. estimated $d = \frac{M_D}{s} = \frac{2}{2} = 1$

23. For a repeated-measures design the same subjects are used in both treatment conditions. In a matched-subjects design, two different sets of subjects are used. However, in a matched-subjects design, each subject in one condition is matched with respect to a specific variable with a subject in the second condition so that the two separate samples are equivalent with respect to the matching variable.

25. For a repeated-measures t statistic, $df = n - 1$ where n is the number of individuals in the sample. If $n - 1 = 10$, then the study requires a sample of $n = 11$. A matched-subjects design would require two samples, each with n participants. The t statistic is the same as for the repeated-measures design and has $df = n - 1$. If $df = 10$, then $n = 11$ and the study would require a total of 22 participants (11 matched pairs). For an independent-measures design, $df = (n_1 - 1) + (n_2 - 1)$. If $(n_1 - 1) + (n_2 - 1) = 10$, then $n_1 + n_2 = 12$ and the study requires a total of 12 participants.

27. **a.** There is not enough information. The problem does not provide the variability of the difference scores D.
   b. Step 1: H_0: $\mu_D \geq 0$, H_1: $\mu_D < 0$

Step 2: With $df = 20$ and $\alpha = .05$, one-tailed, the critical value is $t = -1.725$.

Step 3: $s_{M_D} = \sqrt{\frac{s^2}{n}} = \sqrt{\frac{5376}{21}} = \sqrt{256} = 16$

$$t = \frac{M_D - \mu_D}{s_{M_D}} = \frac{-32 - 0}{16.00} = -2.00$$

Step 4: Reject the null hypothesis and conclude that exposure to blue light significantly decreased the delay to respond.

CHAPTER 12 Introduction to Analysis of Variance

1. With three or more treatment conditions you need three or more t tests to evaluate all the mean differences. Each test involves a risk of Type I error. The more tests you do, the more risk there is of a Type I error occurring in any of the tests. The ANOVA performs all of the tests simultaneously with a single, fixed level for α.

3. When there is no treatment effect, the numerator and the denominator of the F-ratio are both measuring the same sources of variability (random and unsystematic differences from sampling error). In this case, the F-ratio is balanced and should have a value near 1.00.

5. Both the F-ratio and the t statistic compare the actual mean differences between sample means (numerator) with the differences that would be expected if there is no treatment effect (the denominator if H_0 is true). If the numerator is sufficiently larger than the denominator, we conclude that there is a significant difference between treatments.

7. $SS_{total} = \Sigma X^2 - \frac{G^2}{N} = 6517 - \frac{395^2}{30} =$

   $6517 - 5200.83 = 1316.17$

   $SS_{within} = \Sigma SS$ within each treatment $= 350.5 +$

   $190.0 + 424 = 964.5$

   $SS_{Between} = SS_{Total} - SS_{Within} = 1316.17 - 964.5 = 351.67$

   or

   $SS_{Between} = \Sigma\frac{T^2}{n} - \frac{G^2}{N} = \frac{105^2}{10} + \frac{180^2}{10} + \frac{110^2}{10} - \frac{395^2}{30} =$

   $1102.5 + 3240 + 1210 - 5200.83 = 5552.5 -$

   $5200.83 = 351.67$

9. **a.** $k = df_{between} + 1 = 3 + 1 = 4$
   b. $N = df_{Within} + k = 40 + 4 = 44$
   c. The critical value for F is equal to 2.84.
   d. The critical value for F is equal to 4.31.

11. **a.** $s^2_{Treatment\ 1} = \frac{SS}{df} = \frac{220}{11} = 20$

   $s^2_{Treatment\ 2} = 22$

   $s^2_{Treatment\ 3} = 18$

b. $df_{within} = N - k = 36 - 3 = 33$

   $SS_{within} = \Sigma SS$ within each treatment

   $= 220 + 242 + 198 = 660$

   $MS_{within} = \frac{SS_{within}}{df_{within}} = \frac{660}{33} = 20$

13.

Source	SS	df	MS	
Between Treatments	32	2	16	$F = 4.00$
Within Treatments	60	15	4	
Total	92	17		

15.

Source	SS	df	MS	
Between Treatments	48	2	24.00	$F = 5.30$
Within Treatments	204	45	4.53	
Total	252	47		

17. **a.** Step 1: H_0: $\mu_1 = \mu_2 = \mu_3$ (The drug has no effect.)

   H_1: At least one of the treatment means is different.

   Step 2: $df_{between} = k - 1 = 3 - 1 = 2$

   $df_{within} = N - k = 15 - 3 = 12$

   The critical value for $\alpha = .05$ is 3.88.

   Step 3:

I − No Drug	II − Medium Dose	III − High Dose
17	29	16
14	21	24
22	27	20
19	23	21
18	25	19
$n = 5$	$n = 5$	$n = 5$
$T_1 = \Sigma X = 90$	$T_2 = \Sigma X = 125$	$T_3 = \Sigma X = 100$
$SS_1 = \Sigma X^2 - \frac{(\Sigma X)^2}{n}$	$SS_2 = 40$	$SS_3 = 34$
$= 1654 - \frac{90^2}{5}$		
$= 1654 - 1620$		
$= 34$		

$G = \Sigma T = 90 + 125 + 100 = 315$

$SS_{total} = \Sigma X^2 - \frac{G^2}{N}$

$= 6853 - \frac{315^2}{15} = 6853 - 6615 = 238$

$SS_{within} = \Sigma SS \text{ within groups} = 34 + 40 + 34 = 108$

$SS_{between} = SS_{total} - SS_{within} = 238 - 108 = 130$

$MS_{within} = \frac{SS_{within}}{df_{within}} = \frac{108}{12} = 9$

$MS_{between} = \frac{SS_{between}}{df_{between}} = \frac{130}{2} = 65$

$F = \frac{65}{9} = 7.22$

Source	SS	df	MS	
Between Treatments	130	2	65	F = 7.22
Within Treatments	108	12	9.0	
Total	238	14		

Step 4: The F value calculated in Step 3 is more extreme than the critical F value. Reject the null hypothesis and conclude that the drug had an effect.

b. The researcher would use a post hoc test.

c. A one-way analysis of variance detected a significant effect of drug dose on performance, $F(2, 12) = 7.22$, $p < .05$.

19. Step 1: $H_0: \mu_1 = \mu_2 = \mu_3$ (The treatment has no effect.)
H_1: At least one of the treatment means is different.

Step 2: With $df = 2, 18$, the critical value for $\alpha = .05$ is 3.55

Step 3: $SS_{total} = \Sigma X^2 - \frac{G^2}{N}$

$= 17035 - \frac{690^2}{21} = 626.95$

$SS_{within} = \Sigma SS \text{ within groups} = 186 + 80 + 168 = 434$

$SS_{between} = SS_{total} - SS_{within} = 626.95 - 434 = 192.95$

Source	SS	df	MS	
Between Treatments	192.95	2	96.48	F = 4.00
Within Treatments	434	18	24.11	
Total	626.95	20		

Step 4: The F value calculated in Step 3 is more extreme than the critical F value. Reject the null hypothesis and conclude that the treatment had an effect.

21. $SS_{Between} = \Sigma \frac{T^2}{n} - \frac{G^2}{N} = \frac{28^2}{7} + \frac{32^2}{8} + \frac{108^2}{9} - \frac{168^2}{24}$

$= 112 + 128 + 1296 - 1176 = 360$

23. a. $k = df_{between} + 1 = 4 + 1 = 5$
b. $N = df_{within} + k = 40 + 5 = 45$
c. The critical F value for $\alpha = .01$ is equal to 3.83. Fail to reject the null hypothesis.

25. a. The F-ratio would decrease because changing the mean from $M = 35$ to $M = 25$ would decrease the variation between groups.
b. The F-ratio would decrease because increasing the SS within a group increases the value of the denominator of the F-ratio.

27. a.

	Treatments		
	I	II	III
	$s^2 = 24$	$s^2 = 26$	$s^2 = 16$

Increasing SS by a factor of two increases variance by a factor of two.

b. Because the variance within treatments was increased, the value of the F-ratio should decrease.

c. As in problem 26, except:

$SS_{within} = \Sigma SS \text{ within groups} = 120 + 130 + 80 = 330$

$SS_{total} = \Sigma X^2 - \frac{G^2}{N} = 576 - \frac{54^2}{18} = 576 - 162 = 414$

Source	SS	df	MS	
Between Treatments	84	2	42	F = 1.91
Within Treatments	330	15	22	
Total	414	17		

Fail to reject the null hypothesis.

29. a. Step 1: $H_0: \mu_1 = \mu_2 = \mu_3$ (The drug has no effect.)
H_1: At least one of the treatment means is different.
Step 2: With df of 2, 12 and $\alpha = .05$, the critical value for F is equal to 3.88.
Step 3: $MS_{within} = \frac{9 + 10 + 11}{3} = 10$

$SS_{total} = \Sigma X^2 - \frac{G^2}{N} = 430 - \frac{60^2}{15} = 190$

$SS_{between} = SS_{total} - SS_{within} = 190 - 120 = 70$

Source	SS	df	MS	
Between Treatments	70	2	35	F = 3.50
Within Treatments	120	12	10	
Total	190	14		

Step 4: The critical F value from Step 2 is more extreme than the F-ratio calculated in Step 3. Fail to reject the null hypothesis.

b. $\eta^2 = \frac{SS_{between}}{SS_{total}} = 0.37$

31. a. The use of a larger sample should increase the F-ratio.

b.

Source	SS	df	MS	
Between Treatments	140	2	70	F = 7.00
Within Treatments	270	27	10	
Total	410	29		

The F-ratio was much larger (see problem 29). For problem 31, with $df = 2, 27$, the critical value equals 3.35. Reject the null hypothesis because the F ratio is greater than the critical value.

c. Increasing the sample size should have little or no effect on η^2. $\eta^2 = \frac{SS_{between}}{SS_{total}} = 0.34$, which is about the same as the value obtained in problem 29.

CHAPTER 13   Two-Factor Analysis of Variance

1. a. In analysis of variance, an independent variable (or a quasi-independent variable) is called a *factor*.
   b. The values of a factor that are used to create the different groups or treatment conditions are called the *levels* of the factor.
   c. A research study with two independent (or quasi-independent) variables is called a *two-factor study*.

3. a. The main effect for treatment is the 6-point difference between the overall column means, $M = 6$ and $M = 12$.
   b. The main effect for age is the 4-point difference between the overall row means, $M = 11$ and $M = 7$.
   c. There is no interaction. The effect of the treatment does not depend on age. With the treatment, scores increase by an average of 6 points for the 3-year old children and also increase by an average of 6 points for the 2-year old children.

5. a. $M = 5$. A_1 row mean is equal to $\frac{3+7}{2} = 5$. No main effect of A would be observed if the A_1 and A_2 row means are equal. Because the given A_2 mean is $M = 5$, the missing mean must also be $M = 5$.
   b. $M = 1$. B_1 column mean is equal to $\frac{3+5}{2} = 4$. No main effect of B would be observed if the B_1 and B_2 column means are equal. Because the given B_2 mean is $M = 7$, the missing mean must be $M = 1$.
   c. $M = 9$. No interaction would occur if the size of the difference between A_1 and A_2 were the same for both levels of B. If $M = 9$, the difference between A_1 and A_2 would be equal to 2 points for both B_1 and B_2.

7. a. $df = 1, 28$. df_A = number of rows -1, thus $df_A = 2 - 1 = 1$.
   $df_{within} = \Sigma df_{each\ treatment} = 7 + 7 + 7 + 7$
   b. $df = 1, 28$. As above, except df_B = number of columns -1.
   c. $df = 1, 28$. As above, except $df_{A \times B} = df_A \times df_B = 1 \times 1 = 1$.

9. a. Step 1: H_0 for factor A: $\mu_{A_1} = \mu_{A_2}$. H_0 for factor B: $\mu_{B_1} = \mu_{B_2}$. H_0 for interaction: The effect of factor A does not depend on the levels of factor B. $\alpha = .05$.
   Step 2: $df_{within} = \Sigma df_{each\ treatment} = 9 + 9 + 9 + 9 = 36$
   df_A = number of rows $- 1$, thus $df_A = 2 - 1 = 1$.
   df_B = number of columns $- 1$, thus $df_B = 2 - 1 = 1$
   $df_{A \times B} = df_A \times df_B = 1 \times 1 = 1$
   The critical F value for all three tests is 4.11.

   Step 3 (Stage 1):

   $SS_{total} = \Sigma X^2 - \frac{G^2}{N} = 640 - \frac{120^2}{40} = 640 - 360 = 280$

   $SS_{within\ treatments} = \Sigma SS_{inside\ each\ treatment}$
   $= 50 + 60 + 30 + 40 = 180$

   $SS_{between\ treatments} = \Sigma \frac{T^2}{n} - \frac{G^2}{N}$
   $= \frac{40^2}{10} + \frac{50^2}{10} + \frac{10^2}{10} + \frac{20^2}{10} - \frac{120^2}{40}$
   $= 160 + 250 + 10 + 40 - 360 = 100$

Step 3 (Stage 2):

$SS_A = \Sigma \frac{T_{ROWS}^2}{n_{ROWS}} - \frac{G^2}{N}$
$= \frac{50^2}{20} + \frac{70^2}{20} - \frac{120^2}{40} = 125 + 245 - 360 = 10$

$SS_B = \Sigma \frac{T_{COLS}^2}{n_{COLS}} - \frac{G^2}{N}$
$= \frac{90^2}{20} + \frac{30^2}{20} - \frac{120^2}{40} = 405 + 45 - 360 = 90$

$SS_{A \times B} = SS_{between\ treatments} - SS_A + SS_B$
$= 100 - 10 + 90 = 0$

$MS_A = \frac{SS_A}{df_A} = \frac{10}{1} = 10$, $MS_B = \frac{SS_B}{df_B} = \frac{90}{1} = 90$, and

$MS_{A \times B} = \frac{SS_{A \times B}}{df_{A \times B}} = \frac{0}{1} = 0$

$MS_{within\ treatments} = \frac{SS_{within\ treatments}}{df_{within\ treatments}} = \frac{180}{36} = 5$

$F_A = \frac{MS_A}{MS_{within\ treatments}} = \frac{10}{5} = 2$, $F_B = \frac{MS_B}{MS_{within\ treatments}} = \frac{90}{5} = 18$,

$F_{A \times B} = \frac{MS_{A \times B}}{MS_{within\ treatments}} = \frac{0}{5} = 0$

Source	SS	df	MS	
Between Treatments	100	3		
Factor A	10	1	10	$F(1, 36) = 2.00$
Factor B	90	1	90	$F(1, 36) = 18.00$
A × B Interaction	0	1	0	$F(1, 36) = 0.00$
Within Treatments	180	36	5	
Total	280	39		

Step 4: The main effect of factor B was significant because the F ratio of $F = 18.00$ was more extreme than the F critical value of $F = 4.11$. Neither the main effect of factor A nor the $A \times B$ interaction was significant.
   b. η^2 for factor $A = \frac{SS_A}{SS_A + SS_{within\ treatments}} = \frac{10}{10 + 180} = .053$
   η^2 for factor $B = \frac{SS_B}{SS_B + SS_{within\ treatments}} = \frac{90}{90 + 180} = .333$
   η^2 for $A \times B = \frac{SS_{A \times B}}{SS_{A \times B} + SS_{within\ treatments}} = \frac{0}{0 + 180} = .000$

11. a.

Source	SS	df	MS	
Between Treatments	340	3		
Factor A	80	1	80	$F(1, 76) = 4.00$
Factor B	180	1	180	$F(1, 76) = 9.00$
A × B Interaction	80	1	80	$F(1, 76) = 4.00$
Within Treatments	1520	76	20	
Total	1860	79		

The critical value for all three F-ratios is 3.98 (using $df = 1, 70$ because 76 degrees of freedom is not listed in the table). Both main effects and the interaction are significant.
   b. For the sport factor, $\eta^2 = 0.050$. For the time factor, $\eta^2 = 0.106$. For the interaction, $\eta^2 = 0.050$.
   c. For the noncontact athletes, there is little or no difference between the beginning of the first season and the end of the second season, but the contact athletes show noticeably lower scores after the second season.

13.

Source	SS	df	MS	
Between Treatments	72	5		
Factor A	12	1	12	$F(1, 42) = 4.00$
Factor B	36	2	18	$F(2, 42) = 6.00$
A × B Interaction	24	2	12	$F(2, 42) = 4.00$
Within Treatments	126	42	3	
Total	198	47		

15. a. Step 1: H_0 for factor A (learning style):

$\mu_{A_{\text{visual style}}} = \mu_{A_{\text{verbal style}}}$. H_0 for factor B (method):

$\mu_{B_{\text{visual method}}} = \mu_{B_{\text{verbal method}}}$. H_0 for interaction: The effect of instructional method does not depend on learning style. $\alpha = .05$.

Step 2: $df_{\text{within}} = \Sigma df_{\text{each treatment}} = 5 + 5 + 5 + 5 = 20$

$df_A = $ number of rows $- 1$, thus

$df_A = 2 - 1 = 1$

$df_B = $ number of columns $- 1$, thus

$df_B = 2 - 1 = 1$

$df_{A \times B} = df_A \times df_B = 1 \times 1 = 1$

The critical F value for all three tests is 4.35.

Step 3: For each cell in the design, compute M, SS, and T, and compute ΣX^2, G and N.

		Factor B: Instructional Method	
		Visual	Verbal
	Visual	M = 18 SS = 240 T = 108	M = 10 SS = 200 T = 60
Factor A: Learning Style	Verbal	M = 16 SS = 360 T = 96	M = 10 SS = 250 T = 60

Stage 1:

$SS_{\text{total}} = \Sigma X^2 - \frac{G^2}{N} = 5730 - \frac{324^2}{24}$

$= 5730 - 4374 = 1356$

$SS_{\text{within treatments}} = \Sigma SS_{\text{inside each treatment}}$

$= 240 + 200 + 360 + 250 = 1050$

$SS_{\text{between treatments}} = \Sigma \frac{T^2}{n} - \frac{G^2}{N}$

$= \frac{108^2}{6} + \frac{60^2}{6} + \frac{96^2}{6} + \frac{60^2}{6} - \frac{324^2}{24}$

$= 1944 + 600 + 1536 + 600 - 4374$

$= 306$

Stage 2:

$SS_{\text{Learning Style}} = \Sigma \frac{T^2_{\text{ROWS}}}{n_{\text{ROWS}}} - \frac{G^2}{N}$

$= \frac{168^2}{12} + \frac{156^2}{12} - \frac{324^2}{24}$

$= 2352 + 2028 - 4374 = 6$

$SS_{\text{Instructional Method}} = \Sigma \frac{T^2_{\text{COLS}}}{n_{\text{COLS}}} - \frac{G^2}{N}$

$= \frac{204^2}{12} + \frac{120^2}{12} - \frac{324^2}{24}$

$= 3468 + 1200 - 4374 = 294$

$SS_{A \times B} = SS_{\text{between treatments}} - SS_A + SS_B$

$= 306 - 6 + 294 = 6$

$MS_{\text{Learning style}} = \frac{SS_{\text{Learning style}}}{df_{\text{Learning style}}} = \frac{6}{1} = 6,$

$MS_{\text{Instructional method}} = \frac{SS_{\text{Instructional method}}}{df_{\text{Instructional method}}} = \frac{294}{1} = 294,$

and $MS_{A \times B} = \frac{SS_{A \times B}}{df_{A \times B}} = \frac{6}{1} = 6$

$MS_{\text{within treatments}} = \frac{SS_{\text{within treatments}}}{df_{\text{within treatments}}} = \frac{1050}{20} = 52.5$

$F_{\text{Learning style}} = \frac{MS_{\text{Learning style}}}{MS_{\text{within treatments}}} = \frac{6}{52.5} = 0.114,$

$F_{\text{Instructional method}} = \frac{MS_{\text{Instructional method}}}{MS_{\text{within treatments}}} = \frac{294}{52.5} = 5.60,$

$F_{A \times B} = \frac{MS_{A \times B}}{MS_{\text{within treatments}}} = \frac{6}{52.5} = 0.114$

Source	SS	df	MS	
Between Treatments	306	3		
Factor A (Learning Style)	6	1	6	$F(1, 20) = 0.114$
Factor B (Instructional Method)	294	1	294	$F(1, 20) = 5.600$
A × B Interaction	6	1	6	$F(1, 20) = 0.114$
Within Treatments	1050	20	52.5	
Total	1356	23		

Step 4: The main effect of instructional method was significant because the F ratio of $F = 5.60$ was more extreme than the F critical value of $F = 4.35$. Thus, we reject the null hypothesis that instructional method does not affect learning. Neither the main effect of learning style nor the interaction between instructional method and learning was significant.

17.

Source	SS	df	MS	
Between Treatments	70	3		
Factor A (Language of Testimony)	20	1	20	$F(1, 16) = 3.333$
Factor B (Instructions)	45	1	45	$F(1, 16) = 7.500$
A × B Interaction	5	1	5	$F(1, 16) = 0.833$
Within Treatments	96	16	6	
Total	166	19		

The critical value for all three F-ratios is 4.49 (using $df = 1, 16$). The main effect of instructions was significant. Neither the main effect of language of testimony nor the interaction was significant.

19. a.

Source	SS	df	MS	
Between Treatments	350	5		
Factor A	120	1	120	$F(1, 24) = 15.00$
Factor B	15	2	7.50	$F(2, 24) = 0.938$
A × B Interaction	215	2	107.50	$F(2, 24) = 13.438$
Within Treatments	192	24	8.00	
Total	542	29		

The critical F-ratio for the main effect of factor A is 4.26 (using $df = 1, 24$). The critical values for the main effect of factor B and the interaction is 3.40 (using $df = 2, 24$). Thus, the main effect of factor A and the interaction is significant. The main effect of factor B is not significant.

b. For B_2

	A_1		A_2	
	$n = 5$		$n = 5$	$N = 10$
	$M = 14$		$M = 3$	$G = 85$
	$T = 70$		$T = 15$	

Step 1: H_0: $\mu_{A1} = \mu_{A2}$ for B_2
Step 2:

$$SS_{\text{Between}} = \Sigma \frac{T^2}{n} - \frac{G^2}{N} = \frac{70^2}{5} + \frac{15^2}{5} - \frac{85^2}{10}$$

$$= 980 + 45 - 722.5 = 302.5$$

$$MS_{\text{Between}} = \frac{SS_{\text{Between}}}{df_{\text{Between}}} = \frac{302.5}{1} = 302.5$$

$$F = \frac{MS_{\text{Between}}}{MS_{\text{Within}}} = \frac{302.5}{8} = 37.813$$

The F-ratio has $df = 1, 24$ and a critical value of 4.26. The simple effect of factor A at level B_2 is significant.

21. a.

Source	SS	df	MS	
Between Treatments	140	2	70	$F(2, 27) = 2.83$
Within Treatments	667.50	27	24.72	
Total	807.50	29		

The critical F-ratio for the effect of the treatment is 3.35 with $df = 2, 27$. The effect of the treatment was not significant.

b.

Source	SS	df	MS	
Between Treatments	327.5	5	65.5	
Factor A (Treatment)	140	2	70	$F(2, 24) = 3.50$
Factor B (Location)	187.5	1	187.5	$F(1, 24) = 9.38$
A × B Interaction	0	2	0	$F(1, 24) = 0.00$
Within Treatments	480	24	20	
Total	807.5	29		

The critical F-ratio for the treatment effect is 3.40 with $df = 2, 24$. The critical F-ratio for the effect of location (online versus in-lab) is 4.26 with $df = 1, 24$. Both the main effect of treatment and the main effect of location are significant. The interaction is not significant.

c. The one-factor ANOVA in part A failed to detect a significant effect of treatment. The two-factor ANOVA in part B detected a significant effect of treatment (factor A). MS for the treatment effect was the same in parts A and B (i.e., $MS = 70$). However, $MS_{\text{within treatment}}$ was smaller in part B ($MS_{\text{within treatment}} = 20.00$) than in part A ($MS_{\text{within treatment}} = 24.72$). In part B, inclusion of location as a factor in the two-factor ANOVA removed variability due to testing location from the denominator of the F-ratio for the treatment effect.

CHAPTER 14 Correlation and Regression

1.

Scores		Deviations		Products
X	Y	$X - M_X$	$Y - M_Y$	$(X - M_X)(Y - M_Y)$
4	8	0	1	0
3	11	−1	4	−4
9	8	5	1	5
0	1	−4	−6	24

$SP = \Sigma(X - M_X)(Y - M_Y) = 0 + (-4) + 5 + 24 = 25$

3. a.

b. There is no linear trend and a straight line would provide a poor description of the relationship between X and Y. The estimated correlation is zero.

c.

Scores		Deviations		Squared Deviations		Products
X	Y	$X - M_X$	$Y - M_Y$	$(X - M_X)^2$	$(Y - M_Y)^2$	$(X - M_X)(Y - M_Y)$
2	5	−3	0	9	0	0
5	6	0	1	0	1	0
4	0	−1	−5	1	25	5
6	3	1	−2	1	4	−2
5	12	0	7	0	49	0
8	4	3	−1	9	1	−3

$$SS_X = \Sigma(X - M_X)^2 = 9 + 0 + 1 + 1 + 0 + 9 = 20$$

$$SS_Y = \Sigma(Y - M_Y)^2 = 0 + 1 + 25 + 4 + 49 + 1 = 80$$

$$SP = \Sigma(X - M_X)(Y - M_Y)$$

$$= 0 + 0 + 5 + (-2) + 0 + (-3) = 0$$

$$r = \frac{SP}{\sqrt{SS_X SS_Y}} = \frac{0}{\sqrt{20(80)}} = 0$$

5. a.

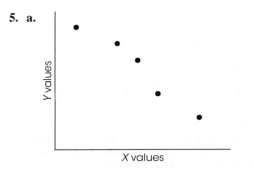

The estimated correlation is strong and negative because the scatterplot suggests that a line pointing down on the right side of the figure would provide a good description of the relationship between X and Y.

b. $SP = -39$, $SS_X = 20$ and $SS_Y = 80$.

$r = \frac{SP}{\sqrt{SS_X SS_Y}} = \frac{-39}{\sqrt{80(20)}} = \frac{-39}{\sqrt{1600}} = \frac{-39}{40} = -0.98$

7. a. $SP = 56$, $SS_X = 36$ and $SS_Y = 100$. SS for both X and Y are identical to problem 6. SP is positive.

b. $r = 0.93$. Notice that the correlation is positive in problem 7 but it was negative in problem 6.

9. a.

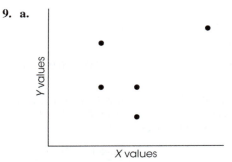

b. $SP = 5$, $SS_X = 6$ and $SS_Y = 24$. $r = 0.42$.

11. a. For the men's weights, $SS = 18$ and for their incomes, $SS = 11,076$. $SP = 330$. The correlation is $r = 0.739$.

b. With $n = 8$, $df = 6$ and the critical value is 0.707. The correlation is significant.

13. a. The critical value is 0.811 with $df = 4$.

b. The critical value is 0.576 with $df = 10$.

c. The critical value is 0.404 with $df = 22$.

15. Not necessarily. Greater social status could have caused increased learning ability or some other, third variable could have caused an increase in both learning ability and social status.

17. a.

Partic-ipant	X	Y (1 = binge-watched, 0 = watched daily)	$X - M_X$	$Y - M_Y$	$(X - M_X)^2$	$(Y - M_Y)^2$	$(X - M_X) \times (Y - M_Y)$
A	87	1	1.5	0.5	2.25	0.25	0.75
B	71	1	-14.5	0.5	210.25	0.25	-7.25
C	73	1	-12.5	0.5	156.25	0.25	-6.25
D	86	1	0.5	0.5	0.25	0.25	0.25
E	78	1	-7.5	0.5	56.25	0.25	-3.75
F	84	0	-1.5	-0.5	2.25	0.25	0.75
G	100	0	14.5	-0.5	210.25	0.25	-7.25
H	87	0	1.5	-0.5	2.25	0.25	-0.75
I	97	0	11.5	-0.5	132.25	0.25	-5.75
J	92	0	6.5	-0.5	42.25	0.25	-3.25

$SP = -32.5$, $SS_X = 814.5$ and $SS_Y = 2.5$. $r = -0.72$.

b. $r^2 = 0.52$

c. $r^2 = \frac{t^2}{t^2 + df} = \frac{-2.94^2}{-2.94^2 + 8} = \frac{8.64}{8.64 + 8} = 0.52$

19. a. $r = \frac{SP}{\sqrt{SS_X SS_Y}} = \frac{20}{\sqrt{16(64)}} = \frac{20}{32} = 0.63$

b. $b = \frac{SP}{SS_X} = \frac{20}{16} = 1.25$

$a = M_Y - bM_X = 8 - 1.25(6) = 8 - 7.5 = 0.50$

$\hat{Y} = 1.25X + 0.50$

21. The standard error of estimate is a measure of the average distance between the predicted Y points, $\hat{Y}$, from the regression equation and the actual Y points in the data.

23. $SP = 15$, $SS_X = 10$ and $SS_Y = 40$. $r = 0.75$.

$b = \frac{SP}{SS_X} = \frac{15}{10} = 1.5$

$a = M_Y - bM_X = 10 - 1.5(2) = 7$

$\hat{Y} = 1.5X + 7$

25. a. $SP = -60$, $SS_X = 40$ and $SS_Y = 160$.

$b = \frac{SP}{SS_X} = \frac{-60}{40} = -1.5$

$a = M_Y - bM_X = 8 - (-1.5)(4) = 14$

$\hat{Y} = -1.5X + 14$

b. $r = -0.75$ and $r^2 = 0.5625$.

$SS_{residual} = (1 - r^2)SS_Y$

$= (1 - 0.5625)160 = 70$

$standard\ error = \sqrt{\frac{SS_{residual}}{df}} = \sqrt{\frac{70}{4}} = 4.18$

27. a. $r^2 = 0.25$, $SS_{residual} = 36$, $standard\ error = 1.5$

b. $standard\ error = 0.75$

29. a. $SS_{regression} = r^2 SS_Y = 0.5625(160) = 90$.

$MS_{regression} = \frac{SS_{regression}}{df_{regression}} = \frac{90}{1} = 90$.

$MS_{residual} = \frac{SS_{residual}}{df_{residual}} = \frac{70}{4} = 17.50$.

$F = \frac{MS_{regression}}{MS_{residual}} = \frac{90}{17.50} = 5.14$. With $df = 1, 4$ the critical F value is 7.71. Fail to reject the null hypothesis.

CHAPTER 15 The Chi-Square Statistic

1. Nonparametric tests make few, if any, assumptions about the populations from which the data are obtained. For example, the populations do not need to form normal distributions, nor is it required that different populations in the same study have equal variances (homogeneity of variance assumption). Parametric tests require data measured on an interval or ratio scale. For nonparametric tests, any scale of measurement is acceptable.

3. **a.** Step 1: The null hypothesis states that there is no preference among the four colors; $p = \frac{1}{4}$ for all categories.

   Step 2: With $df = 3$, the critical value is 7.81.

   Step 3: The expected frequencies are $f_e = pn = .25(80) = 20$ for all categories.

   $$\chi^2 = \Sigma \frac{(f_o - f_e)^2}{f_e}$$

   $$\chi^2 = \frac{(30 - 20)^2}{20} + \frac{(13 - 20)^2}{20} + \frac{(23 - 20)^2}{20} + \frac{(14 - 20)^2}{20}$$

   $$= 5.00 + 2.45 + 0.45 + 1.80 = 9.70$$

   Step 4: the χ^2 statistic computed in Step 3 is more extreme than the critical value. Reject the null hypothesis and conclude that there are significant preferences.

   b. The results indicate that there are significant preferences among the four colors, $\chi^2(3, N = 80) = 9.70, p < .05$.

5. **a.** Step 1: The null hypothesis states that there is no preference among the four positions; $p = \frac{1}{4}$ for all categories.

   Step 2: With $df = 3$, the critical value is 7.81.

   Step 3: The expected frequencies are $f_e = pn = .25(60) = 15$ for all categories.

   $$\chi^2 = \Sigma \frac{(f_o - f_e)^2}{f_e}$$

   $$\chi^2 = \frac{(20 - 15)^2}{15} + \frac{(20 - 15)^2}{15} + \frac{(10 - 15)^2}{15} + \frac{(10 - 15)^2}{15}$$

   $\chi^2 = 1.67 + 1.67 + 1.67 + 1.67 = 6.67$ (note that you might have come up with 6.68 if you rounded to two decimal places for each cell in the analysis).

   Step 4: The χ^2 value calculated in Step 3 is less extreme than the critical value. Fail to reject the null hypothesis.

7. **a.** H_0 states that the distribution of automobile accidents is the same as the distribution of registered drivers: 16% under age 20, 28% ages 20 to 29, and 56% age 30 or older. With $df = 2$, the critical value is 5.99. The expected frequencies for these three categories are 48, 84, and 168. Chi-square $= 13.76$. Reject H_0 and conclude that the distribution of automobile accidents is not identical to the distribution of registered drivers.

b. $w = \sqrt{\Sigma \frac{(p_o - p_e)^2}{p_e}}$

   $w = \sqrt{\frac{(0.227 - 0.16)^2}{0.16} + \frac{(0.307 - 0.28)^2}{0.28} + \frac{(0.467 - 0.56)^2}{0.56}}$

   $w = \sqrt{0.028 + 0.003 + 0.015}$

   Cohen's $w = 0.214$

c. The chi-square test shows that the age distribution for people in automobile accidents is significantly different from the age distribution of licensed drivers, $\chi^2(2, n = 300) = 13.76, p < .05, w = 0.214$.

9. The null hypothesis states that the distribution of preferences is the same for both groups (same proportions). With $df = 2$, the critical value is 5.99. The expected frequencies are:

	Design 1	Design 2	Design 3
Students	30	20	10
Older Adults	30	20	10

$$\chi^2 = \Sigma \frac{(f_o - f_e)^2}{f_e}$$

$$\chi^2 = \frac{(40 - 30)^2}{30} + \frac{(20 - 30)^2}{30} + \frac{(10 - 20)^2}{20} + \frac{(30 - 20)^2}{20} + \frac{(10 - 10)^2}{10} + \frac{(10 - 10)^2}{10}$$

$\chi^2 = 3.33 + 3.33 + 5.00 + 5.00 + 0.00 + 0.00 = 16.67$

Reject H_0.

11. The null hypothesis states that there is no relationship between happiness and living longer. With $df = 1$, the critical value is 3.84. The expected frequencies are:

	Lived	Died	
Happy Most of the Time	384	16	400
Unhappy Most of the Time	192	8	200
	576	24	

a. Chi-square $= 0.78$. Fail to reject H_0.

b. $\phi = \sqrt{\frac{\chi^2}{n}} = \sqrt{\frac{0.78}{600}} = \sqrt{.0013} = .036$

13. **a.** The null hypothesis states that the proportion who falsely recall seeing broken glass should be the same for all three groups. The expected frequency of saying yes is 9.67 for all groups, and the expected frequency for saying no is 40.33 for all groups. With $df = 2$, the critical value is 5.99. For these data, chi-square $= 7.78$. Reject the null hypothesis and conclude that the likelihood of recalling broken glass depends on the question that the participants were asked.

b. $V = \sqrt{\frac{\chi^2}{n(df^*)}} = \sqrt{\frac{7.78}{150(1)}} = 0.228$.

c. Participants who were asked about the speed with which the cars "smashed into" each other were more than two times more likely to falsely recall seeing broken glass.

d. The results of the chi-square test indicate that the phrasing of the question had a significant effect on the participants' recall of the accident, $\chi^2(2, N = 150) = 7.78, p < .05, V = 0.228$.

15. a. Step 1: The null hypothesis states that there is no preference among the three sounds; $p = \frac{1}{3}$ for all categories.
Step 2: With $df = 2$, the critical value is 5.99.
Step 3: The expected frequencies are $f_e = pn = \frac{1}{3}(300) = 100$ for all categories. Notice that you should use multiply n by the fraction ($\frac{1}{3}$) instead of a decimal rounded to two decimal places. $\chi^2 = 156.48$
Step 4: The χ^2 statistic computed in Step 3 is more extreme than the critical value. Reject the null hypothesis and conclude that there are significant preferences.

b. The results indicate that there are significant preferences among the four colors, $\chi^2(2, N = 300) = 156.48, p < .05$.

17. a. The null hypothesis states that there is no relationship between the personalities of the participants and the personalities of the avatars they create. With $df = 1$

and $\alpha = .05$, the critical value is 3.84. The expected frequencies are:

	Participant Personality		
	Introverted	Extroverted	
Introverted Avatar	17.1	27.9	45
Extroverted Avatar	20.9	34.1	55
	38	62	

The chi-square statistic is 4.12. Reject H_0.
b. The phi-coefficient is 0.203.

19. The null hypothesis states that there is no relationship between IQ and volunteering. With $df = 2$ and $\alpha = .05$, the critical value is 5.99. The expected frequencies are:

	IQ		
	High	Medium	Low
Volunteer	37.5	75	37.5
Not Volunteer	12.5	25	12.5

The chi-square statistic is 4.75. Fail to reject H_0 with $\alpha = .05$ and $df = 2$.

Solution to the Matchstick Puzzle

There are several possible solutions to the matchstick puzzle presented in the Chapter 10 end of chapter problems but all involve destroying two of the existing squares. One square is destroyed by removing two matchsticks from one of the corners and a second square is destroyed by removing one matchstick.

The three removed matchsticks are then used to build a new square using a line that already exists in the figure as the fourth side. One solution is shown in the figure below. The red lines in the top panel of the figure are moved to the positions marked by red lines in the bottom panel of the figure.

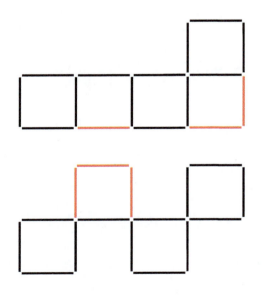

General Instructions for Using SPSS®

The Statistical Package for the Social Sciences, commonly known as SPSS, is a computer program that performs statistical calculations and is widely available on college campuses. Detailed instructions for using SPSS for specific statistical calculations (such as computing sample variance or performing an independent-measures *t* test) are presented at the end of the appropriate chapter in the text. Look for the SPSS section near the end of each chapter. In this appendix, we provide a general overview of the SPSS program.

■ SPSS Layout

After you open SPSS, you are prompted to either open an existing dataset or create a **New Dataset**. For all examples in this textbook, you should create a new dataset by clicking on **New Dataset** and then **Open**. SPSS consists of three basic components: the **Data View** of the **Data Editor**, the **Variable View** of the **Data Editor**, and a set of statistical commands.

■ SPSS Data View

After you create a new dataset, you will see the Data View of the **data editor**, which is a huge matrix of numbered rows and columns and information about variables in your analysis. To begin any analysis, you must enter information about variables in your analysis and type your data into the data editor. To enter data into the editor, the **Data View** tab must be set at the bottom left of the screen. Typically, scores are entered into columns of the editor in the Data View. Before scores are entered, each of the columns is labeled "var." After scores are entered, the first column becomes VAR00001, the second column becomes VAR00002, and so on.

■ SPSS Variable View

To enter information about the variables in your analysis, click on the **Variable View** tab at the bottom of the data editor. You will get a description of each variable in the editor. Use the **Variable View** to enter information about the variables in your analysis. The **Name** field allows you to change the name of your variable to something descriptive. **Type** can be used to select the type of variable you are analyzing. For example, most of the variables you will analyze will be simple numeric variables. However, you might occasionally use a string to enter text labels for nominal data. **Width** controls the number of characters for the score that should be displayed in Data View (the default width of eight characters is usually acceptable). **Decimals** allows you to specify the number of places after a decimal to be displayed. **Label** allows a user to assign a long, descriptive title to a variable. This title will be displayed in the results of statistical analyses. The **Values** field will apply labels to specific values of the variable. For example, when using an ordinal scale, you might want a value of "1" to be labeled "first," a value of "2" to be labeled "second," and so on. **Missing** is a collection of settings for how the program should identify missing values. **Columns** is the width of the variable's column in the Data View. **Align** controls whether values for the

Statistics Organizer: Finding the Right Statistics for Your Data

Overview: Three Basic Data Structures

After students have completed a statistics course, they occasionally are confronted with situations in which they have to apply the statistics they have learned. For example, in the context of a research methods course, or while working as a research assistant, students are presented with the results from a study and asked to do the appropriate statistical analysis. The problem is that many of these students have no idea where to begin. Although they have learned the individual statistics, they cannot match the statistical procedures to a specific set of data. The Statistics Organizer attempts to help you find the right statistics by providing an organized overview for most of the statistical procedures presented in this book.

We assume that you know (or can anticipate) what your data look like. Therefore, we begin by presenting some basic categories of data so you can find the one that matches your own data. For each data category, we then present the potential statistical procedures and identify the factors that determine which are appropriate for you based on the specific characteristics of your data. Most research data can be classified in one of three basic categories.

Category 1: A single group of participants with one score per participant.

Category 2: A single group of participants with two variables measured for each participant.

Category 3: Two (or more) groups of scores with each score a measurement of the same variable.

In this section we present examples of each structure. Once you match your own data to one of the examples, you can proceed to the section of the chapter in which we describe the statistical procedures that apply to that example.

■ Scales of Measurement

Before we begin discussion of the three categories of data, there is one other factor that differentiates data within each category and helps to determine which statistics are appropriate. In Chapter 1 we introduced four scales of measurement and noted that different measurement scales allow different kinds of mathematical manipulation, which result in different statistics. For most statistical applications, however, ratio and interval scales are equivalent, so we group them together for the following review.

ranks or ordinal categories (for example, small, medium, and large). In either case, the median is appropriate for describing central tendency for ordinal measurements and proportions can be used to describe the distribution of individuals across categories. For example, a researcher might report that 60% of the students were in the high self-esteem category, 30% in the moderate self-esteem category, and only 10% in the low self-esteem category.

Inferential Statistics   If there is a basis for a null hypothesis specifying the proportions in each ordinal category for the population from which the scores were obtained, then a chi-square test for goodness of fit (Chapter 15) can be used to evaluate the hypothesis. For example, it may be reasonable to hypothesize that the categories occur equally often (equal proportions) in the population and the test would determine whether the sample proportions are significantly different.

■ Scores from a Nominal Scale

For these data, the scores simply indicate the nominal category for each individual. For example, individuals could be classified as Republican or Democrat or grouped into different occupational categories.

Descriptive Statistics   The only descriptive statistics available for these data are the mode (Chapter 3) for describing central tendency or using proportions (or percentages) to describe the distribution across categories.

Inferential Statistics   If there is a basis for a null hypothesis specifying the proportions in each category for the population from which the scores were obtained, then a chi-square test for goodness of fit (Chapter 15) can be used to evaluate the hypothesis. For example, it may be reasonable to hypothesize that the categories occur equally often (equal proportions) in the population. If proportions are known for a comparison population or for a previous time, the null hypothesis could specify that the proportions are the same for the population from which the scores were obtained. For example, if it is known that 35% of the adults in the United States get a flu shot each season, then a researcher could select a sample of college students and count how many got a shot and how many did not [see the data in Table 1(c)]. The null hypothesis for the chi-square test would state that the distribution for college students is not different from the distribution for the general population.

Figure 1 summarizes the statistical procedures used for data in Category 1.

Section II: Statistical Procedures for Data from a Single Group of Participants with Two Variables Measured for Each Participant

The goal of the statistical analysis for data in this category is to describe and evaluate the relationships between variables, typically focusing on two variables at a time. With only two variables, the appropriate statistics are correlations and regression (Chapter 14), and the chi-square test for independence (Chapter 15).

■ Two Numerical Variables from Interval or Ratio Scales

The Pearson correlation measures the degree and direction of linear relationship between the two variables (see Example 14.3 on page 485). Linear regression determines the

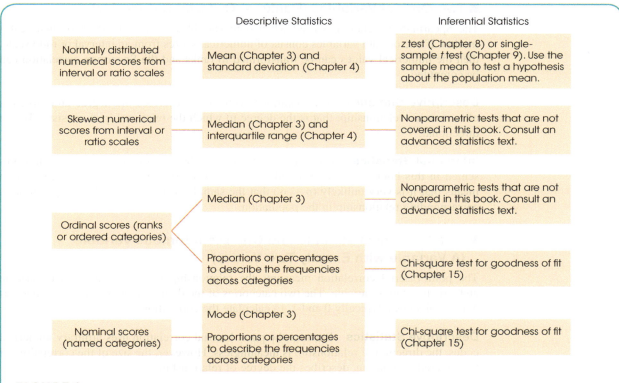

FIGURE 1

Statistics for Category 1 data. A single group of participants with one score per participant. The goal is to describe the variable as it exists naturally.

equation for the straight line that gives the best fit to the data points. For each X value in the data, the equation produces a predicted Y value on the line so that the squared distances between the actual Y values and the predicted Y values are minimized.

Descriptive Statistics The Pearson correlation serves as its own descriptive statistic. Specifically, the sign and magnitude of the correlation describe the linear relationship between the two variables. The squared correlation is often used to describe the strength of the relationship. The linear regression equation provides a mathematical description of the relationship between X values and Y. The slope constant describes the amount that Y changes each time the X value is increased by 1 point. The constant (Y intercept) value describes the value of Y when X is equal to zero.

Inferential Statistics The statistical significance of the Pearson correlation is evaluated with a t statistic or by comparing the sample correlation with critical values listed in Table B6 in Appendix B. A significant correlation means that it is very unlikely ($p < \alpha$) that the sample correlation would occur without a corresponding relationship in the population. Analysis of regression is a hypothesis-testing procedure that evaluates the significance of the regression equation. Statistical significance means that the equation predicts more of the variance in the Y scores than would be reasonable to expect if there were not a real underlying relationship between X and Y.

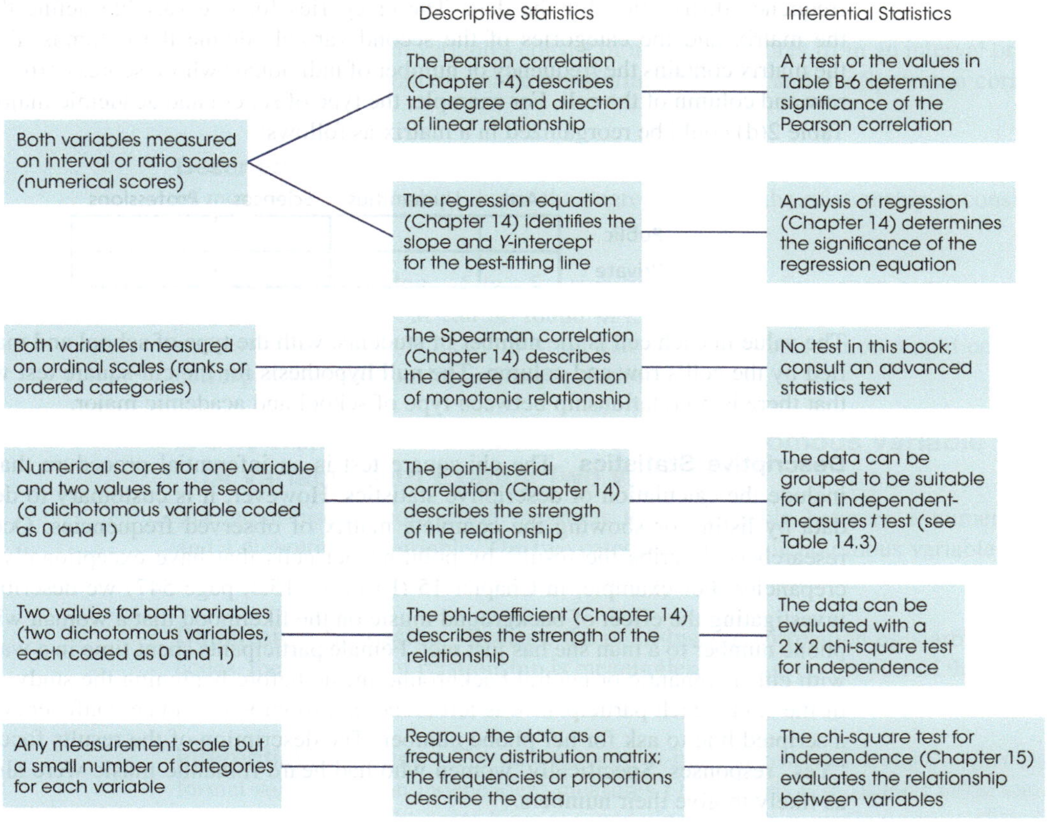

Descriptive Statistics   Inferential Statistics

Both variables measured on interval or ratio scales (numerical scores)	The Pearson correlation (Chapter 14) describes the degree and direction of linear relationship	A *t* test or the values in Table B-6 determine significance of the Pearson correlation
	The regression equation (Chapter 14) identifies the slope and *Y*-intercept for the best-fitting line	Analysis of regression (Chapter 14) determines the significance of the regression equation
Both variables measured on ordinal scales (ranks or ordered categories)	The Spearman correlation (Chapter 14) describes the degree and direction of monotonic relationship	No test in this book; consult an advanced statistics text
Numerical scores for one variable and two values for the second (a dichotomous variable coded as 0 and 1)	The point-biserial correlation (Chapter 14) describes the strength of the relationship	The data can be grouped to be suitable for an independent-measures *t* test (see Table 14.3)
Two values for both variables (two dichotomous variables, each coded as 0 and 1)	The phi-coefficient (Chapter 14) describes the strength of the relationship	The data can be evaluated with a 2 x 2 chi-square test for independence
Any measurement scale but a small number of categories for each variable	Regroup the data as a frequency distribution matrix; the frequencies or proportions describe the data	The chi-square test for independence (Chapter 15) evaluates the relationship between variables

FIGURE 2

Statistics for Category 2 data. One group of participants with two (or more) variables measured for each participant. The goal is to describe and evaluate the relationship between variables.

The goal for a single-factor research design is to demonstrate a relationship between the two variables by showing consistent differences between groups. The scores in each group can be numerical values measured on interval or ratio scales.

Other considerations regarding inferential statistics are the assumptions of the statistical test. Recall that the *t* test (Chapters 9, 10, and 11) and analysis of variance (Chapters 12 and 13) assume that the population from which samples are selected are normally distributed. This assumption is less important with very large samples. However, with small samples the validity of the test might be compromised when the assumption has been violated, as would be the case with a skewed distribution.

■ Scores from Interval or Ratio Scales: Assumption of Normality Satisfied

Descriptive Statistics   When the scores in each group are numerical values, the standard procedure is to compute the mean (Chapter 3) and the standard deviation (Chapter 4) as descriptive statistics to summarize and describe each group. For a repeated-measures study comparing exactly two groups, it also is common to compute the difference between the two scores for each participant and then report the mean and the standard deviation for the difference scores.

Inferential Statistics   Analysis of variance (ANOVA) and t tests are used to evaluate the statistical significance of the mean differences between the groups of scores. With only two groups, the two tests are equivalent and either may be used. With more than two groups, mean differences are evaluated with an ANOVA. For independent-measures designs (between-subjects designs), the independent-measures t (Chapter 10) and independent-measures ANOVA (Chapter 12) are appropriate. For repeated-measures designs consisting of two treatments, the repeated-measures t (Chapter 11) is used. When the repeated-measures study has more than two treatments, a repeated-measures ANOVA is used. (This analysis is not covered in this text. See an advanced statistics text for a description of the repeated-measures ANOVA.) For all tests, a significant result indicates that the sample mean differences in the data are very unlikely ($p < \alpha$) to occur if there are not corresponding mean differences in the population. For an ANOVA comparing more than two means, a significant F-ratio indicates that post-tests such as Scheffé or Tukey (Chapter 12) are necessary to determine exactly which sample means are significantly different. Significant results from a t test should be accompanied by a measure of effect size such as Cohen's d or r^2. For ANOVA, effect size is measured by computing the percentage of variance accounted for, η^2.

■ Two-Factor Designs with Scores from Interval or Ratio Scales: Assumption of Normality Satisfied

Research designs with two independent (or quasi-independent) variables are known as two-factor designs. These designs can be presented as a matrix with the levels of one factor defining the rows and the levels of the second factor defining the columns. A third variable (the dependent variable) is measured to obtain a group of scores in each cell of the matrix (see Table 13.6 on page 448).

Descriptive Statistics   When the scores in each group are numerical values, the standard procedure is to compute the mean (Chapter 3) and the standard deviation (Chapter 4) as descriptive statistics to summarize and describe each group.

Inferential Statistics   A two-factor ANOVA is used to evaluate the significance of the mean differences between cells. The ANOVA separates the mean differences into three categories and conducts three separate hypothesis tests:

1. The main effect for factor A evaluates the overall mean differences for the first factor; that is, the mean differences between rows in the data matrix.

2. The main effect for factor B evaluates the overall mean differences for the second factor; that is, the mean differences between columns in the data matrix.

3. The interaction between factors evaluates the mean differences between cells that are not accounted for by the main effects.

For each test, a significant result indicates that the sample mean differences in the data are very unlikely ($p < \alpha$) to occur if there are not corresponding mean differences in the population. For each of the three tests, effect size is measured by computing the percentage of variance accounted for, η^2.

Figure 3 summarizes the statistical procedures used for normally distributed interval or ratio data in Category 3.

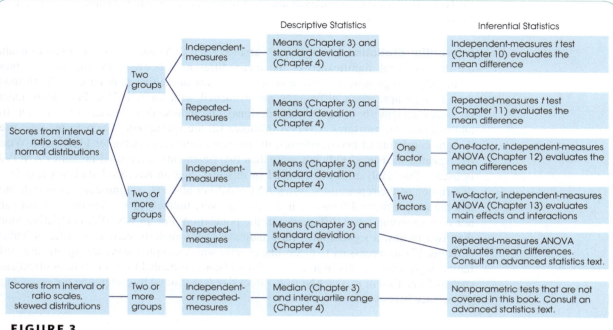

FIGURE 3

Statistics for Category 3 Data. Two or more groups of scores that consist of measurements of the same variable on a ratio or interval scale of measurement. Options for descriptive and inferential procedures are shown when the assumption of a normally distributed population has been satisfied or violated. For skewed distributions an advanced text on nonparametric statistical tests should be consulted.

■ Scores from Interval or Ratio Scales: Skewed Distributions

Descriptive Statistics   You can compute the mean (M) and standard deviation (s) for scores in skewed distributions that are on a ratio or an interval scale. However, the presence of extreme values in one tail of the distribution will distort the values you obtain for the mean and standard deviation. This distortion can make them relatively poor descriptive statistics for skewed distributions. We saw that just a few extreme scores can draw the mean in the direction of the tail (Chapter 3) and the value for the standard deviation will be inflated (Chapter 4). For these reasons the preferred measures of central tendency and variability for skewed distributions are the median and interquartile range (IQR). They provide descriptive measures that are less influenced by extreme scores.

Inferential Statistics   Inferential statistical procedures for data consisting of scores on an interval or ratio scale require that certain assumptions are satisfied. These hypothesis tests, for example the t test (Chapters 9, 10, 11) and analysis of variance (Chapters 12, 13), assume that the population from which samples are selected are normally distributed. If the distribution is skewed, then the results of these statistical tests might not be valid, especially if the size of the samples is small. In these instances, one should use nonparametric statistical tests. This text does not cover the nonparametric tests that should be used when the assumptions for a parametric test have not been satisfied. An advanced text should be consulted.

Summary of Statistics Formulas

The Mean

Population: $\mu = \dfrac{\Sigma X}{N}$

Sample: $M = \dfrac{\Sigma X}{n}$

The Weighted Mean

$$M = \dfrac{\Sigma X_1 + \Sigma X_2}{n_1 + n_2}$$

The Median (for Tied Scores in Center of Distribution)

$$\text{median} = X_{\text{LRL}} + \left(\dfrac{0.5N - f_{\text{BELOW LRL}}}{f_{\text{TIED}}} \right)$$

Sum of Squares

Definitional: $SS = \Sigma(X - \mu)^2$ or for a sample, $SS = \Sigma(X - M)^2$

Computational: $SS = \Sigma X^2 - \dfrac{(\Sigma X)^2}{N}$ or for a sample, $SS = \Sigma X^2 - \dfrac{(\Sigma X)^2}{n}$

Variance

Population: $\sigma^2 = \dfrac{SS}{N}$

Sample: $s^2 = \dfrac{SS}{n - 1} = \dfrac{SS}{df}$

Standard Deviation

Population: $\sigma = \sqrt{\dfrac{SS}{N}}$

Sample: $s = \sqrt{\dfrac{SS}{n - 1}} = \sqrt{\dfrac{SS}{df}}$

z-Score (for Locating an X Value)

$$z = \frac{X - \mu}{\sigma}$$

z-Score (for Locating a Sample Mean)

$$z = \frac{M - \mu}{\sigma_M} \text{ where } \sigma_M = \frac{\sigma}{\sqrt{n}} = \sqrt{\frac{\sigma^2}{n}}$$

t Statistic (Single Sample)

$$t = \frac{M - \mu}{s_M} \text{ where } s_M = \frac{s}{\sqrt{n}} = \sqrt{\frac{s^2}{n}}$$

t Statistic (Independent Measures)

$$t = \frac{(M_1 - M_2) - (\mu_1 - \mu_2)}{s_{(M_1 - M_2)}}$$

$$\text{where } s_{(M_1 - M_2)} = \sqrt{\frac{s_p^2}{n_1} + \frac{s_p^2}{n_2}} \text{ and } s_p^2 = \frac{SS_1 + SS_2}{df_1 + df_2} = \frac{df_1 s_1^2 + df_2 s_2^2}{df_1 + df_2}$$

t Statistic (Repeated Measures, Two Related Samples)

$$t = \frac{M_D - \mu_D}{s_{M_D}} \text{ where } s_{M_D} = \sqrt{\frac{s^2}{n}}, M_D = \frac{\Sigma D}{n}, \text{ and } D = X_2 - X_1$$

Independent-Measures Analysis of Variance

$$SS_{total} = \Sigma X^2 - \frac{G^2}{N} \qquad df_{total} = N - 1$$

$$SS_{between} = \Sigma \frac{T^2}{n} - \frac{G^2}{N} \qquad df_{between} = k - 1$$

$$SS_{within} = \Sigma SS_{inside \ each \ treatment} \qquad df_{within} = \Sigma df_{each \ treatment} = N - k$$

$$MS_{between} = \frac{SS_{between}}{df_{between}} \qquad MS_{within} = \frac{SS_{within}}{df_{within}}$$

$$F = \frac{MS_{between}}{MS_{within}}$$

Two-Factor Analysis of Variance

$$SS_{total} = \Sigma X^2 - \frac{G^2}{N} \qquad df_{total} = N - 1$$

$$SS_{between \ treatments} = \Sigma \frac{T^2}{n} - \frac{G^2}{N} \qquad df_{between \ treatments} = \text{number of cells} - 1$$

$$SS_{\text{within treatments}} = \Sigma SS_{\text{each treatment}} \qquad df_{\text{within treatments}} = \Sigma df_{\text{each treatment}}$$

$$SS_A = \Sigma \frac{T^2_{ROW}}{n_{ROW}} - \frac{G^2}{N} \qquad df_A = \text{number of rows} - 1$$

$$SS_B = \Sigma \frac{T^2_{COL}}{n_{COL}} - \frac{G^2}{N} \qquad df_B = \text{number of columns} - 1$$

$$SS_{A \times B} = SS_{\text{between treatments}} - SS_A - SS_B \qquad df_{A \times B} = df_{\text{between treatments}} - df_A - df_B$$

$$MS_A = \frac{SS_A}{df_A} \qquad MS_B = \frac{SS_B}{df_B} \qquad MS_{A \times B} = \frac{SS_{A \times B}}{df_{A \times B}} \qquad MS_{\text{within}} = \frac{SS_{\text{within}}}{df_{\text{within}}}$$

$$F_A = \frac{MS_A}{MS_{\text{within}}} \qquad F_B = \frac{MS_B}{MS_{\text{within}}} \qquad F_{A \times B} = \frac{MS_{A \times B}}{MS_{\text{within}}}$$

Pearson Correlation

$$r = \frac{SP}{\sqrt{SS_X SS_Y}}$$

where $SP = \Sigma(X - M_X)(Y - M_Y) = \Sigma XY - \dfrac{(\Sigma X)(\Sigma Y)}{n}$

Spearman Correlation

$$r_s = 1 - \frac{6 \Sigma D^2}{n(n^2 - 1)}$$

where D is the difference between the X rank and the Y rank for each individual

Regression

$$\hat{Y} = bX + a \text{ where } b = \frac{SP}{SS_X} \text{ and } a = M_Y - bM_X$$

Chi-Square Statistic

$$\chi^2 = \Sigma \frac{(f_o - f_e)^2}{f_e}$$

where f_e for the goodness-of-fit test is $f_e = pn$ and $df = C - 1$

and f_e for the test of independence is $f_e = \dfrac{f_c f_r}{n}$ and $df = (C - 1)(R - 1)$

Effect Size Measures and Confidence Intervals

For the z test, Cohen's $d = \dfrac{\text{mean difference}}{\text{standard deviation}} = \dfrac{\mu_{\text{treatment}} - \mu_{\text{no treatment}}}{\sigma}$

For the single-sample t test, estimated $d = \dfrac{M - \mu}{s}$

For the independent-measures t test, estimated $d = \dfrac{M_1 - M_2}{\sqrt{s_P^2}}$

For the repeated-measures t test, estimated $d = \dfrac{M_D}{s}$

r^2 and η^2 (percentage of variance accounted for by effect or relationship)

$$r^2 = \frac{t^2}{t^2 + df} \qquad \text{(for } t \text{ tests)}$$

$$\eta^2 = \frac{SS_{\text{between treatments}}}{SS_{\text{total}}} \qquad \text{(for independent-measures analysis of variance)}$$

$$\eta^2 = \frac{SS_A}{SS_A + SS_{\text{within treatments}}} \qquad \text{(for main effect of } A \text{ in two factor analysis of variance)}$$

$$\eta^2 = \frac{SS_B}{SS_B + SS_{\text{within treatments}}} \qquad \text{(for main effect of } B \text{ in two factor analysis of variance)}$$

$$\eta^2 = \frac{SS_{A \times B}}{SS_{A \times B} + SS_{\text{within treatments}}} \qquad \text{(for interaction in two factor analysis of variance)}$$

Confidence interval (for single-sample t statistic)

$$\mu = M \pm t(s_M)$$

Confidence interval (for independent-measures t statistic)

$$\mu_1 - \mu_2 = M_1 - M_2 \pm ts_{(M_1 - M_2)}$$

Confidence interval (for repeated-measures t statistic)

$$\mu_D = M_D \pm ts_{M_D}$$

References

Ackerman, P. L., & Beier, M. E. (2007). Further explorations of perceptual speed abilities in the context of assessment methods, cognitive abilities, and individual differences during skill acquisition. *Journal of Experimental Psychology: Applied, 13*(4), 249–272.

Ackerman, R., & Goldsmith, M. (2011). Metacognitive regulation of text learning: On screen versus on paper. *Journal of Experimental Psychology: Applied, 17,* 18–32. doi:10.1037/a0022086

American Pet Products Association. (n.d.). APPA Survey 2017–2018. In *The Humane Society of the United States, Pets by the Numbers.* Retrieved from https://www.animalsheltering.org/page/pets-by-the-numbers

American Psychological Association. (2010). *Publication manual of the American Psychological Association* (6th ed.). Washington, DC: American Psychological Association.

American Veterinary Medical Association. (2012). *2012 American Veterinary Medical Association Sourcebook.* Retrieved from https://www.animalsheltering.org/page/pets-by-the-numbers

Anderson, D. R., Huston, A. C., Wright, J. C., & Collins, P. A. (1998). Initial findings on the long-term impact of Sesame Street and educational television for children: The recontact study. In R. Noll and M. Price (Eds.), *A communication cornucopia: Markle Foundation essays on information policy* (pp. 279–296). Washington, DC: Brookings Institution.

Anderson, N. D. (1999). The attentional demands of encoding and retrieval in younger and older adults: II. Evidence from secondary task reaction time distributions. *Psychology and Aging, 14*(4), 645–655.

ASPCA. (n.d.). *ASPCA policy and position statements. Position statement on pit bulls.* Retrieved from https://www.aspca.org/about-us/aspca-policy-and-position-statements/position-statement-pit-bulls

Bakhshi, S., Kanuparthy, P., & Gilbert, E. (2014). *Demographics, weather and online reviews: A study of restaurant recommendations.* International World Wide Web Conference Committee, IW3C2 (published online April 7–11, 2014). Retrieved from http://dx.doi.org/10.1145/2566486.2568021

Bar-Hillel, M. (1980). The base-rate fallacy in probability judgments. *Acta Psychologica, 44,* 211–233.

Beaven, C. M., & Ekstrom, J. (2013). A comparison of blue light and caffeine effects on cognitive function and alertness in humans. *PLoS ONE, 8*(10), 1–7.

Bélisle, J., & Bodur, H. O. (2010). Avatars as information: Perception of consumers based on the avatars in virtual worlds. *Psychology & Marketing, 27,* 741–765.

Bjornsen, C. A., & Archer, K. J. (2015). Relations between college students' cell phone use during class and grades. *Scholarship of Teaching and Learning in Psychology, 1,* 326–336.

Blum, J. (1978). *Pseudoscience and mental ability.* New York: Monthly Review Press.

Boehm, J. K., Winning, A., Segerstrom, S., & Kubzansky, L. D. (2015). Variability modifies life satisfaction's association with mortality risk in older adults. *Psychological Science, 26*(7), 1063–1070.

Boogert, N. J., Reader, S. M., & Laland, K. N. (2006). The relation between social rank, neophobia, and individual learning in starlings. *Animal Behaviour, 72*(6), 1229–1239.

Bowden, V. K., Loft, S., Tatasciore, M., & Visser, T. A. (2017). Lowering thresholds for speed limit enforcement impairs peripheral object detection and increases driver subjective workload. *Accident Analysis & Prevention, 98,* 118–122.

Boysen, G. A, & Vogel, D. L. (2009). Bias in the classroom: Types, frequencies, and responses. *Teaching of Psychology, 36,* 12–17. doi:10.1080/00986280802529038

Bransford, J. D., & Johnson, M. K. (1972). Contextual prerequisites for understanding: Some investigations of comprehension and recall. *Journal of Verbal Learning and Verbal Behavior, 11,* 717–726.

Callahan, L. F. (2009). Physical activity programs for chronic arthritis. *Current Opinion in Rheumatology, 21,* 177–182.

Cameron, J., Pierce, W. D., & So, S. (2004). Rewards, task difficulty, and intrinsic motivation: A test of learned industriousness theory. *Alberta Journal of Educational Research, 50*(3), 317–320.

Centers for Disease Control and Prevention. (2016). *Mortality multiple cause data files.* Retrieved from https://www.cdc.gov/nchs/data_access/vitalstatsonline.htm

Centers for Disease Control and Prevention, National Center for Health Statistics. (2016). Anthropometric reference data for children and adults: United States, 2011–2014. *Vital and Health Statistics*, Series 3(39). Retrieved from https://www.cdc.gov/nchs/data/series/sr_03/sr03_039.pdf

Cepeda, N. J., Vul, E., Rohrer, D., Wixted, J. T., & Pashler, H. (2008). Spacing effects in learning: A temporal ridgeline of optimal retention. *Psychological Science*, *19*, 1095–1102.

Chang, A., Aeschbach, D., Duffy, J. F., & Czeisler, C. A. (2015). Evening use of light-emitting eReaders negatively affects sleep, circadian timing, and next-morning alertness. *Proceedings of the National Academy of Science of the United States, 112.* doi:10.1073/pnas.1418490112

Cheng, R. K., MacDonald, C. J., & Meck, W. H. (2006). Differential effects of cocaine and ketamine on time estimation: Implications for neurobiological models of interval timing. *Pharmacology, Biochemistry, and Behavior*, *85*, 114–122.

Cobb-Clark, D., & Schurer, S. (2012). The stability of big-five personality traits. *Economic Letters, 115*(1), 11–15.

Cohen, J. (1988). *Statistical power analysis for the behavioral sciences*. Hillsdale, NJ: Lawrence Erlbaum Associates.

Cohen, J. (1990). Things I have learned (so far). *American Psychologist, 45*, 1304–1312.

Cohen, J. (1992). A power primer. *Psychological Bulletin, 112*, 155–159.

Cohen, R. (2013, August). Sugar love: A not so sweet tale. *National Geographic Magazine*. Retrieved from http://ngm.nationalgeographic.com/2013/08/sugar/cohen-text

Cowles, M., & Davis, C. (1982). On the origins of the .05 level of statistical significance. *American Psychologist, 37*, 553–558.

Derks, D., Bos, A. E. R., & von Grumbkow, J. (2007). Emoticons and social interaction on the Internet: The importance of social context. *Computers in Human Behavior, 23*, 842–849.

Deters, F. G., & Mehl, M. R. (2013). Does posting Facebook status updates increase or decrease loneliness? An online social networking experiment. *Social Psychological and Personality Science, 4*(5), 1–13.

Dickey, B. (2016). *Pit bull: The battle over an American icon*. New York: Knopf.

Doebel, S., & Munakata, Y. (2018). Group influences on engaging self-control: Children delay gratification and value it more when their in-group delays and their out-group doesn't. *Psychological Science, 29*(5), 738–748.

Drèze, X., & Nunes, J. (2006). The endowed progress effect: How artificial advancement increases effort. *Journal of Consumer Research, 32*, 504–512.

Dunning, D., Johnson, K., Ehrlinger, J., & Kruger, J. (2003). Why people fail to recognize their own incompetence. *Psychological Science, 12*(3), 83–87.

Dwyer, M. D., Figueroa, J., Gasalla, P., & Lopez, M. (2018). Reward adaptation and the mechanisms of learning: Contrast changes reward value in rats and drives learning. *Psychological Science, 29*(2), 219–227.

Elbel, B., Gyamfi, J., & Kersh, R. (2011). Child and adolescent fast-food choice and the influence of calorie labeling: A natural experiment. *International Journal of Obesity, 35*, 493–500. doi:10.1038/ijo.2011.4

Elliot, A. J., & Niesta, D. (2008). Romantic red: Red enhances men's attraction to women. *Journal of Personality and Social Psychology, 95*, 1150–1164.

Environmental Protection Agency. (2019). *Fuel economy data*. Retrieved from https://www.fueleconomy.gov/feg/download.shtml

Evans, S. W., Pelham, W. E., Smith, B. H., Bukstein, O., Gnagy, E. M., Greiner, A. R. ... Baron-Myak, C. (2001). Dose-response effects of methylphenidate on ecologically valid measures of academic performance and classroom behavior in adolescents with ADHD. *Experimental and Clinical Psychopharmacology, 9*, 163–175.

Federal Aviation Administration. (2017). *Air traffic by the numbers*. Retrieved from https://www.faa.gov/air_traffic/by_the_numbers/

Flynn, J. R. (1984). The mean IQ of Americans: Massive gains 1932 to 1978. *Psychological Bulletin, 95*, 29–51.

Flynn, J. R. (1999). Searching for justice: The discovery of IQ gains over time. *American Psychologist, 54*, 5–20.

Ford, A. M., & Torok, D. (2008). Motivational signage increases physical activity on a college campus. *Journal of American College Health, 57*, 242–244.

Gentile, D. A., Lynch, P. J., Linder, J. R., & Walsh, D. A. (2004). The effects of video game habits on adolescent hostility, aggressive behaviors, and school performance. *Journal of Adolescence, 27*, 5–22. doi:10.1016/j.adolescence.2003.10.002

Gillam, B., & Chambers, D. (1985). Size and position are incongruous: Measurements on the Müller-Lyer figure. *Perception & Psychophysics, 37*, 549–556.

Gillen-O'Neel, C., Huynh, V. W., & Fuligni, A. J. (2013). To study or to sleep? The academic costs of extra studying at the expense of sleep. *Child Development, 84*, 133–142. doi:10/1111/j.1467-8624.2012.01834.x

Gino, F., & Ariely, D. (2012). The dark side of creativity: Original thinkers can be more dishonest. *Journal of Personality and Social Psychology, 102*, 445–459. doi:10.1037/a0026406

Gorant, J. (2010). *The lost dogs. Michael Vick's dogs and their tale of rescue and redemption.* New York: Gotham Books.

Greenlees, I. A., Eynon, M., & Thelwell, R. C. (2013). Color of soccer goalkeepers' uniforms influences the outcome of penalty kicks. *Perceptual and Motor Skills: Exercise and Sport,* 117, 1–10.

Guéguen, N., Jacob, C., & Lamy, L. (2010). "Love is in the air": Effects of songs with romantic lyrics on compliance with a courtship request. *Psychology of Music,* 38, 303–307.

Guéguen, N., & Jacob, C. (2012, April). Clothing color and tipping: Gentlemen patrons give more tips to waitresses with red clothes. *Journal of Hospitality & Tourism Research.* doi:10.1177/1096348012442546

Guidry, K. (2017). Delivery versus time devoted to assignments: The effect on course performance. *Journal of Instructional Pedagogies, 19,* 2–9.

Gunter, L. M., Barber, R. T., & Wynne, C. D. L (2016). What's in a name? Effect of breed perceptions & labeling on attractiveness, adoptions and length of stay for pit-bull-type dogs. *PLoS ONE, 11,* e0136857. doi:10.1371/journal.pone.0146857

Gutierrez-Cebollada, J., de la Torre, R., Ortuno, J., Garces, J. M., & Cami, J. (1994). Psychotropic drug consumption and other factors associated with heroin overdose. *Drug and Alcohol Dependence, 35,* 169–174.

Harman, B. A., & Sato, T. (2011). Cell phone use and grade point average among undergraduate university students. *College Student Journal, 45,* 544–549.

Horvath, J. C., Horton, A. J., Lodge, J. M., & Hattie, J. A. (2017). The impact of binge watching on memory and perceived comprehension. *First Monday, 22*(9).

Humane Society of the United States. (2018). American Pet Products Association Survey 2017–2018. *Pets by the Numbers.* Retrieved from https://www.animalsheltering .org/page/pets-by-the-numbers

Hunter, J. E. (1997). Needed: A ban on the significance test. *Psychological Science, 8,* 3–7.

Indiana University. (2018). *National Survey of Student Engagement.* Retrieved from http://nsse.indiana.edu/2018 _institutional_report/pdf/Frequencies/FreqSex.pdf

Insurance Information Institute. (2015). *Facts and statistics: Mortality risks.* Retrieved from https://www.iii.org/fact -statistic/facts-statistics-mortality-risk

Jacobsen, W. C., & Forste, R. (2011). The wired generation: Academic and social outcomes of electronic media use among university students. *Cyberpsychology, Behaviour, and Social Networking, 14,* 275–280. doi:10.1089/cyber. 2010.0135

Jena, A. B., Jain, A., & Hicks, T. R. (2018, February 3). Do "Fast and Furious" movies cause a rise in speeding? *The Record* [Kitchener, Ontario]. pD4.

Johnston, J. J. (1975). Sticking with first responses on multiple-choice exams: For better or worse? *Teaching of Psychology, 2,* 178–179.

Judge, T. A., & Cable, D. M. (2010). When it comes to pay, do the thin win? The effect of weight on pay for men and women. *Journal of Applied Psychology, 96,* 95–112. doi:10.1037/a0020860

Junco, R. (2015). Student class standing, Facebook use, and academic performance. *Journal of Applied Developmental Psychology, 36,* 18–29. doi:10.1016/j .appdev.2014.11.001

Katona, G. (1940). *Organizing and memorizing.* New York, NY: Columbia University Press.

Killeen, P. R. (2005). An alternative to null-hypothesis significance tests. *Psychological Science, 16,* 345–353.

Kornell, N. (2009). Optimising learning using flashcards: Spacing is more effective than cramming. *Applied Cognitive Psychology, 23,* 1297–1317.

Kuo, M., Adlaf, E. M., Lee, H., Gliksman, L., Demers, A., & Wechsler, H. (2002). More Canadian students drink but American students drink more: Comparing college alcohol use in two countries. *Addiction, 97,* 1583–1592.

Kurinec, C. A., & Weaver, C. A. (2018). Do memory-focused jury instructions moderate the influence of eyewitness word choice? *Applied Psychology in Criminal Justice, 14,* 55–69.

Lay, C. H. (1986). At last, my research article on procrastination. *Journal of Research in Personality, 20,* 474–495.

Lepp, A., Barkley, J. E., & Karpinski, A. C. (2014). The relationship between cell phone use, academic performance, anxiety, and satisfaction with life in college students. *Computers in Human Behavior, 31,* 343–350. doi:10.10116/j.chb.2013.10.049

Lepp, A., Barkley, J. E., & Karpinski, A. C. (2015). The relationship between cell phone use and academic performance in a sample of U.S. college students. *SAGEOpen, 5.* doi:10.1177/2158244015573169

Li, A. (2008). Experiencing visuo-motor plasticity by prism adaptation in a classroom setting. *Journal of Undergraduate Neuroscience Education, 7*(1), A13–A18.

Li, L., Chen, R., & Chen, J. (2016). Playing action video games improves visuomotor control. *Psychological Science, 27*(8), 1092–1108.

Liu, B., Floud, S., Pirie, K., Green, J., Peto, R., & Beral, V. (2015). Does happiness itself directly affect mortality? The prospective UK million women study. *The Lancet* (published online Dec. 9, 2015). Retrieved from http:// dx,doi.org/10.1016/S0140-6736(15)01087-9

Liu, M., Huang, Y., & Zhang, D. (2017). Gamification's impact on manufacturing: Enhancing motivation, satisfaction and operational performance with smartphone-based gamified job design. *Human Factors and Ergonomics in Manufacturing & Service Industries, 28*(1), 38–51.

Live Science Staff. (2011, June 22). These states are really lightning rods. *Live Science.* Retrieved from

https://www.livescience.com/14714-lightning-prone
-states-110620.html

Loftus, E. F., & Palmer, J. C. (1974). Reconstruction of automobile destruction: An example of the interaction between language and memory. *Journal of Verbal Learning & Verbal Behavior, 13*, 585–589.

Loftus, G. R. (1996). Psychology will be a much better science when we change the way we analyze data. *Current Directions in Psychological Science, 5*, 161–171.

Loke, W. H. (1988). Effects of caffeine on mood and memory. *Physiology & Behavior, 44*, 367–372.

Luhmann, M., Schimmack, U., & Eid, M. (2011). Stability and variability in the relationship between subjective well-being and income. *Journal of Research in Personality, 45*(2), 186–197.

Mackay, G. J., & Neill, J. T. (2010). The effect of "green exercise" on state anxiety and the role of exercise duration, intensity, and greenness: A quasi-experimental study. *Psychology of Sport and Exercise, 11*(3), 238–245.

Marchewka, A., Zurawski, L., Jenorog, K., & Grabowska, A. (2014). The Nencki affective picture system (NAPS): Introduction to a novel, standardized, wide-range, high-quality, realistic picture database. *Behavioral Research Methods, 46*(2), 596–620.

Massa, L. J., & Mayer, R. E. (2006). Testing the ATI hypothesis: Should multimedia instruction accommodate verbalizer-visualizer cognitive style? *Learning and Individual Differences, 16*, 321–336.

McAllister, T. W., Flashman, L. A., Maerlender, A., Greenwald, R. M., Beckwith, J. G., Tosteson, T. D.,... Turco, J. H. (2012). Cognitive effects of one season of head impacts in a cohort of collegiate contact sport athletes. *Neurology, 78*, 1777–1784. doi:10.1212/WNL.0b013e3182582fe7

McGee, R., Williams, S., Howden-Chapman, P., Martin, J., & Kawachi, I. (2006). Participation in clubs and groups from childhood to adolescence and its effects on attachment and self-esteem. *Journal of Adolescence, 29*, 1–17.

McMorris, B. J., Catalano, R. F., Kim, M. J., Toumbourou, J. W., & Hemphill, S. A. (2011). Influence of family factors and supervised alcohol use on adolescent alcohol use and harms: Similarities between youth in different alcohol policy contexts. *Journal of Studies on Alcohol and Drugs, 72*, 418–428.

Neuenschwander, L. M., Abbott, A., & Mobley, A. R. (2012). Assessment of low-income adults' access to technology: Implications for nutrition education. *Journal of Nutrition Education and Behavior, 44*, 60–65.

Nitzschner, M., Melis, A. P., Kaminski, J., & Tomasello, M. (2012). Dogs (*Canis familiaris*) evaluate humans on the basis of direct experiences only. *PLoS ONE, 7*(October), Issue 10, e46880 Retrieved from https://doi.org/10.1371/journal.pone.0046880

Noland, S. A. & The Society for the Teaching of Psychology Statistical Literacy Taskforce. (2012). *Statistical literacy in the undergraduate psychology curriculum.* Retrieved from http://www.teachpsych.org/Resources/Documents/otrp/resources/statistics/STP_Statistical_Literacy_Psychology_Major_Learning_Goals_4-2014.pdf

Oishi, S., & Schimmack, U. (2010). Residential mobility, well-being, and mortality. *Journal of Personality and Social Psychology, 98*, 980–994.

Oreg, S., & Berson, Y. (2018). The impact of top leaders' personalities: The process through which organizations become reflections of their leaders. *Current Directions in Psychological Science, 27*, 241–248.

Otto, A. R., Fleming, S. M., & Glimcher, P. W. (2016). Unexpected but incidental positive outcomes predict real-world gambling. *Psychological Science, 27*(3), 299–311.

Pashler, H., McDaniel, M., Rohrer, D., & Bjork, R. (2009). Learning styles: Concepts and evidence. *Psychological Science in the Public Interest, 9*, 105–119.

Piff, P. K., Kraus, M. W., Cote, S., Cheng, B. H., & Keltner, D. (2010). Having less, giving more: The influence of social class on prosocial behavior. *Journal of Personality and Social Psychology, 99*, 771–784.

Polman, H., de Castro, B. O., & van Aken, M. A. G. (2008). Experimental study of the differential effects of playing versus watching violent video games on children's aggressive behavior. *Aggressive Behavior, 34*, 256–264. doi:10.1002/ab.20245

Popham, J., Lee, M., Sublette, M., Kent, T., & Carswell, C. M. (2017). Graphic vs. text-only résumés: Effects of design elements on simulated employment decisions. *Proceedings of the Human Factors and Ergonomics Society Annual Meeting, 61*, 1242–1246.

Ramirez, M. (2006). "My dog's just like me": Dog ownership as a gender display. *Symb Interact, 29*, 373–391.

Rello, L., & Bigham, J. P. (2017). Good background colors for readers: A study of people with and without dyslexia. In *Proceedings of the 19th International ACM SIGACCESS Conference on Computers and Accessibility* (pp. 72–80). Baltimore, MD: ACM. doi:10.1145/3132525.3132546

Remmers, C., & Zander, T. (2018). Why you don't see the forest for the trees when you are anxious: Anxiety impairs intuitive decision making. *Clinical Psychological Science, 6*(1), 48–62.

Robbins, L., & Margulis, S. W. (2014). The effects of auditory enrichment on gorillas. *Zoo Biology, 2014*, 197–203.

Roberts, J. A., Yaya, L. H. P., & Manolis, C. (2014). The invisible addiction: Cell-phone activities and addiction among male and female college students. *Journal of Behavioral Addictions, 3*, 354–265. doi:10.1556/JBA.3.2014.015

Rosati, A. G., & Santos, L. R. (2016). Spontaneous meta-cognition in rhesus monkeys. *Psychological Science, 27*, 1181–1191.

Rosenberg, M. (1965). *Society and the adolescent self-image*. Princeton, NJ: Princeton University Press.

Rosenberg, M. (1989). *Society and the adolescent self-image*. Revised edition. Middletown, CT: Wesleyan University Press.

Sachs, B. D., Ni, J. R., & Caron, M. G. (2014). Brain 5-HT deficiency increases stress vulnerability and impairs antidepressant responses following psychosocial stress. *Proceedings of the National Academy of Sciences, 112*, 2557–2562.

Schachter, S. (1968). Obesity and eating. *Science, 161*, 751–756.

Scharf, R., Demmer, R., & DeBoer, M. M. (2013). Longitudinal evaluation of milk type consumed and weight status in preschoolers. *Archives of Disease in Childhood, 98*, 335–340. doi:10.1136/archdis-child-2012-302941

Sibbald, T. (2014). Occurrence of bimodal classroom achievement in Ontario, Alberta. *Journal of Educational Research, 60*, 221–225.

Simon, N. W., & Moghaddam, B. (2016). Methylphenidate has nonlinear effects on cued response inhibition in adults but not adolescents. *Brain Research, 1654*, 171–176.

Singh, S. (2006). Impact of color on marketing. *Management Decision, 44*(6), 783–789. doi:10.1108/0025170610673322

So, W. C., Ching, T. H-W., Lim, P. E., Cheng, X., & Ip, K. Y. (2014). Producing gestures facilitates route learning. *PLoS ONE, 9*(11): e112543. doi:10.1371/journal.pone.0112543

Starling, M. J., Branson, N., Thomson, P. C., & McGreevy, P. D. (2013). "Boldness" in the domestic dog differs among breeds and breed groups. *Behavioural Processes, 97*, 53–62.

Stephens, R., Atkins, J., & Kingston, A. (2009). Swearing as a response to pain. *NeuroReport: For Rapid Communication of Neuroscience Research, 20*, 1056–1060. doi:10.1097/WNR.0b013e32832e64b1

Taylor, A. M., Reby, D., & McComb, K. (2011). Cross modal perception of body size in domestic dogs (Canis familiaris). *PLoS ONE, 6*(2): e17069. doi:10.1371/journal.pone.0017069

Telles, S., Singh, N., & Balkrishna, A. (2012). Finger dexterity and visual discrimination following two yoga breathing practices. *International Journal of Yoga, 5*, 37–41.

Tolman, E. C. (1948). Cognitive maps in rats and men. *Psychological Review, 55*(4), 189–208.

Troyer, A. K., Leach, L., & Strauss, E. (2006). Computerized Victoria Stroop Test in adult unipolar depressed patients and healthy subjects: Influence of age and gender. *Aging, Neuropsychology, and Cognition, 13*, 20–35.

Tukey, J. W. (1977). *Exploratory data analysis*. Reading, MA: Addison-Wesley.

Tversky, A., & Kahneman, D. (1974). Judgments under uncertainty: Heuristics and biases. *Science, 185*, 1124–1131.

Twenge, J. M., Joiner, T. E., Rogers, M. L., & Martin, G. N. (2018). Increases in depressive symptoms, suicide-related outcomes, and suicide rates among U.S. adolescents after 2010 and links to increased new media screen time. *Clinical Psychological Science, 6*, 3–17.

US Census Bureau. (2017). Commuting times, median rents and language other than English use in the home on the rise. *Newsroom*. Retrieved from https://www.census.gov/newsroom/press-releases/2017/acs-5yr.html

United States Department of Labor, Bureau of Labor Statistics. (2017). *Occupational outlook handbook*. Retrieved from https://www.bls.gov/ooh/

von Hippel, P. T. (2005). Mean, median, and skew: Correcting a textbook rule. *Journal of Statistics Education, 13*. Retrieved from http://www.amstat.org/publications/jse/v13n2/vonhippel.html

Weinberg, G. H., Schumaker, J. A., & Oltman, D. (1981). *Statistics: An intuitive approach*. Belmont, CA: Wadsworth.

Weinstein, Y., McDermott, K. B., & Roediger, H. L. III (2010). A comparison of study strategies for passages: Rereading, answering questions, and generating questions. *Journal of Experimental Psychology: Applied, 16*, 308–316.

Weiss, E., Miller, K., Mohan-Gibbons, H., & Vela, C. (2012). Why did you choose this pet? Adopters and pet selection preferences in five animal shelters in the United States. *Animals, 2*, 144–159. doi:10.3390/ani2020144

West, G. L., Zendel, B. R., Konishi, K., Benady-Chorney, J., Bohbot, V. D., Peretz, I., Belleville, S. (2017). Playing Super Mario 64 increases hippocampal grey matter in older adults. *PLoS One, 12*(12).

Wilcox, K., & Stephen, A. T. (2013). Are close friends the enemy? Online social networks, self-esteem, and self-control. *Journal of Consumer Research, 40*, 90–103. doi:10.1086/668794

Wilkinson, L., and the Task Force on Statistical Inference. (1999). Statistical methods in psychology journals. *American Psychologist, 54*, 594–604.

Williamson, S., Block, L. G., & Keller, P. A. (2016). Of waste and waists: The effect of plate material on food consumption and waste. *Journal of the Association for Consumer Research, 1*(1), 147–160.

Yan, D., & Sengupta, J. (2011). Effects of construal level on the price-quality relationship. *Journal of Consumer Research, 38,* 376–389. doi:10.1086/659755

Zagorcheve, L., Meyer, C., Stehle, T., Wenzel, F., Young, S., Peters, J., … McAllister, T. (2016). Differences in regional brain volumes two months and one year after mild traumatic brain injury. *Journal of Neurotrauma, 33*(1), 29–34.

Zezulka, L. A., & Seigfried-Spellar, K. C. (2016). Differentiating cyberbullies and internet trolls by personality characteristics and self-esteem. *Journal of Digital Forensics, Security, and Law, 11*(3).

Zhong, C., Bohns, V. K., & Gino, F. (2010). Good lamps are the best police: Darkness increases dishonesty and self-interested behavior. *Psychological Science, 21,* 311–314.

Name Index

Subject Index